Metals

Metalloids

Nonmetals

10 VIII	11 IB	12 II B	13 IIIA	14 IVA	15 VA	16 VIA	17 VIIA	18 VIIIA
								2 **He** 4.00 helium
			5 **B** 10.81 boron	6 **C** 12.01 carbon	7 **N** 14.01 nitrogen	8 **O** 16.00 oxygen	9 **F** 19.00 fluorine	10 **Ne** 20.18 neon
			13 **Al** 26.98 aluminum	14 **Si** 28.09 silicon	15 **P** 30.97 phosphorus	16 **S** 32.07 sulfur	17 **Cl** 35.45 chlorine	18 **Ar** 39.95 argon
28 **Ni** 58.69 nickel	29 **Cu** 63.55 copper	30 **Zn** 65.39 zinc	31 **Ga** 69.72 gallium	32 **Ge** 72.61 germanium	33 **As** 74.92 arsenic	34 **Se** 78.96 selenium	35 **Br** 79.90 bromine	36 **Kr** 83.80 krypton
46 **Pd** 106.42 palladium	47 **Ag** 107.87 silver	48 **Cd** 112.41 cadmium	49 **In** 114.82 indium	50 **Sn** 118.71 tin	51 **Sb** 121.75 antimony	52 **Te** 127.60 tellurium	53 **I** 126.90 iodine	54 **Xe** 131.29 xenon
78 **Pt** 195.08 platinum	79 **Au** 196.97 gold	80 **Hg** 200.59 mercury	81 **Tl** 204.38 thallium	82 **Pb** 207.2 lead	83 **Bi** 208.98 bismuth	84 **Po** (209) polonium	85 **At** (210) astatine	86 **Rn** (222) radon
110 — (269)	111 — (272)	112 — (277)	113	114 — (285)	115	116 — (289)	117	118

64 **Gd** 157.25 gadolinium	65 **Tb** 158.93 terbium	66 **Dy** 162.50 dysprosium	67 **Ho** 164.93 holmium	68 **Er** 167.26 erbium	69 **Tm** 168.93 thulium	70 **Yb** 173.04 ytterbium	71 **Lu** 174.97 lutetium
96 **Cm** (247) curium	97 **Bk** (247) berkelium	98 **Cf** (251) californium	99 **Es** (252) einsteinium	100 **Fm** (257) fermium	101 **Md** (258) mendelevium	102 **No** (259) nobelium	103 **Lr** (260) lawrencium

Introductory Chemistry Essentials

Introductory Chemistry Essentials

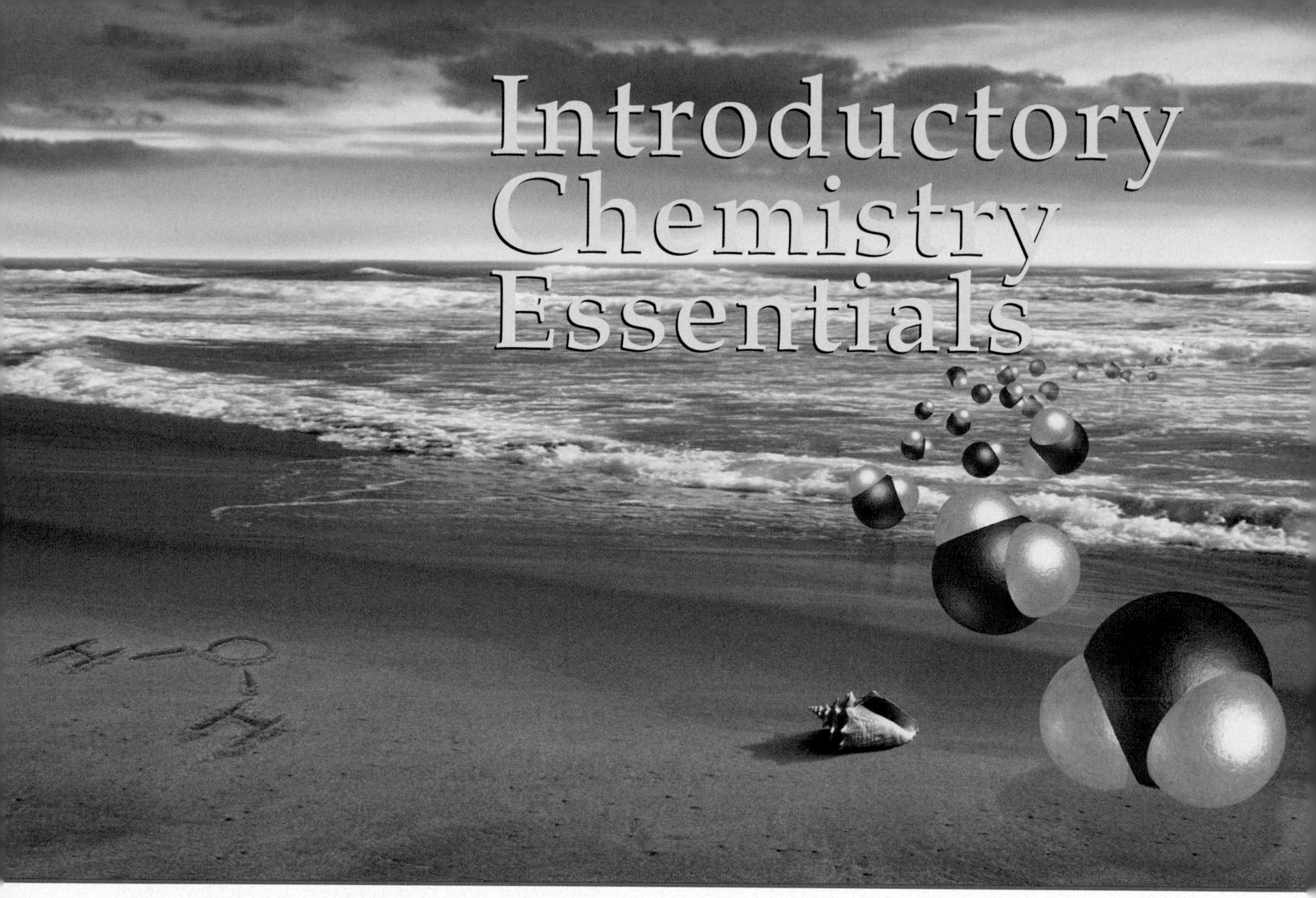

Nivaldo J. Tro

WESTMONT COLLEGE

PEARSON EDUCATION, INC.
Upper Saddle River, New Jersey 07458

Library of Congress Cataloging-in-Publication Data

Tro, Nivaldo J.
Introductory chemistry essentials/Nivaldo J. Tro.
p. cm.
Includes index.
ISBN 0-13-111903-6
1. Chemistry I. Title.

QD33.2 .T763 2003
540--dc21 2002042502

CIP

Senior Editor: Kent Porter Hamann
Editor in Chief, Physical Sciences: John Challice
Development Editor: Donald Gecewicz
Editor in Chief, Development: Ray Mullaney
Vice President of Production and Manufacturing: David W. Riccardi
Executive Managing Editor: Kathleen Schiaparelli
Assistant Managing Editor: Beth Sweeten
Senior Marketing Manager: Steve Satori
Assistant Managing Editor, Science Media: Nicole Bush
Assistant Managing Editor, Science Supplements: Dinah Thong
Media Editor: Michael J. Richards
Project Manager: Kristen Kaiser
Director of Creative Services: Paul Belfanti
Director of Design: Carole Anson
Art Editor: Connie Long
Art Direction and Cover Design: Kenny Beck
Interior Design: Laura C. Ierardi
Cover and Chapter Opening Illustrations: Quade Paul
Manufacturing Manager: Trudy Pisciotti
Assistant Manufacturing Manager: Michael Bell
Photo Researcher: Kathy Ringrose
Photo Editor: Nancy Seise
Art Studio: Artworks
Senior Manager: Patricia Burns
Production Manager: Ronda Whitson
Manager, Production Technologies: Matthew Haas
Illustrators: Royce Copenheaver, Jay McElroy, Mark Landis
Art Quality Assurance: Timothy Nguyen, Stacy Smith, Pamela Taylor
Editorial Assistants: Nancy Bauer, Jacquelyn Howard
Production Supervision/Composition: Lithokraft II

Pearson Education, Inc.
Upper Saddle River, New Jersey 07458

Printed in the United States of America

10 9 8 7 6 5 4 3 2 1

ISBN 0-13-111903-6

Pearson Education Ltd., *London*
Pearson Education Australia Pty. Limited, *Sydney*
Pearson Education Singapore Pte. Ltd.
Pearson Education North Asia Ltd., *Hong Kong*
Pearson Education Canada, Ltd., *Toronto*
Pearson Educación de Mexico, S.A. de C.V
Pearson Education—Japan, *Tokyo*
Pearson Education Malaysia, Pte. Ltd.

TO ANNIE

ABOUT THE AUTHOR

Professor Tro has been a faculty member at Westmont College in Santa Barbara, California, since 1990. He received his B.A. degree in chemistry from Westmont College in 1985 and his Ph.D. from Stanford University in 1989. He performed post-doctoral research at the University of California at Berkeley. He was honored as Westmont's outstanding teacher of the year in 1994 and again in 2001. He was also honored as Westmont's outstanding researcher of the year in 1996.

Professor Tro lives in Santa Barbara with his wife, Ann, and their three children, Michael, Alicia, and Kyle. For leisure, he enjoys snowboarding, camping, biking, and wakeboarding with his family.

BRIEF CONTENTS

CONTENTS

6 Chemical Composition 163

7 Chemical Reactions 203

TRO
Introductory Chemistry

13 Solutions 443

14 Acids and Bases 483

TO THE STUDENT

This book is for you, and every text feature has you in mind. I have two main goals for you in this course: to see chemistry like you never have before, and to develop the problem-solving skills you need to succeed in introductory chemistry.

I want you to experience chemistry in a new way. Each chapter of this book is written to show you that chemistry is not just something that happens in a laboratory, but that chemistry surrounds you at every moment. I have worked with several outstanding artists to develop photographs and art that help you visualize the molecular world. From the opening example to the closing chapter, you will see chemistry. I hope that, when you finish this course, you think differently about your world because you understand the molecular interactions that lurk beneath everything around you.

I also want you to develop problem-solving skills. No one succeeds in chemistry—or in life, really—without the ability to solve problems. I can't give you a formula for problem solving, but I can give you strategies and help you to develop the chemical intuition you need to understand chemical reasoning. Look for several recurring structures throughout this book designed to help you master problem solving. The two most important ones are the solution map, a visual aid that helps you navigate your way through a problem, and the multi-column examples, where you learn a problem-solving procedure as you see it applied to two different examples.

Lastly, know that chemistry is not reserved only for those with some special talent or special capability. With the right amount of effort and some clear guidance, anyone can master chemistry, including you.

Sincerely,

Nivaldo J. Tro
tro@westmont.edu

TO THE INSTRUCTOR

Introductory Chemistry Essentials is designed for a one-semester, college-level, introductory or preparatory chemistry course. The goal of the book is to teach chemical principles and build chemical skills in a truly relevant context. Students taking this course need to develop problem-solving skills—but they also must see *why* these skills are important to them and to their world. *Introductory Chemistry Essentials* extends chemistry from the laboratory to the student's world. It motivates students to learn chemistry by demonstrating how chemistry plays out in their everyday lives.

This is a visual book. Today's students often learn by seeing, so wherever possible, I have used images to help communicate the subject. For example, in developing problem-solving skills, I use a solution map to help students see the general logic of working through a multi-step problem. I have also developed multi-column examples, where students learn a problem-solving procedure as they see it applied to two different examples at once. In developing chemical principles, I have worked with several artists to develop multi-part images that show the connection between everyday processes visible to the eye and the molecular interactions responsible for these processes. This book is designed to meet the students where they are and bring them to the level that they need to be.

I hope that you find as much joy in using this book as I have in teaching students the fundamental principles of the chemistry that surrounds them. It is a worthwhile cause, even though it requires constant effort. Please feel free to email me (tro@westmont.edu) with any questions or comments you might have. I look forward to hearing from you as you use this book in your course.

Sincerely,

Nivaldo J. Tro
tro@westmont.edu

PREFACE

The design and features of this book represent a conscious, deliberate, and sustained effort to achieve, in our students, the goals of the book—teaching chemical principles, and building chemical skills in the context of relevance. Students must understand chemical concepts, solve chemical problems, and understand why they are important.

Chemical Principles

The understanding of basic chemical principles and concepts in topics such as atomic structure, chemical bonding, chemical reactions, and gas laws is critical to the success of the introductory chemistry student. Students should understand the principles they need to succeed in the normal general chemistry sequence. The book integrates qualitative and quantitative material and proceeds from concrete concepts to more abstract ones.

The main divergence in topic ordering among instructors teaching preparatory chemistry courses is the placement of electronic structure and chemical bonding. Should these topics come early, at the point where models for the atom are being discussed? Or should they come later, after the student is exposed to chemical compounds and chemical reactions? Early placement gives the student a theoretical framework out of which they can understand compounds and reactions. However, it might also present students with abstract models before they understand why they are necessary. I have chosen a later placement for the following reasons:

1) *A later placement seems more flexible.* An instructor who wants to cover atomic theory and bonding earlier can simply cover Chapters 9 and 10 after Chapter 4. However, if atomic theory and bonding were placed earlier, it would be more difficult for the instructor to skip these chapters and come back to them later.
2) *A later placement allows earlier coverage of topics that students can more easily visualize.* Coverage of abstract topics too early in a course can lose some students. Chemical compounds and chemical reactions can be more tangible than atomic orbitals, and the relevance of these is easier to demonstrate to the beginning student.
3) *A later placement gives students a reason to learn an abstract theory.* Once students learn about compounds and reactions, they are more easily motivated to learn a theory that explains the underlying causes behind them.
4) *A later placement follows the scientific method.* In science, we normally make observations, form laws, and then build models or theories that explain our observations and laws. A later placement reflects this ordering.

Nonetheless, I know that every course is unique and that each instructor chooses to cover topics in his or her own way. Consequently, I have written each chapter for maximum flexibility in topic ordering. In addition, the book is offered in two formats. ***Introductory Chemistry***, the full version, contains 19 chapters and includes organic chemistry and biochemistry. Since some courses do not cover these two topics, we offer ***Introductory Chemistry Essentials***, which contains 17 chapters and omits these topics.

Problem-Solving Skills ▶

The development of problem-solving skills is the other main goal of this text; it is often the primary reason that students take introductory/preparatory chemistry. To this end, *Introductory Chemistry* develops a systematic approach to problem solving. Problem-solving skills are emphasized throughout each chapter, developed through many in-chapter examples, reinforced with *skillbuilder* exercises immediately following each example, reviewed in unique chapter summaries, and practiced and synthesized in end-of-chapter exercises.

EXAMPLE 2.11 Solving Multistep Unit Conversion Problems

A running track measures 0.250 mi per lap. To run 10.0 km, how many laps should you run?

We set up the problem in the standard way.

Given: 10.0 km

Find: laps

Conversion Factors: 1 lap = 0.250 mi

0.6214 mi = 1 km

We then build our solution map, focusing on the units and how to get from km to laps.

Solution Map:

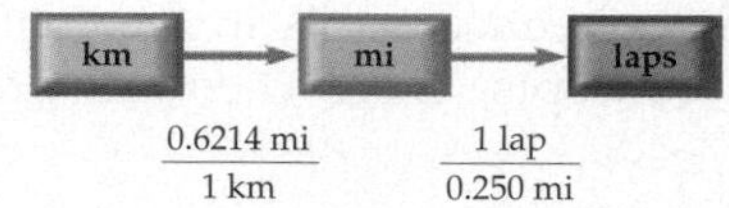

$$\frac{0.6214\ \text{mi}}{1\ \text{km}} \qquad \frac{1\ \text{lap}}{0.250\ \text{mi}}$$

We then follow the solution map to solve the problem.

Solution:

$$10.0\ \text{km} \times \frac{0.6214\ \text{mi}}{1\ \text{km}} \times \frac{1\ \text{lap}}{0.250\ \text{mi}} = 24.856\ \text{laps} = 24.9\ \text{laps}$$

The intermediate answer (blue) in blue is rounded to three significant figures as limited by the given quantity, 10.0 km. The units of the answer are correct and the value of the answer makes sense—a lap is shorter than a km, so the value in laps should be larger than the value in km.

SKILLBUILDER 2.11 Solving Multistep Unit Conversion Problems

A running track measures 1056 ft per lap. To run 15.0 km, how many laps should you run? 1 mi = 5280 ft

SKILLBUILDER PLUS

Convert 5.72 nautical mi to meters. A nautical mile is equal to 1.151 mi.

Solution Maps

Many problems in this course can be solved using dimensional analysis, so this is an emphasis of the early chapters. The text presents students with a basic procedure for solving most chemical problems. Part of this procedure uses a unique visual approach in which students draw a *solution map* to the problem. In this map, students outline the steps—using conversion factors and equations—that are required to get from the information they are given to the information they are trying to find. They can follow their map to solve the problem. The map is a good way for students to get an overview of how the problem is solved. ▼

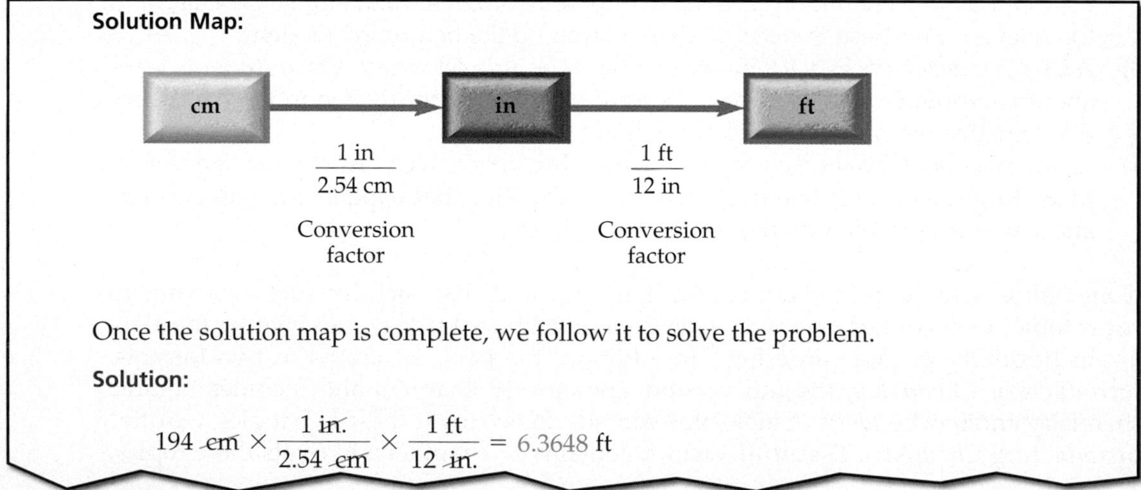

Solution Map:
cm
in
ft
1 in / 2.54 cm
1 ft / 12 in
Conversion factor
Conversion factor
Once the solution map is complete, we follow it to solve the problem.
Solution:
194 cm × 1 in / 2.54 cm × 1 ft / 12 in = 6.3648 ft

Multi-Column Examples

Unique to this book are the multi-column examples, where students learn a problem-solving procedure as they see it applied to two different examples simultaneously. The student can then use the same procedure to solve the two accompanying skillbuilder exercises. ▼

Solving Unit Conversion Problems	**EXAMPLE 2.8** **Unit Conversion** Convert 7.8 km to miles.	**EXAMPLE 2.9** **Unit Conversion** Convert 0.825 m to millimeters.
1. Write down the *given* quantity and its unit(s).	**Given:** 7.8 km	**Given:** 0.825 m
2. Write down the quantity that you are asked to *find* and its unit(s).	**Find:** mi	**Find:** mm
3. Write down the appropriate *conversion factor(s)*. Some of these will be given in the problem. Others you find in tables within the text.	**Conversion Factors:** 1 km = 0.6214 mi (This conversion factor is from Table 2.3.)	**Conversion Factors:** 1 mm = 10^{-3} m (This conversion factor is from Table 2.2.)
4. Write a *solution map* for the problem. Begin with the *given* quantity and draw an arrow symbolizing each conversion step. Below each arrow, write the appropriate conversion factor for that step. Focus on the units. The solution map should end at the *find* quantity.	**Solution Map:** 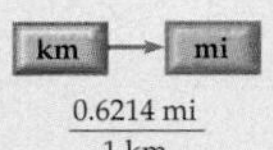 $\frac{0.6214 \text{ mi}}{1 \text{ km}}$ The conversion factor is written so that km, the unit we are converting *from*, is on the bottom and mi, the unit we are converting *to*, is on the top.	**Solution Map:** 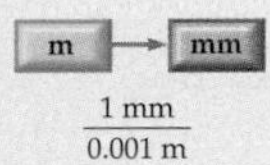 $\frac{1 \text{ mm}}{0.001 \text{ m}}$ The conversion factor is written so that m, the unit we are converting *from*, is on the bottom and mm, the unit we are converting *to*, is on the top.
5. Follow the solution map to solve the problem. Begin with the *given* quantity and its units. Multiply by the appropriate conversion factor(s), canceling units, to arrive at the *find* quantity.	**Solution:** $7.8 \cancel{\text{km}} \times \frac{0.6214 \text{ mi}}{1 \cancel{\text{km}}} =$ 4.84692 mi	**Solution:** $0.825 \cancel{\text{m}} \times \frac{1 \text{ mm}}{10^{-3} \cancel{\text{m}}} =$ 825 mm
6. Round the answer to the correct number of significant figures. Follow the significant figure rules in Sections 2.3 and 2.4. Remember that exact conversion factors do not limit the number of significant figures in your answer.	4.84692 mi = 4.8 mi We round to two significant figures since the quantity given has two significant figures. (If possible, obtain conversion factors to enough significant figures so that they do not limit the number of significant figures in the answer.)	825 mm = 825 mm We leave the answer with three significant figures since the quantity given has three significant figures and the conversion factor is a definition, which therefore does not limit the number of significant figures in the answer.
7. Check your answer. Make certain that the units are correct and that the magnitude of the answer makes physical sense.	The units, mi, are correct. The magnitude of the answer is reasonable. A mi is longer than a km, so the value in mi should be smaller than the value in km.	The units, mm, are correct and the magnitude is reasonable. A mm is shorter than a m, so the value in m should be smaller than the value in mm.
	SKILLBUILDER 2.8 **Unit Conversion** Convert 56.0 cm to inches.	**SKILLBUILDER 2.9** **Unit Conversion** Convert 5,678 m to kilometers.

Molecular Art

Many chemical principles involve making a connection between atoms and molecules and the properties of the substances they compose. For example, electronic structure shows how the reactivity of elements is related to the electron configuration of their atoms. The gas laws show how pressure, volume, and temperature are related to gas particles and their motions and collisions. The art program helps students visualize this connection between the molecular world and the macroscopic world. Many concepts are portrayed using a two-part visual image. One part of the image is a photograph of a real world object or process, such as an inflated balloon, for example. The second part, either superimposed on the photograph or shown as a magnification window, shows what the molecules magnified by many orders of magnification are doing in the photograph. For the balloon, molecules are superimposed on the photograph to show how their collisions with the balloon's walls keep it distended. Many molecular formulas will be portrayed, not only with structural formulas, but also with space-filling drawings of the molecule; the idea is to portray the beauty and form of the molecular world. ▼

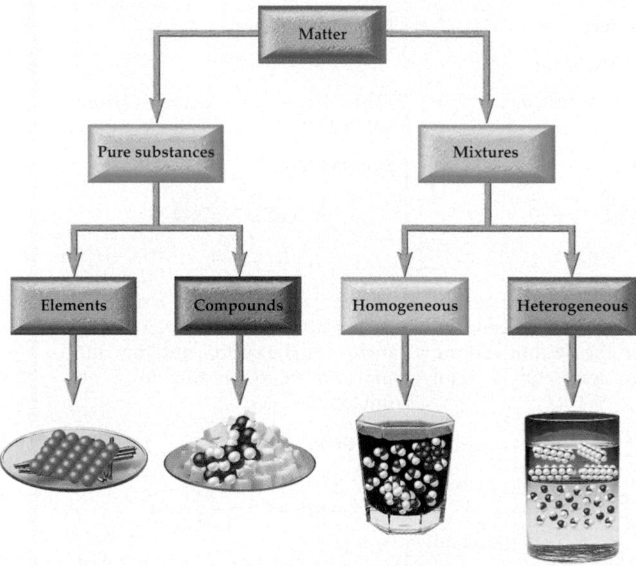

Figure 3.8 Classification of matter. Matter may be a pure substance or it may be a mixture. A pure substance may either be an element or a compound, and a mixture may be either homogenous or heterogeneous.

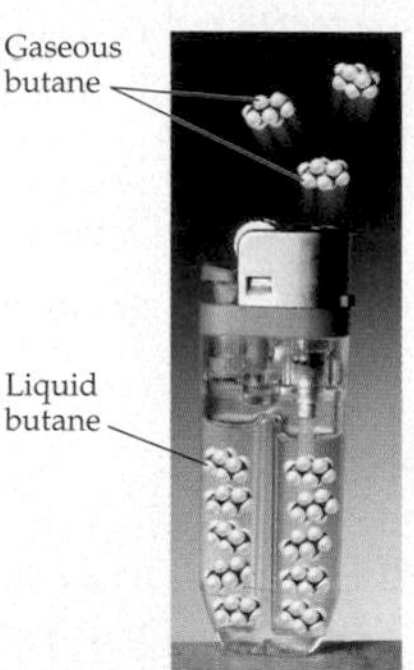

Figure 3.11 If you push the button on a lighter without turning the flint, some of the liquid butane vaporizes to gaseous butane. Since the liquid butane and the gaseous butane are both composed of butane molecules, this is a physical change.

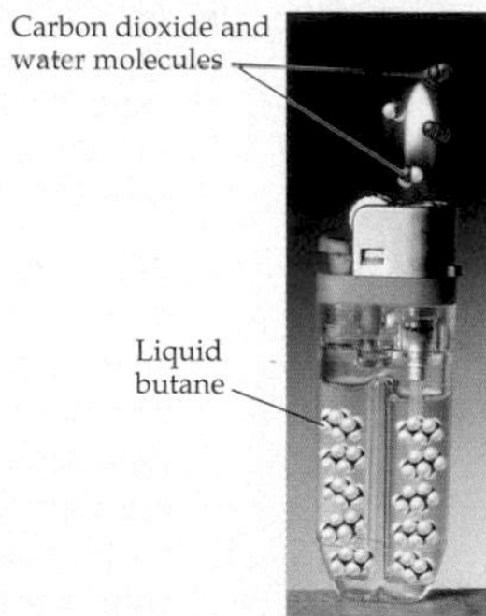

Figure 3.12 If you push the button *and* turn the flint to create a spark, you produce a flame. The butane molecules react with oxygen molecules in air to form new molecules, carbon dioxide and water. This is a chemical change.

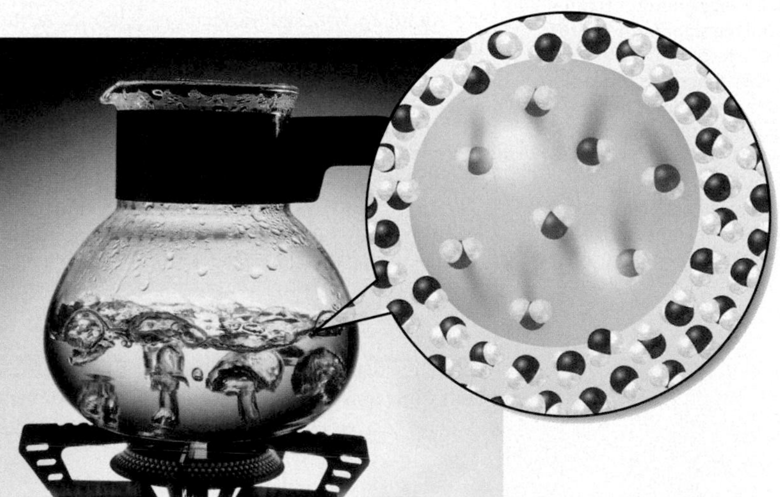

Figure 12.13 During boiling, thermal energy is enough to cause water molecules in the interior of the liquid to become gaseous, forming bubbles containing gaseous water molecules.

Generating Interest in Chemistry

Interest in the field of chemistry and in the topic under study is generated using two recurring features: the chapter openers and interest boxes.

Chapter Openers

The first feature in the opening section of every chapter presents a description of something practical and applied that clearly demonstrates the need for the material covered in that chapter. These openers often involve consumer, environmental, or societal issues. For example, the chapter on stoichiometry opens by making the connection between how much gasoline is burned each year and how much carbon dioxide, the most significant greenhouse gas, is emitted into the atmosphere. The chapter on chemical bonding begins with a description of how bonding theories helped scientists develop AIDS drugs. These narrative chapter introductions are accompanied by a striking piece of art by Quade Paul that portrays visually the chemistry underlying an everyday event. Chapter openings such as these give students a clear reason for why they are learning the current topic. The topics in preparatory chemistry are foundational to so many important things—we simply need to show students examples so that they can make a connection between the principles and skills they are learning and the real world. ▼

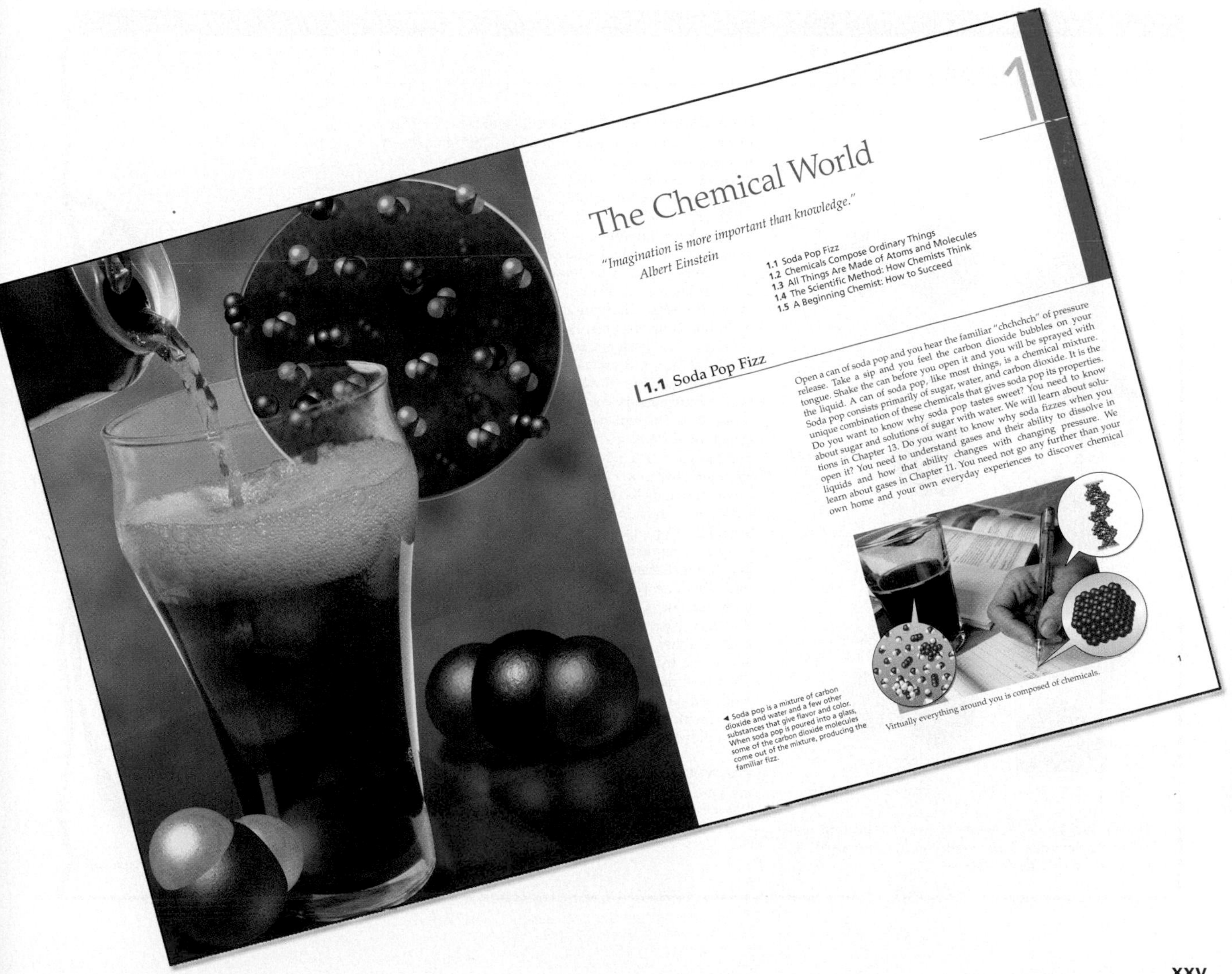

1

The Chemical World

"Imagination is more important than knowledge."
Albert Einstein

1.1 Soda Pop Fizz
1.2 Chemicals Compose Ordinary Things
1.3 All Things Are Made of Atoms and Molecules
1.4 The Scientific Method: How Chemists Think
1.5 A Beginning Chemist: How to Succeed

1.1 Soda Pop Fizz

Open a can of soda pop and you hear the familiar "chchchch" of pressure release. Take a sip and you feel the carbon dioxide bubbles on your tongue. Shake the can before you open it and you will be sprayed with the liquid. A can of soda pop, like most things, is a chemical mixture. Soda pop consists primarily of sugar, water, and carbon dioxide. It is the unique combination of these chemicals that gives soda pop its properties. Do you want to know why soda pop tastes sweet? You need to know about sugar and solutions of sugar with water. We will learn about solutions in Chapter 13. Do you want to know why soda fizzes when you open it? You need to understand gases and their ability to dissolve in liquids and how that ability changes with changing pressure. We learn about gases in Chapter 11. You need not go any further than your own home and your own everyday experiences to discover chemical

◀ Soda pop is a mixture of carbon dioxide and water and a few other substances that give flavor and color. When soda pop is poured into a glass, some of the carbon dioxide molecules come out of the mixture, producing the familiar fizz.

Virtually everything around you is composed of chemicals.

1

Interest Boxes

The second feature is the frequent use of interest boxes. ***Introductory Chemistry*** has four types of interest boxes:

- Everyday Chemistry
- Chemistry in the Media
- Chemistry in the Environment
- Chemistry and Health

The *Everyday Chemistry* boxes describe what is happening with molecules and atoms in common, everyday processes. For example, a number of recent children's toys and clothing will change color based on temperature changes. A bowl changes from green to yellow when filled with warm oatmeal, or a shirt becomes two-toned because of body warmth. The *Everyday Chemistry* box describes, in chemical terms, what molecules do that explain these phenomena. *Chemistry in the Media* boxes describe chemical topics that have gained recent media attention. For example, a discussion on limiting reagent has a *Chemistry in the Media* box discussing the controversy over oxygenated fuels and MTBE. *Chemistry in the Environment* boxes describe environmental issues relevant to the subject under study. *Chemistry and Health* boxes describe chemistry and health related topics. Often interest boxes such as these are contrived, or they show little relevance to the topic under study. The interest boxes in ***Introductory Chemistry*** all contain questions that relate directly to the chapter material. The students benefit from reading the box because they can apply what they have just learned to something that is clearly relevant. ▼

CHEMISTRY AND HEALTH

Hydrogen Bonding in DNA

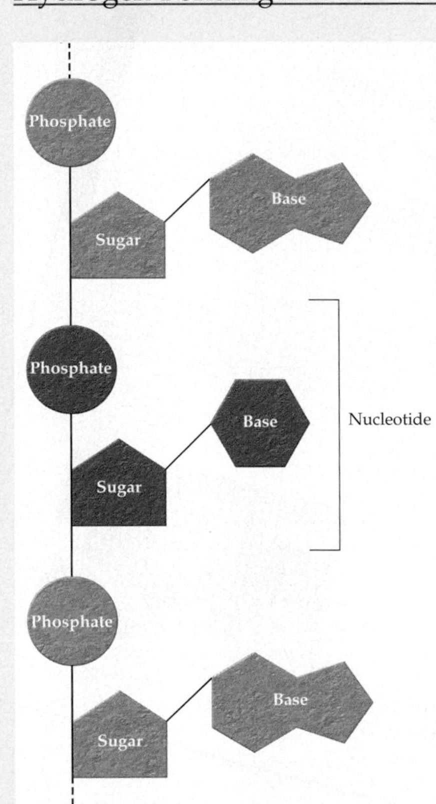

Figure 12.21 DNA is composed of repeating units called nucleotides. Each nucleotide is composed of a sugar, a phosphate, and a base.

DNA is a long chain-like molecule that acts as a blueprint for living organisms. Copies of DNA are passed from parent to offspring, which is why we inherit traits from our parents. A DNA molecule is composed of thousands of repeating units called *nucleotides* (Figure 12.21). Each nucleotide contains one of four different bases: adenine, thymine, cytosine, and guanine (abbreviated *A, T, C,* and *G*). The order of these bases along DNA contains the code that specifies how proteins—the workhorse molecules in living organisms—are made. Proteins determine many human characteristics including how we look, how we fight infections, and even how we behave. Consequently, human DNA is a blueprint for how humans are made.

Each human cell actually contains two complete and complementary copies of DNA, both necessary for replication. The replicating mechanism is related to the structure of DNA, discovered in 1953 by James Watson and Francis Crick. DNA consists of two complementary strands wrapped around each other in the now famous double helix. Each strand is held to the other by hydrogen bonds that occur between the bases on each strand. DNA replicates because each base (A, T, C, and G) has a complementary partner with which it hydrogen bonds (Figure 12.22). Adenine (A) hydrogen bonds with thymine (T) and cytosine (C) hydrogen bonds with guanine (G). The hydrogen bonds are so specific that each base will pair only with its complementary partner. When a cell is going to divide, the DNA unzips across the hydrogen bonds that run along its length. Then new bases, complementary to the bases in each half, add along each of the halves, forming hydrogen bonds with their complement. The result is two identical copies of the original DNA (See Chapter 19).

CAN YOU ANSWER THIS? *Why would dispersion forces not work as a way to hold the two halves of DNA together? Why would covalent bonds not work?*

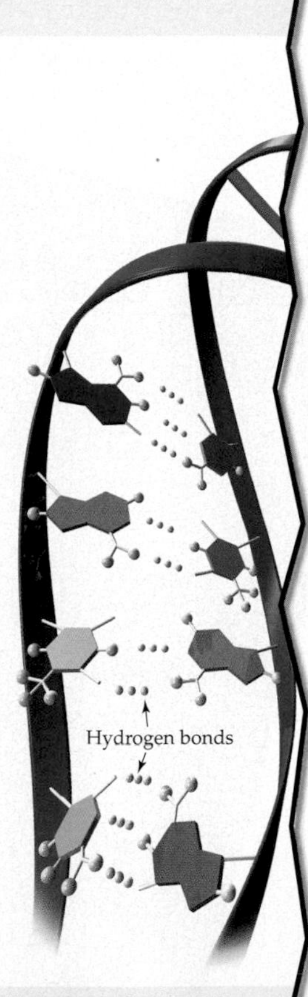

Figure 12.22 The two halves of th

End-of-Chapter Review and Assessment

Chapter in Review

Each chapter ends with a review consisting of two sections. The first section reviews chemical principles and the second one reviews chemical skills. Each section is itself divided into two columns. In the chemical principles review section, one column summarizes the principle, and the other column tells why it is important. ▼

CHAPTER IN REVIEW

Chemical Principles	Relevance
Uncertainty: Measured quantities are reported so that the number of digits reflects the certainty in the measurement. Write measured quantities so that every digit is certain except the last, which is estimated.	**Uncertainty:** Measurement is a hallmark of science and the precision of a measurement must be communicated with the measurement so that others know how reliable the measurement is. When you write or manipulate measured quantities you must show and retain the precision with which the measurement was made.

In the chemical skills section, one column describes the skill, while the other column shows a worked example. The last Chapter in Review section has a list of key terms whose definitions can be found in the chapter. ▼

Chemical Skills / Examples

Scientific Notation (Section 2.2)

To express a number in scientific notation:

- Move the decimal point to obtain a number between 1 and 10.
- Write the decimal part multiplied by 10 raised to the number of places you moved the decimal point.
- The exponent is positive if you moved the decimal point to the left and negative if you moved the decimal point to the right.

EXAMPLE 2.18 Scientific Notation

Express the number 45,000,000 in scientific notation.

45,000,000
7 6 5 4 3 2 1

4.5×10^7

Reporting Measured Quantities to the Right Number of Digits (Section 2.3)

Report measured quantities so that every digit is certain, except the last, which is estimated.

EXAMPLE 2.19 Reporting Measured Quantities to the Right Number of Digits

Record the volume of liquid in the graduated cylinder to the correct number of digits. Laboratory glassware should be read from the bottom of the meniscus (See figure).

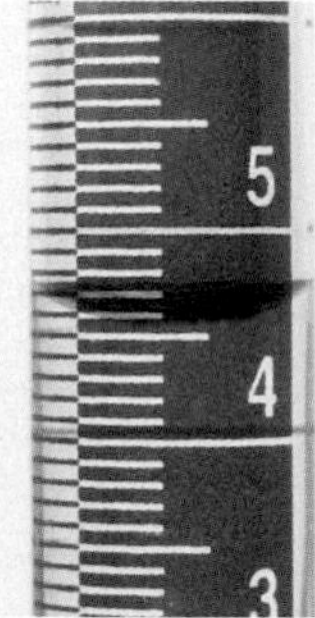

Since the graduated cylinder has markings every 0.1 mL, the measurement should be recorded to the nearest 0.01 mL. In this case, that is 4.57 mL.

Student Exercises

All chapters contain exercises divided into four types: Questions, Problems, Cumulative Problems, and Highlight Problems. The *Questions* are qualitative and require the student to summarize important chapter concepts. The *Problems* section is the longest, containing quantitative problems arranged in pairs and divided with subheadings into the major categories of the chapter. ▼

Problems

(Note: The exercises in this section of Problems are paired and the answers in the odd-numbered exercises appear in Appendix 3.)

Scientific Notation

27. Express each of the following numbers in scientific notation.
a) 32,667,000 (population of California)
b) 1,193,000 (population of Hawaii)
c) 18,175,000 (population of New York)
d) 481,000 (population of Wyoming)

28. Express each of the following numbers in scientific notation.
a) 5,926,467,000 (population of the world)
b) 1,236,915,000 (population of China)
c) 11,051,000 (population of Cuba)
d) 3,619,000 (population of Ireland)

Cumulative Problems, also arranged in pairs, require students to synthesize several of the skills they have learned in the chapter and in previous chapters to solve the problem. ▼

Cumulative Problems

89. A thief uses a bag of sand to replace a gold vase that sits on a weight-sensitive, alarmed pedestal. The bag of sand and the vase are exactly the same volume, 1.75 L.
a) Calculate the mass of each object. (density of gold = $19.3\ g/cm^3$, density of sand = $3.00\ g/cm^3$)
b) Did the thief set off the alarm? Explain.

90. One of the particles that compose atoms is called the proton. The proton has a radius of approximately 1.0×10^{-13} cm and a mass of 1.7×10^{-24} g. Determine the density of a proton. For a sphere $V = 4/3\pi r^3$;

93. The density of aluminum is $2.7\ g/cm^3$. What is its density in pounds per cubic inch?

94. The density of platinum is $21.4\ g/cm^3$. What is its density in pounds per cubic inch?

95. A typical backyard swimming pool holds 150 yd^3 of water. What is the mass in pounds of the water?

96. An iceberg has a volume of 8975 ft^3. What is the mass of the ice (in kilograms) composing the iceberg?

97. Honda has recently produced a hybrid electric car called the Honda Insight. The Insight has both a gasoline-powered engine and an electric motor and

All paired *Problems* and *Cumulative Problems* have one problem answered in the back of the book and another similar problem without an answer. All answers are available in the instructor's full Solutions Manual. *Highlight Problems* are within a relevant or well-known context that will often include photographs or art. ▼

Highlight Problems

99. In 1999, NASA lost a $94-million orbiter because one group of engineers used metric units in their calculations while another group used English units. Consequently, the orbiter pushed too far into the Martian atmosphere and burned up. Suppose that the orbiter was to have established orbit at 155 km and that one group of engineers specified this distance as 1.55×10^5 m. Suppose further that a second group of engineers programmed the orbiter to go to 1.55×10^5 ft. What was the difference (in kilometers) between the two altitudes? How low did the probe go?

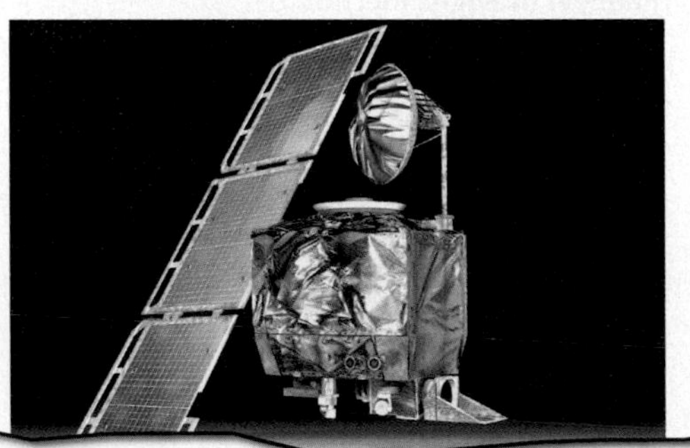

For the Instructor

Instructor's Resource & Full Solutions Manual (0-13-100202-3) by Mark Ott, Jackson Community College and Matthew Johll, St. Norbert College, features lecture outlines with presentation suggestions, teaching tips, suggested in-class demonstrations, and topics for classroom discussion.

Test Item File (0-13-100213-9) by Matthew Johll, St. Norbert College, is a test bank of over 1200 questions.

TestGen-EQ (0-13-100206-6) The computerized version of the Test Item File is available on a dual-platform CD-ROM. The software available with this database allows you to create and tailor exams to your specific needs.

Transparency Pack (0-13-100203-1) This set contains 150 full-color transparencies chosen from the text. This select group of transparencies put principles into visual perspective and save you time while you are preparing your lectures.

Instructor's Resource CD-ROM (0-13-100205-8) An Instructor CD-ROM that contains almost all of the art from the text. Using the included MediaPortfolio software, instructors can browse for figures and other media elements by thumbnail and description as well as search by keyword or title. This CD also contains two pre-built PowerPoint™ Presentations for each chapter; one contains the illustrations, one per slide, for easy incorporation into your lecture. The other contains a complete lecture outline including selections from the available media assets. Also included are PDF files of each image as well as the Word™ files for the Instructor's Resource Manual.

For the Student

Study Guide (0-13-141192-6) This book assists students through the text material with chapter overviews and practice problems applied to each major concept in the text, followed by two or three self-tests with answers at the end of each chapter.

Selected Solutions Manual (0-13-100201-5) by Matthew Johll, St. Norbert College, provides solutions only to those problems that have a short answer in the text's Selected Answer Appendix (problems numbered in red in the text).

Math Review Toolkit (0-13-100202-3) by Gary L. Long, Virginia Polytechnic University. This free book reinforces the skills necessary to succeed in chemistry. It is keyed specifically to chapters in ***Introductory Chemistry***, and includes additional mathematics review, problem-solving tools and examples.

Companion Website http://chem.prenhall.com/trointro
Built to complement ***Introductory Chemistry*** as part of an integrated course package, the easy to use Companion Website features the following modules for each chapter:

- a gallery of animated *student tutorials,*
- a collection of interactive *live examples,*
- a selection of *3D molecular models,*
- multiple choice and true/false *practice quizzes,*
- a *math tutorial,*
- a library of chemistry-related *web links,* and
- a useful *reference center* consisting of an interactive periodic table and a conversion table.

Student Accelerator CD-ROM (0-13-100204-X)
The accelerator CD-ROM contains all of the Student Tutorials and Live Examples from the Companion Website, so that the student can view these files via an Internet connection or the CD-ROM. This allows for maximum flexibility for the student.

Course Management Options Prentice Hall provides support for the course management systems that are most popular at institutions today, including WebCT®, BlackBoard®, and Pearson Education's own CourseCompass® (powered by BlackBoard®). Course management systems allow complete course administration, including roster and gradebook management, distribution of course materials, setup and maintenance of bulletin boards and announcements, and other tasks.

Acknowledgments

This book has been a group effort. I am grateful to my colleagues, Allan Nishimura and David Marten, who have supported me in my department while I worked on this book. I am also grateful to my Provost, Shirley Mullen, who gives me the freedom to be who I am and encourages me to develop in ways unique to myself. She is an outstanding faculty leader and an inspiration to me. This book bears many hours of labor from my editor, Kent Porter Hamann, whose wisdom and experience have been invaluable. Thanks also to my developmental editor, Don Gecewicz; to my copy editor, Jennifer Murtoff; to my media editors, Paul Draper and Michael Richards; to my ancillaries editor, Kristen Kaiser; to my senior marketing manager, Steve Sartori; to the chemistry team assistant, Jackie Howard; to my production editor, Marty Sopher; and to the rest of the Prentice Hall team—they are a first-class operation. This book would only be a shadow of itself without their talents. Thanks to my students Katie Pointer, Daniel Arnold, and Andy Ribbens who put many hours into reading my manuscript, working problems, and giving me feedback from a student's perspective. Thanks also to Quade Paul who made my ideas come alive in his art.

I am grateful to those who have supported me personally while writing this book. On the top of that list is my wife, Ann. Her patience and love for me are beyond description. Thanks also to my children, Michael, Ali, and Kyle—they are my reasons for reason. I come from a large Cuban family, whose closeness and support most people would envy. Thanks to my parents, Nivaldo and Sara; my siblings, Sarita, Mary, and Jorge; my siblings-in-law Jeff, Nachy, Karen, and John; my nephews and nieces, Germain, Danny, Lisette, Sara, and Kenny. These are the people with whom I celebrate life.

Lastly, I am grateful to the many reviewers, listed below, whose ideas are scattered throughout this book. They have corrected me, inspired me, and sharpened my thinking on how to best teach this subject we call chemistry. I am particularly grateful to Kim Summerhays, Michael McCallum, Rill Ann Reuter, Roy Kennedy, and Connie Roberts. This core group of reviewers shaped many of the ideas and features in these pages, and I am incredibly grateful for their commitment to this project.

Reviewers

Lori Allen
University of Wisconsin—Parkside

Laura Andersson
Big Bend Community College

Danny R. Bedgood
Arizona State University

Christine V. Bilicki
Pasadena City College

Warren Bosch
Elgin Community College

Bryan E. Breyfogle
Southwest Missouri State University

Carl J. Carrano
Southwest Texas State University

Donald C. Davis
College of Lake County

Donna G. Friedman
St. Louis Community College at Florissant Valley

Leslie Wo-Mei Fung
Loyola University of Chicago

Dwayne Gergens
San Diego Mesa College

George Goth
Skyline College

Jan Gryko
Jacksonville State University

Roy Kennedy
Massachusetts Bay Community College

C. Michael McCallum
University of the Pacific

Kathy Mitchell
St. Petersburg Junior College

Bill Nickels
Schoolcraft College

Bob Perkins
Kwantlen University College

Mark Porter
Texas Tech University

Caryn Prudenté
University of Southern Maine

Connie M. Roberts
Henderson State University

Rill Ann Reuter
Winona State University

Jeffery A. Schneider
SUNY–Oswego

Kim D. Summerhays
University of San Francisco

Ronald H. Takata
Honolulu Community College

Calvin D. Tormanen
Central Michigan University

Eric L. Trump
Emporia State University

The Chemical World

1

"Imagination is more important than knowledge."
Albert Einstein

1.1 Soda Pop Fizz

Open a can of soda pop and you hear the familiar "chchchch" of pressure release. Take a sip and you feel the carbon dioxide bubbles on your tongue. Shake the can before you open it and you will be sprayed with the liquid. A can of soda pop, like most things, is a chemical mixture. Soda pop consists primarily of sugar, water, and carbon dioxide. It is the unique combination of these chemicals that gives soda pop its properties. Do you want to know why soda pop tastes sweet? You need to know about sugar and solutions of sugar with water. We will learn about solutions in Chapter 13. Do you want to know why soda fizzes when you open it? You need to understand gases and their ability to dissolve in liquids and how that ability changes with changing pressure. We learn about gases in Chapter 11. You need not go any further than your own home and your own everyday experiences to discover chemical

◀ Soda pop is a mixture of carbon dioxide and water and a few other substances that give flavor and color. When soda pop is poured into a glass, some of the carbon dioxide molecules come out of the mixture, producing the familiar fizz.

Virtually everything around you is composed of chemicals.

questions. Chemicals compose virtually everything: the soda; this book; your pencil; indeed, even your own body.

We will define atoms and molecules more carefully in Chapters 4 and 5—for now, think of them as tiny particles that compose all matter.

Chemical bonds will be defined carefully later; for now, think of them as the attachments that hold atoms together.

Chemists are particularly interested in the connections between the properties of substances and the particles that compose them. For example, why does soda pop fizz? Like all substances, soda pop is composed of tiny particles called atoms. Atoms are so small that a single drop of soda pop contains about one billion trillion of them. In soda pop, as in most substances, these atoms are bound together to form several different types of molecules. The molecules important to fizzing are carbon dioxide and water. Carbon dioxide molecules consist of three atoms—one carbon and two oxygen—held together in a straight line by chemical bonds.

Carbon dioxide molecule

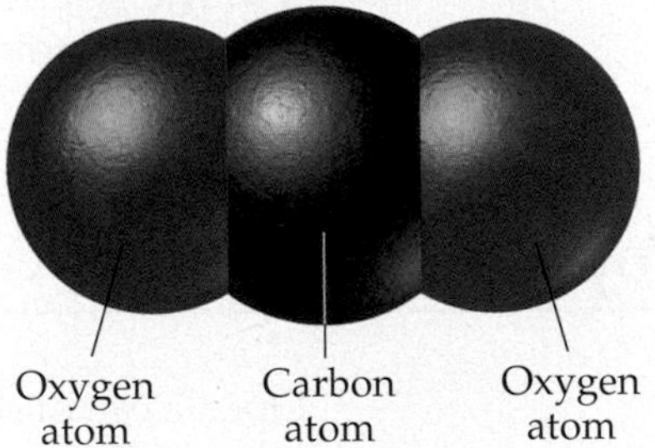

Water molecules also consist of three atoms—one oxygen and two hydrogen—bonded together, but rather than being straight like carbon dioxide, the water molecule is bent.

Water molecule

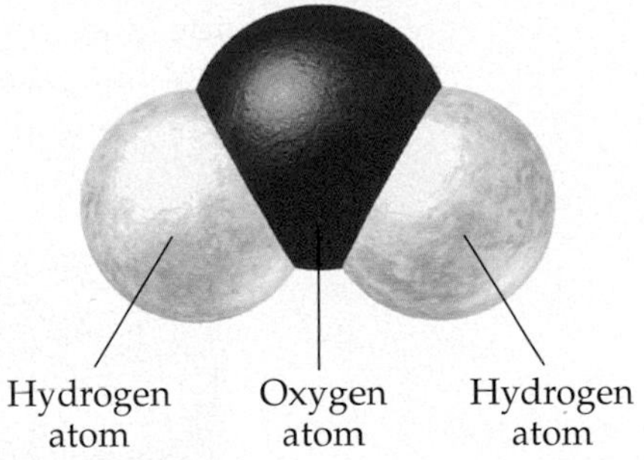

The details of how atoms bond together to form a molecule—straight, bent, or some other shape—as well as the type of atoms in the molecule, determine everything about the substance that the molecule composes. The characteristics of water molecules make water a liquid at room temperature. The characteristics of carbon dioxide molecules make carbon dioxide a gas at room temperature.

The makers of soda pop use pressure to force gaseous carbon dioxide molecules to mix with liquid water molecules. As long as the can of soda is sealed, the carbon dioxide molecules remain mixed with the water molecules, held there by pressure. When the can is opened, the pressure is released and carbon dioxide molecules escape out of the soda mixture (Figure 1.1). As they do, they create bubbles—the familiar fizz of soda pop.

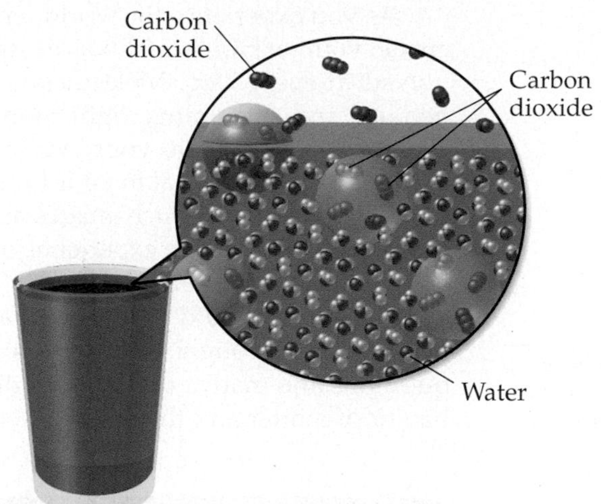

Figure 1.1 Bubbles in soda pop are pockets of carbon dioxide gas molecules escaping out of the liquid water.

1.2 Chemicals Compose Ordinary Things

Is soda pop composed of chemicals? In the broad definition of chemicals, yes. In fact, there is nothing you can hold or touch that is *not* made of chemicals. Unfortunately, when most people think of chemicals, they envision a can of paint thinner in their garage marked with a skull and crossbones, or they see bottles of hazardous acids on a dusty shelf. But chemicals compose more than just these things—they compose ordinary things, too. Chemicals compose the air we breathe and the water we drink. They compose toothpaste, Tylenol, and toilet paper. Chemicals make up virtually everything we come into contact with. This broader notion of chemicals is what chemists use. Chemistry explains the properties and behavior of chemicals, in the broadest sense, by helping us understand the molecules that compose them.

The public often has a very narrow view of chemicals, thinking of them only as dangerous poisons or pollutants.

As you experience the world around you, molecules are interacting to create your experience. Imagine watching a sunset. Molecules are involved in every step. Molecules in air interact with light from the sun, bending red and orange light more than blue and green light to create the color. Molecules in your eyes then interact with the bent light, rearranging to send a signal to your brain. Molecules in your brain then interpret the signal to produce images and emotions. All of this—mediated by molecules—creates the experience of seeing a sunset.

Chemists are interested in why ordinary things are the way they are. Why is water a liquid? Why is salt a solid? Why does soda fizz? Why is a sunset red? Throughout this book you will learn the answers to these questions and many others. You will learn the connection between the behavior of matter and the behavior of the particles that compose it.

Matter is defined thoroughly in Chapter 3. For now, think of matter as anything that has mass and occupies space.

Chemists are interested in knowing why ordinary things, such as water, are the way they are. When a chemist sees a pitcher of water, she sees beyond the surface to the molecules that lie within.

1.3 All Things Are Made of Atoms and Molecules

Professor Richard Feynman, in a lecture to first-year physics students at the California Institute of Technology, said that the most important idea in all human knowledge is that *all things are made of atoms*. Since atoms are usually bound together to form molecules, however, a chemist might add the concept of *molecules* to Feynman's bold assertion. This simple idea—that all things are made of atoms and molecules—explains much about our world and our experience of it. Atoms and molecules determine how matter behaves—if they were different, matter would be different. Water molecules, for example, determine how water behaves. Sugar molecules determine how sugar behaves, and the molecules that compose humans determine how we behave.

Richard Feynman (1918–1988), Nobel-Prize-winning physicist and popular professor at California Institute of Technology.

There is a direct connection between the world of atoms and molecules and the world you and I experience everyday. Chemists explore this connection. They seek to understand it. A good, simple definition of **chemistry** is the science that tries to understand what matter does by studying what atoms and molecules do.

Chemistry – the science that seeks to understand what matter does by studying what atoms and molecules do.

1.4 The Scientific Method: How Chemists Think

Chemists are scientists and use the **scientific method**—a way of learning that emphasizes observation and experimentation—to understand the world. The scientific method stands in contrast to ancient Greek philosophies that emphasized *reason* as the way to understand the world. The first step in the scientific method (Figure 1.2) is the **observation** or measurement of some aspect of nature. Some observations are simple, requiring nothing more than the naked eye, while others rely on the use of increasingly sensitive instrumentation. An observation must measure or describe something about the physical or natural world. For example, **Antoine Lavoisier** (1743–1794), a French chemist who studied combustion, made careful measurements on the mass of objects before and after burning them in closed containers. He noticed that there was no change in the mass during combustion. Lavoisier made an *observation* about the physical world.

The mass of an object is a measure of the quantity of matter within it.

Painting of French chemist Antoine Lavoisier and his wife, Marie, who helped him in his work by illustrating his experiments, recording results, and translating scientific articles from English.

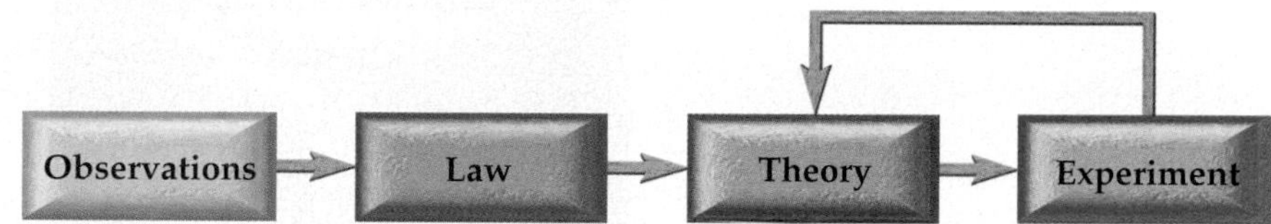

Figure 1.2 The scientific method.

A number of similar observations can be generalized in a **scientific law**, a brief statement that summarizes past observations and predicts future ones. For example, Lavoisier summarized his observations on combustion with the **law of conservation of mass** that states, "In a chemical reaction matter is neither created nor destroyed." This statement summarized Lavoisier's observations and predicted the outcome of similar experiments on any chemical reaction.

A number of related laws and observations may lead to a scientific **theory**, which is sometimes called a **hypothesis** before it is well established. Theories describe the underlying reasons for observations and laws. They are models for the way nature is, and they predict behavior that extends well beyond the observations and laws from which they are formed. A good example of a theory is **John Dalton's** (1766–1844) **atomic theory**. Dalton explained the law of conservation of mass, as well as other laws and observations of the time, by proposing that all matter was composed of small, indestructible particles called atoms. Dalton's theory was a model for the physical world—it went beyond the laws and observations of the time to explain why these laws and observations existed.

Scientific theories are also called *models*.

John Dalton, English chemist who formulated the atomic theory.

Theories are validated by **experiments**, tests that look for other observable predictions of a theory. Notice that the scientific method begins with observation, forms laws and theories based on those observations, and then returns to observation to determine if the laws and theories are valid. If an experiment is inconsistent with a given law or theory, the law or theory must be revised and new experiments must be conducted to test the revisions. Over time, poor theories are eliminated, and good theories—those consistent with experiments—remain. Established theories with strong experimental support are the most powerful pieces of scientific knowledge. People unfamiliar with science sometimes say, "That is just a theory," as if theories were easily discardable. However, well-established theories are as close to truth as we get in science. For example, the idea that all matter is made of atoms is "just a theory," but it is a theory with two hundred years of experimental evidence to support it, including the recent imaging of atoms themselves (Figure 1.3). Established theories should not be taken lightly—they are the pinnacle of scientific understanding.

Figure 1.3 The atomic theory has two hundred years of experimental evidence to support it including recent images, such as this one, of atoms themselves. This image shows the Kanji characters for "atom" written with individual iron atoms on top of a copper surface.

EVERYDAY CHEMISTRY

Combustion and the Scientific Method

Early chemical theories attempted to explain common phenomena such as combustion. Why did things burn? What was happening to a substance when it burned? Could something that was burned be unburned? Early chemists burned different substances and made observations to try to answer these questions. They observed that things would stop burning if placed in a closed container. They found that many metals would burn to form a white powder that they called a *calx* (now we know these white powders are oxides of the metal), and that the metal could be recovered from the calx, or unburned, by combining it with charcoal and heating.

Chemists in the first part of the eighteenth century formed a theory about combustion to explain these observations. In this theory, combustion involved a fundamental substance that they called *phlogiston*. Phlogiston was present in anything that burned and was released during combustion. Flammable objects were flammable because they contained phlogiston. When things were burned in a closed container, they didn't burn for very long because the space within the container became saturated with phlogiston. When things burned in the open, they continued to burn until all of the phlogiston within them was gone. This theory also explained how metals that had burned could be unburned. Charcoal was a phlogiston-rich material—they knew this because it burned so well—and when it was combined with a calx, which was a metal that had been emptied of its phlogiston, it transferred some of its phlogiston back into the calx, converting it back into the unburned form of the metal. The phlogiston theory was consistent with all of the observations and was widely accepted as valid.

Like any theory, the phlogiston theory had to be tested continually by experiment. One set of experiments, conducted in the mid-eighteenth century by Louis-Bernard Guyton de Morveau (1737–1816), consisted of weighing metals before and after burning them. In every case the metals *gained* weight when they were burned. However, the phlogiston theory predicted that they should *lose* weight because phlogiston was supposed to be lost during combustion. The phlogiston theory needed modification.

The first modification was to suppose that phlogiston was a very light substance so that it actually "buoyed up" the materials that contained it. When phlogiston was released, the material actually became heavier. Such a modification seemed to fit the observations but also seemed far fetched. Antoine Lavoisier developed a more

1.5 A Beginning Chemist: How to Succeed

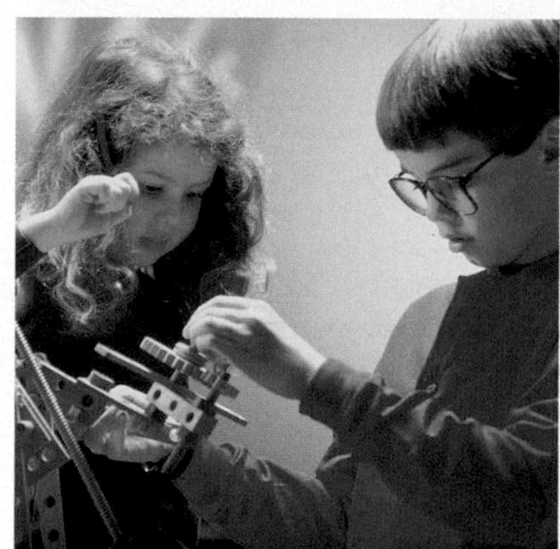

To succeed as a scientist, you must have the curiosity of a child.

You are a beginning chemist. This may be your first chemistry course, but it will probably not be your last. To succeed as a beginning chemist, keep these things in mind. First, chemistry requires curiosity. If you are content knowing that the sky is blue, but don't care *why* it is blue, then you may have to rediscover your curiosity. I say "rediscover" because even children—or better said, *especially* children—have this kind of curiosity. To succeed as a chemist, you must have the curiosity of a child—you must want to know the *why* of things.

Second, chemistry requires calculation. Throughout this book, I will ask you to calculate things and to quantify information. *Quantification* is the assigning of a number to an observation—it is one of the most important tools in science. Quantification allows us to go beyond merely saying that this object is hot and that one is cold or that this one is large and that one is small. It allows us to specify the difference precisely. For example, two samples of water may feel equally hot to our hands, but when we measure their temperatures, we find that one is 40 °C and the other is 44 °C. By assigning numbers to the water temperatures, we can differentiate between small differences. So assigning numbers to observations and manipulating those numbers becomes very important in chemistry.

likely explanation by devising a completely new theory of combustion. According to Lavoisier, when a substance burned, it actually took something *out* of the air, and when it unburned, it released something back into the air. Lavoisier said that burning objects "fixed" the air and that the "fixed" air was released when unburning. In a confirming experiment (Figure 1.4), Lavoisier roasted a mixture of calx and charcoal with the aid of sunlight focused by a giant burning lens—a huge volume of "fixed air" was released in the process. The scientific method had worked. The phlogiston theory was proven wrong, and a new theory of combustion took its place—a theory that, with a few refinements, is still valid today.

CAN YOU ANSWER THIS? *What is the difference between a law and theory? How does the preceding story demonstrate this difference?*

Figure 1.4 The great burning lens belonging to the Academy of Sciences. Lavoisier used a similar lens to show that a mixture of *calx* (metal oxide) and charcoal released a large volume of *fixed air* (oxygen) when heated.

Lastly, chemistry requires commitment. To succeed in this course, you must commit yourself to learning chemistry. Roald Hoffman, winner of the 1981 Nobel Prize for chemistry, said,

> *I like the idea that human beings can do anything they want to. They need to be trained sometimes. They need a teacher to awaken the intelligence within them. But to be a chemist requires no special talent, I'm glad to say. Anyone can do it, with hard work.*

Professor Hoffman is right. The key to success in this course is hard work, and that requires commitment. You must do your work regularly and carefully. If you do, you will succeed, and you will be rewarded by seeing a whole new world—the world of molecules and atoms. This world exists beneath the surface of nearly everything you encounter. I welcome you to this world and consider it a privilege, together with your professor, to be your guide.

CHAPTER IN REVIEW

Chemical Principles

Matter and Molecules: Chemists are interested in all matter, even ordinary matter such as water or air. You need not go to a chemical storeroom to find chemical questions because chemicals are all around you. Chemistry is the science that tries to understand what matter does by understanding what molecules do.

The Scientific Method: Chemists use the scientific method, consisting of observation, laws, theories, and experiments. Observations involve measuring or observing some aspect of nature. Laws summarize the results of a large number of observations, and theories are models that give the underlying causes for observations and laws. Theories are tested and validated by experiment. If an experimental result is inconsistent with a theory, the theory is revised.

Success as a Beginning Chemist: To succeed as a beginning chemist, you must be curious, be willing to do calculations, and be committed to learning the material.

Relevance

Matter and Molecules: Chemists want to understand matter for several reasons. First, chemists are simply curious—they want to know why. Why are some substances reactive and others not? Why are some substances gases, some liquids, and others solids? Chemists are also practical; they want to understand matter so that they can control it and produce substances that are useful to society and to humankind.

The Scientific Method: The scientific method is important because it works as a way to understand the world. Since its inception, knowledge about the natural world and corresponding technologies using that knowledge have grown rapidly. The scientific method has increased living standards throughout the world with advances such as increased food production, rapid transportation, unparalleled access to information, and longer life spans.

Success as a Beginning Chemist: Understanding chemistry will give you a deeper appreciation for the world in which you live, and if you choose science as a career, it will be a foundation upon which you will continue to build.

KEY TERMS

atomic theory [1.4]
chemistry [1.3]
John Dalton [1.4]
experiment [1.4]
hypothesis [1.4]
Antoine Lavoisier [1.4]
law of conservation of mass [1.4]
observation [1.4]
scientific law [1.4]
scientific method [1.4]
theory [1.4]

EXERCISES

Questions

The answers to all exercises numbered in blue appear in Appendix 5.

1. Why does soda fizz?
2. What are chemicals? Give some examples.
3. What do chemists try to do? How do they understand the natural world?
4. What is meant by the statement, "matter does what molecules do"? Give an example.
5. Define chemistry.
6. How is chemistry connected to everyday life? Is chemistry only relevant in the chemistry laboratory?
7. Explain the scientific method.
8. Give an example of how the scientific method works.
9. What is the difference between a law and a theory?
10. What is the difference between a hypothesis and a theory?
11. What is wrong with the statement, "It is just a theory"?
12. What is the law of conservation of mass and who discovered it?
13. What is the atomic theory and who discovered it?
14. What are three things you need to do to succeed in this course?

Problems

(Note: The exercises in this section of Problems are paired and the answers to the odd-numbered exercises appear in Appendix 5.)

15. Classify each of the following as an observation, a law, or a theory.
a) When a metal is burned in a closed container, the mass of the container and its contents does not change.
b) Matter is made of atoms.
c) Matter is conserved in chemical reactions.
d) When wood is burned in a closed container, its mass does not change.

16. Classify each of the following as an observation, a law, or a theory.
a) The closest star to the earth is moving away from the earth at high speed.
b) A body in motion stays in motion unless acted upon by a force.
c) The universe began as a cosmic explosion called the big bang.
d) A stone dropped from an altitude of 450 meters falls to the ground in 9.6 seconds.

17. A chemist in an imaginary universe does an experiment that attempts to correlate the size of an atom with its chemical reactivity. The results are as follows.

Size of Atom	*Chemical Reactivity*
small	low
medium	intermediate
large	high

a) Can you formulate a law from this data?
b) Can you formulate a theory to explain the law?

18. A chemist decomposes several samples of water into hydrogen and oxygen and weighs (or more correctly "obtains the mass of") the hydrogen and the oxygen. The results are as follows.

Sample Number	*Grams of Hydrogen*	*Grams of Oxygen*
1	1	8
2	2	16
3	3	24

a) Can you summarize these observations in a short statement?

Next, the chemist decomposes several samples of carbon dioxide into carbon and oxygen. The results are as follows.

Sample Number	*Grams of Carbon*	*Grams of Oxygen*
1	0.3	0.8
2	0.6	1.6
3	0.9	2.4

b) Can you summarize these observations in a short statement?
c) Can you formulate a law from the observations in **a)** and **b)**?
d) Can you formulate a theory that might explain your law in **c)**?

Weights and Measures
SCALE
Weighs 25 pounds
by ounces

Measurement and Problem Solving

2

"The important thing in science is not so much to obtain new facts as to discover new ways of thinking about them."

Sir William Lawrence Bragg 1890–1971

2.1 Measuring Global Temperatures

As scientists grow increasingly concerned about changes in global climate, global warming has become a household term. Average global temperatures affect everything from agriculture to weather and ocean levels. The media have reported that global temperatures are increasing. These reports are based on the work of scientists who—after analyzing records from thousands of temperature-measuring stations around the world—concluded that average global temperatures have risen by 0.6 °C in the last century.

Notice how the scientists reported their results. What if they had reported a temperature increase of simply 0.6 without any units? The result would be unclear. Units are extremely important in reporting and working with scientific measurements, and they must always be written down and reported. What if the scientists reported additional zeroes with their results, for example 0.60 °C? or 0.600 °C, or what if they reported the number their computer displayed after averaging many measurements, something like 0.58759824 °C?

A unit is a standard, agreed-upon quantity by which other quantities are measured.

Scientists agree to a standard way of reporting measured quantities in which the number of reported digits reflects the precision in the measurement—more digits, more precision, fewer digits, less precision. Numbers are usually written so that the uncertainty is in the last reported digit. For example, by reporting a temperature increase of 0.6 °C, the scientists mean 0.6 ± 0.1 °C ($\pm$ means plus or minus). The temperature rise could be as much as 0.7 °C or as little as 0.5 °C, but it is not 1.0 °C.

◀ Measurement is part of our daily lives. In this illustration, a pumpkin is weighed for pricing. **Question:** Can you read the weight on the scale?

2.2 Scientific Notation: Writing Big and Small Numbers

Lasers such as this one can measure time periods as short as 1×10^{-15} seconds.

Science has constantly pushed the boundaries of the very big and the very small. We can now measure time periods as short as 0.000000000000001 seconds and distances as great as 14,000,000,000 light years. The many zeroes in these numbers are cumbersome to write, so scientists often use **scientific notation** to write them more compactly. In scientific notation, 0.000000000000001 becomes 1×10^{-15}, and 14,000,000,000 becomes 1.4×10^{10}. a number written in scientific notation consists of a **decimal part**, a number that is usually between 1 and 10, and an **exponential part**, 10 raised to an **exponent**, n.

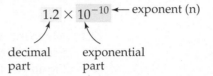

$$1.2 \times 10^{-10}$$

A positive exponent means 1 multiplied by 10 n times.

$$10^0 = 1$$
$$10^1 = 1 \times 10$$
$$10^2 = 1 \times 10 \times 10 = 100$$
$$10^3 = 1 \times 10 \times 10 \times 10 = 1{,}000$$

A negative exponent $(-n)$ means 1 divided by 10 n times.

$$10^{-1} = \frac{1}{10} = 0.1$$
$$10^{-2} = \frac{1}{10 \times 10} = 0.01$$
$$10^{-3} = \frac{1}{10 \times 10 \times 10} = 0.001$$

To convert a number to scientific notation, we move the decimal point to obtain a number between 1 and 10 and then multiply by 10 raised to the appropriate power. For example, to write 5983 in scientific notation, we move the decimal point to the left three places to get 5.983 (a number between 1 and 10) and then multiply by 1000 to make up for moving the decimal point.

$$5983 = 5.983 \times 1000$$

Since 1000 is 10^3, we write:

$$= 5.983 \times 10^3$$

We can do this in one step by counting how many places we move the decimal point to obtain a number between 1 and 10 and then writing the decimal part multiplied by 10 raised to the number of places we moved the decimal point.

$$5983 = 5.983 \times 10^3$$

3 2 1

If the decimal point is moved to the left, as in the previous example, the exponent is positive. If the decimal is moved to the right, the exponent is negative.

$$0.00034 = 3.4 \times 10^{-4}$$

1234

To express a number in scientific notation:

1. Move the decimal point to obtain a number between 1 and 10.
2. Write the result from Step 1 multiplied by 10 raised to the number of places you moved the decimal point.
   - *The exponent is positive if you moved the decimal point to the left.*
   - *The exponent is negative if you moved the decimal point to the right.*

EXAMPLE 2.1 Scientific Notation

The U.S. population in the year 2000 was 274,520,000 people. Express this number in scientific notation.

Solution:

To obtain a number between 1 and 10, move the decimal point to the left 8 decimal places; therefore the exponent is 8. Since we moved the decimal point to the left, the sign of the exponent is positive. So we write:

$$274{,}520{,}000 \text{ people} = 2.7452 \times 10^{8} \text{ people}$$

SKILLBUILDER 2.1 Scientific Notation

The total federal debt in the year 2000 was approximately $5,479,000,000,000. Express this number in scientific notation.

EXAMPLE 2.2 Scientific Notation

The radius of a carbon atom is approximately 0.000000000070 meters. Express this number in scientific notation.

Solution:

To obtain a number between 1 and 10, we move the decimal point to the right 11 decimal places; therefore the exponent is 11. Since we moved the decimal to the right, the sign of the exponent is negative. So we write:

$$0.000000000070 \text{ m} = 7.0 \times 10^{-11} \text{ m}$$

SKILLBUILDER 2.2 Scientific Notation

Express the number 0.000038 in scientific notation.

2.3 Significant Figures: Writing Numbers to Reflect Precision

If you tell someone you have seven pennies, the meaning is clear. Pennies generally come in whole numbers, and seven pennies means seven whole pennies—it is unlikely that you would have 7.4 pennies. On the other hand, if you tell someone that you have 1 gram (g) of gold powder, the meaning is unclear. The actual amount of gold powder depends on how precisely you measured it, which in turn depends on the scale or balance you used to make the measurement. As we just learned, measured quantities are written in a way that reflects the uncertainty in the measurement. If the gold measurement was rough, you might write "1 g of gold". If however, you used a more precise balance, you would write "1.0 g." An even more precise measurement would be reported as "1.00 g."

Figure 2.1 This balance has markings every 1 g, so we estimate to the tenths place. To estimate between markings, mentally divide the space into ten equal spaces and estimate the last digit. This reading is 1.2 g.

Since pennies come in whole numbers, 7 pennies means 7.00000 pennies. It is an exact number and therefore never limits significant figures in calculations.

The amount of gold in 1 g of gold powder depends on how precisely it was measured.

Scientific numbers are reported so that every digit is certain except the last, which is estimated.

For example, suppose a reported number is:

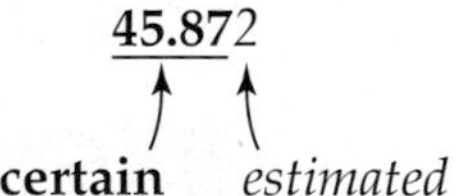

The first four digits are certain; the last digit is estimated.

Suppose that we weigh an object on a balance with marks at every 1 g, and suppose that the pointer is between the 1-g mark and the 2-g mark (Figure 2.1) but much closer to the 1-g mark. We mentally divide the space between the 1- and 2-g marks into ten equal spaces and estimate that the pointer is at about 1.2 g. We then write the measurement as 1.2 g indicating that we are sure of the "1" but have estimated the ".2."

A balance with marks every tenth of a gram requires us to write the result with more digits. For example, suppose that on this more precise balance the pointer is between the 1.2-g mark and the 1.3-g mark (Figure 2.2). We again divide the space between the two marks into ten equal spaces and estimate the third digit. For the figure shown, we report 1.26 g.

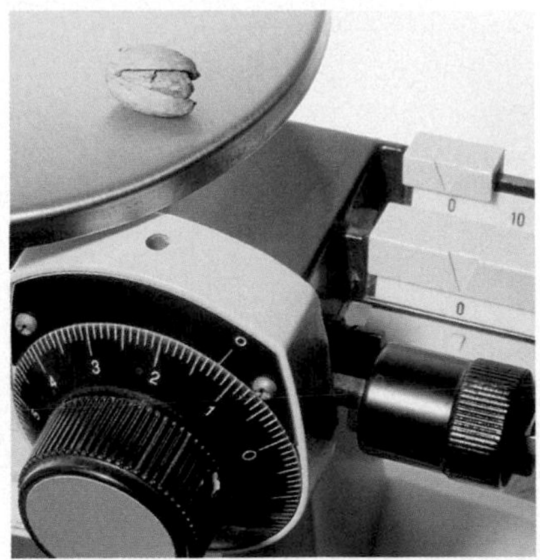

Figure 2.2 Since this scale has markings every 0.1 g, we estimate to the hundredths place. The correct reading is 1.26 g.

Figure 2.3

EXAMPLE 2.3 Reporting the Right Number of Digits

The bathroom scale in Figure 2.3 has markings at every 1 lb. Report the reading to the correct number of digits.

Solution:

Since the pointer is between the 147- and 148-lb markings, we mentally divide the space between the markings into ten equal spaces and estimate the next digit. In this case, the result should be reported as:

147.7 lb

What if you estimated a little differently and wrote 147.6 lb? In general, one unit of difference in the last digit is acceptable because the last digit is estimated and different people might estimate it slightly differently. However, if you wrote 147.2 lb, you would clearly be wrong.

SKILLBUILDER 2.3 Reporting the Right Number of Digits

A thermometer is used to measure the temperature of a backyard hot tub, and the reading is shown in Figure 2.4. Write the temperature reading to the right number of digits.

Figure 2.4

Counting Significant Figures

The non-place-holding digits in a reported measurement are called **significant figures** (or **significant digits**) and, as we have seen, represent the precision of a measured quantity. The greater the number of significant figures, the greater the precision of the measurement. We can determine the number of significant figures in a written number fairly easily; however, if the number contains zeroes, we must distinguish between those zeroes that are significant and those that simply mark the decimal place. In the number 0.002, for example, the leading zeroes simply mark the decimal place; they *do not* add to the precision of the measurement. In the number 0.00200, the trailing zeroes *do* add to the precision of the measurement.

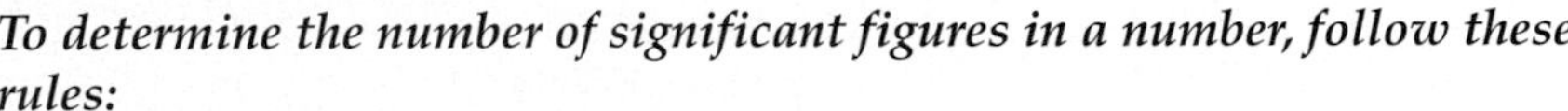

To determine the number of significant figures in a number, follow these rules:

1. All nonzero digits are significant.

   1.05 0.0110

2. Interior zeroes (zeroes between two numbers) are significant.

   4.0208 50.1

3. Trailing zeroes (zeroes after a decimal point) are significant.

   5.10 3.00

4. Leading zeroes (zeroes to the left of the of the first nonzero number) are not significant. They only serve to locate the decimal point.

   Thus, the number 0.0005 has only one significant digit.

5. Zeroes at the end of a number but before a decimal point are ambiguous and should be avoided by using scientific notation.

   For example, does 350 have two or three significant figures? Write the number as 3.5×10^2 to indicate two significant figures or as 3.50×10^2 to indicate three.

Exact Numbers

Exact numbers have an unlimited number of significant figures. Exact numbers originate from measurements where the objects are countable, such as the number of pennies for example: 7 pennies means 7.000000.... pennies. Exact numbers also originate from defined quantities, such as the number of centimeters (cm) in 1 meter(m). One hundred centimeters is defined as 1 m; therefore:

$$100 \text{ cm} = 1 \text{ m means} \quad 100.00000\ldots \text{cm} = 1.0000000\ldots \text{m}$$

EXAMPLE 2.4 Determining the Number of Significant Figures in a Number

How many significant figures are in each of the following numbers?

a) 0.0035
b) 1.080
c) 2371
d) 2.97×10^5
e) 0.000004
f) 100,000

Solution:

a) 0.0035
The 3 and the 5 are significant. The leading zeroes only mark the decimal place and are not significant. Therefore the number has two significant figures.
b) 1.080
The interior zero and the trailing zero are significant as are the 1 and the 8 for a total of 4 significant figures.
c) 2371
All digits are significant for a total of 4 significant figures.
d) 2.97×10^5
All digits in the decimal part are significant for a total of 3 significant figures.
e) 0.000004
The leading zeroes only mark the decimal place—they are not significant. The 4 is the only significant digit for a total of 1 significant digit.
f) 100,000
This number is ambiguous. Write as 1×10^5 to indicate one significant figure or as 1.00000×10^5 to indicate six significant figures.

SKILLBUILDER 2.4 Determining the Number of Significant Figures in a Number

How many significant figures are in each of the following numbers?

a) 58.31
b) 0.00250
c) 2.7×10^3
d) 7.007
e) 0.500
f) 2100

CHEMISTRY IN THE MEDIA

The COBE Satellite and Very Precise Measurements That Illuminate Our Cosmic Past

Since the earliest times, humans have wondered about the origins of our planet. Science has slowly probed this question and has developed theories for how the universe and the earth began. The most accepted theory today for the origin of the universe is the big bang theory. According to the big bang theory, the universe began in a tremendous expansion about 16 billion years ago and has been expanding ever since. A measurable prediction of this theory is the presence of a remnant "background radiation" from the expansion itself. That remnant is characteristic of the current temperature of the universe. When the big bang occurred, the temperature of the universe was very hot and the associated radiation very bright. Today, 16 billion years later, the temperature of the universe is very cold and the background radiation very faint.

In the early 1960's, Robert H. Dicke, P.J.E. Peebles, and their coworkers at Princeton University began to build a device that might measure this background radiation and thus take a direct look into the cosmological past and provide evidence for the big bang theory. At about the same time, quite by accident, Arno Penzias and Robert Wilson of Bell Telephone Laboratories measured excess radio noise on one of their communications satellites. As it turned out, this noise was the background radiation that the Princeton scientists were looking for. The two groups published papers together in 1965 reporting their findings along with the corresponding temperature of the universe, about 3 degrees above absolute zero (3 kelvin (K)).

In 1989, the Cosmic Background Explorer (COBE) satellite was developed by NASA's Goddard Space Flight Center to measure the background radiation more precisely. It was launched in late 1989 and determined that the background radiation corresponded to a universe with a temperature of 2.735 K. (Notice the difference in significant figures from the previous measurement). The satellite went on to measure tiny fluctuations in the background radiation that amount to temperature differences of 1 part in 100,000. These fluctuations, though small, are an important prediction of the big bang theory. Scientists announced that the Cosmic Background Explorer (COBE) satellite had produced the strongest evidence yet for the big bang theory of the creation of the universe. This is the way that science works. Measurement, and precision in measurement, are important to understanding the world—so important that we dedicate most of this chapter just to the concept of measurement.

Note: We will define temperature measurement scales in Chapter 3. For now, know that 3 K is an extremely low temperature, several hundred degrees below zero on the Fahrenheit scale.

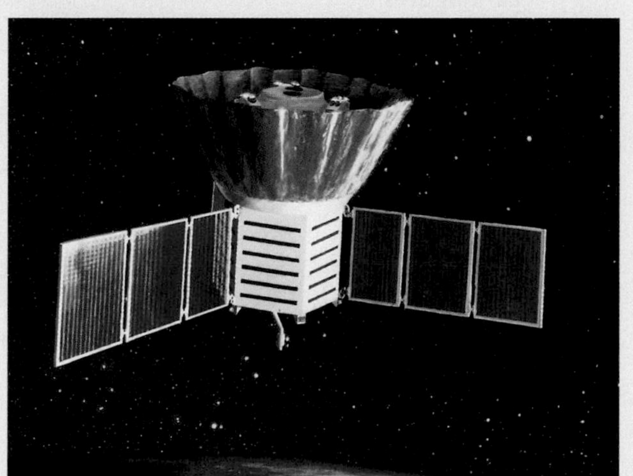

COBE Satellite, launched in 1989 to measure background radiation. Background radiation is a remnant of the big bang that is believed to have formed the universe.

CAN YOU ANSWER THIS? *How many significant figures are there in each of the preceding temperature measurements (3 K, 2.735 K)?*

2.4 Significant Figures in Calculations

When we use measured quantities in calculations, the results of the calculation must reflect the precision of the measured quantities. We should not lose or gain precision during mathematical operations.

Multiplication and Division

In multiplication or division, the result carries the same number of significant figures as the factor with the fewest significant figures.

For example:

5.02	×	89.665	×	0.10	= 45.0118 =	45
(3 sig. figures)		(5 sig. figures)		(2 sig. figures)		(2 sig. figures)

The intermediate result (in blue) is rounded to two significant figures to reflect the least precisely known number (0.10), which has two significant figures.

For division, we follow the same rule.

5.892	÷	6.10	= 0.96590 =	0.966
(4 sig. figures)		(3 sig. figures)		(3 sig. figures)

The intermediate result (in blue) is rounded to three significant figures to reflect the least precisely known number (6.10), which has three significant figures.

Rounding

When rounding to the correct number of significant figures,

round down if the last (or left-most) digit dropped is four or less;
round up if the last (or left-most) digit dropped is five or more.

For example, consider rounding each of the following to 2 significant figures.

2.33 rounds to 2.3
2.37 rounds to 2.4
2.34 rounds to 2.3
2.35 rounds to 2.4

Be certain to use only the *last (or left-most) digit being dropped* to decide in which direction to round—ignore all digits to the right of it. For example, to round 2.349 to two significant figures, only the 4 in the hundredths place (2.349) determines which direction to round—the 9 is irrelevant.

2.349 rounds to 2.3

For calculations involving multiple steps, round only the final answer—do not round off intermediate steps. This prevents small rounding errors from affecting your final answer.

EXAMPLE 2.5 Significant Figures in Multiplication and Division

Perform the following calculations to the correct number of significant figures.

a) $1.01 \times 0.12 \times 53.51 \div 96$
b) $56.55 \times 0.920 \div 34.2585$

Solution:

a) $1.01 \times 0.12 \times 53.51 \div 96 = 0.067556 = 0.068$
We round the intermediate result (in blue) to 2 significant figures to reflect the 2 significant figures in the least precisely known quantities (0.12 and 96).
b) $56.55 \times 0.920 \div 34.2585 = 1.51863 = 1.52$

We round the intermediate result (in blue) to 3 significant figures to reflect the 3 significant figures in the least precisely known quantity (0.920).

SKILLBUILDER 2.5 Significant Figures in Multiplication and Division

Perform the following calculations to the correct number of significant figures.

a) $1.10 \times 0.512 \times 1.301 \times 0.005 \div 3.4$
b) $4.562 \times 3.99870 \div 89.5$

Addition and Subtraction

In addition or subtraction the result carries the same number of decimal places as the quantity carrying the fewest decimal places.

For example:

$$\begin{array}{r} 5.74 \\ 0.823 \\ +2.651 \\ \hline 9.214 \end{array} = 9.21$$

It is sometimes helpful to draw a vertical line directly to the right of the number with the fewest decimal places. The line shows the number of decimal places that should be in the answer.

We round the intermediate answer in blue to two decimal places because the quantity with the fewest decimal places (5.74), has two decimal places.

For subtraction, we follow the same rule. For example:

$$\begin{array}{r} 4.8 \\ -3.965 \\ \hline 0.835 \end{array} = 0.8$$

We round the intermediate answer in blue to one decimal place because the quantity with the fewest decimal places (4.8), has one decimal place. *Remember, for multiplication and division, the quantity with the fewest* significant figures *determines the number of* significant figures *in the answer, but for addition and subtraction, the quantity with the fewest* decimal places *determines the number of* decimal places *in the answer*. In multiplication and division, we focus on significant figures, but in addition and subtraction we focus on decimal places.

EXAMPLE 2.6 **Significant Figures in Addition and Subtraction**

Perform the following calculations to the correct number of significant figures.

a)
$$\begin{array}{r} 0.987 \\ +125.1 \\ -1.22 \\ \hline \end{array}$$

b)
$$\begin{array}{r} 0.765 \\ -3.449 \\ -5.98 \\ \hline \end{array}$$

Solution:

a)
$$\begin{array}{r} 0.987 \\ +125.1 \\ -1.22 \\ \hline 124.867 \end{array} = 124.9$$

We round the intermediate answer in blue to one decimal place to reflect the quantity with the fewest decimal places (125.1). Notice that 125.1 is not the quantity with the fewest significant figures—it has four while the other quantities only have three—but because it has the fewest decimal places, it determines the number of decimal places in the answer.

b)
$$\begin{array}{r} 0.765 \\ -3.449 \\ -5.98 \\ \hline -8.664 \end{array} = -8.66$$

We round the intermediate answer (in parentheses) to two decimal places to reflect that the quantity with the fewest decimal places (5.98).

SKILLBUILDER 2.6 **Significant Figures in Addition and Subtraction**

Perform the following calculations to the correct number of significant figures.

a)
$$\begin{array}{r} 2.18 \\ +\ 5.621 \\ +\ 1.5870 \\ -\ 1.8 \\ \hline \end{array}$$

b)
$$\begin{array}{r} 7.876 \\ -0.56 \\ +123.792 \\ \hline \end{array}$$

Calculations Involving Both Multiplication/Division and Addition/Subtraction

In calculations involving both multiplication/division and addition/subtraction, do the steps in parentheses first; determine the number of significant figures in the intermediate answer; then do the remaining steps.

For example:

$$3.489 \times (5.67 - 2.3)$$

We do the subtraction step first.

$$5.67 - 2.3 = 3.37$$

We use the subtraction rule to determine that the intermediate answer (3.37) has only one significant decimal place. To avoid small errors, it is best not to round at this point; instead, underline the least significant figure as a reminder.

$$= 3.489 \times 3.\underline{3}7$$

We then do the multiplication step.

$$3.489 \times 3.\underline{3}7 = 11.758$$
$$= 12$$

We use the multiplication rule to determine that the intermediate answer (11.758) rounds to two significant figures (12) limited by the two significant figures in $3.\underline{3}7$.

EXAMPLE 2.7 Significant Figures in Calculations Involving Both Multiplication/Division and Addition/Subtraction

a) $6.78 \times 5.903 \times (5.489 - 5.01)$
b) $19.667 - (5.4 \times 0.916)$

Solution:

a) $6.78 \times 5.903 \times (5.489 - 5.01)$

We do the step in parentheses first.

$$= 6.78 \times 5.903 \times (0.4790)$$

We use the subtraction rule to mark 0.4790 to two decimal places since 5.01, the number with the least number of decimal places, has two.

$$= 6.78 \times 5.903 \times 0.4\underline{7}90$$

We then perform the multiplication and round the answer to two significant figures since the number with the least number of significant figures ($0.4\underline{7}90$) has two.

$$6.78 \times 5.903 \times 0.4\underline{7}90 = 19.1707$$
$$= 19$$

b) $19.667 - (5.4 \times 0.916)$

We do the step in parentheses first.

$$= 19.667 - (4.9464)$$

The number with the least number of significant figures (5.4) has two, so we mark the answer to two significant figures.

$$= 19.667 - 4.\underline{9}464$$

We then perform the subtraction and round the answer to one decimal place since the number with the least number of decimal places has one.

$$19.667 - 4.\underline{9}464 = 14.7206$$
$$= 14.7$$

SKILLBUILDER 2.7 **Significant Figures in Calculations Involving Both Multiplication/Division and Addition/Subtraction**

a) $3.897 \times (782.3 - 451.88)$

b) $(4.58 \div 1.239) - 0.578$

2.5 The Basic Units of Measurement

By themselves, numbers have little meaning. Read this sentence: When my son was 7 he walked 3, and when he was 4 he threw his baseball 8 and said his school was 5 away. The sentence is confusing because we don't know what the numbers mean—the **units** are missing. The meaning becomes clear, however, when we add the missing units to the numbers: When my son was 7 *months old* he walked 3 *steps*, and when he was 4 *years old* he threw his baseball 8 *feet* and said his school was 5 *minutes* away. Units make all the difference. In chemistry, units are critical. Never write a number by itself; always use its associated units—otherwise your work will be as confusing as the preceding sentence.

The two most common unit systems are the **English system**, used in the United States, and the **metric system**, used in most of the rest of the world. The English system uses units such as inches, yards, and pounds, while the metric system uses centimeters, meters, and kilograms. The most convenient system for science measurements is based on the metric system and is called the **International System** of units or **SI units**. SI units are a set of standard units agreed upon by scientists throughout the world.

The abbreviation *SI* comes from the French, *le Systéme International.*

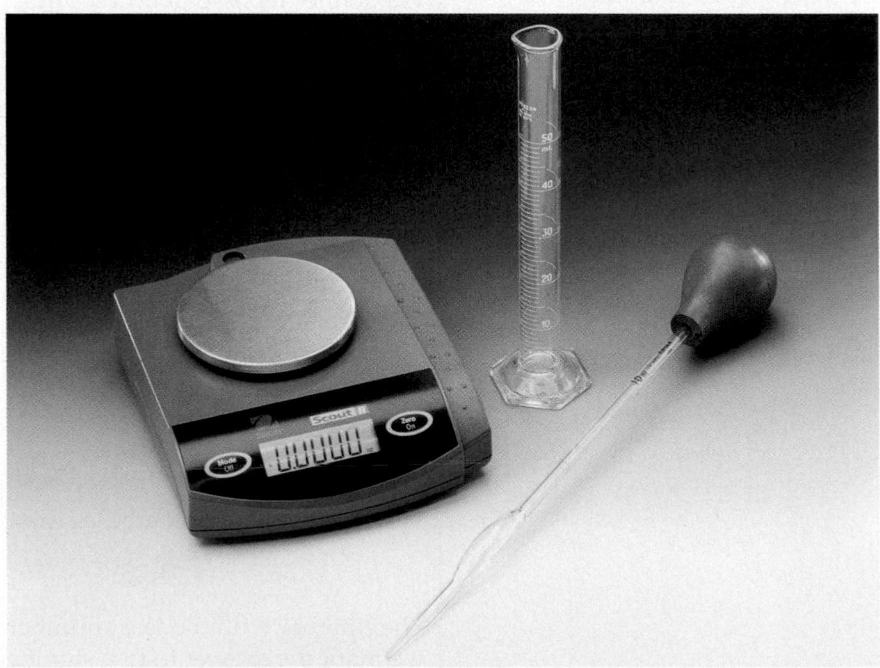

Science uses instruments to make measurements. Every instrument is calibrated in a particular unit without which the measurements would be meaningless.

TABLE 2.1

Important SI Standard Units

Quantity	*Unit*	*Symbol*
length	meter	m
mass	kilogram	kg
time	second	s
temperature*	kelvin	K

*Temperature units are discussed in Chapter 3.

The Standard Units

The standard units in the SI system are shown in Table 2.1. They include the **meter (m)** as the standard unit of length; the **kilogram (kg)** as the standard unit of mass; and the **second (s)** as the standard unit of time. Each of these standard units is precisely defined. The meter is defined as the distance light travels in a certain period of time, 1/299,792,458 s (Figure 2.5). The kilogram is defined as the mass of a block of metal kept at the International Bureau of Weights and Measures at Sèvres, France (Figure 2.6), and the second is defined using an atomic standard (Figure 2.7).

The velocity of light is 3.0×10^8 m/s.

More precisely, the second is defined as the duration of 9,192,631,770 periods of the radiation emitted from a certain transition in a cesium-133 atom.

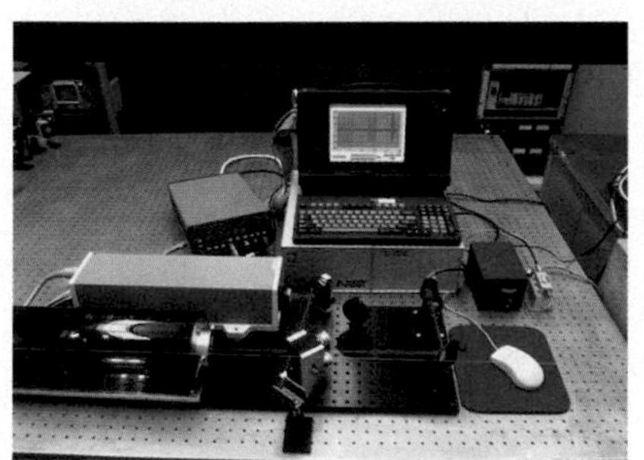

Figure 2.5 The definition of a meter, established by international agreement in 1983, is the distance that light travels in vacuum in 1/299792458 s. **Question:** What reasons can you think of for why such a precise standard is necessary?

Figure 2.6 A duplicate of the international standard kilogram, called kilogram 20, is kept at the National Institute of Standards and Technology near Washington D.C.

Figure 2.7 The standard of time, the second, is defined using a cesium clock.

Most people are familiar with the SI standard unit of time, the second. However, if you live in the United States, you may be less familiar with the meter and the kilogram. The meter is slightly longer than a yard (a yard is 36 inches while a meter is 39.37 inches). Thus, a 100-yd football field measures only 91.4 m. The kilogram is a measure of mass, which is different from weight. The **mass** of an object is a measure of the quantity of matter within it, while the weight of an object is a measure of the gravitational pull on that matter. Consequently, weight depends on gravity while mass does not. If you weigh yourself on Jupiter, for example, the greater gravity pulls you toward the scale more than earth's gravity, resulting in a greater weight. A 150-lb person on earth weighs 351 lb on Jupiter. However, the person's mass, the quantity of matter in his body, remains the same. A kilogram of mass is the equivalent of 2.205 lb of weight on Earth, so if we express mass in kilograms, a 150-lb person on Earth has a mass of approximately 68 kg. A second common unit of mass is the gram (g) defined as:

$$1000 \text{ g} = 10^3 \text{ g} = 1 \text{ kg}$$

A nickel (5¢) has a mass of about 5 g.

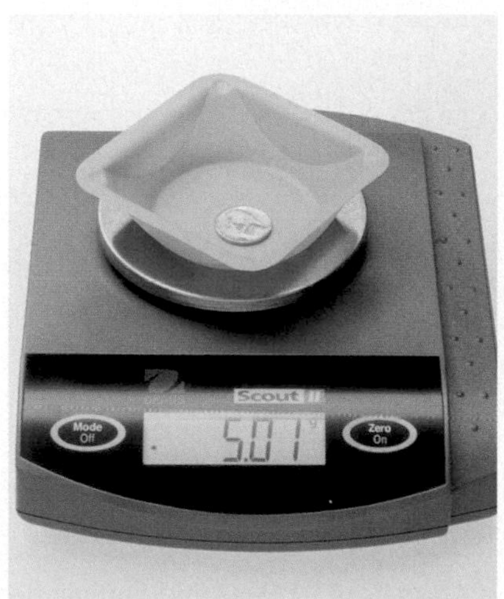

A nickel has a mass of about 5 g.

Prefix Multipliers

The international system of units uses **prefix multipliers** (Table 2.2) with the standard units. These multipliers change the value of the unit by powers of 10. For example, the kilometer (km) has the prefix *kilo-* meaning 1000 or 10^3. Therefore:

$$1 \text{ km} = 1000 \text{ m} = 10^3 \text{ m}$$

Similarly, the millisecond (ms) has the prefix *milli-* meaning 0.001 or 10^{-3}.

$$1 \text{ (ms)} = 0.001 \text{ s} = 10^{-3} \text{ s}$$

TABLE 2.2

SI Prefix Multipliers

Prefix	*Symbol*	*Multiplier*	
tera-	T	1,000,000,000,000	(10^{12})
giga-	G	1,000,000,000	(10^{9})
mega-	M	1,000,000	(10^{6})
kilo-	k	1,000	(10^{3})
deci-	d	0.1	(10^{-1})
centi-	c	0.01	(10^{-2})
milli-	m	0.001	(10^{-3})
micro-	μ	0.000001	(10^{-6})
nano-	n	0.000000001	(10^{-9})
pico-	p	0.000000000001	(10^{-12})
femto-	f	0.000000000000001	(10^{-15})

The diameter of a jelly donut is about 8 cm. **Question:** Why would you *not* use meters to make this measurement?

TABLE 2.3

Some Common Units and Their Equivalents
Length
1 kilometer (km) = 0.6214 mile (mi)
1 meter (m) = 39.37 inches (in.)
= 1.094 yards (yd)
1 foot (ft) = 30.48 centimeters (cm)
1 inch (in.) = 2.54 centimeters (cm) (exact)
Mass
1 kilogram (kg) = 2.205 pounds (lb)
1 pound (lb) = 453.59 grams (g)
1 ounce (oz) = 28.35 grams (g)
Volume
1 liter (L) = 1000 milliliters (mL)
= 1000 cubic centimeters (cm^3)
1 liter (L) = 1.057 quarts (qt)
1 U.S. gallon (gal) = 3.785 liters (L)

The prefix multipliers allow us to express a wide range of measurements in units that are similar in size to the quantity we are measuring. Choose the prefix multiplier that is most convenient for a particular measurement. For example, to measure the size of a jelly doughnut, use centimeters because jelly doughnuts have diameters of about 8 cm. A centimeter is a common metric unit and is about equivalent to the width of your pinky finger (2.54 cm = 1 in). You could have also chosen the decimeter to express the diameter of a jelly doughnut; then the doughnut would measure 0.8 dm, but you should not choose the kilometer since, in that unit, the diameter is 0.00008 km. Pick a unit similar in size to (or smaller than) the quantity you are measuring. Consider expressing the length of a chemical bond, about 1.2×10^{-10} m. Which prefix multiplier should you use? The most convenient one is probably the picometer (pico = 10^{-12}). Chemical bonds measure about 120 pm.

Derived Units

A derived unit is formed from other units. A common derived unit is for **volume**, which is a measure of space. Any unit of length, when cubed (raised to the third power), becomes a unit of volume. Thus, cubic meters (m^3), cubic centimeters (cm^3), and cubic millimeters (mm^3) are all units of volume. In these units, a three-bedroom house has a volume of about 630 m^3, a can of soda pop has a volume of about 350 cm^3, and a rice grain has a volume of about 3 mm^3. We will also use the **liter (L)** and milliliter (mL) to express volume. A gallon of gasoline is equal to 3.785 L. A milliliter is equivalent to 1 cm^3. Table 2.3 lists some common units and their equivalents.

2.6 Converting From One Unit to Another

Using units as a guide to solving problems is often called *dimensional analysis.*

Units are critical in calculations. Knowing how to work with and manipulate units in calculations is one of the most important skills you will learn in this course. In calculations, units help to determine correctness. Units should always be included in calculations, and we will think of many calculations as converting from one unit to another. Units are multiplied, divided, and canceled like any other algebraic quantity. Remember:

1. **Always write every number with its associated unit. Never ignore units; they are critical.**
2. **Always include units in your calculations, dividing them and multiplying them as if they were algebraic quantities. Do not let units magically appear or disappear in calculations. Units must flow logically from beginning to end.**

Consider converting 17.6 inches to centimeters. We know from Table 2.3 that 1 in = 2.54 cm. How many centimeters are in 17.6 inches? We perform the conversion as follows:

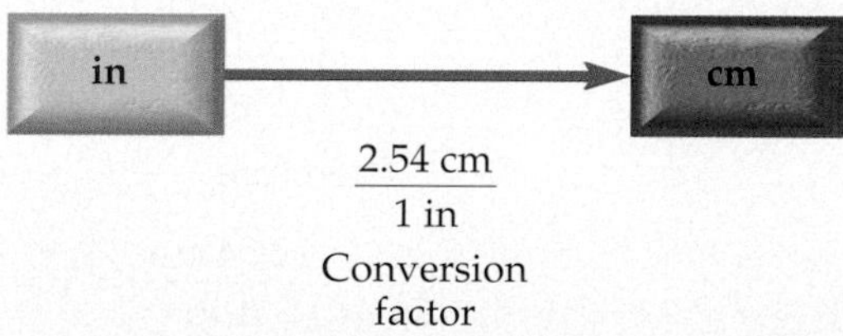

$$17.6\,\cancel{\text{in}} \times \frac{2.54\ \text{cm}}{1\,\cancel{\text{in}}} = 44.7\ \text{cm}$$

The unit *in* cancels and we are left with *cm* as our final unit. The quantity $\frac{2.54\ \text{cm}}{1\ \text{in}}$ is a **conversion factor** between in and cm—it is a fraction with *cm* on top and *in* on bottom.

For most conversion problems, we are given information in some units and asked to convert it to another unit. These calculations take the form of:

information given × conversion factor(s) = information sought

Conversion factors are constructed from any two quantities known to be equivalent. In our example, 2.54 cm = 1 in, so we construct the conversion factor by dividing both sides of the equality by 1 inch and cancelling the units.

$$2.54\ \text{cm} = 1\ \text{in}$$

$$\frac{2.54\ \text{cm}}{1\ \text{in}} = \frac{1\,\cancel{\text{in}}}{1\,\cancel{\text{in}}}$$

$$\frac{2.54\ \text{cm}}{1\ \text{in}} = 1$$

The quantity $\frac{2.54\ \text{cm}}{1\ \text{in}}$ is equal to 1 and can be used to convert between in and cm.

Suppose we want to perform the conversion the other way, from cm to in. If we try to use the same conversion factor, the units do not cancel correctly.

$$44.7\ \text{cm} \times \frac{2.54\ \text{cm}}{1\ \text{in}} = \frac{114\ \text{cm}^2}{\text{in}}$$

The units in the answer, as well as the value of the answer, are incorrect. The unit cm^2/in is not correct, and, based on our knowledge of cm and in, we know that 44.7 cm cannot be equivalent to 114 in. In performing every problem, always look at the final units to see if they are correct and also look at the magnitude of the answer to see if it makes sense. In this case, our mistake was in how we used the conversion factor. It must be inverted.

$$44.7\,\cancel{\text{cm}} \times \frac{1\ \text{in}}{2.54\,\cancel{\text{cm}}} = 17.6\ \text{in}$$

Conversion factors can be inverted because they are equal to 1 and the inverse of 1 is 1.

$$\frac{1}{1} = 1$$

Therefore,

$$\frac{2.54\ \text{cm}}{1\ \text{in}} = 1 = \frac{1\ \text{in}}{2.54\ \text{cm}}$$

In this book, we diagram conversions such as the preceding ones using a **solution map**. A solution map is a visual outline that shows the strategic route required to solve a problem. For unit conversion, the solution map focuses on units and how to convert from one unit to another. The solution map for converting from inches to centimeters is:

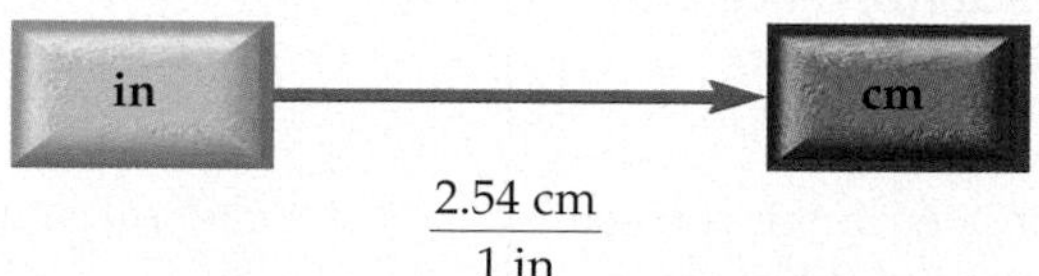

$$\frac{2.54\text{ cm}}{1\text{ in}}$$

The solution map for converting from centimeters to inches is:

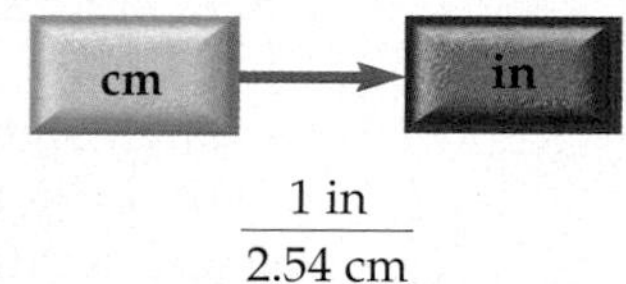

$$\frac{1\text{ in}}{2.54\text{ cm}}$$

Each arrow in a solution map for a unit conversion has an associated conversion factor with the units of the previous step in the denominator and the units of the following step in the numerator. For one-step problems such as these, the solution map is only moderately helpful, but for multistep problems, it becomes a powerful way to develop a problem-solving strategy. The procedure for solving unit conversion problems using the solution map is given in the left column of the following box. The middle and right columns demonstrate examples of how to use the procedure.

Solving Unit Conversion Problems	**EXAMPLE 2.8** **Unit Conversion** Convert 7.8 km to miles.	**EXAMPLE 2.9** **Unit Conversion** Convert 0.825 m to millimeters.
1. Write down the *given* quantity and its unit(s).	**Given:** 7.8 km	**Given:** 0.825 m
2. Write down the quantity that you are asked to *find* and its unit(s).	**Find:** mi	**Find:** mm
3. Write down the appropriate *conversion factor(s)*. Some of these will be given in the problem. Others you find in tables within the text.	**Conversion Factors:** $1\text{ km} = 0.6214\text{ mi}$ (This conversion factor is from Table 2.3.)	**Conversion Factors:** $1\text{ mm} = 10^{-3}\text{ m}$ (This conversion factor is from Table 2.2.)
4. Write a *solution map* for the problem. Begin with the *given* quantity and draw an arrow symbolizing each conversion step. Below each arrow, write the appropriate conversion factor for that step. Focus on the units. The solution map should end at the *find* quantity.	**Solution Map:** 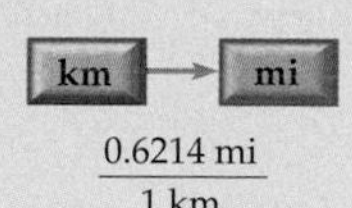 $\frac{0.6214\text{ mi}}{1\text{ km}}$ The conversion factor is written so that km, the unit we are converting *from*, is on the bottom and mi, the unit we are converting *to*, is on the top.	**Solution Map:** 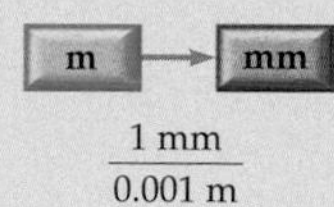 $\frac{1\text{ mm}}{0.001\text{ m}}$ The conversion factor is written so that m, the unit we are converting *from*, is on the bottom and mm, the unit we are converting *to*, is on the top.

5. Follow the solution map to solve the problem. Begin with the *given* quantity and its units. Multiply by the appropriate conversion factor(s), canceling units, to arrive at the *find* quantity.	**Solution:** $7.8\ \cancel{\text{km}} \times \frac{0.6214\ \text{mi}}{1\ \cancel{\text{km}}} =$ $4.84692\ \text{mi}$	**Solution:** $0.825\ \cancel{\text{m}} \times \frac{1\ \text{mm}}{10^{-3}\ \cancel{\text{m}}} =$ $825\ \text{mm}$
6. Round the answer to the correct number of significant figures. Follow the significant figure rules in Sections 2.3 and 2.4. Remember that exact conversion factors do not limit the number of significant figures in your answer.	$4.84692\ \text{mi} = 4.8\ \text{mi}$ We round to two significant figures since the quantity given has two significant figures. (If possible, obtain conversion factors to enough significant figures so that they do not limit the number of significant figures in the answer.)	$825\ \text{mm} = 825\ \text{mm}$ We leave the answer with three significant figures since the quantity given has three significant figures and the conversion factor is a definition, which therefore does not limit the number of significant figures in the answer.
7. Check your answer. Make certain that the units are correct and that the magnitude of the answer makes physical sense.	The units, mi, are correct. The magnitude of the answer is reasonable. A mi is longer than a km, so the value in mi should be smaller than the value in km. **SKILLBUILDER 2.8** **Unit Conversion** Convert 56.0 cm to inches.	The units, mm, are correct and the magnitude is reasonable. A mm is shorter than a m, so the value in m should be smaller than the value in mm. **SKILLBUILDER 2.9** **Unit Conversion** Convert 5,678 m to kilometers.

2.7 Solving Multistep Conversion Problems

When solving multistep conversion problems, we follow the preceding procedure but simply add more steps to the solution map. Every step in the solution map must have a conversion factor with the units of the previous step in the denominator and the units of the following step in the numerator. For example, suppose we want to convert 194 cm to feet. We set up the problem as we outlined earlier.

Given: 194 cm

Find: ft

Conversion Factors:

$2.54\ \text{cm} = 1\ \text{in.}$
$12\ \text{in.} = 1\ \text{ft}$

Now we can build our solution map. We start with the quantity given and focus on the units. We use the first conversion factor to convert centimeters to inches and the second to convert inches to feet.

Solution Map:

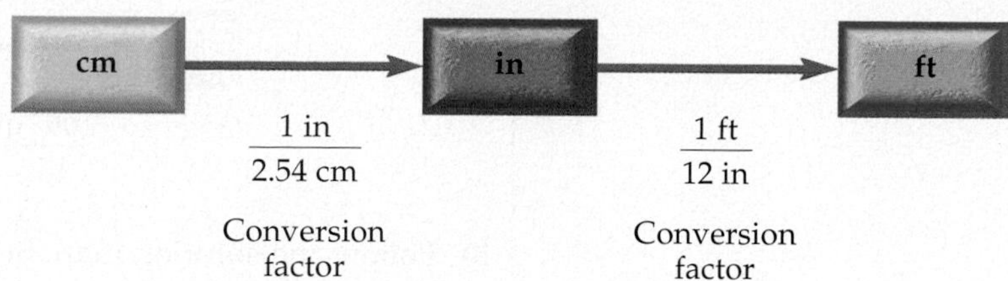

$\frac{1 \text{ in}}{2.54 \text{ cm}}$ Conversion factor

$\frac{1 \text{ ft}}{12 \text{ in}}$ Conversion factor

Once the solution map is complete, we follow it to solve the problem.

Solution:

$$194 \cancel{\text{cm}} \times \frac{1 \cancel{\text{in}}}{2.54 \cancel{\text{cm}}} \times \frac{1 \text{ ft}}{12 \cancel{\text{in}}} = 6.3648 \text{ ft}$$

We then round to the correct number of significant figures, in this case, three (from 194 cm, which has three significant figures).

Since one foot is *defined* as 12 in., it does not limit significant figures.

$$6.3648 \text{ ft} = 6.36 \text{ ft}$$

Finally, we check the answer. The units of the answer, ft, are the correct ones and the magnitude seems about right. Since ft is a larger unit than cm, it is reasonable that the value in ft is smaller than the value in cm.

EXAMPLE 2.10 Solving Multistep Unit Conversion Problems

An Italian recipe for making creamy pasta sauce calls for 0.75 L of cream. Your measuring cup only measures in cups. How many cups of cream should you use?

4 cups = 1 quart

1) Write down the quantity that is *given* and its unit(s).

Given: 0.75 L

2) Write down the quantity that you are asked to *find* and its unit(s).

Find: cu

3) Write down the appropriate *conversion factor(s).*

Conversion Factors: 4 cu = 1 qt

1.057 qt = 1 L

The second conversion factor is from Table 2.3.

Note: It is okay if you don't have all the necessary conversion factors at this point. You will discover what you need as you work through the problem.

4) Write a *solution map.*

Focus on the units, finding the appropriate conversion factors with the units of the previous step in the denominator and the units of the next step in the numerator.

Solution Map:

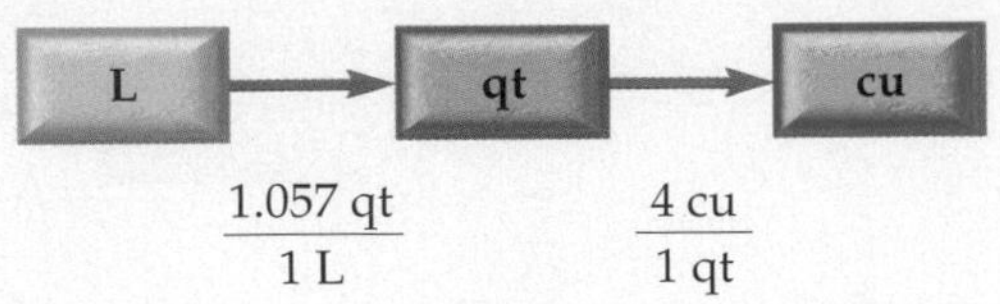

5) Follow the solution map. Starting with the quantity that is given and its units, multiply by the appropriate conversion factors, canceling units, to arrive at the quantity that you are trying to find in the desired units.

Solution:

$$0.75\,\cancel{\text{L}} \times \frac{1.057\,\cancel{\text{qt}}}{1\,\cancel{\text{L}}} \times \frac{4\text{ cu}}{1\text{ qt}} = 3.171\text{ cu}$$

6) Round the final answer.

$$3.171\text{ cu} = 3.2\text{ cu}$$

We round to two significant figures since the quantity given has two significant figures.

7) Check your answer. The answer has the right units (cu) and seems reasonable. We know that cu is smaller than L, so the value in cu should be larger than the value in L.

SKILLBUILDER 2.10 **Solving Multistep Unit Conversion Problems**

Suppose a recipe calls for 1.2 cu of oil. How many liters is this?

EXAMPLE 2.11 **Solving Multistep Unit Conversion Problems**

A running track measures 0.250 mi per lap. To run 10.0 km, how many laps should you run?

We set up the problem in the standard way.

Given: 10.0 km

Find: laps

Conversion Factors: 1 lap = 0.250 mi

0.6214 mi = 1 km

We then build our solution map, focusing on the units and how to get from km to laps.

Solution Map:

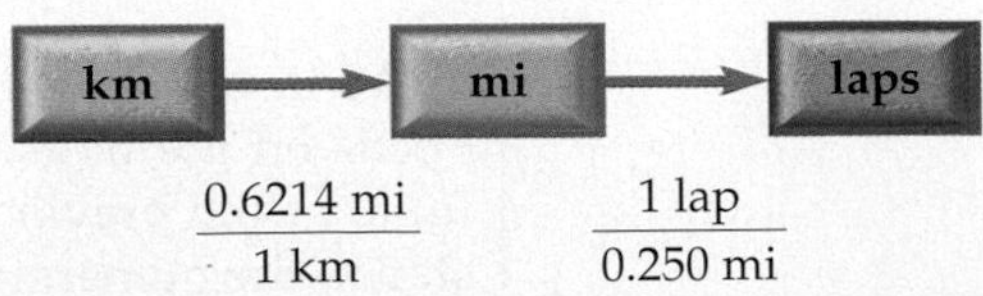

We then follow the solution map to solve the problem.

Solution:

$$10.0\ \cancel{\text{km}} \times \frac{0.6214\ \cancel{\text{mi}}}{1\ \cancel{\text{km}}} \times \frac{1\ \text{lap}}{0.250\ \cancel{\text{mi}}} = 24.856\ \text{laps} = 24.9\ \text{laps}$$

The intermediate answer in blue is rounded to three significant figures as limited by the given quantity, 10.0 km. The units of the answer are correct and the value of the answer makes sense—a lap is shorter than a km, so the value in laps should be larger than the value in km.

SKILLBUILDER 2.11 **Solving Multistep Unit Conversion Problems**

A running track measures 1056 ft per lap. To run 15.0 km, how many laps should you run? 1 mi = 5280 ft

SKILLBUILDER PLUS

Convert 5.72 nautical mi to meters. A nautical mile is equal to 1.151 mi.

2.8 Units Raised to a Power

The unit cm^3 is often abbreviated as c.c.

When converting quantities with units raised to a power, such as cubic centimeters (cm^3), the conversion factor must also be raised to that power. For example, suppose we want to convert the size of a motorcycle engine reported as 1255 cm^3 into cubic inches. We know that:

$$2.54\ \text{cm} = 1\ \text{in}$$

Most tables of conversion factors will not include conversions between cubic units, but we can derive them from the conversion factors for the basic units. We cube both sides of the preceding equality to obtain the proper conversion factor.

$$(2.54\ \text{cm})^3 = (1\ \text{in})^3$$
$$16.387\ \text{cm}^3 = 1\ \text{in}^3$$

(Note: 2.54 cm = 1 in. is an exact conversion factor. After cubing, we retain five significant figures so that the conversion factor does not limit the four significant figures of our original quantity (1255 cm^3).

We can do the same thing in fractional form.

$$\frac{1\ \text{in}}{2.54\ \text{cm}} = \frac{(1\ \text{in})^3}{(2.54\ \text{cm})^3} = \frac{1\ \text{in}^3}{16.387\ \text{cm}^3}$$

We then proceed with the conversion in the normal way.

Solution Map:

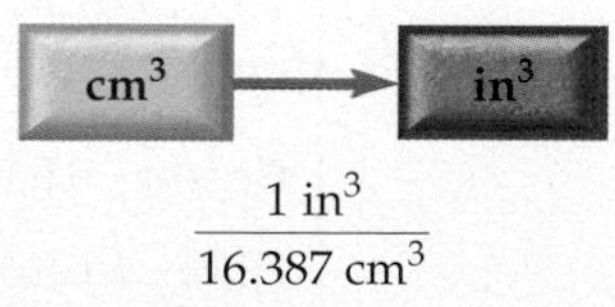

$$\frac{1\ \text{in}^3}{16.387\ \text{cm}^3}$$

Solution:

$$1255 \text{ cm}^3 \times \frac{1 \text{ in}^3}{16.387 \text{ cm}^3} = 76.5851 \text{ in}^3 = 76.59 \text{ in}^3$$

EXAMPLE 2.12 Converting Quantities Involving Units Raised to a Power

A circle has an area of 28 in^2. What is its area in square centimeters? We set up the problem in the standard way.

Given: 28 in^2
Find: cm^2

Conversion Factor: 2.54 cm = 1 in

Solution Map:

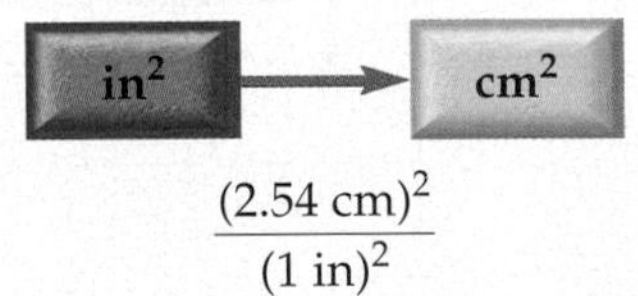

$$\frac{(2.54 \text{ cm})^2}{(1 \text{ in})^2}$$

We then follow the solution map to solve the problem.

Solution:

$$28 \text{ in}^2 \times \frac{(2.54 \text{ cm})^2}{(1 \text{ in})^2}$$

$$= 28 \text{ in}^2 \times \frac{6.45 \text{ cm}^2}{1 \text{ in}^2}$$

$$= 1.80645 \times 10^2 \text{ cm}^2$$

$$= 1.8 \times 10^2 \text{ cm}^2$$

The units of the answer are correct, and the magnitude makes physical sense. A cm^2 is smaller than a in^2 so the value in cm^2 should be larger than the value in in^2.

SKILLBUILDER 2.12 Converting Quantities Involving Units Raised to a Power

An automobile engine has a displacement (a measure of the size of the engine) of 289.7 in^3. What is its displacement in cubic centimeters?

EXAMPLE 2.13 Solving Multistep Problems Involving Units Raised to a Power

The average annual per person oil consumption in the United States is 15,615 dm^3. What is this value in cubic inches?
Set up the problem in the standard way.

Given: 15,615 dm^3

Find: volume in in^3

Conversion Factors: 1 dm = 0.1 m
1 cm = 0.01 m
2.54 cm = 1 in

We must convert from dm^3 to in^3. Each of these conversion factors must be cubed since the quantities involve cubic units.

Solution Map:

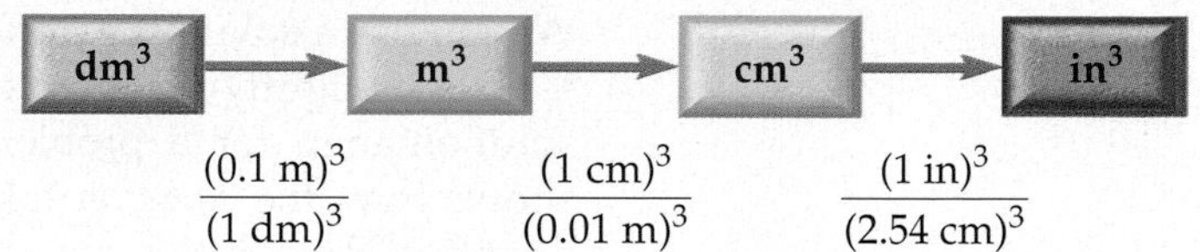

$$\frac{(0.1\text{ m})^3}{(1\text{ dm})^3} \qquad \frac{(1\text{ cm})^3}{(0.01\text{ m})^3} \qquad \frac{(1\text{ in})^3}{(2.54\text{ cm})^3}$$

We then use the solution map to solve the problem.

Solution:

$$15{,}615\text{ dm}^3 \times \frac{(0.1\text{ m})^3}{(1\text{ dm})^3} \times \frac{(1\text{ cm})^3}{(0.01\text{ m})^3} \times \frac{(1\text{ in.})^3}{(2.54\text{ cm})^3}$$
$$= 9.5289 \times 10^5\text{ in}^3$$

We round our answer to five significant figures to reflect the five significant figures in the least precisely known quantity (15,615 dm^3). These conversion factors are all exact and therefore do not limit the number of significant figures. The units of the answer are correct and the magnitude makes sense. A in^3 is a smaller unit than a dm^3, so the value in in^3 should be larger than the value in dm^3.

SKILLBUILDER 2.13 **Solving Multistep Problems Involving Units Raised to a Power**

How many cubic inches are there in 3.25 yd^3?

2.9 Density

Why do some people pay over three thousand dollars for a bicycle made of titanium? A steel frame would be just as strong for a fraction of the cost. The difference between the two, of course, is mass. The titanium bike is lighter than the steel bike because titanium is less dense than steel; that is, for a given volume of metal, titanium has less mass. The **density** of a substance is the ratio of its mass to its volume.

$$\text{Density} = \frac{\text{Mass}}{\text{Volume}} \quad \text{or} \quad d = \frac{m}{V}$$

Top-end bicycle frames are made of titanium because of its low density and high relative strength. Titanium has a density of 4.50 g/cm^3, while iron, for example, has a density of 7.86 g/cm^3.

Density is a fundamental property of materials and differs from one substance to another. The units of density are those of mass divided by those of volume, most conveniently expressed in grams per cubic centimeter or grams per milliliter. See Table 2.4 for a list of the densities of same common substances. Aluminum is among the least dense structural metals with a density of 2.70 g/cm^3, while platinum is among the densest with a density of 21.4 g/cm^3. Titanium has a density of 4.50 g/cm^3.

Remember that cubic centimeters and milliliters are equivalent units.

Calculating Density

The density of a substance is calculated by dividing the mass of a given amount of the substance by its volume. For example, a sample of liquid has a volume of 22.5 mL and a mass of 27.2 g. To find its density, we divide the mass by the volume.

$$d = \frac{m}{V} = \frac{27.2\ \text{g}}{22.5\ \text{mL}} = 1.21\ \text{g/mL}$$

We can use a solution map for solving problems involving equations, but it takes a slightly different form than for pure conversion problems. In a solution map for a problem involving an equation, the solution map shows how the *equation* takes you from the *given* quantities to the *find* quantity. The solution map for this problem is:

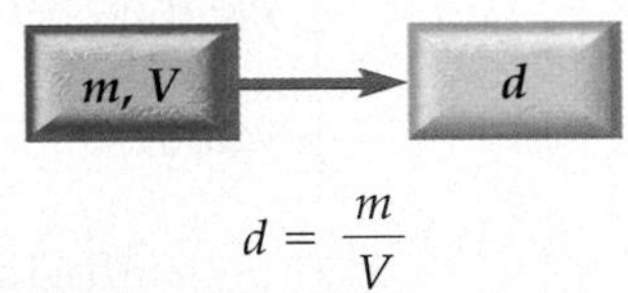

$$d = \frac{m}{V}$$

The solution map shows how the values of m and V, when substituted into the equation, $d = \frac{m}{V}$, give the desired result, d.

TABLE 2.4

The Density of Some Common Substances

Substance	*Density* (g/cm^3)
water	1.0
ice	0.92
ethanol	0.789
lead	11.4
copper	8.96
gold	19.3
aluminum	2.7
platinum	21.4
iron	7.86
titanium	4.50
charcoal, oak	0.57
glass	2.6

EXAMPLE 2.14 Calculating Density

Platinum has become a popular metal for fine jewelry. A man gives a woman an engagement ring and tells her that it is made of platinum. Noting that the ring felt a little light, the woman decides to perform a test to determine the ring's density before giving him an answer about marriage. She places the ring without the diamond on a balance and finds it has a mass of 5.84 g. She then finds that the ring displaces 0.556 cm^3 of water. Is the ring made of platinum? (The density of platinum is 21.4 g/cm^3.)

Given: $m = 5.84$ g

$V = 0.556\ \text{cm}^3$

Find: density in g/cm^3

If the density matches that of platinum, then the ring is probably real; if it does not, it is a fake.

Equation: $d = \frac{m}{V}$

Solution Map:

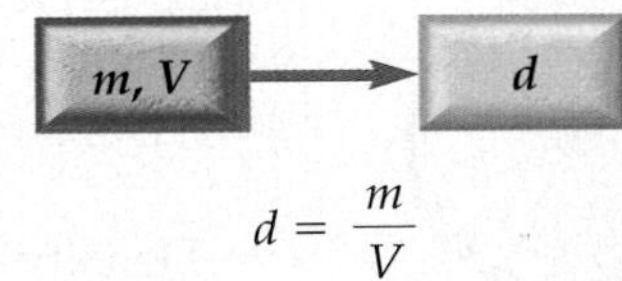

$$d = \frac{m}{V}$$

Solution:

$$d = \frac{m}{V} = \frac{5.84\ \text{g}}{0.556\ \text{cm}^3} = 10.5\ \text{g/cm}^3$$

The density of the ring is much too low to be platinum, and therefore the ring is a fake.

Note: The displacement of water is a common way to measure the volume of irregularly shaped objects. To say that an object *displaces* 0.556 cm^3 of water means that when the object is submerged into water, the water level rises by 0.556 cm^3. Therefore, the volume of the object is 0.556 cm^3.

SKILLBUILDER 2.14 **Calculating Density**

The man takes the ring back to the jewelry shop where he is met with endless apologies. They accidentally had made the ring out of silver rather than platinum. They give him a new ring that they promise is platinum. This time he checks the density himself and finds the mass of the ring to be 9.67 g and its volume to be 0.452 cm^3. Is this ring genuine?

Density as a Conversion Factor

Density can also be thought of as a conversion factor between mass and volume. We may know the mass of an object and want to calculate its volume or vice versa. For example, suppose we need 68.4 g of a liquid with a density of 1.32 g/cm^3 and only have a graduated cylinder to measure its volume. How much volume should we measure? We can set up this problem like a standard conversion problem.

Given: 68.4 g

Find: volume in mL

Conversion Factors: $1.32\ g/cm^3$

$1\ cm^3 = 1\ mL$

Density is a conversion factor between mass and volume. To convert from g to mL we need to invert the density because we want g, the unit we are converting from, to be on the bottom and cm^3, the unit we are converting to, on the top. Our solution map is:

Solution Map:

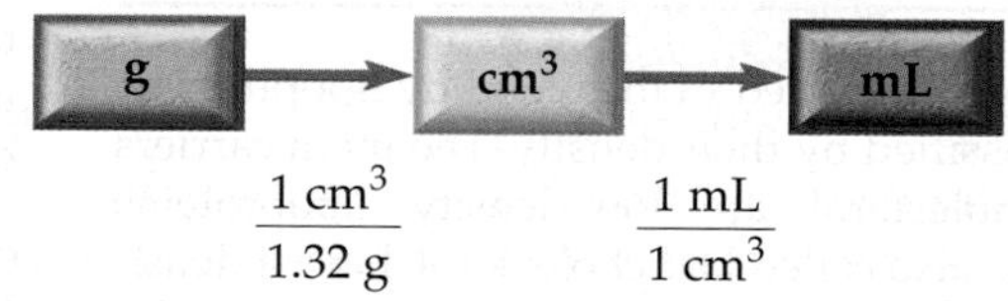

$$\frac{1\ cm^3}{1.32\ g} \qquad \frac{1\ mL}{1\ cm^3}$$

Solution:

$$68.4\ g \times \frac{1\ cm^3}{1.32\ g} \times \frac{1\ mL}{1\ cm^3} = 51.8\ mL$$

We know we must measure 51.8 mL to obtain 68.4 g of the liquid.

EXAMPLE 2.15 **Density as a Conversion Factor**

The gasoline in an automobile gas tank has a mass of 60.0 kg and a density of 0.752 g/cm^3. What is its volume in cm^3?

Given: 60.0 kg

Find: volume in cm^3

Conversion Factors: $0.752\ g/cm^3$

$1000\ g = 1\ kg$

Chemical Skills

Examples

Scientific Notation (Section 2.2)

To express a number in scientific notation:

- Move the decimal point to obtain a number between 1 and 10.
- Write the decimal part multiplied by 10 raised to the number of places you moved the decimal point.
- The exponent is positive if you moved the decimal point to the left and negative if you moved the decimal point to the right.

EXAMPLE 2.18 Scientific Notation

Express the number 45,000,000 in scientific notation.

45,000,000
7 6 5 4 3 2 1

4.5×10^7

Reporting Measured Quantities to the Right Number of Digits (Section 2.3)

Report measured quantities so that every digit is certain, except the last, which is estimated.

EXAMPLE 2.19 Reporting Measured Quantities to the Right Number of Digits

Record the volume of liquid in the graduated cylinder to the correct number of digits. Laboratory glassware should be read from the bottom of the meniscus (See figure).

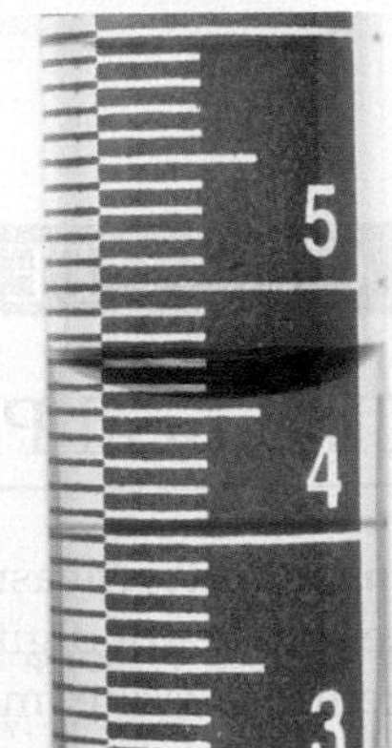

Since the graduated cylinder has markings every 0.1 mL, the measurement should be recorded to the nearest 0.01 mL. In this case, that is 4.57 mL.

Counting Significant Digits (Section 2.3)

The following digits should always be counted as significant:

- nonzero digits
- interior zeroes
- trailing zeroes after a decimal point

The following digits should never be counted as significant:

- zeroes to the left of the first nonzero number

The following digits are ambiguous and should be avoided by using scientific notation:

- zeroes at the end of a number, but before a decimal point.

EXAMPLE 2.20 Counting Significant Digits

How many significant figures in the following numbers?

1.0050	five significant figures
0.00870	three significant figures
5,400	Write as 5.4×10^3, 5.40×10^3, or 5.400×10^3 depending on the number of significant figures intended.

Rounding (Section 2.4)

When rounding numbers to the correct number of significant figures, round down if the last digit dropped is four or less; round up if the last digit dropped is five or more.

EXAMPLE 2.21 Rounding

Round 6.442 and 6.456 to two significant figures.

6.442 rounds to 6.4
6.456 rounds to 6.5

Significant Figures in Multiplication and Division (Section 2.4)

The result of a multiplication or division should carry the same number of significant figures as the factor with the least number of significant figures.

EXAMPLE 2.22 Significant Figures in Multiplication and Division

Perform the following calculation and report the answer to the correct number of significant figures.

$$8.54 \times 3.589 \div 4.2$$
$$= 7.297$$
$$= 7.3$$

Round the intermediate result to two significant figures to reflect the two significant figures in the factor with the least number of significant figures (4.2).

Significant Figures in Addition and Subtraction (Section 2.4)

The result of an addition or subtraction should carry the same number of decimal places as the quantity carrying the least number of decimal places.

EXAMPLE 2.23 Significant Figures in Addition and Subtraction

Perform the following operation and report the answer to the correct number of significant figures.

$$\begin{array}{r} 3.098 \\ 0.67 \\ -0.9452 \\ \hline 2.8228 \end{array} = 2.82$$

Round the final result to two decimal places to reflect the two decimal places in the quantity with the least number of decimal places (0.67).

Significant Figures in Calculations Involving Both Addition/Subtraction and Multiplication/Division (Section 2.4)

In calculations involving both addition/subtraction and multiplication/division, do the steps in parentheses first, keeping track of how many significant digits in the answer by underlining the least significant digit, then do the remaining steps. It is best to not round off until the very end.

EXAMPLE 2.24 Significant Figures in Calculations Involving both Addition/Subtraction and Multiplication/Division

Perform the following operation and report the answer to the correct number of significant figures.

$$8.16 \times (5.4323 - 5.411)$$
$$= 8.16 \times 0.02\underline{1}3$$
$$= 0.1738$$
$$= 0.17$$

Unit Conversion (Section 2.6, 2.7)

Solve unit conversion problems by following these steps.

- Write down the given quantity and its units.
- Write down the quantity to find and/or its units.
- Write down the appropriate conversion factor(s).
- Draw a solution map.
- Follow the solution map. Starting with the given quantity and its units, multiply by the appropriate conversion factor(s), canceling units, to arrive at the quantity to find in the desired units.
- Round the final answer to the correct number of significant figures.
- Check the answer.

EXAMPLE 2.25 Unit Conversion

Convert 108 ft to meters.

Given: 108 ft

Find: m

Conversion Factors: 1 m = 39.37 in
1 ft = 12 in

Solution Map:

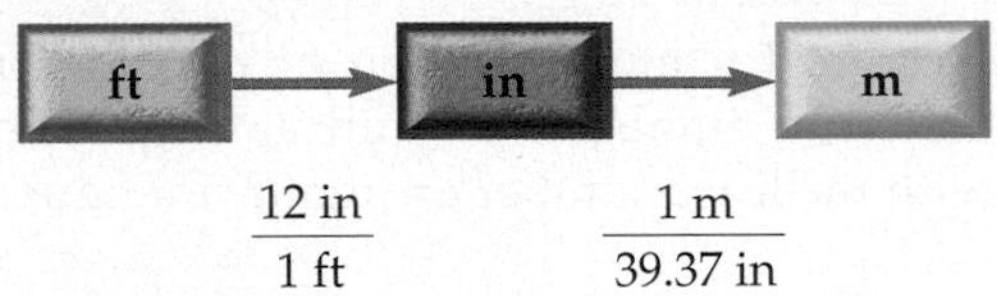

$$\frac{12\text{ in}}{1\text{ ft}} \qquad \frac{1\text{ m}}{39.37\text{ in}}$$

Solution:

$$108\text{ ft} \times \frac{12\text{ in}}{1\text{ ft}} \times \frac{1\text{ m}}{39.37\text{ in}}$$
$$= 32.918\text{ m}$$
$$= 32.9\text{ m}$$

Unit Conversion Involving Units Raised to a Power (Section 2.8)

When working problems involving units raised to a power, raise the conversion factors to the same power.

EXAMPLE 2.26 Unit Conversion Involving Units Raised to a Power

How many square meters in 1.0 km^2?

Given: 1.0 km^2

Find: m^2

Conversion Factors: 1 km = 1000 m

Solution Map:

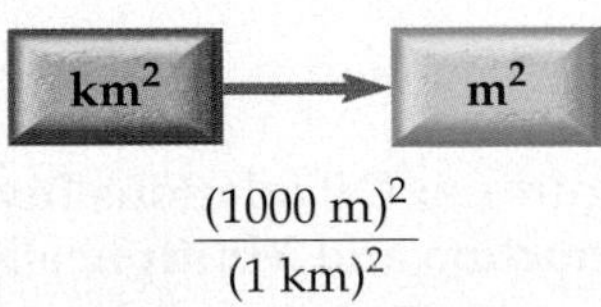

$$\frac{(1000\text{ m})^2}{(1\text{ km})^2}$$

Solution:

$$1.0\text{ km}^2 \times \frac{(1000\text{ m})^2}{(1\text{ km})^2}$$
$$= 1.0\text{ km}^2 \times \frac{1 \times 10^6\text{ m}^2}{1\text{ km}^2}$$
$$= 1.0 \times 10^6\text{ m}^2$$

Calculating Density (Section 2.10)

The density of an object or substance is its mass divided by its volume.

$$d = \frac{m}{V}$$

EXAMPLE 2.27 Calculating Density

An object has a mass of 23.4 g and displaces 5.7 mL of water. What is its density in grams per milliliter?

Given: $m = 23.4$ g
$V = 5.7$ mL

Find: density in g/mL

Solution Map:

m, V → d

$$d = \frac{m}{V}$$

Solution:

$$d = \frac{m}{V}$$
$$= \frac{23.4 \text{ g}}{5.7 \text{ mL}}$$
$$= 4.11 \text{ g/mL}$$
$$= 4.1 \text{ g/mL}$$

Density as a Conversion Factor (Section 2.10)

Density can be used as a conversion factor from mass to volume or from volume to mass. To convert between volume and mass, use density directly. To convert between mass and volume, you must invert the density.

EXAMPLE 2.28 Density as a Conversion Factor

What is the volume (in liters) of 321 g of a liquid with a density of 0.84 g/mL?

Given: 321 g

Find: volume in L

Conversion Factors: 0.84 g/mL
1 L = 1000 mL

Solution Map:

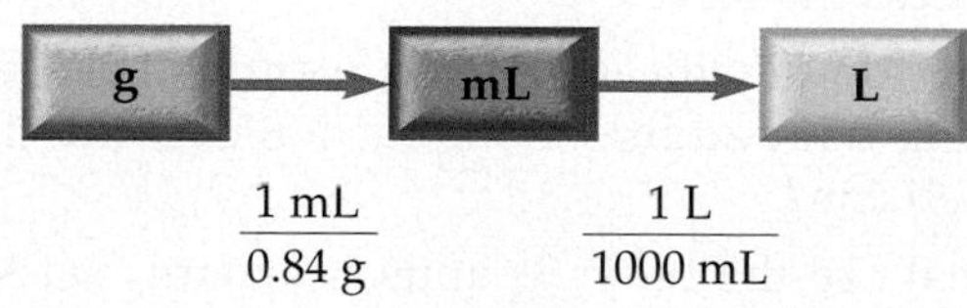

Solution:

$$321 \text{ g} \times \frac{1 \text{ mL}}{0.84 \text{ g}} \times \frac{1 \text{ L}}{1000 \text{ mL}}$$
$$= 0.382 \text{ L}$$
$$= 0.38 \text{ L}$$

KEY TERMS

conversion factor [2.6]
decimal part [2.2]
density [2.9]
English system [2.5]
exponent [2.2]
exponential part [2.2]
International System [2.5]
kilogram (kg) [2.5]
liter (L) [2.5]
mass [2.5]
meter (m) [2.5]
metric system [2.5]
prefix multipliers [2.5]
scientific notation [2.2]
second (s) [2.5]
SI units [2.5]
significant figures (digits) [2.3]
solution map [2.6]
units [2.5]
volume [2.5]

EXERCISES

Questions

The answers to all exercises numbered in blue appear in Appendix 5.

1. Why is it important to report units with scientific measurements?

2. Why is it important to report a certain number of digits with scientific measurements?

3. Why is scientific notation necessary?

4. If a measured quantity is written correctly, which digits are certain? Which are uncertain?

5. Explain when zeroes count as significant digits and when they do not.

6. How many significant digits are there in exact numbers? What kinds of numbers are exact?

7. What limits the number of significant digits in a calculation involving only multiplication and division?

8. What limits the number of significant digits in a calculation involving only addition and subtraction?

9. What are the rules for rounding numbers? What happens if the last digit being dropped is five or more? Less than five?

10. How do significant figures work in calculations involving both addition/subtraction and multiplication/division?

11. What are the basic SI units of length, mass, and time?

12. What are common units of volume?

13. Suppose you are trying to measure the diameter of a Frisbee. What unit and prefix multiplier would you use?

14. What is the difference between mass and weight?

15. Obtain a metric ruler and measure the following objects to the correct number of significant figures.
a) quarter (diameter)
b) dime (diameter)
c) notebook paper (width)
d) this book (width)

16. Obtain a stop watch and measure each of the following to the correct number of significant figures.
a) time between your heart beats
b) time it takes you to do the next problem
c) time between your breaths

17. Explain why units are important in calculations.

18. How are units treated in a calculation?

19. What is a conversion factor?

20. Write the conversion factor that converts a measurement in inches to feet.

21. Draw a solution map to convert a measurement in grams to pounds.

22. Draw a solution map to convert a measurement in milliliters to gallons.

23. Draw a solution map to convert a measurement in meters to feet.

24. Draw a solution map to convert a measurement in ounces to grams (1 lb = 16 oz)

25. What is density? Explain why density can work as a conversion factor. Between what quantities does it convert?

26. Explain how you would calculate the density of a substance. Include a solution map in your explanation.

Problems

(Note: The exercises in this section of Problems are paired and the answers to the odd-numbered exercises appear in Appendix 5.)

Scientific Notation

27. Express each of the following numbers in scientific notation.
a) 32,667,000 (population of California)
b) 1,193,000 (population of Hawaii)
c) 18,175,000 (population of New York)
d) 481,000 (population of Wyoming)

28. Express each of the following numbers in scientific notation.
a) 5,926,467,000 (population of the world)
b) 1,236,915,000 (population of China)
c) 11,051,000 (population of Cuba)
d) 3,619,000 (population of Ireland)

29. Express each of the following numbers in scientific notation.
a) 0.00000000007461 m (length of a hydrogen-hydrogen chemical bond)
b) 0.0000158 mi (number of miles in an inch)
c) 0.000000632 m (wavelength of red light)
d) 0.000015 m (diameter of human hair)

30. Express each of the following numbers in scientific notation.
a) 0.000000001 s (time it takes light to travel one foot)
b) 0.143 s (time it takes light to travel around the world)
c) 0.000000000001 s (time it takes a chemical bond to undergo one vibration)
d) 0.000001 m (approximate size of a dust particle)

31. Complete the following table.

Decimal Notation	*Scientific Notation*
0.000874	____
____	3.456×10^{10}
100,000,000	____
9.8×10^{12}	____

32. Complete the following table.

Decimal Notation	*Scientific Notation*
45,678,000	____
____	3.2×10^{-8}
____	9.8×10^{-4}
589,000	____

33. Express each of the following in decimal notation.
a) 6.022×10^{23} (number of carbon atoms in 12.01 g of carbon)
b) 1.6×10^{-19} c (charge of a proton in coulombs)
c) 2.99×10^{8} m/s (speed of light)
d) 3.44×10^{2} m/s (speed of sound)

34. Express each of the following in decimal notation.
a) 450×10^{-9} m (wavelength of blue light)
b) 16×10^{9} years (approximate age of the universe)
c) 5×10^{9} years (approximate age of planet earth)
d) 37 years (approximate age of this author)

35. Express each of the following in decimal notation.
a) 3.89×10^{9}
b) 5.9×10^{-4}
c) 8.68×10^{12}
d) 7.86×10^{-8}

36. Express each of the following in decimal notation.
a) 1.90×10^{8}
b) 1.4×10^{-2}
c) 3.2×10^{3}
d) 4.53×10^{-14}

Significant Figures

37. Read each of the following to the correct number of significant figures. Laboratory glassware should always be read from the bottom of the meniscus.

a)

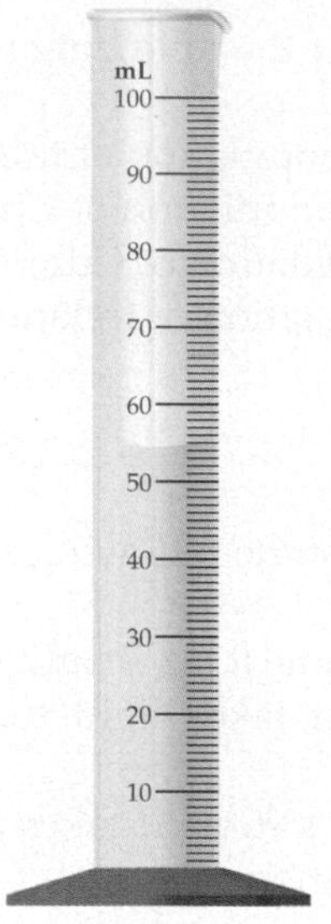

b)

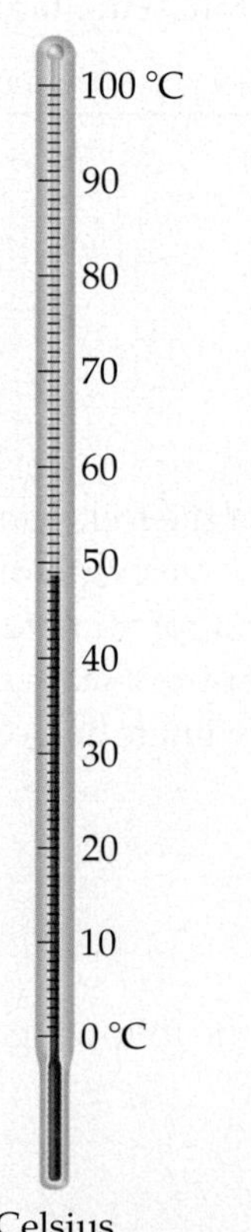

Celsius

38. Read each of the following to the correct number of significant figures. Laboratory glassware should always be read from the bottom of the meniscus.

a)

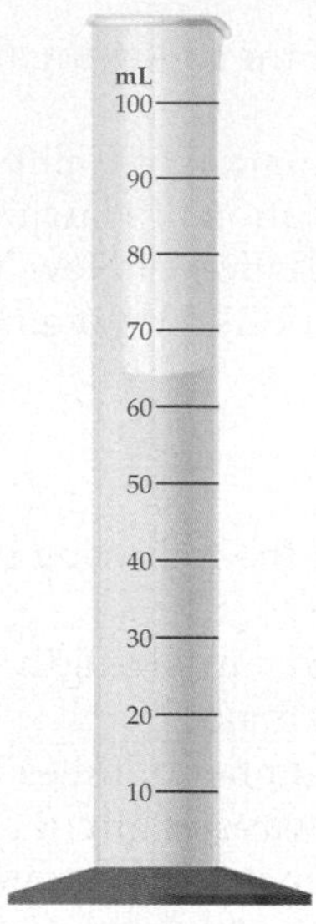

b)

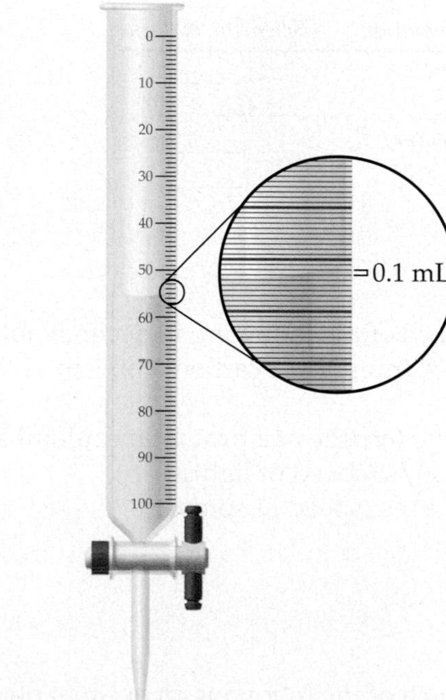

Note: A burette reads from the top down.

37. (continued)

c)

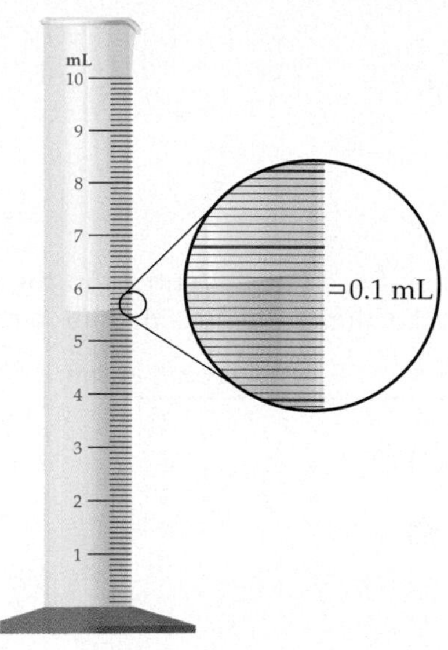

d)

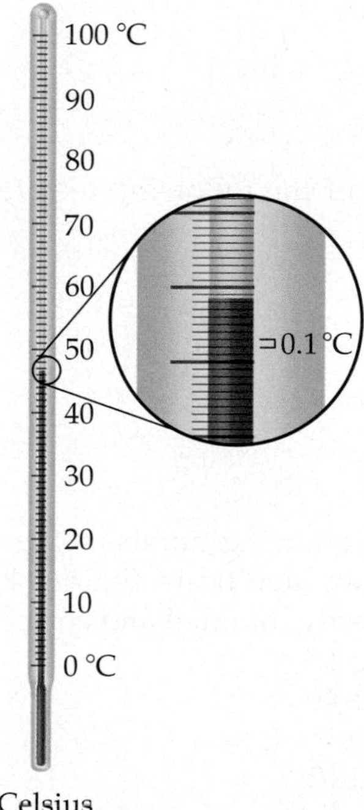

Celsius

38. (continued)

c)

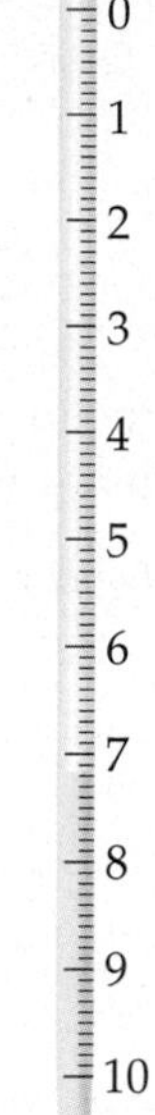

Note: A pipette reads from the top down.

d)

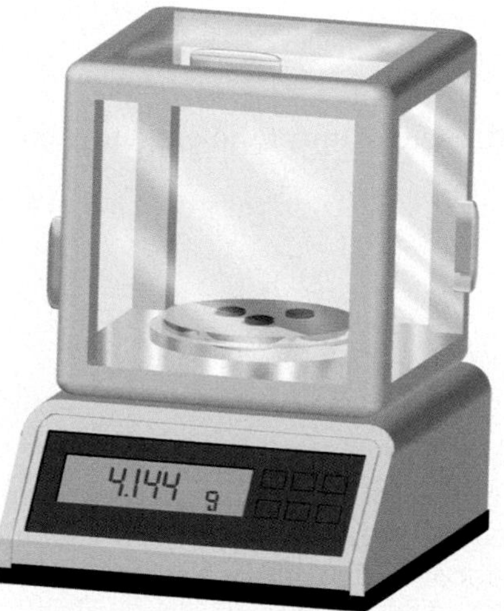

Note: Digital balances normally display mass to the correct number of significant figures for that particular balance.

39. For each of the following numbers, underline the zeroes that are significant and draw an x through the zeroes that are not.

a) 0.00608

b) 450,802

c) 0.0000000000000005

d) 0.009090

40. For each of the following numbers, underline the zeroes that are significant and draw an x through the zeroes that are not.

a) 0.09045

b) 709,980,761

c) 0.00008760

d) 0.00090220

41. How many significant figures are in each of the following numbers?
a) 0.9987
b) 0.009987
c) 99870
d) 9.98700×10^4

42. How many significant figures are in each of the following numbers?
a) 0.00072
b) 72000
c) 7.20×10^4
d) 72387

43. Determine if each of the entries in the following table is correct. Correct the ones that are wrong.

Number	*Significant Figures*
a) 895675	6
b) 0.000869	6
c) 0.5672100	5
d) 6.022×10^{23}	4

44. Determine if each of the entries in the following table is correct. Correct the ones that are wrong.

Number	*Significant Figures*
a) 24	2
b) 5.6×10^{-17}	3
c) 3.14	3
d) 0.00383	5

Rounding

45. The following numbers are displayed on a calculator. Round each to four significant figures.
a) 342.985318
b) 0.009650901
c) $3.52569241 \times 10^{-8}$
d) 1,127,436,092

46. The following numbers are displayed on a calculator. Round each to three significant figures.
a) 9,845.87492
b) $1.548937483 \times 10^{12}$
c) 2.3499999995
d) 0.000045389

47. Round each of the following to two significant figures.
a) 2.34
b) 2.35
c) 2.349
d) 2.359

48. Round each of the following to three significant figures.
a) 65.74
b) 65.749
c) 65.75
d) 65.750

49. Each of the following numbers was supposed to be rounded to three significant figures. Find the ones that were incorrectly rounded and correct them.
a) 42.3492 to 42.4
b) 56.9971 to 57.0
c) 231.904 to 232
d) 0.04555 to 0.046

50. Each of the following numbers was supposed to be rounded to two significant figures. Find the ones that were incorrectly rounded and correct them.
a) 1.249×10^3 to 1.3×10^3
b) 3.999×10^2 to 40
c) 56.21 to 56.2
d) 0.009964 to 0.010

Significant Figures in Calculations

51. Perform the following calculations to the correct number of significant figures.
a) $4.5 \times 0.03060 \times 0.391$
b) $5.55 \div 8.97$
c) $7.890 \times 10^{12} \div 6.7 \times 10^4$
d) $67.8 \times 9.8 \div 100.04$

52. Perform the following calculations to the correct number of significant figures.
a) $89.3 \times 77.0 \times 0.08$
b) $5.01 \times 10^5 \div 7.8 \times 10^2$
c) $4.005 \times 74 \times 0.007$
d) $453 \div 2.031$

53. Determine if each of the following calculations is performed to the correct number of significant figures. If it is not, correct it.
a) $34.00 \times 567 \div 4.564 = 4.2239 \times 10^3$
b) $79.3 \div 0.004 \times 35.4 = 7 \times 10^5$
c) $89.763 \div 22.4581 = 3.997$
d) $4.32 \times 10^{12} \div 3.1 \times 10^{-4} = 1.4 \times 10^{16}$

54. Determine if each of the following calculations is performed to the correct number of significant figures. If it is not, correct it.
a) $45.3254 \times 89.00205 = 4034.05$
b) $0.00740 \times 45.0901 = 0.334$
c) $49857 \div 904875 = 0.05510$
d) $0.009090 \times 6007.2 = 54.605$

55. Perform the following calculations to the correct number of significant figures.
a) $87.6 + 9.888 + 2.3 + 100.77$
b) $43.7 - 2.341$
c) $89.6 + 98.33 - 4.674$
d) $6.99 - 5.772$

56. Perform the following calculations to the correct number of significant figures.
a) $1459.3 + 9.77 + 4.32$
b) $0.004 + 0.09879$
c) $432 + 7.3 - 28.523$
d) $2.4 - 1.777$

57. Determine if each of the following calculations is performed to the correct number of significant figures. If not, correct it.
a) $3.8 \times 10^5 - 8.45 \times 10^5 = -4.7 \times 10^5$
b) $0.00456 + 1.0936 = 1.10$
c) $8475.45 - 34.899 = 8440.55$
d) $908.87 - 905.34095 = 3.5291$

58. Determine if each of the following calculations is performed to the correct number of significant figures. If not, correct it.
a) $78.9 + 890.43 - 23 = 9.5 \times 10^2$
b) $9354 - 3489.56 + 34.3 = 5898.74$
c) $0.00407 + 0.0943 = 0.0984$
d) $0.00896 - 0.007 = 0.00196$

59. Perform each of the following calculations to the correct number of significant figures.
a) $(78.4 - 44.889) \div 0.0087$
b) $(34.6784 \times 5.38) + 445.56$
c) $(78.7 \times 10^5 \div 88.529) + 356.99$
d) $(892 \div 986.7) + 5.44$

60. Perform each of the following calculations to the correct number of significant figures.
a) $(1.7 \times 10^6 \div 2.63 \times 10^5) + 7.33$
b) $(568.99 - 232.1) \div 5.3$
c) $(9443 + 45 - 9.9) \times 8.1 \times 10^6$
d) $(3.14 \times 2.4367) - 2.34$

61. Determine if each of the following calculations is performed to the correct number of significant figures. If not, correct it.
a) $(78.56 - 9.44) \times 45.6 = 3152$
b) $(8.9 \times 10^5 \div 2.348 \times 10^2) + 121 = 3.9 \times 10^3$
c) $(45.8 \div 3.2) - 12.3 = 2$
d) $(4.5 \times 10^3 - 1.53 \times 10^3) \div 34.5 = 86$

62. Determine if each of the following calculation is performed to the correct number of significant figures. If not, correct it.
a) $(908.4 - 3.4) \div 3.52 \times 10^4 = 0.026$
b) $(1206.7 - 0.904) \times 89 = 1.07 \times 10^5$
c) $(876.90 + 98.1) \div 56.998 = 17.11$
d) $(455 \div 407859) + 1.00098 = 1.00210$

Unit Conversion

63. Perform each of the following conversions within the metric system.
a) 2.14 kg to g
b) 6172 mm to m
c) 1316 mg to kg
d) 0.0256 L to mL

64. Perform each of the following conversions within the metric system.
a) 8.95 cm to m
b) 1298.4 g to kg
c) 129 cm to mm
d) 2452 mL to L

65. Perform the following conversions between the English and metric systems.
a) 22.5 in. to cm
b) 126 ft to m
c) 825 yd to km
d) 2.4 in. to mm

66. Perform the following conversions between the English and metric systems.
a) 78.3 in. to cm
b) 445 yd to m
c) 336 ft to cm
d) 45.3 in. to mm

67. Perform the following conversions between the metric and English systems.
a) 40 cm to in.
b) 27.8 m to ft
c) 10 km to mi
d) 3845 kg to lb

68. Perform the following conversions between the metric and English systems.
a) 254 cm to in.
b) 89 mm to in.
c) 7.5 L to qt
d) 122 kg to lb

69. A student loses 2.4 lb in one month. How many grams did she lose?

70. A student gains 1.6 lb in two weeks. How many grams did he gain?

71. A runner wants to run 10.0 km. She knows that her running pace is 7.5 mi/h. How many minutes must she run? (Hint: Use 7.5 mi/h as a conversion factor between distance and time.)

72. A cyclist rides at an average speed of 24 mi/h. If she wants to bike 195 km, how long (in hours) must she ride?

73. A baking recipe calls for 5.0 qt of milk. What is this quantity in cubic centimeters?

74. A gas can holds 2.0 gal of gasoline. What is this quantity in cubic centimeters?

Units Raised to a Power

75. Fill in the blanks.
a) 1.0 km^2 = ____ m^2
b) 1.0 cm^3 = ____ m^3
c) 1.0 mm^3 = ____ m^3

76. Fill in the blanks.
a) 1.0 ft^2 = ____ $in.^2$
b) 1.0 yd^2 = ____ ft^2
c) 1.0 m^2 = ____ yd^2

77. A modest-sized house has an area of 215 m^2. What is its area in:
a) km^2
b) dm^2
c) cm^2

78. A classroom has a volume of 285 m^3. What is its volume in:
a) km^3
b) dm^3
c) cm^3

79. Total U.S. farmland occupies 954 million acres. How many square miles is this? (1 acre = 43,560 ft^2, 1 mi = 5280 ft)

80. The average U.S. farm occupies 435 acres. How many square miles is this? (1 acre = 43,560 ft^2, 1 mi = 5280 ft)

Density

81. A sample of an unknown metal has a mass of 35.4 g and a volume of 3.11 cm^3. Calculate its density and identify the metal by comparison to Table 2.4.

82. A new penny has a mass of 2.49 g and a volume of 0.349 cm^3. Is the penny made of pure copper?

83. Glycerol is a syrupy liquid often used in cosmetics and soaps. A 2.50-L sample of pure glycerol has a mass of 3.15×10^3 g. What is the density of glycerol in g/cm^3?

84. An aluminum engine block has a volume of 4.77 L and a mass of 12.88 kg. What is the density of the aluminum in g/cm^3?

85. A supposedly gold crown is tested to determine its density. It is found to displace 10.7 mL of water and has a mass of 206 g. Could the crown be made of gold?

86. A vase is said to be solid platinum. It is found to displace 18.65 mL of water and has a mass of 157 g. Could the vase be solid platinum?

87. Ethylene glycol (antifreeze) has a density of 1.11 g/cm^3.
a) What is the mass in grams of 417 mL of this liquid?
b) What is the volume in liters of 4.1 kg of this liquid?

88. Acetone (nail polish remover) has a density of 0.7857 g/cm^3.
a) What is the mass, in grams, of 28.56 mL of acetone?
b) What is the volume, in milliliters, of 6.54 g of acetone?

Cumulative Problems

89. A thief uses a bag of sand to replace a gold vase that sits on a weight-sensitive, alarmed pedestal. The bag of sand and the vase are exactly the same volume, 1.75 L.
a) Calculate the mass of each object. (density of gold = 19.3 g/cm^3, density of sand = 3.00 g/cm^3)
b) Did the thief set off the alarm? Explain.

90. One of the particles that compose atoms is called the proton. The proton has a radius of approximately 1.0×10^{-13} cm and a mass of 1.7×10^{-24} g. Determine the density of a proton. For a sphere $V = 4/3\pi r^3$; $\pi = 3.14$

91. A block of metal has a volume of 22.4 $in.^3$ and weighs 8.56 lb. What is its density in grams per cubic centimeter?

92. A log is known to be either oak or pine. It displaces 2.7 gal of water and weighs 19.8 lb. Is the log oak or pine? (Oak has a density of 0.9 g/cm^3 while pine has a density of 0.4 g/cm^3).

93. The density of aluminum is 2.7 g/cm^3. What is its density in pounds per cubic inch?

94. The density of platinum is 21.4 g/cm^3. What is its density in pounds per cubic inch?

95. A typical backyard swimming pool holds 150 yd^3 of water. What is the mass in pounds of the water?

96. An iceberg has a volume of 8975 ft^3. What is the mass of the ice (in kilograms) composing the iceberg?

97. Honda has recently produced a hybrid electric car called the Honda Insight. The Insight has both a gasoline-powered engine and an electric motor and has an EPA gas mileage rating of 70 mi/gal on the highway. What is the Insight's rating in kilometers per liter?

98. You rent a car in Germany with a gas mileage rating of 10.3 km/L. What is its rating in miles per gallon?

Highlight Problems

99. In 1999, NASA lost a \$94-million orbiter because one group of engineers used metric units in their calculations while another group used English units. Consequently, the orbiter pushed too far into the Martian atmosphere and burned up. Suppose that the orbiter was to have established orbit at 155 km and that one group of engineers specified this distance as 1.55×10^5 m. Suppose further that a second group of engineers programmed the orbiter to go to 1.55×10^5 ft. What was the difference (in kilometers) between the two altitudes? How low did the probe go?

The \$94-million Mars Climate Orbiter was lost in the Martian atmosphere in 1999 because two groups of engineers failed to communicate to each other the units that they used for their calculations.

100. A NASA satellite showed that the 1999 ozone hole was smaller than it was in 1998. The 1998 hole measured $1.05 \times 10^7\ mi^2$ while the 1999 hole measured $9.8 \times 10^6\ mi^2$. Calculate the difference in size between the two holes in square meters.

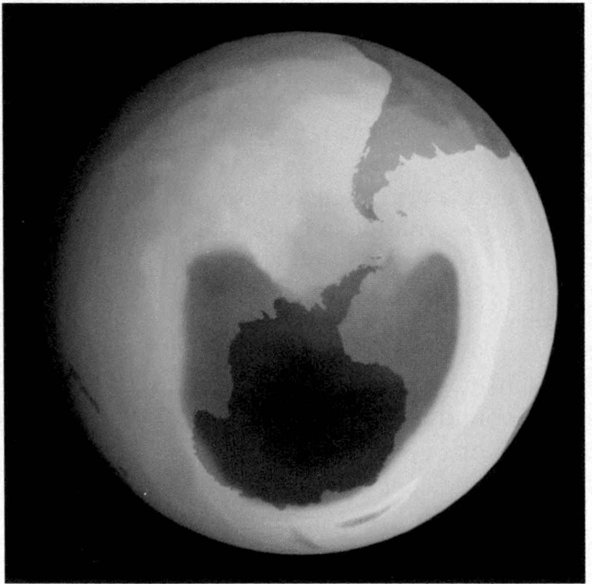

The 1999 Antarctic ozone hole was smaller than in 1998.

101. In 1999, scientists discovered a new class of black holes with masses 100 to 10,000 times the mass of our sun, but occupying less space than our moon. Suppose that one of these black holes has a mass of 1×10^3 suns and a radius equal to one half the radius of our moon. What is its density in grams per cubic centimeter? The mass of the sun is 2.0×10^{30} kg, and the radius of the moon is 2.16×10^3 mi. For a sphere $V = 4/3\pi r^3$.

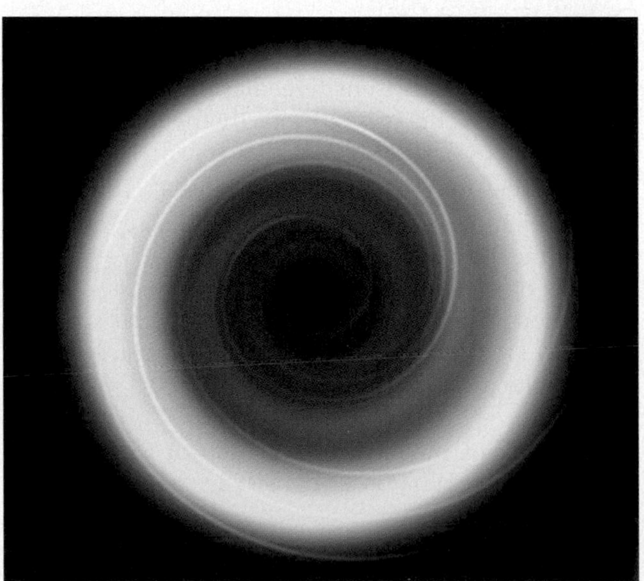

M82 is a nearby galaxy now thought to harbor a middleweight black hole in its nucleus.

102. A titanium bicycle frame contains the same amount of titanium as a titanium cube measuring 6.8 cm on a side. Use the density of titanium to calculate the mass (in kilograms) of titanium in the frame. What would be the mass of a similar frame composed of iron?

A titanium bicycle frame contains the same amount of titanium as a titanium cube measuring 6.8 cm on a side.

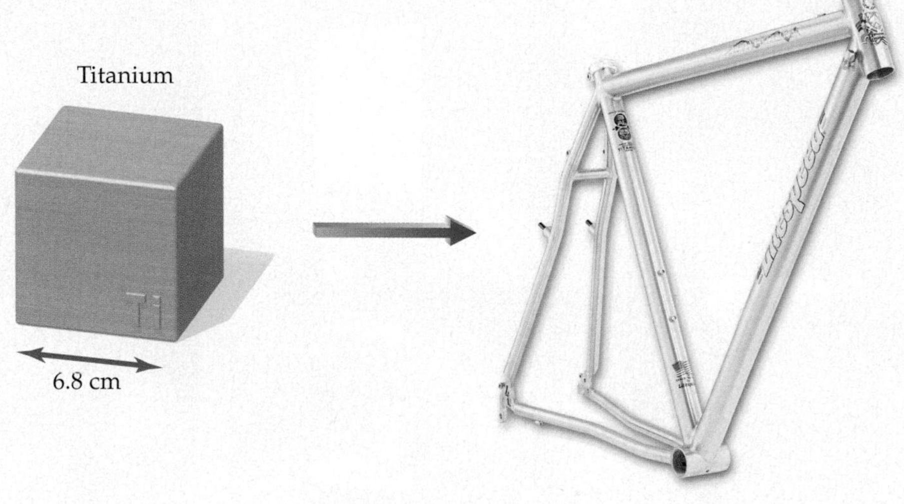

Answers to Skillbuilder Exercises

Skillbuilder 2.1 \$5.479 × 10^{12}
Skillbuilder 2.2 3.8 × 10^{-5}
Skillbuilder 2.3 103.4 °F
Skillbuilder 2.4 **a)** 4 significant figures **b)** 3 significant figures **c)** 2 significant figures **d)** 4 significant figures **e)** 3 significant figures **f)** ambiguous
Skillbuilder 2.5 **a)** 0.001 or 1 × 10^{-3} **b)** 0.204
Skillbuilder 2.6 **a)** 7.6 **b)** 131.11
Skillbuilder 2.7 **a)** 1288 **b)** 3.12
Skillbuilder 2.8 22.0 in.
Skillbuilder 2.9 5.678 km
Skillbuilder 2.10 0.28 L
Skillbuilder 2.11 46.6 laps
Skillbuilder Plus, p. 31 1.06 × 10^4 m
Skillbuilder 2.12 4747 cm^3
Skillbuilder 2.13 1.52 × 10^5 $in.^3$
Skillbuilder 2.14 Yes, the density is 21.4 g/cm^3 and matches that of platinum.
Skillbuilder 2.15 4.4 × 10^{-2} cm^3
Skillbuilder Plus, p. 37 1.95 kg
Skillbuilder 2.16 16 kg
Skillbuilder 2.17 d = 19.3 g/cm^3; yes, it is gold.

Matter and Energy

3

"Thus, the task is, not so much to see what no one has yet seen; but to think what nobody has yet thought, about that which everybody sees."
Erwin Schrödinger

3.1 In Your Room

Look around your room—what do you see? You might see your desk, your bed, or a glass of water. Maybe you have a window and can see trees, grass, or mountains. You can certainly see this book and possibly the table it sits on. What are these things made of? They are made of matter, which we will define more carefully shortly. For now, know that all you see is matter—your desk, your bed, the glass of water, the trees, the mountains, and this book. Some of what you don't see is matter as well. For example, you are constantly breathing air, which is also matter, in and out of your lungs. You feel the matter in air when you feel wind on your skin. Virtually everything is made of matter.

Think about the differences between different kinds of matter. Air is different from water, and water is different from wood. One of our first tasks as we learn about matter is to identify the similarities and differences among kinds of matter. How are sugar and salt similar? How are air and water different? Why are they different? Why is a mixture of sugar and water similar to a mixture of salt and water but different from a mixture of sand and water? As students of chemistry, we are particularly interested in the connection between the similarities and differences in matter and the similarities and differences in the atoms and molecules that compose it. We want to understand the connection between the macroscopic world and the microscopic one.

◀ Everything that you can see in this room is made of matter. As chemists, we are interested in how the differences between different kinds of matter are related to the differences between the molecules and atoms that compose it. The molecular structures shown here are water molecules on the left and carbon atoms in graphite on the right.

3.2 What Is Matter?

Matter is defined as anything that occupies space and has mass. Matter is often visible—such as steel, water, wood, and plastic—but is sometimes not visible to our eyes, such as air or microscopic dust. Matter may sometimes appear smooth and continuous, but actually it is not. Matter is ultimately composed of **atoms,** tiny particles too small to see. In many cases, these atoms are bonded together to form **molecules,** two or more atoms grouped in well-defined clusters (Figure 3.1). Recent advances in microscopy have allowed us to see the atoms (Figure 3.2) and molecules (Figure 3.3) that compose matter, sometimes with stunning clarity.

(a)

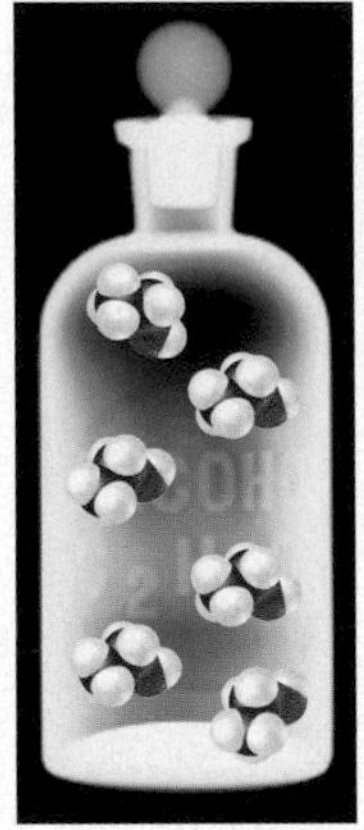

(b)

Figure 3.1 All matter is ultimately composed of atoms. **a)** In some cases, such as aluminum, the atoms exist as independent particles. **b)** In other cases, such as in alcohol, several atoms bond together to form well-defined clusters called molecules.

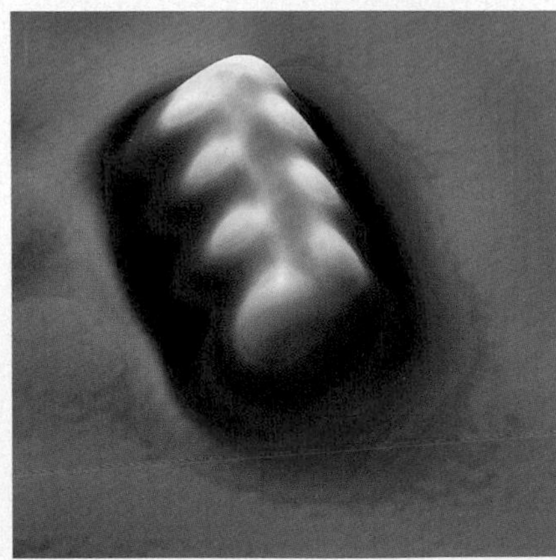

Figure 3.3 A scanning tunneling microscope image of a cesium iodide "molecule." (Cesium iodide is an ionic compound and does not normally form molecules. However, for this STM image, scientists bonded 8 cesium atoms and 8 iodine atoms together to make an ionic "molecule.")

Figure 3.2 A scanning tunneling microscope image of nickel atoms.

3.3 Classifying Matter According to Its State: Solid, Liquid, and Gas

There are three different **states of matter: solid, liquid**, and **gas** (Figure 3.4). In solid matter, atoms or molecules pack close to each other in fixed locations. Although neighboring atoms or molecules may vibrate or oscillate,

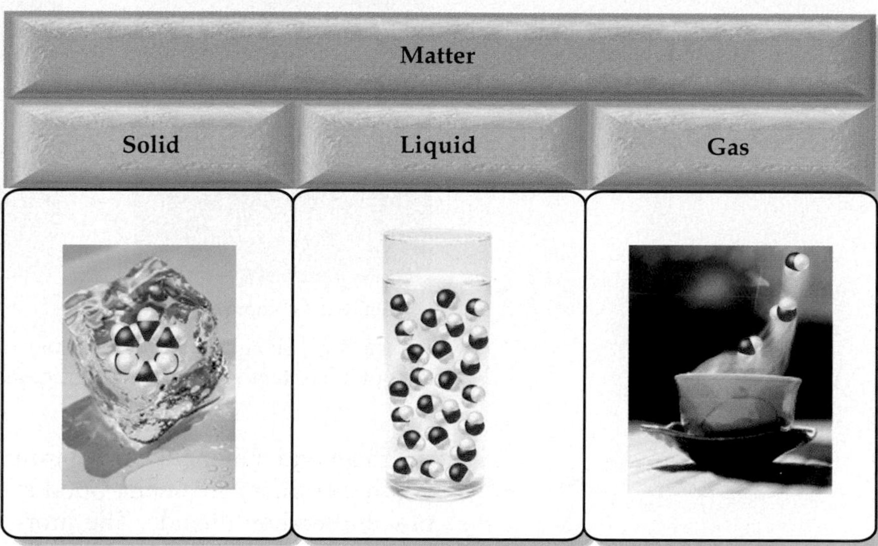

Figure 3.4 Water exists as ice (solid), water (liquid), and steam (gas). In ice, the water molecules are closely spaced and do not move relative to one another. In liquid water, the water molecules are closely spaced but are free to move around and past each other. In steam, water molecules are separated by large distances and do not interact with one another.

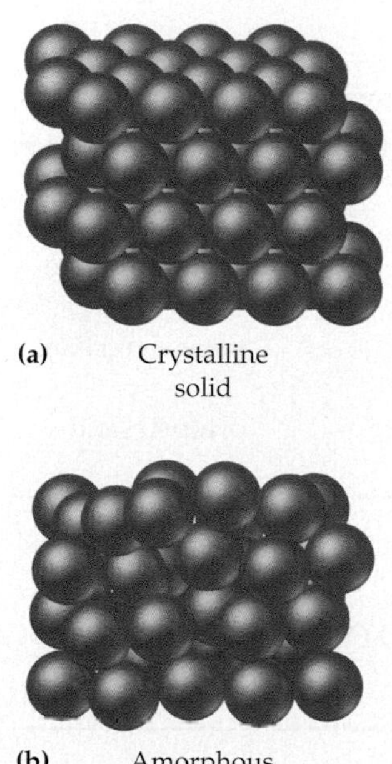

Figure 3.5 **a)** In a crystalline solid, atoms or molecules assume well-ordered, three-dimensional structure. **b)** In an amorphous solid, atoms do not have any long-range order.

they do not move around each other, giving solids the fixed volume and rigid shape we are accustomed to seeing. Ice, diamond, quartz, and iron are all good examples of solid matter. Solid matter may be **crystalline**, in which case its atoms or molecules arrange in geometric patterns with long-range, repeating order (Figure 3.5a), or it may be **amorphous**, in which case its atoms or molecules do not have long-range order (Figure 3.5b). Good examples of *crystalline* solids include salt (Figure 3.6) and diamond; the well-ordered, geometric shapes of salt and diamond crystals reflect the well-ordered geometric arrangement of their atoms. Good examples of *amorphous* solids include glass and plastic.

In liquid matter, atoms or molecules are packed close to each other (about as closely as in a solid) but are free to move around and by each other. Like solids, liquids have a fixed volume because their atoms or molecules are in close contact. Unlike solids, however, liquids assume the shape of their container because the atoms or molecules are free to move relative to one another. Water, gasoline, alcohol, and mercury are all good examples of liquid matter.

In gaseous matter, atoms or molecules are separated by large distances and are free to move relative to one another. Since the atoms or molecules that compose gases are not in contact with one another, gases are **compressible** (Figure 3.7). When you inflate a bicycle tire, for example, you push more atoms and molecules into the same space, compress-

Figure 3.6 Sodium chloride is an example of a crystalline solid. The well-ordered, cubic shape of salt crystals is due to the well-ordered, cubic arrangement of its atoms.

Figure 3.7 Since the atoms or molecules that compose gases are not in contact with one another, gases are compressible.

ing them and making the tire harder. Gases always assume the shape and volume of their containers. Good examples of gases include oxygen, helium, and carbon dioxide. The properties of solids, liquids, and gases are summarized in Table 3.1.

TABLE 3.1

Properties of Liquids, Solids, and Gases

State	*Atomic/Molecular Motion*	*Atomic/Molecular Spacing*	*Shape*	*Volume*	*Compressibility*
solid	oscillation/vibration about fixed point	close together	definite	definite	incompressible
liquid	free to move relative to one another	close together	indefinite	definite	incompressible
gas	free to move relative to one another	far apart	indefinite	indefinite	compressible

3.4 Classifying Matter According to Its Composition: Elements, Compounds, and Mixtures

In addition to classifying matter according to its state, we can classify it according to its composition (Figure 3.8). Matter may be either a **pure substance**, composed of only one type of atom or molecule, or it may be a **mixture**, composed of two or more different types of atoms or molecules combined in variable proportions.

The majority of matter that we encounter is in the form of mixtures. A glass of apple juice, a flame, salad dressing, and soil are all examples of mixtures; they contain several substances with proportions that vary from one sample to another. Other common mixtures include air, seawater, and

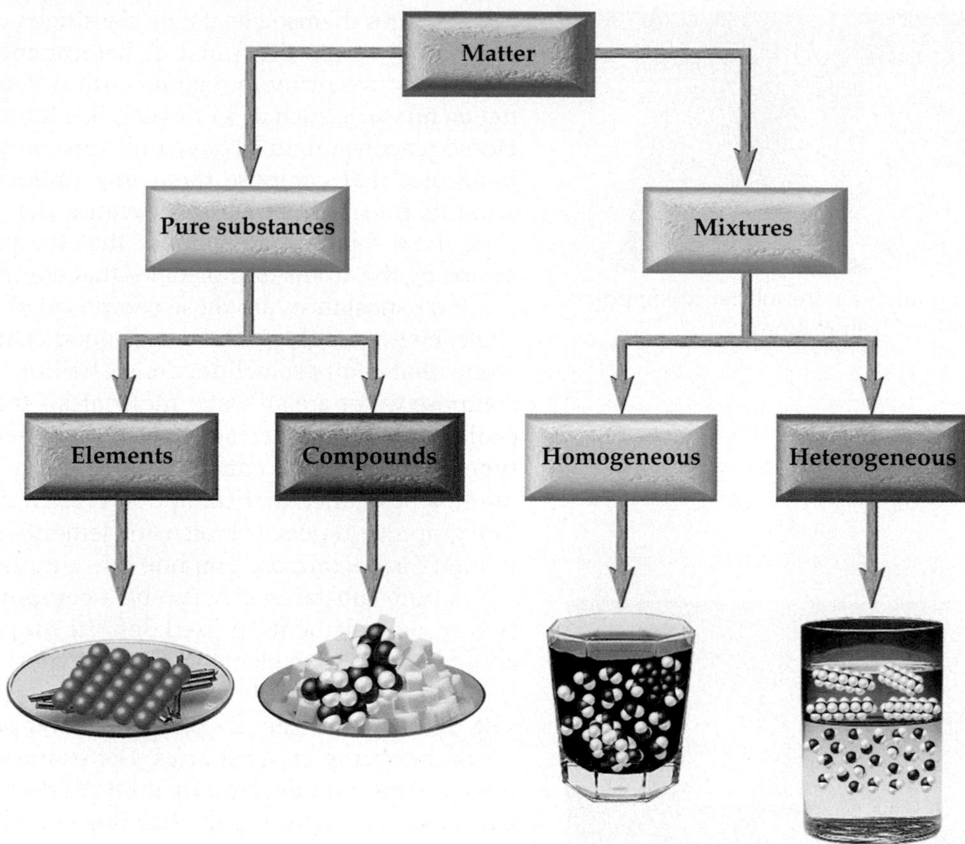

Figure 3.8 Classification of matter. Matter may be a pure substance or it may be a mixture. A pure substance may either be an element or a compound, and a mixture may be either homogenous or heterogeneous.

brass. Air is a mixture composed primarily of nitrogen and oxygen gas. Seawater is a mixture composed primarily of salt and water, and brass is a mixture composed of copper and zinc. Each of these mixtures can have different proportions of its constituent components. For example, metallurgists vary the relative amounts of copper and zinc in brass to tailor the metal's properties to its intended use—the higher the zinc content relative to the copper content, the more brittle the brass.

Air and seawater are good examples of mixtures. Air contains primarily nitrogen and oxygen. Seawater contains primarily salt and water.

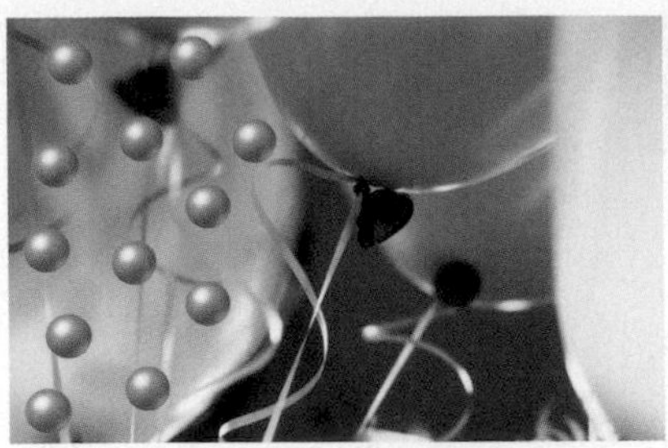

Helium is a pure substance composed only of helium atoms.

Mixtures themselves can be classified according to how uniformly the substances within them mix. A **heterogeneous mixture**, such as oil and water, has two or more regions with different compositions. A **homogeneous mixture**, such as salt water, has the same composition throughout. Homogeneous mixtures have uniform compositions because the atoms or molecules that compose them mix uniformly. Heterogeneous mixtures separate into distinct regions because the atoms or molecules that compose them separate. Remember that the properties of matter are determined by the atoms or molecules that compose it.

Pure substances are those composed of only one type of atom or molecule. Helium and water are both good examples of pure substances. The atoms that compose helium are all helium atoms, and the molecules that compose water are all water molecules—there are no other atoms or molecules mixed in. Pure substances can themselves be divided into two types: elements and compounds. Helium is a good example of an **element**, a substance that cannot be broken down into simpler substances. The graphite in pencils is also an element—carbon. No chemical transformation can decompose graphite into simpler substances; it is pure carbon.

Water is a pure substance composed only of water molecules.

A pure substance can also be a **compound**, a substance composed of two or more elements in fixed definite proportions. Compounds are more common than pure elements because most elements are chemically reactive and combine with other elements to form compounds. Water, table salt, and sugar are good examples of compounds; they can all be decomposed into simpler substances. For example, if you heat sugar on a pan over a flame, you decompose it into carbon and gaseous water. The black substance left on your pan after burning is the carbon; the water escapes into the air as steam.

To summarize, as shown in Figure 3.8:

- ***Matter may be a pure substance or it may be a mixture.***
- ***A pure substance may either be an element or a compound.***
- ***A mixture may be either homogenous or heterogeneous.***
- ***Mixtures may be composed of two or more elements, two or more compounds, or a combination of both.***

EXAMPLE 3.1 Classifying Matter

Classify each of the following as a pure substance or a mixture. If it is a pure substance, classify it as an element or a compound; if it is a mixture, classify it as homogeneous or heterogeneous.

a) Lead fishing weight
b) Seawater
c) Distilled water
d) Italian salad dressing

Solution:

Begin by examining the alphabetical listing of pure elements in the back of this text. If the substance appears in that table, it is a pure substance and an element. If it is not in the table, but is a pure substance, then it is a compound. If the substance is not a pure substance, then it is a mixture. Use your knowledge about the mixture to determine whether it is homogenous or heterogeneous.

a) Lead is listed in the table of elements. It is a pure substance and an element.
b) Seawater is composed of several substances including salt and water; it is a mixture. It has a uniform composition, so it is a homogeneous mixture.
c) Distilled water is not listed in the table of elements, but it is a pure substance (water), therefore it is a compound.
d) Italian salad dressing contains a number of substances and is therefore a mixture. It usually separates into at least two distinct regions with different composition and is therefore a heterogeneous mixture.

SKILLBUILDER 3.1 **Classifying Matter**

Classify each of the following as a pure substance or a mixture. If it is a pure substance, classify it as an element or a compound. If it is a mixture, classify it as homogeneous or heterogeneous.

a) Mercury in a thermometer
b) Exhaled air
c) Minestrone soup
d) Sugar

3.5 How We Tell Matter Apart: Physical and Chemical Properties

In our daily life, we distinguish one substance from another based on the substance's properties. For example, we distinguish water from alcohol based on their different smells, or we distinguish catsup from mustard based on their different colors or flavors. The characteristics we use to distinguish one substance from another are called **properties**. Every unique substance has unique properties that characterize it and distinguish it from other substances.

Figure 3.9 The boiling point of water is a physical property, and boiling is a physical change. When water boils, it turns into a gas, but the water molecules are the same in both the liquid water and the gaseous steam.

In chemistry, we differentiate between **physical properties**, those that a substance displays without changing its composition, and **chemical properties**, those that a substance displays only through changing its composition. For example, the characteristic odor of gasoline is a physical property—gasoline does not change its composition when it exhibits its odor. On the other hand, the flammability of gasoline is a chemical property—gasoline does change its composition when it burns.

The atoms or molecules that compose a substance do not change when the substance displays its physical properties. For example, the boiling point of water—a physical property—is 100°C. When water boils, it changes from a liquid to a gas, but the gas is still water (Figure 3.9). On the other hand, the susceptibility of iron to rust is a chemical property—iron must change into iron oxide to display this property (Figure 3.10). Physical properties include odor, taste, color, appearance, melting point, boiling point, and density. Chemical properties include flammability and other chemical activities.

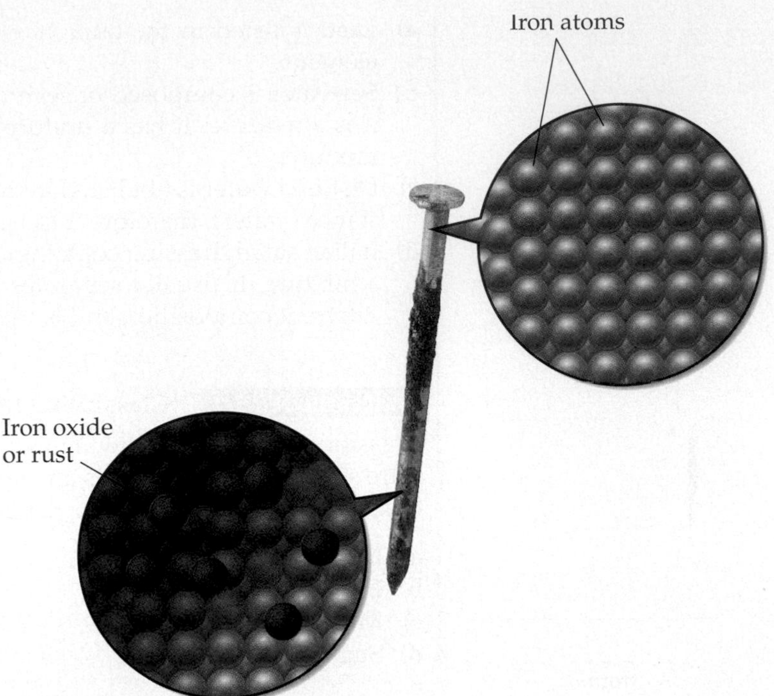

Figure 3.10 The susceptibility of iron to rusting is a chemical property, and rusting is a chemical change. When iron rusts, it turns from iron to iron oxide.

EXAMPLE 3.2 Physical and Chemical Properties

Determine whether each of the following is a physical or chemical property.

a) The tendency of copper to turn green when exposed to air
b) The tendency of automobile paint to dull over time
c) The tendency of gasoline to evaporate quickly when spilled
d) The low mass (for a given volume) of aluminum relative to other metals

Solution:

a) Copper turns green because it reacts with gases in air to form compounds; this is a chemical property.
b) Automobile paint dulls over time because it reacts with oxygen in air; this is a chemical property.
c) Gasoline evaporates quickly because it has a low boiling point; this is a physical property.
d) Aluminum's low mass (for a given volume) relative to other metals is due to its low density; this is a physical property.

SKILLBUILDER 3.2 Physical and Chemical Properties

Determine whether each of the following is a physical or chemical property.

a) The explosiveness of hydrogen gas
b) The bronze color of copper
c) The shiny appearance of silver
d) The ability of dry ice to vaporize without melting

The transition from a solid directly into a gas is called sublimation. We will further study this transformation in Chapter 12.

3.6 How Matter Changes: Physical and Chemical Changes

In a **physical change**, matter does not change its composition even though its appearance might change. For example, when ice melts, it looks different—water looks different from ice—but its composition is the same. Solid ice and liquid water are both composed of water molecules, so melting is a physical change. Similarly, when glass shatters, it looks different, but its composition remains the same—it is still glass and this is again a physical change. On the other hand, in a **chemical change**, matter does change its composition. For example, when copper turns green upon continued exposure to air, it is because it has reacted with gases in air to form new compounds. This is a chemical change.

The differences between physical and chemical changes are not always apparent. Only chemical examination of the substances before and after the change can verify whether the change is physical or chemical. For many cases, however, we can identify chemical and physical changes based on what we know about them. Phase changes, such as melting or boiling, or changes that are merely in appearance, such as cutting or crushing, are always physical changes. Changes involving chemical reactions—often evidenced by heat exchange or color changes—are chemical changes.

Phase changes are transformations from one state of matter (such as solid or liquid) to another.

The main difference between chemical and physical changes is related to the changes at the molecular and atomic level. In physical changes, the atoms or molecules that compose the matter *do not* change their identity, even though the matter may change its appearance. In chemical changes, the atoms and molecules that compose the matter *do* change their identity.

Consider physical and chemical changes in liquid butane, the substance used to fuel butane lighters. In many lighters, you can see the liquid butane through the plastic case of the lighter. If you push the fuel button on the lighter without turning the flint, some of the liquid butane vaporizes to gaseous butane—you can usually hear hissing as it leaks out. (See Figure 3.11.) Since the liquid butane and the gaseous butane are both composed of butane molecules, this is a physical change. On the other hand, if you push the button *and* turn the flint to create a spark, a chemical change occurs. The butane molecules react with oxygen molecules in air to form new molecules, carbon dioxide and water. (See Figure 3.12.) The change is chemical because the molecules that compose the butane have changed into different molecules.

EXAMPLE 3.3 Physical and Chemical Changes

Determine whether each of the following is a physical or chemical change.

a) The rusting of iron
b) The evaporation of fingernail-polish remover (acetone) from the skin
c) The burning of coal
d) The fading of a carpet upon repeated exposure to sunlight

Solution:

a) Iron rusts because it reacts with oxygen in air to form iron oxide; therefore this is a chemical change.

b) When fingernail-polish remover (acetone) evaporates, it changes from liquid to gas, but it remains acetone; therefore, this is a physical change.
c) Coal burns because it reacts with oxygen in air to form carbon dioxide; this is a chemical change.
d) A carpet fades on repeated exposure to sunlight because the molecules that give the carpet its color are decomposed by sunlight; this is a chemical change.

SKILLBUILDER 3.3 **Physical and Chemical Changes**

Determine whether each of the following is a physical or chemical change.

a) Copper metal forming a blue solution when it is dropped into colorless nitric acid
b) A passing train flattening a penny placed on a railroad track
c) Ice melting into liquid water
d) A match igniting a firework

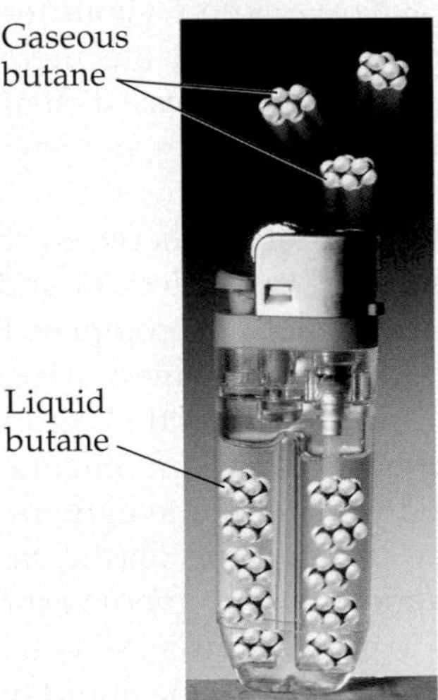

Figure 3.11 If you push the button on a lighter without turning the flint, some of the liquid butane vaporizes to gaseous butane. Since the liquid butane and the gaseous butane are both composed of butane molecules, this is a physical change.

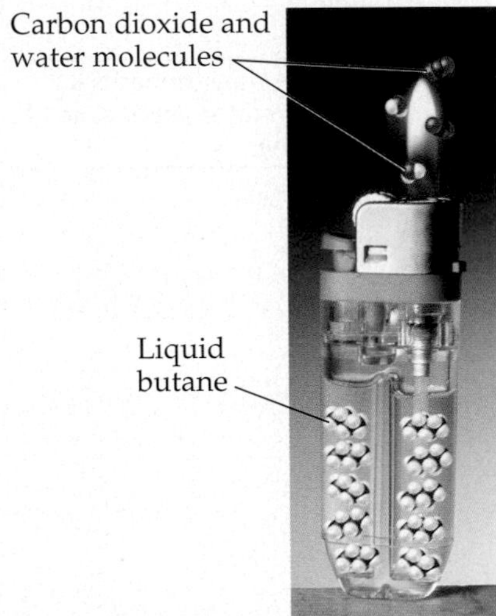

Figure 3.12 If you push the button *and* turn the flint to create a spark, you produce a flame. The butane molecules react with oxygen molecules in air to form new molecules, carbon dioxide and water. This is a chemical change.

3.7 Conservation of Mass: There Is No New Matter

As we have seen, our planet, our air, and even our own bodies are composed of matter. Chemical changes do not destroy matter, nor do they create new matter. Recall from Chapter 1 that Antoine Lavoisier, by studying combustion, established the law of conservation of mass, which states

Matter is neither created nor destroyed in a chemical reaction.

During chemical changes, the total amount of matter remains constant. How does this happen? When we burn butane in a lighter, the butane slowly disappears. Where does it go? It combines with oxygen to form carbon dioxide and water that go into the surrounding air. The mass of the carbon dioxide and water that form, however, must exactly equal the mass of the butane and oxygen that combined.

For example, 58 g of butane will react with 208 g of oxygen to form 176 g of carbon dioxide and 90 g of water.

$$\underbrace{\begin{matrix}\text{Butane} & + & \text{Oxygen}\\ 58\text{g} & + & 208\text{g}\end{matrix}}_{266\text{g}} \rightarrow \underbrace{\begin{matrix}\text{Carbon dioxide} & + & \text{Water}\\ 176\text{g} & + & 90\text{ g}\end{matrix}}_{266\text{g}}$$

We examine the quantitative relationships in chemical reactions in Chapter 8.

The sum of the masses of the butane and oxygen, 266 g, is equal to the sum of the masses of the carbon dioxide and water, which is also 266 g. Matter is conserved.

EXAMPLE 3.4 Conservation of Mass

A chemist forms 16.6 g of potassium iodide by combining 3.9 g of potassium with 12.7 g of iodine. Show that these results are consistent with the law of conservation of mass.

Solution:

The sum of the masses of the potassium and iodine is:

$$3.9\text{ g} + 12.7\text{ g} = 16.6\text{ g}$$

The sum of the masses of potassium and iodine equals the mass of the product, potassium iodide. The results are consistent with the law of conservation of mass.

SKILLBUILDER 3.4 Conservation of Mass

Suppose 12 g of natural gas combines with 48 g of oxygen in a flame. The chemical change produces 33 g of carbon dioxide and how many grams of water?

3.8 Energy

Water behind a dam contains potential energy. As the water flows through the dam, the potential energy is converted to electrical energy.

Matter is one of the two major components of our universe. The other major component is **energy**, defined as *the capacity to do work*. Like matter, energy is conserved. The **law of conservation of energy** states

Energy can neither be created nor destroyed.

The total energy is constant and cannot change; it can be transferred but not created.

An object possessing energy can do work on another object—it can cause the other object to move. For example, a moving billiard ball contains **kinetic energy**, energy associated with its motion. It can collide with another billiard ball and cause it to move. Water behind a dam contains **potential energy**, energy associated with its position. The water can flow from a higher position to a lower position through the dam, turn a turbine, and produce electrical energy. **Electrical energy** is the energy associated with the flow of electrical charge. Chemical systems contain **chemical energy**, energy associated with potential chemical changes, which can move or heat other objects. When we drive a car, we use chemical energy stored in gasoline to move the car forward. When we heat a home, we use chemical energy stored in natural gas to produce heat and warm the air in the house.

Units of Energy

Since energy can be converted from one form to another, all forms of energy have the same units. The SI unit of energy is the joule (J) named after the English scientist **James Joule** (1818–1889), who demonstrated that energy could be converted from one type to another as long as the total energy was conserved. A second unit of energy in common use is the **calorie (cal)**, the amount of energy required to raise the temperature of 1 g of water by 1°C. A calorie is a larger unit than a joule with the conversion 1 cal = 4.184 J. A related energy unit is the nutritional or capital *C* **Calorie**

CHEMISTRY IN THE MEDIA

Perpetual Motion

The law of conservation of energy has significant implications for energy use. The best we can do with energy is break even; we can't continually draw energy from a device without putting energy into it. A device that continuously produces energy without the need for energy input is called a **perpetual motion machine** (Figure 3.13) and, according to the law of conservation of energy, cannot exist. Occasionally, the media reports or speculates on the discovery of a system that appears to produce more energy than it consumes. For example, I once heard a radio talk show on the subject of energy and gasoline costs. The reporter suggested that we simply design an electric car that recharges itself while you drive. The battery in the electric car would charge during operation in the same way that the battery in a conventional car recharges, except the electric car would run with energy from the battery. People have dreamed of perpetual motion machines such as this for decades. However, such ideas violate the law of conservation of energy because they produce energy without any energy input. In the case of the perpetually moving electric car, the fault lies in the idea that driving the electric car can recharge the battery—it can't.

The reason that the battery in a conventional car recharges is that energy from gasoline combustion is converted into electrical energy that then charges the battery. The electric car needs energy to move forward and the battery would eventually discharge. Some new hybrid cars (electric and gasoline powered) can capture energy from braking and use that energy to recharge the battery. However, they could never run indefinitely without adding fuel. Our society has a continual need for energy, and as our current energy resources dwindle, new energy sources will be required. Unfortunately, those sources must also follow the law of conservation of energy—energy must be conserved.

Figure 3.13 A perpetual motion machine. The rolling balls supposedly keep the wheel perpetually spinning. **Question:** Can you explain why this would not work?

CAN YOU ANSWER THIS? *A friend asks you to invest in a new flashlight he invented that never needs batteries. What kind of questions should you ask before writing a check?*

(Cal), equivalent to 1000 little *c* calories. Electricity bills usually come in yet another energy unit, the **kilowatt-hour (kWh)**. Electricity costs about $0.15 per kWh. Table 3.2 shows various energy units and their conversion factors. Table 3.3 shows the amount of energy required for various processes in each of these units.

TABLE 3.2

Energy Conversion Factors

1 calorie (cal)	=	4.184 joules (J)
1 Calorie (Cal)	=	1000 calories (cal)
1 kilowatt-hour (kWh)	=	3.60×10^6 joules (J)

TABLE 3.3

Energy Use in Various Units

Unit	*Energy Required to Raise Temperature of 1 g of Water by 1 °C*	*Energy Required to Light 100-W Bulb for 1 Hour*	*Energy Used by Average U.S. Citizen in 1 Day*
joule (J)	4.18	3.6×10^5	9.0×10^8
calorie (cal)	1	8.6×10^4	2.15×10^8
Calorie (Cal)	0.001	86	215,000
kWh	1.1×10^{-6}	0.10	250

EXAMPLE 3.5 **Conversion of Energy Units**

A candy bar contains 225 Cal of nutritional energy. How many joules does it contain?

We solve this problem using the procedure to solve numerical problems from Chapter 2 (Section 2.10).

Given: 225 Cal

Find: J

Conversion Factors:

1000 calories = 1 Cal
4.184 J = 1 cal

Solution Map:

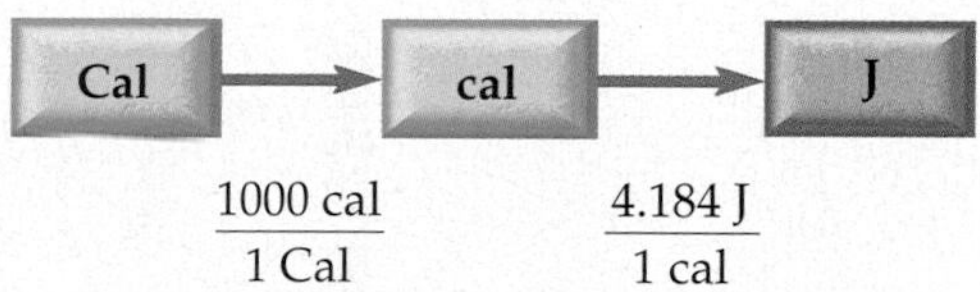

$$\frac{1000 \text{ cal}}{1 \text{ Cal}} \qquad \frac{4.184 \text{ J}}{1 \text{ cal}}$$

Solution:

We then follow the solution map to solve the problem and round to the correct number of significant digits.

$$225 \cancel{\text{Cal}} \times \frac{1000 \cancel{\text{cal}}}{1 \cancel{\text{Cal}}} \times \frac{4.184 \text{ J}}{1 \cancel{\text{cal}}} = 9.41 \times 10^5 \text{ J}$$

SKILLBUILDER 3.5 **Conversion of Energy Units**

The complete combustion of a small wooden match produces approximately 512 cal of heat. How many kilojoules are produced?

SKILLBUILDER PLUS

Remember that *kilo* means 1000.

Convert 2.75×10^4 kJ to cal.

3.9 Temperature: Random Molecular and Atomic Motion

The temperature of matter is related to the random motion of the atoms and molecules that compose it. The hotter an object is, the greater the motion and the higher the temperature. There are three different temperature scales in common use. The most familiar in the United States is the **Fahrenheit (°F)** scale. On the Fahrenheit scale, water freezes at 32 °F and boils at 212 °F. Room temperature is approximately 75 °F. The Fahrenheit scale was initially set up by assigning 0°F to the freezing point of a concentrated saltwater solution and 100 °F to body temperature.

The scale often used by scientists is the **Celsius (°C)** scale. On this scale, water freezes at 0 °C and boils at 100 °C. Room temperature is approximately 25 °C. The Fahrenheit and Celsius scales differ both in the size of their respective degrees and the temperature each calls "zero" (Figure 3.14). Both the Fahrenheit and Celsius scales contain negative temperatures. However, a third temperature scale, called the **Kelvin (K)** scale, avoids negative temperatures by assigning 0 K to the coldest temperature possible, absolute zero. Absolute zero (−273 °C or −459 °F) is the

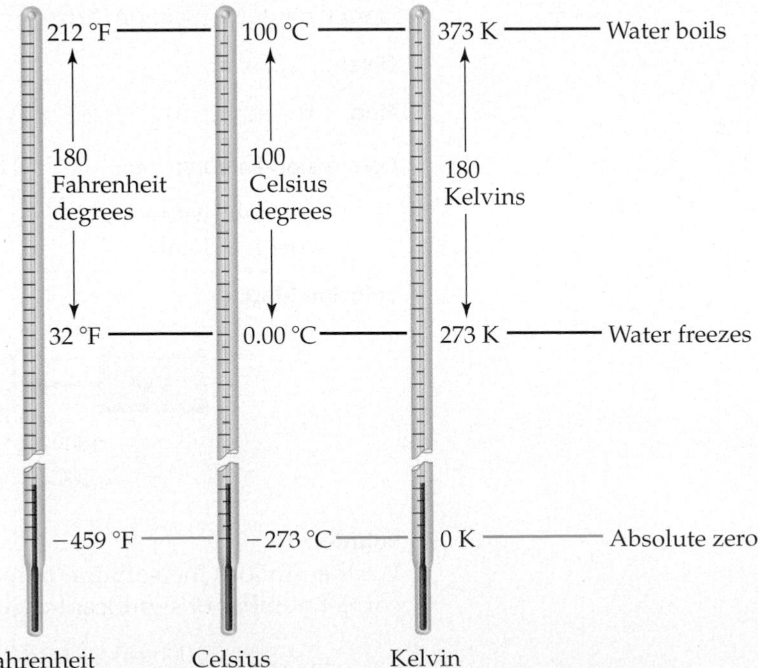

Figure 3.14 A comparison of the Fahrenheit, Celsius, and Kelvin temperature scales. The Fahrenheit degree is five-ninths the size of a Celsius degree. The Celsius degree and the kelvin degree are the same size.

temperature at which molecular motion stops. There is no lower temperature. The size of the kelvin degree is identical to the Celsius degree—the only difference is the temperature that each calls zero.

Conversions between these temperature scales are achieved using the following formulas.

$$K = °C + 273$$

$$°C = \frac{(°F - 32)}{1.8}$$

For example, suppose we want to convert 212 K to Celsius. Following the procedure for solving numerical problems (Section 2.10), we write:

Given: 212 K

Find: °C

Equation:

The equation that relates the given quantity (K) to the find quantity (°C) is:

$$K = °C + 273$$

Solution Map:

In a solution map involving a formula, the formula establishes the relationship between the variables. However, the formula under the arrow is not necessarily solved for the correct variable until later, as is the case here.

We then build a solution map.

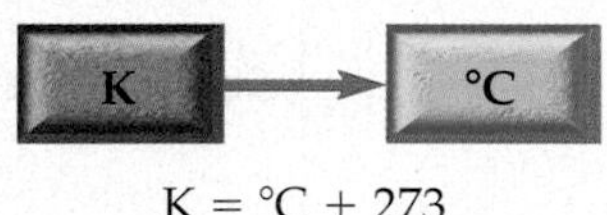

K = °C + 273

Solution:

The equation below the arrow shows the relationship between K and °C, but it is not in the correct form. We must solve the equation for °C.

$$K = °C + 273$$
$$°C = K - 273$$

Finally, we substitute the given value for K and compute the answer to the correct number of significant figures.

$$°C = 212 - 273$$
$$= -61\ °C$$

EXAMPLE 3.6 Converting between Celsius and Kelvin Temperature Scales

Convert −25 °C to kelvin.

Given: −25 °C

Find: K

Equation:

We find the equation that shows the relationship between the given quantity (°C) and the find quantity (K).

$$K = °C + 273$$

Solution Map:

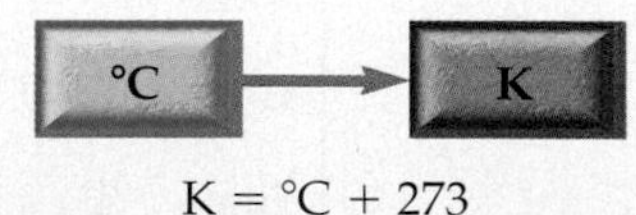

K = °C + 273

Solution:

We then follow the solution map to solve the problem by substituting the correct value for °C and computing the answer to the correct number of significant figures.

$$K = °C + 273$$

$$K = -25\,°C + 273 = 248\text{ K}$$

SKILLBUILDER 3.6 Converting Between Celsius and Kelvin Temperature Scales

Convert 358 K to Celsius.

EXAMPLE 3.7 Converting Between Fahrenheit and Celsius Temperature Scales

Convert 55 °F to Celsius.

Given: 55 °F

Find: °C

Equation: $°C = \frac{(°F - 32)}{1.8}$

Solution Map:

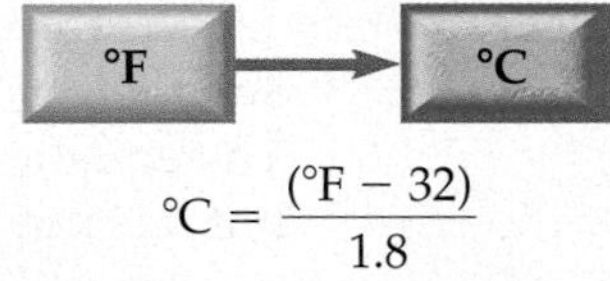

$$°C = \frac{(°F - 32)}{1.8}$$

Solution:

The equation beneath the arrow in the solution map shows the relationship between the find quantity (°C) and the given quantity (°F). We can now substitute into this equation and compute the answer to the correct number of significant figures.

$$°C = \frac{(°F - 32)}{1.8}$$

$$°C = \frac{(55 - 32)}{1.8} = 12.778\,°C = 13\,°C$$

SKILLBUILDER 3.7 Converting Between Fahrenheit and Celsius Temperature Scales

Convert 139 °C to Fahrenheit.

EXAMPLE 3.8 Converting Between Fahrenheit and Kelvin Temperature Scales

Convert 310 K to Fahrenheit.

Given: 310 K

Find: °F

Equations:

This problem requires two equations: one relating K and °C and the other relating °C and °F.

$$K = °C + 273$$

$$°C = \frac{(°F - 32)}{1.8}$$

Solution Map:

This conversion requires two steps, one to convert from K to °C and one to convert from °C to °F.

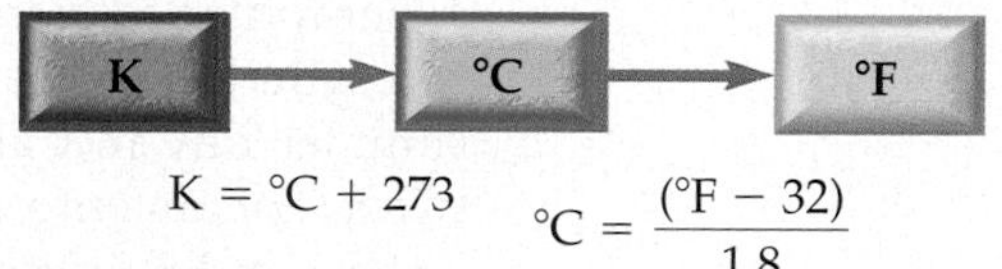

$$K = °C + 273 \qquad °C = \frac{(°F - 32)}{1.8}$$

Solution:

The first equation must be solved for °C.

$$K = °C + 273$$
$$°C = K - 273$$

We now substitute into this equation to convert from K to °C.

$$°C = K - 273$$
$$°C = 310 - 273 = 37\ °C$$

The second equation must be solved for °F.

$$°C = \frac{(°F - 32)}{1.8}$$
$$1.8\ °C = (°F - 32)$$
$$°F = 1.8\ °C + 32$$

We then substitute into this equation to convert from °C to °F and compute the answer to the correct number of significant digits.

$$°F = 1.8(°C) + 32$$
$$°F = 1.8\,(37) + 32 = 98.6\ °F = 99\ °F$$

SKILLBUILDER 3.8 **Converting Between Fahrenheit and Kelvin Temperature Scales**

Convert −321 °F to kelvin.

3.10 Temperature Changes: Heat Capacity

All substances change temperature when they are heated, but how much they change for a given amount of heat varies significantly from one substance to another. For example, if you put a steel skillet on a flame, its temperature rises rapidly. However, if you put some water in the skillet, the temperature increase is much slower. Why? The first reason is that when you add water, the same amount of heat energy must warm more matter, so the temperature rises more slowly. The second and more significant reason is that water is more resistant to temperature change than steel. This is because water has a higher **heat capacity**. The heat capacity of a substance is the quantity of heat energy (usually in joules) required to change the temperature of a given amount of the substance by 1 °C. When the amount of the substance is expressed in grams, the heat capacity is called the **specific heat capacity** (or simply the **specific heat**) and has the units of joules per gram degree Celsius (J/g °C). Table 3.4 shows the values of the specific heat capacity for several substances.

Notice that water has the highest heat capacity on the list—changing its temperature requires a lot of heat. If you have traveled from an inland geographical region to a coastal one and have felt the drop in temperature, you have experienced the effects of water's high heat capacity. On a summer day in California, for example, the temperature difference between Sacramento (inland city) and San Francisco (coastal city) can be 30 °F; San Francisco enjoys a cool 68 °F, while Sacramento bakes at near 100 °F. Yet the intensity of sunlight falling on these two cities is the same. Why the large temperature difference? The difference between the two locations is the presence of the Pacific Ocean, which practically engulfs San Francisco. Water, with its high heat capacity, absorbs much of the sun's heat without undergoing a large increase in temperature, keeping San Francisco cool. The low heat-capacity land surrounding Sacramento, on the other hand, cannot absorb a lot of heat without a large increase in temperature—it has a lower *capacity* to absorb heat without a large temperature increase.

Similarly, only two U.S. states have never recorded a temperature above 100 °F. One of them is obvious, Alaska. It is too far north to get that hot. The other one, however, may come as a surprise. It is Hawaii. The high heat-capacity water that surrounds the only island state moderates the temperature, keeping Hawaii from ever getting too hot.

TABLE 3.4

Specific Heat Capacities of Some Common Substances

Substance	*Specfic Heat Capacity (J/g °C)*
lead	0.128
gold	0.128
silver	0.235
copper	0.385
iron	0.449
aluminum	0.903
ethanol	2.42
water	4.18

It is simply by coincidence that lead and gold have the same value of specific heat capacity.

San Francisco enjoys cool weather even in summer months because of the high heat capacity of the surrounding ocean.

EVERYDAY CHEMISTRY

Coolers, Camping, and the Heat Capacity of Water

Have you ever loaded a cooler with ice and then added room-temperature drinks? If you have, you know that the ice quickly melts. In contrast, if you load your cooler with prechilled drinks, the ice lasts for hours. Why the difference? The answer is related to the high heat capacity of water within the drinks. As we just learned, water must absorb a lot of heat to raise its temperature, but it must also release a lot of heat to lower its temperature. When the warm drinks are placed into the ice, they release heat, which then melts the ice. The prechilled drinks, on the other hand, are already cold, so they do not release much heat. It is always better to load your cooler with prechilled drinks—that way, the ice will last the rest of the day.

The ice in a cooler loaded with cold drinks lasts much longer than the ice in a cooler loaded with warm drinks. **Question:** Can you explain why?

CAN YOU ANSWER THIS? *Suppose you are cold-weather camping and decide to heat some objects to bring into your sleeping bag for added warmth. You place a large water jug and a rock of equal mass close to the fire. Over time, both the rock and the water jug warm to about 38 °C (100 °F). If you could bring only one into your sleeping bag, which one should you bring to keep you the warmest? Why?*

3.11 Energy and Heat Capacity Calculations

The specific heat capacity of a substance can be used to quantify the relationship between the amount of heat added to a given amount of the substance and the corresponding temperature increase. The equation that relates these quantities is:

$$\text{Heat} = \text{Mass} \times \text{Heat capacity} \times \text{Temperature change}$$
$$q = m \times C \times \Delta T$$

where q is the amount of heat in joules, m is the mass of the substance in grams, C is the heat capacity in joules per gram degree Celsius, and ΔT is the temperature change in Celsius. The symbol Δ means *the change in*, so ΔT means *the change in temperature*. For example, suppose you are making a cup of tea and want to know how much heat energy will warm 235 g of water (about 8 oz) from 25 °C to 100.0 °C (boiling). We set up this problem as follows:

ΔT in °C is equal to ΔT in K but is not equal to ΔT in °F.

Given: 235 g water
25 °C initial temperature (T_i)
100.0 °C final temperature (T_f)

Find: amount of heat needed, q

Equation:

The equation that relates the given and find quantities is:

$$q = m \cdot C \cdot \Delta T$$

Solution Map:

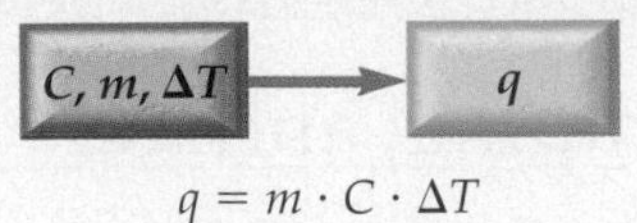

$$q = m \cdot C \cdot \Delta T$$

In addition to m and ΔT, the equation requires C, the heat capacity of water. Our next step is to gather all of the required quantities for the equation (C, m, and ΔT) in the correct units. These are:

$$C = 4.18 \text{ J/g °C}$$
$$m = 235 \text{ g}$$

The other required quantity is ΔT. The change in temperature is the difference between the final temperature (T_f) and the initial temperature (T_i).

$$\Delta T = T_f - T_i$$
$$= 100.0 \text{ °C} - 25 \text{ °C} = 75 \text{ °C}$$

Solution:
Finally, we substitute the correct values into the equation and compute the answer to the correct number of significant figures.

$$q = m \times C \times \Delta T$$
$$= 235 \cancel{\text{g}} \times 4.18 \frac{\text{J}}{\cancel{\text{g}} \cancel{\text{°C}}} \times 75 \cancel{\text{°C}}$$
$$= 7.367 \times 10^4 \text{ J} = 7.4 \times 10^4 \text{ J}$$

It is critical that you substitute each of the correct variables into the equation in the correct units and cancel units as you compute the answer. If, during this process, you learn that one of your variables is not in the correct units, convert it to the correct units using the skills you learned in Chapter 2. Notice that the sign of q is positive (+) if the substance is increasing in temperature (heat going into the substance) and negative (−) if the substance is decreasing in temperature (heat going out of the substance).

EXAMPLE 3.9 Relating Heat Energy to Temperature Changes

Gallium is a solid metal at room temperature but melts at 29.9 °C. If you hold gallium in your hand, it melts from body heat. How much heat must 2.5 g of gallium absorb from your hand to raise its temperature from 25.0 °C to 29.9 °C? The heat capacity of gallium is 0.372 J/g °C.

Given: 2.5 g gallium

$T_i = 25.0$ °C

$T_f = 29.9$ °C

$C = 0.372$ J/g °C

Find: q

Equation: $q = m \cdot C \cdot \Delta T$

Solution Map:

C, m, ΔT → q

$$q = m \cdot C \cdot \Delta T$$

The necessary quantities for the equation are C, m, and ΔT. The values of these are:

$$C = 0.372\ \text{J/g °C}$$
$$m = 2.5\ \text{g}$$
$$\Delta T = 29.9\ \text{°C} - 25.0\ \text{°C} = 4.9\ \text{°C}$$

Solution:
We then substitute the correct variables into the equation, canceling units, and compute the answer to the right number of significant figures.

$$q = m \times C \times \Delta T$$
$$= 2.5\ \cancel{\text{g}} \times 0.372 \frac{\text{J}}{\cancel{\text{g}}\ \cancel{\text{°C}}} \times 4.9\ \cancel{\text{°C}} = 4.557\ \text{J} = 4.6\ \text{J}^{*}$$

SKILLBUILDER 3.9 **Relating Heat Energy to Temperature Changes**

You find a copper penny (pre-1982) in the snow and pick it up. How much heat is absorbed by the penny as it warms from the temperature of the snow, −5.0 °C, to the temperature of your body, 37.0 °C? Assume the penny is pure copper and has a mass of 3.10 g. You can find the heat capacity of copper in Table 3.4.

SKILLBUILDER PLUS

The temperature of a lead fishing weight rises from 26 °C to 38 °C as it absorbs 11.3 J of heat. What is the mass of the fishing weight in grams?

*Note: This is the amount of heat required to raise the temperature to melting point. Actually melting the gallium would require additional heat.

EXAMPLE 3.10 **Relating Heat Capacity to Temperature Changes**

A chemistry student finds a shiny rock that she suspects is gold. She weighs the rock on a balance and obtains the mass, 14.3 g. She then finds that the temperature of the rock rises from from 25 °C to 52 °C upon absorption of 174 J of heat. Find the heat capacity of the rock and determine if the value is consistent with the heat capacity of gold.

Given: 14.3 g "gold" rock

174 J of heat absorbed

$T_i = 25$ °C

$T_f = 52$ °C

Find: C

Equation: $q = m \cdot C \cdot \Delta T$

Solution Map:

$m, q, \Delta T$ → C

$$q = m \cdot C \cdot \Delta T$$

Solution:

In this case, the required quantities for the equation are m, q, and ΔT.

$$m = 14.3 \text{ g}$$
$$q = 174 \text{ J}$$
$$\Delta T = 52\ °\text{C} - 25\ °\text{C} = 27\ °\text{C}$$

Before we substitute the variables into the equation, we solve the equation for C.

$$q = m \cdot C \cdot \Delta T$$
$$C = \frac{q}{m\Delta T}$$

We then substitute the variables into the equation and compute C to the correct number of significant digits.

$$C = \frac{174 \text{ J}}{14.3 \text{ g} \times 27\ °\text{C}} = 0.4506 \frac{\text{J}}{\text{g}\ °\text{C}} = 0.45 \frac{\text{J}}{\text{g}\ °\text{C}}$$

By comparing the computed value of the heat capacity (0.45 J/g °C) with the heat capacity of gold from Table 3.4 (0.128 J/g °C), we conclude that the rock is not gold.

SKILLBUILDER 3.10 **Relating Heat Capacity to Temperature Changes**

A 328 g sample of water absorbs 5.78×10^3 J of heat. Find the change in temperature for the water. If the water is initially at 25.0 °C, what is its final temperature?

CHAPTER IN REVIEW

Chemical Principles	Relevance
Matter: Matter is anything that occupies space and has mass. It is composed of atoms, which are often bonded together as molecules. Matter can exist as a solid, a liquid, or a gas. Solid matter can either be amorphous or crystalline.	**Matter:** Understanding matter is relevant because everything is made of matter—you, me, the chair you sit on, and the air we breathe. The physical universe basically contains only two things: matter and energy. So we begin our study of chemistry by defining and classifying these two building blocks of the universe.
Classification of Matter: Matter can be classified according to its composition. Pure matter is composed of only one type of substance; that substance may be an element, a substance that cannot be decomposed into simpler substances, or it may be a compound, a substance composed of two or more elements in fixed definite proportions. Mixtures are composed of two or more different substances whose proportions may vary from one sample to the next. Mixtures can be either homogenous, having the same composition throughout, or they may be heterogeneous, having two or more regions with different compositions.	**Classification of Matter:** Since ancient times, humans have tried to understand matter and harness it for their purposes. The earliest humans shaped matter into tools and used the transformation of matter—especially fire—to keep warm and to cook food. To manipulate matter, we must understand it. Fundamental to this understanding is the connection between the properties of matter and the molecules and atoms that compose it. If matter is composed of a single type of atom or molecule, it is a pure substance. If it is composed of two or more types of atoms or molecules, it is a mixture.

Properties and Changes of Matter: The properties of matter can be divided into two types, physical and chemical. The physical properties of matter are those that are displayed without a change in composition. The chemical properties of matter can only be displayed with a change in composition. The changes in matter can themselves be divided into physical and chemical. In a physical change, the appearance of matter may change but its composition does not. In a chemical change, the composition of matter changes.

Properties and Changes of Matter: The physical and chemical properties of matter make the world around us the way it is. For example, a physical property of water is its boiling point at sea level—100 °C. If water boiled at a different temperature at sea level, it would be a different substance. The physical properties of water—and all matter—are determined by the atoms and molecules that compose it. If water molecules were different—even slightly different—water would boil at a different temperature. Imagine a world where water boiled at room temperature.

Conservation of Mass: Whether the changes in matter are chemical or physical, matter is always conserved. In a chemical change the masses of the matter undergoing the chemical change must equal the sum of the masses of matter resulting from the chemical change.

Conservation of Mass: The conservation of matter is relevant to, for example, pollution. We often think that we create pollution, but actually, we are powerless to create anything. Matter cannot be created. So pollution is simply misplaced matter—matter that has been put into places it does not belong.

Energy: Besides matter, energy is the other major component of our universe. Like matter, energy is conserved—it can neither be created nor destroyed. Energy comes in various different types, and these can be converted from one to another. Some common units of energy are the joule (J), the calorie (cal), the nutritional Calorie (Cal), and the kilowatt-hour (kWh).

Energy: Our society's energy sources will not last forever because as we burn fossil fuels—our primary energy source—we convert chemical energy, stored in molecules, to kinetic and thermal energy. The kinetic and thermal energy is not readily available to be used again. Consequently, our energy resources are dwindling, and the conservation of energy implies that we will not be able to simply create new energy—it must come from somewhere.

Temperature: The temperature of matter is related to the random motions of the molecules and atoms that compose it—the greater the motion, the higher the temperature. Temperature is commonly measured on three scales: Fahrenheit (°F), Celsius (°C), and Kelvin (K).

Temperature: The temperature of matter and its measurement is relevant to many everyday phenomena. Humans are constantly interested in the weather, and air temperature is a fundamental part of weather. We use body temperature as one measure of human health and global temperature as one measure of the planet's health.

Heat Capacity: The temperature change that a sample of matter undergoes upon absorption of a given amount of heat is related to the heat capacity for the substance composing the matter. Water has one of the highest heat capacities, meaning that it is most resistant to rapid temperature changes.

Heat Capacity: The heat capacity of water explains why it is cooler in coastal areas, which are near large bodies of high heat-capacity water, than in inland areas, which are surrounded by low heat-capacity land. It also explains why it takes longer to cool a refrigerator filled with liquids than an empty one.

Chemical Skills

Classifying Matter (Sections 3.3, 3.4)

Begin by examining the alphabetical listing of elements in the back of this text. If the substance is listed in that table, it is a pure substance and an element.

If the substance is not listed in that table, use your knowledge about the substance to determine if it is a pure substance. If it is a pure substance not listed in the table, then it is a compound.

If it is not a pure substance, then it is a mixture. Use your knowledge about the mixture to determine if it has uniform composition throughout (homogeneous) or non-uniform composition (heterogeneous).

Examples

EXAMPLE 3.11 Classifying Matter

Classify each of the following as a pure substance or a mixture. If it is a pure substance, classify it as an element or compound. If it is a mixture, classify it as homogeneous or heterogeneous.

a) Pure silver
b) Swimming pool water
c) Dry ice (solid carbon dioxide)
d) Blueberry muffin

Solution:

a) Pure element; silver appears in the element table.
b) Homogeneous mixture; pool water contains at least water and chlorine in variable proportions, and it is uniform throughout.
c) Compound; dry ice is a pure substance (carbon dioxide), but it is not listed in the table.
d) Heterogeneous mixture; a blueberry muffin is a mixture of several things and has nonuniform composition.

Physical and Chemical Properties (Section 3.5)

To distinguish between physical and chemical properties, ask whether or not the substance changes composition while displaying the property. If it *does not* change composition, the property is physical; if it *does*, the property is chemical.

EXAMPLE 3.12 Physical and Chemical Properties

Determine whether each of the following is a physical or chemical property.

a) The tendency for platinum jewelry to scratch easily
b) The ability of sulfuric acid to burn the skin
c) The ability of hydrogen peroxide to bleach hair
d) The density of lead relative to other metals

Solution:

a) Physical; scratched platinum is still platinum.
b) Chemical; the acid chemically reacts with the skin to produce the burn.
c) Chemical; the hydrogen peroxide chemically reacts with hair to bleach it.
d) Physical; the heaviness can be felt without changing the lead into anything else.

Chemical and Physical Changes (Section 3.6)

To distinguish between physical and chemical changes, ask whether or not the substance changes composition during the change. If it *does not* change composition the change is physical; if it *does*, the change is chemical.

EXAMPLE 3.13 Chemical and Physical Changes

Determine whether each of the following is a physical or chemical change.

a) The explosion of gunpowder in the barrel of a gun
b) The melting of gold in a furnace
c) The bubbling that occurs upon mixing baking soda and vinegar
d) The bubbling that occurs when water boils

Solution:

a) Chemical; the gunpowder reacts with oxygen during the explosion.
b) Physical; the liquid gold is still gold.
c) Chemical; the bubbling is a result of a chemical reaction between the two substances to form new substances, one of which is carbon dioxide released as bubbles.
d) Physical; the bubbling is due to liquid water turning into gaseous water, but it is still water.

Conservation of Mass (Section 3.7)

The sum of the masses of the substances involved in a chemical change must be the same before and after the change.

EXAMPLE 3.14 Conservation of Mass

An automobile runs for 10 minutes and burns 47 g of gasoline. The gasoline combined with oxygen from air and formed 132 g of carbon dioxide and 34 g of water. How much oxygen was consumed in the process?

Solution:

The total mass after the chemical change is:

$$132 \text{ g} + 34 \text{ g} = 166 \text{ g}$$

The total mass before the change must also be 166 g.

$$47 \text{ g} + \text{g oxygen} = 166 \text{ g}$$

So, the mass of oxygen consumed is the total mass (166 g) minus the mass of gasoline (47 g).

$$\text{g oxygen} = 166 \text{ g} - 47 \text{ g} = 119 \text{ g}$$

Conversion of Energy Units (Section 3.8)

Solve energy units conversion problems using the problem solving procedure that we learned in Section 2.10.

- Write down the given quantity and its units.
- Write down the quantity to find and/or its units.
- Write down the appropriate conversion factor(s).
- Write a solution map.

EXAMPLE 3.15 Conversion of Energy Units

Convert 1.7×10^3 kWh (the amount of energy used by the average U.S. citizen in one week) into calories.

Given: 1.7×10^3 kWh

Find: cal

Conversion Factors: $1 \text{ kWh} = 3.60 \times 10^6 \text{ J}$

$1 \text{ cal} = 4.18 \text{ J}$

Solution Map:

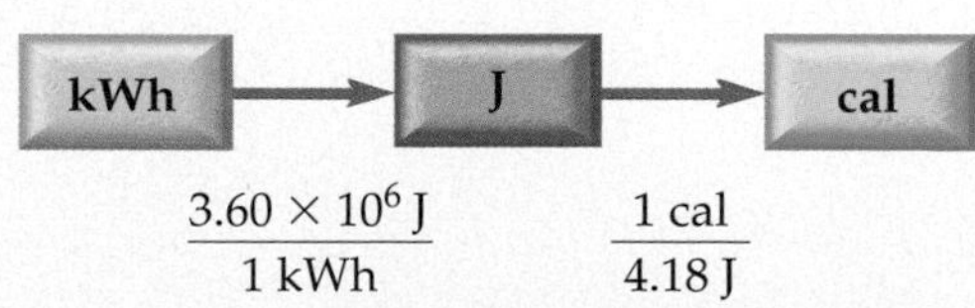

$$\frac{3.60 \times 10^6 \text{ J}}{1 \text{ kWh}} \qquad \frac{1 \text{ cal}}{4.18 \text{ J}}$$

- Starting with the given quantity and its units, multiply by the appropriate conversion factor(s), canceling units, to arrive at the quantity to find in the desired units.
- Round the final answer to the correct number of significant figures.
- Check the answer.

Solution:

$$1.7 \times 10^3 \cancel{\text{kWh}} \times \frac{3.60 \times 10^6 \cancel{\text{J}}}{1 \cancel{\text{kWh}}} \times \frac{1 \text{ cal}}{4.18 \cancel{\text{J}}}$$

$$= 1.464 \times 10^9 \text{ cal}$$

$$1.464 \times 10^9 \text{ cal} = 1.5 \times 10^9 \text{ cal}$$

The unit of the answer, cal, is correct. The magnitude of the answer makes sense since cal is a smaller unit than kWh.

Converting Between Celsius and Kelvin Temperature Scales (Section 3.9)

Solve temperature conversion problems using the problem solving procedure that we learned in Section 2.10. Take the steps appropriate for equations.

- Write down the given and find quantities.
- Write down the appropriate equation that relates K and °C.
- Draw a solution map. Use the appropriate equation that relates the given quantities to the find quantities.
- Solve the equation for the find quantity (°C).
- Substitute the appropriate quantity into the equation and compute the answer to the correct number of significant figures.
- Check the answer.

EXAMPLE 3.16 Converting Between Celsius and Kelvin Temperature Scales

Convert 257 K to Celsius

Given: 257 K

Find: °C

Equation: K = °C + 273

Solution Map:

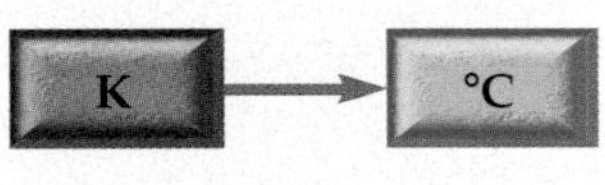

K = °C + 273

Solution:

$$\text{K} = {}^\circ\text{C} + 273$$

$$^\circ\text{C} = \text{K} - 273$$

$$^\circ\text{C} = 257 - 273 = -16\ ^\circ\text{C}$$

The answer has the correct unit, and its magnitude seems correct (See Figure 3.14).

Converting Between Fahrenheit and Celsius Temperature Scales (Section 3.9)

- Write down the given and find quantities.
- Write down the appropriate equation that relates °C and °F.
- Draw a solution map.

EXAMPLE 3.17 Converting Between Fahrenheit and Celsius Temperature Scales

Convert 62.0 °C to °F.

Given: 62.0 °C

Find: °F

Equation: $^\circ\text{C} = \dfrac{(^\circ\text{F} - 32)}{1.8}$

Solution Map:

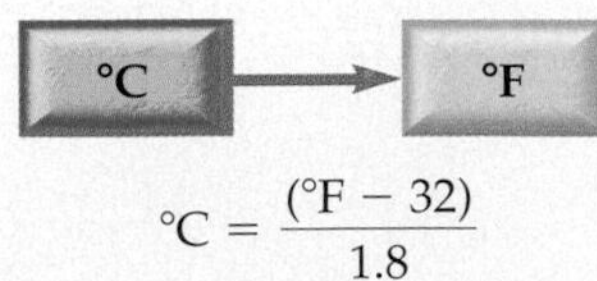

- Solve the equation for the find quantity (°F).

Solution:

$$°C = \frac{(°F - 32)}{1.8}$$

$$1.8\,(°C) = °F - 32$$

$$°F = 1.8\,(°C) + 32$$

- Substitute the appropriate quantity into the equation and compute the answer to the correct number of significant figures.

$$°F = 1.8\,(62.0) + 32 = 143.60\ °F = 144\ °F$$

- Check the answer.

The answer has the correct unit, and its magnitude seems correct (See Figure 3.14).

Energy, Temperature Change, and Heat Capacity Calculations (Sections 3.10, 3.11)

EXAMPLE 3.18 Energy, Temperature Change, and Heat Capacity Calculations

What is the temperature change in 355 mL of water upon absorption of 34 kJ of heat?

- Write down the given and find quantities.

Given: 355 mL water; 34 kJ of heat

Find: ΔT

- Write down the equation that relates the given and find quantities.

Equation: $q = m \cdot C \cdot \Delta T$

- We build a solution map writing the appropriate equation—the one showing the relationship between the given and find quantities—below the arrow.

Solution Map:

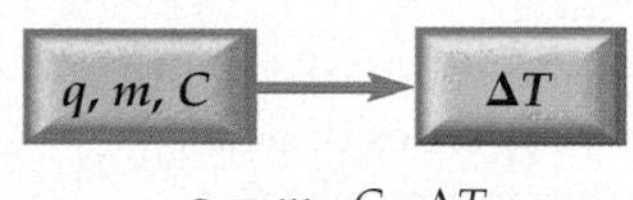

$$q = m \cdot C \cdot \Delta T$$

- Collect each of the quantities required in the equation in the correct units.

The value for q must be converted from kJ to J:

$$q = 34\ \text{kJ} \times \frac{1000\ \text{J}}{1\ \text{kJ}} = 3.4 \times 10^4\ \text{J}$$

The value for m must converted from mL to g; use the density of water, 1.0 g/mL, to convert mL to g.

$$m = 355\ \text{mL} \times \frac{1.0\ \text{g}}{1\ \text{mL}} = 355\ \text{g}$$

$$C = 4.18\ \text{J/g °C}$$

- Solve the equation for the find quantity (ΔT).

Solution:

$$q = m \times C \times \Delta T$$

$$\Delta T = \frac{q}{mC}$$

- Substitute the appropriate quantities into the equation and compute the answer to the correct number of significant figures.

$$\Delta T = \frac{3.4 \times 10^4\ \text{J}}{355\ \text{g} \times 4.18\ \text{J/g °C}}$$

$$= 22.91\ °C = 23\ °C$$

- Check the answer.

The answer has the correct units and the magnitude seems correct. If the magnitude of the answer were a huge number, such as 3×10^6 for example, we would go back and look for a mistake. If water were to go above 100 °C, it would boil, so such a large answer would be unlikely.

KEY TERMS

amorphous [3.3]
atoms [3.2]
calorie (c) [3.8]
Calorie (C) [3.8]
Celsius (°C) [3.9]
chemical changes [3.6]
chemical energy [3.8]
chemical properties [3.5]
compound [3.4]
compressible [3.3]
crystalline [3.3]
electrical energy [3.8]
element [3.4]
energy [3.8]
Fahrenheit (°F) [3.9]
gas [3.3]
specific heat [3.10]
heat capacity [3.10]
heterogeneous mixture [3.4]
homogeneous mixture [3.4]
James Joule [3.8]
kelvin (K) [3.9]
kilowatt-hour (kWh) [3.8]
kinetic energy [3.8]
law of conservation of energy [3.8]
liquid [3.3]
matter [3.2]
mixture [3.4]
molecules [3.2]
perpetual motion machine [3.8]
physical changes [3.6]
physical properties [3.5]
potential energy [3.8]
properties [3.5]
pure substance [3.4]
solid [3.3]
specific heat [3.10]
state of matter [3.3]

EXERCISES

Questions

The answers to all exercises numbered in blue appear in Appendix 5.

1. Define matter and give some examples.
2. What is matter composed of?
3. What are the three states of matter?
4. What are the properties of a solid?
5. What is the difference between a crystalline solid and an amorphous solid?
6. What are the properties of a liquid?
7. What are the properties of a gas?
8. Why are gases compressible?
9. What is a mixture?
10. What is the difference between a homogeneous mixture and a heterogeneous mixture?
11. What is a pure substance?
12. What is an element? A compound?
13. What is the difference between a mixture and a compound?
14. What is a physical property? What is a chemical property?
15. What is the difference between a physical change and a chemical change?
16. What is the law of conservation of mass?
17. What is the definition of energy?
18. What is the law of conservation of energy?
19. Name some different kinds of energy.
20. What are three common units for energy?
21. What are three common units for measuring temperature?
22. How do the three temperature scales differ?
23. What is heat capacity?
24. Why are coastal geographic regions normally cooler in the summer than inland geographic regions?
25. The following equation can be used to convert temperature in Fahrenheit to temperature in Celsius.

$$°C = \frac{(°F - 32)}{1.8}$$

Use algebra to change the equation to convert temperature in Celsius to temperature in Fahrenheit.

26. The following equation can be used to convert temperature in Celsius to temperature in kelvin.

$$K = °C + 273$$

Use algebra to change the equation to convert temperature in kelvin to temperature in Celsius.

Problems

(Note: The exercises in this section of Problems are paired and the answers to the odd-numbered exercises appear in Appendix 5.)

Classifying Matter

27. Classify each of the following pure substances as an element or a compound.
a) aluminum
b) sulfur
c) methane
d) acetone

28. Classify each of the following pure substances as an element or a compound.
a) carbon
b) baking soda (sodium bicarbonate)
c) nickel
d) gold

29. Classify each of the following mixtures as homogeneous or heterogeneous.
a) coffee
b) chocolate sundae
c) apple juice
d) gasoline

30. Classify each of the following mixtures as homogeneous or heterogeneous.
a) baby oil
b) chocolate chip cookie
c) water and gasoline
d) wine

31. Classify each of the following as a pure substance or a mixture. If it is a pure substance, classify it as an element or a compound. If it is a mixture, classify it as homogeneous or heterogeneous.
a) helium gas
b) clean air
c) rocky road ice cream
d) concrete

32. Classify each of the following as a pure substance or a mixture. If it is a pure substance, classify it as an element or a compound. If it is a mixture, classify it as homogeneous or heterogeneous.
a) urine
b) pure water
c) Snickers bar
d) soil

Physical and Chemical Properties and Physical and Chemical Changes

33. Classify each of the following properties as physical or chemical.
a) the tendency of silver to tarnish
b) the shine of chrome
c) the color of gold
d) the flammability of propane gas

34. Classify each of the following properties as physical or chemical.
a) the boiling point of ethyl alcohol
b) the temperature at which dry ice evaporates
c) the flammability of ethyl alcohol
d) the smell of perfume

35. The following list contains several properties of ethylene (a ripening agent for bananas). Which are physical properties and which are chemical?
- colorless
- odorless
- flammable
- gas at room temperature
- one liter has a mass of 1.260 g under standard conditions
- mixes with acetone
- polymerizes to form polyethylene

36. The following list contains several properties of ozone (a pollutant in the lower atmosphere but part of a protective shield against UV light in the upper atmosphere). Which are physical and which are chemical?
- bluish color
- pungent odor
- very reactive
- decomposes on exposure to ultraviolet light
- gas at room temperature

37. Determine whether each of the following changes is physical or chemical.
a) A balloon filled with hydrogen gas explodes upon contact with a spark.
b) The liquid propane in a barbecue evaporates away because the user left the valve open.
c) The liquid propane in a barbecue ignites upon contact with a spark.
d) Copper metal turns green on exposure to air and water.

38. Determine whether each of the following changes is physical or chemical.
a) Sugar dissolves in hot water.
b) Sugar burns in a pot.
c) A metal surface becomes dull because of continued abrasion.
d) A metal surface becomes dull on exposure to air.

39. A block of aluminum is a) ground into aluminum powder and then b) ignited. It then emits flames and smoke. Classify a) and b) as chemical or physical changes.

40. Several pieces of graphite from a mechanical pencil are a) broken into tiny pieces. Then the pile of graphite is b) ignited with a hot flame. Classify **a)** and **b)** as chemical or physical changes.

The Conservation of Mass

41. An automobile gasoline tank holds 42 kg of gasoline. When the gasoline burns, 168 kg of oxygen are consumed and carbon dioxide and water are produced. What is the total combined mass of carbon dioxide and water that is produced?

42. In the explosion of a hydrogen-filled balloon, 0.50 g of hydrogen reacted with 4.0 g of oxygen to form how many grams of water vapor? (Water vapor is the only product.)

43. Are the following data sets on chemical changes consistent with the law of conservation of mass?
a) A 6-g sample of hydrogen react with 48 g of oxygen to form 54 g of water.
b) A 55-g sample of gasoline react with 221 g of oxygen to form 187 g of carbon dioxide and 80 g of water.

44. Are the following data sets on chemical changes consistent with the law of conservation of mass?
a) An 8-g sample of natural gas react with 32 g of oxygen gas to form 17 g of carbon dioxide and 16 g of water.
b) An 11.4-g sample of sodium react with 17.8 g of chlorine to form 29.2 g of sodium chloride.

45. In a butane lighter, 9.7 g of butane combine with 34.7 g of oxygen to form 29.3 g carbon dioxide and how many grams of water?

46. A 56-g sample of iron reacts with 24 g of oxygen to form how many grams of iron oxide?

Conversion of Energy Units

47. Perform each of the following conversions.
a) 32.5 J to cal
b) 562 cal to J
c) 35.7 kJ to cal
d) 287 cal to kJ

48. Perform each of the following conversions.
a) 456 cal to J
b) 22.3 J to Cal
c) 234 kJ to Cal
d) 12.4 Cal to J

49. Perform each of the following conversions.
a) 25 kWh to J
b) 249 cal to Cal
c) 113 cal to kWh
d) 44 kJ to cal

50. Perform each of the following conversions.
a) 345 Cal to kWh
b) 23 J to cal
c) 5.7×10^3 J to kJ
d) 326 kJ to J

51. An energy bill indicates that the customer used 955 kWh in July. How many joules did the customer use?

52. A television uses 25 kWh of energy per year. How many joules does it use?

53. An adult eats food whose nutritional energy totals approximately 2.0×10^3 Cal per day. How many joules is that?

54. How many joules of nutritional energy are in a bag of chips whose label reads 225 Cal?

Converting between Temperature Scales

55. Perform each of the following temperature conversions.
a) 212 °F to Celsius (temperature of boiling water)
b) 77 K to Fahrenheit (temperature of liquid nitrogen)
c) 25 °C to kelvin (room temperature)
d) 98.6 °F to kelvin (body temperature)

56. Perform each of the following temperature conversions.
a) 102 °F to Celsius
b) 0 K to Fahrenheit
c) −48 °C to Fahrenheit
d) 273 K to Celsius

57. The coldest temperature ever measured in the USA is −80 °F on January 23, 1971, in Prospect Creek, Alaska. Convert that temperature to Celsius and kelvin (assume that −80 °F is accurate to two significant figures.).

58. The warmest temperature ever measured in the USA is 134 °F on July 10, 1913, in Death Valley, California. Convert that temperature to Celsius and kelvin.

59. Vodka will not freeze in the freezer because it contains a high percentage of ethanol. The freezing point of pure ethanol is −114 °C. Convert that temperature to Fahrenheit and kelvin.

60. Liquid helium boils at 4.2 K. Convert this temperature to Fahrenheit and kelvin.

Energy, Heat Capacity, and Temperature Changes

61. Calculate the amount of heat required to raise the temperature of a 97-g sample of water from 42 °C to 67 °C.

62. Calculate the amount of heat required to raise the temperature of a 27-g sample of water from 5 °C to 22 °C.

63. Calculate the amount of heat required to heat a 45-kg sample of ethanol from 11 °C to 19 °C.

64. Calculate the amount of heat required to heat a 3.5-kg gold bar from 21 °C to 67 °C.

65. If 89 J of heat are added to a pure gold coin with a mass of 12 g, what is its temperature change?

66. If 57 J of heat are added to an aluminum can with a mass of 17.1 g, what is its temperature change?

67. An iron nail with a mass of 12 g absorbs 15 J of heat. If the nail was initially at 28 °C, what is its final temperature?

68. A 45-kg sample of water absorbs 345 kJ of heat. If the water was initially at 22.1 °C, what is its final temperature?

69. Calculate the temperature change that occurs when 248 cal of heat are added to 24 g of water.

70. A lead fishing weight with a mass of 57 g absorbs 146 cal of heat. If its initial temperature is 47 °C, what is its final temperature?

71. An unknown metal with a mass of 28 g absorbs 58 J of heat. Its temperature rises from 31.1 °C to 39.9 °C. Calculate the heat capacity of the metal and identify it using Table 3.4.

72. An unknown metal is suspected to be gold. When 2.8 J of heat are added to 5.6 g of the metal, its temperature rises by 3.9 °C. Are these data consistent with the metal being gold?

73. When 56 J of heat are added to 11 g of a liquid, its temperature rises from 10.4 °C to 12.7 °C. What is the heat capacity of the liquid?

74. When 47.5 J of heat are added to 13.2 g of a liquid, its temperature rises by 1.72 °C. What is the heat capacity of the liquid?

Cumulative Problems

75. Calculate the final temperature of 245 mL of water initially at 32 °C upon absorption of 17 kJ of heat.

76. Calculate the final temperature of 32 mL of ethanol initially at 11 °C upon absorption of 562 J of heat. (Density of ethanol is 0.789 g/mL.)

77. A pure gold ring with a volume of 1.57 cm^3 is initially at 11.4 °C. When it is put on, it warms to 29.5 °C. How much heat did the ring absorb? (Density of gold = 19.3 g/cm^3)

78. A block of aluminum with a volume of 98.5 cm^3 absorbs 67.4 J of heat. If its initial temperature was 32.5 °C, what is its final temperature? (Density of aluminum is 2.70 g/cm^3.)

79. How much heat (in kilojoules) is required to heat 56 L of water from 85 °F to 212 °F?

80. How much heat (in joules) is required to heat a 43 g sample of aluminum from 72 °F to 145 °F.

81. What is the temperature change (in Celsius) when 29.5 L of water absorbs 2.3 kWh of heat?

82. If 1.45 L of water is initially at 25.0 °C, what temperature will it be after absorption of 9.4×10^{-2} kWh of heat?

83. A water heater contains 55 gal of water. How many kilowatt-hours of energy are necessary to heat the water in the water heater by 25 °C?

84. A room contains 48 kg of air. How many kilowatt-hours of energy are necessary to heat the air in the house from 7 °C to 28 °C? The heat capacity of air is 1.03 J/g °C.

85. A backpacker wants to carry enough fuel to heat 2.5 kg of water from 25 °C to 100.0 °C. If the fuel he carries produces 36 kJ of heat per gram, how much fuel should he carry? (For the sake of simplicity, assume that the transfer of heat is 100% efficient.)

86. A cook wants to heat 1.35 kg of water from 32.0 °C to 100.0 °C. If he uses natural gas to heat the water, how much natural gas will he need to burn? Natural gas produces 49.3 kJ of heat per gram. (For the sake of simplicity, assume that the transfer of heat is 100% efficient.)

Highlight Problems

87. Classify each of the following molecular pictures as a pure substance or a mixture.

a)

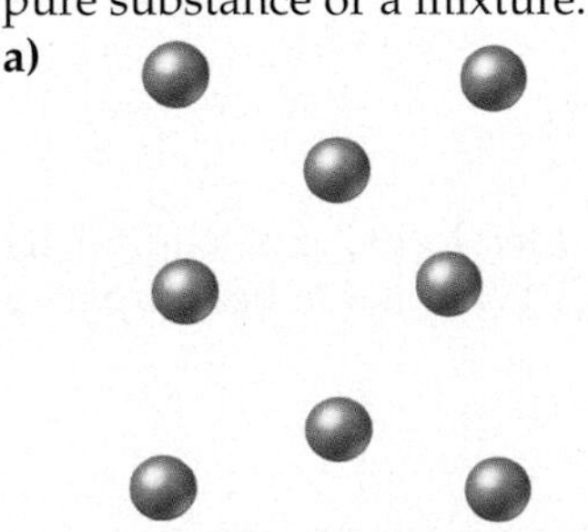

b)

c)

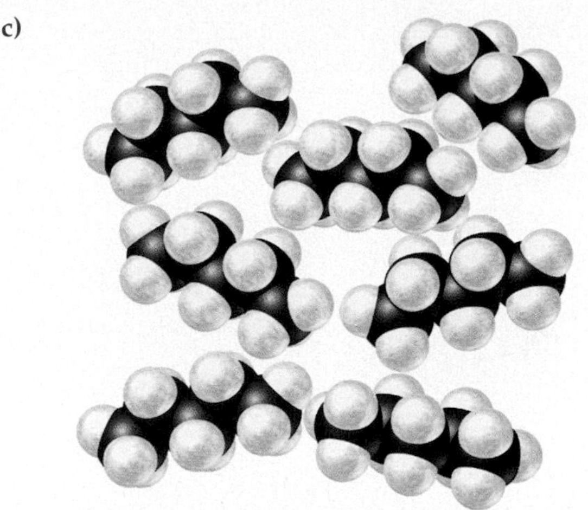

d)

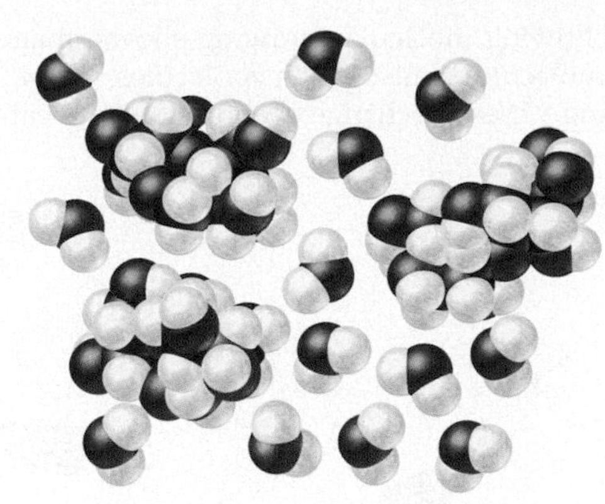

88. Classify each of the following molecular pictures as a pure substance or a mixture. If it is a pure substance, classify it as an element or a compound. If it is a mixture, classify it as homogeneous or heterogeneous.

a)

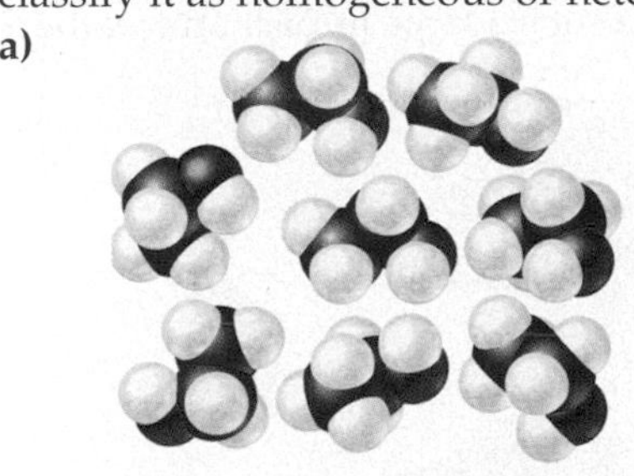

b)

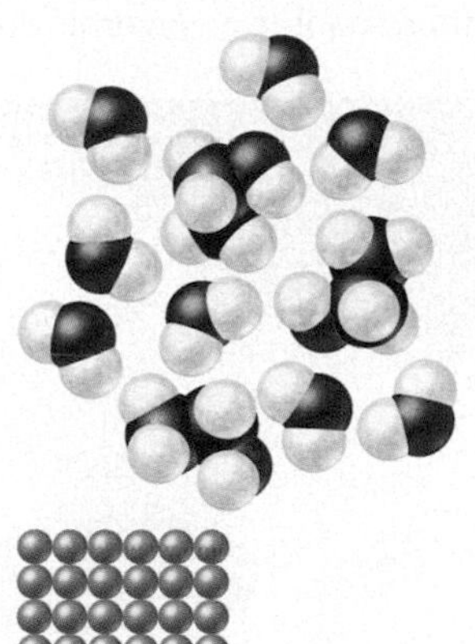

c)

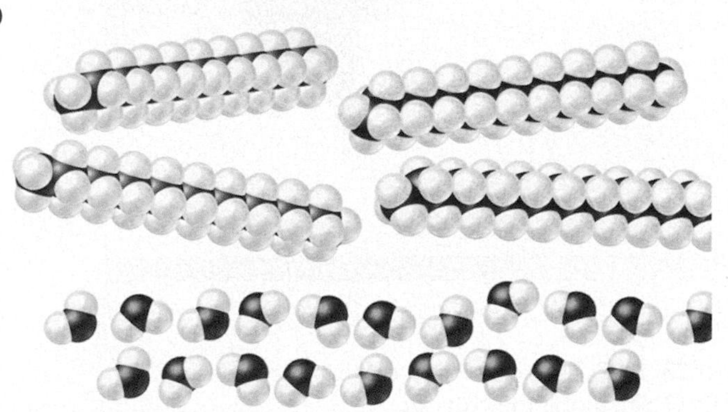

d)

89. The following molecular drawing shows images of acetone molecules before and after a change. Was the change chemical or physical?

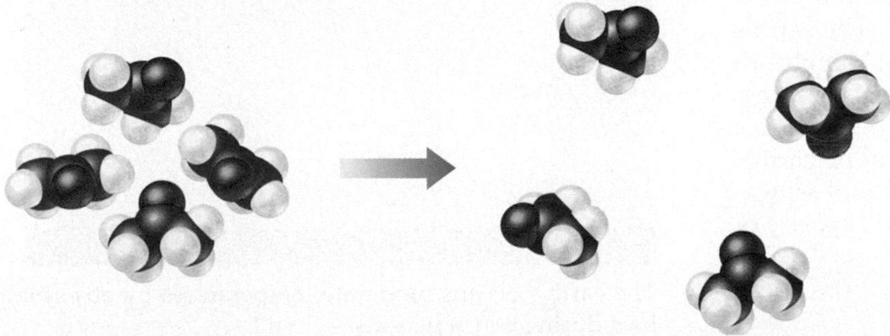

90. The following molecular drawing shows images of methane molecules and oxygen molecules before and after a change. Was the change chemical or physical?

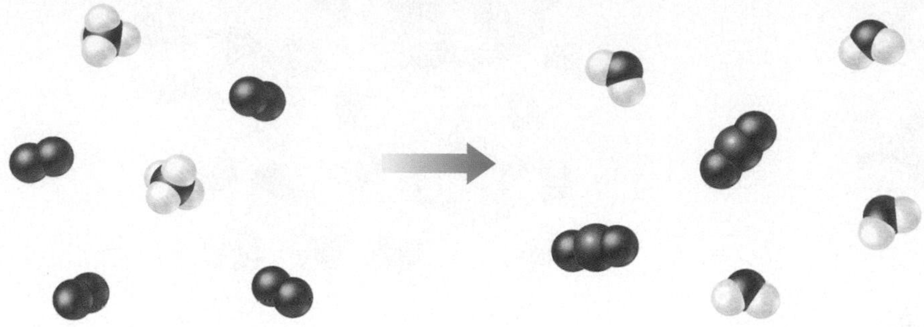

91. A major event affecting global climate is the El Niño/La Niña cycle. In this cycle, equatorial Pacific Ocean waters warm by several degrees Celsius above normal (El Niño) and then cool by several degrees Celsius below normal (La Niña). This cycle affects weather not only in North and South America, but as far away as Africa. Why does a seemingly small change in ocean temperature have such a large impact on weather?

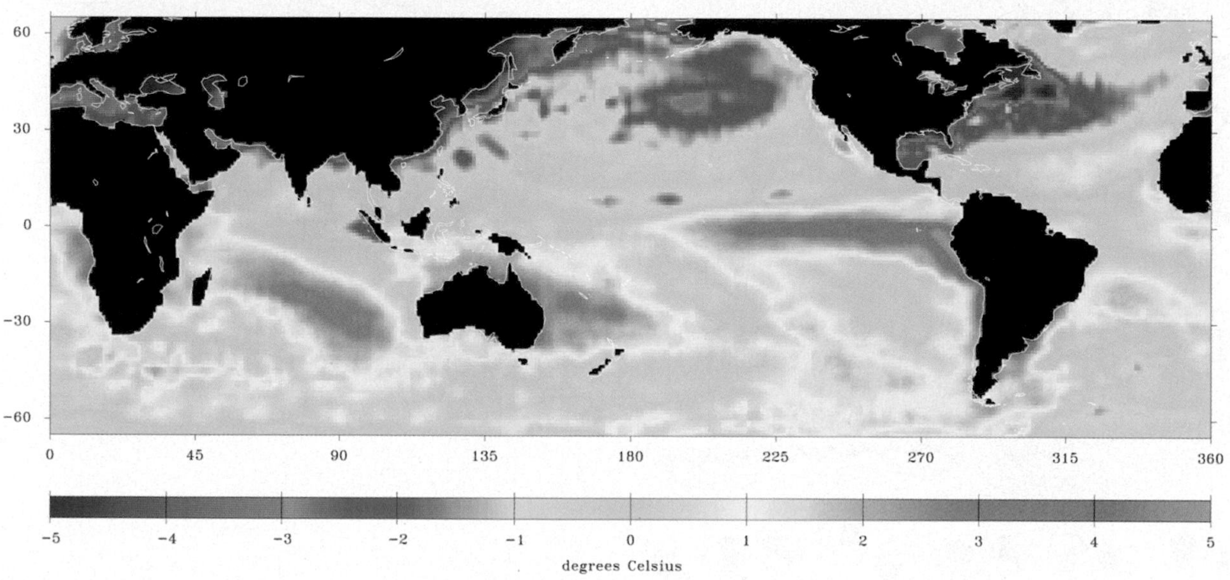

Temperature anomaly plot of the world's oceans for December 23, 1997. The red section off the western coast of South America is the El Niño effect, a warming of the Pacific Ocean along the equator.

92. Global warming refers to the increase in average global temperature due to the increase of certain gases, called greenhouse gases, in our atmosphere. The earth's oceans, because of their high heat capacity, can absorb heat and therefore act to slow down global warming. How much heat would be required to warm the earth's oceans by 1.0 °C? Assume that the volume of earth's oceans is 137×10^7 km^3 and that the density of seawater is 1.03 g/cm^3. Also assume that the heat capacity of seawater is the same as water.

The earth's oceans moderate temperatures by absorbing heat during warm periods.

93. Examine the following data for the maximum and minimum average temperatures of San Francisco and Sacramento in the summer and in the winter.

San Francisco (Coastal City)

January		*August*	
High	*Low*	*High*	*Low*
57.4 °F	43.8 °F	64.4 °F	54.5 °F

Sacramento (Inland City)

January		*August*	
High	*Low*	*High*	*Low*
53.2 °F	37.7 °F	91.5 °F	57.7 °F

a) Notice the difference between the August high in San Francisco and Sacramento. Why is it much hotter in the summer in Sacramento?
b) Notice the difference between the January low in San Francisco and Sacramento. How might the heat capacity of the ocean contribute to this difference?

Answers to Skillbuilder Exercises

Skillbuilder 3.1 **a)** pure substance, element **b)** mixture, homogeneous **c)** mixture, heterogeneous **d)** pure substance, compound
Skillbuilder 3.2 **a)** chemical **b)** physical **c)** physical **d)** physical
Skillbuilder 3.3 **a)** chemical **b)** physical **c)** physical **d)** chemical
Skillbuilder 3.4 27 g
Skillbuilder 3.5 2.14 kJ
Skillbuilder Plus, p. 16 6.58×10^6 cal
Skillbuilder 3.6 85 °C
Skillbuilder 3.7 282 °F
Skillbuilder 3.8 77 K
Skillbuilder 3.9 50.1 J
Skillbuilder Plus, p. 23 7.4 g
Skillbuilder 3.10 $\Delta T = 4.22$ °C; $T_f = 29.2$ °C

Figure 4.1 If every atom within a pebble were the size of the pebble itself, then the pebble would be larger than Mt. Everest.

There are about ninety-one different elements in nature, and consequently about ninety-one different kinds of atoms. In addition, scientists have succeeded in making about twenty synthetic elements (not found in nature).

The exact number of naturally occurring elements is controversial because there is a possibility that some elements previously considered only synthetic may be naturally occurring.

In this chapter, we learn about atoms: what they are made of; how they differ from one another; and how they are structured. We also learn about the elements that atoms compose and some of the properties of those elements. Elements are organized in the periodic table of elements shown in the inside front cover of this book.

4.2 Not to Cut: The Atomic Theory

If we simply examine matter, even under a microscope, it is not obvious that matter is composed of tiny particles. In fact, it appears to be just the opposite. If we divide a sample of matter into smaller and smaller pieces, it seems that we could divide it forever. From our perspective, matter seems continuous. The first person recorded as thinking otherwise was **Democritus** (460–370 B.C.), a Greek philosopher who theorized that matter was ultimately composed of small, indivisible particles he called *atomos* or "atoms" meaning "not to cut." Democritus suggested that if you divided matter into smaller and smaller pieces, you would eventually end up with tiny, indestructible particles—atoms.

Image of Democritus with Diogenes as captured in a woodcut. Democritus is the first person on record to have postulated that matter was composed of atoms.

Democritus' ideas were not widely accepted, and it was not until the early nineteenth century that John Dalton formalized a theory of atoms that gained broad acceptance. Dalton's atomic theory has three parts.

1) Each element is composed of tiny indestructible particles called atoms.
2) All atoms of a given element have the same mass and other properties that distinguish them from the atoms of other elements.
3) Atoms combine in simple, whole-number ratios to form compounds.

Today, the evidence for the atomic theory is overwhelming. Recent advances in microscopy have allowed scientists not only to image individual atoms but also to pick them up and move them (Figure 4.2). Matter is indeed composed of atoms.

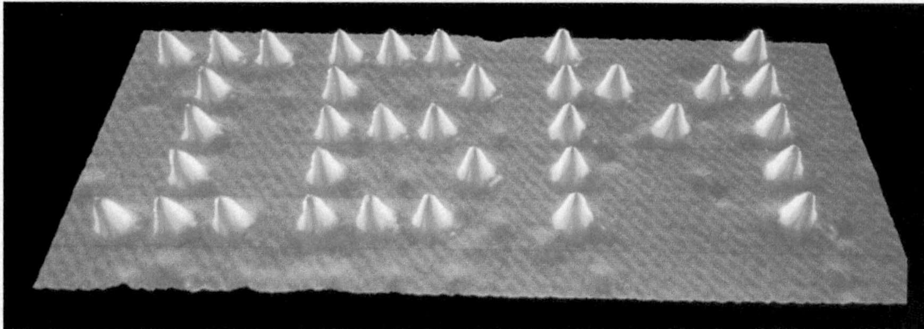

Figure 4.2 Scientists at IBM used a special microscope, called a scanning tunneling microscope (STM), to move xenon atoms to form the letters IBM. The cone-shape of these atoms is due to the peculiarities of the instrumentation. Atoms are, in general, spherical in shape.

EVERYDAY CHEMISTRY

Atoms and Humans

All matter is composed of atoms. What does that mean? What does it imply? It means that everything before you is composed of tiny particles too small to see. It means that even you and I are composed of these same particles. We acquired those particles from the food we have eaten over the years. The average carbon atom in our own bodies has been used by twenty other living organisms before we got it and will be used by other organisms after we die.

The idea that all matter is composed of atoms has far-reaching implications. It implies that our bodies, our hearts, and even our brains are composed of atoms acting according to the laws of chemistry and physics. Some have viewed this as a devaluation of human life. We have always wanted to distinguish ourselves from everything else, and the idea that we are made of the same basic particles as all other matter takes something away from that distinction.... or does it?

CAN YOU ANSWER THIS? *Do you find the idea that you are made of atoms disturbing? Why or why not?*

4.3 The Nuclear Atom

By the end of the nineteenth century, scientists were convinced that matter was composed of atoms, the permanent, indestructible building blocks from which all substances are constructed. However, an English physicist named **J.J. Thomson** (1856–1940), complicated the picture by his discovery of an even smaller and more fundamental particle called the **electron**. Thomson discovered that electrons were negatively charged, that they were much smaller and lighter than atoms, and that they were uniformly present in many different kinds of substances. The indestructible building block called the atom could apparently be "chipped."

Charge is more fully defined in section 4.4. For now think of it as an inherent property of electrons that causes them to interact with other charged particles.

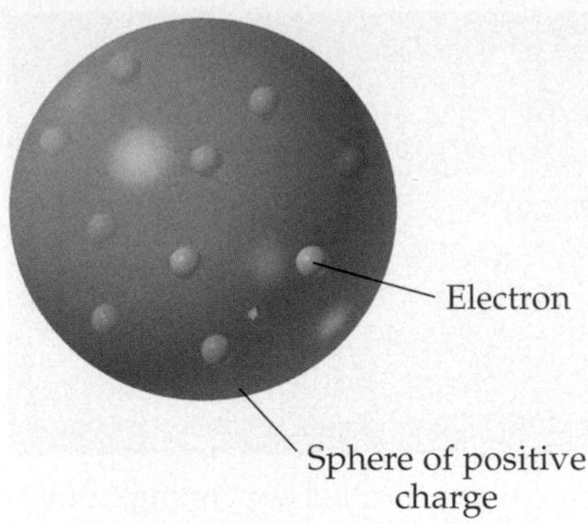

Figure 4.3 Plum-pudding model of the atom as suggested by J.J. Thomson. In this model, negatively charged electrons (yellow) were held in a sphere of positive charge (red).

Alpha-particles or α-particles are discussed more fully in Chapter 17.

The discovery of negatively charged particles within atoms raised the question of a neutralizing positive charge. Atoms were known to be charge neutral, so they must contain positive charge that neutralized the negative charge of electrons. But how did the positive and negative charges within the atom fit together? Were atoms just a jumble of even more fundamental particles? Where they solid spheres? Did they have some internal structure? J.J. Thomson proposed that the negatively charged electrons were small particles held within a positively charged sphere. This model, the most popular of the time, became known as the plum-pudding model (plum-pudding was an English dessert) (Figure 4.3). The picture suggested by Thomson was—to those of us not familiar with plum-pudding—like a blueberry muffin, where the blueberries are the electrons and the muffin is the positively charged sphere.

In 1909, **Ernest Rutherford** (1871–1937), who had worked under Thomson and adhered to his plum-pudding model, performed an experiment in an attempt to confirm it. His experiment proved it wrong instead. In his experiment, Rutherford directed tiny, positively charged particles—called alpha-particles—at an ultra-thin sheet of gold foil (Figure 4.4). These particles were to act as probes of the gold atoms' structure. If the gold atoms were indeed like blueberry muffins or plum pudding—with their mass and charge spread throughout the entire volume of the atom—these speeding probes should pass right through the gold foil with minimum deflection. Rutherford performed the experiment, but the results were not as he expected. A majority of the particles did pass directly through the foil, but some particles were deflected, and some (1 in 20,000) even bounced back. The results puzzled Rutherford who said, "[this is]

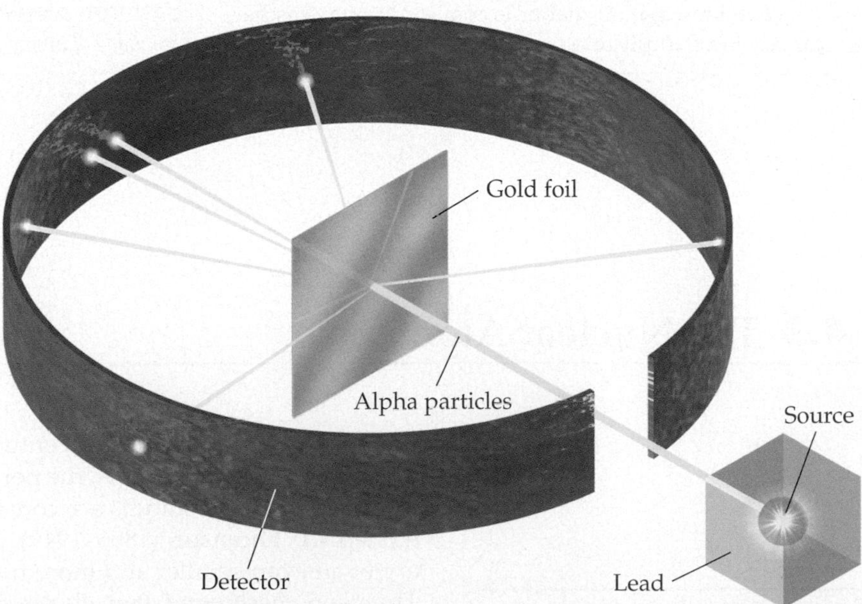

Figure 4.4 Rutherford's gold foil experiment. Tiny particles, called α-particles, were directed at a thin sheet of gold foil. Most of the particles passed directly through the foil, but some were deflected.

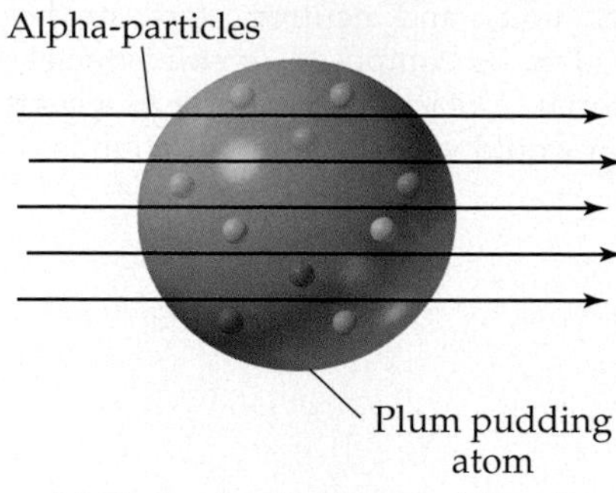

(a) Rutherford's Expected Result

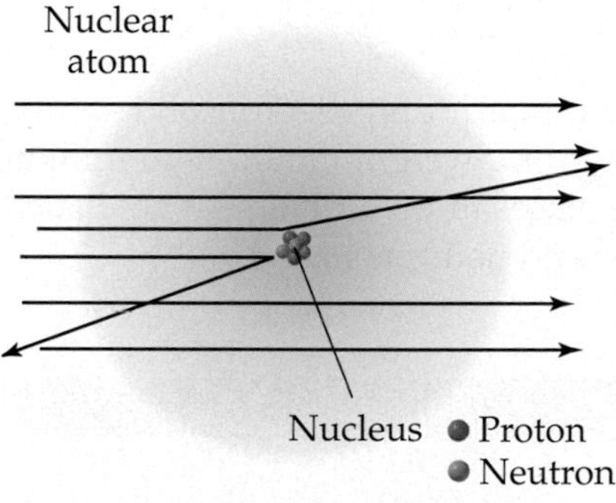

(b) Rutherford's Actual Result

Figure 4.5 **a)** Expected result of Rutherford's gold foil experiment. If the plum-pudding model were correct, the α-particles should pass right through the gold foil with minimal deflection. **b)** Actual result of Rutherford's gold foil experiment. A small number of α-particles were deflected or bounced back. The only way to explain the deflections was to suggest that most of the mass and all of the positive charge of an atom must be concentrated in a space much smaller than the size of the atom itself.

about as credible as if you had fired a 15-inch shell at a piece of tissue paper and it came back and hit you." What must the structure of the atom be in order to explain this odd behavior?

Rutherford created a new model to explain his results (Figure 4.5). He concluded that matter must not be as uniform as it appears. It must contain large regions of empty space accompanied by small regions of very dense matter. In order to explain the deflections he observed, the mass and positive charge of an atom must all be concentrated in a space much smaller than the size of the atom itself. Using this idea, he proposed the **nuclear theory of the atom**, which has three basic parts.

1) Most of the atom's mass and all of its positive charge are contained in a small core called the **nucleus**.
2) Most of the volume of the atom is empty space occupied by tiny, negatively charged electrons.
3) There are as many negatively charged electrons outside the nucleus as there are positively charged particles (**protons**) inside the nucleus, so that the atom is electrically neutral.

Later work by Rutherford and others demonstrated that the atom's nucleus contains both positively charged protons and neutral particles called **neutrons**. The dense nucleus makes up over 99.9% of the mass of the atom, but electrons take up most of its volume. For now, we can think of these electrons like the water droplets that make up a cloud—they occupy much volume but don't have much mass (Figure 4.6).

Rutherford's nuclear theory was a success and is still valid today. The revolutionary part of this theory is the idea that matter—at its core—is much less uniform than it appears. If the nucleus of the atom were the size of this dot •, the average electron would be about ten meters away. Yet the dot would contain almost the entire mass of the atom. Imagine what matter would be like if atomic structure broke down. What if matter were composed of atomic nuclei piled on top of each other like marbles? Such matter

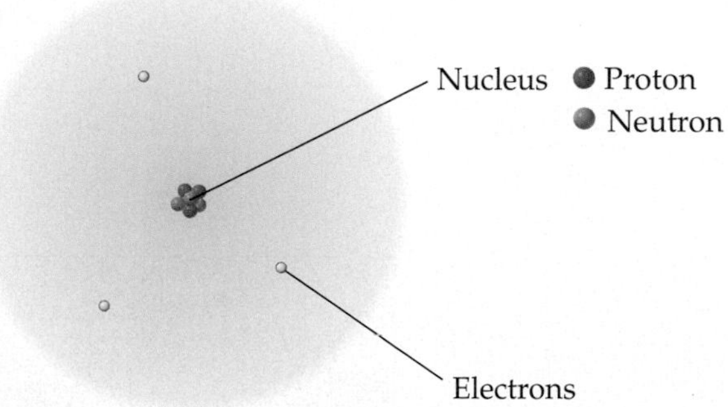

Figure 4.6 The nuclear atom. In this model, 99.9% of the atom's mass is concentrated in a small, dense, nucleus that contains protons and neutrons. The rest of the volume of the atom is mostly empty space occupied by negatively charged electrons. The number of electrons outside the nucleus is equal to the number of protons inside the nucleus. In this image, the nucleus is greatly enlarged. Electrons do not really exist as particles as shown here, but exist in quantum mechanical orbitals as covered in Chapter 9.

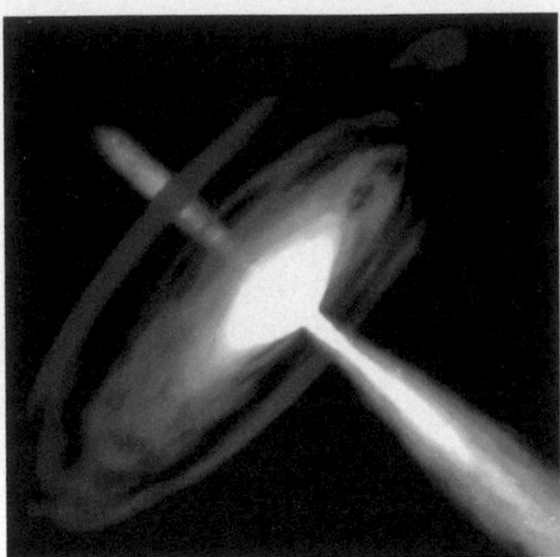

Black holes and neutron stars are believed to be composed of solid nuclear material. A black hole the size of a sand grain would weigh 10 million pounds.

would be incredibly dense; a single grain of sand composed of solid atomic nuclei would have a mass of 5 million kg (or a weight of about 10 million lb). Astronomers believe there are some places in the universe where such matter exists—they are called black holes and neutron stars.

EVERYDAY CHEMISTRY

Solid Matter?

If matter really is mostly empty space as Rutherford suggested, then why does it appear so solid? Why can I tap my knuckles on the table and feel a solid thump? Matter appears solid because the variation in the density is on such a small scale that our eyes can't see it. Imagine a jungle gym one hundred stories high and the size of a football field. It is mostly empty space. Yet if you viewed it from an airplane, it would appear as a solid mass. Matter is similar. When you tap your knuckle on the table, it is much like one giant jungle gym (your finger) crashing into another (the table). Even though they are both primarily empty space, one does not fall into the other.

CAN YOU ANSWER THIS? *Use the jungle gym analogy to explain why most of Rutherford's alpha-particles went right through the gold foil and why a few bounced back. Remember that his gold foil was extremely thin.*

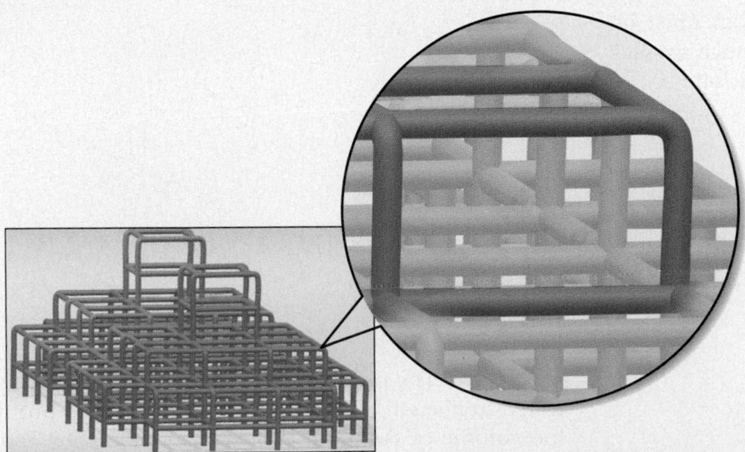

Matter appears solid and uniform because the variation in density is on a scale too small for our eyes to see. Just as this scaffolding appears solid at a distance, so matter appears solid to us.

4.4 The Properties of Protons, Neutrons, and Electrons

If a proton had the mass of a baseball, an electron would have the mass of a rice grain.

The mass of an electron is actually 0.00055 amu.

Protons and neutrons have very similar masses. In SI units, the mass of the proton is 1.67262×10^{-27} kg, and the mass of the neutron is a close 1.67493×10^{-27} kg. A more common unit to express these masses, however, is called the **atomic mass unit (amu)**, defined as one-twelfth of the mass of a carbon atom containing six protons and six neutrons. In this unit, protons and neutrons each have a mass of 1 amu. Electrons, in contrast, have an almost negligible mass of 0.0009×10^{-27} kg or approximately 0 amu.

The proton and the electron both have electrical **charge.** The proton's charge is +1 and the electron's charge is −1. The charge of the proton and the electron are equal in magnitude but opposite in sign so that when the two particles are paired the charges exactly cancel. The neutron has no charge.

What is electrical charge? Electrical charge is a fundamental property of protons and electrons just as mass is a fundamental property of matter. Most matter is charge neutral because protons and electrons occur together and their charges cancel. However, you have probably experienced excess electrical charge when brushing your hair on a dry day. The brushing action causes the accumulation of charged particles in your hair, which repel each other, causing your hair to stand on end.

We can summarize the nature of electrical charge as follows (Figure 4.7):

- ***Electrical charge is a fundamental property of protons and electrons.***
- ***Positive and negative electrical charges attract each other.***
- ***Positive–positive or negative–negative charges repel each other.***
- ***Positive and negative charges cancel each other so that a proton and an electron, when paired, are charge neutral.***

Notice that matter is usually charge neutral due to the canceling effect of protons and electrons. When matter does acquire charge imbalances, these imbalances usually equalize quickly, often in dramatic ways. For example, the shock you receive when touching a doorknob during dry weather is the equalization of a charge imbalance that developed as you walked across the carpet. Lightning is an equalization of charge imbalances that develop during electrical storms.

Positive (red) and negative (yellow) charges attract.

Positive-positive or negative-negative charges repel.

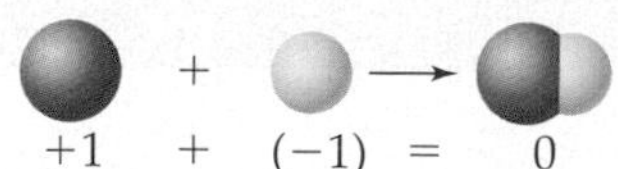

Positive and negative charges cancel.

Figure 4.7 The nature of electrical charge.

Matter is normally charge neutral, having equal numbers of positive and negative charges that exactly cancel. When the charge balance of matter is disturbed, as in an electrical storm, it quickly rebalances, often in dramatic ways such as lightning.

If you had a sample of matter—even a tiny sample, such as a sand grain—that was composed of only protons or only electrons, the forces around that matter would be extraordinary, and the matter would be unstable. Luckily, that is not the way matter is—protons and electrons exist together, canceling each other's charge and making matter charge neutral. The properties of protons, neutrons, and electrons are summarized in Table 4.1.

TABLE 4.1

Subatomic Particles

	Mass (kg)	*Mass (amu)*	*Charge*
proton	1.67262×10^{-27}	1	+1
neutron	1.67493×10^{-27}	1	0
electron	0.00091×10^{-27}	0.00055	−1

4.5 Elements: Defined by Their Numbers of Protons

We have seen that atoms are composed of protons, neutrons, and electrons. However, it is the number of protons in the nucleus of an atom that identifies it as a particular element. For example, atoms with 2 protons in their nucleus are helium atoms, atoms with 13 protons in their nucleus are aluminum atoms, and atoms with 92 protons in their nucleus are uranium

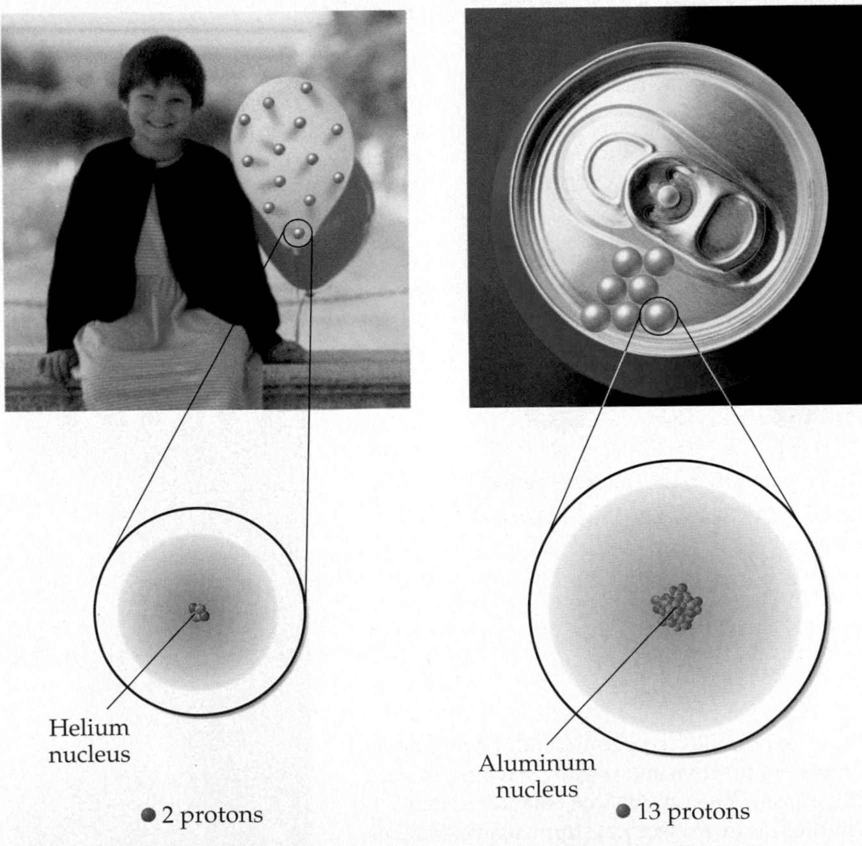

Figure 4.8 The number of protons in the nucleus defines the element.

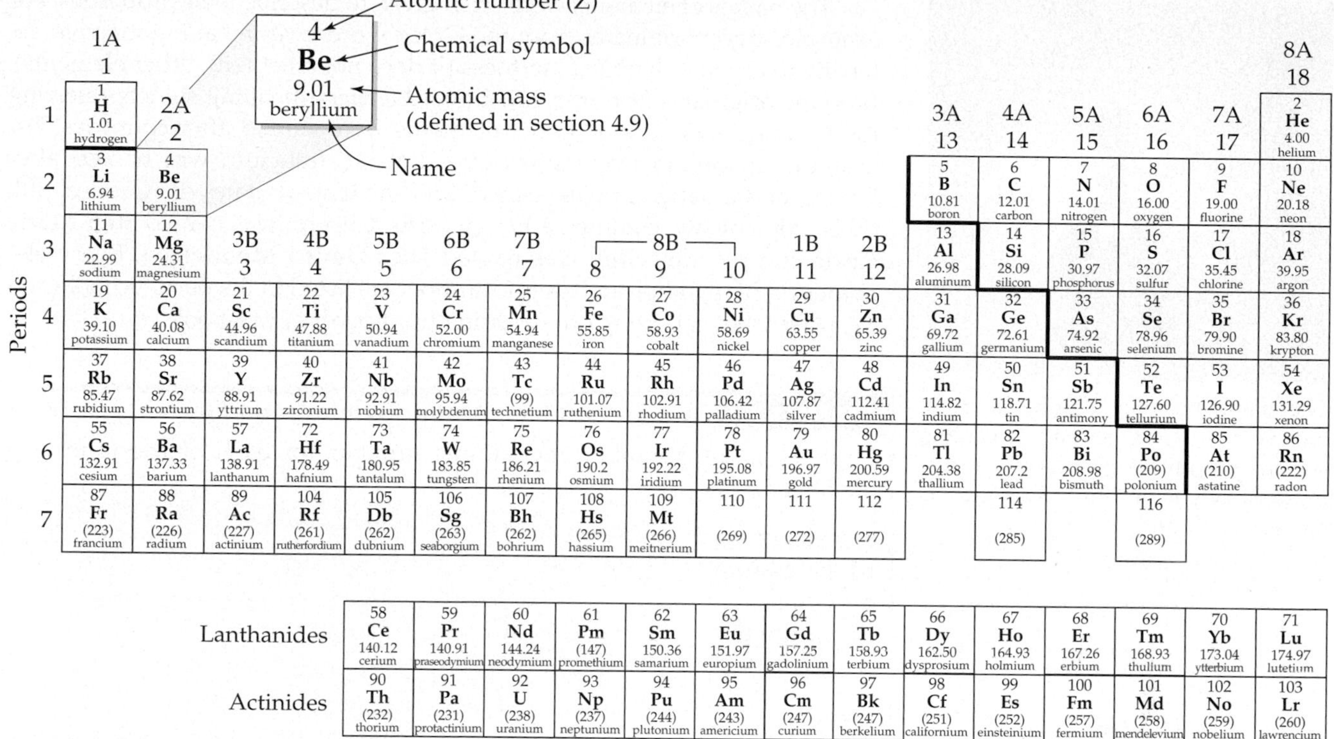

Figure 4.9 The periodic table of the elements.

atoms. The number of protons in an atom's nucleus defines the element (Figure 4.8). Every aluminum atom has 13 protons in its nucleus—if it had a different number of protons, it would be a different element. The number of protons in the nucleus of an atom is called the **atomic number** and is given the symbol **Z**.

The periodic table of the elements (Figure 4.9) lists all known elements according to their atomic number. Each element, identified by a unique atomic number, is represented with a unique **chemical symbol**, a one or two letter abbreviation for the element that is listed directly below its atomic number on the periodic table. The chemical symbol for helium is He, for aluminum is Al, and for uranium is U. The chemical symbol and the atomic number always go together. If the atomic number is 13, the chemical symbol must be Al. If the atomic number is 92, the chemical symbol must be U. This is just another way of saying that the number of protons defines the element.

Most chemical symbols are based on the English name of the element. For example, the symbol for carbon is C, for silicon is Si, and for bromine is Br. Some elements, however, have symbols based on their Latin names. For example, the symbol for potassium is K from the Latin *kalium*, and the symbol for sodium is Na from the Latin *natrium*. Other elements with symbols based on their Greek or Latin names include:

lead	Pb	*plumbum*	silver	Ag	*argentum*
mercury	Hg	*hydrargyrum*	tin	Sn	*stannum*
iron	Fe	*ferrum*	copper	Cu	*cuprum*

The name bromine originates from the Greek word *bromos,* meaning stench. Bromine vapor, seen as the red-brown gas in this photograph, has a strong odor.

The names of elements were often given to describe their properties. For example, argon originates from the Greek word *argos* meaning inactive, referring to argon's chemical inertness (it does not react with other elements). Bromine originates from the Greek word *bromos,* meaning stench, referring to bromine's strong odor. Other elements were named after countries. For example, polonium was named after Poland, francium was named after France, and americium was named after the United States of America. Still other elements were named after scientists. Curium was named after **Marie Curie** and mendelevium was named after **Dmitri Mendeleev**. Every element's name, symbol, and atomic number are listed in the periodic table and in an alphabetical listing on the inside front cover of this book.

Curium
96
Cm
(247)

Curium is named after Marie Curie, a chemist who helped discover radioactivity and also discovered two new elements. Curie won two Nobel prizes for her work.

EXAMPLE 4.1 Atomic Number, Atomic Symbol, and Element Name

Find the atomic symbol and atomic number for each of the following elements.

a) Silicon
b) Potassium
c) Gold
d) Antimony

Solution:

As you become familiar with the periodic table, you will be able to quickly locate elements on it. For now, it might be easier to find them in the alphabetical listing on the inside back cover of this book, but you should also find their position in the periodic table.

Element	*Symbol*	*Atomic Number*
silicon	Si	14
potassium	K	19
gold	Au	79
antimony	Sb	51

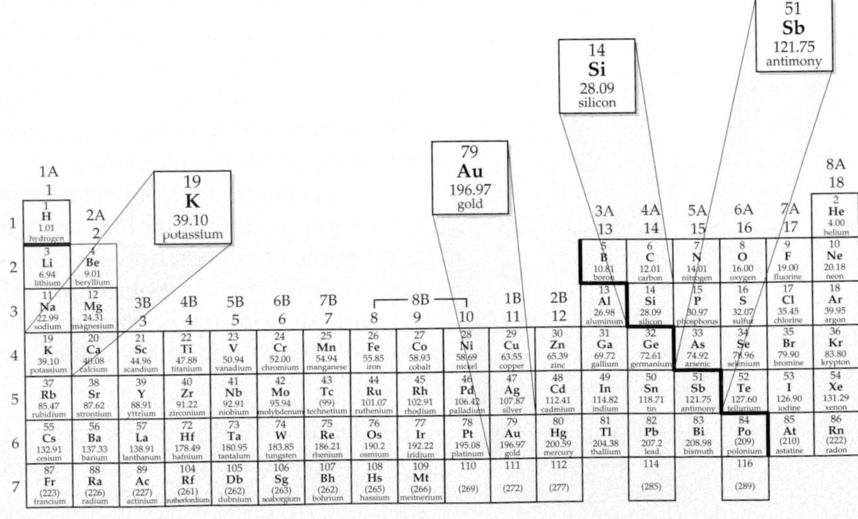

SKILLBUILDER 4.1 **Atomic Number, Atomic Symbol, and Element Name**

Find the name and atomic number for each of the following elements.

a) Na
b) Ni
c) P
d) Ta

4.6 Looking for Patterns: The Periodic Law and the Periodic Table

Dmitri Mendeleev, a Russian chemistry professor who proposed the periodic law and arranged early versions of the periodic table.

Periodic means repeating pattern. The properties of the elements, when listed in order of increasing relative mass, formed a *repeating pattern*.

The organization of the periodic table has its origins in the work of Dmitri Mendeleev (1834–1907), a nineteenth-century Russian chemistry professor. In his time, about sixty-five different elements had been discovered. Through the work of a number of chemists, much was known about each of these elements including their relative masses, chemical activity, and some of their physical properties. However, there was no systematic way of organizing them.

In 1869, Mendeleev noticed that certain groups of elements had similar properties. Mendeleev found that by listing elements in order of increasing relative mass, their properties recurred in a periodic pattern (Figure 4.10). Mendeleev summarized these observations in the **periodic law** that states, "When the elements are arranged in order of increasing relative mass, certain sets of properties recur periodically." Mendeleev then organized all the known elements in a table in which relative mass increased from left to right and elements with similar properties aligned in the same vertical columns (Figure 4.11). Since many elements had not

1	2	3	4	5	6	7	8	9	10	11	12	13	14	15	16	17	18	19	20
H	He	Li	Be	B	C	N	O	F	Ne	Na	Mg	Al	Si	P	S	Cl	Ar	K	Ca

Figure 4.10 The above elements are listed in order of increasing atomic number (Mendeleev used relative mass, which is similar). The color of each element represents its properties. Notice that the properties (colors) of these elements forms a repeating pattern.

1 H							2 He
3 Li	4 Be	5 B	6 C	7 N	8 O	9 F	10 Ne
11 Na	12 Mg	13 Al	14 Si	15 P	16 S	17 Cl	18 Ar
19 K	20 Ca						

Figure 4.11 If we place the above elements from Figure 4.10 in a table in which atomic number increases from left to right, elements with similar properties align in the same vertical columns. This is similar to Mendeleev's first periodic table.

yet been discovered, Mendeleev's table contained some gaps, which allowed him to predict the existence of yet-undiscovered elements. For example, Mendeleev predicted the existence of an element he called eka-silicon, which fell below silicon on the table and between gallium and arsenic. In 1886, eka-silicon was discovered by German chemist **Clemens Winkler** (1838–1904) who named it germanium, after his home country.

Mendeleev's original listing has evolved into the modern periodic table. In the modern table, elements are listed in order of increasing atomic number rather than increasing relative mass. The modern periodic table also contains more elements than Mendeleev's original table because many more have been discovered since his time.

Mendeleev's periodic law was based on observation. Like all scientific laws, the periodic law summarized many observations but did not give the underlying reason for the observation—only theories do that. For now, we accept the periodic law as it is, but in Chapter 9 we examine a powerful theory that explains the law and gives the underlying reasons for it.

The elements in the periodic table can be broadly classified as metals, nonmetals, and metalloids (Figure 4.12). **Metals** tend toward the left of the periodic table and have similar properties: they are good conductors of heat and electricity; they can be pounded into flat sheets (malleability); they can be drawn into wires (ductility); they are often shiny; and they tend to lose electrons when they undergo chemical changes. Good examples of metals include iron, magnesium, chromium, and sodium.

Nonmetals tend toward the upper right side of the periodic table. The dividing line between metals and nonmetals is the zigzag diagonal line running from boron to astatine. Nonmetals have more varied

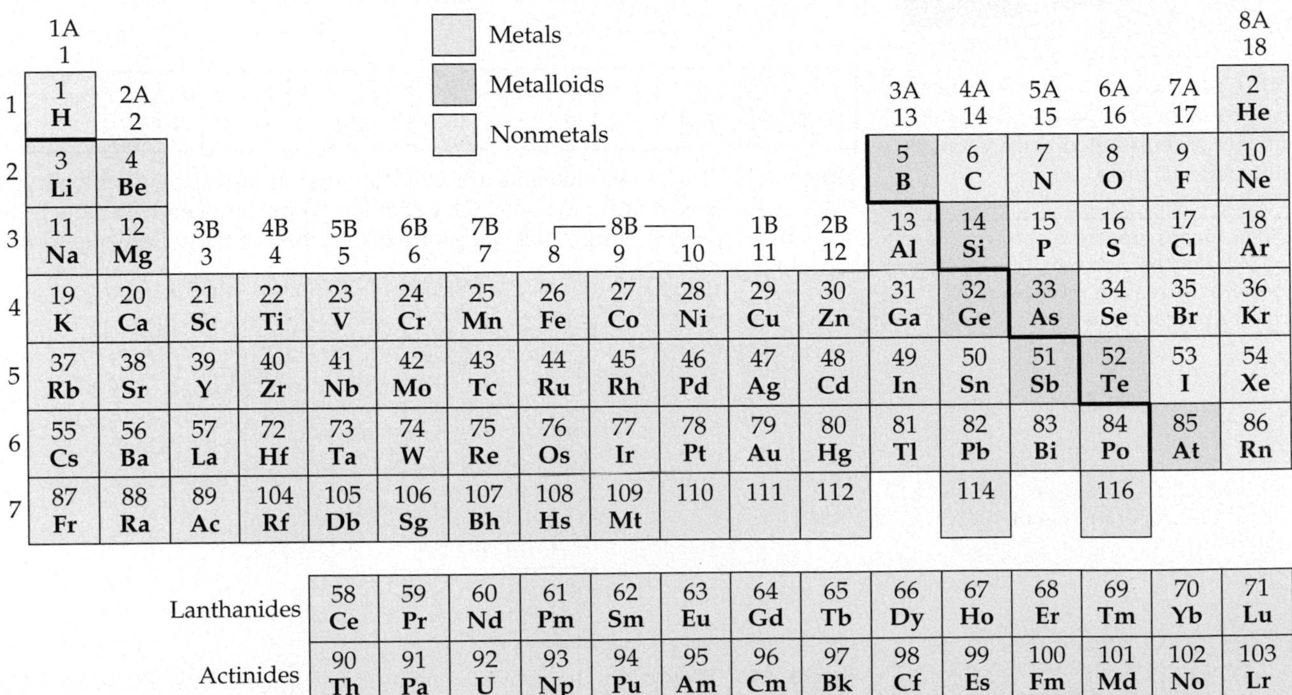

Figure 4.12 The elements in the periodic table can be broadly classified as metals, nonmetals, and metalloids.

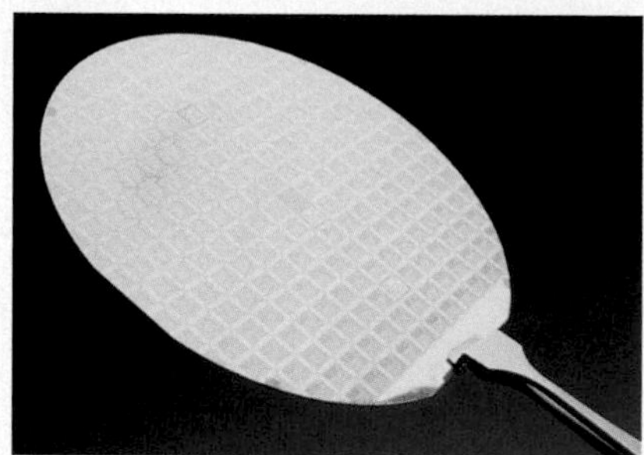

Silicon, a metalloid used extensively in the computer and electronics industries.

Metalloids are sometimes called semimetals.

properties—some are solids at room temperature, others are gases—but as a whole they tend to be poor conductors of heat and electricity and they all tend to gain electrons when they undergo chemical changes. Good examples of nonmetals include oxygen, nitrogen, chlorine, and iodine. The elements that fall in the middle-right of the periodic table—on the zigzag diagonal line that divides metals and nonmetals—are called **metalloids** and show mixed properties. Metalloids are also called **semiconductors** because of their intermediate electrical conductivity, which can be changed and controlled. This property makes semiconductors useful in the manufacture of the electronic devices central to computers, cellular telephones, and many other modern gadgets. Good examples of metalloids include silicon, arsenic, and germanium.

EXAMPLE 4.2 **Classifying Elements as Metals, Nonmetals or Metalloids**

Classify each of the following elements as a metal, nonmetal, or metalloid.

a) Ba
b) I
c) O
d) Te

Solution:

a) Barium is on the left hand side of the periodic table; it is a metal.
b) Iodine is on the right hand side of the periodic table; it is a nonmetal.
c) Oxygen is on the right hand side of the periodic table; it is a nonmetal.
d) Tellurium is in the middle-right section of the periodic table in the area of the metalloids; it is a metalloid.

SKILLBUILDER 4.2 **Classifying Elements as Metals, Nonmetals or Metalloids**

Classify each of the following elements as a metal, nonmetal, or metalloid.

a) S
b) Cl
c) Ti
d) Sb

Main-group elements are in columns labeled with a number and the letter A. Transition elements are in columns labeled with a number and the letter B.

A competing numbering system does not use letters, but only the numbers 1–18. Both numbering systems are shown in the periodic table in this book.

The noble gases are inert (or unreactive) compared to other elements. However, some noble gases, especially the heavier ones, do form compounds with other elements.

The periodic table can also be broadly divided into **main-group** elements, whose properties tend to be more predictable based on their position in the periodic table, and **transition elements** or **transition metals** whose properties tend to be less predictable based simply on their position in the periodic table (Figure 4.13). Each column within the main-group elements in the periodic table is called a **family** or **group** of elements and is designated with a number and a letter printed directly above the column.

The elements within a group usually have similar properties. For example, the Group 8A elements, called the **noble gases**, are chemically inert gases. The most familiar noble gas is probably helium, used to fill buoyant balloons. Helium, like the other noble gases, is chemically stable—it won't combine with other elements to form compounds—and is therefore safe to put into balloons. Other noble gases include neon, often used in neon signs; argon, which makes up a small percentage of our atmosphere; krypton;

Noble gases

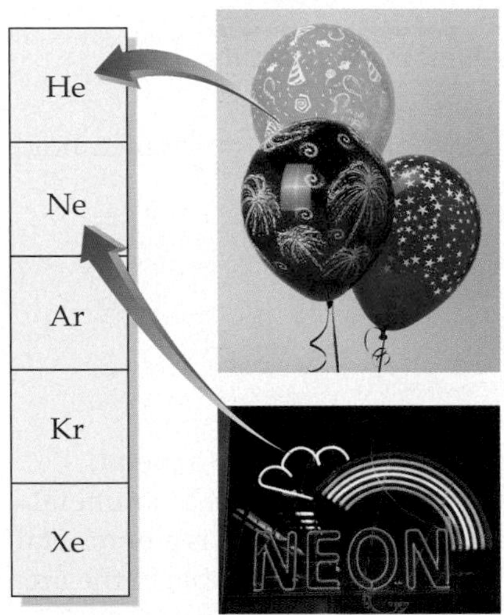

The noble gases include helium (used in balloons), neon (found in neon signs), argon, krypton, and xenon.

Alkali metals

The alkali metals include lithium (shown in the first photo), sodium (shown in the second photo reacting with water), potassium, rubidium and cesium.

Main group elements | Transition elements | Main group elements

Group number

Periods	1A	2A	3B	4B	5B	6B	7B	8B	8B	8B	1B	2B	3A	4A	5A	6A	7A	8A
1	1 H																	2 He
2	3 Li	4 Be											5 B	6 C	7 N	8 O	9 F	10 Ne
3	11 Na	12 Mg											13 Al	14 Si	15 P	16 S	17 Cl	18 Ar
4	19 K	20 Ca	21 Sc	22 Ti	23 V	24 Cr	25 Mn	26 Fe	27 Co	28 Ni	29 Cu	30 Zn	31 Ga	32 Ge	33 As	34 Se	35 Br	36 Kr
5	37 Rb	38 Sr	39 Y	40 Zr	41 Nb	42 Mo	43 Tc	44 Ru	45 Rh	46 Pd	47 Ag	48 Cd	49 In	50 Sn	51 Sb	52 Te	53 I	54 Xe
6	55 Cs	56 Ba	57 La	72 Hf	73 Ta	74 W	75 Re	76 Os	77 Ir	78 Pt	79 Au	80 Hg	81 Tl	82 Pb	83 Bi	84 Po	85 At	86 Rn
7	87 Fr	88 Ra	89 Ac	104 Rf	105 Db	106 Sg	107 Bh	108 Hs	109 Mt	110	111	112		114		116		

Figure 4.13 The periodic table can be broadly divided into main group elements, those whose properties tend to be more predictable based on their position, and transition elements, those whose properties tend to be less predictable based on their position.

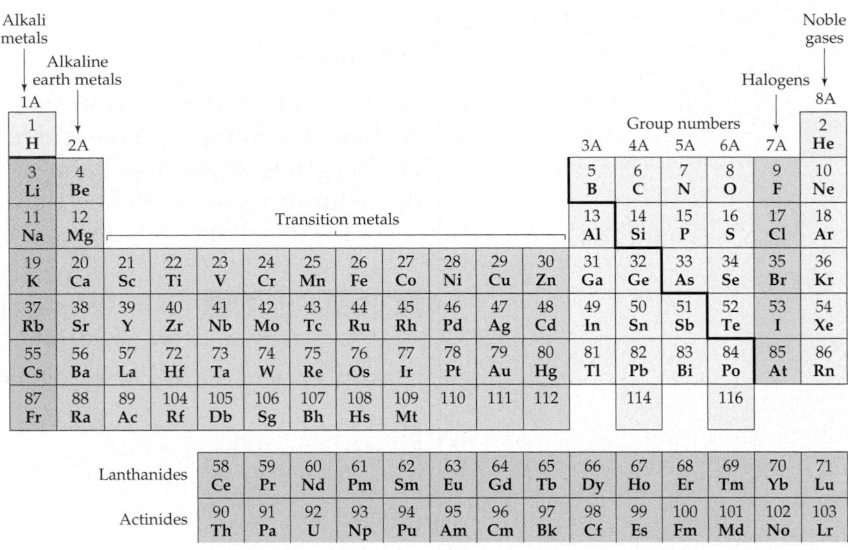

and xenon. The Group 1A elements, called the **alkali metals**, are all very reactive metals. A marble-sized piece of sodium explodes violently when dropped in water. Other alkali metals include lithium, potassium, and rubidium. The Group 2A elements, called the **alkaline earth metals**, are also fairly reactive, although not quite as reactive as the alkali metals. Calcium for example, reacts fairly vigorously when dropped in water but will not explode as easily as sodium. Other alkaline earth metals include magnesium, a common low-density structural metal; strontium; and barium. The Group 7A elements, called the **halogens**, are very reactive nonmetals. The most familiar halogen is probably chlorine, a greenish yellow gas with a pungent odor. Because of its reactivity, chlorine is often used as a sterilizing and disinfecting agent. Other halogens include bromine, a red-brown liquid that easily evaporates into a gas; iodine, a purple solid; and fluorine, a pale yellow gas.

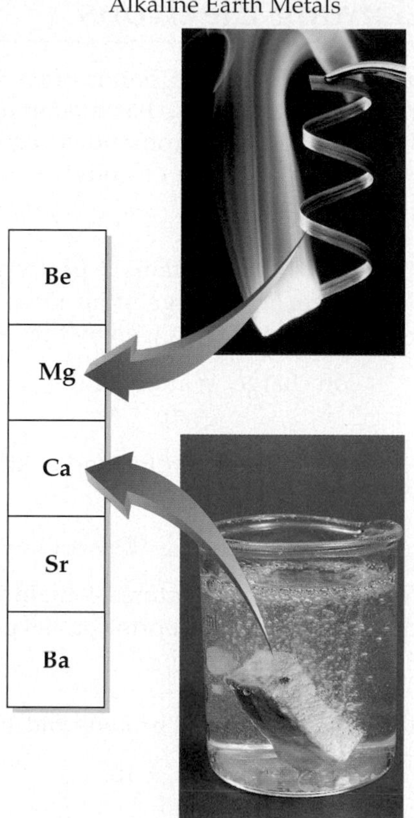

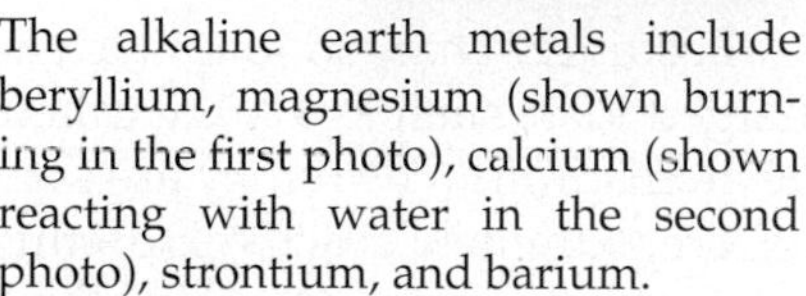
The alkaline earth metals include beryllium, magnesium (shown burning in the first photo), calcium (shown reacting with water in the second photo), strontium, and barium.

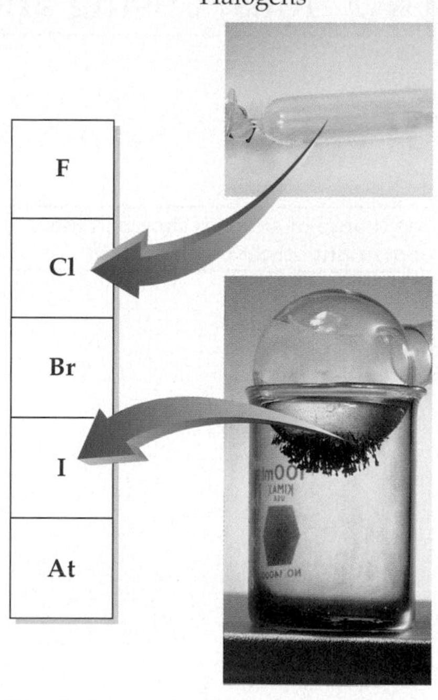

The halogens include fluorine, chlorine (shown in the first photo), bromine, iodine (shown in the second photo), and astatine.

EXAMPLE 4.3 Groups and Families of Elements

To which group or family of elements does each of the following elements belong?

a) Mg
b) N
c) K
d) Br

Solution:

a) Mg is in Group 2A; it is an alkaline earth metal.
b) N is in Group 5A.
c) K is in Group 1A; it is an alkali metal.
d) Br is in Group 7A; it is a halogen.

SKILLBUILDER 4.3 Groups and Families of Elements

To which group or family of elements does each of the following elements belong?

a) Li
b) B
c) I
d) Ar

4.7 Ions: Losing and Gaining Electrons

In chemical reactions, atoms often lose or gain electrons to form charged particles called **ions**. For example, neutral lithium (Li) atoms contain 3 protons and 3 electrons; however, in its reactions, lithium atoms lose one electron (e^-) to form Li^+ ions.

The charge of an ion is shown in the upper right corner of the symbol.

$$Li \rightarrow Li^+ + e^-$$

The Li^+ *ion* contains 3 protons but only 2 electrons, resulting in a +1 charge. The charge of an ion depends on how many electrons were gained or lost and is given by the following formula.

$$\begin{aligned}\text{Ion charge} &= \text{number of protons} - \text{number of electrons} \\ &= \#p - \#e^-\end{aligned}$$

where *p* stands for proton and e^- stands for electron.
For the Li^+ ion:

$$\text{Ion charge} = 3 - 2 = +1$$

Neutral fluorine (F) atoms contain 9 protons and 9 electrons; however, in chemical reactions fluorine atoms gain one electron to form F^- ions.

$$F + e^- \rightarrow F^-$$

The F^- *ion* contains 9 protons and 10 electrons resulting in a −1 charge.

$$\begin{aligned}\text{Ion charge} &= 9 - 10 \\ &= -1\end{aligned}$$

Positively charged ions, such as Li^+, are called **cations** and negatively charged ions, such as F^-, are called **anions**. Ions act very differently than the atoms from which they are formed. Neutral sodium atoms, for example, are extremely reactive, reacting violently with most things they contact. Sodium cations (Na^+) on the other hand are relatively inert—we eat them all the time in sodium chloride (table salt). In nature, cations and anions always occur together so that, again, matter is charge neutral.

EXAMPLE 4.4 Determining Ion Charge from Numbers of Protons and Electrons

Determine the charge of each of the following ions.

a) A magnesium ion with 10 electrons
b) A sulfur ion with 18 electrons
c) An iron ion with 23 electrons

Solution:
To determine the charge of each ion, we use the ion charge equation.

$$\text{Ion charge} = \#p - \#e^-$$

The number of electrons is given in the problem. The number of protons is obtained from the element's atomic number in the periodic table.

a) magnesium with atomic number 12

$$\text{Ion charge} = 12 - 10 = +2 \quad (Mg^{2+})$$

b) sulfur with atomic number 16

$$\text{Ion charge} = 16 - 18 = -2 \quad (S^{2-})$$

c) iron with atomic number 26

$$\text{Ion charge} = 26 - 23 = +3 \quad (Fe^{3+})$$

SKILLBUILDER 4.4 **Determining Ion Charge from Numbers of Protons and Electrons**

Determine the charge of each of the following ions.

a) A nickel ion with 26 electrons
b) A bromine ion with 36 electrons
c) A phosphorus ion with 18 electrons

EXAMPLE 4.5 **Determining the Number of Protons and Electrons in an Ion**

Find the number of protons and electrons in the Ca^{2+} ion.

Solution:

From the periodic table, we find that the atomic number for calcium is 20, so calcium has 20 protons. The number of electrons can be found using the ion charge equation.

$$\text{Ion charge} = \#p - \#e^-$$
$$+2 = 20 - \#e^-$$
$$\#e^- = 20 - 2 = 18$$

Therefore the number of electrons is 18. The Ca^{2+} ion has 20 protons and 18 electrons.

SKILLBUILDER 4.5 **Determining the Number of Protons and Electrons in an Ion**

Find the number of protons and electrons in the S^{2-} ion.

Ions and the Periodic Table

We have learned that in chemical reactions metals have a tendency to lose electrons and nonmetals have a tendency to gain them. For many main group elements, we can use the periodic table to predict how many electrons tend to be lost or gained. In other words, many main-group elements fall into groups that tend to form ions with the same charge. For example, the alkali metals (Group 1A) tend to lose 1 electron and therefore form +1 ions, while the alkaline earth metals (Group 2A) tend to lose 2 electrons and therefore form +2 ions. The halogens (Group 7A) tend to

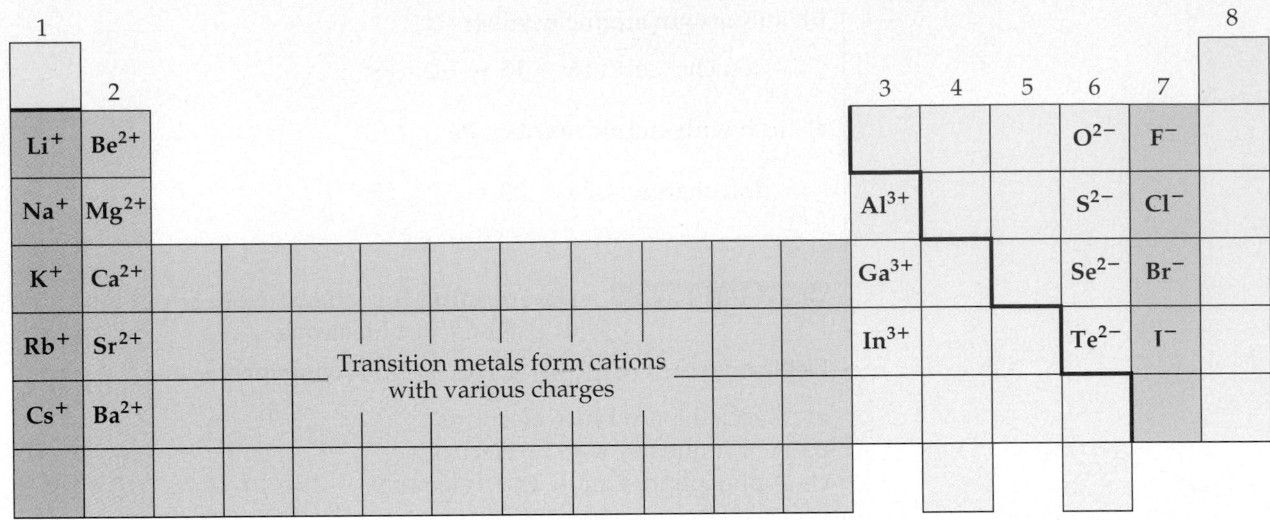

Figure 4.14 Elements that form predictable ions.

gain 1 electron and therefore form −1 ions. The groups in the periodic table that form predictable ions are shown in Figure 4.14. For now, memorize these groups and the ions they form. In Chapter 9, we examine a theory that explains why these groups form ions as they do.

EXAMPLE 4.6 **Charge of Ions from Position in Periodic Table**

Based on their position in the periodic table, what ions will barium and iodine tend to form?

Solution:

Since barium is in Group 2A, it will tend to form a cation with a +2 charge (Ba^{2+}). Since iodine is in Group 7A, it will tend to form an anion with a −1 charge (I^{-}).

SKILLBUILDER 4.6 **Charge of Ions from Position in Periodic Table**

Based on their position in the periodic table, what ions will potassium and sulfur tend to form?

4.8 Isotopes: When the Number of Neutrons Varies

All atoms of a given element have the same number of protons; however, they do not necessarily have the same number of neutrons. Since neutrons have nearly the same mass as protons (1 amu), this means that—contrary to what John Dalton originally proposed in his atomic theory—all atoms of a given element *do not* have the same mass. For example, all neon atoms in nature contain 10 protons, but they may have 10, 11, or 12 neutrons (Figure 4.15). All three types of neon atoms exist, and each has a

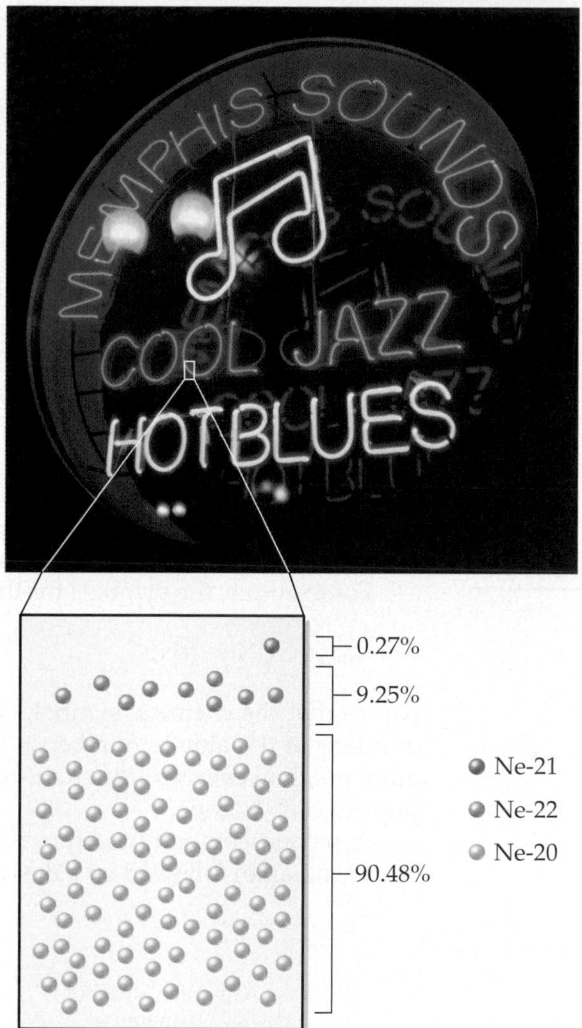

Figure 4.15 Naturally occurring neon contains three different isotopes, Ne-20 (with 10 neutrons), Ne-21 (with 11 neutrons), and Ne-22 (with 12 neutrons).

slightly different mass. Atoms with the same number of protons but different numbers of neutrons are called **isotopes**. Some elements, such as beryllium (Be) and aluminum (Al), have only one naturally occurring isotope, while other elements, such as neon (Ne) and chlorine (Cl), have two or more.

Fortunately, the relative amount of each different isotope in a naturally occurring sample of a given element is always the same. For example, in any natural sample of neon atoms, 90.48% of them are the isotope with 10 neutrons, 0.27% are the isotope with 11 neutrons, and 9.25% are the isotope with 12 neutrons. This means that out of 10,000 neon atoms, for example, 9048 have 10 neutrons, 27 have 11 neutrons, and 925 have 12 neutrons. These percentages are called the **percent natural abundance** of the isotopes. The preceding numbers are for neon, but all elements have their own unique percent natural abundance of isotopes.

Percent means *per hundred.* 90.48% means that 90.48 atoms out of 100 are the isotope with 10 neutrons.

The sum of the number of neutrons and protons in an atom is called the **mass number** and is given the symbol **A**.

$$A = \text{Number of protons} + \text{Number of neutrons}$$

For neon, with 10 protons, the mass numbers of the three different naturally occurring isotopes are 20, 21, and 22 corresponding to 10, 11, and 12 neutrons, respectively.

Isotopes are often symbolized in the following way.

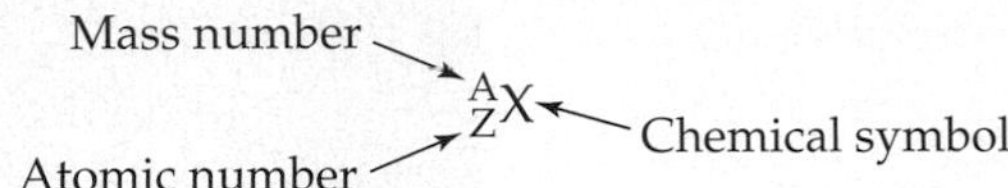

In general, mass number roughly increases with increasing atomic number.

where X is the chemical symbol, A is the mass number, and Z is the atomic number.

For example, the symbols for the neon isotopes are:

$^{20}_{10}Ne$ $^{21}_{10}Ne$ $^{22}_{10}Ne$

Notice that the chemical symbol, Ne, and the atomic number, 10, are redundant: if the atomic number is 10, the symbol must be Ne. The mass numbers, however, are different, reflecting the different number of neutrons in each isotope.

A second common notation for isotopes is the chemical symbol (or chemical name) followed by a dash and the mass number of the isotope.

X—A

Chemical symbol or name → X; Mass number → A

In this notation, the neon isotopes are:

Ne-20	Ne-21	Ne-22
neon-20	neon-21	neon-22

We can summarize what we have learned about the neon isotopes in Table 4.2.

TABLE 4.2

Neon Isotopes

Symbol	*Number of Protons*	*Number of Neutrons*	*A (Mass Number)*	*Natural Abundance*
Ne-20 or $^{20}_{10}Ne$	10	10	20	90.48%
Ne-21 or $^{21}_{10}Ne$	10	11	21	0.27%
Ne-22 or $^{22}_{10}Ne$	10	12	22	9.25%

Notice that all isotopes of a given element have the same number of protons (otherwise they would be a different element). Notice also that the mass number is the *sum* of the number of protons and the number of neutrons. The number of neutrons in an isotope is the difference between the mass number and the atomic number.

EXAMPLE 4.7 Atomic Numbers, Mass Numbers, and Isotope Symbols

What are the atomic number (Z), mass number (A), and symbols of the carbon isotope with 7 neutrons?

Solution:

From the periodic table, we find that the atomic number (Z) of carbon is 6, so carbon atoms have 6 protons. The mass number (A) for the isotope with 7 neutrons is the sum of the number of protons and the number of neutrons.

$$A = 6 + 7 = 13$$

So, Z = 6, A = 13 and the symbols for the isotope are C-13 or ${}^{13}_{6}C$.

SKILLBUILDER 4.7 Atomic Numbers, Mass Numbers, and Isotope Symbols

What are the atomic number, mass number, and symbols for the chlorine isotope with 18 neutrons?

EXAMPLE 4.8 Numbers of Protons and Neutrons from Isotope Symbols

How many protons and neutrons are in the following chromium isotope?

$${}^{52}_{24}Cr$$

Solution:

The number of protons is equal to Z (lower left number).

$$\#p = Z = 24$$

The number of neutrons is equal to A (upper left number) −Z (lower left number).

$$\#n = 52 - 24 = 28$$

SKILLBUILDER 4.8 Numbers of Protons and Neutrons from Isotope Symbols

How many protons and neutrons are in:

$${}^{39}_{19}K$$

CHEMISTRY IN THE ENVIRONMENT

Radioactive Isotopes at Hanford, Washington

Nuclei of the isotopes of a given element are not all equally stable. For example, naturally occurring lead is composed primarily of Pb-206, Pb-207, and Pb-208. Other isotopes of lead also exist, but their nuclei are unstable. Scientists can make some of these other isotopes, such as Pb-185 for example, in the laboratory. However, within seconds of their synthesis, Pb-185 atoms emit a few energetic, subatomic particles from their nuclei and change into different isotopes of different elements (which are themselves unstable). These emitted subatomic particles are called **nuclear radiation**. Nuclear radiation, always associated with unstable nuclei, can be harmful to humans and other living organisms because the energetic particles interact with and damage biological molecules. For Pb-185, the radiation is short-lived; however, for some isotopes, the radiation can last a long time—in some cases millions or even billions of years.

The nuclear power and nuclear weapons industries both produce byproducts containing unstable isotopes of several different elements. Many of these isotopes emit nuclear radiation for a long time, and their disposal is an environmental problem. For example, in Hanford, Washington, which for fifty years produced fuel for nuclear weapons, 177 underground storage tanks contain 55 million gallons of high-level nuclear waste. Certain isotopes within that waste have unstable nuclei that will produce nuclear radiation for the foreseeable future. Unfortunately, some of the underground storage tanks are aging and leaks have allowed some of the waste to seep into the environment. While the danger from short-term external exposure to this waste is minimal, ingestion of this waste through contamination of drinking water or food supplies would pose significant health risks. Consequently, Hanford is now the site of the largest environmental clean-up project in U.S. history. The U.S. government expects the project to take over twenty years and cost about $10 billion.

Radioactive isotopes are not always harmful, however, and many have beneficial uses. For example, technetium-99 (Tc-99) is often given to patients to diagnose disease. The radiation emitted by Tc-99 helps doctors to image internal organs or detect infection.

CAN YOU ANSWER THIS? *Give the number of neutrons in each of the isotopes mentioned here (Pb-206, Pb-207, Pb-208, Pb-185, Tc-99).*

Storage tanks at Hanford, Washington contain 55 million gallons of high-level nuclear waste. Each tank pictured here holds 1 million gallons.

4.9 Atomic Mass

Some books call this *average atomic mass* instead of simply *atomic mass*.

An important part of Dalton's atomic theory was that all atoms of a given element have the same mass. However, we just learned that because of isotopes, the atoms of a given element may have different masses, so Dalton was not completely correct. We can, however, calculate an average mass—called the **atomic mass**—for each element. The atomic mass of each element is listed in the periodic table directly beneath the element's symbol and represents the average mass of the atoms that compose that element. For example, the periodic table lists the atomic mass of chlorine as 35.45 amu. Naturally occurring chlorine consists of 75.77% chlorine-35 (mass 34.97 amu) and 24.23% chlorine-37 (mass 36.97 amu). Its atomic mass is:

$$\begin{aligned}\text{Atomic mass} &= (0.7577 \times 34.97\text{ amu}) + (0.2423 \times 36.97\text{ amu}) \\ &= 35.45\text{ amu}\end{aligned}$$

Notice that the atomic mass of chlorine is closer to 35 than 37. Naturally occurring chlorine contains more chlorine-35 atoms than chlorine-37 atoms. Notice also that when percentages are used in calculations, they must always be converted to their decimal value. To convert a percentage to its decimal value, simply divide by 100. For example:

$$\begin{aligned}75.77\% &= 75.77/100 = 0.7577 \\ 24.23\% &= 24.23/100 = 0.2423\end{aligned}$$

In general, the atomic mass is calculated according to the following equation.

$$\begin{aligned}\text{atomic mass} = {} & (\text{fraction of isotope 1} \times \text{mass of isotope 1}) + \\ & (\text{fraction of isotope 2} \times \text{mass of isotope 2}) + \\ & (\text{fraction of isotope 3} \times \text{mass of isotope 3}) + \cdots\cdots\end{aligned}$$

where the fractions of each isotope are the percent natural abundances converted to their decimal values. Atomic mass is useful because it allows us to assign a characteristic mass to each element, and as we will see in Chapter 6, it allows us to quantify the number of atoms in a sample of that element.

EXAMPLE 4.9 Calculating Atomic Mass

Gallium has two naturally occurring isotopes: Ga-69 with mass 68.9256 amu and a natural abundance of 60.11% and Ga-71 with mass 70.9247 amu and a natural abundance of 39.89%. Calculate the atomic mass of gallium.

Solution:

Convert the percent natural abundances into decimal form by dividing by 100.

$$\text{Fraction Ga-69} = \frac{60.11}{100} = 0.6011$$

$$\text{Fraction Ga-71} = \frac{39.89}{100} = 0.3989$$

The atomic mass is:

$$\begin{aligned}\text{Atomic mass} &= (0.6011 \times 68.9256\text{ amu}) \\ &\quad + (0.3989 \times 70.9247\text{ amu}) \\ &= 41.4\overline{3}12\text{ amu} + 28.2\overline{9}19\text{ amu} \\ &= 69.7\overline{2}31 = 69.72\text{ amu}\end{aligned}$$

SKILLBUILDER 4.9 **Calculating Atomic Mass**

Magnesium has three naturally occurring isotopes with masses of 23.99, 24.99, 25.98 amu and natural abundances of 78.99%, 10.00%, and 11.01%. Calculate the atomic mass of magnesium.

CHAPTER IN REVIEW

Chemical Principles	Relevance
The Atomic Theory: Democritus, an ancient Greek philosopher, is the first person on record to assert that matter is ultimately composed of small, indestructible particles. It was not until 2000 years later, however, that John Dalton introduced a formal atomic theory stating that matter is composed of atoms; atoms of a given element have unique properties that distinguish it from atoms of other elements; and atoms combine in simple, whole-number ratios to form compounds.	**The Atomic Theory:** The concept of atoms is important because it explains the physical world. You and everything you see are made of atoms. To understand the physical world, we must begin by understanding atoms. Atoms are the key concept—they determine the properties of matter.
Discovery of the Atom's Nucleus: Rutherford's gold foil experiment probed atomic structure and his results led to the nuclear model of the atom, which, with minor modifications to accommodate neutrons, is still valid today. In this model, the atom is composed of protons and neutrons—which compose most of the atom's mass and are grouped together in a dense nucleus—and electrons, which compose most of the atom's volume. Protons and neutrons have similar masses (1 amu) while electrons have a much smaller mass (≈ 0 amu).	**Discovery of the Atom's Nucleus:** We can see why this is relevant by asking the question, what if it were otherwise? What if matter was *not* mostly empty space? While we cannot know for certain, it seems probable that such matter would not form the diversity of substances required for life and then, of course, we would not be around to ask the question.
Charge: Protons and electrons both have electrical charge; the charge of the proton is $+1$ and the charge of the electron is -1. The neutron is charge neutral. When protons and electrons combine in atoms, their charges cancel.	**Charge:** Electrical charge is relevant to so much of our modern world. Many of the machines and computers we depend on are powered by electricity, which is simply the movement of electrical charge through a wire.
The Periodic Table: The periodic table tabulates all known elements in order of increasing atomic number. The periodic table is arranged so that similar elements are grouped together. Elements on the left side of the periodic table are metals and tend to lose electrons in their chemical changes. Elements on the upper right side of the periodic table are nonmetals and tend to gain electrons in their chemical changes. Elements between the two are called metalloids. Columns of elements in the periodic table have similar properties and are called groups or families.	**The Periodic Table:** The periodic table helps us organize the elements in ways that allow us to predict their properties. Have you ever breathed helium from a balloon and talked in a high voice as a result? Breathing helium is not dangerous (unless you breathe too much and suffocate) because helium is an inert gas—it does not react with anything and goes into and out of your lungs without changing any of the molecules that compose you. The gases in the column below it on the periodic table are also inert gases and form a family or group of elements called the noble gases. By tabulating the elements

Atomic Number: The most fundamental quantity of an atom is the number of protons in its nucleus; this number is called the atomic number (Z) and defines the element.

Ions: When an atom gains or loses electrons it becomes an ion. Positively charged ions are called cations and negatively charged ions are called anions. Cations and anions occur together so that matter, as usual, is charge neutral.

and grouping similar ones together, we begin to understand their properties.

Isotopes: While all atoms of a given element have the same number of protons, they do not necessarily have the same number of neutrons. Atoms of the same element with different numbers of neutrons are called isotopes. Isotopes are characterized by their mass number (A), the sum of the number of protons and the number of neutrons in their nucleus.

Each naturally occurring sample of an element has the same percent natural abundance of each isotope. These percentages, together with the mass of each isotope, are used to compute the atomic mass of the element, a weighted average of the masses of the individual isotopes.

Isotopes: Isotopes are relevant because they influence tabulated atomic masses. To understand these masses, we must understand the presence and abundance of isotopes. In nuclear processes—processes in which the nuclei of atoms actually change—the presence of different isotopes becomes even more important. Some isotopes are not stable, they lose subatomic particles, and are transformed into other elements. The emission of subatomic particles by unstable nuclei is called nuclear radiation. In many cases, such as in diagnosing and treating certain diseases, nuclear radiation is extremely useful. In other cases, such as in the disposal of radioactive waste, it can pose environmental problems.

Chemical Skills

Determining Ion Charge from Numbers of Protons and Electrons (Section 4.7)

- From the periodic table or from the alphabetical list of elements, find the atomic number of the element; this number is equal to the number of protons.
- Use the ion charge equation to compute charge.

$$\text{Ion charge} = \#\text{p} - \#\text{e}^-$$

Determining the Number of Protons and Electrons in an Ion (Section 4.7)

- From the periodic table or from the alphabetical list of elements, find the atomic number of the element; this number is equal to the number of protons.
- Use the ion charge equation and substitute in the known values.

$$\text{Ion charge} = \#\text{p} - \#\text{e}^-$$

- Solve the equation for the number of electrons.

Examples

EXAMPLE 4.10 Determining Ion Charge from Numbers of Protons and Electrons

Determine the charge of a selenium ion with 36 electrons.

Solution:

Selenium is atomic number 34; therefore it has 34 protons.

$$\text{Ion charge} = 34 - 36 = -2$$

EXAMPLE 4.11 Determining the Number of Protons and Electrons in an Ion

Find the number of protons and electrons in the O^{2-} ion.

Solution:

The atomic number of O is 8, therefore there are 8 protons.

$$\text{Ion charge} = \#\text{p} - \#\text{e}^-$$
$$-2 = 8 - \#\text{e}$$
$$\#\text{e} = 8 + 2 = 10$$

The ion has 8 protons and 10 electrons.

Determining Atomic Numbers, Mass Numbers, and Isotope Symbols for an Isotope (Section 4.8)

- From the periodic table or from the alphabetical list of elements, find the atomic number of the element.
- The mass number (A) is equal to the atomic number plus the number of neutrons.
- Write the symbol for the isotope by writing the symbol for the element with the mass number in the upper left corner and the atomic number in the lower left corner.
- The other way to write the symbol for the isotope is simply the chemical symbol followed by a dash and the mass number.

EXAMPLE 4.12 Determining Atomic Numbers, Mass Numbers, and Isotope Symbols for an Isotope

What are the atomic number (Z), mass number (A), and symbols for the iron isotope with 30 neutrons?

Solution:

The atomic number of iron is 26.

$$A = 26 + 30 = 56.$$

The mass number is 56.

$^{56}_{26}\text{Fe}$

Fe-56

Number of Protons and Neutrons from Isotope Symbols (Section 4.8)

- The number of protons is equal to Z (lower left number).
- The number of neutrons is equal to A (upper left number) −Z (lower left number).

EXAMPLE 4.13 Number of Protons and Neutrons from Isotope Symbols

How many protons and neutrons are in $^{62}_{28}\text{Ni}$?

Solution:

28 protons

$$\#n = 62 - 28 = 34 \text{ neutrons}$$

Calculating Atomic Mass from Percent Natural Abundances and Isotopic Masses (Section 4.9)

- Convert the natural abundances from percent to decimal values by dividing by 100.
- Find the atomic mass by multiplying the fractions of each isotope by their respective masses and adding.
- Round to the correct number of significant figures
- Check your work.

EXAMPLE 4.14 Calculating Atomic Mass from Percent Natural Abundances and Isotopic Masses

Copper has two naturally occurring isotopes, Cu-63 with mass 62.9395 amu and a natural abundance of 69.17%, and Cu-65 with mass 64.9278 amu and a natural abundance of 30.83%. Calculate the atomic mass of copper.

Solution:

$$\text{Fraction Cu-63} = \frac{69.17}{100} = 0.6917$$

$$\text{Fraction Cu-65} = \frac{30.83}{100} = 0.3083$$

$$\begin{aligned}\text{Atomic mass} &= (0.6917 \times 62.9395 \text{ amu}) \\ &\quad + (0.3082 \times 64.9278 \text{ amu}) \\ &= 43.5353 \text{ amu} + 20.0107 \text{ amu} \\ &= 63.5460 \text{ amu} \\ &= 63.55 \text{ amu}\end{aligned}$$

KEY TERMS

alkali metals [4.6]
alkaline earth metals [4.6]
anions [4.7]
atomic mass [4.9]
atomic mass unit (amu) [4.4]
atomic number (Z) [4.5]
cations [4.7]
charge [4.4]
chemical symbol [4.5]
Marie Curie [4.4]
Democritus [4.2]
electron [4.3]
family (of elements) [4.6]
group (of elements) [4.6]
halogens [4.6]
ions [4.7]
isotopes [4.8]
main-group elements [4.6]
mass number (A) [4.8]
Dmitri Mendeleev [4.5]
metalloids [4.6]
metals [4.6]
neutrons [4.3]
noble gases [4.6]
nonmetals [4.6]
nuclear radiation [4.8]
nuclear theory of the atom [4.3]
nucleus [4.3]
percent natural abundance [4.8]
periodic law [4.6]
periodic table [4.1]
protons [4.3]
Ernest Rutherford [4.3]
semiconductor [4.6]
J.J. Thomson [4.3]
transition elements [4.6]
transition metals [4.6]
Clemens Winkler [4.6]

EXERCISES

Questions

1. Why is it important to understand atoms?
2. What is an atom?
3. What did Democritus contribute to our modern understanding of matter?
4. What are three main ideas in Dalton's atomic theory?
5. Describe Rutherford's gold foil experiment and the results of that experiment. How did these results contradict the plum-pudding model of the atom?
6. What are the main ideas in the nuclear theory of the atom?
7. List the three subatomic particles and their properties.
8. What is electrical charge?
9. Is matter usually charge neutral? What if it wasn't?
10. What does the atomic number of an element specify?
11. What is a chemical symbol?
12. Give some examples of how elements got their names.
13. What was Dmitri Mendeleev's main contribution to our modern understanding of chemistry?
14. What is the main idea in the periodic law?
15. How is the periodic table organized?
16. What are the properties of metals? Where are metals found on the periodic table?
17. What are the properties of nonmetals? Where are nonmetals found on the periodic table?
18. Where on the periodic table do you find metalloids?
19. What is a family or group of elements?
20. Locate each of the following on the periodic table and give their group number.
a) alkali metals
b) alkaline earth metals
c) halogens
d) noble gases
21. What is an ion?
22. What is an anion? Cation?
23. Locate each of the following groups on the periodic table and list the charge of the ions they tend to form.
a) Group 1A
b) Group 2A
c) Group 3A
d) Group 6A
e) Group 7A
24. What are isotopes?
25. What is the percent natural abundance of isotopes?
26. What is the mass number of an isotope?
27. What notations are commonly used to specify isotopes? What do each of the numbers in these symbols mean?
28. What is the atomic mass of an element?

Problems

Atomic and Nuclear Theory

29. Which of the following statements are *inconsistent* with Dalton's atomic theory as it was originally stated. Why?
a) All carbon atoms are identical.
b) Helium atoms can be split into two hydrogen atoms.
c) An oxygen atom combines with 1.5 hydrogen atoms to form water molecules.
d) Two oxygen atoms combine with a carbon atom to form carbon dioxide molecules.

30. Which of the following statements are *consistent* with Dalton's atomic theory as it was originally stated? Why?
a) Calcium and titanium atoms have the same mass.
b) Neon and argon atoms are the same.
c) All cobalt atoms are identical.
d) Sodium and chlorine atoms combine in a 1:1 ratio to form sodium chloride.

31. Which of the following statements are *inconsistent* with Rutherford's nuclear theory as it was originally stated? Why?
a) Helium atoms have two protons in their nucleus and two electrons outside of their nucleus.
b) Most of the volume of hydrogen atoms is due to the nucleus.
c) Aluminum atoms have 13 protons in their nucleus and 22 electrons outside of their nucleus.
d) The majority of the mass of nitrogen atoms is due to their 7 electrons.

32. Which of the following statements are *consistent* with Rutherford's nuclear theory as it was originally stated? Why?
a) Atomic nuclei are small compared to the size of atoms.
b) The volume of an atom is mostly empty space.
c) Neutral potassium atoms contain more protons than electrons.
d) Neutral potassium atoms contain more neutrons than protons.

Protons, Neutrons, and Electrons

33. Which of the following statements about electrons are true?
a) Electrons repel each other.
b) Electrons are attracted to protons.
c) Some electrons have a charge of -1 and some have no charge.
d) Electrons are much lighter than neutrons.

34. Which of the following statements about electrons are false?
a) Most atoms have more electrons than protons.
b) Electrons have a charge of -1.
c) If an atom has an equal number of protons and electrons, it will be charge neutral.
d) Electrons experience an attraction to protons.

35. Which of the following statements about protons are true?
a) Protons have twice the mass of neutrons.
b) Protons have the same magnitude of charge as electrons, but opposite in sign.
c) Most atoms have more protons than electrons.
d) Protons have a charge of $+1$.

36. Which of the following statements about protons are false?
a) Protons have about the same mass as neutrons.
b) Protons have about the same mass as electrons.
c) Some atoms don't have any protons.
d) Protons have the magnitude of charge as neutrons, but opposite in sign.

37. How many electrons would it take to the equal the mass of a proton?

38. A helium nucleus has two protons and two neutrons. How many electrons would it take to equal the mass of a helium nucleus?

Elements, Symbols, and Names

39. Find the atomic number (Z) for each of the following elements.
a) Co
b) Ir
c) U
d) Si
e) Be

40. Find the atomic number (Z) for each of the following elements.
a) Al
b) Sn
c) Ca
d) Fr
e) Xe

41. How many protons are in the nucleus of each of the following atoms?
a) Mn
b) Ag
c) Au
d) Pb
e) S

42. How many protons are in the nucleus of each of the following atoms?
a) Y
b) N
c) Ne
d) K
e) Mo

43. Give the symbol and atomic number corresponding to each of the following elements.
a) carbon
b) nitrogen
c) sodium
d) potassium
e) copper

44. Give the symbol and atomic number corresponding to each of the following elements.
a) boron
b) neon
c) silver
d) mercury
e) curium

45. Give the name and the atomic number corresponding to the symbol for each of the following elements.
a) Au
b) Si
c) Ni
d) Zn
e) W

46. Give the name and the atomic number corresponding to the symbol for each of the following elements.
a) Ta
b) H
c) S
d) As
e) Br

47. Fill in the blanks to complete the following table.

Element Name	*Element Symbol*	*Atomic Number*
_____	Ag	47
Gallium	_____	_____
_____	As	_____
Astatine	_____	85

48. Fill in the blanks to complete the following table.

Element Name	*Element Symbol*	*Atomic Number*
_____	Pt	78
Strontium	_____	_____
Xenon	_____	_____
_____	Rn	86

The Periodic Table

49. Classify each of the following elements as a metal, nonmetal, or metalloid.
a) Sr
b) Mg
c) F
d) N
e) As

50. Classify each of the following elements as a metal, nonmetal, or metalloid.
a) Na
b) Ge
c) Si
d) Br
e) Ag

51. Which of the following elements would you expect to lose electrons in chemical changes?
a) potassium
b) sulfur
c) fluorine
d) barium
e) copper

52. Which of the following elements would you expect to gain electrons in chemical changes?
a) nitrogen
b) iodine
c) tungsten
d) strontium
e) gold

53. Which of the following elements are main-group elements?
a) Te
b) K
c) V
d) Re
e) Ag

54. Which of the following elements are *not* main-group elements?
a) Al
b) Br
c) Mo
d) Cs
e) Pb

55. Which of the following elements is an alkaline earth metal?
a) sodium
b) aluminum
c) calcium
d) barium
e) lithium

56. Which of the following elements is an alkaline earth metal?
a) rubidium
b) tungsten
c) magnesium
d) cesium
e) beryllium

57. Which of the following elements is an alkali metal?
a) sodium
b) potassium
c) silver
d) magnesium
e) cesium

58. Which of the following elements is an alkali metal?
a) barium
b) niobium
c) lithium
d) rubidium
e) calcium

59. Classify each of the following as a halogen, a noble gas, or neither.
a) Cl
b) Kr
c) F
d) Ga
e) He

60. Classify each of the following as a halogen, a noble gas, or neither.
a) Ne
b) Br
c) S
d) Xe
e) I

61. To what group number does each of the following elements belong?
a) oxygen
b) aluminum
c) silicon
d) tin
e) phosphorus

62. To what group number does each of the following elements belong?
a) germanium
b) nitrogen
c) sulfur
d) carbon
e) boron

63. Which of the following elements do you expect to be most like sulfur? Why?
a) nitrogen
b) oxygen
c) fluorine
d) lithium
e) potassium

64. Which of the following elements do you expect to be most like magnesium? Why?
a) potassium
b) silver
c) bromine
d) calcium
e) lead

65. Which of the following pairs of elements do you expect to be most similar? Why?
a) Si and P
b) Cl and F
c) Na and Mg
d) Mo and Sn
e) N and Ni

66. Which of the following pairs of elements do you expect to be most similar? Why?
a) Ti and Ga
b) N and O
c) Li and Na
d) Ar and Br
e) Ge and Ga

67. Fill in the blanks to complete the following table.

Chemical Symbol	*Group Number*	*Group Name*	*Metal or Nonmetal*
Na	_____	_____	metal
He	8A	_____	_____
I	_____	halogens	_____
Sr	_____	_____	metal

68. Fill in the blanks to complete the following table.

Chemical Symbol	*Group Number*	*Group Name*	*Metal or Nonmetal*
Rn	_____	_____	nonmetal
F	7A	_____	_____
Ba	_____	_____	metal
K	_____	alkali metal	_____

Ions

69. Complete each of the following.
a) $Na \rightarrow Na^+ + \_$
b) $O + 2e^- \rightarrow \_$
c) $Ca \rightarrow Ca^{2+} + \_$
d) $Cl + e^- \rightarrow \_$

70. Complete each of the following.
a) $Mg \rightarrow \_ + 2e^-$
b) $Ba \rightarrow Ba^{2+} + \_$
c) $I + e^- \rightarrow \_$
d) $Al \rightarrow \_ + 3e^-$

71. Determine the charge of each of the following ions.
a) oxygen ion with 10 electrons
b) aluminum ion with 10 electrons
c) titanium ion with 18 electrons
d) iodine ion with 54 electrons

72. Determine the charge of each of the following ions.
a) tungsten ion with 68 electrons
b) tellurium ion with 54 electrons
c) nitrogen ion with 10 electrons
d) barium ion with 54 electrons

73. Determine the number of protons and electrons in each of the following ions.
a) K^+
b) S^{2-}
c) Sr^{2+}
d) Cr^{3+}

74. Determine the number of protons and electrons in each of the following.
a) Ga^{3+}
b) Se^{2-}
c) Al^{3+}
d) Br^-

75. Determine if each of the following is true or false. If false, correct it.
a) The Ti^{2+} ion contains 22 protons and 24 electrons.
b) The I^- ion contains 53 protons and 54 electrons.
c) The Mg^{2+} ion contains 14 protons and 12 electrons.
d) The O^{2-} ion contains 8 protons and 10 electrons.

76. Determine if each of the following is true or false. If false, correct it.
a) The Fe^{3+} ion contains 29 protons and 26 electrons.
b) The Cs^+ ion contains 55 protons and 56 electrons.
c) The Se^{2-} ion contains 32 protons and 34 electrons.
d) The Li^+ ion contains 3 protons and 2 electrons.

77. Predict the ion formed by each of the following:
a) Rb
b) K
c) Al
d) O

78. Predict the ion formed by each of the following:
a) F
b) N
c) Mg
d) Na

79. Predict how many electrons will most likely be gained or lost by each of the following:
a) Ga
b) Li
c) Br
d) S

80. Predict how many electrons will most likely be gained or lost by each of the following:
a) I
b) Ba
c) Cs
d) Se

81. Fill in the blanks to complete the following table.

Symbol	Ion Formed	Number of Electrons in Ion	Number of Protons in Ion
Ca	Ca^{2+}	_____	_____
Te	_____	54	_____
In	_____	_____	49
_____	Sr^{2+}	_____	38

82. Fill in the blanks to complete the following table.

Symbol	Ion Formed	Number of Electrons in Ion	Number of Protons in Ion
Cl	_____	_____	17
_____	Be^{2+}	2	_____
Al	_____	_____	13
Br	Br^-	_____	_____

Isotopes

83. What are the atomic number and mass number for each of the following isotopes?
a) the hydrogen isotope with 1 neutron
b) the chromium ion with 27 neutrons
c) the calcium isotope with 20 neutrons
d) the tantalum isotope with 108 neutrons

84. How many neutrons are in an atom with the following atomic numbers and mass numbers?
a) $Z = 28, A = 58$
b) $Z = 92, A = 238$
c) $Z = 21, A = 45$
d) $Z = 18, A = 40$

85. Write isotopic symbols of the form $^{A}_{Z}X$ for each of the following isotopes.
a) the oxygen isotope with 8 neutrons
b) the fluorine isotope with 10 neutrons
c) the sodium isotope with 12 neutrons
d) the aluminum isotope with 14 neutrons

86. Write isotopic symbols of the form X-A (eg. C-13) for each of the following isotopes.
a) the iodine isotope with 74 neutrons
b) the phosphorus isotope with 16 neutrons
c) the uranium isotope with 234 neutrons
d) the argon isotope with 22 neutrons

87. Write the symbols of each of the following in the form $^{A}_{Z}X$.
a) cobalt-60
b) neon-22
c) iodine-131
d) plutonium-244

88. Write the symbols of each of the following in the form $^{A}_{Z}X$.
a) U-235
b) V-52
c) P-32
d) Xe-144

89. Determine the number of protons, and neutrons in each of the following:
a) $^{23}_{11}Na$
b) $^{266}_{88}Ra$
c) $^{208}_{82}Pb$
d) $^{14}_{7}N$

90. Determine the number of protons, and neutrons in each of the following:
a) $^{33}_{15}P$
b) $^{40}_{19}K$
c) $^{222}_{86}Rn$
d) $^{99}_{43}Tc$

91. Carbon-14, present within living organisms and substances derived from living organisms, is often used to establish the age of fossils and artifacts. Determine the number of protons and neutrons in a carbon-14 isotope and write its symbol in the form $^{A}_{Z}X$.

92. Plutonium-239 is used in nuclear bombs. Determine the number or protons and neutrons in plutonium-239 and write its symbol in the form $^{A}_{Z}X$.

Atomic Mass

93. Rubidium has two naturally occurring isotopes: Rb-85 with mass 84.9118 amu and a natural abundance of 72.17%; and Rb-87 with mass 86.9092 amu and a natural abundance of 27.83%. Calculate the atomic mass of rubidium.

94. Silicon has three naturally occurring isotopes: Si-28 with mass 27.9769 amu and a natural abundance of 92.2%; Si-29 with mass 28.9765 amu and a natural abundance of 4.67% and Si-30 with mass 29.9737 amu and a natural abundance of 3.10%. Calculate the atomic mass of silicon.

95. Bromine has two naturally occurring isotopes (Br-79 and Br-81) and an atomic mass of 79.904 amu.
a) If the natural abundance of Br-79 is 50.69%, what is the natural abundance of Br-81?
b) If the mass of Br-81 is 80.9163 amu, what is the mass of Br-79?

96. Silver has two naturally occurring isotopes (Ag-107 and Ag-109).
a) Use the periodic table to find the atomic mass of silver.
b) If the natural abundance of Ag-107 is 51.84%, what is the natural abundance of isotope 2?
c) If the mass of Ag-107 is 106.905, what is the mass of Ag-109?

97. An element has two naturally occurring isotopes. Isotope one has a mass of 120.9038 amu and a relative abundance of 57.4% and isotope two has a mass of 122.9042 amu and a relative abundance of 42.6%. Find the atomic mass of this element and by comparison to the periodic table, identify it.

98. Copper has two naturally occurring isotopes. Cu-63 has a mass of 62.939 amu and relative abundance of 69.17%. Use the atomic weight of copper to determine the mass of the second copper isotope.

Cumulative Problems

99. Electrical charge is sometimes reported in coulombs (C). On this scale, one electron has a charge of -1.6×10^{-19}C. Suppose your body acquires -125 mC (millicoulombs) of charge on a dry day. How many excess electrons has it acquired? (Hint: Use the charge of an electron in C as a conversion factor between charge and electrons.)

100. How many excess protons are in a positively charged object with a charge of $+398$ mC (millicoulombs)? The charge of one proton is $+1.6 \times 10^{-19}$C.

101. Prepare a table such as Table 4.2 for the 4 different isotopes of Sr that have the following natural abundances and masses.

Sr-84	0.56%	83.9134 amu
Sr-86	9.86%	85.9093 amu
Sr-87	7.00%	86.9089 amu
Sr-88	82.58%	87.9056 amu

Use your table and the preceding atomic masses to calculate the atomic mass of strontium.

102. Determine the number of protons and neutrons in each of the following isotopes of chromium and use the following natural abundances and masses to calculate its atomic mass.

Cr-50	4.345%	49.9460 amu
Cr-52	83.89%	51.9405 amu
Cr-53	9.50%	52.9407 amu
Cr-54	2.365%	53.9389 amu

103. Fill in the blanks to complete the following table.

Symbol	*Z*	*A*	*Number of Protons*	*Number of Electrons*	*Number of Neutrons*	*Charge*
Fe	___	58	___	24	___	___
Cu	___	___	___	___	24	2+
___	15	___	___	15	16	___
O	___	16	___	___	___	2−

104. Fill in the blanks to complete the following table.

Symbol	*Z*	*A*	*Number of Protons*	*Number of Electrons*	*Number of Neutrons*	*Charge*
___	16	___	___	___	16	2−
Mg	___	25	___	___	24	2+
Ni	___	61	___	26	___	___
N	___	14	___	10	___	___

105. Europium has two naturally occurring isotopes: Eu-151 with a mass of 150.9198 and a natural abundance of 47.8% and Eu-153. Use the atomic mass of europium to find the mass and natural abundance of Eu-153.

106. Rhenium has two naturally occurring isotopes: Re-185 with a natural abundance of 37.40% and Re-187 with a natural abundance of 62.60%. The sum of the masses of the two isotopes is 371.9087 amu. Find the masses of the individual isotopes.

Highlight Problems

107. Below is a representation of 50 atoms of a fictitious element with the symbol Nt and atomic number 120. Nt has three isotopes represented by the following colors: Nt-304 (red), Nt-305 (blue), and Nt-306 (green).

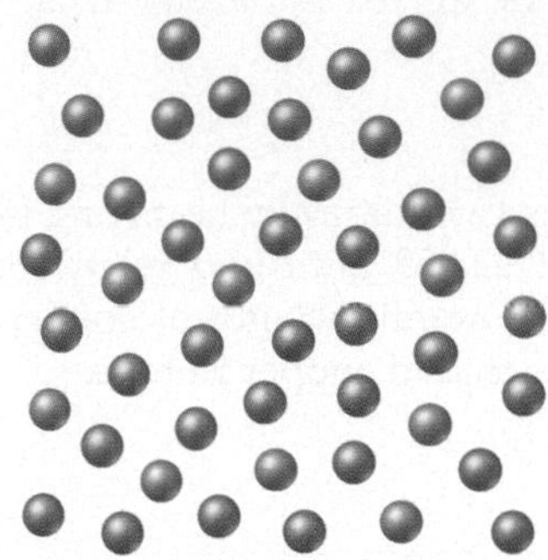

a) Assuming that the preceding diagram is statistically representative of naturally occurring Nt, what is the percent natural abundance of each Nt isotope?
b) Use the following masses of each isotope to calculate the atomic mass of Nt. Then draw a box for the element similar to the boxes for each element shown in the periodic table in the inside front cover of this book. Make sure your box includes the atomic number, symbol, and atomic mass. (Assume that the percentages from part a are good to four significant figures)

Nt-304	303.956 amu
Nt-305	304.962 amu
Nt-306	305.978 amu

108. Neutron stars are believed to be composed of solid nuclear matter, primarily neutrons.
a) If the radius of a neutron is 1.0×10^{-13} cm, calculate the density of a neutron. (Hint: For a sphere $V = \frac{4}{3}\pi r^3$)
b) Assuming that a neutron star has the same density as a neutron, calculate the mass (in kilograms) of a small piece of a neutron star the size of a spherical pebble with a radius of 0.10 mm.

Neutron stars are believed to be composed of solid nuclear matter.

Answers to Skillbuilder Exercises

Skillbuilder 4.1 **a)** sodium, 11 **b)** nickel, 28 **c)** phosphorus, 15 **d)** tantalum, 73
Skillbuilder 4.2 **a)** nonmetal **b)** nonmetal **c)** metal **d)** metalloid
Skillbuilder 4.3 **a)** alkali metal **b)** Group 3A **c)** halogen **d)** noble gas
Skillbuilder 4.4 **a)** +2 **b)** −1 **c)** −3
Skillbuilder 4.5 16 protons, 18 electrons
Skillbuilder 4.6 K^+ and S^{2-}
Skillbuilder 4.7 Z = 17, A = 35, Cl-35 or $^{35}_{17}Cl$
Skillbuilder 4.8 19 protons, 20 neutrons
Skillbuilder 4.9 24.31 amu

Across
1. A type of sugar
16. Element N
26. Thirsty?
42. Part of water
51. Not real sugar
66. ______composition

Molecules and Compounds 5

"Chemistry is a science, the object of which is to recognize the nature and properties of all substances by analyzing them and their compounds"

Pierre Joseph Macquer (1718–1784)

5.1 Sugar and Salt

Figure 5.1 Sodium is a reactive metal that dulls almost instantly upon exposure to air.

◀ Ordinary table sugar is a compound called sucrose. A sucrose molecule, such as the one shown here, contains carbon, hydrogen, and oxygen atoms. However, the properties of sucrose are much different than those of carbon, hydrogen, and oxygen. The properties of a compound are much different than the properties of the elements that compose it.

Sodium, a shiny metal (Figure 5.1) that dulls almost instantly upon exposure to air, is extremely reactive and poisonous. If you were to consume any appreciable amount of elemental sodium, you would need immediate medical help. Chlorine, a pale yellow gas (Figure 5.2), is equally reactive and poisonous. Yet the compound formed between these two elements, sodium chloride, is the relatively harmless flavor enhancer that we call table salt (Figure 5.3). When elements combine to form compounds, their properties change completely.

Consider ordinary sugar. Sugar is a compound composed of carbon, hydrogen, and oxygen. Each of these elements has its own unique properties. Carbon is most familiar to us as the graphite found in pencils or the diamonds in jewelry. Hydrogen is an extremely flammable gas used as a fuel for the space shuttle, and oxygen is one of the gases that compose air. When these three elements combine to form sugar, however, a sweet, white, crystalline solid results.

In Chapter 4, we learned how protons, neutrons, and electrons compose different elements, each with its own properties and its own chemistry, each different from the other. In this chapter, we learn how these elements combine with each other to form different compounds, each with its own properties and its own chemistry, each different from the other and different from the elements that compose it. Here is the great wonder of nature, how from such simplicity—protons, neutrons, and electrons—we get such great complexity. It is exactly this complexity, however, that makes life possible. Life could not exist with just

Figure 5.2 Chlorine is a yellow gas with a pungent odor. It is highly reactive and poisonous.

Figure 5.3 The compound formed by sodium and chlorine is table salt.

ninety-one different elements if they did not combine to form compounds. It takes compounds in all of their diversity to make living organisms.

5.2 Compounds Display Constant Composition

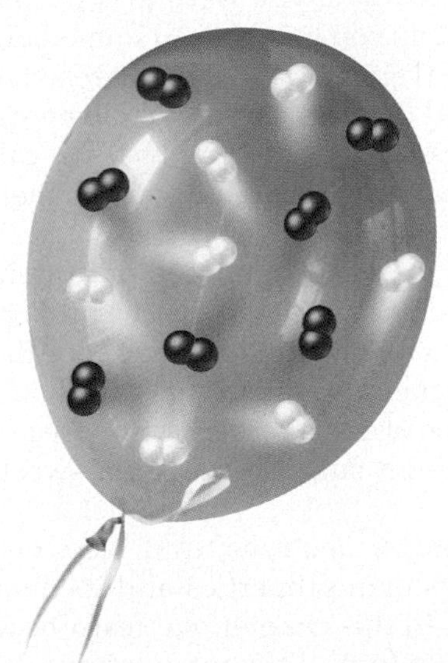

Figure 5.4 This balloon is filled with a mixture of hydrogen and oxygen gas. The relative amounts of hydrogen and oxygen are variable. We could easily add either more hydrogen or more oxygen to the balloon.

Although some of the substances we encounter in every day life are elements, most are not—they are compounds. Free atoms are rare in nature. As we learned in Chapter 3, a compound is different from a mixture of elements. In a compound, the elements combine in fixed, definite proportions, while in a mixture, they can have any proportions whatsoever. For example, consider the difference between a mixture of hydrogen and oxygen gas (Figure 5.4), and the compound water (Figure 5.5). A mixture of hydrogen and oxygen gas can have any proportions of hydrogen and oxygen. Water, on the other hand, is composed of water molecules that must contain 2 hydrogen atoms to every 1 oxygen atom. Consequently, water has a definite proportion of hydrogen to oxygen.

The first chemist to formally state the idea that elements combine in fixed proportions to form compounds was **Joseph Proust** (1754–1826) in the **law of constant composition**, which states:

> All samples of a given compound have the same proportions of their constituent elements.

For example, if we decompose water, we find 16.0 g of oxygen to every 2.0 g of hydrogen, or an oxygen-to-hydrogen mass ratio of:

$$\text{Mass ratio} = \frac{16.0\text{ g O}}{2.0\text{ g H}} = 8.0$$

Figure 5.5 This balloon is filled with water, composed of molecules which have a fixed ratio of hydrogen to oxygen.

Two H atoms (○) to every 1 O atom (●)

Two H atoms (○) to every 1 O atom (●)

Figure 5.6 Water will always have a constant ratio of hydrogen to oxygen, no matter what its source.

This is true of any sample of pure water, no matter what its origin (Figure 5.6). The law of constant composition applies not only to water, but to every compound. Consider ammonia, a compound composed of nitrogen and hydrogen. Ammonia contains 14.0 g of nitrogen to every 3.0 g of hydrogen or a nitrogen-to-hydrogen mass ratio of:

$$\text{Mass ratio} = \frac{14.0 \text{ g N}}{3.0 \text{ g H}} = 4.7$$

Even though atoms combine in whole number ratios, their mass ratios are not necessarily whole numbers.

Again, this ratio is the same for every sample of ammonia—the composition of each compound is constant.

EXAMPLE 5.1 Constant Composition of Compounds

Two samples of carbon dioxide, obtained from different sources, were decomposed into their constituent elements. One sample produced 4.8 g of oxygen and 1.8 g of carbon, and the other sample produced 17.1 g of oxygen and 6.4 g of carbon. Show that these results are consistent with the law of constant composition.

Solution:

To show this, compute the mass ratio of one element to the other for both samples by dividing the larger mass by the smaller one.

For the first sample:

$$\frac{\text{Mass oxygen}}{\text{Mass carbon}} = \frac{4.8 \text{ g}}{1.8 \text{ g}} = 2.7$$

For the second sample:

$$\frac{\text{Mass oxygen}}{\text{Mass carbon}} = \frac{17.1 \text{ g}}{6.4 \text{ g}} = 2.7$$

Since the ratios are the same for the two samples, these results are consistent with the law of constant composition.

SKILLBUILDER 5.1 **Constant Composition of Compounds**

Two samples of carbon monoxide, obtained from different sources, were decomposed into their constituent elements. One sample produced 4.3 g of oxygen and 3.2 g of carbon, and the other sample produced 7.5 g of oxygen and 5.6 g of carbon. Are these results consistent with the law of constant composition?

5.3 Chemical Formulas: How to Represent Compounds

We represent a compound with a **chemical formula**, which indicates the elements present in the compound and the relative number of atoms of each. For example, H_2O is the chemical formula for water; it indicates that water consists of hydrogen and oxygen atoms in a two-to-one ratio. The formula contains the symbol for each element and a subscript indicating the number of atoms of the element. When the subscript is 1, it is omitted.

The reason that compounds have constant composition with respect to mass (as we learned in the previous section) is because they are composed of atoms in fixed ratios.

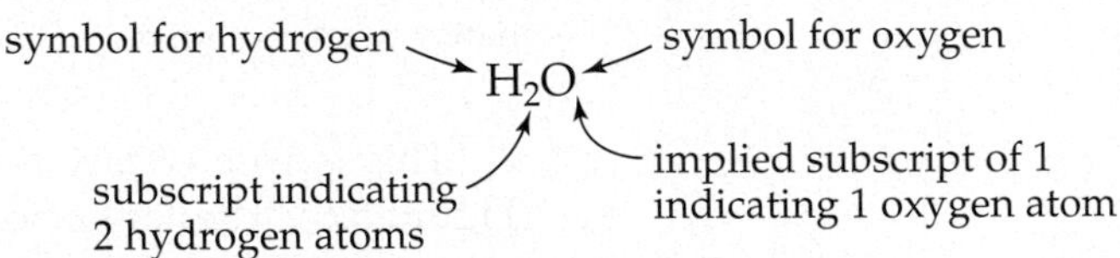

Some other common chemical formulas include NaCl for table salt indicating sodium and chlorine atoms in a one-to-one ratio; CO_2 for carbon dioxide indicating carbon and oxygen atoms in a one-to-two ratio; and $C_{12}H_{22}O_{11}$ for table sugar (sucrose) indicating carbon, hydrogen and oxygen atoms in a 12:22:11 ratio.

The subscripts in a chemical formula are part of the compound's definition—if they change, the compound changes. For example, CO is the chemical formula for carbon monoxide, an air pollutant with adverse health effects on humans. When inhaled, carbon monoxide interferes with the blood's ability to carry oxygen and can cause death. CO is the primary substance responsible for deaths of people who inhale too much automobile exhaust. Changing the subscript on O in CO from a 1 to a 2, however, makes a totally different compound.

The subscripts in a chemical formula are part of what define the compound—if you change a subscript, you change the compound.

CO_2 is the chemical formula for carbon dioxide, the relatively harmless product of combustion and human respiration. We breathe small amounts of CO_2 all the time with no harmful effects. So, remember:

> The subscripts in a chemical formula represent the relative number of each type of atom in a chemical compound; they do not change for a given compound.

Chemical formulas normally list the most metallic elements first. Therefore, the formula for table salt is NaCl, not ClNa. For compounds that do not include a metal, the more metal-like element is listed first. Recall from Chapter 4 that metals are found on the left side of the periodic table and nonmetals on the upper right side. Therefore, among nonmetals, those to the left in the periodic table are more metal-like than

TABLE 5.1

Order of Listing Nonmetal Elements in a Chemical Formula*

C	P	N	H	S	I	Br	Cl	O	F

Elements on the left are generally listed before elements on the right.

*There are a few historical exceptions to this rule such as the hydroxide ion, which is written as OH^-.

those to the right and are normally listed first. Therefore we write CO_2 and NO, not O_2C and ON. Within a single column in the periodic table, elements toward the bottom are more metal-like than elements towards the top. Therefore we write SO_2 not O_2S. The specific order for listing nonmetal elements in a chemical formula is shown in Table 5.1.

EXAMPLE 5.2 Writing Chemical Formulas

Write a chemical formula for each of the following:

a) The compound containing 2 aluminum atoms to every 3 oxygen atoms
b) The compound containing 3 oxygen atoms to every 1 sulfur atom
c) The compound containing 4 chlorine atoms to every 1 carbon atom

Solution:

a) Since aluminum is the metal, it is listed first. The formula is Al_2O_3.
b) Since sulfur is below oxygen on the periodic table and since it occurs before oxygen in Table 5.1, it is listed first. The formula is SO_3.
c) Since carbon is to the left of chlorine on the periodic table and since it occurs before chlorine in Table 5.1, it is listed first. The formula is CCl_4.

SKILLBUILDER 5.2 Writing Chemical Formulas

Write a chemical formula for each of the following:

a) The compound containing two silver atoms to every sulfur atom.
b) The compound containing two nitrogen atoms to every oxygen atom.
c) The compound containing two oxygen atoms to every titanium atom.

Some chemical formulas contain several identical groupings of atoms. These groupings are set off in parentheses and given a subscript indicating how many of that group are present. For example, $Mg(NO_3)_2$ indicates a compound containing 1 magnesium atom and two NO_3^- groups.

Many times these groups have a charge associated with them and are called *polyatomic ions*. Polyatomic ions are described in more detail in Section 5.7.

symbol for NO_3^- group

symbol for magnesium → $Mg(NO_3)_2$ ← subscript indicating 2 NO_3^- groups

implied subscript indicating 1 magnesium atom

subscript indicating 3 oxygen atoms per NO_3^- group

implied subscript indicating 1 nitrogen atom per NO_3^- group

To determine the total number of each type of atom within the parentheses, multiply the subscript outside of the parenthesis by the subscript for each atom inside the parentheses. Therefore, the preceding formula has the following numbers of each type of atom.

Mg: 1 Mg

N: $1 \times 2 = 2$ N (implied 1 inside parentheses times 2 outside of parentheses)

O: $3 \times 2 = 6$ O (3 inside parentheses times 2 outside parentheses)

EXAMPLE 5.3 Total Number of Each Type of Atom in a Chemical Formula

Determine the number of each type of atom in $Mg_3(PO_4)_2$.

Solution:

Mg: There are 3 Mg atoms as indicated by the subscript 3.

P: There are 2 P atoms as indicated by multiplying the subscript outside of the parentheses (2) by the subscript for P inside the parentheses, which is 1 (implied).

O: There are 8 O atoms as indicated by muliplying the subscript outside of the parentheses (2) by the subscript for O inside the parentheses (4).

SKILLBUILDER 5.3 Total Number of Each Type of Atom in a Chemical Formula

Determine the number of each type of atom in K_2SO_4.

SKILLBUILDER PLUS

Determine the number of each type of atom in $Al_2(SO_4)_3$.

5.4 A Molecular View of Elements and Compounds

In Chapter 3, we learned that pure substances could be divided into elements or compounds. We can further subdivide elements and compounds according to the basic units that compose them (Figure 5.7). Pure substances may be elements, or they may be compounds. Elements may be either atomic or molecular. Compounds may be either molecular or ionic.

Atomic Elements

Atomic elements are those that exist in nature with single atoms as their basic units. Most elements fall into this category. For example, helium is composed of helium atoms, copper is composed of copper atoms, and mercury is composed of mercury atoms (Figure 5.8).

TABLE 5.2

Elements That Occur as Diatomic Molecules

Name of Element	*Formula of Basic Unit*
hydrogen	H_2
nitrogen	N_2
oxygen	O_2
fluorine	F_2
chlorine	Cl_2
bromine	Br_2
iodine	I_2

Molecular Elements

Molecular elements do not normally exist in nature with single atoms as their basic units. Instead, these elements exist as diatomic molecules—two atoms of that element bonded together—as their basic units. For example, hydrogen is composed of H_2 molecules, oxygen is composed of O_2 molecules, and chlorine is composed of Cl_2 molecules (Figure 5.9). Elements that exist as diatomic molecules are shown in Table 5.2 and in Figure 5.10.

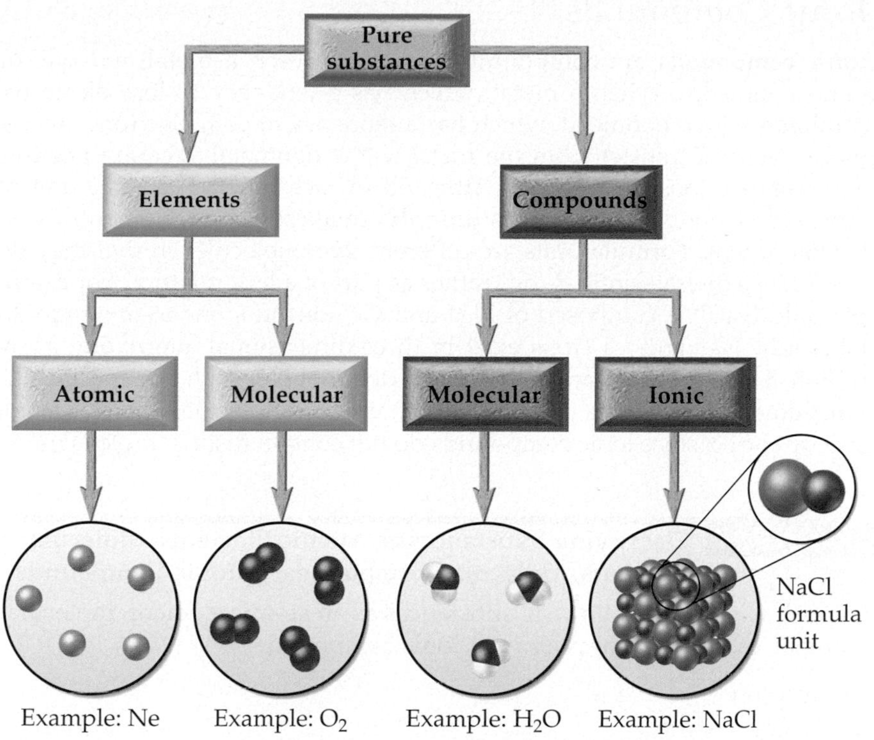

Figure 5.7 A molecular view of elements and compounds.

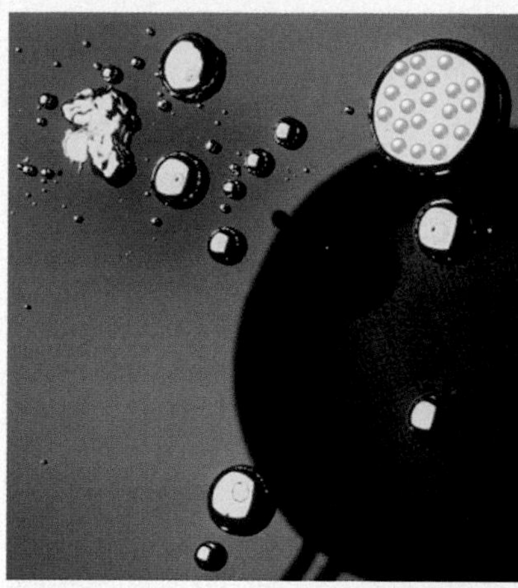

Figure 5.8 The basic units that compose mercury, an atomic element and a metal, are single mercury atoms.

Molecular Compounds

Molecular compounds are compounds formed between two or more nonmetals. The basic units of molecular compounds are molecules composed of the constituent atoms. For example, water is composed of H_2O molecules, dry ice is composed of CO_2 molecules, and acetone (nail-polish remover) is composed of C_3H_6O molecules (Figure 5.11).

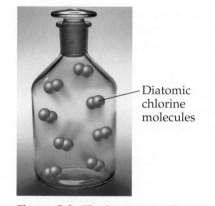

Diatomic chlorine molecules

Figure 5.9 The basic units that compose chlorine, a molecular element, are diatomic chlorine molecules, each composed of two chlorine atoms.

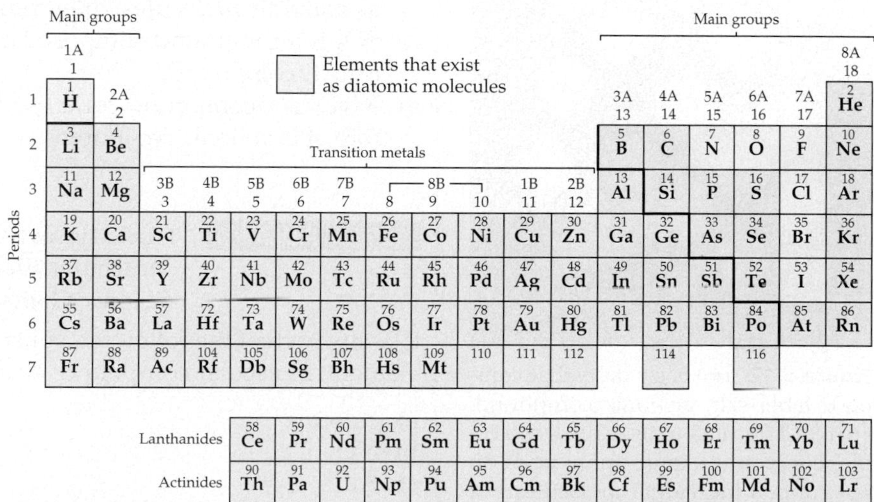

Figure 5.10

Remember that nonmetals tend toward the upper right side of the periodic table.

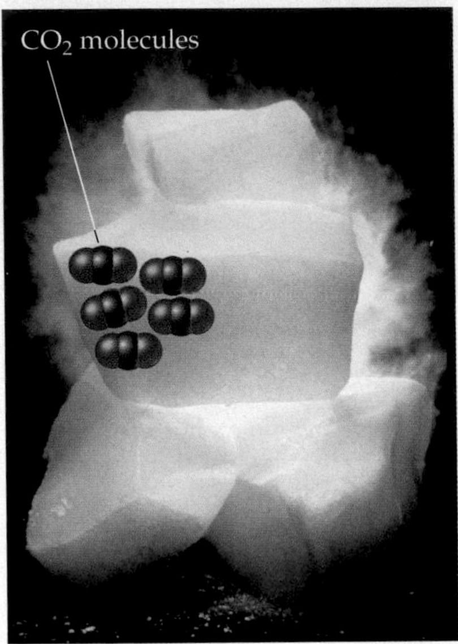

Figure 5.11 The basic units that compose dry ice, a molecular compound, are CO_2 molecules.

Figure 5.12 The basic units that compose table salt, an ionic compound, are NaCl formula units. Unlike molecular compounds, ionic compounds do not contain individual molecules but rather sodium and chloride ions in an alternating three-dimensional array.

Ionic Compounds

Ionic compounds are compounds formed between a metal and one or more nonmetals. When a metal, which has a tendency to lose electrons, combines with a nonmetal, which has a tendency to gain electrons, one or more electrons transfer from the metal to the nonmetal, creating positive and negative ions that are then attracted to each other. The basic unit of ionic compounds is the **formula unit**, the smallest electrically neutral collection of ions. Formula units are different than molecules in that they do not exist as discrete entities, but rather as part of a larger lattice. For example, salt (NaCl) is composed of Na^+ and Cl^- ions in a one-to-one ratio. In table salt, Na^+ and Cl^- ions exist in three-dimensional alternating array (Figure 5.12). However, any one Na^+ ion does not pair with one specific Cl^- ion. Sometimes chemists refer to formula units as molecules, but this is not strictly correct since ionic compounds do not contain distinct molecules.

EXAMPLE 5.4 Classifying Substances as Atomic Elements, Molecular Elements, Molecular Compounds, or Ionic Compounds

Classify each of the following substances as an atomic element, molecular element, molecular compound, or ionic compound.

a) krypton
b) $CoCl_2$
c) nitrogen
d) SO_2
e) KNO_3

Solution:

a) Krypton is an element and it is not listed on the table of elements that exist as diatomic molecules; therefore, it is an atomic element.
b) $CoCl_2$ is a compound composed of a metal (left side of periodic table) and nonmetal (right side of the periodic table); therefore, it is an ionic compound.
c) Nitrogen is an element that is listed on the table of elements that exist as diatomic molecules; therefore, it is a molecular element.
d) SO_2 is a compound composed of two nonmetals; therefore, it is a molecular compound.
e) KNO_3 is a compound composed of a metal and two nonmetals; therefore, it is an ionic compound.

SKILLBUILDER 5.4 Classifying Substances as Atomic Elements, Molecular Elements, Molecular Compounds, or Ionic Compounds

Classify each of the following substances as an atomic element, molecular element, molecular compound, or ionic compound.

a) chlorine
b) NO
c) Au
d) Na_2O
e) $CrCl_3$

5.5 Writing Formulas for Ionic Compounds

Review Section 4.7 and Figure 4.14 to learn the elements that form ions with a predictable charge

Since ionic compounds must be charge neutral and since many elements form only one type of ion with a predictable charge, the formulas for many ionic compounds can be determined based on their constituent elements. For example, the formula for the ionic compound composed of sodium and chlorine must be NaCl and not anything else because in compounds Na always forms +1 cations and Cl always forms −1 anions. In order for the compound to be charge neutral, it must contain one Na^+ cation to every one Cl^- anion. The formula for the ionic compound composed of magnesium and chlorine, however, must be $MgCl_2$ because Mg always forms +2 cations and Cl always forms −1 anions. In order for the compound to be charge neutral, it must contain one Mg^{2+} cation to every two Cl^- anions. In general:

- Ionic compounds always contain positive and negative ions.
- In the chemical formula, the sum of the charges of the positive ions (cations) must always equal the sum of the charges of the negative ions (anions).

To write the formula for an ionic compound, follow the procedure in the left column of the following box. Two examples of how to apply the procedure are provided in the center and right columns.

Writing Formulas for Ionic Compounds	**EXAMPLE 5.5** Write a formula for the ionic compound that forms between aluminum and oxygen.	**EXAMPLE 5.6** Write a formula for the ionic compound that forms between magnesium and oxygen.
1. Write the symbol for the metal and its charge followed by the symbol of the nonmetal and its charge. For many elements, these charges can be determined from their group number in the periodic table (refer to Figure 4.14).	Al^{3+} O^{2-}	Mg^{2+} O^{2-}
2. Make the magnitude of the charge on each ion (without the sign) become the subscript for the other ion.	Al^{3+} O^{2-} ↓ Al_2O_3	Mg^{2+} O^{2-} ↓ Mg_2O_2
3. Reduce the subscripts to give a ratio with the smallest whole numbers.	In this case, the numbers cannot be reduced any further; the correct formula is Al_2O_3.	To reduce the subscripts, divide both subscripts by 2. $Mg_2O_2 \div 2 = MgO$
4. Check that the sum of the charges of the cations exactly cancels the sum of the charges of the anions.	Cations: $+3+3=+6$ Anions: $-2-2-2=-6$ The charges cancel.	Cations: $+2$ Anions: -2 The charges cancel.
	SKILLBUILDER 5.5 Write a formula for the compound formed between cesium and oxygen.	**SKILLBUILDER 5.6** Write a formula for the compound formed between aluminum and nitrogen.

EXAMPLE 5.7 Writing Formulas for Ionic Compounds

Write a formula for the compound that forms between potassium and oxygen.

Solution:

We first write the symbol for each ion along with its appropriate charge from its group number in the periodic table.

K^+ O^{2-}

We then make the magnitude of each ion's charge become the subscript for the other ion.

K^+ O^{2-} becomes K_2O

There is no reducing of subscripts necessary in this case. Finally, we check to see that the sum of the charges of the cations $(+1 + 1 = +2)$ exactly cancels the sum of the charges of the anion (-2). The correct formula is K_2O.

SKILLBUILDER 5.7 Writing Formulas for Ionic Compounds

Write a formula for the compound that forms between calcium and bromine.

5.6 Nomenclature: Naming Compounds

Since there are so many different compounds, chemists have developed systematic ways to name them. If you learn these, you can simply examine a compound's formula to determine its name or vice versa. Many compounds, however, also have a common name. For example, H_2O has the common name *water* and the systematic name *dihydrogen monoxide*. A common name is like a nickname for a compound, used by those who are familiar with it. Since water is such a familiar compound, everyone uses its common name and not its systematic name. In the sections that follow, we learn how to systematically name simple ionic and molecular compounds. Keep in mind, however, that some compounds also have common names that are often used instead of the systematic name. Common names can only be learned through familiarity.

5.7 Naming Ionic Compounds

The first step in naming an ionic compound is identifying it as one. Remember, *ionic compounds are formed between metals and nonmetals*; any time you see a metal and one or more nonmetals together in a chemical formula, it is an ionic compound. Ionic compounds can be divided into two types (Figure 5.13) depending on the metal in the compound. If the metal, in all of its different compounds, always forms a cation with the same charge, then the charge is implied and the compound is a **Type I compound**. Sodium, for instance, has a +1 charge in all of its compounds and therefore forms Type I compounds. Some examples of Type I metals are

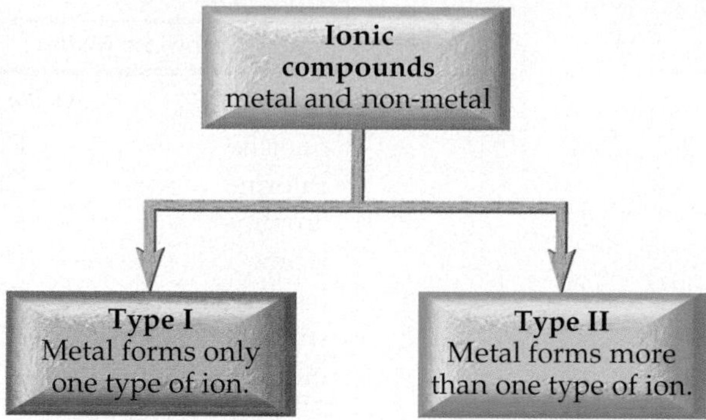

Figure 5.13 Ionic compounds can be divided into two types, depending on the metal in the compound. If the metal, in all of its compounds, always forms an ion with the same charge, it is a Type I ionic compound. If the metal, in all of its compounds, forms ions with different charges, it is a Type II ionic compound.

listed in Table 5.3. For most of these metals, their charge can be inferred from their group number in the periodic table (See Figure 4.14).

If, on the other hand, the metal forms compounds in which its charge is *not* always the same, then the charge is not implied and the compound is a **Type II compound**. Iron, for instance, has a +2 charge in some of its compounds and a +3 charge in others. These metals are often found in the section of the periodic table known as the **transition metals** (Figure 5.14).

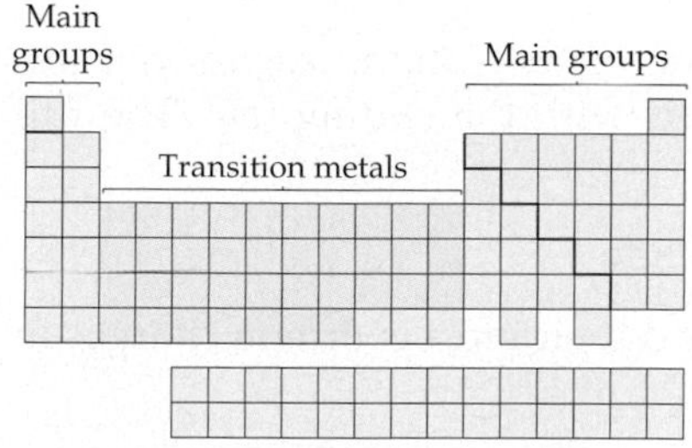

Figure 5.14 Metals that form Type II ionic compounds are usually transition metals.

Some examples of metals that form Type II ionic compounds are shown in Table 5.5.

Naming Type I Binary Ionic Compounds

Binary compounds are those containing only two different kinds of elements. The names for binary Type I ionic compounds have the following form.

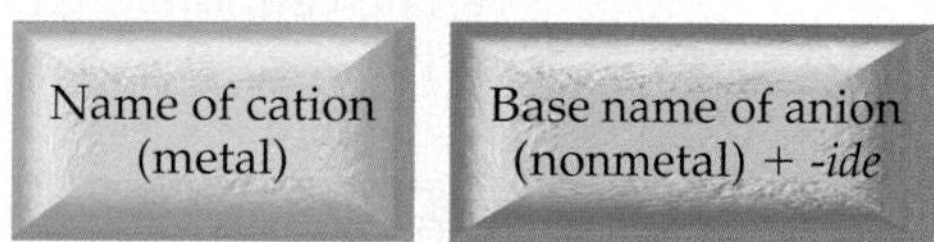

TABLE 5.3

Metals That Form Type I Ionic Compounds

Metal	*Ion*	*Name*	*Group Number*
Li	Li^+	lithium	1A
Na	Na^+	sodium	1A
K	K^+	potassium	1A
Rb	Rb^+	rubidium	1A
Cs	Cs^+	cesium	1A
Be	Be^{2+}	beryllium	2A
Mg	Mg^{2+}	magnesium	2A
Ca	Ca^{2+}	calcium	2A
Sr	Sr^{2+}	strontium	2A
Ba	Ba^{2+}	barium	2A
Al	Al^{3+}	aluminum	3A
Zn	Zn^{2+}	zinc	*
Ag	Ag^+	silver	*

*The charge of these metals cannot be inferred from their group number.

TABLE 5.4

Some Common Anions

Nonmetal	*Symbol for Ion*	*Base Name*	*Anion Name*
fluorine	F^-	fluor-	fluoride
chlorine	Cl^-	chlor-	chloride
bromine	Br^-	brom-	bromide
iodine	I^-	iod-	iodide
oxygen	O^{2-}	ox-	oxide
sulfur	S^{2-}	sulf-	sulfide
nitrogen	N^{3-}	nitr-	nitride

For example, the name for NaCl consists of the name of the cation, *sodium*, followed by the base name of the anion, *chlor*, with the ending *-ide*. The full name is *sodium chloride*.

NaCl Sodium chloride

The name for MgO consists of the name of the cation, *magnesium*, followed by the base name of the anion, *ox*, with the ending *-ide*. The full name is *magnesium oxide*.

MgO Magnesium oxide

The base names for various nonmetals and their most common charges in ionic compounds are shown in Table 5.4.

EXAMPLE 5.8 Naming Type I Ionic Compounds

Give the name for the compound MgF_2.

Solution:

The cation is magnesium. The anion is fluorine, which becomes *fluoride*. The correct name is *magnesium fluoride*.

SKILLBUILDER 5.8 Naming Type I Ionic Compounds

Give the name for the compound KBr.

SKILLBUILDER PLUS

Give the name for the compound Zn_3N_2.

Naming Type II Binary Ionic Compounds

The names for binary (two-element) Type II ionic compounds have the following form.

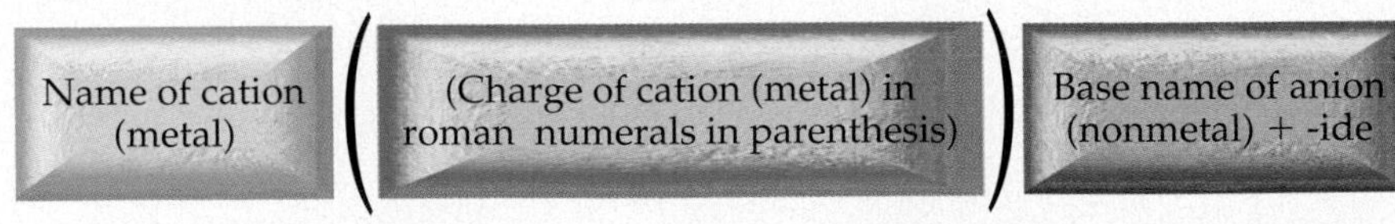

See Table 5.5 for some examples of metals that form Type II ionic compounds. The charge of the metal cation is obtained by inference from the sum of the charges of the nonmetal anions—remember that the sum of all the charges must be zero. For example, the name for $FeCl_3$ consists of the

TABLE 5.5

Some Metals That Form Type II Ionic Compounds and Their Common Charges

Metal	*Symbol Ion*	*Name*	*Older Name**
chromium	Cr^{2+}	chromium(II)	chromous
	Cr^{3+}	chromium(III)	chromic
iron	Fe^{2+}	iron(II)	ferrous
	Fe^{3+}	iron(III)	ferric
cobalt	Co^{2+}	cobalt(II)	cobaltous
	Co^{3+}	cobalt(III)	cobaltic
copper	Cu^{+}	copper(I)	cuprous
	Cu^{2+}	copper(II)	cupric
tin	Sn^{2+}	tin(II)	stannous
	Sn^{4+}	tin(IV)	stannic
mercury	Hg_2^{2+}	mercury(I)	mercurous
	Hg^{2+}	mercury(II)	mercuric
lead	Pb^{2+}	lead(II)	plumbous
	Pb^{4+}	lead(IV)	plumbic

*An older naming system substitutes the names found in this column for the name of the metal and its charge. Under this system, chromium(II) oxide is named chromous oxide. We will *not* use this older system in this text.

name of the cation, *iron*, followed by the charge of the cation in parentheses *(III)*, followed by the base name of the anion, *chlor*, with the ending *-ide*. The full name is *iron(III) chloride*.

$FeCl_3$ Iron(III) chloride

The charge of iron must be +3 in order for the compound to be charge neutral with three Cl^- anions. Likewise, the name for CrO consists of the name of the cation, *chromium*, followed by the charge of the cation in parentheses *(II)*, followed by the base name of the anion, *ox-*, with the ending *-ide*. The full name is *chromium(II) oxide*.

CrO Chromium(II) oxide

The charge of chromium must be +2 in order for the compound to be charge neutral with one O^{2-} anion.

EXAMPLE 5.9 Naming Type II Ionic Compounds

Give the name for the compound $PbCl_4$.

Solution:

The name for $PbCl_4$ consists of the name of the cation, *lead*, followed by the charge of the cation in parenthesis *(IV)*, followed by the base name of the anion, *chlor-*, with the ending *-ide*. The full name is *lead(IV) chloride*. We know the charge on Pb is +4 because the charge on Cl is −1. Since there are 4 Cl^- anions, the Pb cation must be Pb^{4+}.

$PbCl_4$ Lead(IV) chloride

SKILLBUILDER 5.9 **Naming Type II Ionic Compounds**

Give the name for the compound PbO.

Naming Ionic Compounds Containing a Polyatomic Ion

Some ionic compounds contain ions that are themselves composed of a group of atoms with an overall charge. These ions are called **polyatomic ions** and are shown in Table 5.6. Ionic compounds containing polyatomic ions are named using the same procedure as all ionic compounds, except that the name of the polyatomic ion is used whenever it occurs. For example, KNO_3 is named according to its cation, K^+, *potassium*, and its polyatomic anion, NO_3^-, *nitrate*. The full name is *potassium nitrate*.

KNO_3 Potassium nitrate

$FeSO_4$ is named according to its cation, *iron*, its charge (II), and its polyatomic ion *sulfate*. The full name is *iron(II) sulfate*.

$FeSO_4$ Iron(II) sulfate

If the compound contains both a polyatomic cation and a polyatomic anion, simply use the names of both polyatomic ions. For example, NH_4NO_3 is named *ammonium nitrate*.

NH_4NO_3 Ammonium nitrate

You must be able to recognize polyatomic ions in a chemical formula, so memorize Table 5.6. Most polyatomic ions are **oxyanions**, anions containing oxygen. Notice that when a series of oxyanions occurs that contain different numbers of oxygen atoms, they are named systematically according to the number of oxygen atoms in the ion. If there are two ions in the series, the one with more oxygen atoms is given the ending *-ate* and the one with fewer is given the ending *-ite*. For example, NO_3^- is called *nitrate* and NO_2^- is called *nitrite*.

TABLE 5.6

Some Common Polyatomic Ions

Name	*Formula*	*Name*	*Formula*
acetate	$C_2H_3O_2^-$	hypochlorite	ClO^-
carbonate	CO_3^{2-}	chlorite	ClO_2^-
hydrogen carbonate (or bicarbonate)	HCO_3^-	chlorate	ClO_3^-
hydroxide	OH^-	perchlorate	ClO_4^-
nitrate	NO_3^-	permanganate	MnO_4^-
nitrite	NO_2^-	sulfate	SO_4^{2-}
chromate	CrO_4^{2-}	sulfite	SO_3^{2-}
dichromate	$Cr_2O_7^{2-}$	hydrogen sulfite (or bisulfite)	HSO_3^-
phosphate	PO_4^{3-}	hydrogen sulfate (or bisulfate)	HSO_4^-
hydrogen phosphate	HPO_4^{2-}		
ammonium	NH_4^+	cyanide	CN^-

EVERYDAY CHEMISTRY

Polyatomic Ions

A glance at the labels of household products reveals the importance of polyatomic ions in everyday compounds. For example, the active ingredient in household bleach is sodium hypochlorite, which acts to destroy color-causing molecules in clothes (bleaching action) and to kill bacteria (disinfectant). A box of baking soda contains sodium bicarbonate, which acts as an antacid when consumed in small amounts and also as a source of carbon dioxide gas in baking. The pockets of carbon dioxide gas make baked goods fluffy rather than flat.

Calcium carbonate is the active ingredient in many antacids such as Tums and Alka-Mints. It neutralizes stomach acids, relieving the symptoms of indigestion and heartburn. Too much calcium carbonate, however, can cause constipation, so Tums should not be overused. Sodium nitrite is a common food additive used to preserve packaged meats such as ham, hot dogs, and bologna. Sodium nitrite inhibits bacterial growth, especially those that cause botulism, an often-fatal type of food poisoning.

CAN YOU ANSWER THIS? *Write a formula for each of these compounds, which contain polyatomic ions: sodium hypochlorite, sodium bicarbonate, calcium carbonate, sodium nitrite.*

Compounds containing polyatomic ions are present in many consumer products.

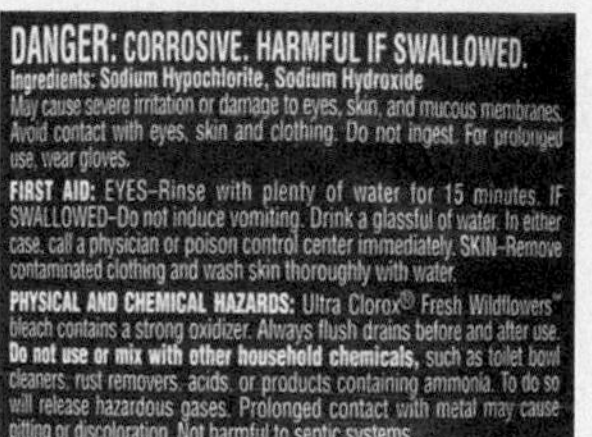

The active ingredient in bleach is sodium hypochlorite.

NO_3^-	Nitrate
NO_2^-	Nitrite

If there are more than two ions in the series then the prefixes *hypo-* meaning "less than" and *per-* meaning "more than" are used. So ClO^- is called *hypochlorite* meaning "less oxygen than chlorite," and ClO_4^- is called *perchlorate* meaning "more oxygen than chlorate."

ClO^-	Hypochlorite
ClO_2^-	Chlorite
ClO_3^-	Chlorate
ClO_4^-	Perchlorate

EXAMPLE 5.10 Naming Ionic Compounds That Contain a Polyatomic Ion

Give the name for the compound K_2CrO_4.

Solution:

The name for K_2CrO_4 consists of the name of the cation, *potassium*, followed by the name of the polyatomic ion *chromate*. The full name is *potassium chromate.*

K_2CrO_4 Potassium chromate

SKILLBUILDER 5.10 Naming Ionic Compounds That Contain a Polyatomic Ion

Give the name for the compound $Mn(NO_3)_2$.

5.8 Naming Molecular Compounds

The first step in naming a molecular compound is identifying it as one. Remember, molecular compounds form between two or more nonmetals. In this section, we learn how to name binary (two-element) molecular compounds. Their names have the following form.

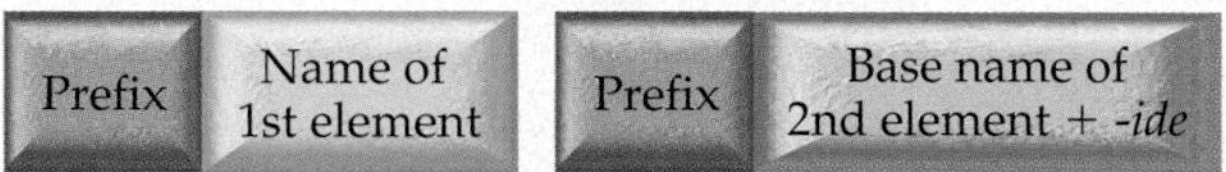

When writing the name of a molecular compound, as when writing the formula, the first element is the more metal-like one (See Table 5.1). The prefixes given to each element indicate the number of atoms present.

mono- 1	penta- 5
di- 2	hexa- 6
tri- 3	hepta- 7
tetra- 4	octa- 8

If there is only one atom of the *first element* in the formula, the prefix *mono-* is normally omitted. For example, CO_2 is named according to the first element, *carbon*, with no prefix because *mono-* is omitted for the first element, followed by the prefix *di-*, to indicate two oxygen atoms, followed by the base name of the second element, *-ox-*, with the ending *ide*.

Carbon di- -ox- -ide

The full name is *carbon dioxide.*

CO_2 Carbon dioxide

The compound N_2O, also called laughing gas, is named according to the first element, *nitrogen*, with the prefix *di-*, to indicate that there are two of them, followed by the base name of the second element, *-ox-*, prefixed by *mono-*, to indicate one, and the suffix *-ide*. Since *mono-* ends with a vowel and *oxide* begins with one, an *o* is dropped and the two are combined as *monoxide*. The entire name is *dinitrogen monoxide*.

N_2O Dinitrogen monoxide

EXAMPLE 5.11 **Naming Molecular Compounds**

Name each of the following: CCl_4 BCl_3 SF_6

Solution:

CCl_4

The name of the compound is the name of the first element, *carbon*, followed by the base name of the second element, *-chlor*, prefixed by *tetra-* to indicate four, and the suffix *-ide*. The entire name is *carbon tetrachloride*.

BCl_3

The name of the compound is the name of the first element, *boron*, followed by the base name of the second element, *-chlor*, prefixed by *tri-* to indicate three, and the suffix *-ide*. The entire name is *boron trichloride*.

SF_6

The name of the compound is the name of the first element, *sulfur*, followed by the base name of the second element, *-fluor*, prefixed by *hexa-* to indicate six, and the suffix *-ide*. The entire name is *sulfur hexafluoride*.

SKILLBUILDER 5.11 **Naming Molecular Compounds**

Name the compound N_2O_4.

5.9 Naming Acids

Acids are molecular compounds that dissolve in water to form H^+ ions. They are composed of hydrogen, usually written first in their formula, and one or more nonmetals, written second. Acids are characterized by their sour taste and their ability to dissolve some metals. For example, HCl is a molecular compound that, when dissolved in water, forms H^+ ions. It is an acid. To emphasize that acids occur in water solutions, HCl in water is sometimes represented as HCl(aq), where (aq) means "aqueous" or "dissolved in water." HCl has a characteristically sour taste. Since HCl is present in stomach fluids, its sour taste becomes painfully obvious during vomiting. HCl also dissolves some metals. For example, if you drop a strip of zinc into a beaker of HCl, it would slowly disappear as the HCl converts the zinc metal into Zn^{2+} cations.

HCl(g) refers to HCl molecules in the gas phase

Acids are present in many foods such as lemons and limes and they are used in some household products such as toilet bowl cleaner and Lime-Away. In this section, we simply learn how to name them, but in Chapter 14 we learn more about their properties. Acids can be divided into two categories: **binary acids**, those containing only hydrogen and a nonmetal, and **oxyacids**, those containing hydrogen, a nonmetal, and oxygen (Figure 5.15).

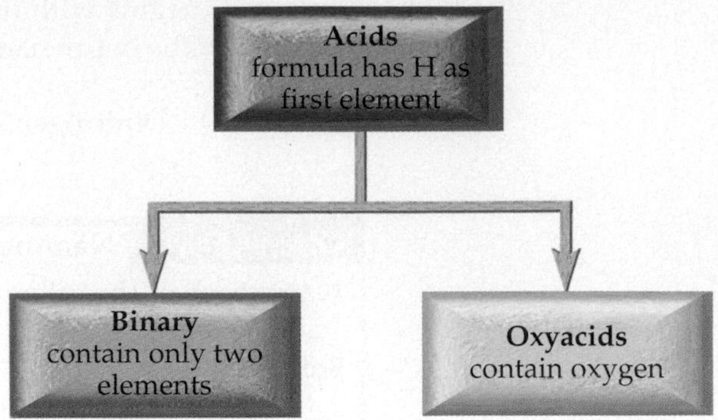

Figure 5.15 Acids can be divided into two types, depending on the number of elements in the acid. If the acid contains only two elements, it is a binary acid. If it contains oxygen, it is an oxyacid.

Naming Binary Acids

Binary acids are composed of hydrogen and a nonmetal. The names for binary acids have the following form.

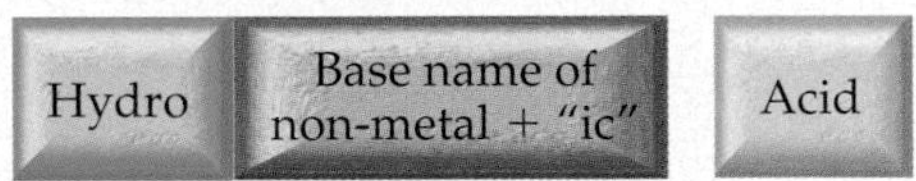

For example, HCl is named hydro*chlor*ic acid and HBr is named hydro*brom*ic acid.

HCl Hydrochloric acid HBr Hydrobromic acid

EXAMPLE 5.12 Naming Binary Acids

Give the name of HF.

Solution:

The base name of F is *-fluor-* so the name is *hydrofluoric acid*.

HF Hydrofluoric acid

SKILLBUILDER 5.12 Naming Binary Acids

Give the name of HI.

Naming Oxyacids

Oxyacids are derived from the oxyanions found in the table of polyatomic ions (Table 5.6). For example, HNO_3 is derived from the nitrate (NO_3^-) ion, H_2SO_3 is derived from the sulfite (SO_3^{2-}) ion, and H_2SO_4 is derived from the sulfate (SO_4^{2-}) ion. Notice that these acids are simply a combination of one or more H^+ ions with an oxyanion. The number of H^+ ions depends on the charge of the oxyanion so that the formula is always charge neutral. The names of oxyacids depend on the ending of the oxyanion (Figure 5.16) and have the following forms.

Oxyanions ending with *-ate*

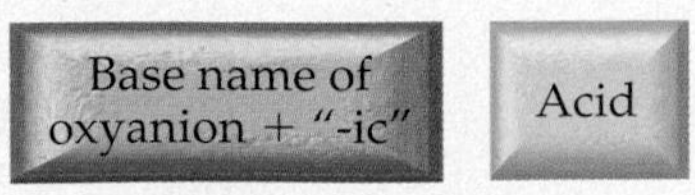

Oxyanions ending with *-ite*

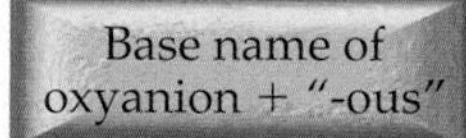

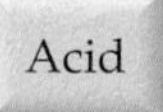

CHEMISTRY IN THE ENVIRONMENT

Acid Rain

Acid rain occurs when rainwater mixes with air pollutants—such as NO, NO_2 and SO_2—that form acids. NO and NO_2, primarily from vehicular emissions, combine with water to form HNO_3. SO_2, primarily from coal-powered electricity generation, combines with water to form H_2SO_4. HNO_3 and H_2SO_4 both cause rainwater to become acidic. The problem is greatest in the northeastern United States where pollutants from midwestern electrical power plants combine with rainwater to produce rain with acid levels that are up to ten times above normal.

When acid rain falls or flows into lakes and streams, it makes them more acidic. Some species of aquatic animals—such as trout, bass, snails, salamanders, and clams—cannot tolerate the increased acidity and die. This then disturbs the ecosystem of the lake, resulting in imbalances that may lead to the death of other aquatic species. Acid rain also weakens trees by dissolving nutrients in the soil and by damaging their leaves. Appalachian red spruce trees have been the hardest hit, with many forests showing significant acid rain damage.

Acid rain also damages building materials. Acids dissolve $CaCO_3$ (limestone), a main component of marble and concrete, and iron, the main component of steel. Consequently, many statues, buildings, and bridges in the northeastern United States show significant acid rain damage. For example, some historical gravestones that are made of limestone are barely legible due to acid rain damage. Although acid rain has been a problem for many years, recent legislation has offered hope for change. In 1990, Congress passed several amendments to the Clean Air Act that included provisions requiring electrical utilities to lower SO_2 emissions. Since then, SO_2 emissions have decreased, and rain in the northeastern United States has become somewhat less acidic. With time, and continued enforcement of the acid rain program, lakes, streams, and forests damaged by acid rain should recover.

CAN YOU ANSWER THIS? *Provide the names for each of the compounds given here as formulas* (NO, NO_2, SO_2, HNO_3, $CaCO_3$).

A forest damaged by acid rain.

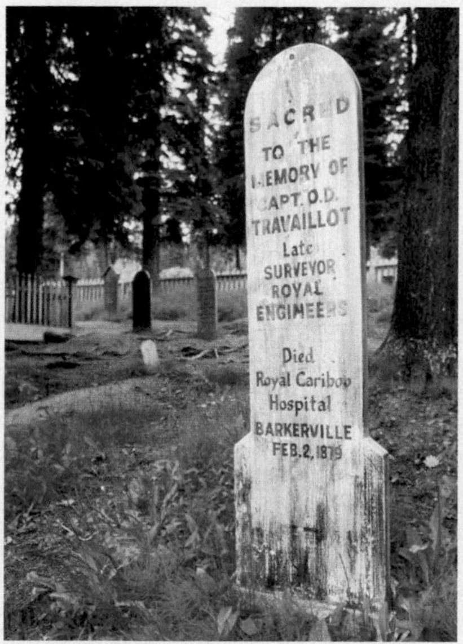

Acid rain damages many materials, including the limestone that composes many gravestones.

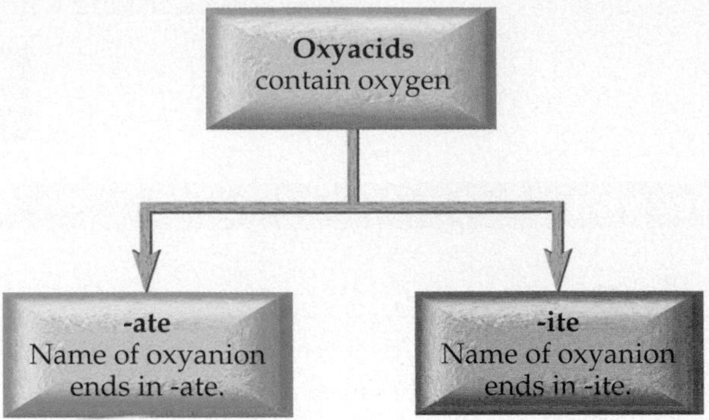

Figure 5.16 Oxyacids can be divided into two types, depending on the ending of the oxyanion from which they are derived.

So HNO_3 is named *nitric acid* (oxyanion is nitrate), and H_2SO_3 is named *sulfurous acid* (oxyanion is sulfite).

HNO_3 Nitric acid H_2SO_3 Sulfurous acid

EXAMPLE 5.13 Naming Oxyacids

Give the name of $HC_2H_3O_2$.

Solution:

The oxyanion is acetate, which ends in *-ate*; therefore, the name of the acid is *acetic acid.*

$HC_2H_3O_2$ Acetic acid

SKILLBUILDER 5.13 Naming Oxyacids

Give the name of HNO_2.

5.10 Nomenclature Summary

Naming compounds requires several steps. The flow chart in Figure 5.17 summarizes the different categories of compounds that we have learned and how to identify and name them. The first step is to decide whether the compound is ionic, molecular, or an acid. You can recognize ionic compounds by the presence of a metal and a nonmetal, molecular compounds by two or more nonmetals, and acids by the presence of hydrogen (written first) and one or more nonmetals.

Acids are technically a subclass of molecular compounds; that is, they are molecular compounds that form H^+ ions when dissolved in water.

Ionic Compounds

For an ionic compound, you must next decide whether it is a Type I or Type II ionic compound. Group 1A (alkali) metals, Group 2A (alkaline earth) metals, and aluminum will always form Type I ionic compounds (Figure 4.14). Most of the transition metals will form Type II ionic compounds. Once you have decided which type of ionic compound it is, name it according to the scheme in the chart. If the ionic compound contains a polyatomic ion—something you must recognize by familiarity—insert the name of the polyatomic ion in place of the metal (positive polyatomic ion) or the nonmetal (negative polyatomic ion).

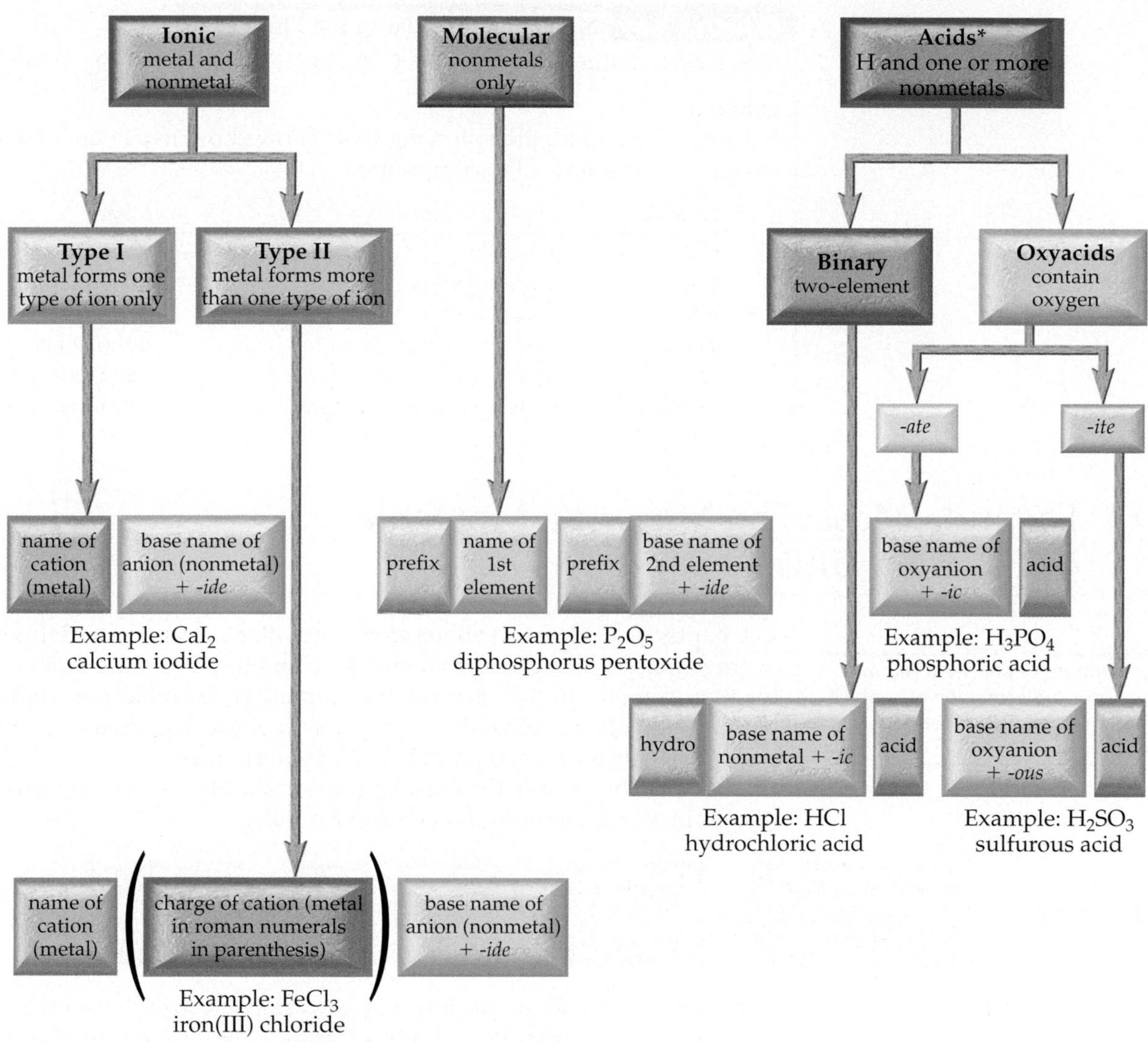

Figure 5.17 Nomenclature Flow Chart

Molecular Compounds

We have only learned how to name one type of molecular compound, the binary (two-element) compound. If you identify a compound as molecular, name it according to the scheme in the chart.

Acids

To name an acid, you must first decide whether it is a binary (two-element) acid, or an oxyacid (an acid containing oxygen). Binary acids are named according to the scheme in the chart. Oxyacids must be further subdivided based on the name of their corresponding oxyanion. If the oxyanion ends in *-ate*, use one scheme; if it ends with *-ite*, use the other.

EXAMPLE 5.14 Nomenclature Using the Flow Chart

Name each of the following: CO, CaF_2, HF, $Fe(NO_3)_3$, $HClO_4$, H_2SO_3

Solution:

For each compound, the following table shows how to use the flow chart to arrive at a name for the compound.

Formula	*Flow Chart Path*	*Name*
CO	molecular	carbon monoxide
CaF_2	Ionic → Type I →	calcium fluoride
HF	Acid → Binary →	hydrofluoric acid
$Fe(NO_3)_3$	Ionic → Type II →	iron(III) nitrate
$HClO_4$	Acids → Oxyacids → *-ate* →	perchloric acid
H_2SO_3	Acids → Oxyacids → *-ite* →	sulfurous acid

5.11 Formula Mass: The Mass of a Molecule or Formula Unit

Also in common use are the terms *molecular mass* or *molecular weight*, which have the same meaning as formula mass.

In Chapter 4, we learned about atoms and elements and we defined the average mass of the atoms that compose an element as the atomic mass for that element. In this chapter, we learned about molecules and compounds, and so we define the average mass of the molecules (or formula units) that compose a compound as the **formula mass**.

For any compound, the formula mass is simply the sum of the atomic masses of all the atoms in its chemical formula.

Formula mass = [# atoms of 1st element in chemical formula] × [Atomic mass of 1st element] +

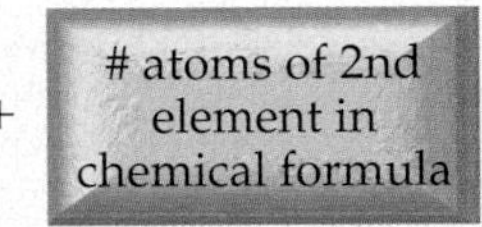

×

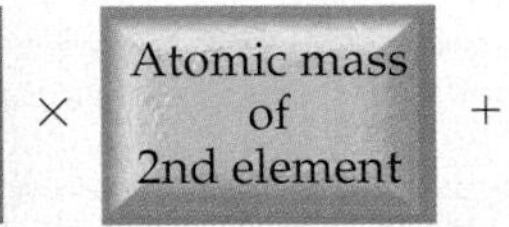

+ . . .

Like atomic mass for atoms, formula mass characterizes the average mass of a molecule or formula unit. For example, the formula mass of water, H_2O, is:

$$\text{Formula mass} = 2(1.01 \text{ amu}) + 16.00 \text{ amu} = 18.02 \text{ amu}$$

and that of sodium chloride, NaCl, is:

$$\text{Formula mass} = 22.99 \text{ amu} + 35.45 \text{ amu} = 58.44 \text{ amu}$$

In addition to giving a characteristic mass to the molecules or formula units of a compound, formula mass—as we will learn in Chapter 6—allows us to quantify the number of molecules or formula units in a sample of a given mass.

EXAMPLE 5.15 Calculating Formula Mass

Calculate the formula mass of carbon tetrachloride, CCl_4.

Solution:

To find the formula mass, we sum the atomic masses of each atom in the chemical formula.

$$\begin{aligned}\text{Formula mass} &= 1 \times (\text{formula mass C}) + 4 \times (\text{formula mass Cl}) \\ &= 12.01 \text{ amu} + 4(35.45 \text{ amu}) \\ &= 153.81 \text{ amu}\end{aligned}$$

SKILLBUILDER 5.15 **Calculating Formula Masses**

Calculate the formula mass of dinitrogen monoxide, N_2O, also called laughing gas.

CHAPTER IN REVIEW

Chemical Principles

Compounds: Matter is ultimately composed of atoms, but those atoms are often combined in compounds. The most important characteristic of a compound is its constant composition. The elements that compose a particular compound are in fixed, definite proportions in all samples of the compound.

Chemical Formulas: Compounds are represented by chemical formulas, which indicate the elements present in the compound and the relative number of atoms of each. These formulas represent the basic units that compose a compound.

Pure substances can be divided according to the basic units that compose them. Elements can be composed of atoms or molecules. Compounds can be molecular, in which case their basic units are molecules, or ionic, in which case their basic units are ions. The formulas for many ionic compounds can be written simply by knowing the elements in the compound.

Chemical Nomenclature: The names of simple ionic compounds, molecular compounds, and acids can all be written by examining their chemical formula. The nomenclature flow chart (Figure 5.17) shows the basic procedure for determining these names.

Formula Mass: The formula mass of a compound is the sum of the atomic masses of all the atoms in the chemical formula for the compound. Like atomic mass for elements, formula mass characterizes the average mass of a molecule or formula unit.

Relevance

Compounds: Most of the matter we encounter is in the form of compounds. Water, salt, and carbon dioxide are all good examples of common simple compounds. More complex compounds include caffeine, aspirin, acetone, and testosterone.

Chemical Formulas: To understand compounds we must understand their composition, which is represented by a chemical formula. The connection between the microscopic world and the macroscopic world hinges on the particles that compose matter. Since most matter is in the form of compounds, the properties of most matter depend on the molecules or ions that compose it. Molecular matter does what its molecules do, ionic matter does what its ions do. The world we see and experience is governed by what these particles are doing.

Chemical Nomenclature: Since there are so many compounds, we need a systematic way to name them. By learning these few simple rules, you will be able to name thousands of different compounds. The next time you look at the label on a consumer product, try to identify as many of the compounds as you can.

Formula Mass: Besides being the characteristic mass of a molecule or formula unit, formula mass is important in many calculations involving the composition of compounds and quantities in chemical reactions.

Chemical Skills

Examples

Constant Composition of Compounds (Section 5.2)

The law of constant composition states that all samples of a given compound should have the same ratio of their constituent elements.

To determine if experimental data is consistent with the law of constant composition, compute the ratios of the masses of each element in all samples. When computing these ratios, it is most convenient to put the larger number in the numerator (top) and the smaller one in the denominator (bottom); that way, the ratio is greater than one. If the ratios are the same, then the data is consistent with the law of constant composition.

EXAMPLE 5.16 Constant Composition of Compounds

Two samples said to be carbon disulfide (CS_2) are decomposed into their constituent elements. One sample produced 8.08 g S and 1.51 g C, while the other produced 31.3 g S and 3.85 g C. Are these results consistent with the law of constant composition?

Solution:

Sample 1

$$\frac{\text{Mass S}}{\text{Mass C}} = \frac{8.08 \text{ g}}{1.51 \text{ g}} = 5.35$$

Sample 2

$$\frac{\text{Mass S}}{\text{Mass C}} = \frac{31.3 \text{ g}}{3.85 \text{ g}} = 8.13$$

These results are not consistent with the law of constant composition and the data given must therefore be in error.

Writing Chemical Formulas (Section 5.3)

Chemical formulas indicate the elements present in a compound and the relative number of atoms of each. When writing formulas, put the more metallic element first.

EXAMPLE 5.17 Writing Chemical Formulas

Write a chemical formula for the compound containing 2 nitrogen atoms to every 1 oxygen atom.

Solution:

N_2O

Total Number of Each Type of Atom in a Chemical Formula (Section 5.3)

The numbers of atoms not enclosed in parentheses are given directly by their subscript.

The numbers of atoms within parentheses are found by multiplying their subscript within the parentheses, by their subscript outside of the parentheses.

EXAMPLE 5.18 Total Number of Each Type of Atom in a Chemical Formula

Determine the number of each type of atom in $Pb(ClO_3)_2$.

Solution:

1 Pb atom

2 Cl atoms
6 O atoms

Classifying Elements as Atomic or Molecular (Section 5.4)

Most elements exist as atomic elements, their basic units in nature being individual atoms. However, several elements (H_2, N_2, O_2, F_2, Cl_2, Br_2, and I_2) exist as molecular elements, their basic units in nature being diatomic molecules.

EXAMPLE 5.19 Classifying Elements as Atomic or Molecular

Classify each of the following elements as atomic or molecular: Na, I, N.

Solution:

Na: atomic
I: molecular (I_2)
N: molecular (N_2)

Classifying Compounds as Ionic or Molecular (Section 5.4)

Compounds containing a metal and a nonmetal are ionic. Metals that form more than one ion, typical of transition metals, are Type II.

Compounds containing a metal and one or more nonmetals are ionic. Metals that form one type of ion—typical of Group I, Group II, and aluminum—are Type I.

Compounds composed of nonmetals are molecular.

EXAMPLE 5.20 Classifying Compounds as Ionic or Molecular

Classify each of the following compounds as ionic or molecular. If they are ionic, classify them as Type I or Type II ionic compounds: $FeCl_3$, K_2SO_4, CCl_4

Solution:

$FeCl_3$: ionic, Type II

K_2SO_4: ionic, Type I

CCl_4 molecular

Writing Formulas for Ionic Compounds (Section 5.5)

Write the symbol for the metal ion followed by the symbol for the nonmetal ion (or polyatomic ion) and their charges. These charges come from the group numbers in the periodic table (In the case of polyatomic ions, the charges come from Table 5.6).

Make the magnitude of the charge on each ion become the subscript for the other ion.

Check to see if the subscripts can be reduced to simpler whole numbers. Subscripts of 1 can be dropped since they are normally implied.

Check that the sum of the charges of the cations exactly cancels the sum of the charges of the anions.

EXAMPLE 5.21 Writing Formulas for Ionic Compounds

Write a formula for the compound that forms between lithium and sulfate ions.

Solution:

Li^+ SO_4^{2-}

$Li_2(SO_4)$

In this case, the subscripts cannot be further reduced.

Li_2SO_4

Cations:

$+1 + 1 = +2$

Anions:

-2

Naming Type I Binary Ionic Compounds (Section 5.7)

The name of the metal is unchanged. The name of the nonmetal is its base name with the ending *-ide*.

EXAMPLE 5.22 Naming Type I Binary Ionic Compounds

Give the name for the compound Al_2O_3.

Solution:

Aluminum oxide

Naming Type II Binary Ionic Compounds (Section 5.7)

Since the name of Type II compounds includes the charge of the metal atom, we must first find that charge. To do this, compute the total charge of the nonmetal atoms.

The total charge of the metal atoms must equal the total charge of the nonmetal atoms, but have the opposite sign.

The name is the name of the metal, followed by the charge of the metal, followed by the base name of the nonmetal + *-ide*.

EXAMPLE 5.23 Naming Type II Binary Ionic Compounds

Give the name for the compound Fe_2S_3.

Solution:

3 sulfur ions × (−2) = −6

2 iron atoms × (?) = +6

? = +3

Charge of each iron atom = +3

Iron(III) sulfide

Naming Compounds Containing a Polyatomic Ion (Section 5.7)

Name Type I and Type II ionic compounds containing a polyatomic ion in the normal way, except substitute the name of the polyatomic ion (from Table 5.6) in place of the nonmetal (This example is Type II).

The charge on the metal must equal the sum of the charges of the polyatomic ions.

The name of the compound is the name of the metal, followed by the charge of the metal, followed by the name of the polyatomic ion.

EXAMPLE 5.24 Naming Compounds Containing a Polyatomic Ion.

Give the name for the compound $Co(ClO_4)_2$.

Solution:

2 perchlorate ions × (−1) = −2

Charge of cobalt ion = +2

Cobalt(II) perchlorate

Naming Molecular Compounds (Section 5.8)

The name consists of a prefix indicating the number of atoms of the first element, followed by the name of the first element, and a prefix for the number of atoms of the second element followed by the base name of the second element plus *-ide*. When *mono-* occurs on the first element, it is normally dropped.

EXAMPLE 5.25 Naming Molecular Compounds

Name the compound NO_2.

Solution:

Nitrogen dioxide

Naming Binary Acids (Section 5.9)

The name begins with *hydro-*, followed by the base name of the nonmetal, plus the ending *-ic* and then the word *acid*.

EXAMPLE 5.26 Naming Binary Acids

Name the acid HI.

Solution:

Hydroiodic acid

Naming Oxyacids with an Oxyanion Ending in *-ate* (Section 5.9)

The name is the base name of the oxyanion + *-ic*, followed by the word *acid* (sulfate violates the rule somewhat, since in strict terms, the base name would be *sulf*).

EXAMPLE 5.27 Naming Oxyacids with an Oxyanion Ending in *-ate*

Name the acid H_2SO_4.

Solution:

The oxyanion is sulfate. The name of the acid is *sulfuric acid*.

Naming Oxyacids with an Oxyanion Ending in *-ite* (Section 5.9)

The name is the base name of the oxyanion + *-ous*, followed by the word *acid*.

EXAMPLE 5.28 Naming Oxyacids with an Oxyanion Ending in *-ite*

Name the acid $HClO_2$.

Solution:

The oxyanion is chlorite. The name of the acid is *chlorous acid*.

Calculating Formula Mass (Section 5.11)

The formula mass is the sum of the atomic masses of all the atoms in the chemical formula. In determining the number of each type of atom, don't forget to multiply subscripts inside of parentheses by subscripts outside of parentheses.

EXAMPLE 5.29 Calculating Formula Mass

Calculate the formula mass of $Mg(NO_3)_2$

Solution:

$$\begin{aligned} \text{Formula mass} &= 24.31 + 2(14.01) + 6(16.00) \\ &= 148.33 \text{ amu} \end{aligned}$$

KEY TERMS

acid [5.9]
atomic element [5.4]
binary acid [5.9]
binary compound [5.7]
chemical formula [5.3]
formula mass [5.11]
formula unit [5.4]
ionic compound [5.4]
law of constant composition [5.2]
molecular compound [5.4]
molecular element [5.4]
oxyacid [5.9]
oxyanion [5.7]
polyatomic ion [5.7]
Joseph Proust [5.2]
Type I compound [5.7]
Type II compound [5.7]

EXERCISES

Questions

1. Do the properties of an element change when it combines with another element to form a compound? Explain.

2. How would the world be different if elements did not combine to form compounds?

3. What is the law of constant composition? Who discovered it?
4. What is a chemical formula? Give some examples.
5. In a chemical formula, which element is listed first?
6. In a chemical formula, how do you calculate the number of atoms of an element within parentheses? Give an example.
7. What is the difference between a molecular element and an atomic element? List the elements that occur as diatomic molecules.
8. What is the difference between an ionic compound and a molecular compound?
9. What is the difference between a common name for a compound and a systematic name?
10. List the metals that form Type I ionic compounds. What are the group numbers of these metals?
11. Find the block in the periodic table of elements that tend to form Type II ionic compounds. What is the name of this block?
12. What is the basic form for the names of Type I ionic compounds?
13. What is the basic form for the names of Type II ionic compounds?
14. Why are numbers needed in the names of Type II ionic compounds?
15. How are compounds containing a polyatomic ion named?
16. What polyatomic ions have a -2 charge? What polyatomic ions have a -3 charge?
17. What is the basic form for the names of molecular compounds?
18. How many atoms does each of the following prefixes specify? *mono-*, *di-*, *tri-*, *tetra-*, *penta-*, *hexa-*.
19. What is the basic form for the names of binary acids?
20. What is the basic form for the name of oxyacids whose oxyanions end with *-ate*?
21. What is the basic form for the name of oxyacids whose oxyanions end with *-ite*?
22. What is the formula mass of a compound?

Problems

Constant Composition of Compounds

23. Two samples of sodium chloride were decomposed into their constituent elements. One sample produced 4.65 g of sodium and 7.16 g of chlorine, and the other sample produced 7.45 g of sodium and 11.5 g of chlorine. Are these results consistent with the law of constant composition? Show why or why not.

24. Two samples of carbon tetrachloride were decomposed into their constituent elements. One sample produced 32.4 g of carbon and 373 g of chlorine, and the other sample produced 12.3 g of carbon and 112 g of chlorine. Are these results consistent with the law of constant composition? Show why or why not.

25. Upon decomposition, one sample of magnesium fluoride produced 1.65 kg of magnesium and 2.57 kg of fluorine. A second sample produced 1.32 kg of magnesium. How much fluorine (in grams) did the second sample produce?

26. The mass ratio of sodium to fluorine in sodium fluoride is 1.21:1. A sample of sodium fluoride produced 34.5 g of sodium upon decomposition. How much fluorine (in grams) was formed?

Chemical Formulas

27. Write a chemical formula for the compound containing one nitrogen atom to every three bromine atoms.

28. Write a chemical formula for the compound containing one carbon atom to every four chlorine atoms.

29. Write chemical formulas for compounds containing each of the following:
a) two iron atoms to every three oxygen atoms
b) one phosphorus atom to every three chlorine atoms
c) one phosphorus atom to every five chlorine atoms
d) two silver atoms to every one oxygen atom

30. Write chemical formulas for compounds containing each of the following:
a) one calcium atom to every two chlorine atoms
b) two nitrogen atoms to every four oxygen atoms
c) one silicon atom to every two oxygen atoms
d) one zinc atom to every two chlorine atoms

31. How many oxygen atoms are in each of the following chemical formulas?
a) H_3PO_4
b) Na_2HPO_4
c) $Ca(HCO_3)_2$
d) $Ba(C_2H_3O_2)_2$

32. How many hydrogen atoms are in each of the formulas in Problem 31?

33. Determine the number of each type of atom in each of the following formulas.
a) $MgCl_2$
b) $NaNO_3$
c) $Ca(NO_2)_2$
d) $Sr(OH)_2$

34. Determine the number of each type of atom in each of the following formulas.
a) NH_4Cl
b) $Mg_3(PO_4)_2$
c) NaCN
d) $Ba(HCO_3)_2$

Molecular View of Elements and Compounds

35. Classify each of the following elements as atomic or molecular.
a) helium
b) chlorine
c) oxygen
d) sodium

36. Which of the following elements have molecules as their basic units?
a) hydrogen
b) argon
c) bromine
d) mercury

37. Classify each of the following compounds as ionic or molecular.
a) CS_2
b) CuO
c) KI
d) PCl_3

38. Classify each of the following compounds as ionic or molecular.
a) PtO_2
b) CF_2Cl_2
c) CO
d) SO_3

39. Match the substances on the left with the basic units that compose them on the right.

helium	molecules
CCl_4	formula units
K_2SO_4	diatomic molecules
bromine	single atoms

40. Match the substances on the left with the basic units that compose them on the right.

NI_3	molecules
copper metal	single atoms
$SrCl_2$	diatomic molecules
nitrogen	formula units

41. What are the basic units—single atoms, molecules, or formula units—that compose each of the following substances?
a) $BaBr_2$
b) Ne
c) I_2
d) CO

42. What are the basic units—single atoms, molecules, or formula units —that compose each of the following substances?
a) Rb_2O
b) N_2
c) $Fe(NO_3)_2$
d) N_2F_4

43. Classify each of the following compounds as ionic or molecular. If it is ionic, classify it as a Type I or Type II ionic compound.
a) KCl
b) CBr_4
c) NO_2
d) $Sn(SO_4)_2$

44. Classify each of the following compounds as ionic or molecular. If it is ionic, classify it as a Type I or Type II ionic compound.
a) $CoCl_2$
b) CF_4
c) $BaSO_4$
d) NO

Writing Formulas for Ionic Compounds

45. Write a formula for the ionic compound that forms between each of the following pairs of elements.
a) strontium and oxygen
b) sodium and oxygen
c) aluminum and sulfur
d) strontium and bromine

46. Write a formula for the ionic compound that forms between each of the following pairs of elements.
a) aluminum and oxygen
b) beryllium and chlorine
c) calcium and sulfur
d) strontium and iodine

47. Write a formula for the compound that forms between potassium and
a) acetate
b) chromate
c) phosphate
d) cyanide

48. Write a formula for the compound that forms between calcium and
a) hydroxide
b) carbonate
c) phosphate
d) hydrogen phosphate

49. Write formulas for the compounds that form between the element on the left and each element on the right.
a) K N, O, F
b) Ba N, O, F
c) Al N, O, F

50. Write formulas for the compounds that form between the element on the left and each polyatomic ion on the right.
a) Rb NO_3^-, SO_4^{2-}, PO_4^{3-}
b) Sr NO_3^-, SO_4^{2-}, PO_4^{3-}
c) In NO_3^-, SO_4^{2-}, PO_4^{3-}

Naming Ionic Compounds

51. Name each of the following Type I ionic compounds.
a) $CsCl$
b) $SrBr_2$
c) K_2O
d) LiF

52. Name each of the following Type I ionic compounds.
a) LiI
b) MgS
c) BaF_2
d) NaF

53. Name each of the following Type II ionic compounds.
a) $CrCl_2$
b) $CrCl_3$
c) SnO_2
d) PbI_2

54. Name each of the following Type II ionic compounds.
a) $HgBr_2$
b) Fe_2O_3
c) CuI_2
d) $SnCl_4$

55. Determine whether each of the following ionic compounds is Type I or Type II and give each an appropriate name.
a) Cr_2O_3
b) NaI
c) $CaBr_2$
d) SnO

56. Determine whether each of the following ionic compounds is Type I or Type II and give each an appropriate name.
a) FeI_3
b) $PbCl_4$
c) SrI_2
d) BaO

57. Name each of the following ionic compounds containing a polyatomic ion.
a) $Ba(NO_3)_2$
b) $Pb(C_2H_3O_2)_2$
c) NH_4I
d) $KClO_3$
e) $CoSO_4$
f) $NaClO_4$

58. Name each of the following ionic compounds containing a polyatomic ion.
a) $Ba(OH)_2$
b) $Fe(OH)_3$
c) $CuNO_2$
d) $PbSO_4$
e) $KClO$
f) $Mg(C_2H_3O_2)_2$

59. Write a formula for each of the following ionic compounds.
a) copper(II) bromide
b) silver(I) nitrate
c) potassium hydroxide
d) sodium sulfate
e) potassium hydrogen sulfate
f) sodium hydrogen carbonate

60. Write a formula for each of the following ionic compounds.
a) copper(I) chlorate
b) potassium permanganate
c) lead(II) chromate
d) calcium fluoride
e) iron(II) phosphate
f) lithium hydrogen sulfite

Naming Molecular Compounds

61. Name each of the following molecular compounds.
a) SO_2
b) NI_3
c) BrF_5
d) NO
e) N_4Se_4

62. Name each of the following molecular compounds.
a) XeF_4
b) PI_3
c) SO_3
d) $SiCl_4$
e) I_2O_5

63. Write a formula for each of the following molecular compounds.
a) carbon monoxide
b) disulfur tetrafluoride
c) dichlorine monoxide
d) phosphorus pentafluoride
e) boron tribromide
f) diphosphorus pentasulfide

64. Write a formula for each of the following molecular compounds.
a) chlorine monoxide
b) xenon tetroxide
c) xenon hexafluoride
d) carbon tetrabromide
e) diboron tetrachloride
f) tetraphosphorus triselenide

65. Determine whether each of the names for the molecular compounds shown is correct. If not, give the compound the correct name.
a) PBr_5 phosphorus(V) pentabromide
b) P_2O_3 phosphorus trioxide
c) SF_4 monosulfur hexafluoride
d) NF_3 nitrogen trifluoride

66. Determine whether each of the names for the molecular compounds shown is correct. If not, give the compound the correct name.
a) NCl_3 nitrogen chloride
b) CI_4 carbon(IV) iodide
c) CO carbon oxide
d) SCl_4 sulfur tetrachloride

Naming Acids

67. Name each of the following acids.
a) $HClO_2$
b) HI
c) H_2SO_4
d) HNO_3

68. Name each of the following acids.
a) H_2CO_3
b) $HC_2H_3O_2$
c) HNO_2
d) HCl

69. Write formulas for each of the following acids.
a) phosphoric acid
b) hydrobromic acid
c) sulfurous acid

70. Write formulas for each of the following acids.
a) hydrofluoric acid
b) hydrocyanic acid
c) chlorous acid

Formula Mass

71. Calculate the formula mass for each of the following compounds.
a) HNO_2
b) $CaCl_2$
c) CCl_4
d) $Mg(NO_3)_2$

72. Calculate the formula mass for each of the following compounds.
a) CO_2
b) $C_6H_{12}O_6$
c) $Fe(NO_3)_3$
d) C_8H_{18}

73. Arrange the following compounds in order of decreasing formula mass.

Ag_2O, PtO_2, $Al(NO_3)_3$, PBr_3

74. Arrange the following compounds in order of decreasing formula mass.

WO_2, Rb_2SO_4, $Pb(C_2H_3O_2)_2$, RbI

Cumulative Problems

75. Determine whether each of the following compounds is ionic, molecular, or an acid and give it a name.
a) $Co(CN)_3$
b) H_2CrO_4
c) KCl
d) N_2H_4

76. Determine whether each of the following compounds is ionic, molecular, or an acid and give it a name.
a) $K_2Cr_2O_7$
b) PbO_2
c) HBr
d) N_2O_5

77. Determine if each of the following names is correct for the given formula. If not, give the correct name.
a) $Ca(NO_2)_2$ calcium nitrate
b) K_2O dipotassium monoxide
c) PCl_3 phosphorus chloride
d) $PbCO_3$ lead(II) carbonate

78. Determine if each of the following names is correct for the given formula. If not, give the correct name.
a) HNO_3 hydrogen nitrate
b) NaClO sodium hypochlorite
c) CaI_2 calcium diiodide
d) $SnCrO_4$ tin chromate

79. For each of the following compounds, give the correct formula and calculate the formula mass.
a) tin(IV) sulfate
b) nitrous acid
c) sodium bicarbonate
d) phosphorus pentafluoride

80. For each of the following compounds, give the correct formula and calculate the formula mass.
a) barium bromide
b) dinitrogen trioxide
c) copper(I) sulfate
d) hydrobromic acid

81. Name each of the following compounds and calculate its formula mass.
a) PtO_2
b) N_2O_5
c) $Al(ClO_3)_3$
d) PBr_5

82. Name each of the following compounds and calculate its formula mass.
a) $Al_2(SO_4)_3$
b) P_2O_3
c) HClO
d) $Cr(C_2H_3O_2)_3$

Highlight Problems

83. Examine each of the following substances and their molecular views and classify each as an atomic element, a molecular element, a molecular compound, or an ionic compound.

a)

b)

c)

d)

84. Molecules can be as small as a few atoms or as large as thousands of atoms. In 1962, Max F. Perutz and John C. Kendrew were awarded the Nobel prize for their discovery of the structure of hemoglobin, a very large molecule found in blood and responsible for transporting oxygen from the lungs to cells within the body. The chemical formula of hemoglobin is $C_{2952}H_{4664}O_{832}N_{812}S_8Fe_4$. Calculate the formula mass of hemoglobin.

In 1962, Max F. Perutz and John C. Kendrew were awarded the Nobel prize for their discovery of the structure of hemoglobin by X-ray diffraction.

85. Examine each of the following consumer product labels. Write chemical formulas for as many of the compounds as possible based on what we have learned in this chapter.

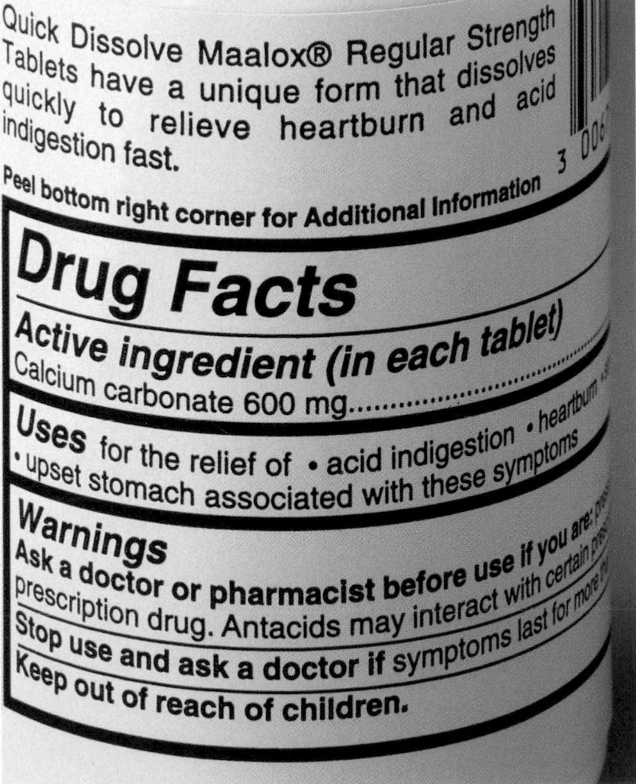

(continued on next page)

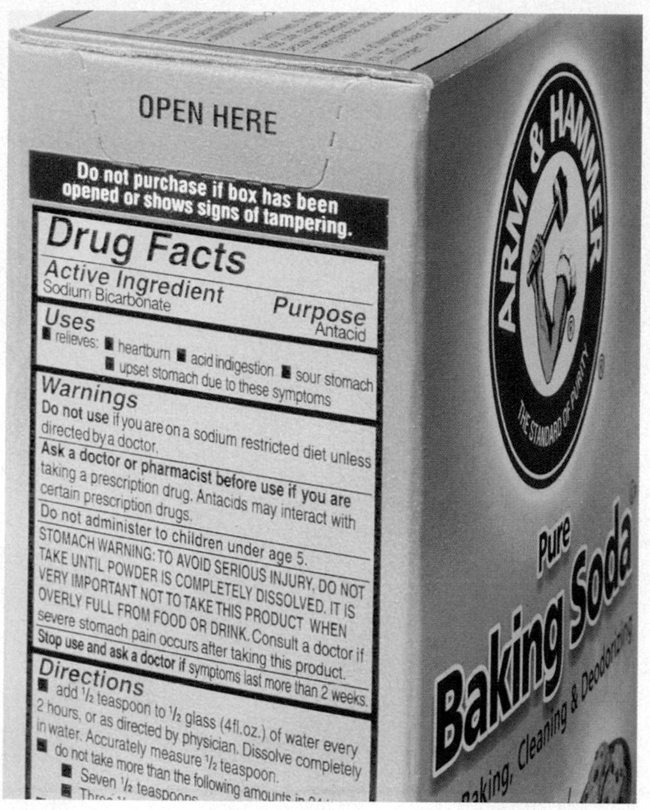

Answers to Skillbuilder Exercises

Skillbuilder 5.1 Yes, because in both cases the

$$\frac{\text{Mass O}}{\text{Mass C}} = 1.3$$

Skillbuilder 5.2 **a)** Ag_2S **b)** N_2O **c)** TiO_2

Skillbuilder 5.3 2 K atoms, 1 S atom, and 4 O atom

Skillbuilder Plus, p. 132 2 Al atoms, 3 S atoms, 12 O atoms

Skillbuilder 5.4 **a)** molecular element **b)** molecular compound **c)** atomic element **d)** ionic compound **e)** ionic compound

Skillbuilder 5.5 Cs_2O

Skillbuilder 5.6 AlN

Skillbuilder 5.7 $CaBr_2$

Skillbuilder 5.8 potassium bromide

Skillbuilder Plus, p. 138 zinc nitride

Skillbuilder 5.9 lead(II) oxide

Skillbuilder 5.10 manganese(II) nitrate

Skillbuilder 5.11 dinitrogen tetroxide

Skillbuilder 5.12 hydroiodic acid

Skillbuilder 5.13 nitrous acid

Skillbuilder 5.15 44.02 amu

Figure 6.1 Estimating the threat of ozone depletion from chlorofluorocarbons requires knowing how much chlorine is in a given amount of a particular chlorofluorocarbon.

example, wants to know how much iron they can recover from a given amount of iron ore, or a company interested in developing hydrogen as a potential fuel wants to know how much hydrogen they can extract from a given amount of water. Many environmental issues also require knowledge of chemical composition. For example, an estimate of the threat of ozone depletion requires knowing how much chlorine is in a given amount of a particular chlorofluorocarbon such as freon-12 (CF_2Cl_2) (Figure 6.1). Answering these questions requires understanding the relationships inherent in a chemical formula and the relationship between numbers of atoms or molecules and their masses. In this chapter, we learn these relationships.

6.2 Counting Nails by the Pound

Some hardware stores sell nails by the pound. This is easier than selling them individually because customers often need hundreds of nails and counting them takes too long. However, a customer may still want to know the number of nails contained in a given weight of nails. This problem is similar to asking how many atoms are in a given mass of an element. With atoms, however, we *must* use their mass as a way to count them because atoms are too small to count individually. Even if you could see atoms and counted them twenty-four hours a day as long as you lived, you would barely begin to count the number of atoms in something as small as a sand grain. However, just as the hardware store customer wants to know how many nails are in a given weight, we want to know the number of atoms in a given mass. How do we do that?

3.4 lbs nails

8.5 grams carbon

Asking how many nails are in a given weight of nails is similar to asking how many atoms are in a given mass of an element. In both cases, we count the objects by weighing them.

Suppose the hardware store customer buys 2.60 lb of medium-sized nails. Suppose further that a dozen nails weigh 0.150 lb. How many nails did the customer buy? This calculation requires two conversions: one between lb and dozen and another between dozen and number of nails. The conversion factor for the first part is the weight per dozen nails.

$$0.150 \text{ lb nails} = 1 \text{ doz nails}$$

The conversion factor for the second part is the number of nails in one dozen.

$$1 \text{ doz nails} = 12 \text{ nails}$$

The solution map for the problem is:

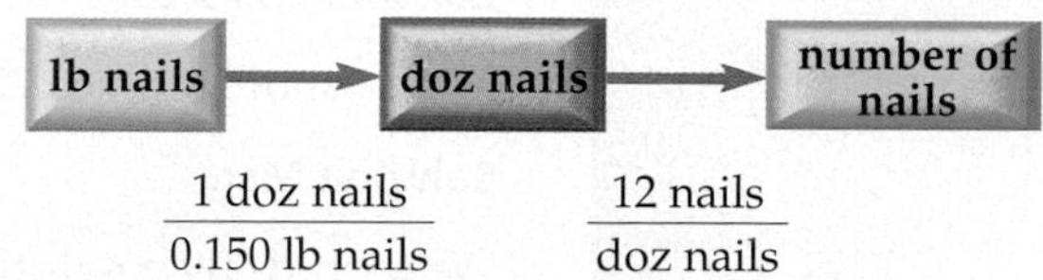

$$\frac{1 \text{ doz nails}}{0.150 \text{ lb nails}} \qquad \frac{12 \text{ nails}}{\text{doz nails}}$$

Beginning with 2.60 lb and using the solution map as a guide we convert from lb to number of nails.

$$2.60 \cancel{\text{lb nails}} \times \frac{1 \cancel{\text{doz nails}}}{0.150 \cancel{\text{lb nails}}} \times \frac{12 \text{ nails}}{1 \cancel{\text{doz nails}}} = 208 \text{ nails}$$

The customer who bought 2.60 lb of nails got 208 nails. He counted the nails by weighing them. If the customer purchased a different size of nail, the first conversion factor—relating lb to doz—would change, but the second conversion factor would not. One dozen corresponds to twelve nails, regardless of their size.

6.3 Counting Atoms by the Gram

Twenty-two copper pennies contain approximately 1 mol of copper atoms.

Beginning in 1982, pennies became almost all zinc with only a copper plating. Prior to 1982, pennies were mostly copper.

Determining the number of atoms in a sample with a certain mass is similar to determining the number of nails in a sample with a certain weight, but the numbers are different. With nails, we used a dozen as a convenient number in our conversions, but a dozen is too small to use with atoms. We need a larger number because atoms are so small. The chemist's "dozen" is called the **mole (mol)** and has a value of 6.022×10^{23}.

$$1 \text{ mol} = 6.022 \times 10^{23}$$

This number is also called Avogadro's number, named after **Amadeo Avogadro** (1776–1856), and is a convenient number to use when dealing with atoms. A mole of atoms makes up objects of reasonable sizes. For example, 22 copper pennies contain approximately 1 mol of copper (Cu) atoms, and a couple of large helium balloons contain approximately 1 mol of helium (He) atoms. There is nothing mysterious about a mole; it is just a certain number of objects (6.022×10^{23}), just as a dozen is a certain number of objects (12).

Two large helium balloons contain approximately one mole of helium atoms.

Recall from Chapter 4 that helium is an atomic element, which means that its fundamental units are individual atoms.

Converting Between Moles and Number of Atoms

Converting between moles and number of atoms is similar to converting between dozens and number of nails. To convert between moles of atoms and number of atoms we simply use the conversion factors:

$$\frac{1 \text{ mol atoms}}{6.022 \times 10^{23} \text{ atoms}} \quad \text{or} \quad \frac{6.022 \times 10^{23} \text{ atoms}}{1 \text{ mol atoms}}$$

For example, suppose we want to convert 3.5 mol of helium to the number of helium atoms. We set up the problem in the standard way.

Given: 3.5 mol He

Find: He atoms

Conversion Factor:

$$1 \text{ mol He} = 6.022 \times 10^{23} \text{ He atoms}$$

Solution Map:

We then draw a solution map showing the conversion from moles of He to He atoms.

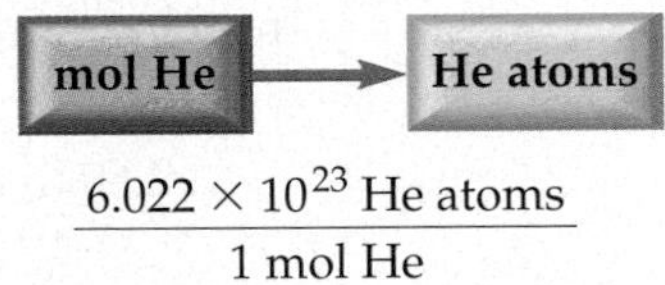

$$\frac{6.022 \times 10^{23} \text{ He atoms}}{1 \text{ mol He}}$$

Solution:

Beginning with 3.5 mol He, we use the conversion factor to get to He atoms.

$$3.5 \cancel{\text{mol He}} \times \frac{6.022 \times 10^{23} \text{ He atoms}}{1 \cancel{\text{mol He}}} = 2.1 \times 10^{24} \text{ He atoms}$$

EXAMPLE 6.1 **Converting Between Moles and Number of Atoms**

A silver ring contains 1.1×10^{22} silver atoms. How many moles of silver are in the ring? We set up the problem in the standard way.

Given:

$$1.1 \times 10^{22} \text{ Ag atoms}$$

Find: mol Ag

Conversion Factor:

$$1 \text{ mol Ag} = 6.022 \times 10^{23} \text{ Ag atoms}$$

Solution Map:

We then draw a solution map showing the conversion from Ag atoms to moles of Ag.

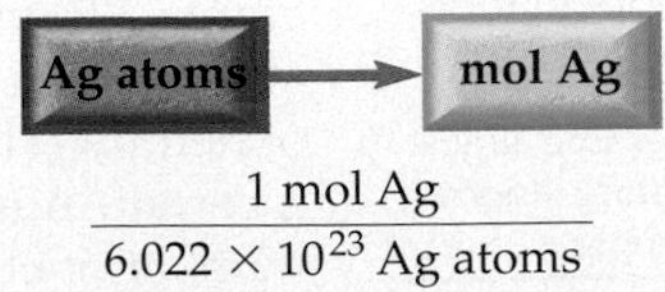

$$\frac{1 \text{ mol Ag}}{6.022 \times 10^{23} \text{ Ag atoms}}$$

Solution:
Beginning with 1.1×10^{22} Ag atoms, we use the conversion factor to get to moles of Ag.

$$1.1 \times 10^{22} \cancel{\text{Ag atoms}} \times \frac{1 \text{ mol Ag}}{6.022 \times 10^{23} \cancel{\text{Ag atoms}}} = 1.8 \times 10^{-2} \text{ mol Ag}$$

SKILLBUILDER 6.1 **Converting Between Moles and Number of Atoms**
How many gold atoms are in a pure gold ring containing 8.83×10^{-2} mol Au?

Converting Between Grams and Moles

We just learned how to convert between moles and number of atoms. As in our nail analogy, we need one more conversion factor to convert from the mass of a sample to the number of atoms in the sample. For nails, we used the weight of one dozen nails; for atoms, we need the mass of one mole of atoms.

> The mass of 1 mol of atoms of an element is called the **molar mass.** The value of an element's molar mass in grams per mole is numerically equivalent to the element's atomic mass in atomic mass units.

For example, copper has an atomic mass of 63.55 amu; one mole of copper atoms has a mass of 63.55 g, and the molar mass is of copper is 63.55 g/mol. Just as the weight of 1 doz nails changes for different nails, so the mass of one mole of atoms changes for different elements: 1 mol of sulfur atoms (sulfur atoms are lighter than copper atoms) has a mass of 32.06 g; 1 mol of carbon atoms (lighter than sulfur) has a mass of 12.01 g; and 1 mol of lithium atoms (lighter yet) has a mass of 6.94 g.

$$32.06 \text{ g sulfur} = 1 \text{ mol sulfur} = 6.022 \times 10^{23} \text{ S atoms}$$
$$12.01 \text{ g carbon} = 1 \text{ mol carbon} = 6.022 \times 10^{23} \text{ C atoms}$$
$$6.94 \text{ g lithium} = 1 \text{ mol lithium} = 6.022 \times 10^{23} \text{ Li atoms}$$

The smaller the atom, the less mass it takes to make one mole (Figure 6.2).

Therefore, the molar mass of any element becomes a conversion factor between grams of that element and moles of that element. For carbon:

$$12.01 \text{ g C} = 1 \text{ mol C} \quad \text{or} \quad \frac{12.01 \text{ g C}}{1 \text{ mol C}} \quad \text{or} \quad \frac{1 \text{ mol C}}{12.01 \text{ g C}}$$

A 0.58 g diamond would be about a three-carat diamond.

Suppose we want to calculate the moles of carbon in a 0.58 g diamond (pure carbon). We set up the problem in the standard way.

Given: 0.58 g C

Find: mol C

Conversion Factor: 12.01 g C = 1 mol C

Part A

1 mole S (32.06 g)

1 mole C (12.011 g)

Part B

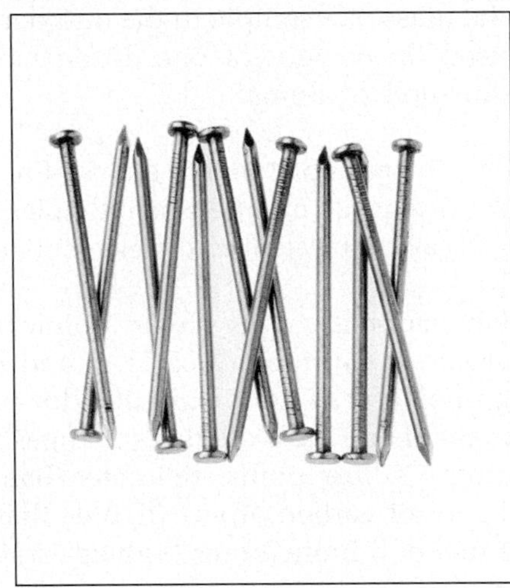

1 dozen large nails

1 dozen small nails

Figure 6.2a Each of these samples has the same number of atoms, 6.022×10^{23}. Since sulfur atoms are heavier and larger, one mole of S atoms has the largest mass and occupies the most space. Carbon atoms are smaller and less massive than sulfur atoms.

Figure 6.2b Each of these pictures shows the same number of nails, twelve. As you can see, twelve large nails have more weight and occupy more space than twelve small nails. The same is true for atoms.

Solution Map:

We then draw a solution map showing the conversion from grams of C to moles of C. The conversion factor is the molar mass of carbon.

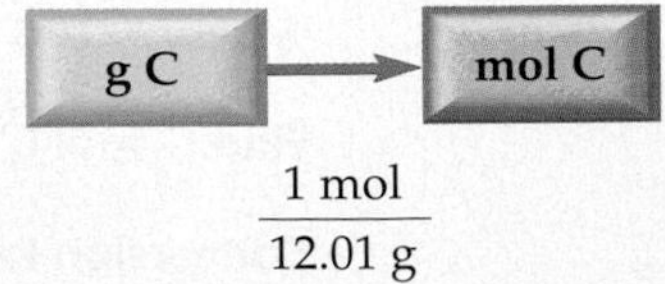

Solution:
Beginning with 0.58 g C we use the conversion factor to get to mol C.

$$0.58 \text{ g C} \times \frac{1 \text{ mol C}}{12.01 \text{ g C}} = 4.8 \times 10^{-2} \text{ mol C}$$

EXAMPLE 6.2 The Mole Concept—Converting Between Grams and Moles

Calculate the moles of sulfur in 57.8 g of sulfur.

Given: 57.8 g S

Find: mol S

Conversion Factor: 32.06 g S = 1 mol S

Solution Map:

Draw a solution map showing the conversion from g S to mol S. The conversion factor is the molar mass of sulfur.

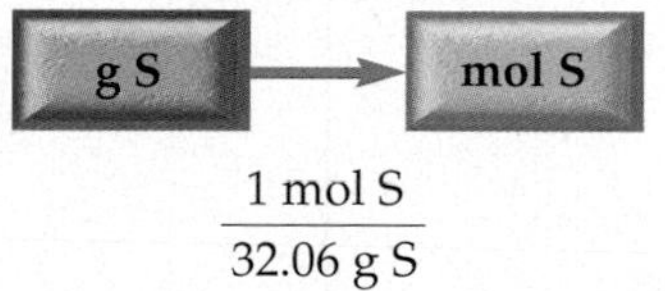

$$\frac{1 \text{ mol S}}{32.06 \text{ g S}}$$

Solution:
Begin with 57.8 g S and use the conversion factor to get to mol S.

$$57.8 \text{ g S} \times \frac{1 \text{ mol S}}{32.06 \text{ g S}} = 1.80 \text{ mol S}$$

SKILLBUILDER 6.2 The Mole Concept—Converting Between Grams and Moles

Calculate the number of grams of sulfur in 2.78 mol of sulfur.

Converting Between Grams and Number of Atoms

Now, suppose we want to know the number of carbon *atoms* in the 0.58 g diamond. We first convert from grams to moles and then from moles to number of atoms. The solution map is:

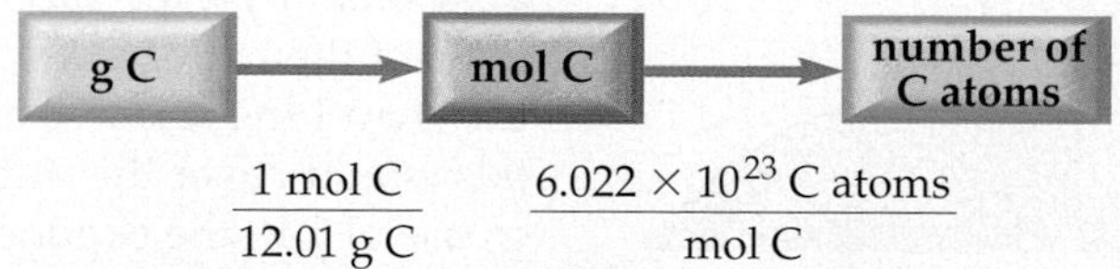

$$\frac{1 \text{ mol C}}{12.01 \text{ g C}} \qquad \frac{6.022 \times 10^{23} \text{ C atoms}}{\text{mol C}}$$

Notice the similarity between this solution map and the one we used for nails.

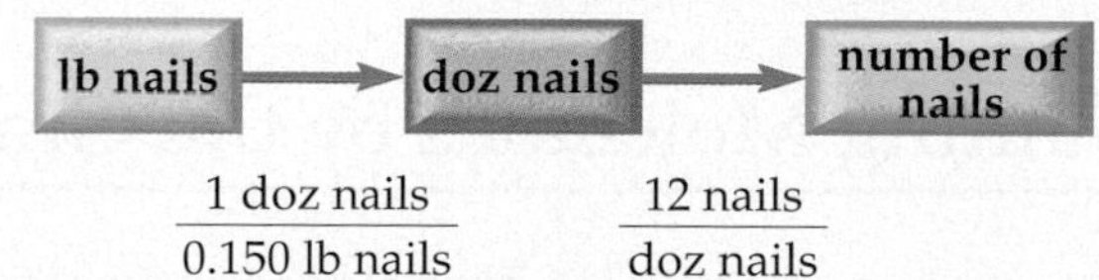

$$\frac{1 \text{ doz nails}}{0.150 \text{ lb nails}} \qquad \frac{12 \text{ nails}}{\text{doz nails}}$$

Beginning with 0.58 g carbon and using the solution map as a guide, we convert to the number of carbon atoms.

$$0.58\ \cancel{\text{g C}} \times \frac{1\ \cancel{\text{mol C}}}{12.01\ \cancel{\text{g C}}} \times \frac{6.022 \times 10^{23}\ \text{C atoms}}{\cancel{\text{mol C}}} = 2.9 \times 10^{22}\ \text{C atoms}$$

EXAMPLE 6.3 The Mole Concept—Converting Between Grams and Number of Atoms

How many aluminum atoms are in an aluminum can with a mass of 16.2 g? We set up the problem in the standard way.

Given: 16.2 g Al

Find: Al atoms

Conversion Factors:

26.98 g Al = 1 mol Al (molar mass of Al)
$6.022 \times 10^{23} = 1$ mol

Solution Map:

The solution map begins with g Al and ends with number of Al atoms.

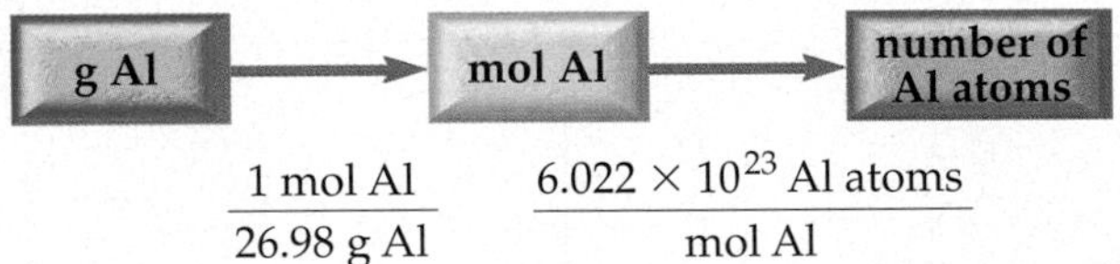

$$\frac{1\ \text{mol Al}}{26.98\ \text{g Al}} \qquad \frac{6.022 \times 10^{23}\ \text{Al atoms}}{\text{mol Al}}$$

Solution:
We then follow the solution map, beginning with 16.2 g Al and multiplying by the appropriate conversion factors, to arrive at Al atoms.

$$16.2\ \cancel{\text{g Al}} \times \frac{1\ \cancel{\text{mol Al}}}{26.98\ \cancel{\text{g Al}}} \times \frac{6.022 \times 10^{23}\ \text{Al atoms}}{1\ \cancel{\text{mol Al}}} = 3.62 \times 10^{23}\ \text{Al atoms}$$

SKILLBUILDER 6.3 The Mole Concept—Converting Between Grams and Number of Atoms

Calculate the mass of 1.23×10^{24} helium atoms.

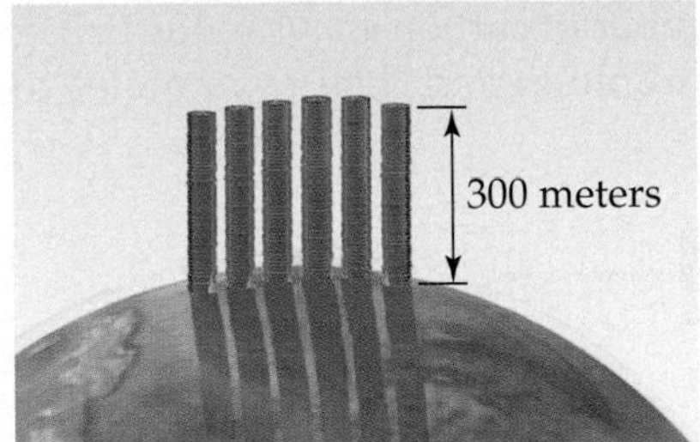

A mole of pennies would cover Earth's surface to depth of 300 meters.

Before we move on, notice that numbers with large exponents, such as 6.022×10^{23}, are deceptively large. Twenty-two copper pennies contain 6.022×10^{23} or one mole of copper atoms, but 6.022×10^{23} pennies would cover the earth's entire surface to a depth of 300 meters. Even objects small by everyday standards occupy a huge space when we have a mole of them. For example, one crystal of granulated sugar has a mass of less than 1 mg and a diameter of less than 0.1 mm, yet 1 mol of sugar crystals would cover the state of Texas to a depth of several feet. For every increase of 1 in the exponent of a number, the number increases by 10. So a number with an exponent of 23 is incredibly large. A mole has to be a large number, however, because atoms are so small.

6.4 Counting Molecules by the Gram

The calculations we just performed for atoms can also be applied to molecules for covalent compounds or formula units for ionic compounds. For elements, the molar mass is the mass of one mole of atoms of that element.

Remember, ionic compounds do not contain individual molecules. In loose language, the smallest electrically neutral collection of ions is sometimes called a molecule but is more correctly called a formula unit.

Remember, the formula mass for a compound is simply the sum of the atomic masses of all of the atoms in a chemical formula.

For compounds, the molar mass is the mass of one mole of molecules or formula units of that compound. The molar mass of a compound in grams per mole is numerically equivalent to the formula mass of the compound in atomic mass units. For example, the formula mass of CO_2 is:

$$\begin{aligned}\text{Formula mass} &= 1 \times (\text{Atomic mass of C}) + 2 \times (\text{Atomic mass of O}) \\ &= 1\,(12.01\ \text{amu}) + 2\,(16.00\ \text{amu}) \\ &= 44.01\ \text{amu}\end{aligned}$$

The molar mass is therefore:

$$\text{Molar mass} = 44.01\ \text{g/mol}$$

The molar mass of CO_2 is a conversion factor between grams of CO_2 and moles of CO_2. Suppose we want to find the number of CO_2 molecules in a sample of dry ice (solid CO_2) with a mass of 22.5 g.

The solution map for the problem is:

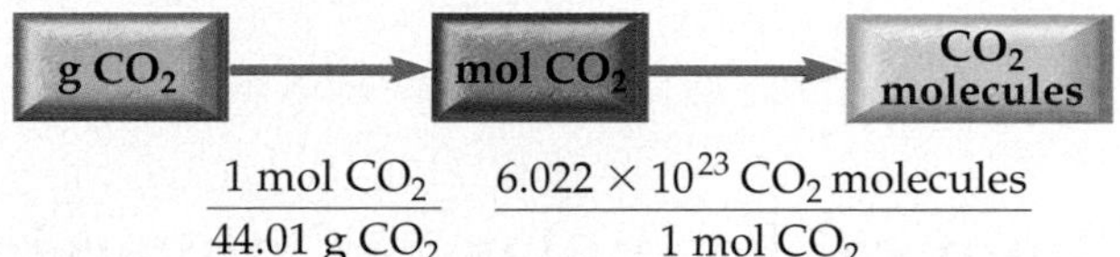

$$\frac{1\ \text{mol}\ CO_2}{44.01\ \text{g}\ CO_2} \qquad \frac{6.022 \times 10^{23}\ CO_2\ \text{molecules}}{1\ \text{mol}\ CO_2}$$

Following the solution map, we get:

$$22.5\ \cancel{\text{g}\ CO_2} \times \frac{1\ \cancel{\text{mol}\ CO_2}}{44.01\ \cancel{\text{g}\ CO_2}} \times \frac{6.022 \times 10^{23}\ CO_2\ \text{molecules}}{1\ \cancel{\text{mol}\ CO_2}} = 3.08 \times 10^{23}\ CO_2\ \text{molecules}$$

EXAMPLE 6.4 The Mole Concept—Converting Between Mass and Number of Molecules

Find the mass of 4.78×10^{24} NO_2 molecules.

We are given a number of NO_2 molecules and asked to find the mass of NO_2.

Given: 4.78×10^{24} NO_2 molecules

Find: g NO_2

Conversion Factor: $6.022 \times 10^{23} = 1$ mol

We also need the molar mass of NO_2 as a conversion factor.

$$\begin{aligned}NO_2\ \text{molar mass} &= 1 \times (\text{Atomic mass N}) + 2 \times (\text{Atomic mass O}) \\ &= 14.01 + 2\,(16.00) \\ &= 46.01\ \text{g/mol}\end{aligned}$$

Solution Map:

Draw a solution map, beginning with the given quantity and using the correct conversion factors to get to g NO_2.

NO_2 molecules → mol NO_2 → g NO_2

$$\frac{1\ \text{mol}\ NO_2}{6.022 \times 10^{23}\ NO_2\ \text{molecules}} \qquad \frac{46.01\ \text{g}\ NO_2}{1\ \text{mol}\ NO_2}$$

Solution:

Using the solution map as a guide, begin with NO_2 molecules and multiply by the appropriate conversion factors to arrive at g NO_2.

$$4.78 \times 10^{24}\ \cancel{NO_2\ molecules} \times \frac{1\ \cancel{mol\ NO_2}}{6.022 \times 10^{23}\ \cancel{NO_2\ molecules}} \times \frac{46.01\ g\ NO_2}{1\ \cancel{mol\ NO_2}} = 365\ g\ NO_2$$

SKILLBUILDER 6.4 **The Mole Concept—Converting Between Mass and Number of Molecules**

How many H_2O molecules are in a sample of water with a mass of 3.64 g?

6.5 Chemical Formulas as Conversion Factors

From our knowledge of clovers, we know that each clover has three leaves. We can express that as an equivalence, 3 leaves $\equiv$ 1 clover.

We are almost ready to address the sodium problem in our opening example. To determine how much of a particular element (such as sodium) is in a given amount of a particular compound (such as sodium chloride), we must understand the numerical relationships inherent in a chemical formula. We can understand these relationships with a simple analogy: asking how much sodium is in a given amount of sodium chloride is much like asking how many leaves are on a given number of clovers. For example, suppose we want to know how many leaves are on 14 clovers. We need a conversion factor between leaves and clovers. For clovers, the conversion factor comes from our knowledge about them—we know that each clover has three leaves. We write:

$$3\ \text{leaves} \equiv 1\ \text{clover}$$

Like other conversion factors, this equivalence gives the relationship between leaves and clovers. We use the equivalence sign ($\equiv$) because while 3 leaves do not equal 1 clover, 3 leaves are equivalent to one clover, meaning that each clover must have 3 leaves to be complete. With this conversion factor, we can easily find the number of leaves in 14 clovers. The solution map is:

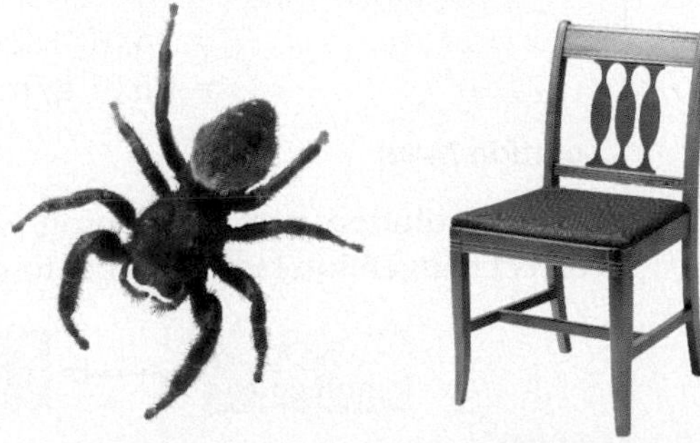

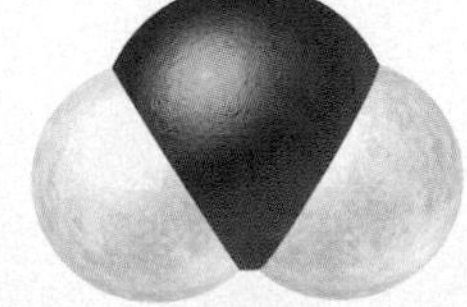

8 legs $\equiv$ 1 spider   4 legs $\equiv$ 1 chair   2 H atoms $\equiv$ 1 H_2O molecule

Each of these shows an equivalence.

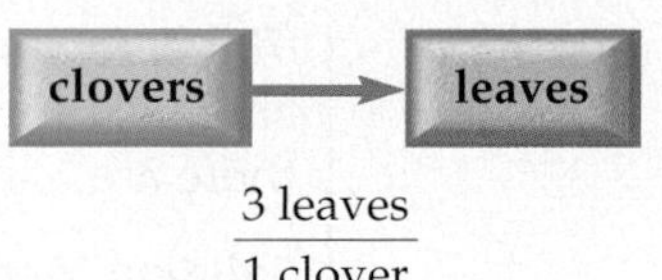

$$\frac{3 \text{ leaves}}{1 \text{ clover}}$$

We solve the problem by beginning with clovers and converting to leaves.

$$14 \text{ clovers} \times \frac{3 \text{ leaves}}{1 \text{ clover}} = 42 \text{ leaves}$$

Similarly, a chemical formula gives us equivalences between elements for a particular compound. For example, the formula for carbon dioxide (CO_2) means there are 2 O atoms per CO_2 molecule. We write this as:

$$2 \text{ O atoms} \equiv 1 \text{ } CO_2 \text{ molecule}$$

Just as *3 leaves* ≡ *1 clover* can also be written as *3 dozen leaves* ≡ *1 dozen clovers*, for molecules we can write:

$$2 \text{ doz O atoms} \equiv 1 \text{ doz } CO_2 \text{ molecules},$$

However, for atoms and molecules, we normally work in moles.

$$2 \text{ mol O} \equiv 1 \text{ mol } CO_2$$

With conversion factors such as these—that come directly from the chemical formula—we can determine the amounts of the constituent elements present in a given amount of a compound.

Chemical formulas are covered in Chapter 5.

Converting Between Moles of a Compound and Moles of a Constituent Element

Suppose we want to know the number of moles of O in 18 mol CO_2. Our solution map is:

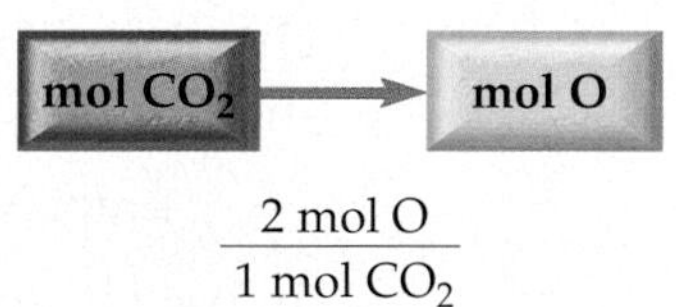

$$\frac{2 \text{ mol O}}{1 \text{ mol } CO_2}$$

We can then calculate the moles of O.

$$18 \text{ mol } CO_2 \times \frac{2 \text{ mol O}}{1 \text{ mol } CO_2} = 36 \text{ mol O}$$

EXAMPLE 6.5 Chemical Formulas as Conversion Factors—Converting Between Moles of a Compound and Moles of a Constituent Element

Determine the moles of O in 1.7 moles of $CaCO_3$.
Set up the problem.

Given: 1.7 mol $CaCO_3$

Find: mol O

Conversion Factor: $3 \text{ mol O} \equiv 1 \text{ mol } CaCO_3$

The conversion factor is obtained from the chemical formula. Since the chemical formula indicates 3 O atoms for every 1 $CaCO_3$, we know that there are 3 mol O for every 1 mol $CaCO_3$.

Solution Map:

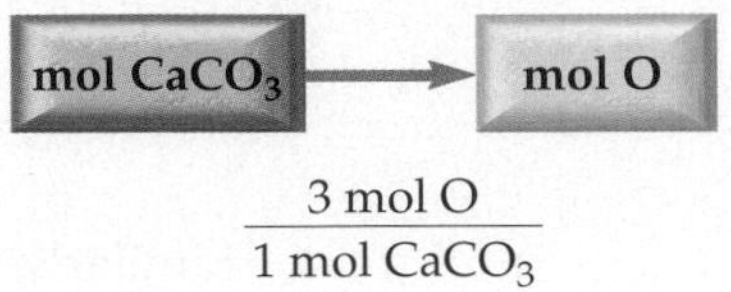

$$\frac{3 \text{ mol O}}{1 \text{ mol } CaCO_3}$$

Solution:

Finally, we perform the required conversions.

$$1.7 \cancel{\text{mol } CaCO_3} \times \frac{3 \text{ mol O}}{1 \cancel{\text{mol } CaCO_3}} = 5.1 \text{ mol O}$$

The subscripts in a chemical formula are exact, so they never limit significant figures.

SKILLBUILDER 6.5 **Chemical Formulas as Conversion Factors—Converting Between Moles of a Compound and Moles of a Constituent Element**

Determine moles of O in 1.4 mol H_2SO_4.

Converting Between Grams of a Compound and Grams of a Constituent Element

Now, we have everything we need to solve our sodium problem. Suppose we want to know how many grams of sodium there are in 15 g NaCl. The chemical formula gives us the relationship between moles of Na and moles of NaCl.

$$1 \text{ mol Na} \equiv 1 \text{ mol NaCl}$$

To use this relationship, we need *mol* NaCl. But, we have *g* NaCl. We can, however, use the *molar mass* of NaCl to convert from g NaCl to mol NaCl. Then we use the conversion factor from the chemical formula to convert to mol Na. Finally, we use the molar mass of Na to convert to g Na. The solution map is:

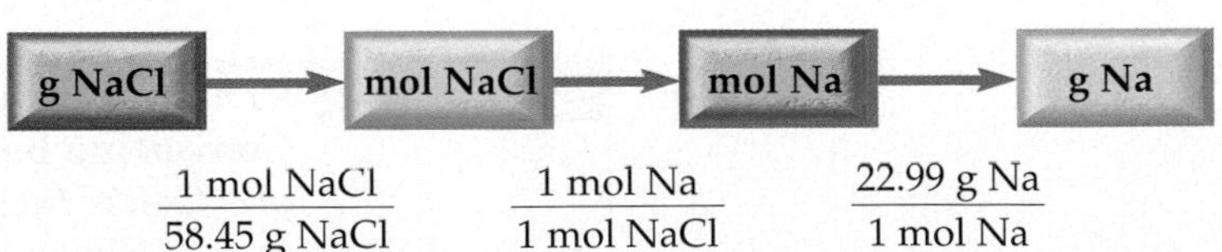

$$\frac{1 \text{ mol NaCl}}{58.45 \text{ g NaCl}} \qquad \frac{1 \text{ mol Na}}{1 \text{ mol NaCl}} \qquad \frac{22.99 \text{ g Na}}{1 \text{ mol Na}}$$

Notice that we must convert from g NaCl to mol NaCl *before* we can use the chemical formula as a conversion factor.

The chemical formula gives us a relationship between moles of substances, not between grams.

We follow the solution map to solve the problem.

$$15 \text{ g NaCl} \times \frac{1 \text{ mol NaCl}}{58.45 \text{ g NaCl}} \times \frac{1 \text{ mol Na}}{1 \text{ mol NaCl}} \times \frac{23.00 \text{ g Na}}{1 \text{ mol Na}} = 5.9 \text{ g Na}$$

The general form for solving problems where you are asked to find the mass of an element present in a given mass of a compound is:

Mass compound → **Moles** compound → **Moles** element → **Mass** element

The conversions between mass and moles are accomplished using the atomic or molar mass, and the conversion between moles and moles is accomplished using the relationships inherent in the chemical formula (Figure 6.3).

1 mol CCl_4 ≡ 4 mol Cl

Figure 6.3 The relationships inherent in a chemical formula allow us to convert between moles of the compound and moles of a constituent element (or vice versa).

EXAMPLE 6.6 Chemical Formulas as Conversion Factors—Converting Between Grams of a Compound and Grams of a Constituent Element

Carvone ($C_{10}H_{14}O$) is the main component of spearmint oil. It has a pleasant aroma and mint flavor. Carvone is often added to chewing gum, liqueurs, soaps, and perfumes. Find the mass of carbon in 55.4 g of carvone. We extract the important information from the problem in the standard way.

Given: 55.4 g $C_{10}H_{14}O$

Find: g C

Conversion Factors:

We need two conversion factors. The first is the molar mass of carvone.

$$\begin{aligned} \text{Molar mass} &= 10\,(12.01) + 14\,(1.01) + 16.00 \\ &= 120.1 + 14.14 + 16.00 \\ &= 150.2 \text{ g/mol} \end{aligned}$$

The second conversion factor is the relationship between mol C and mol carvone. This comes from the molecular formula.

$$10 \text{ mol C} \equiv 1 \text{ mol } C_{10}H_{14}O$$

Solution Map:

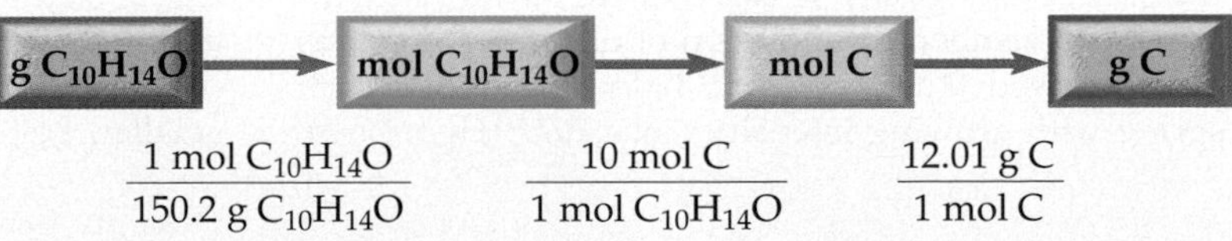

The solution map is based on Grams → Mole → Mole → Grams. Remember, the conversion factor obtained from the chemical formula (10 mol C ≡ 1 mol $C_{10}H_{14}O$) applies only to mol; since we are given g of carvone, we must first convert from g to mol.

Solution:

We then follow the solution map, beginning with g $C_{10}H_{14}O$ and multiplying by the appropriate conversion factors to arrive at g C.

$$55.4\ \cancel{\text{g } C_{10}H_{14}O} \times \frac{1\ \cancel{\text{mol } C_{10}H_{14}O}}{150.2\ \cancel{\text{g } C_{10}H_{14}O}} \times \frac{10\ \cancel{\text{mol C}}}{1\ \cancel{\text{mol } C_{10}H_{14}O}} \times \frac{12.01\text{ g C}}{1\ \cancel{\text{mol C}}} = 44.3\text{ g C}$$

SKILLBUILDER 6.6 **Chemical Formulas as Conversion Factors—Converting Between Grams of a Compound and Grams of a Constituent Element**

Determine the mass of oxygen in a 5.8 g sample of sodium bicarbonate ($NaHCO_3$).

SKILLBUILDER PLUS

Determine the mass of oxygen in a 7.20 g sample of $Al_2(SO_4)_3$.

CHEMISTRY IN THE ENVIRONMENT

Chlorine in Chlorofluorocarbons

About 25 years ago, scientists began to suspect that synthetic compounds known as chlorofluorocarbons (or CFCs) were destroying vital molecules, called ozone (O_3), in Earth's upper atmosphere. Upper atmospheric ozone is important because it acts as a shield to protect life on earth from harmful ultraviolet light (Figure 6.4). CFCs are chemically inert molecules—they do not readily react with other substances—used primarily as refrigerants and industrial solvents. Their inertness, however, has allowed them to leak into the atmosphere and stay there for many years. In the upper atmosphere sunlight breaks bonds within CFCs that results in the release of chlorine atoms. The chlorine atoms then react with ozone and destroy it, converting it into O_2.

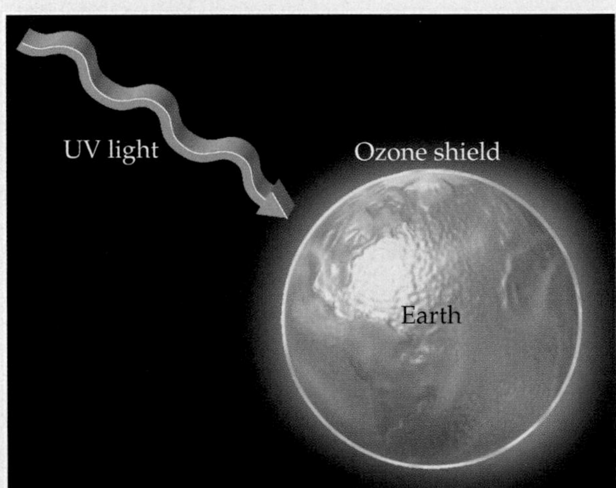

Figure 6.4 Atmospheric ozone shields life on earth from harmful ultraviolet light.

In 1985, scientists discovered a large hole in the ozone layer over Antarctica that has since been attributed to CFCs. The amount of ozone depletion over Antarctica was a startling 50 percent. The ozone hole is transient, existing only in the Antarctic spring, late August to November. Examination of data from previous years showed that this ozone hole had formed each spring with growing intensity since 1977 (Figure 6.5), and it continues to form today. A similar hole has been observed during some years over the North Pole, and a smaller, but still significant, drop in ozone has been

6.6 Mass Percent Composition of Compounds

Another way to express how much of an element is in a given compound is to use the element's mass percent composition for that compound. The **mass percent composition** or simply **mass percent** of an element is the element's percentage of the total mass of the compound. For example, the mass percent composition of sodium in sodium chloride is 39%; a 100 g sample of sodium chloride contains 39 g of sodium. The mass percent composition can be determined from experimental data using the following formula.

$$\text{Mass percent of element } X = \frac{\text{Mass of } X \text{ in a sample of the compound}}{\text{Mass of the sample of the compound}} \times 100\%$$

For example, suppose a 0.358 g sample of chromium reacts with oxygen to form 0.523 g of the metal oxide. Then the mass percent of chromium is:

$$\text{Mass percent Cr} = \frac{\text{Mass Cr}}{\text{Mass metal oxide}} \times 100\%$$

$$= \frac{0.358 \text{ g}}{0.523 \text{ g}} \times 100\% = 68.5\%$$

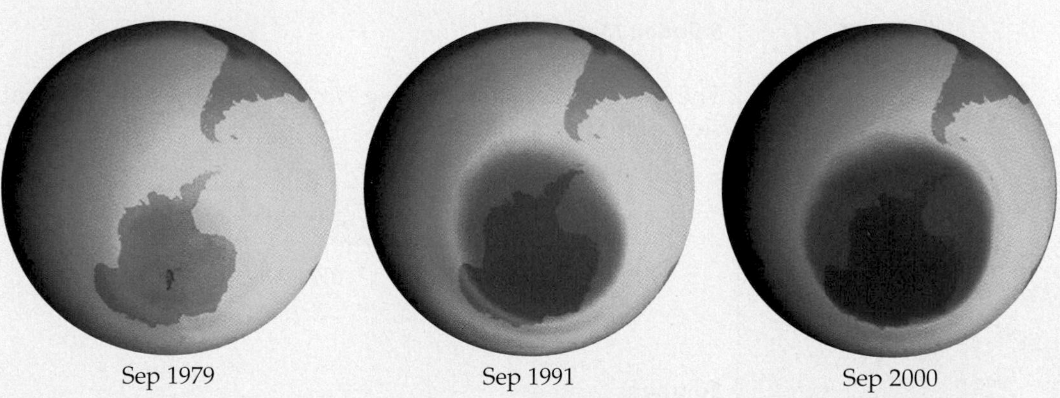

observed over more populated areas such as the northern United States and Canada. The thinning of ozone over these areas is dangerous because ultraviolet light can harm living things and induce skin cancer in humans. Based on this evidence, most developed nations banned the production of CFCs on January 1, 1996. However, CFCs still lurk in most older refrigerators and air conditioning units and can leak into the atmosphere and destroy ozone.

Figure 6.5 Antarctic ozone levels on three Septembers from 1979 to 2000. The blue colors indicate low ozone levels.

CAN YOU ANSWER THIS? *Suppose a car air conditioner contains 2.5 kg of freon-12 (CCl_2F_2), a CFC. How many kilograms of Cl are contained within the freon?*

Mass percent composition can be used as a conversion factor between grams of a constituent element and grams of the compound. For example, we saw that the mass percent composition of sodium in sodium chloride is 39%. This can be written as:

$$39 \text{ g sodium} \equiv 100 \text{ g sodium chloride}$$

or in fractional form:

$$\frac{39 \text{ g Na}}{100 \text{ g NaCl}} \quad \text{or} \quad \frac{100 \text{ g NaCl}}{39 \text{ g Na}}$$

These fractions are conversion factors between g Na and g NaCl.

EXAMPLE 6.7 Using Mass Percent Composition as a Conversion Factor

The FDA recommends that you consume less than 2.4 g of sodium per day. How many grams of sodium chloride can you consume and still be within the FDA guidelines? Sodium chloride is 39% sodium by mass.

We set up the problem in the standard way.

Given: 2.4 g Na

Find: g NaCl

Conversion Factor:

Our conversion factor is the mass percent of sodium in sodium chloride.

$$39 \text{ g Na} \equiv 100 \text{ g NaCl}$$

Solution Map:

The solution map starts with g Na and uses the mass percent as a conversion factor to get to g NaCl.

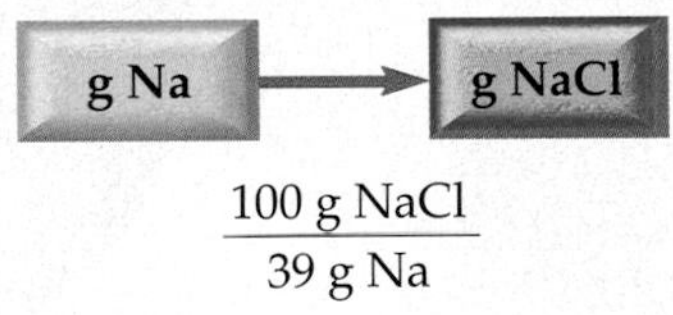

$$\frac{100 \text{ g NaCl}}{39 \text{ g Na}}$$

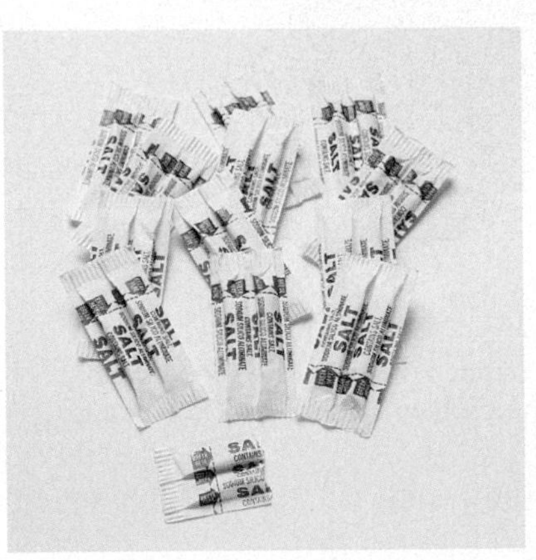

Twelve and a half salt packets contain 6.2 g NaCl.

Solution:

$$2.4 \cancel{\text{g Na}} \times \frac{100 \text{ g NaCl}}{39 \cancel{\text{g Na}}} = 6.2 \text{ g NaCl}$$

So, you can consume 6.2 g NaCl and still be within the FDA guideline.

SKILLBUILDER 6.7 Using Mass Percent Composition as a Conversion Factor

If someone consumes 22 g of sodium chloride, how much sodium does that person consume? Sodium chloride is 39% sodium by mass.

6.7 Mass Percent Composition From a Chemical Formula

The mass percent of any element in a compound can also be computed from the chemical formula. Based on the chemical formula, the mass percent of element X in a compound is:

$$\text{Mass percent of element } X = \frac{\text{Mass of element } X \text{ in 1 mol of compound}}{\text{Mass of 1 mol of the compound}} \times 100\%$$

Suppose we want to calculate the mass percent composition of Cl in CCl_2F_2. The mass percent Cl is given by:

$$\text{Mass percent Cl} = \frac{\text{Mass of Cl in 1 mol of } CCl_2F_2}{\text{Mass of 1 mol of } CCl_2F_2} \times 100\%$$

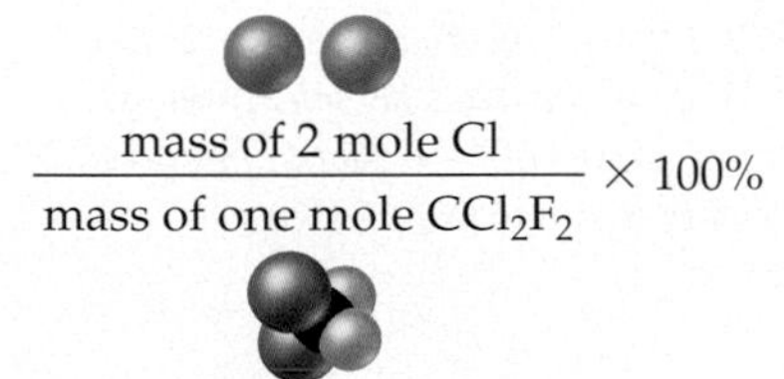

The mass percent of Cl in CCl_2F_2 is equal to the mass of 2 mol Cl divided by the mass of 1 mol CCl_2F_2 times 100%.

To get the mass of Cl in one mole of CCl_2F_2, we convert from mol CCl_2F_2 to mol Cl to mass of Cl. The solution map is:

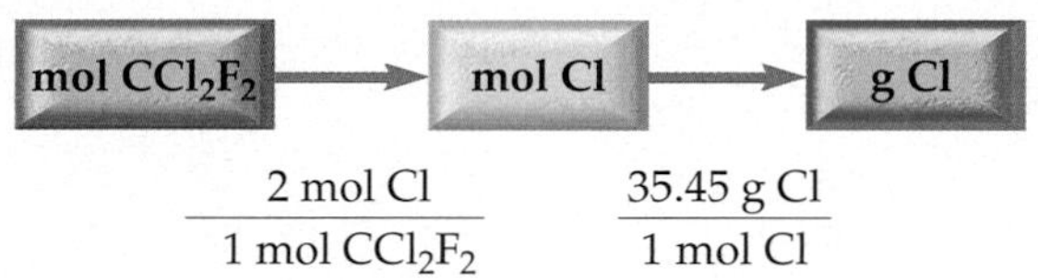

$$\frac{2 \text{ mol Cl}}{1 \text{ mol } CCl_2F_2} \qquad \frac{35.45 \text{ g Cl}}{1 \text{ mol Cl}}$$

The first conversion factor comes from the chemical formula; the second one is the molar mass of Cl. Beginning with 1 mol CCl_2F_2 and converting:

$$1 \cancel{\text{mol } CCl_2F_2} \times \frac{2 \cancel{\text{mol Cl}}}{1 \cancel{\text{mol } CCl_2F_2}} \times \frac{35.45 \text{ g Cl}}{1 \cancel{\text{mol Cl}}} = 70.90 \text{ g Cl}$$

So 1 mol of CCl_2F_2 contains 70.90 g Cl. The molar mass of CCl_2F_2 is:

$$\begin{aligned}\text{Molar mass} &= 12.01 + 2(35.45) + 2(19.00) \\ &= 120.91 \text{ g/mol}\end{aligned}$$

A mole of CCl_2F_2 has a mass of 120.91 g. Now we use our mass percent equation to find the mass percent of Cl.

$$\begin{aligned}\text{Mass percent Cl} &= \frac{\text{Mass of Cl in one mole of } CCl_2F_2}{\text{Mass of one mole of } CCl_2F_2} \times 100\% \\ &= \frac{70.90 \text{ g}}{120.91 \text{ g}} \times 100\% \\ &= 58.64\%\end{aligned}$$

CHEMISTRY AND HEALTH

Fluoridation of Drinking Water

In the early 1900s, scientists discovered that people whose drinking water naturally contained fluoride (F^-) ions had fewer cavities than people whose water did not. At the proper levels, fluoride acts to strengthen tooth enamel, which prevents tooth decay. In an effort to improve public health, fluoride has been artificially added to drinking water supplies since 1945. In the United States today, about 62% of the population drinks artificially fluoridated drinking water. The American Dental Association and public health agencies estimate that water fluoridation reduces tooth decay by 40 to 65%.

The fluoridation of public drinking water, however, is often controversial. Some opponents argue that fluoride is available from other sources—such as toothpaste, mouthwash, drops, or pills—and therefore should not be added to drinking water. Anyone who wants fluoride can get it from these optional sources, and the government should not impose fluoride on the general population, they argue. Other opponents argue that the risks associated with fluoridation are too great. Indeed, too much fluoride can cause teeth to become brown and spotted, a condition known as dental fluorosis. Extremely high levels can lead to skeletal fluorosis, a condition in which the bones become brittle and arthritic.

The scientific consensus is that, like many minerals, fluoride shows some health benefits at certain levels—about 1–4 mg/day for adults—but can have detrimental effects at higher levels. Consequently, most major cities fluoridate their drinking water at a level of about 1 mg/L. Since adults drink between 1 and 2 L of water per day, they should receive the beneficial amounts of fluorine from the water.

CAN YOU ANSWER THIS? *Fluoride is often added to water as sodium fluoride (NaF). What is the mass percent composition of* F^- *in NaF? How many grams of NaF should be added to 1500 L of water to fluoridate it at a level of 1.0 mg* F^-/L?

EXAMPLE 6.8 **Mass Percent Composition**

Calculate the mass percent Cl in freon-114 ($C_2Cl_4F_2$).

Given: $C_2Cl_4F_2$

Find: mass percent Cl

Equation:

$$\text{Mass percent of element } X = \frac{\text{Mass of element } X \text{ in 1 mol of compound}}{\text{Mass of 1 mol of the compound}} \times 100\%$$

Conversion Factors:

To compute the quantities that go into the preceding equation, we need three conversion factors. The first is the relationship between mol Cl and mol $C_2Cl_4F_2$.

$$4 \text{ mol Cl} \equiv 1 \text{ mol } C_2Cl_4F_2$$

The second is the molar mass of Cl.

$$\frac{35.45 \text{ g Cl}}{1 \text{ mol Cl}}$$

The third is the molar mass of $C_2Cl_4F_2$.

$$\begin{aligned} C_2Cl_4F_2 \text{ molar mass} &= 2(12.01) + 4(35.45) + 2(19.00) \\ &= 24.02 + 141.8 + 38.00 \\ &= 203.82 \\ &= 203.8 \text{ g/mol} \end{aligned}$$

To solve this problem, we must find the two quantities required for the mass fraction equation. The first is the mass of Cl in 1 mol of $C_2Cl_4F_2$.

Solution Map:

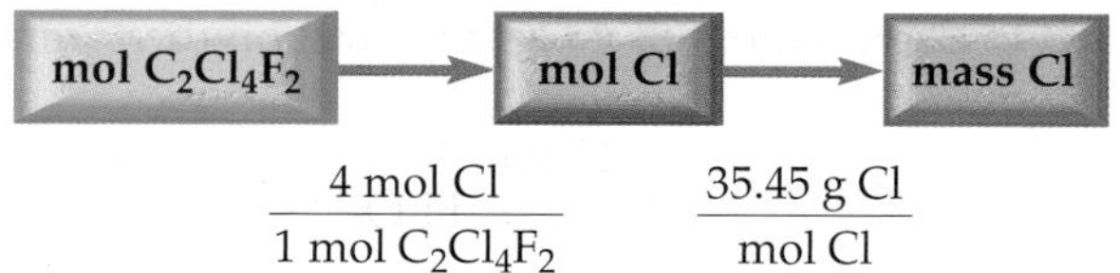

$$\frac{4 \text{ mol Cl}}{1 \text{ mol } C_2Cl_4F_2} \qquad \frac{35.45 \text{ g Cl}}{\text{mol Cl}}$$

Solution:

$$1 \text{ mol } C_2Cl_4F_2 \times \frac{4 \text{ mol Cl}}{1 \text{ mol } C_2Cl_4F_2} \times \frac{35.45 \text{ g Cl}}{1 \text{ mol Cl}} = 141.8 \text{ g Cl}$$

The second quantity is the mass of 1 mol of $C_2Cl_4F_2$, which is simply its molar mass, 203.8 g. We now substitute these two quantities into the mass percent equation.

$$\begin{aligned} \text{Mass percent of Cl} &= \frac{\text{Mass of Cl in 1 mol of } C_2Cl_4F_2}{\text{Mass of 1 mol of } C_2Cl_4F_2} \times 100\% \\ &= \frac{141.8 \text{ g Cl}}{203.8 \text{ g } C_2Cl_4F_2} \times 100\% \\ &= 69.58\% \end{aligned}$$

SKILLBUILDER 6.8 **Mass Percent Composition**

Acetic acid ($HC_2H_3O_2$) is the active ingredient in vinegar. Calculate the mass percent composition of O in acetic acid.

6.8 Calculating Empirical Formulas for Compounds

In the previous section we learned how to calculate mass percent composition from a chemical formula. But can we go the other way? Can we calculate a chemical formula from mass percent composition? This is important because laboratory analyses of compounds do not often give chemical formulas directly, but rather, they give the relative masses of each element present in a compound. For example, if we decompose water into hydrogen and oxygen in the laboratory, we could measure the masses of hydrogen and oxygen produced. Can we get a chemical formula from this kind of data?

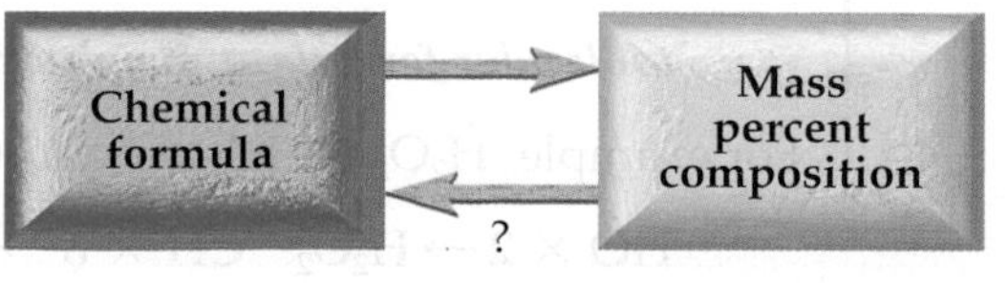

We just learned how to go from the chemical formula of a compound to its mass percent composition. Can we also go the other way?

Fructose, a sugar found in fruit.

180.2 g/mol. We know that the molecular formula is a whole-number multiple of CH_2O.

$$\text{Molecular formula} = CH_2O \times n$$

We also know that the molar mass is a whole-number multiple of the **empirical formula molar mass**, the sum of the masses of all the atoms in the empirical formula.

$$\text{Molar mass} = \text{Empirical formula molar mass} \times n$$

For a particular compound, the value of n in both cases is the same. Therefore, we can find n by computing the ratio of the molar mass to the empirical formula molar mass.

$$n = \frac{\text{Molar mass}}{\text{Empirical formula mass}}$$

For fructose, the empirical formula molar mass is:

$$\text{Empirical formula molar mass} = 12.01 + 2(1.01) + 16.00$$
$$= 30.03 \text{ g/mol}$$

therefore n is:

$$n = \frac{180.2 \text{ amu}}{30.03 \text{ amu}} = 6$$

We can then use this value of n to find the molecular formula.

$$\text{Molecular formula} = CH_2O \times 6 = C_6H_{12}O_6$$

EXAMPLE 6.12 Calculating Molecular Formula From Empirical Formula and Molar Mass

Naphthalene is a carbon and hydrogen containing compound often used in moth balls. Its empirical formula is C_5H_4 and its molar mass is 128.16 g/mol. Find its molecular formula.

Given: empirical formula $= C_5H_4$; molar mass $=$ 128.16 g/mol

Find: molecular formula

Solution:
The molecular formula is n times the empirical formula. To find n, divide the molar mass by the empirical formula molar mass.

$$\text{Empirical formula molar mass} = 5(12.01) + 4(1.01)$$
$$= 64.09 \text{ g/mol}$$

$$n = \frac{\text{Molar mass}}{\text{Empirical formula mass}} = \frac{128.16}{64.09} = 2$$

Therefore, the molecular formula is 2 times the empirical formula.

$$\text{Molecular formula} = C_5H_4 \times 2 = C_{10}H_8$$

SKILLBUILDER 6.12 **Calculating Molecular Formula From Empirical Formula and Molar Mass**

Butane is a compound containing carbon and hydrogen that is used as a fuel in butane lighters. Its empirical formula is C_2H_5 and its molar mass is 58.12 g/mol. Find its molecular formula.

SKILLBUILDER PLUS

A compound with the following percent composition has a molar mass of 60.10 g/mol. Find its molecular formula.

C: 39.97% H: 13.41% N: 46.62%

CHAPTER IN REVIEW

Chemical Principles

The Mole Concept: The mole is a convenient number, 6.022×10^{23}, to use when dealing with atoms or molecules. A mole of atoms or molecules composes objects of reasonable size. A mole of any element has a mass equivalent to its atomic mass in grams and a mole of any compound has a mass equivalent to its formula mass in grams. The mass of one mole of an element or compound is called the *molar mass*.

Chemical Formulas and Chemical Composition: Chemical formulas give us the relative number of each kind of element in a compound. These numbers are based on atoms or moles. By using molar masses, we can use the information in a chemical formula to determine the relative masses of each kind of element in a compound. We can then relate the mass of a sample of a compound to the masses of the elements contained in the compound.

Empirical and Molecular Formulas From Laboratory Data: The relative masses of the elements within a compound can be used to determine the empirical formula of the compound. If the chemist also knows the molar mass of the compound, he or she can also determine its molecular formula.

Relevance

The Mole Concept: The mole concept allows us to determine the number of atoms or molecules in a sample from its mass. Just as a hardware-store customer wants to know the number of nails in a certain weight of nails, so we want to know the number of atoms in a certain mass of atoms. Since atoms are too small to count, we use their mass.

Chemical Formulas and Chemical Composition: The chemical composition of compounds is important because it lets us determine how much of a particular element is contained within a particular compound. For example, a patient monitoring sodium intake probably wants to know how much sodium is contained within the amount of sodium chloride (table salt) consumed each day.

Empirical and Molecular Formulas From Laboratory Data: The first thing a chemist wants to know about an unknown compound is its chemical formula, because the formula reveals the compound's composition. The way chemists often get formulas is by analyzing the compounds in the laboratory—either by decomposing them or by synthesizing them—to determine the relative masses of the elements they contain.

Chemical Skills

Converting Between Moles and Number of Atoms (Section 6.3)

To convert between moles and number of atoms, use Avogadro's number, 6.022×10^{23} atoms = 1 mol, as a conversion factor.

Examples

EXAMPLE 6.13 Converting Between Moles and Number of Atoms

Calculate the number of atoms in 4.8 mol of copper.

Given: 4.8 mol Cu

Find: Cu atoms

Conversion Factor:

$$1 \text{ mol Cu} = 6.022 \times 10^{23} \text{ Cu atoms}$$

Solution Map:

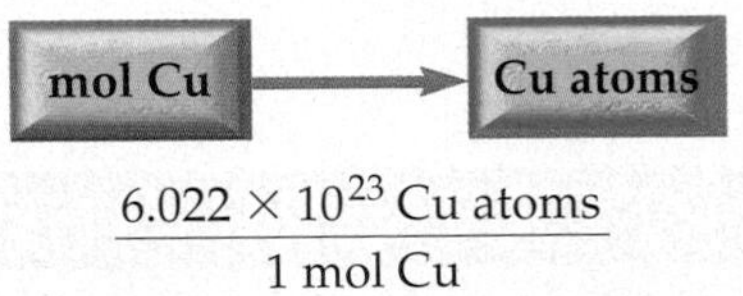

$$\frac{6.022 \times 10^{23} \text{ Cu atoms}}{1 \text{ mol Cu}}$$

Solution:

$$4.8 \cancel{\text{mol Cu}} \times \frac{6.022 \times 10^{23} \text{ Cu atoms}}{1 \cancel{\text{mol Cu}}} = 2.9 \times 10^{24} \text{ Cu atoms}$$

Converting Between Grams and Moles (Section 6.3)

To convert between grams of an element and moles of an element, use that element's molar mass. The molar mass of any element in atomic mass units is its atomic mass with the units of grams per mole. Therefore the molar mass is a conversion factor between grams and moles. To convert between grams of a *compound* and moles of a *compound*, use the compound's molar mass.

EXAMPLE 6.14 Converting Between Grams and Moles

Calculate the mass of aluminum (in grams) of 6.73 moles of aluminum.

Given: 6.73 mol Al

Find: g Al

Conversion Factor:

$$26.98 \text{ g Al} = 1 \text{ mol Al}$$

Solution Map:

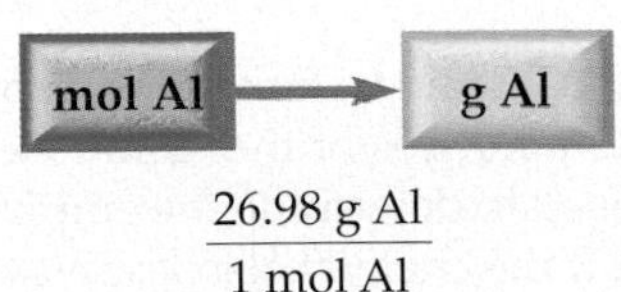

$$\frac{26.98 \text{ g Al}}{1 \text{ mol Al}}$$

Solution:

$$6.73 \cancel{\text{mol Al}} \times \frac{26.98 \text{ g Al}}{1 \cancel{\text{mol Al}}} = 182 \text{ g Al}$$

Converting Between Grams and Number of Atoms or Molecules (Section 6.4)

To convert from grams of an element to the number of atoms, first use the molar mass of the element to convert from grams to moles, and then use Avogadro's number to convert moles to number of atoms.

To convert from the number of atoms of an element to grams of the element, first use Avogadro's number to convert from number of atoms to moles, and then use the molar mass to convert from moles to grams.

For compounds the molar mass is simply the formula mass with the units of grams per mole.

EXAMPLE 6.15 Converting Between Grams and Number of Atoms or Molecules

Determine the number of atoms in a 48.3-g sample of zinc.

Given: 48.3 g Zn

Find: Zn atoms

Conversion Factors:

65.38 g Zn = 1 mol Zn

1 mol = 6.022×10^{23} atoms

Solution Map:

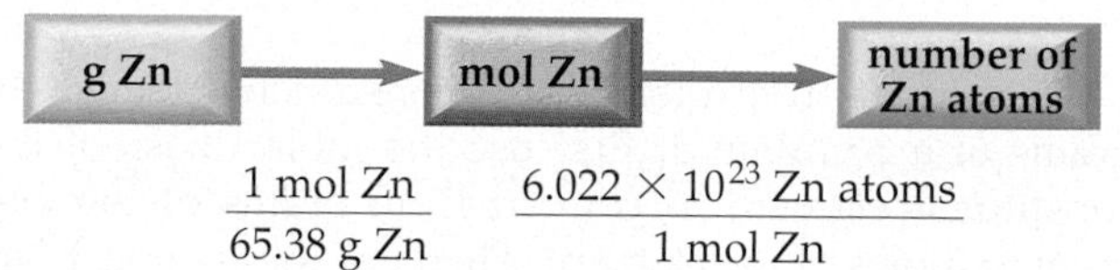

$$\frac{1 \text{ mol Zn}}{65.38 \text{ g Zn}} \qquad \frac{6.022 \times 10^{23} \text{ Zn atoms}}{1 \text{ mol Zn}}$$

Solution:

$$48.3 \text{ g Zn} \times \frac{1 \text{ mol Zn}}{65.38 \text{ g Zn}} \times \frac{6.022 \times 10^{23} \text{ Zn ato}}{1 \text{ mol Zn}}$$

$$= 4.45 \times 10^{23} \text{ Zn atoms}$$

Converting Between Moles of a Compound and Moles of a Constituent Element (Section 6.5)

To convert between moles of a compound and moles of a constituent element, use the chemical formula of the compound as a conversion factor. The subscripts of each element in the formula specify the number of moles of that element that is equivalent to one mole of the compound.

EXAMPLE 6.16 Converting Between Moles of a Compound and Moles of a Constituent Element

Determine the number of moles of oxygen in 7.2 mol of H_2SO_4.

Given: 7.2 mol H_2SO_4

Find: mol O

Conversion Factor:

4 mol O ≡ 1 mol H_2SO_4

Solution Map:

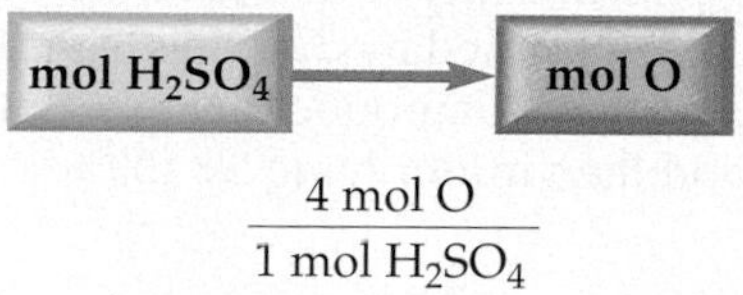

$$\frac{4 \text{ mol O}}{1 \text{ mol } H_2SO_4}$$

Solution:

$$7.2 \text{ mol } H_2SO_4 \times \frac{4 \text{ mole O}}{1 \text{ mol } H_2SO_4} = 28.8 \text{ mol O}$$

Converting Between Grams of a Compound and Grams of a Constituent Element (Section 6.5)

To convert from grams of a compound to grams of a constituent element, first use the molar mass of the compound to convert from grams of the compound to moles of the compound. Then use the chemical formula to obtain a conversion factor to convert from moles of the compound to moles of the constituent element. Finally, use the molar mass of the constituent element to convert from moles of the element to grams of the element.

To convert from grams of a constituent element to grams of a compound, first use the molar mass of the constituent element to convert from grams of the element to moles of the element. Then use the chemical formula to obtain a conversion factor to convert from moles of the constituent element to moles of the compound. Finally, use the molar mass of the compound to convert from moles of the compound to grams of the compound.

EXAMPLE 6.17 Converting Between Grams of a Compound and Grams of a Constituent Element

Find the grams of iron in 79.2 g of Fe_2O_3

Given: 79.2 g Fe_2O_3

Find: g Fe

Conversion Factors:

$$\text{Molar mass } Fe_2O_3 = 2(55.85) + 3(16.00) = 159.70 \text{ g/mol}$$

$$2 \text{ mol Fe} \equiv 1 \text{ mol } Fe_2O_3$$

Solution Map:

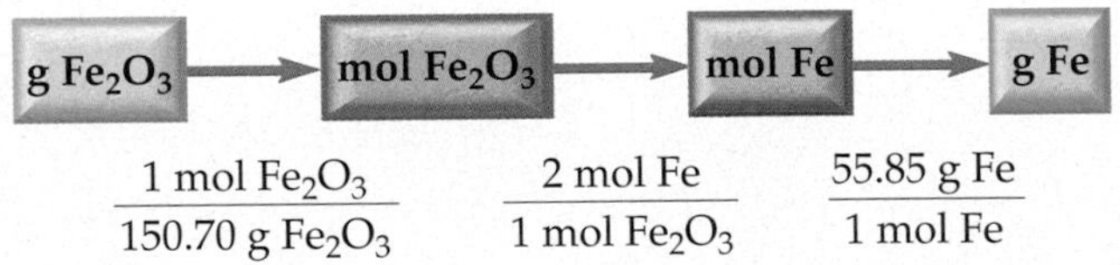

Solution:

$$79.2 \text{ g } Fe_2O_3 \times \frac{1 \text{ mol } Fe_2O_3}{150.7 \text{ g } Fe_2O_3} \times \frac{2 \text{ mol Fe}}{1 \text{ mol } Fe_2O_3} \times \frac{55.85 \text{ g Fe}}{1 \text{ mol Fe}} = 55.4 \text{ g Fe}$$

Calculating Mass Percent Composition From Experimental Data (Section 6.6)

The mass percent composition of an element in a compound can be determined by dividing the mass of the element within the compound by the mass of the compound and then multiplying by 100%.

EXAMPLE 6.18 Calculating Mass Percent Composition From Experimental Data

A 3.52 g sample of chromium reacts with fluorine to produce 7.38 g of the metal fluoride. What is the mass percent composition of chromium in the fluoride?

Given: 3.52 g Cr, 7.38 g metal fluoride

Find: Cr mass percent

Solution:

$$\% \text{ Cr} = \frac{\text{Mass Cr}}{\text{Mass metal fluoride}} \times 100\%$$

$$= \frac{3.52 \text{ g}}{7.38 \text{ g}} \times 100\% = 47.7\%$$

Using Mass Percent Composition as a Conversion Factor (Section 6.6)

The mass percent composition of an element in a compound can be used as conversion factor between grams of the element and grams of the compound. Since percent means per hundred, the mass percent composition of an element in a compound is equal to the number of grams of that element per 100 g of the compound.

EXAMPLE 6.19 Using Mass Percent Composition as a Conversion Factor

Determine the mass of titanium in 57.2 g of titanium(IV) oxide. The mass percent of titanium in titanium(IV) oxide is 59.9%.

Given: 57.2 g TiO_2

Find: g Ti

Conversion Factor:

$$59.9 \text{ g Ti} \equiv 100 \text{ g TiO}_2$$

Solution Map:

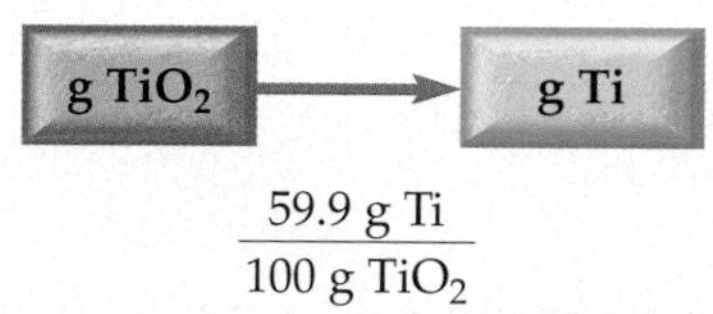

$$\frac{59.9 \text{ g Ti}}{100 \text{ g TiO}_2}$$

Solution:

$$57.2 \cancel{\text{g TiO}_2} \times \frac{59.9 \text{ g Ti}}{100 \cancel{\text{g TiO}_2}} = 34.3 \text{ g Ti}$$

Determining Mass Percent Composition From a Chemical Formula (Section 6.7)

The mass percent of element X in a compound is:

mass% element X

$$= \frac{\text{Mass of element } X \text{ in 1 mol of the compound}}{\text{Mass of 1 mol of the compond}} \times 100\%$$

EXAMPLE 6.20 Determining Mass Percent Composition From a Chemical Formula

Calculate the mass percent composition of potassium in potassium oxide (K_2O).

Given: K_2O

Find: mass percent composition of K.

Conversion Factors:

$$2 \text{ mol K} = 1 \text{ mol K}_2\text{O}$$

$$39.10 \text{ g K} = 1 \text{ mol K}$$

Equation:

$$\text{Mass\% K} = \frac{\text{Mass of K in 1 mol K}_2\text{O}}{\text{Mass of 1 mol K}_2\text{O}} \times 100\%$$

To calculate the mass of the element in one mole of the compound, use the chemical formula to obtain a conversion factor and convert from moles of the compound to moles of the element. Then use the molar mass of the element to convert to mass of the element.

Solution Map: (to find mass K in 1 mol K)

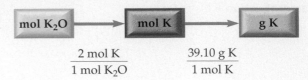

mol K_2O → mol K → g K

$$\frac{2 \text{ mol K}}{1 \text{ mol } K_2O} \qquad \frac{39.10 \text{ g K}}{1 \text{ mol K}}$$

Solution:

$$1 \cancel{\text{mol } K_2O} \times \frac{2 \cancel{\text{mol K}}}{1 \cancel{\text{mol } K_2O}} \times \frac{39.10 \text{ g K}}{1 \cancel{\text{mol K}}} = 78.20 \text{ g K}$$

Finally, substitute the quantities into the mass percent equation and compute the answer.

$$\text{Molar mass } K_2O = 2(39.10) + 16.00 = 94.20 \text{ g/mol}$$

$$\text{Mass\% K} = \frac{78.20 \text{ g K}}{94.20 \text{ g } K_2O} \times 100\% = 83.01\%$$

Determining an Empirical Formula From Experimental Data (Section 6.8)

To determine an empirical formula from experimental data, follow these steps.

EXAMPLE 6.21 Determining an Empirical Formula From Experimental Data

A laboratory analysis of vanillin, the flavoring agent in vanilla, determined the following mass percent composition: C 63.15% H 5.30% O 31.55% Determine the empirical formula of vanillin.

1) Write down (or compute) as **given** the masses of each element present in a sample of the compound. If you are given mass percent composition, assume a 100 g sample and compute the masses of each element from the given percentages.

Given: In a 100 g sample, we have 63.15 g C, 5.30 g H, and 31.55 g O

Find: empirical formula

Solution:

2) Convert each of the masses in Step 1 to moles by using the appropriate molar mass for each element as a conversion factor.

$$63.15 \cancel{\text{g C}} \times \frac{1 \text{ mol C}}{12.01 \cancel{\text{g C}}} = 5.258 \text{ mol C}$$

$$5.30 \cancel{\text{g H}} \times \frac{1 \text{ mol H}}{1.01 \cancel{\text{g H}}} = 5.25 \text{ mol H}$$

$$31.55 \cancel{\text{g O}} \times \frac{1 \text{ mol O}}{16.00 \cancel{\text{g O}}} = 1.972 \text{ mol O}$$

3) Write down a pseudoformula for the compound using the moles of each element (from Step 2) as subscripts.

$$C_{5.258}H_{5.25}O_{1.972}$$

4) Divide all the subscripts in the formula by the smallest subscript.

$$C_{\frac{5.258}{1.972}}H_{\frac{5.25}{1.972}}O_{\frac{1.972}{1.972}} \rightarrow C_{2.67}H_{2.66}O_1$$

5) If the subscripts are not whole numbers, multiply all the subscripts by a small whole number to get whole-number subscripts.

$$C_{2.67}H_{2.66}O_1 \times 3 \rightarrow C_8H_8O_3$$

The correct empirical formula is $C_8H_8O_3$.

Calculating a Molecular Formula From an Empirical Formula and Molar Mass (Section 6.9)

To calculate the molecular formula from the empirical formula and the molar mass, first compute the empirical formula molar mass, which is the sum of the masses of the all the atoms in the empirical formula.

Next, find n, the ratio of the molar mass to empirical mass.

Finally, multiply the empirical formula by n to get the molecular formula.

EXAMPLE 6.22 Calculating a Molecular Formula From an Empirical Formula and Molar Mass

Acetylene, a gas often used in welding torches, has the empirical formula CH and a molar mass of 26.04 amu. Find its molecular formula.

Given: empirical formula = CH

molar mass = 26.04 g/mol

Find: molecular formula

Solution:

Empirical formula molar mass
$$= 12.01 + 1.01$$
$$= 13.02\ \text{g/mol}$$

$$n = \frac{\text{Molar mass}}{\text{Empirical formula molar mass}}$$

$$= \frac{26.04\ \text{g/mol}}{13.02\ \text{g/mol}} = 2$$

$$\text{Molecular formula} = \text{CH} \times 2 \longrightarrow C_2H_2$$

KEY TERMS

Amadeo Avogadro [6.3]
empirical formula [6.8]
molar mass [6.9]
mass percent (composition) [6.6]
molar mass [6.3]
mole (mol) [6.3]
molecular formula [6.8]

EXERCISES

Questions

1. Why is chemical composition important?
2. How can you determine the number of atoms in a sample of an element? Why is counting them not an option?
3. How many atoms are in one mole of atoms?
4. How many molecules are in one mole of molecules?
5. What is the mass of one mole of atoms for an element?
6. What is the mass of one mole of molecules for a compound?
7. What is the mass of one mole of atoms of each of the following elements?
   a) P
   b) Pt
   c) C
   d) Cr
8. What is the mass of one mole of molecules of each of the following compounds?
   a) CO_2
   b) CH_2Cl_2
   c) $C_{12}H_{22}O_{11}$
   d) SO_2
9. The subscripts in a chemical formula give relationships between moles of the constituent elements and moles of the compound. Explain why these subscripts DO NOT give relationships between grams of the constituent elements and grams of the compound.
10. Write the conversion factors between each constituent element and the compound for $C_{12}H_{22}O_{11}$.

11. Mass percent composition can be used as a conversion factor between grams of a constituent element and grams of the compound. Write the conversion factor (including units) inherent in each of the following percent compositions.
a) Water is 11.19% hydrogen by mass.
b) Fructose, also known as fruit sugar, is 53.29% oxygen by mass.
c) Octane, a component of gasoline, is 84.12% carbon by mass.
d) Ethanol, the alcohol in alcoholic beverages, is 52.14% carbon by mass.

12. What is the mathematical formula to compute mass percent composition from a chemical formula?
13. How are the empirical formula and the molecular formula of a compound related?
14. Why is it important to be able to calculate an empirical formula from experimental data?
15. What is the empirical formula mass of a compound?
16. How are the molar mass and empirical formula mass for a compound related?

Problems

The Mole Concept

17. How many mercury atoms are in 5.9 mol of mercury?

18. How many moles of gold atoms do 5.8×10^{24} gold atoms constitute?

19. How many atoms are in each of the following?
a) 3.4 mol Cu
b) 9.7×10^{-3} mol C
c) 22.9 mol Hg
d) 0.215 mol Na

20. How many moles of atoms are in each of the following?
a) 4.6×10^{24} Pb atoms
b) 2.87×10^{22} He atoms
c) 7.91×10^{23} K atoms
d) 4.41×10^{21} Ca atoms

21. Consider the following definitions.

1 doz = 12
1 gross = 144
1 ream = 500
1 mol = 6.022×10^{23}

Suppose you have 872 sheets of paper. How many _____ of paper do you have?
a) dozens
b) gross
c) reams
d) moles

22. A pure copper penny contains approximately 3.0×10^{22} copper atoms. Use the definitions in Problem 21 to determine how many _____ of copper atoms in a penny.
a) dozens
b) gross
c) reams
d) moles

23. How many moles of tin atoms are in a pure tin cup with a mass of 42.3 g?

24. A lead fishing weight contains 0.12 mol of lead atoms. What is its mass?

25. How many moles of atoms are in each of the following?
a) 1.54 g Zn
b) 22.8 g Ar
c) 86.2 g Ta
d) 0.034 g Li

26. What is the mass, in grams, of each of the following?
a) 7.8 mol W
b) 0.943 mol Ba
c) 43.9 mol Xe
d) 1.8 mol S

27. A pure silver ring contains 0.0134 mmol (millimol) Ag. How many silver atoms does it contain?

28. A pure gold ring contains 0.0102 mmol (millimol) Au. How many gold atoms does it contain?

29. How many aluminum atoms are in 3.78 g of aluminum?

30. What is the mass of 4.91×10^{21} platinum atoms?

31. How many atoms are in each of the following?
a) 12.8 g Sr
b) 45.2 g Fe
c) 9.87 g Bi
d) 36.1 g P

32. Calculate the mass, in grams, of each of the following:
a) 7.9×10^{21} uranium atoms
b) 3.82×10^{22} zinc atoms
c) 5.8×10^{23} lead atoms
d) 2.1×10^{24} silicon atoms

33. How many carbon atoms are in a diamond (pure carbon) with a mass of 52 mg?

34. How many helium atoms are in a helium blimp containing 536 kg of helium?

35. How many titanium atoms are in a pure titanium bicycle frame with a mass of 1.28 kg?

36. How many copper atoms are in a pure copper statue with a mass of 133 kg?

37. A mothball, composed of naphthalene ($C_{10}H_8$), has mass of 1.32 g. How many naphthalene molecules does it contain?

38. Calculate the mass (in grams) of a single water molecule.

39. How many molecules are in each of the following?
a) 3.5 g H_2O
b) 56.1 g N_2
c) 89 g CCl_4
d) 19 g $C_6H_{12}O_6$

40. Calculate the mass (in grams) of each of the following:
a) 5.94×10^{20} H_2O_2 molecules
b) 2.8×10^{22} SO_2 molecules
c) 4.5×10^{25} O_3 molecules
d) 9.85×10^{19} CH_4 molecules

41. A sugar crystal contains approximately 1.8×10^{17} sucrose ($C_{12}H_{22}O_{11}$) molecules. What is its mass in milligrams?

42. A salt crystal has a mass of 0.12 mg. How many NaCl formula units does it contain?

Chemical Formulas as Conversion Factors

43. Determine the number of moles of Cl in 4.7 mol $CaCl_2$.

44. How many moles of O are in 14.8 mol $Fe(NO_3)_3$?

45. Which of the following contains the greatest number of moles of O?
a) 2.3 mol H_2O
b) 1.2 mol H_2O_2
c) 0.9 mol $NaNO_3$
d) 0.5 mol $Ca(NO_3)_2$

46. Which of the following contains the greatest number of moles of Cl?
a) 3.8 mol HCl
b) 1.7 mol CH_2Cl_2
c) 4.2 mol $NaClO_3$
d) 2.2 mol $Mg(ClO_4)_2$

47. Determine the number of moles of C in each of the following:
a) 3.8 mol CH_4
b) 0.273 mol C_2H_6
c) 4.89 mol C_4H_{10}
d) 22.9 mol C_8H_{18}

48. Determine the number of moles of hydrogen in each of the following:
a) 3.15 mol H_2O
b) 9.88 mol NH_3
c) 0.0737 mol N_2H_4
d) 54.1 mol $C_{10}H_{22}$

49. How many grams of Cl are in 55 g of each of the following chlorofluorocarbons (CFCs)?
a) CF_2Cl_2
b) $CFCl_3$
c) $C_2F_3Cl_3$
d) CF_3Cl

50. Calculate the number of grams of sodium in 5.0 grams of each of the following sodium-containing food additives.
a) NaCl (table salt)
b) Na_3PO_4 (sodium phosphate)
c) $NaC_7H_5O_2$ (sodium benzoate)
d) $Na_2C_6H_6O_7$ (sodium hydrogen citrate)

51. Iron is found in the earth's crust as several iron compounds often called iron ores. Calculate the mass (in kilograms) of each of the following iron compounds that contain 1.0×10^3 kg of iron.
a) Fe_2O_3 (hematite)
b) Fe_3O_4 (magnetite)
c) $FeCO_3$ (siderite)

52. Lead is often found in the earth's crust as several lead compounds called lead ores. Calculate the mass (in kilograms) of each of the following lead compounds that contain 1.0×10^3 kg of lead.
a) PbS (galena)
b) $PbCO_3$ (cerussite)
c) $PbSO_4$ (anglesite)

Mass Percent Composition

53. A 2.45 g sample of strontium completely reacts with oxygen to form 2.89 g of strontium oxide. Use these data to calculate the mass percent composition of strontium in strontium oxide.

54. A 4.78 g sample of aluminum completely reacts with oxygen to form 6.67 g of aluminum oxide. Use these data to calculate the mass percent composition of aluminum in aluminum oxide.

55. A 1.912 g sample of calcium chloride is decomposed into its constituent elements and found to contain 0.690 g Ca and 1.222 g Cl. Calculate the mass percent composition of Ca and Cl in calcium chloride.

56. A 0.45 g sample of aspirin is decomposed into its constituent elements and found to contain 0.27 g C, 0.020 g H, and 0.16 g O. Calculate the mass percent composition of C, H, and O in aspirin.

57. Copper(II) fluoride contains 37.42% F by mass. Use this percentage to calculate the mass of fluorine (in grams) contained in 28.5 g of copper(II) fluoride.

58. Silver chloride, often used in silver plating, contains 75.27% Ag. Calculate the mass of silver chloride required to make 4.8 g of silver plating.

59. In small amounts, the fluoride ion (often consumed as NaF) prevents tooth decay. According to the American Dental Association, an adult female should consume 3.0 mg of fluorine per day. Calculate the amount of sodium fluoride (45.24% F) that should be consumed to get the recommended amount of fluoride.

60. The iodide ion, usually consumed as potassium iodide, is a dietary mineral essential to good nutrition. In countries where potassium iodide is added to salt, iodine deficiency or goiter has been almost completely eliminated. The recommended daily allowance (RDA) for iodine is $150 \mu g/day$. How much potassium iodide (76.45% I) should be consumed to meet the RDA?

Mass Percent Composition From Chemical Formula

61. Calculate the mass percent composition of nitrogen in each of the following nitrogen compounds.
a) N_2O
b) NO
c) NO_2
d) N_2O_5

62. Calculate the mass percent composition of carbon in each the following carbon compounds.
a) C_2H_2
b) C_3H_6
c) C_2H_6
d) C_2H_6O

63. Calculate the mass percent composition of each element in each of the following compounds.
a) $C_2H_4O_2$
b) CH_2O_2
c) C_3H_9N
d) $C_4H_{12}N_2$

64. Calculate the mass percent composition of each element in each of the following compounds.
a) $FeCl_3$
b) TiO_2
c) H_3PO_4
d) HNO_3

65. Iron ores have different amounts of iron per kilogram of ore. Calculate the mass percent composition of iron for the following iron ores: Fe_2O_3 (hematite), Fe_3O_4 (magnetite), $FeCO_3$ (siderite). Which ore has the highest iron content?

66. Plants need nitrogen to grow, so many fertilizers consist of nitrogen-containing compounds. Calculate the mass percent composition of nitrogen in each of the following nitrogen-containing compounds used as fertilizers: NH_3, $CO(NH_2)_2$, NH_4NO_3, $(NH_4)_2SO_4$. Which fertilizer has the highest nitrogen content?

Calculating Empirical Formulas

67. A nitrogen- and oxygen-containing compound is decomposed in the laboratory and produces 1.78 g nitrogen and 4.05 g oxygen. Calculate the empirical formula of the compound.

68. A selenium- and fluorine-containing compound is decomposed in the laboratory and produces 2.231 g selenium and 3.221 g fluorine. Calculate the empirical formula of the compound.

69. Samples of several compounds are decomposed and the following are the masses of their constituent elements. Calculate the empirical formula for each compound.
a) 1.245 g Ni, 5.381 g I
b) 1.443 g Se, 5.841 g Br
c) 2.128 g Be, 7.557 g S, 15.107 g O

70. Samples of several compounds are decomposed and the following are the masses of their constituent elements. Calculate the empirical formula for each compound.
a) 2.677 g Ba, 3.115 g Br
b) 1.651 g Ag, 0.1224 g O
c) 0.672 g Co, 0.569 g As, 0.486 g O

71. The rotten smell of a decaying animal carcass is partially due to a nitrogen-containing compound called putrescine. Elemental analysis of putrescine showed that it consisted of C 54.50%, H 13.73%, and N 31.77%. Calculate the empirical formula of putrescine.

72. Citric acid, the compound responsible for the sour taste of lemons, has the following elemental composition: C 37.51%, H 4.20%, O 58.29%. Calculate the empirical formula of citric acid.

73. Calculate the empirical formula for each of the following compounds often found in many natural flavors and smells.
a) ethyl butyrate (pineapple oil): C 62.04%, H 10.41%, O 27.55%
b) methyl butyrate (apple flavor): C 58.80%, H 9.87%, O 31.33%
c) benzyl acetate (oil of jasmine): C 71.98%, H 6.71%, O 21.31%

74. Calculate the empirical formula for each of the following over-the-counter pain-relievers.
a) acetaminophen (Tylenol): C 63.56%, H 6.00%, N 9.27%, O 21.17%
b) naproxen (Aleve): C 73.03%, H 6.13%, O 20.84%

75. A 1.45 g sample of phosphorus burns in air and forms 2.57 g of a phosphorus oxide. Calculate the empirical formula of the oxide.

76. A 2.241 g sample of nickel reacts with oxygen to form 2.852 g of the metal oxide. Calculate the empirical formula of the oxide.

77. A 0.77 mg sample of nitrogen reacts with chlorine to form 6.61 mg of the chloride. What is the empirical formula of the nitrogen chloride?

78. A 45.2 mg sample of phosphorus reacts with selenium to form 131.6 mg of the selenide. What is the empirical formula of the phosphorus selenide?

Calculating Molecular Formulas

79. A carbon and hydrogen containing compound has a molar mass of 56.11 g/mol and an empirical formula of CH_2. Find its molecular formula.

80. A phosphorus and oxygen containing compound has a molar mass of 219.9 and an empirical formula of P_2O_3. Find its molecular formula.

81. The following are the molar masses and empirical formulas of several carbon- and chlorine-containing compounds. Find the molecular formula of each compound.
a) 284.77 g/mol, CCl
b) 131.39 g/mol, C_2HCl_3
c) 181.44 g/mol, C_2HCl

82. The following are the molar masses and empirical formulas of several carbon- and nitrogen-containing compounds. Find the molecular formula of each compound.
a) 163.26 g/mol, $C_{11}H_{17}N$
b) 186.24 g/mol, C_6H_7N
c) 312.29 g/mol, C_3H_2N

Cumulative Problems

83. A pure copper cube has an edge length of 1.42 cm. How many copper atoms does it contain? The volume of a cube is its edge length cubed and copper has a density of 8.96 g/cm^3.

84. A pure silver sphere has a radius 0.886 cm. How many silver atoms does it contain? The volume of a sphere is $\frac{4}{3}\pi r^3$ and silver has a density of 10.5 g/cm^3.

85. A drop of water has a volume of approximately 0.05 mL. How many water molecules does it contain? The density of water is 1.0 g/cm^3.

86. Nail polish remover is primarily acetone (C_3H_6O). How many acetone molecules are in a bottle of acetone with a volume of 325 mL? The density of acetone is 0.788 g/cm^3.

87. Determine the chemical formula of each of the following compounds and then use it to calculate the mass percent composition of each constituent element.
a) copper(II) iodide
b) sodium nitrate
c) lead(II) sulfate
d) calcium fluoride

88. Determine the chemical formula of each of the following compounds and then use it to calculate the mass percent composition of each constituent element.
a) nitrogen triiodide
b) xenon tetrafluoride
c) phosphorus trichloride
d) carbon monoxide

89. The rock in a particular iron ore deposit contains 78% Fe_2O_3 by mass. How many kg of the rock must be processed to obtain 1.0×10^3 kg of iron?

90. The rock in a lead ore deposit contains 84% PbS by mass. How many kg of the rock must be processed to obtain 1.0 kg of Pb?

91. A leak in the air conditioning system of a corporate building releases 12 kg of CHF_2Cl per month. If the leak were allowed to continue, how many kg of Cl would be emitted into the atmosphere each year?

92. A leak in an automotive air conditioning system of an older car releases 55 g of CF_2Cl_2 per month. How much Cl is emitted into the atmosphere each year by this car?

93. Hydrogen is a possible future fuel. However, elemental hydrogen is rare, so it must be obtained from a hydrogen-containing compound. One candidate is water. If hydrogen were obtained from water, how much hydrogen, in grams, could be obtained from 1.0 L of water? Assume that the density of water is 1.0 g/cm^3.

94. Hydrogen, a possible future fuel mentioned in Problem 93, can also be obtained from other compounds such as ethanol. Ethanol can be made from the fermentation of crops such as corn. How much hydrogen, in grams, can be obtained from 1.0 kg of ethanol (C_2H_6O)? The density of ethanol is 0.789g/cm^3.

95. Butanedione, a component of butter and body odor, has a cheesy smell. Elemental analysis of butanedione gave the following mass percent composition: C 55.80%, H 7.03%, O 37.17%. The molar mass of butanedione is 86.09 g/mol. Find the molecular formula of butanedione.

96. Caffeine, a stimulant found in coffee and soda pop, has the following mass percent composition: C 49.48%, H 5.19%, N 28.85%, O 16.48%. The molar mass of caffeine is 194.19 g/mol. Find the molecular formula of caffeine.

97. Nicotine, a stimulant found in tobacco, has the following mass percent composition: C 74.03%, H 8.70%, N 17.27%. The molar mass of nicotine is 162.23 g/mol. Find the molecular formula of nicotine.

98. Estradiol is a female sexual hormone that causes maturation and maintenance of the female reproductive system. Elemental analysis of estradiol gave the following mass percent composition: C 79.37%, H 8.88%, O 11.75%. The molar mass of estradiol is 272.37. Find the molecular formula of estradiol.

Highlight Problems

99. You can use the concepts in this chapter to obtain an estimate of the number of atoms in the universe. The following steps will guide you through this calculation.
a) Begin by calculating the number of atoms in the sun. Assume that the sun is pure hydrogen with a density of 1.4 g/cm^3. The radius of the sun is 7×10^8m and the volume of a sphere is given by $V = \frac{4}{3}\pi r^3$.

b) Since the sun is an average sized star, and since stars are believed to compose most of the mass of the visible universe (planets are so small they can be ignored), we can estimate the number of atoms in a galaxy by assuming that every star in the galaxy has the same number of atoms as our sun. The Milky Way galaxy is believed to contain 1×10^{11} stars. Use your answer from part a to calculate the number of atoms in the Milky Way galaxy.

c) The universe is estimated to contain approximately 1×10^9 galaxies. If each of these galaxies contains the same number of atoms as the Milky Way galaxy, what is the total number of atoms in the universe?

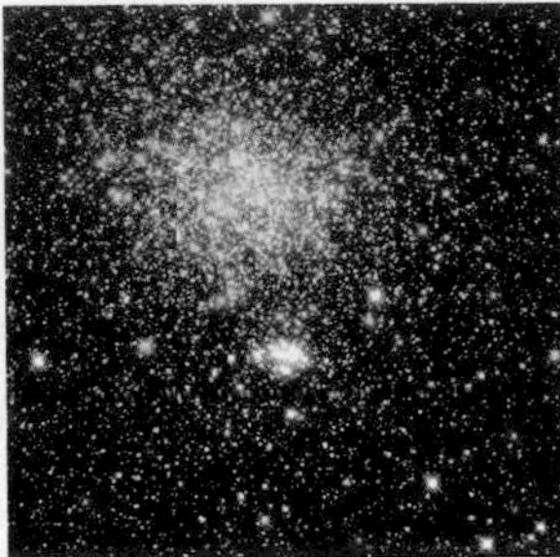

Our sun is one of the 100 billion stars in the Milky Way galaxy. The universe is estimated to contain about 1 billion galaxies.

100. Because of increasing evidence of damage to the ozone layer, chlorofluorocarbon (CFC) production was banned in 1996. However, there are about 100 million auto air conditioners that still use CFC-12 (CF_2Cl_2). These air conditioners are recharged from stockpiled supplies of CFC-12. If each of the 100 million automobiles contains 1.1 kg of CFC-12 and leaks 25% of its CFC-12 into the atmosphere per year, how much Cl, in kilograms, is added to the atmosphere each year due to auto air conditioners? (Assume two significant figures in your calculations.)

101. In 1996, the media reported that possible evidence of life on Mars was found on a meteorite called Allan Hills 84001 (AH 84001). The meteorite was discovered in Antarctica in 1984 and is believed to have originated on Mars. Elemental analysis of substances within its crevices revealed carbon-containing compounds that normally derive only from living organisms. Suppose that one of those compounds had a molar mass of 202.23 g/mol and the following mass percent composition: C 95.02%, H 4.98%. What is the molecular formula for the carbon-containing compound?

EP/TOMS total ozone for Oct. 1, 2001

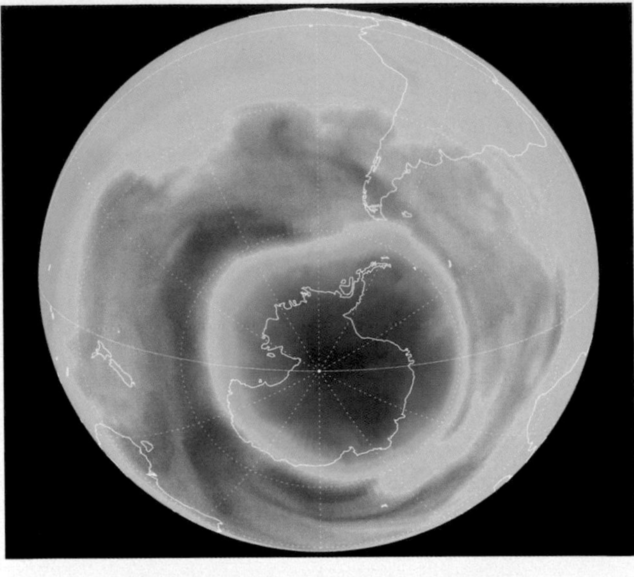

$<$ 60 140 220 300 380 460 $>$

Dobson units

dark gray $<$ 60, red $>$ 460

Source: http://jwocky.gsfc.nasa.gov/

The ozone hole, October 1, 2001.

Allan Hills 84001 meteorite. Elemental analysis of the substances within the crevices of this meteorite revealed carbon-containing compounds that normally originate from living organisms.

Answers to Skillbuilder Exercises

Skillbuilder 6.1 5.32×10^{22} Au atoms
Skillbuilder 6.2 89.1 g
Skillbuilder 6.3 8.17 g He
Skillbuilder 6.4 1.22×10^{23} H_2O molecules
Skillbuilder 6.5 5.6 mol O
Skillbuilder 6.6 3.3 g O
Skillbuilder Plus, p. 176 4.04 g O
Skillbuilder 6.7 8.6 g Na
Skillbuilder 6.8 53.29%
Skillbuilder 6.9 CH_2O
Skillbuilder 6.10 $C_{13}H_{18}O_2$
Skillbuilder 6.11 CuO
Skillbuilder 6.12 C_4H_{10}
Skillbuilder Plus, p. 187 $C_2H_8N_2$

Chemical Reactions

7

"Science, at bottom, is really anti-intellectual. It always distrusts pure reason and demands the production of objective fact."

Henry Louis Mencken 1880–1956

7.1 Kindergarten Volcanoes, Automobiles, and Laundry Detergents

Did you ever make a clay volcano in kindergarten that—when filled with vinegar, baking soda, and red food coloring for effect—erupted? Have you pushed the gas pedal of a car and felt the acceleration as the car moved forward? Have you wondered why laundry detergents work better than normal soap to clean your clothes? Each of these processes depends on a **chemical reaction**—the change of one or more substances into different substances—to occur.

In the classic kindergarten volcano, the baking soda (which is sodium bicarbonate) reacts with acetic acid in the vinegar to form carbon dioxide gas, water, and sodium acetate. The newly formed carbon dioxide gas bubbles out of the mixture, causing the eruption. Reactions that occur in liquids and form a gas are called **gas evolution reactions**. A similar reaction causes the fizzing in antacids such as Alka-Seltzer.

Hydrocarbons are covered in detail in Chapter 18.

◀ In the space shuttle's main engines, hydrogen and oxygen (which are stored in the central fuel tank) react to form water. The reaction emits the energy that helps propel the shuttle into space.

When you drive your car, hydrocarbons such as octane (in gasoline) react with oxygen from the air to form carbon dioxide gas and water (Figure 7.1). This reaction produces heat, which is used to expand the air in the car's cylinders and accelerate the car forward. Reactions such as this one—

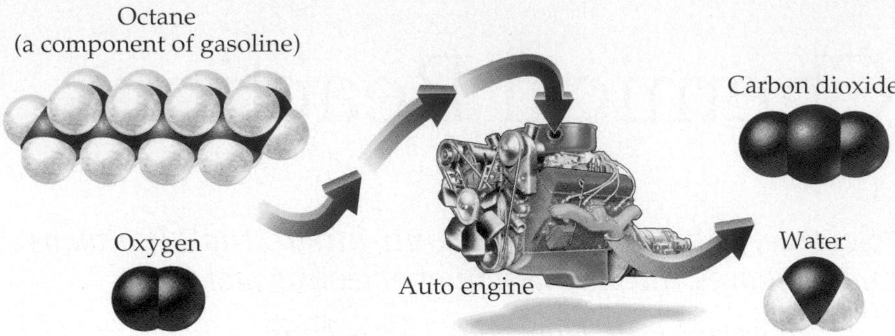

Figure 7.1 In an automobile engine hydrocarbons, such as octane (C_8H_{18}) from gasoline, combine with oxygen from the air and react to form carbon dioxide and water.

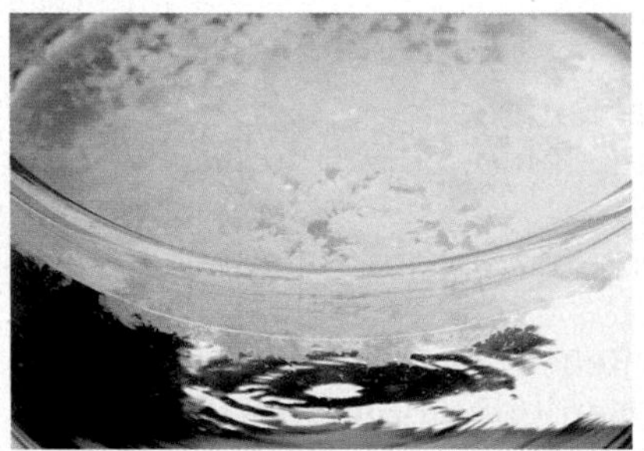

Figure 7.2 Hard water ions react with soap to form curd, a gray, slimy substance that adheres to clothes.

To dissociate means to separate.

in which a substance reacts with oxygen, emitting heat, and forming one or more oxygen containing compounds—are called **combustion reactions**. Combustion reactions are a subcategory of **oxidation–reduction reactions**, in which electrons are transferred from one substance to another. The formation of rust and the dulling of automobile paint are other examples of oxidation–reduction reactions.

One reason that laundry detergents work better than soap to wash clothes is that they contain substances that soften hard water. Hard water contains dissolved calcium (Ca^{2+}) and magnesium (Mg^{2+}) ions. These ions interfere with soap action by reacting with it to form a gray, slimy substance called *curd* (Figure 7.2). If you have ever washed your clothes in ordinary soap, you may have noticed the curd as a gray residue on your clothes.

Laundry detergents inhibit curd formation because they contain substances such as sodium carbonate (Na_2CO_3) that remove calcium and magnesium ions from the water. When sodium carbonate dissolves in water, it dissociates into sodium ions (Na^+) and carbonate ions (CO_3^{2-}). The dissolved carbonate ions then react with calcium and magnesium ions in the hard water to form solid calcium carbonate ($CaCO_3$) and solid magnesium carbonate ($MgCO_3$). These solids simply settle to the bottom of the laundry mixture, resulting in the removal of the ions from the water. In other words, laundry detergents contain substances that react with the ions in hard water to immobilize them. Reactions such as these—that form solid substances in water—are called **precipitation reactions**. Precipitation reactions are also used to remove dissolved toxic metals in industrial wastes.

Chemical reactions are all around us and even inside us. They are involved in many of the products we use daily and in many of our experiences. Chemical reactions can be relatively simple, such as the combination of hydrogen and oxygen to form water, or they can be complex, such as the synthesis of a protein molecule from thousands of simpler molecules. In some cases, chemical reactions are not noticeable to the naked eye, such as the neutralization reaction that occurs in a swimming pool when acid is added to adjust the water's acidity level. In other cases, chemical reactions are obvious, such as the combustion reaction that produces a pillar of smoke and fire under the space shuttle during lift-off. In all cases, however, chemical reactions produce changes in the arrangements of the molecules and atoms that compose matter. Many times, these molecular changes cause macroscopic changes that we experience.

7.2 Evidence of a Chemical Reaction

A child's temperature sensitive spoon changes color upon warming due to a reaction induced by the higher temperature.

If we could see the atoms and molecules that compose matter, we could easily identify a chemical reaction. Are atoms combining with other atoms to form compounds? Are new molecules forming? Are the original molecules decomposing? Are atoms in one molecule changing places with atoms in another? If one or more of these are happening, a chemical reaction is occurring. Of course, we don't normally see atoms and molecules, so we need other ways to identify a chemical reaction.

Fortunately, many chemical reactions produce easily detectable changes when they occur. For example, when the color-causing molecules in a brightly colored shirt decompose with repeated exposure to sunlight, the color of the shirt fades. Similarly, when the molecules imbedded in the plastic of a child's temperature sensitive spoon transform upon warming, the color of the spoon changes. These **color changes** are evidence that a chemical reaction has occurred.

Other changes that identify chemical reactions include the **formation of a solid** in a previously clear solution (Figure 7.3) or the **formation of a gas** (Figure 7.4). Dropping Alka-Seltzer tablets into water or combining baking soda and vinegar (as in our opening example of the kindergarten volcano), are both good examples of chemical reactions that produce a gas—the gas is visible as bubbles in the liquid.

Heat absorption or **emission**, as well as **light emission**, is also evidence for reactions. For example, a natural gas flame produces heat and light. A chemical cold pack becomes cold when the plastic barrier separating two substances is broken. Both of these changes suggest that a chemical reaction is occurring.

Figure 7.3 The formation of a solid in a previously clear solution is evidence of a chemical reaction.

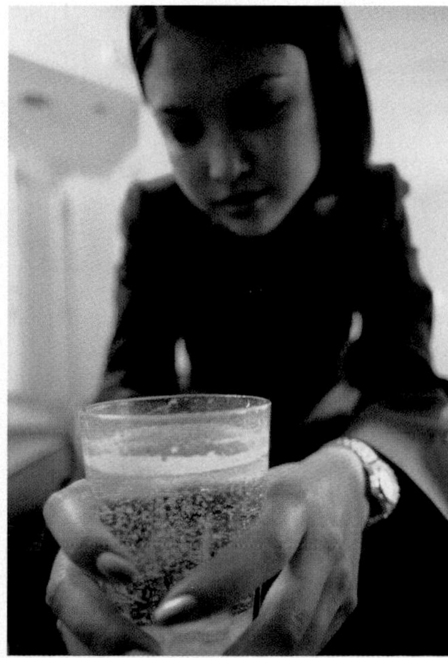

Figure 7.4 The formation of a gas is evidence of a chemical reaction.

In summary, each of the following provides **evidence of a chemical reaction.**

- *a* **color change**

- *the* **formation of a solid** *in a previously clear solution*

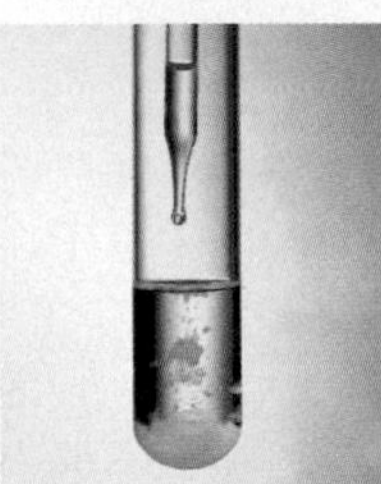

- *the* **formation of a gas** *when you add a substance to a solution*

- *the* **emission of light**

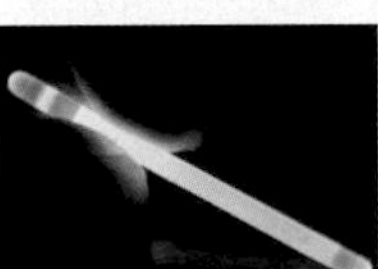

- *the* **emission** *or* **absorption of heat**

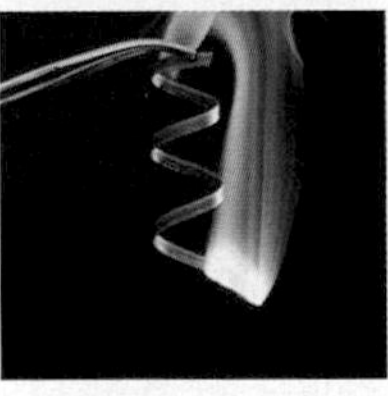

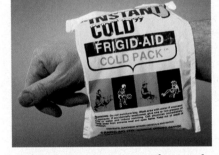

A change in temperature due to absorption or emission of heat is evidence of a chemical reaction. This chemical cold pack becomes cold when the barrier separating two substances is broken.

While these changes provide evidence of a chemical reaction, they are not *definitive* evidence. Only chemical analysis showing that the initial substances have changed into other substances conclusively proves that a chemical reaction has occurred. It is possible to be fooled. For example, when water boils, bubbles form, but no chemical reaction has occurred. Boiling water forms gaseous steam, but both water and steam are composed of water molecules—no chemical change has occurred (Figure 7.5). On the other hand, chemical reactions may occur without any obvious signs, yet chemical analysis may show that a reaction has indeed occurred. It is the changes occurring at the atomic and molecular level that determine if a chemical reaction has occurred. Fortunately, much of the time these molecular changes also result in changes that we can perceive.

Figure 7.5 When water boils, bubbles are formed and a gas is evolved. However, no chemical change has occurred because the gas, like the liquid water, is also composed of water molecules.

EXAMPLE 7.1 Evidence of a Chemical Reaction

Which of the following is a chemical reaction? Why?

a) Ice melting upon warming
b) Water forming hydrogen and oxygen gas bubbles as an electric current is passed through it
c) Iron rusting
d) Bubbles forming when a soda can is opened

Solution:

a) Not a chemical reaction; melting ice forms water, but both the ice and water are composed of water molecules.
b) Chemical reaction; water decomposes into hydrogen and oxygen as evidenced by the bubbling.
c) Chemical reaction; iron changes into iron oxide, changing color in the process.
d) Not a chemical reaction; even though there is bubbling, it is just carbon dioxide coming out of the liquid.

SKILLBUILDER 7.1 Evidence of a Chemical Reaction

Which of the following is a chemical reaction? Why?

a) Butane burning in a butane lighter
b) Butane evaporating out of a butane lighter
c) Wood burning
d) Dry ice evaporating

7.3 The Chemical Equation

TABLE 7.1

Abbreviations Indicating the States of Reactants and Products in Chemical Equations

Abbreviation	*State*
(g)	gas
(l)	liquid
(s)	solid
(aq)	aqueous (water solution)*

*The (aq) designation stands for *aqueous*, which means a substance dissolved in water. When a substance dissolves in water, the mixture is called a *solution* (See Section 7.5).

Chemical reactions are represented with **chemical equations**. For example, the reaction occurring in a natural gas flame, such as the flame on your kitchen stove, is methane (CH_4) reacting with oxygen (O_2) to form carbon dioxide (CO_2) and water (H_2O). This reaction is represented by the following equation.

$$\underbrace{CH_4 + O_2}_{\text{reactants}} \rightarrow \underbrace{CO_2 + H_2O}_{\text{products}}$$

The substances on the left side of the equation are called the **reactants** and the substances on the right side are called the **products**. We often specify the states of each reactant or product in parentheses next to the formula. If we add these to our equation, it becomes:

$$CH_4(g) + O_2(g) \rightarrow CO_2(g) + H_2O(g)$$

The (g) indicates that these substances are gases in the reaction. The common states of reactants and products and their symbols used in chemical reactions are summarized in Table 7.1.

Let's look more closely at our equation for the burning of natural gas. How many oxygen atoms are on each side of the equation?

$$CH_4(g) + O_2(g) \longrightarrow CO_2(g) + H_2O(l)$$

2 O atoms — 2 O atoms + 1 O atom = 3 O atoms

There are two oxygen atoms on the left side and three on the right. Where did this additional oxygen atom on the right side come from? Since chemical equations represent real chemical reactions, atoms cannot simply appear or disappear because, as we know, atoms don't simply appear or disappear in nature. Notice also that there are four hydrogen atoms on the left and only two on the right.

In chemical equations, atoms cannot change from one type to another—hydrogen atoms cannot change into oxygen atoms.

$$CH_4(g) + O_2(g) \longrightarrow CO_2(g) + H_2O(l)$$

2 O atoms — 2 O atoms + 1 O atom = 3 O atoms

To correct these problems, we must **balance** the equation; that is, we must add coefficients—not subscripts—to ensure that the number of each type of atom on both sides of the equation is equal. New atoms do not form during a reaction, nor do atoms vanish—matter must be conserved.

We balance chemical equations by adding coefficients as needed to the reactants and products. This changes the number of molecules in the equation, but it does not change the *kind* of molecules. To balance the preceding equation, for example, we put the coefficient 2 before O_2 in the reactants, and the coefficient 2 before H_2O in the products.

$$CH_4(g) + \mathbf{2}O_2(g) \rightarrow CO_2(g) + \mathbf{2}H_2O(g)$$

The equation is now balanced because the numbers of each type of atom on either side of the equation are equal. We can verify this by summing the number of each type of atom.

The number of a particular type of atom within a chemical formula embedded in an equation is obtained by multiplying the subscript for the atom by the coefficient for the chemical formula.

If there is no coefficient or subscript, a 1 is implied. So, for the combustion of natural gas we have:

$$CH_4(g) + 2O_2(g) \rightarrow CO_2(g) + 2H_2O(g)$$

Reactants	*Products*
1 C atom ($1 \times \underline{C}H_4$)	1 C atom ($1 \times \underline{C}O_2$)
4 H atoms ($1 \times C\underline{H_4}$)	4 H atoms ($2 \times \underline{H_2}O$)
4 O atoms ($2 \times \underline{O_2}$)	4 O atoms ($1 \times C\underline{O_2} + 2 \times H_2\underline{O}$)

The numbers of each type of atom on both sides of the equation are equal—the equation is balanced.

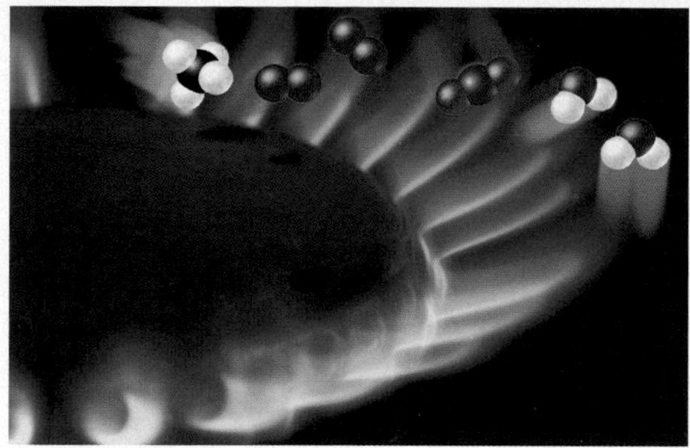

$CH_4(g) + 2\,O_2(g) \longrightarrow CO_2(g) + 2\,H_2O(g)$

A balanced chemical equation represents a chemical reaction. In this photograph, methane molecules combine with oxygen to form carbon dioxide and water.

7.4 How to Write Balanced Chemical Equations

The following procedure box details the steps for writing balanced chemical equations. As in other procedures, we show the steps in the left column and examples of applying each step in the center and right columns. Remember, change only the *coefficients* to balance a chemical equation, *never the subscripts*.

Writing Balanced Chemical Equations	**EXAMPLE 7.2** Write a balanced equation for the reaction between solid silicon dioxide and solid carbon to produce solid silicon carbide and carbon monoxide gas.	**EXAMPLE 7.3** Write a balanced equation for the combustion of liquid octane (C_8H_{18}), a component of gasoline, in which it combines with gaseous oxygen to form gaseous carbon dioxide and gaseous water.
1. Write a skeletal equation by writing chemical formulas for each of the reactants and products. Review Chapter 5 for nomenclature rules. (If a skeletal equation is provided, go to Step 2).	**Solution:** $SiO_2(s) + C(s) \rightarrow SiC(s) + CO(g)$	**Solution:** $C_8H_{18}(l) + O_2(g) \rightarrow CO_2(g) + H_2O(g)$
2. If an element occurs in only one compound on both sides of the equation, balance it first. If there is more than one such element, balance metals before nonmetals.	**Begin with Si:** $SiO_2(s) + C(s) \rightarrow SiC(s) + CO(g)$ **1 Si atom → 1 Si atom** Si is already balanced. **Balance O next:** $SiO_2(s) + C(s) \rightarrow SiC(s) + CO(g)$ **2 O atoms → 1 O atom**	**Begin with C:** $C_8H_{18}(l) + O_2(g) \rightarrow CO_2(g) + H_2O(g)$ **8 C atoms → 1 C atom** To balance C, put an 8 before $CO_2(g)$. $C_8H_{18}(l) + O_2(g) \rightarrow 8CO_2(g) + H_2O(g)$ **8 C atoms → 8 C atoms**

To balance O, put a 2 before $CO(g)$.

$$SiO_2(s) + C(s) \rightarrow SiC(s) + \mathbf{2}CO(g)$$

2 O atoms → 2 O atoms

Balance H next:

$$C_8H_{18}(l) + O_2(g) \rightarrow 8\,CO_2(g) + H_2O(g)$$

18 H atoms → 2 H atoms

To balance H, put a 9 before $H_2O(g)$.

$$C_8H_{18}(l) + O_2(g) \rightarrow 8\,CO_2(g) + \mathbf{9}H_2O(g)$$

18 H atoms → 18 H atoms

3. If an element occurs as a free element on either side of the chemical equation, balance it last. Always balance free elements by adjusting the coefficient *on the free element*.

Balance C:

$$SiO_2(s) + C(s) \rightarrow SiC(s) + 2\,CO(g)$$

1 C atom → 1 C + 2 C = **3 C atoms**

To balance C, put a 3 before $C(s)$.

$$SiO_2(s) + \mathbf{3}\,C(s) \rightarrow SiC(s) + 2\,CO(g)$$

3 C atoms → 1 C + 2 C = **3 C atoms**

Balance O:

$$C_8H_{18}(l) + O_2(g) \rightarrow 8\,CO_2(g) + 9\,H_2O(g)$$

2 O atoms → 16 O + 9 O = **25 O atoms**

To balance O, put a $\frac{25}{2}$ before $O_2(g)$.

$$C_8H_{18}(l) + \tfrac{25}{2}\,O_2(g) \rightarrow 8\,CO_2(g) + 9\,H_2O(g)$$

25 O atoms → 16 O + 9 O = **25 O atoms**

4. If the balanced equation contains coefficient fractions, clear these by multiplying the entire equation by the appropriate factor.

This step is not necessary in this example. Proceed to Step 5.

$$[C_8H_{18}(l) + \tfrac{25}{2}\,O_2(g) \rightarrow 8\,CO_2(g) + 9\,H_2O(g)] \times 2$$

$$2\,C_8H_{18}(l) + 25\,O_2(g) \rightarrow 16\,CO_2(g) + 18\,H_2O(g)$$

5. Check to make certain the equation is balanced by summing the total number of each type of atom on both sides of the equation.

$$SiO_2(s) + 3\,C(s) \rightarrow SiC(s) + 2\,CO(g)$$

Reactants		*Products*
1 Si atom	→	1 Si atom
2 O atoms	→	2 O atoms
3 C atoms	→	3 C atoms

The equation is balanced.

$$2\,C_8H_{18}(l) + 25\,O_2(g) \rightarrow 16\,CO_2(g) + 18\,H_2O(g)$$

Reactants		*Products*
16 C atoms	→	16 C atoms
36 H atoms	→	36 H atoms
50 O atoms	→	50 O atoms

The equation is balanced.

SKILLBUILDER 7.2

Write a balanced equation for the reaction between solid chromium(III) oxide and solid carbon to produce solid chromium and dioxide gas.

SKILLBUILDER 7.3

Write a balanced equation for the combustion of C_4H_{10} in which it combines with gaseous oxygen to form gaseous carbon dioxide and gaseous water.

EXAMPLE 7.4 Balancing Chemical Equations

Write a balanced equation for the reaction of solid aluminum with aqueous sulfuric acid to form aqueous aluminum sulfate and hydrogen gas.

Solution:

We first use our knowledge of chemical nomenclature from Chapter 5 to write a skeletal equation containing formulas for each of the reactants and products.

$$Al(s) + H_2SO_4(aq) \rightarrow Al_2(SO_4)_3(aq) + H_2(g)$$

The formulas for each compound MUST BE CORRECT before we begin to balance the equation. Since both aluminum and hydrogen occur as pure elements, we will balance those last. Sulfur and oxygen occur in only one compound on both sides of the equation, so we balance these first. Sulfur and oxygen are also part of a polyatomic ion that stays intact on both sides of the equation. *Balance polyatomic ions such as these as a group*. There are 3 SO_4^{2-} ions on the right hand side of the equation, so we need 3 on the left. Place a 3 in front of H_2SO_4 to balance the SO_4^{2-} ion.

$$Al(s) + \mathbf{3}H_2SO_4(aq) \rightarrow Al_2(SO_4)_3(aq) + H_2(g)$$

We balance Al next. Since there are 2 Al atoms on the right side of the equation, we place a 2 in front of Al on the left side of the equation.

$$\mathbf{2}Al(s) + 3\,H_2SO_4(aq) \rightarrow Al_2(SO_4)_3(aq) + H_2(g)$$

We balance H next. Since there are 6 H atoms on the left side, we place a 3 in front of $H_2(g)$ on the right side.

$$2\,Al(s) + 3\,H_2SO_4(aq) \rightarrow Al_2(SO_4)_3(aq) + \mathbf{3}\,H_2(g)$$

Finally we sum the number of atoms on each side to make sure that the equation is balanced.

$$2\,Al(s) + 3\,H_2SO_4(aq) \rightarrow Al_2(SO_4)_3(aq) + 3\,H_2(g)$$

Reactants		*Products*
2 Al atoms	⟶	2 Al atoms
6 H atoms	⟶	6 H atoms
3 S atoms	⟶	3 S atoms
12 O atoms	⟶	12 O atoms

The equation is balanced.

SKILLBUILDER 7.4 Balancing Chemical Equations

Write a balanced equation for the reaction of aqueous lead(II) acetate with aqueous potassium iodide to form solid lead(II) iodide and aqueous potassium acetate.

EXAMPLE 7.5 Balancing Chemical Equations

Balance the following chemical equation.

$$Al(s) + HCl(aq) \rightarrow AlCl_3(aq) + H_2(g)$$

Solution:

Since Cl occurs in only one compound on each side of the equation, we balance it first. There is 1 Cl atom on the left hand side of the equation and

3 Cl atoms on the right hand side. We balance Cl by placing a 3 in front of HCl.

$$Al(s) + 3HCl(aq) \rightarrow AlCl_3(aq) + H_2(g)$$

Since H and Al occur as free elements, we balance them last. There is 1 Al atom on the left side of the equation and 1 Al atom on the right, so Al is balanced. There are 3 H atoms on the left and 2 H atoms on the right. We balance H by adjusting the coefficient on H_2 (that way we don't alter other elements that are already balanced). Balance H by placing a $\frac{3}{2}$ in front of H_2.

$$Al(s) + 3\,HCl(aq) \rightarrow AlCl_3(aq) + \frac{3}{2}H_2(g)$$

Since the equation contains a coefficient fraction we clear it by multiplying the entire equation by 2.

$$[Al(s) + 3\,HCl(aq) \rightarrow AlCl_3(aq) + \frac{3}{2}H_2(g)] \times 2$$

$$2\,Al(s) + 6\,HCl(aq) \rightarrow 2\,AlCl_3(aq) + 3\,H_2(g)$$

Finally, we sum the number of atoms on each side to check that the equation is balanced.

$$2\,Al(s) + 6\,HCl(aq) \rightarrow 2\,AlCl_3(aq) + 3\,H_2(g)$$

Reactants		*Products*
2 Al atoms	⟶	2 Al atoms
6 Cl atoms	⟶	6 Cl atoms
6 H atoms	⟶	6 H atoms

The equation is balanced.

SKILLBUILDER 7.5 **Balancing Chemical Equations**

Balance the following chemical equation.

$$HCl(g) + O_2(g) \rightarrow H_2O(l) + Cl_2(g)$$

7.5 Aqueous Solutions and Solubility: Compounds Dissolved in Water

NaCl solutions are often called *saline solutions* or *brine*.

Reactions occurring in aqueous solution are among the most common and important. An **aqueous solution** is a mixture of a substance with water. For example, sodium chloride (NaCl) solutions, common in both the oceans and in living cells, are composed of sodium chloride dissolved in water. You may have formed a NaCl solution yourself by adding table salt to water. As you stir the NaCl into the water, it seems to disappear. However, you know the NaCl is still there because you can taste its saltiness in the water. How does NaCl dissolve in water?

When ionic compounds such as NaCl dissolve in water they usually dissociate into their component ions. A NaCl solution, represented as NaCl(*aq*), does not contain any NaCl units, but rather dissolved Na^+ ions and Cl^- ions.

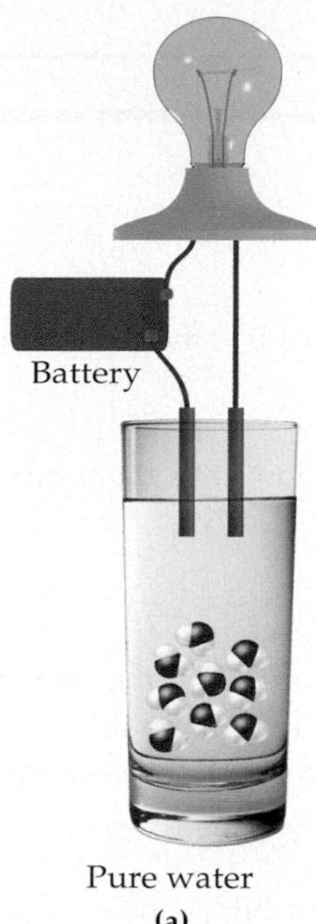

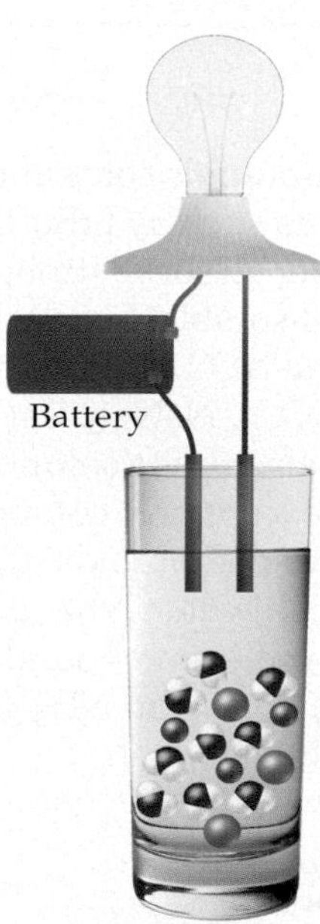

Figure 7.6 **a)** Pure water will not conduct electricity. **b)** Ions in a sodium chloride solution conduct electricity, causing the bulb to light. Solutions such as these are called strong electrolyte solutions.

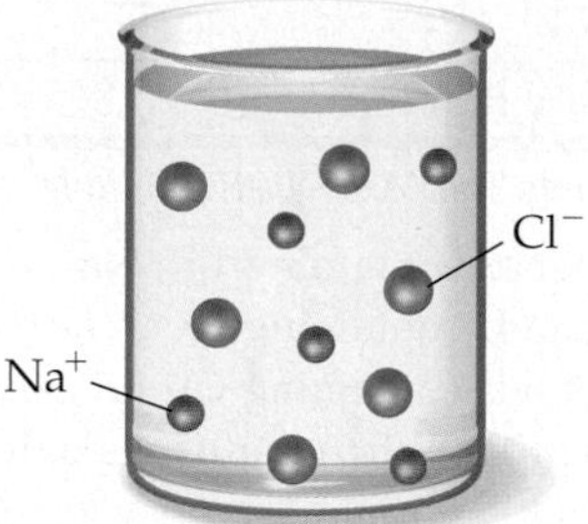

A NaCl solution contains independent Na^+ and Cl^- ions.

We know that NaCl is present as independent sodium and chloride ions in solution because sodium chloride solutions conduct electricity, which requires the presence of freely moving charged particles. Solutions that contain dissolved ionic compounds, such as NaCl, are called **strong electrolyte solutions** (Figure 7.6). Similarly, a $AgNO_3$ solution, represented as $AgNO_3(aq)$, does not contain any $AgNO_3$ units, but rather dissolved Ag^+ ions and NO_3^- ions.

A $AgNO_3$ solution contains independent Ag^+ and NO_3^- ions.

When compounds containing polyatomic ions such as NO_3^- dissolve, the polyatomic ions usually dissolve as intact units.

Not all ionic compounds, however, dissolve in water. AgCl, for example, does not dissolve in water. If we add AgCl to water, it remains as solid AgCl and appears as a white powder at the bottom of the water.

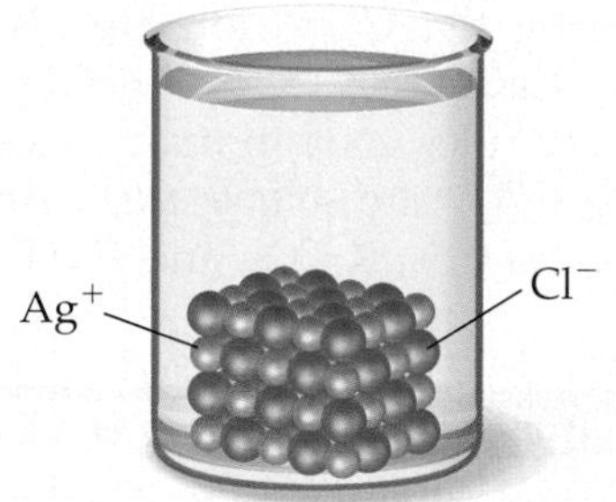

When AgCl is added to water, it remains as solid AgCl—it does not dissolve into independent ions.

Solubility

A compound is **soluble** if it dissolves in water. A compound is **insoluble** if it does not dissolve in water. NaCl, for example, is soluble. If we mix solid sodium chloride into water, it dissolves and forms a strong electrolyte solution. AgCl, on the other hand, is insoluble. If we mix solid silver chloride into water, it remains as a solid within the liquid water.

There is no easy way to tell if a particular compound will be soluble or insoluble. For ionic compounds, however, there are empirical rules that have been deduced from observations on many compounds. These are called **solubility rules** and are summarized in Table 7.2.

TABLE 7.2

Solubility Rules

Compounds That Are Mostly Soluble

Compounds containing Li^+, Na^+, K^+ or NH_4^+
Compounds containing NO_3^- and $C_2H_3O_2^-$
Compounds containing Cl^-,Br^-, and I^-
(Exception: When these ions pair with Ag^+, Hg_2^{2+} or Pb^{2+}, the compounds are insoluble.)
Compounds containing SO_4^{2-}
(Exception: When SO_4^{2-} pairs with Sr^{2+}, Ba^{2+}, Pb^{2+}, and Ca^{2+}, the compounds are insoluble.)

Compounds That Are Mostly Insoluble

Compounds containing OH^- and S^{2-}
(Exceptions: When these ions pair with Li^+, Na^+, K^+ or NH_4^+, the compounds are soluble. When S^{2-} pairs with Ca^{2+}, Sr^{2+}, or Ba^{2+} the compounds are soluble. When OH^- pairs with Ca^{2+}, Sr^{2+}, Ba^{2+} the compounds are slightly soluble*.)
Compounds containing CO_3^{2-} and PO_4^{3-}
(Exception: When these ions pair with Li^+, Na^+, K^+ or NH_4^+, the compounds are soluble.)

*For many purposes, these can be considered insoluble.

For example, the solubility rules state that compounds containing the lithium ion are *soluble*. That means that compounds such as LiBr, $LiNO_3$, Li_2SO_4, LiOH, and Li_2CO_3 will all dissolve in water to form strong electrolyte solutions. If a compound contains Li^+, it is soluble. Similarly, the solubility rules state that compounds containing the NO_3^- ion are soluble. That means that compounds such as $AgNO_3$, $Pb(NO_3)_2$, $NaNO_3$, $Ca(NO_3)_2$ and $Sr(NO_3)_2$ all dissolve in water to form strong electrolyte solutions.

The solubility rules also state that, with some exceptions, compounds containing the CO_3^{2-} ion are *insoluble*. Therefore, compounds such as $CuCO_3$, $CaCO_3$, $SrCO_3$, and $FeCO_3$ do not dissolve in water. Note that the solubility rules contain many exceptions. For example, compounds containing CO_3^{2-} are *soluble when paired with* Li^+, Na^+, K^+, or NH_4^+. Thus Li_2CO_3, Na_2CO_3, K_2CO_3, and $(NH_4)_2CO_3$ are all soluble.

EXAMPLE 7.6 Determining if a Compound Is Soluble

Determine whether each of the following compounds is soluble or insoluble.

a) AgBr
b) $CaCl_2$
c) $Pb(NO_3)_2$
d) $PbSO_4$

Solution:

a) Insoluble; compounds containing Br^- are normally soluble, but Ag^+ is an exception.
b) Soluble; compounds containing Cl^- are normally soluble and Ca^{2+} is not an exception.

c) Soluble; compounds containing NO_3^- are always soluble.
d) Insoluble; compounds containing SO_4^{2-} are normally soluble, but Pb^{2+} is an exception.

SKILLBUILDER 7.6 **Determining if a Compound Is Soluble**

Determine whether each of the following compounds is soluble or insoluble.

a) CuS
b) $FeSO_4$
c) $PbCO_3$
d) NH_4Cl

7.6 Precipitation Reactions: Reactions in Aqueous Solution That Form a Solid

In our opening example, we learned how sodium carbonate is added to laundry detergent to react with dissolved Mg^{2+} and Ca^{2+} ions to form solids that precipitate (or come out of) solution. These reactions are examples of **precipitation reactions**, reactions that form a solid or **precipitate** upon mixing two aqueous solutions.

Precipitation reactions are common in chemistry. Potassium iodide and lead nitrate, for example, both form colorless, strong electrolyte solutions when dissolved in water (see the solubility rules). When the two solutions are combined, however, a brilliant yellow precipitate forms (Figure 7.7). This precipitation reaction can be described with the following chemical equation.

$$2\ KI(aq) + Pb(NO_3)_2(aq) \rightarrow PbI_2(s) + 2\ KNO_3(aq)$$

Precipitation reactions do not always occur when mixing two aqueous solutions. For example, if solutions of KI(*aq*) and NaCl(*aq*) are combined, nothing happens (Figure 7.8).

$$KI(aq) + NaCl(aq) \rightarrow \text{NO REACTION}$$

Figure 7.7 When a potassium iodide solution is mixed with a lead(II) nitrate solution, a brilliant yellow precipitate of $PbI_2(s)$ forms.

Figure 7.8 When a potassium iodide solution is mixed with a sodium chloride solution, no reaction occurs.

Predicting Precipitation Reactions

The key to predicting precipitation reactions is to understand that *only insoluble compounds form precipitates.* In a precipitation reaction, two solutions containing soluble compounds combine and an insoluble compound precipitates. For example, consider the precipitation reaction from Figure 7.7.

$$\underset{\text{soluble}}{2\,KI(aq)} + \underset{\text{soluble}}{Pb(NO_3)_2(aq)} \rightarrow \underset{\text{insoluble}}{PbI_2(s)} + \underset{\text{soluble}}{2\,KNO_3(aq)}$$

KI and $Pb(NO_3)_2$ are both soluble, but the precipitate, PbI_2, is *insoluble.* Before mixing, KI(*aq*) and $Pb(NO_3)_2(aq)$ are both dissociated in their respective solutions.

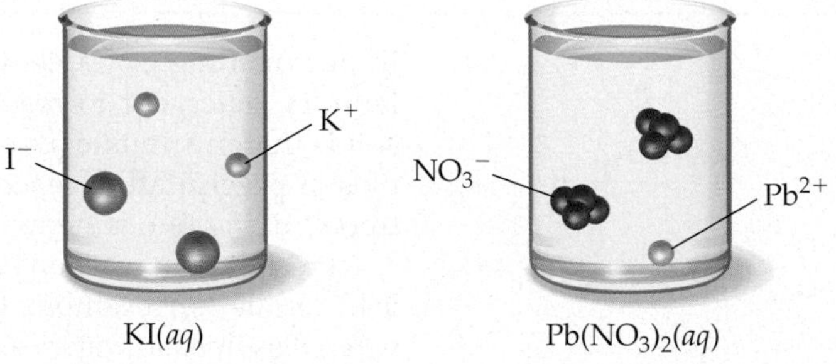

The instant that the solutions are mixed, all four ions are present.

However, new compounds—potentially insoluble ones—are now possible. Specifically, the cation from one compound can now pair with the anion from the other compound to form potentially insoluble products.

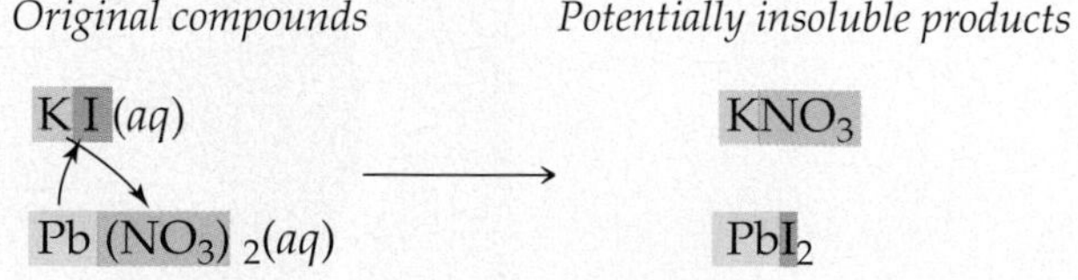

If the *potentially insoluble* products are both *soluble*, then no reaction occurs. If, on the other hand, one or both of the potentially insoluble products are *indeed insoluble*, a precipitation reaction occurs. In this case, KNO_3 is soluble, but PbI_2 is insoluble. Consequently, PbI_2 precipitates.

$PbI_2(s)$ and $KNO_3(aq)$

To predict whether a precipitation reaction will occur when two solutions are mixed and to write an equation for the reaction, follow the steps in the procedure box. As usual, the steps are shown in the left column, and two examples of applying the procedure are shown in the center and right column.

Writing Equations for Precipitation Reactions	EXAMPLE 7.7	EXAMPLE 7.8
	Write an equation for the precipitation reaction that occurs (if any) when solutions of sodium carbonate and copper(II) chloride are mixed.	Write an equation for the precipitation reaction that occurs (if any) when solutions of lithium nitrate and sodium sulfate are mixed.
	Solution:	**Solution:**
1. Write the formulas of the two compounds being mixed as reactants in a chemical equation.	$Na_2CO_3(aq) + CuCl_2(aq) \longrightarrow$	$LiNO_3(aq) + Na_2SO_4(aq) \longrightarrow$
2. Below the equation, write the formulas of the potentially insoluble products that could form from the reactants. Obtain these by combining the cation from one reactant with the anion from the other. Make sure to write correct formulas for these ionic compounds as described in Section 5.5.	$Na_2CO_3(aq) + CuCl_2(aq) \longrightarrow$ **Potentially Insoluble Products** NaCl $CuCO_3$	$LiNO_3(aq) + Na_2SO_4(aq) \longrightarrow$ **Potentially Insoluble Products** $NaNO_3$ Li_2SO_4
3. Use the solubility rules to determine if any of the potentially insoluble products are indeed insoluble.	NaCl is *soluble* (compounds containing Cl^- are usually soluble and Na^+ is not an exception). $CuCO_3$ is *insoluble* (compounds containing CO_3^{2-} are usually insoluble and Cu^{2+} is not an exception).	$NaNO_3$ is *soluble* (compounds containing NO_3^- are soluble and Na^+ is not an exception). Li_2SO_4 is *soluble* (compounds containing SO_4^{2-} are soluble and Li^+ is not an exception).
4. If all of the potentially insoluble products are soluble, there will be no precipitate. Write *NO REACTION* next to the arrow.	Since this example has an insoluble product, we proceed to the next step.	$LiNO_3(aq) + Na_2SO_4(aq) \rightarrow$ NO REACTION

5. If one or both of the potentially insoluble products are indeed insoluble, write their formula(s) as the product(s) of the reaction using (s) to indicate solid. Write any soluble products with *(aq)* to indicate aqueous.

$Na_2CO_3(aq) + CuCl_2(aq) \rightarrow CuCO_3(s) + NaCl(aq)$

6. Balance the equation. Remember to only adjust coefficients here, not subscripts.

$Na_2CO_3(aq) + CuCl_2(aq) \rightarrow CuCO_3(s) + \mathbf{2}\,NaCl(aq)$

SKILLBUILDER 7.7

Write an equation for the precipitation reaction that occurs (if any) when solutions of potassium hydroxide and nickel(II) bromide are mixed.

SKILLBUILDER 7.8

Write an equation for the precipitation reaction that occurs (if any) when solutions of ammonium chloride and iron(III) nitrate are mixed.

EXAMPLE 7.9 Predicting and Writing Equations for Precipitation Reactions

Write an equation for the precipitation reaction that occurs (if any) when solutions of lead (II) acetate and sodium sulfate are mixed. If no reaction occurs, write *NO REACTION*.

Solution:

Following the procedure for writing precipitation reactions:

1) Write the formulas of the two compounds being mixed as reactants in a chemical equation.

$$Pb(C_2H_3O_2)_2(aq) + Na_2SO_4(aq) \rightarrow$$

2) Below the equation, write the formulas of the potentially insoluble products that could form from the reactants. These are obtained by combining the cation from one reactant with the anion from the other. Make sure to adjust the subscripts so that all formulas are charge neutral.

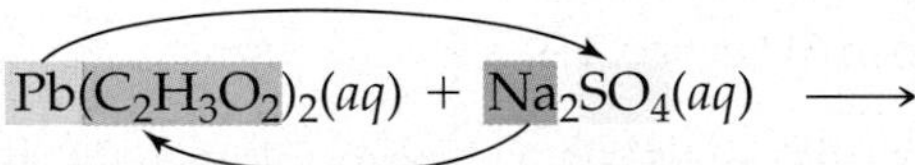

Potentially insoluble products

$NaC_2H_3O_2$ $\quad$ $PbSO_4$

3) Use the solubility rules to determine if any of the potentially insoluble products are indeed insoluble.

$NaC_2H_3O_2$ is *soluble* (compounds containing Na^+ are always soluble).
$PbSO_4$ is *insoluble* (compounds containing SO_4^{2-} are normally soluble, but Pb^{2+} is an exception).

4) If all of the potentially insoluble products are soluble, there will be no precipitate. Write *NO REACTION* next to the arrow.

Since we have an insoluble product, we proceed to the next step.

5) If one or both of the potentially insoluble products are indeed insoluble, write their formula(s) as the product(s) of the reaction using *(s)* to indicate solid. Write any soluble products with *(aq)* to indicate aqueous.

$$Pb(C_2H_3O_2)_2(aq) + Na_2SO_4(aq) \rightarrow PbSO_4(s) + NaC_2H_3O_2(aq)$$

6) Balance the equation.

$$Pb(C_2H_3O_2)_2(aq) + Na_2SO_4(aq) \rightarrow PbSO_4(s) + \mathbf{2}\,NaC_2H_3O_2(aq)$$

SKILLBUILDER 7.9 **Predicting and Writing Equations for Precipitation Reactions**

Write an equation for the precipitation reaction that occurs (if any) when solutions of potassium sulfate and strontium nitrate are mixed. If no reaction occurs, write *NO REACTION*.

7.7 Writing Chemical Equations for Reactions in Solution: Molecular, Complete Ionic, and Net Ionic Equations

Consider the following equation for a precipitation reaction.

$$AgNO_3(aq) + NaCl(aq) \rightarrow AgCl(s) + NaNO_3(aq)$$

This equation is written as a **molecular equation**, an equation showing the complete neutral formulas for every compound in the reaction. However, equations for reactions occurring in aqueous solution may be written to show that aqueous ionic compounds normally dissociate in solution. For example, the previous equation can be written as follows:

$$Ag^+(aq) + NO_3^-(aq) + Na^+(aq) + Cl^-(aq) \rightarrow AgCl(s) + Na^+(aq) + NO_3^-(aq)$$

Equations such as this one, showing the reactants and products as they are actually present in solution, are called **complete ionic equations**.

Notice that in the complete ionic equation, some of the ions in solution appear unchanged on both sides of the equation. These ions are called **spectator ions** because they do not participate in the reaction.

$$Ag^+(aq) + NO_3^-(aq) + Na^+(aq) + Cl^-(aq) \longrightarrow AgCl(s) + Na^+(aq) + NO_3^-(aq)$$

Spectator ions

To simplify the equation, and to more clearly show what is happening, spectator ions can be omitted.

$$Ag^+(aq) + Cl^-(aq) \rightarrow AgCl(s)$$

Species refers to a kind or sort of thing. In this case, the species are the reactants or products.

Equations such as this one, that show only the species that actually change during the reaction, are called **net ionic equations**.

As another example, consider the following reaction between $HCl(aq)$ and $NaOH(aq)$.

$$HCl(aq) + NaOH(aq) \rightarrow H_2O(l) + NaCl(aq)$$

Since HCl, NaOH, and NaCl are all aqueous ionic compounds, they exist in solution as independent ions. The complete ionic equation for this reaction is:

$$H^+(aq) + Cl^-(aq) + Na^+(aq) + OH^-(aq) \rightarrow H_2O(l) + Na^+(aq) + Cl^-(aq)$$

To write the net ionic equation, we remove the spectator ions, those that are unchanged on both sides of the equation.

$$H^+(aq) + \cancel{Cl^-}(aq) + \cancel{Na^+}(aq) + OH^-(aq) \longrightarrow H_2O(l) + \cancel{Na^+}(aq) + \cancel{Cl^-}(aq)$$

Spectator ions

The net ionic equation is $H^+(aq) + OH^-(aq) \rightarrow H_2O(l)$

Summary of equations for reactions in solution.

To Summarize:

- **A molecular equation** ***is a chemical equation showing the complete, neutral formulas for every compound in a reaction.***
- **A complete ionic equation** ***is a chemical equation showing all of the species as they are actually present in solution.***
- **A net ionic equation** ***is an equation showing only the species that actually change during the reaction.***

EXAMPLE 7.10 Writing Complete Ionic and Net Ionic Equations

Consider the following precipitation reaction occurring in aqueous solution.

$$Pb(NO_3)_2(aq) + 2\,LiCl(aq) \rightarrow PbCl_2(s) + 2\,LiNO_3(aq)$$

Write a complete ionic equation and a net ionic equation for this reaction.

Solution:

We write the complete ionic equation by separating aqueous ionic compounds into their constituent ions. The $PbCl_2(s)$ remains as one unit.

Complete ionic equation:

$$Pb^{2+}(aq) + \cancel{2\,NO_3^-(aq)} + \cancel{2\,Li^+(aq)} + 2\,Cl^-(aq) \rightarrow PbCl_2(s) + \cancel{2\,Li^+(aq)} + \cancel{2\,NO_3^-(aq)}$$

The net ionic equation eliminates the spectator ions, those that are not changing during the reaction.

Net ionic equation:

$$Pb^{2+}(aq) + 2\,Cl^-(aq) \rightarrow PbCl_2(s)$$

SKILLBUILDER 7.10 **Writing Molecular, Ionic, and Net Ionic Equations**

Consider the following reaction occurring in aqueous solution.

$$2\ HBr(aq) + Ca(OH)_2(aq) \rightarrow 2\ H_2O(l) + CaBr_2(aq)$$

Write a complete ionic equation and net ionic equation for this reaction.

7.8 Acid–Base and Gas Evolution Reactions

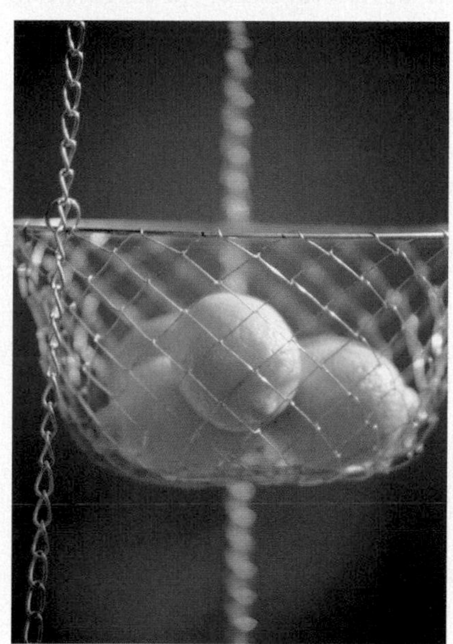

Foods such as lemons, limes, and vinegar contain acids.

Even though coffee itself is acidic overall, it contains some naturally occurring bases (such as caffeine) that give it a bitter taste.

Milk of magnesia is basic and tastes bitter.

Two other kinds of reactions that occur in solution are acid–base reactions, reactions that form water upon mixing an acid and a base, and gas evolution reactions, reactions that evolve a gas. Like precipitation reactions, these reactions occur when the cation of one reactant combines with the anion of another. As we will see in the next section, many gas evolution reactions also happen to be acid–base reactions.

Acid–Base Reactions

We learned in Chapter 5 that an acid is a compound characterized by its sour taste, its ability to dissolve some metals, and its tendency to form H^+ ions in solution. Some common acids are listed in Table 7.3. A base is a compound characterized by its bitter taste, its slippery feel, and its tendency to form OH^- ions in solution. Some common bases are listed in Table 7.3. Acids and bases are also found in many everyday substances. Foods such as lemons, limes, and vinegar contain acids. Soap, coffee, and milk of magnesia all contain bases.

When an acid and base are mixed, the $H^+(aq)$ from the acid combines with the $OH^-(aq)$ from the base to form $H_2O(l)$. For example, consider the reaction between hydrochloric acid and sodium hydroxide mentioned earlier.

$$\underset{\text{Acid}}{HCl(aq)} + \underset{\text{Base}}{NaOH(aq)} \longrightarrow \underset{\text{Water}}{H_2O(l)} + \underset{\text{Salt}}{NaCl(aq)}$$

Acid–base reactions generally form water and an ionic compound—called a **salt**—that usually remains dissolved in the solution. The net ionic equation for many acid–base reactions is:

$$H^+(aq) + OH^-(aq) \rightarrow H_2O(l)$$

TABLE 7.3

Some Common Acids and Bases

Acid	*Formula*	*Base*	*Formula*
hydrochloric acid	HCl	sodium hydroxide	$NaOH$
hydrobromic acid	HBr	lithium hydroxide	$LiOH$
nitric acid	HNO_3	potassium hydroxide	KOH
sulfuric acid	H_2SO_4	calcium hydroxide	$Ca(OH)_2$
perchloric acid	$HClO_4$	barium hydroxide	$Ba(OH)_2$
acetic acid	$HC_2H_3O_2$		

Another example of an acid–base reaction is that between sulfuric acid and potassium hydroxide.

$$\underset{\text{acid}}{H_2SO_4(aq)} + \underset{\text{base}}{2\,KOH} \rightarrow \underset{\text{water}}{2\,H_2O(l)} + \underset{\text{salt}}{K_2SO_4(aq)}$$

Again, notice the pattern of acid and base reacting to form water and a salt.

$$\text{Acid} + \text{Base} \rightarrow \text{Water} + \text{Salt} \qquad \text{(acid–base reactions)}$$

When writing equations for acid–base reactions, write the formula of the salt using the procedure for writing formulas of ionic compounds given in Section 5.5.

EXAMPLE 7.11 Writing Equations for Acid–Base Reactions

Write a molecular and net ionic equation for the reaction between aqueous HNO_3 and aqueous $Ca(OH)_2$.

Solution:

We must recognize these substances as an acid and a base. Acid–base reactions generally form water and an ionic compound (or salt) that usually remains dissolved in the solution. We first write the skeletal reaction following this general pattern.

$$\underset{\text{acid}}{HNO_3(aq)} + \underset{\text{base}}{Ca(OH)_2(aq)} \rightarrow \underset{\text{water}}{H_2O(l)} + \underset{\text{salt}}{Ca(NO_3)_2(aq)}$$

We then balance the equation.

$$\mathbf{2}\,HNO_3(aq) + Ca(OH)_2(aq) \rightarrow \mathbf{2}\,H_2O(l) + Ca(NO_3)_2(aq)$$

The net ionic equation is:

$$2\,H^+(aq) + 2\,OH^-(aq) \rightarrow 2\,H_2O(l)$$
$$\text{or simply } H^+(aq) + OH^-(aq) \rightarrow H_2O(l)$$

SKILLBUILDER 7.11 Writing Equations for Acid–Base Reactions

Write a molecular and net ionic equation for the reaction that occurs between aqueous H_2SO_4 and aqueous KOH.

Gas Evolution Reactions

Some aqueous reactions form a gas as a product. These reactions, as we learned in the opening section of this chapter, are called gas evolution reactions. Some gas evolution reactions form a gaseous product directly when the cation of one reactant reacts with the anion of the other. For example, when sulfuric acid reacts with lithium sulfide, dihydrogen sulfide gas is formed.

$$H_2SO_4(aq) + Li_2S(aq) \longrightarrow \underset{\text{Gas}}{H_2S(g)} + Li_2SO_4(aq)$$

Other gas evolution reactions form an intermediate product that then decomposes into a gas. For example, when aqueous hydrochloric acid is mixed with aqueous sodium bicarbonate the following reaction occurs.

$$HCl(aq) + NaHCO_3(aq) \longrightarrow H_2CO_3(aq) + NaCl(aq) \longrightarrow H_2O(l) + \underset{\text{Gas}}{CO_2(g)} + NaCl(aq)$$

Many gas evolution reactions such as this one are also acid–base reactions. In Chapter 14 we learn how ions such as CO_3^{2-} act as bases in aqueous solution.

The intermediate product, H_2CO_3, is not stable and decomposes to form H_2O and gaseous CO_2. This reaction is almost identical to the reaction in the kindergarten volcano of Section 7.1, which involves the mixing of acetic acid and sodium bicarbonate.

$$HC_2H_3O_2(aq) + NaHCO_3(aq) \rightarrow H_2CO_3(aq) + NaC_2H_3O_2(aq) \rightarrow H_2O(l) + CO_2(g) + NaC_2H_3O_2(aq)$$

A gas-evolution reaction.

The bubbling is caused by the newly formed carbon dioxide gas. Other important gas-evolution reactions form either H_2SO_3 or NH_4OH as intermediate products.

$$HCl(aq) + NaHSO_3(aq) \rightarrow H_2SO_3(aq) + NaCl(aq) \rightarrow H_2O(l) + SO_2(g) + NaCl(aq)$$

$$NH_4Cl(aq) + NaOH(aq) \rightarrow NH_4OH(aq) + NaCl(aq) \rightarrow H_2O(l) + NH_3(g) + NaCl(aq)$$

The main types of compounds that form gases in aqueous reactions, as well as the gases that they form, are listed in Table 7.4.

TABLE 7.4

Types of Compounds That Undergo Gas Evolution Reactions

Reactant Type	*Intermediate Product*	*Gas Evolved*	*Example*
sulfides	none	H_2S	$2\,HCl(aq) + K_2S(aq) \rightarrow H_2S(g) + 2\,KCl(aq)$
carbonates and bicarbonates	H_2CO_3	CO_2	$2\,HCl(aq) + K_2CO_3(aq) \rightarrow H_2O(l) + CO_2(g) + 2\,KCl(aq)$
sulfites and bisulfites	H_2SO_3	SO_2	$2\,HCl(aq) + K_2SO_3(aq) \rightarrow H_2O(l) + SO_2(g) + 2\,KCl(aq)$
Ammonium	NH_4OH	NH_3	$NH_4Cl(aq) + KOH(aq) \rightarrow H_2O(l) + NH_3(g) + KCl(aq)$

EXAMPLE 7.12 Writing Equations for Gas Evolution Reactions

Write a molecular equation for the gas evolution reaction that occurs when you mix aqueous nitric acid and aqueous sodium carbonate.

Solution:

We begin by writing a skeletal equation that includes the reactants and products that form when the cation of each reactant reacts with the anion of the other.

$$HNO_3(aq) + Na_2CO_3(aq) \longrightarrow H_2CO_3(aq) + NaNO_3(aq)$$

We recognize that $H_2CO_3(aq)$ decomposes into $H_2O(l)$ and $CO_2(g)$ and write the corresponding equation.

$$HNO_3(aq) + Na_2CO_3(aq) \rightarrow H_2O(l) + CO_2(g) + NaNO_3(aq)$$

Finally, we balance the equation.

$$\mathbf{2}\,HNO_3(aq) + Na_2CO_3(aq) \rightarrow H_2O(l) + CO_2(g) + \mathbf{2}\,NaNO_3(aq)$$

SKILLBUILDER 7.12 **Writing Equations for Gas Evolution Reactions**

Write a molecular equation for the gas evolution reaction that occurs when you mix aqueous hydrobromic acid and aqueous potassium sulfite.

SKILLBUILDER PLUS

Write a net ionic equation for the previous reaction.

CHEMISTRY AND HEALTH

Neutralizing Excess Stomach Acid

Your stomach normally contains acids that are involved in food digestion. Certain foods and stress, however, can increase the acidity of your stomach to uncomfortable levels, causing acid stomach or heartburn. Antacids are over-the-counter medicines that work by reacting with and neutralizing stomach acid. Antacids employ different bases as neutralizing agents. Tums, for example, contains $CaCO_3$; milk of magnesia contains $Mg(OH)_2$; and Mylanta contains $Al(OH)_3$. They all, however, have the same effect of neutralizing stomach acid and relieving heartburn.

CAN YOU ANSWER THIS? *Assume that stomach acid is HCl and write equations showing how each of these antacids neutralizes stomach acid.*

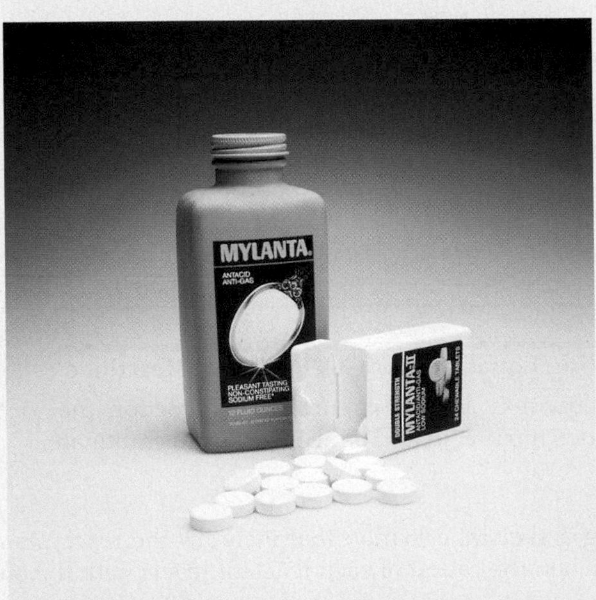

Antacids contain bases such $Mg(OH)_2$, $Al(OH)_3$, and $NaHCO_3$.

The base in an antacid neutralizes excess stomach acid, relieving heartburn and acid stomach.

7.9 Oxidation–Reduction Reactions

We cover oxidation–reduction reactions more thoroughly in Chapter 16.

Reactions involving the transfer of electrons are called oxidation–reduction or redox reactions. Redox reactions are responsible for the rusting of iron, the bleaching of hair, and the production of electricity in batteries. Many redox reactions involve the reaction of a substance with oxygen.

$$2\,H_2(g) + O_2(g) \rightarrow 2\,H_2O(g)$$
(reaction that powers the space shuttle)

$$4\,Fe(s) + 3\,O_2(g) \rightarrow 2\,Fe_2O_3(s)$$
(rusting of iron)

$$CH_4(g) + 2\,O_2(g) \rightarrow CO_2(g) + 2\,H_2O(g)$$
(combustion of natural gas)

However, redox reactions need not involve oxygen. Consider, for example, the reaction between sodium and chlorine to form table salt (NaCl).

$$2\,Na(s) + Cl_2(g) \rightarrow 2\,NaCl(s)$$

This reaction is similar to the reaction between sodium and oxygen to form sodium oxide.

$$4\,Na(s) + O_2(g) \rightarrow 2\,Na_2O(s)$$

Helpful mnemonics:
O I L R I G–**O**xidation **I**s **L**oss; **R**eduction **I**s **G**ain.
L E O G E R–**L**ose **E**lectrons **O**xidation; **G**ain **E**lectrons **R**eduction

What do these two reactions have in common? In both cases, sodium (a metal with a tendency to lose electrons) reacts with a nonmetal (that has a tendency to gain electrons). In both cases, sodium atoms lose electrons to nonmetal atoms. A fundamental definition of oxidation is *the loss of electrons*, and a fundamental definition of reduction is *the gain of electrons*.

Notice that oxidation and reduction must occur together. If one substance loses electrons (oxidation) then another substance must gain electrons (reduction). For now, be able to identify redox reactions.

A reaction can be classified as a redox reaction if it meets **any one** of these requirements.

Redox reactions are those in which:

- ***A substance reacts with elemental oxygen.***
- ***A metal reacts with a nonmetal.***
- ***One substance transfers electrons to another substance.***

EXAMPLE 7.13 Identifying Redox Reactions

Which of the following is a redox reaction?

a) $2\,Mg(s) + O_2(g) \rightarrow 2\,MgO(s)$
b) $2\,HBr(aq) + Ca(OH)_2(aq) \rightarrow 2\,H_2O(l) + CaBr_2(aq)$
c) $Ca(s) + Cl_2(g) \rightarrow CaCl_2(s)$
d) $Zn(s) + Fe^{2+}(aq) \rightarrow Zn^{2+}(aq) + Fe(s)$

Solution:

a) Redox reaction; Mg reacts with elemental oxygen.
b) Not a redox reaction; it is an acid–base reaction.
c) Redox reaction; a metal reacts with a nonmetal.
d) Redox reaction; Zn transfers two electrons to Fe^{2+}.

SKILLBUILDER 7.13 Identifying Redox Reactions

Which of the following is a redox reaction?

a) $2\,Li(s) + Cl_2(g) \rightarrow 2\,LiCl(s)$
b) $2\,Al(s) + 3\,Sn^{2+}(aq) \rightarrow 2\,Al^{3+}(aq) + 3\,Sn(s)$
c) $Pb(NO_3)_2(aq) + 2\,LiCl(aq) \rightarrow PbCl_2(s) + 2\,LiNO_3(aq)$
d) $C(s) + O_2(g) \rightarrow CO_2(g)$

Combustion Reactions

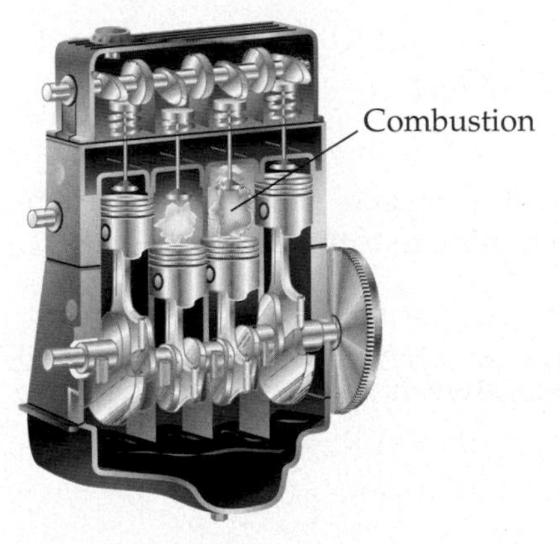

Combustion reactions are a type of redox reaction. They are important because most of our society's energy is derived from combustion reactions. Combustion reactions are characterized by the reaction of a substance with O_2 to form one or more oxygen-containing compounds, often including water. Combustion reactions also emit heat. For example, as we saw in Section 7.3, natural gas (CH_4) reacts with oxygen to form carbon dioxide and water.

$$CH_4(g) + 2\,O_2(g) \rightarrow CO_2(g) + 2\,H_2O(g)$$

As we learned in the opening section of this chapter, combustion reactions power automobiles. For example, octane, a component of gasoline, reacts with oxygen to form carbon dioxide and water.

$$2\,C_8H_{18}(l) + 25\,O_2(g) \rightarrow 16\,CO_2(g) + 18\,H_2O(g)$$

The water formed in combustion reactions may be gaseous (*g*) or liquid (*l*) depending on the reaction conditions.

Ethanol, the alcohol in alcoholic beverages, also reacts with oxygen in a combustion reaction to form carbon dioxide and water.

$$C_2H_6O(l) + 3\,O_2(g) \rightarrow 2\,CO_2(g) + 3\,H_2O(g)$$

Compounds containing carbon and hydrogen—or carbon, hydrogen and oxygen—always form carbon dioxide and water upon combustion. Other combustion reactions include the reaction of carbon with oxygen to form carbon dioxide:

$$C(s) + O_2(g) \rightarrow CO_2(g)$$

and the reaction of hydrogen with oxygen to form water.

$$2\,H_2(g) + O_2(g) \rightarrow 2\,H_2O(g)$$

EXAMPLE 7.14 Writing Combustion Reactions

Write a balanced equation for the combustion of liquid methyl alcohol (CH_3OH).

Solution:

Begin by writing a skeletal equation showing the reaction of CH_3OH with O_2 to form CO_2 and H_2O.

$$CH_3OH(l) + O_2(g) \rightarrow CO_2(g) + H_2O(g)$$

Balance the skeletal equation using the rules in Section 7.4.

$$2\,CH_3OH(l) + 3\,O_2(g) \rightarrow 2\,CO_2(g) + 4\,H_2O(g)$$

SKILLBUILDER 7.14 **Writing Combustion Reactions**

Write a balanced equation for the combustion of liquid pentane (C_5H_{12}), a component of gasoline.

SKILLBUILDER PLUS

Write a balanced equation for the combustion of liquid ethanol (C_2H_5OH).

7.10 Classifying Chemical Reactions

Throughout this chapter, we have examined different types of chemical reactions. We have seen examples of precipitation reactions, acid–base reactions, gas evolution reactions, oxidation–reduction reactions, and combustion reactions. We can organize these different types of reactions with the following flow chart.

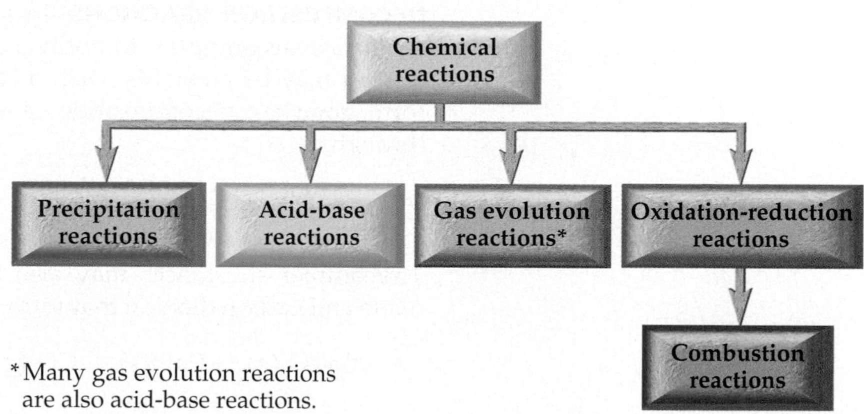

This classification scheme focuses on the type of chemistry or phenomenon that is occurring during the reaction (such as the formation of a precipitate or the transfer of electrons). However, another way to classify chemical reactions is by what atoms or groups of atoms do during the reaction.

Classifying Chemical Reactions by What Atoms Do

Many chemical reactions can be classified into one of the following four categories. In this classification scheme, the letters (A, B, C, D) represent atoms or groups of atoms.

Type of Reaction	*Generic Equation*
synthesis or combination	$A + B \rightarrow AB$
decomposition	$AB \rightarrow A + B$
displacement	$A + BC \rightarrow AC + B$
double-displacement	$AB + CD \rightarrow AD + CB$

SYNTHESIS OR COMBINATION REACTIONS. In a **synthesis** or **combination reaction**, simpler substances combine to form more complex substances. The simpler substances may be elements, such as sodium and chlorine combining to form sodium chloride.

$$2\,Na(s) + Cl_2(g) \rightarrow 2\,NaCl(s)$$

The simpler substances may also be compounds, such as calcium oxide and carbon dioxide combining to form calcium carbonate.

$$CaO(s) + CO_2(g) \rightarrow CaCO_3(s)$$

In either case, a synthesis reaction follows the general equation:

$$A + B \rightarrow AB$$

Other examples of synthesis reactions include:

Note that the first two of these reactions are also redox reactions.

$$2\,H_2(g) + O_2(g) \rightarrow 2\,H_2O(l)$$
$$2\,Mg(s) + O_2(g) \rightarrow 2\,MgO(s)$$
$$SO_3(g) + H_2O(l) \rightarrow H_2SO_4(aq)$$

DECOMPOSITION REACTIONS. In a **decomposition reaction**, a complex substance decomposes to form simpler substances. The simpler substances may be elements, such as the hydrogen and oxygen gases that form upon the decomposition of water when electrical current passes through it.

$$2\,H_2O(l) \xrightarrow[\text{electrical current}]{} 2\,H_2(g) + O_2(g)$$

The simpler substances may also be compounds, such as the calcium oxide and carbon dioxide that form upon heating calcium carbonate.

$$CaCO_3(s) \xrightarrow[\text{heat}]{} CaO(s) + CO_2(g)$$

$$2\,Na(s) + Cl_2(g) \longrightarrow 2\,NaCl(s)$$

In a synthesis reaction, two simpler substances combine to make a more complex substance. In this series of photographs we see sodium metal and chlorine gas. When they combine, a chemical reaction occurs that forms sodium chloride.

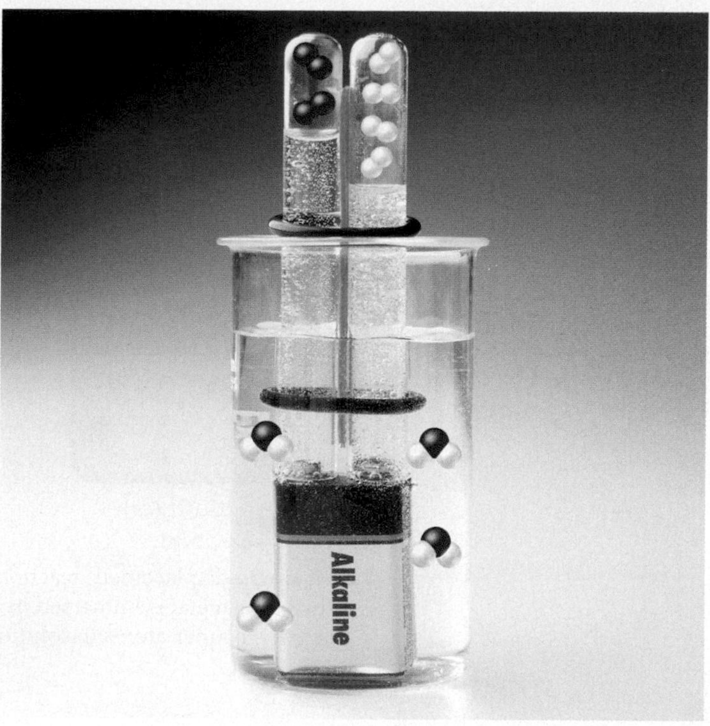

$$2\,H_2O(l) \longrightarrow 2\,H_2(g) + O_2(g)$$

When electrical current is passed through water, the water undergoes a decomposition reaction to form hydrogen gas and oxygen gas.

In either case, a decomposition reaction follows the general equation:

$$AB \rightarrow A + B$$

Other examples of decomposition reactions include:

$$2\,HgO(s) \xrightarrow[\text{heat}]{} 2\,Hg(l) + O_2(g)$$

$$2\,KClO_3(s) \xrightarrow[\text{heat}]{} 2\,KCl(s) + 3\,O_2(g)$$

$$CH_3I(g) \xrightarrow[\text{light}]{} CH_3(g) + I(g)$$

Notice that most decomposition reactions require energy in the form of heat, electrical current, or light to make them happen. This is because compounds are normally stable and energy must be used to decompose them. **Ultraviolet** or **UV light** is light in the ultraviolet region of the spectrum. UV light carries more energy than visible light and can therefore initiate the decomposition of many compounds.

Light in general, and ultraviolet light in particular, is covered in more detail in Chapter 9.

DISPLACEMENT REACTIONS. In a **displacement** or **single-displacement reaction**, one element displaces another in a compound. For example, when metallic zinc is added to a solution of copper(II) chloride, the zinc replaces the copper.

$$Zn(s) + CuCl_2(aq) \rightarrow ZnCl_2(aq) + Cu(s)$$

(a)

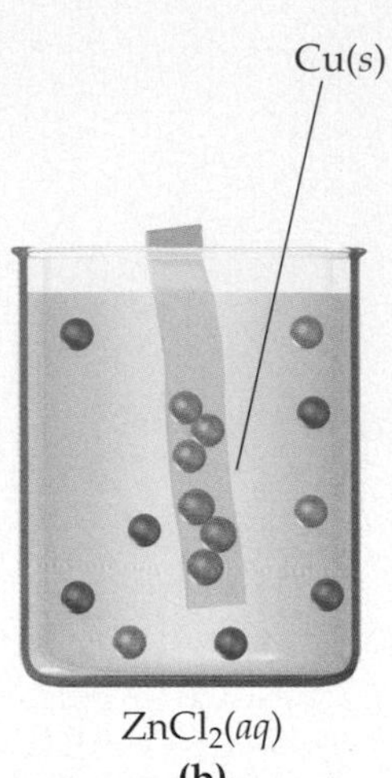

(b)

In a single-displacement reaction, one element displaces another in a compound. When zinc metal is immersed in a copper(II) chloride solution, the zinc atoms displace the copper atoms in solution.

A displacement reaction follows the general equation:

$$A + BC \rightarrow AC + B$$

Other examples of displacement reactions include:

$$Mg(s) + 2\,HCl(aq) \rightarrow MgCl_2(aq) + H_2(g)$$
$$2\,Na(s) + 2\,H_2O(l) \rightarrow 2\,NaOH(aq) + H_2(g)$$

The last reaction can be seen more easily if we write water as HOH(l).

$$2\,Na(s) + 2\,HOH(l) \rightarrow 2\,NaOH(aq) + H_2(g)$$

DOUBLE-DISPLACEMENT REACTIONS. In a **double-displacement reaction**, two elements or groups of elements in two different compounds exchange places to form two new compounds. For example, in aqueous solution, the silver in silver nitrate exchanges with the sodium in sodium chloride to form solid silver chloride and aqueous sodium nitrate.

This double-displacement reaction is also a precipitation reaction.

$$AgNO_3(aq) + NaCl(aq) \rightarrow AgCl(s) + NaNO_3(aq)$$

A double-displacement reaction follows the general form:

$$AB + CD \rightarrow AD + CB$$

Other examples of double-displacement reactions include:

These double-displacement reactions are also acid–base reactions.

$$HCl(aq) + NaOH(aq) \rightarrow H_2O(l) + NaCl(aq)$$
$$2\,HCl(aq) + Na_2CO_3(aq) \rightarrow 2\,NaCl(aq) + H_2CO_3(aq)$$

This double-displacement reaction is also a gas evolution reaction and an acid–base reaction.

As we learned in Section 7.7, $H_2CO_3(aq)$ is not stable and decomposes to form $H_2O(l) + CO_2(g)$, so the overall equation is:

$$2\,HCl(aq) + Na_2CO_3(aq) \rightarrow 2\,NaCl(aq) + H_2O(l) + CO_2(g)$$

Classification Flow Chart

A flow chart for this classification scheme of chemical reactions is as follows:

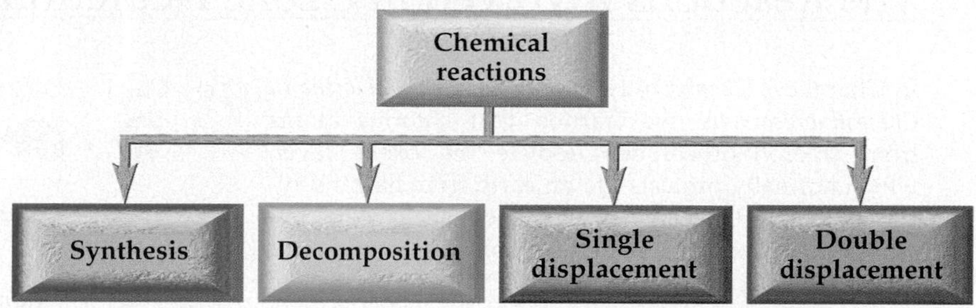

Of course no single classification scheme is perfect because all chemical reactions are unique in some sense. However, both classification schemes—one that focuses on the type of chemistry occurring and the other that focuses on what atoms or groups of atoms are doing—are helpful because they help us see differences and similarities among chemical reactions.

EXAMPLE 7.15 Classifying Chemical Reactions According to What Atoms Do

Classify each of the following reactions as a synthesis, decomposition, single-displacement, or double-displacement reaction.

a) $Na_2O(s) + H_2O(l) \rightarrow 2\,NaOH(aq)$
b) $Ba(NO_3)_2(aq) + K_2SO_4(aq) \rightarrow BaSO_4(s) + 2\,KNO_3(aq)$
c) $2\,Al(s) + Fe_2O_3(s) \rightarrow Al_2O_3(s) + 2\,Fe(l)$
d) $2\,H_2O_2(aq) \rightarrow 2\,H_2O(l) + O_2(g)$
e) $Ca(s) + Cl_2(g) \rightarrow CaCl_2(s)$

Solution:

a) Synthesis; a more complex substance forms from two simpler ones
b) Double-displacment; Ba and K switch places to form two new compounds.
c) Single-displacement; Al displaces Fe in Al_2O_3.
d) Decomposition; a complex substance decomposes into simpler ones.
e) Synthesis; a more complex substance forms from two simpler ones

SKILLBUILDER 7.15 Classifying Chemical Reactions According to What Atoms Do

Classify each of the following reactions as a synthesis, decomposition, single-displacement, or double-displacement reaction.

a) $2\,Al(s) + 2\,H_3PO_4(aq) \rightarrow 2\,AlPO_4(aq) + 3\,H_2(g)$
b) $CuSO_4(aq) + 2\,KOH(aq) \rightarrow Cu(OH)_2(s) + K_2SO_4(aq)$
c) $2\,K(s) + Br_2(l) \rightarrow 2\,KBr(s)$
d) $CuCl_2(aq) \xrightarrow[\text{electrical current}]{} Cu(s) + Cl_2(g)$

CHEMISTRY IN THE ENVIRONMENT

The Reactions Involved in Ozone Depletion

In Chapter 6, *Chemistry in the Environment: Chlorine in Chlorofluorocarbons*, we learned that chlorine atoms from chlorofluorocarbons deplete the ozone layer, which normally protects life on earth from harmful ultraviolet light. Through research, chemists have discovered the reactions by which this depletion occurs.

Ozone normally forms in the upper atmosphere according to the following reaction.

a) $O_2(g) + O(g) \rightarrow O_3(g)$

When chlorofluorocarbons drift to the upper atmosphere, they are exposed to ultraviolet light and undergo the following reaction.

b) $CF_2Cl_2(g) \xrightarrow{\text{uv light}} CF_2Cl(g) + Cl(g)$

UV light

Atomic chlorine then reacts with and depletes ozone according to the following cycle of reactions.

c) $Cl(g) + O_3(g) \rightarrow ClO(g) + O_2(g)$

d) $O_3(g) \xrightarrow{\text{uv light}} O(g) + O_2(g)$

UV light

e) $O(g) + ClO(g) \rightarrow O_2(g) + Cl(g)$

Notice that in the final reaction, atomic chlorine is regenerated and can go through the cycle again to deplete more ozone. Through this cycle of reactions, a single chlorofluorocarbon molecule can deplete thousands of ozone molecules.

CAN YOU ANSWER THIS? *Classify each of these reactions (a–e) as a synthesis, decomposition, single-displacement, or double-displacement reaction.*

CHAPTER IN REVIEW

Chemical Principles	Relevance
Chemical Reactions: In a chemical reaction, one or more substances—either elements or compounds—change into different substances.	**Chemical Reactions:** Chemical reactions are central to many processes including transportation, energy generation, household products, vision, and even life.
Evidence of a Chemical Reaction: The only absolute evidence for a chemical reaction is chemical analysis showing that one or more substances change into other substances. However, one or more of the following often accompany a chemical reaction: a color change; the formation of a solid or precipitate; the formation of a gas; the emission of light; and the emission or absorption of heat.	**Evidence of a Chemical Reaction:** We can often perceive the changes that accompany chemical reactions. In fact, we often use chemical reactions for the changes they produce. For example, we use the heat emitted by the combustion of fossil fuels to heat our homes, drive our cars, and generate electricity.
Chemical Equations: Chemical equations represent chemical reactions. They include formulas for the reactants, the substances present before the reaction, and formulas for the products, the new substances formed by the reaction. Chemical equations must be balanced to reflect the conservation of matter in nature; atoms do not spontaneously appear or disappear.	**Chemical Equations:** Chemical equations allow us to represent and understand chemical reactions. For example, the equations for the combustion reactions of fossil fuels let us see that carbon dioxide, a gas that contributes to global warming, is one of the products of these reactions.

Aqueous Solutions and Solubility: Aqueous solutions are mixtures of a substance dissolved in water. If a substance dissolves in water it is soluble. Otherwise, it is insoluble.

Aqueous Solutions and Solubility: Aqueous solutions are common. For example, oceans, lakes, and most of the fluids in our bodies are aqueous solutions.

Some Specific Types of Reactions

Precipitation reaction: A solid or precipitate forms upon mixing two aqueous solutions.
Acid–base reaction: Water forms upon mixing an acid and a base.
Gas evolution reaction: A gas forms upon mixing two aqueous solutions.
Redox reaction: Electrons are transferred from one substance to another.
Combustion reaction: A substance reacts with oxygen, emitting heat, and forming an oxygen-containing compound and, many times, water.

Some Specific Types of Reactions: Many of the specific types of reactions discussed here occur in aqueous solutions and are therefore important to living organisms. Acid–base reactions, for example, constantly occur in the blood of living organisms to maintain constant blood acidity levels. In humans, a small change in blood acidity levels would result in death, so the body carries out chemical reactions to prevent this. Combustion reactions are important because they are the main energy source for our society.

Classifying Chemical Reactions: Many chemical reactions can be classified into one of the following four categories according to what atoms or groups of atoms do: synthesis $(A + B \rightarrow AB)$, decomposition $(AB \rightarrow A + B)$, single-displacement $(A + BC \rightarrow AC + B)$, and double-displacement $(AB + CD \rightarrow AD + CB)$.

Classifying Chemical Reactions: We classify chemical reactions to better understand them and to better see similarities and differences among reactions.

Chemical Skills

Identifying a Chemical Reaction (Section 7.2)

To identify a chemical reaction, determine if one or more of the initial substances changed into different substances. If so, a chemical reaction occurred. One or more of the following often accompanies a chemical reaction: a color change; the formation of a solid or precipitate; the formation of a gas; the emission of light; and the emission or absorption of heat.

Examples

EXAMPLE 7.16 Identifying a Chemical Reaction

Which of the following is a chemical reaction?

a) Copper turns green on exposure to air.
b) When sodium bicarbonate is combined with hydrochloric acid, bubbling is observed.
c) Liquid water freezes to form solid ice.
d) A pure copper penny forms bubbles of a dark brown gas when dropped into nitric acid. The nitric acid solution turns blue.

Solution:

a) Chemical reaction as evidenced by the color change
b) Chemical reaction as evidenced by the evolution of a gas
c) Not a chemical reaction; solid ice is still water
d) Chemical reaction as evidenced by the evolution of a gas and by a color change

Writing Balanced Chemical Equations (Sections 7.3, 7.4)

To write balanced chemical equations, follow these steps.

1) Write a skeletal equation by writing chemical formulas for each of the reactants and products. (If a skeletal equation is provided, proceed to Step 2).
2) If an element occurs in only one compound on both sides of the equation, balance that element first. If there is more than one such element, and the equation contains both metals and nonmetals, balance metals before nonmetals.
3) If an element occurs as a free element on either side of the chemical equation, balance that element last.
4) If the balanced equation contains coefficient fractions, clear these by multiplying the entire equation by the appropriate factor.
5) Check to make certain the equation is balanced by summing the total number of each type of atom on both sides of the equation.

Reminders

- Change only the *coefficients* to balance a chemical equation, *never the subscripts*. Changing the subscripts would change the compound itself.
- If the equation contains polyatomic ions that stay intact on both sides of the equation, balance the polyatomic ions as a group.

EXAMPLE 7.17 Writing Balanced Chemical Equations

Write a balanced chemical equation for the reaction of solid vanadium(V) oxide with hydrogen gas to form solid vanadium(III) oxide and liquid water.

$$V_2O_5(s) + H_2(g) \rightarrow V_2O_3(s) + H_2O(l)$$

Vanadium occurs in only one compound on both sides of the equation. However, it is balanced, so we proceed to balance oxygen by placing a 2 in front of H_2O on the right hand side.

$$V_2O_5(s) + H_2(g) \rightarrow V_2O_3(s) + \mathbf{2}\,H_2O(l)$$

Hydrogen occurs as a free element, so we balance it last by placing a 2 in front of H_2 on the left hand side.

$$V_2O_5(s) + \mathbf{2}\,H_2(g) \rightarrow V_2O_3(s) + 2\,H_2O(l)$$

Finally, we check the equation.

$$V_2O_5(s) + 2\,H_2(g) \rightarrow V_2O_3(s) + 2\,H_2O(l)$$

Reactants	*Products*
2 V atoms	2 V atoms
5 O atoms	5 O atoms
4 H atoms	4 H atoms

Determining if a Compound Is Soluble (Section 7.5)

To determine if a compound is soluble, use the solubility rules in Table 7.2. It is easiest to begin by looking for those ions that always form soluble compounds (Li^+, Na^+, K^+, NH_4^+, NO_3^-, and $C_2H_3O_2^-$). If a compound contains one of those, it is soluble. If it does not, look at the anion and determine if it is mostly soluble (Cl^-, Br^-, I^- or SO_4^{2-}) or mostly insoluble (OH^-, S^{2-}, CO_3^{2-} or PO_4^{3-}). Look also at the cation to determine if it is one of the exceptions.

EXAMPLE 7.18 Determining if a Compound Is Soluble

Determine if the following compounds are soluble.

a) $CuCO_3$
b) $BaSO_4$
c) $Fe(NO_3)_3$

Solution:

a) Insoluble; compounds containing CO_3^{2-} are insoluble, and Cu^{2+} is not an exception.
b) Insoluble; compounds containing SO_4^{2-} are usually soluble, but Ba^{2+} is an exception.
c) Soluble; all compounds containing NO_3^- are soluble.

Predicting Precipitation Reactions (Section 7.6)

To predict whether a precipitation reaction occurs when two solutions are mixed and to write an equation for the reaction, follow these steps.

1) Write the formulas of the two compounds being mixed as reactants in a chemical equation.
2) Below the equation, write the formulas of the potentially insoluble products that could form from the reactants. These are obtained by combining the cation from one reactant with the anion from the other. Make sure to adjust the subscripts so that all formulas are charge neutral.
3) Use the solubility rules to determine if any of the potentially insoluble products are indeed insoluble.
4) If all of the potentially insoluble products are soluble, there will be no precipitate. Write *NO REACTION* next to the arrow.
5) If one or both of the potentially insoluble products are indeed insoluble, write their formula(s) as the product(s) of the reaction using (*s*) to indicate solid. Write any soluble products with (*aq*) to indicate aqueous.
6) Balance the equation.

EXAMPLE 7.19 Predicting Precipitation Reactions

Write an equation for the precipitation reaction that occurs, if any, when solutions of sodium phosphate and cobalt(II) chloride are mixed.

Solution:

$$Na_3PO_4(aq) + CoCl_2(aq) \rightarrow$$

Potentially Insoluble Products

$NaCl$ $\quad Co_3(PO_4)_2$

$NaCl$ is soluble.

$Co_3(PO_4)_2$ is insoluble.

$$Na_3PO_4(aq) + CoCl_2(aq) \rightarrow Co_3(PO_4)_2(s) + NaCl(aq)$$

$$\mathbf{2}\,Na_3PO_4(aq) + \mathbf{3}\,CoCl_2(aq) \rightarrow Co_3(PO_4)_2(s) + \mathbf{6}\,NaCl(aq)$$

Writing Complete Ionic and Net Ionic Equations (Section 7.7)

To write a complete ionic equation from a molecular equation, separate all aqueous ionic compounds into independent ions. Do not separate solid, liquid, or gaseous compounds.

To write a net ionic equation from a complete ionic equation, eliminate all species that do not change (spectator ions) in the course of the reaction.

EXAMPLE 7.20 Writing Complete Ionic and Net Ionic Equations

Write a complete ionic and net ionic equation for the following reaction.

$$2\,NH_4Cl(aq) + Hg_2(NO_3)_2(aq) \rightarrow Hg_2Cl_2(s) + 2\,NH_4NO_3(aq)$$

Complete ionic equation:

$$2\,NH_4^+(aq) + 2\,Cl^-(aq) + Hg_2^{2+}(aq) + 2\,NO_3^-(aq) \rightarrow Hg_2Cl_2(s) + 2\,NH_4^-(aq) + 2\,NO_3^-(aq)$$

Net ionic equation:

$$2\,Cl^-(aq) + Hg_2^{2+}(aq) \rightarrow Hg_2Cl_2(s)$$

Writing Equations for Acid–Base Reactions (Section 7.8)

When you see an acid and a base (see Table 7.3) as reactants in an equation, write a reaction in which the acid and the base react to form water and a salt.

EXAMPLE 7.21 Writing Equations for Acid–Base Reactions

Write an equation for the reaction that occurs when aqueous hydroiodic acid is mixed with aqueous barium hydroxide.

$$\underset{\text{acid}}{2\,HI(aq)} + \underset{\text{base}}{Ba(OH)_2(aq)} \rightarrow \underset{\text{water}}{2\,H_2O(l)} + \underset{\text{salt}}{BaI_2(aq)}$$

Writing Equations for Gas Evolution Reactions (Section 7.8)

See Table 7.4 to identify gas evolution reactions.

EXAMPLE 7.21 Writing Equations for Gas Evolution Reactions

Write an equation for the reaction that occurs when you mix aqueous hydrobromic acid with aqueous potassium bisulfite.

$$HBr(aq) + KHSO_3(aq) \rightarrow H_2SO_3(aq) + KBr(aq) \rightarrow H_2O(l) + SO_2(g) + KBr(aq)$$

Identifying Redox Reactions (Section 7.9)

Redox reactions are those in which:

- a substance reacts with elemental oxygen
- a metal reacts with a nonmetal
- one substance transfers electrons to another substance

(any one of these)

EXAMPLE 7.21 Identifying Redox Reactions

Which of the following is a redox reaction?

a) $4\,Fe(s) + 3\,O_2(g) \rightarrow 2\,Fe_2O_3(s)$
b) $CaO(s) + CO_2(g) \rightarrow CaCO_3(s)$
c) $AgNO_3(aq) + NaCl(aq) \rightarrow AgCl(s) + NaNO_3(aq)$

Solution:
Only **a)** is a redox reaction.

Writing Equations for Combustion Reactions (Section 7.9)

In a combustion reaction, a substance reacts with O_2 to form one or more oxygen-containing compounds and in many cases water.

EXAMPLE 7.22 Writing Equations for Combustion Reactions

Write a balanced equation for the combustion of gaseous ethane (C_2H_6), a minority component of natural gas.

The skeletal equation is:

$$C_2H_6(g) + O_2(g) \rightarrow CO_2(g) + H_2O(g)$$

The balanced equation is:

$$2\,C_2H_6(g) + 7\,O_2(g) \rightarrow 4\,CO_2(g) + 6\,H_2O(g)$$

Classifying Chemical Reactions (Section 7.10)

Chemical reactions can be classified by inspection. The four major categories are:

Synthesis or combination

$A + B \rightarrow AB$

Decomposition

$AB \rightarrow A + B$

Single-displacement

$A + BC \rightarrow AC + B$

Double-displacement

$AB + CD \rightarrow AD + CB$

EXAMPLE 7.23 Classifying Chemical Reactions

Classify each of the following chemical reactions as a synthesis, decomposition, single-displacement, or double-displacement reaction.

a) $2K(s) + Br_2(g) \rightarrow 2\,KBr(s)$
b) $Fe(s) + 2\,AgNO_3(aq) \rightarrow Fe(NO_3)_2(aq) + 2\,Ag(s)$
c) $CaSO_3(s) \xrightarrow{heat} CaO(s) + SO_2(g)$
d) $CaCl_2(aq) + Li_2SO_4(aq) \rightarrow CaSO_4(s) + 2\,LiCl(aq)$

Solution:

a) Synthesis; KBr, a more complex substance is formed from simpler substances.
b) Single-displacement; Fe displaces Ag in $AgNO_3$
c) Decomposition; $CaSO_3$ decomposes into simpler substances
d) Double-displacement; Ca and Li switch places to form new compounds.

KEY TERMS

acid-base reaction [7.8]
aqueous solution [7.5]
balanced equation [7.3]
chemical equation [7.3]
chemical reaction [7.1]
color change [7.2]
combination reaction [7.10]
combustion reaction [7.1]
complete ionic equation [7.7]
decomposition reaction [7.10]
double-displacement reaction [7.10]
formation of a gas [7.1]
formation of a solid [7.2]
gas evolution reaction [7.1]
heat absorption [7.2]
heat emission [7.2]
insoluble [7.5]
light emission [7.2]
molecular equation [7.7]
net ionic equation [7.7]
oxidation-reduction (redox) reaction [7.1]
precipitate [7.6]
precipitation reaction [7.6]
products [7.3]
reactants [7.3]
salt [7.8]
single-displacement reaction [7.10]
solubility rules [7.5]
soluble [7.5]
spectator ions [7.7]
strong electrolyte solution [7.5]
synthesis reaction [7.10]
ultraviolet (UV) light [7.10]

EXERCISES

Questions

1. What is a chemical reaction? Give some examples.
2. If you could see atoms and molecules, what would you look for as conclusive evidence of a chemical reaction?
3. What are the main evidences of a chemical reaction?
4. What is a chemical equation? Give an example and identify the reactants and products.
5. What do each of the following abbreviations, often used in chemical equations, represent?
a) (g)
b) (l)
c) (s)
d) (aq)
6. To balance a chemical equation, adjust the _____ as necessary to make the numbers of each type of atom on both sides of the equation equal. Never adjust the _____ to balance a chemical equation.
7. List the number of each type of atom on both sides of each of the following equations. Are the equations balanced?
a) $2\,Ag_2O(s) + C(s) \rightarrow CO_2(g) + 4\,Ag(s)$
b) $Pb(NO_3)_2(aq) + 2\,NaCl(aq) \rightarrow PbCl_2(s) + 2\,NaNO_3(aq)$
c) $C_3H_8(g) + O_2(g) \rightarrow 3\,CO_2(g) + 4\,H_2O(g)$
8. What is an aqueous solution? Give two examples.
9. What does it mean for a compound to be soluble? Insoluble?
10. How do ionic compounds dissolve in water?
11. How do polyatomic ions within ionic compounds dissolve in water?
12. What is a strong electrolyte solution?
13. What are the solubility rules and how are they useful?
14. What is a precipitation reaction? Give an example and identify the precipitate.
15. Will the precipitate in a precipitation reaction always be a compound that is soluble or insoluble? Explain.
16. Describe the differences between a molecular equation and a complete ionic equation. Give an example to illustrate the difference.
17. Describe the differences between a complete ionic and net ionic equation. Give an example to illustrate the difference.
18. What is an acid–base reaction? Give an example and identify the acid and the base.
19. What are the properties of acids and bases?
20. What is a gas evolution reaction? Give an example.
21. What is a redox reaction? Give an example.
22. What is a combustion reaction? Give an example.
23. What are two different ways to classify chemical reactions presented in Section 7.10? Explain the differences between them.
24. Describe each of the following general types of chemical reactions and write the generic equation for each.
a) synthesis
b) decomposition
c) single-displacement
d) double-displacement
25. Give two examples of a synthesis reaction.
26. Give two examples of a decomposition reaction.
27. Give two examples of a single-displacement reaction.
28. Give two examples of a double-displacement reaction.

Problems

Evidence of Chemical Reactions

29. Which of the following is a chemical reaction? Why?
a) Solid copper deposits on a piece of aluminum foil when the foil is placed in a blue copper nitrate solution. The blue color of the solution fades.
b) Liquid ethyl alcohol turns into a solid when placed in a low-temperature freezer.
c) A white precipitate forms when solutions of barium nitrate and sodium sulfate are mixed.
d) A mixture of sugar and water bubbles when yeasts are added. After several days, the sugar is gone and ethyl alcohol is found in the water.

30. Which of the following is a chemical reaction? Why?
a) Propane forms a flame and emits heat as it burns.
b) Acetone feels cold as it evaporates from the skin.
c) Bubbling is observed when potassium carbonate and hydrochloric acid solutions are mixed.
d) Heat is felt when a warm object is placed in your hand.

31. Vinegar forms bubbles when it is added to the calcium deposits on a faucet. Has a chemical reaction occurred? Why or why not?

32. When a chemical drain opener is added to a clogged sink, bubbles form and the water warms. Has a chemical reaction occurred? Why or why not?

33. When a commercial hair bleaching mixture is applied to brown hair, it turns blonde. Has a chemical reaction occurred? Why or why not?

34. When water is boiled in a pot, it bubbles. Has a chemical reaction occurred? Why or why not?

Writing and Balancing Chemical Equations

35. Write a balanced chemical equation for each of the following:
a) Solid copper reacts with solid sulfur to form solid copper(I) sulfide.
b) Sulfur dioxide gas reacts with oxygen gas to form sulfur trioxide gas.
c) Aqueous hydrochloric acid reacts with solid manganese(IV) oxide to form aqueous manganese(II) chloride, liquid water, and chlorine gas.
d) Liquid benzene (C_6H_6) reacts with gaseous oxygen to form carbon dioxide and liquid water.

36. Write a balanced chemical equation for each of the following:
a) Solid lead(II) sulfide reacts with aqueous hydrochloric acid to form solid lead(II) chloride and dihydrogen sulfide gas.
b) Gaseous carbon monoxide reacts with hydrogen gas to form gaseous methane (CH_4) and liquid water.
c) Solid iron(III) oxide reacts with hydrogen gas to form solid iron and liquid water.
d) Gaseous ammonia (NH_3) reacts with gaseous oxygen to form gaseous nitrogen monoxide and gaseous water.

37. Write a balanced chemical equation for each of the following:
a) Solid magnesium reacts with aqueous copper(I) nitrate to form aqueous magnesium nitrate and solid copper.
b) Gaseous dinitrogen pentoxide decomposes to form nitrogen dioxide and oxygen gas.
c) Solid calcium reacts with aqueous nitric acid to form aqueous calcium nitrate and hydrogen gas.
d) Liquid methanol (CH_3OH) reacts with oxygen gas to form gaseous carbon dioxide and gaseous water.

38. Write a balanced chemical equation for each of the following:
a) Gaseous acetylene (C_2H_2) reacts with oxygen gas to form gaseous carbon dioxide and gaseous water.
b) Chlorine gas reacts with aqueous potassium iodide to form solid iodine and aqueous potassium chloride.
c) Solid lithium oxide reacts with liquid water to form aqueous lithium hydroxide.
d) Gaseous carbon monoxide reacts with oxygen gas to form carbon dioxide gas.

39. When solid sodium is added to liquid water, it reacts with the water to produce hydrogen gas and aqueous sodium hydroxide. Write a balanced chemical equation for this reaction.

40. When iron rusts, solid iron reacts with gaseous oxygen to form solid iron(III) oxide. Write a balanced chemical equation for this reaction.

41. Sulfuric acid in acid rain forms when gaseous sulfur dioxide pollutant reacts with gaseous oxygen and liquid water to form aqueous sulfuric acid. Write a balanced chemical equation for this reaction.

42. Nitric acid in acid rain forms when gaseous nitrogen dioxide pollutant reacts with gaseous oxygen and liquid water to form aqueous nitric acid. Write a balanced chemical equation for this reaction.

43. Write a balanced chemical equation for the reaction of solid vanadium(V) oxide with hydrogen gas to form solid vanadium(III) oxide and liquid water.

44. Write a balanced chemical equation for the reaction of gaseous nitrogen dioxide with hydrogen gas to form gaseous ammonia and liquid water.

45. Write a balanced chemical equation for the fermentation of sugar ($C_{12}H_{22}O_{11}$) by yeasts in which the aqueous sugar reacts with water to form aqueous ethyl alcohol (C_2H_5OH) and carbon dioxide gas.

46. Write a balanced chemical equation for the photosynthesis reaction in which gaseous carbon dioxide and liquid water react in the presence of chlorophyll to produce aqueous glucose ($C_6H_{12}O_6$) and oxygen gas.

47. Balance each of the following chemical equations.
a) $N_2H_4(l) \rightarrow NH_3(g) + N_2(g)$
b) $H_2(g) + N_2(g) \rightarrow NH_3(g)$
c) $Cu_2O(s) + C(s) \rightarrow Cu(s) + CO(g)$
d) $H_2(g) + Cl_2(g) \rightarrow HCl(g)$

48. Balance each of the following chemical equations.
a) $Na_2S(aq) + Cu(NO_3)_2(aq) \rightarrow NaNO_3(aq) + CuS(s)$
b) $HCl(aq) + O_2(g) \rightarrow H_2O(l) + Cl_2(g)$
c) $H_2(g) + O_2(g) \rightarrow H_2O(l)$
d) $FeS(s) + HCl(aq) \rightarrow FeCl_2(aq) + H_2S(g)$

49. Balance each of the following chemical equations.
a) $BaO_2(s) + H_2SO_4(aq) \rightarrow BaSO_4(s) + H_2O_2(aq)$
b) $Co(NO_3)_3(aq) + (NH_4)_2S(aq) \rightarrow Co_2S_3(s) + NH_4NO_3(aq)$
c) $Li_2O(s) + H_2O(l) \rightarrow LiOH(aq)$
d) $Hg_2(C_2H_3O_2)_2(aq) + KCl(aq) \rightarrow Hg_2Cl_2(s) + KC_2H_3O_2(aq)$

50. Balance each of the following chemical equations.
a) $MnO_2(s) + HCl(aq) \rightarrow Cl_2(g) + MnCl_2(aq) + H_2O(l)$
b) $CO_2(g) + CaSiO_3(s) + H_2O(l) \rightarrow SiO_2(s) + Ca(HCO_3)_2(aq)$
c) $Fe(s) + S(l) \rightarrow Fe_2S_3(s)$
d) $NO_2(g) + H_2O(l) \rightarrow HNO_3(aq) + NO(g)$

51. Determine if each of the following chemical equations is correctly balanced. If not, correct it.
a) $2\,Al(s) + 3\,O_2(g) \rightarrow Al_2O_3(s)$
b) $2\,N_2H_4(g) + N_2O_4(g) \rightarrow 3\,N_2(g) + 4\,H_2O(g)$
c) $NiS(s) + O_2(g) \rightarrow NiO(s) + SO_2(g)$
d) $Fe_2O_3(s) + CO(g) \rightarrow 2\,Fe(s) + CO_2(g)$

52. Determine if each of the following chemical equations is correctly balanced. If not, correct it.
a) $Al_2S_3(s) + H_2O(l) \rightarrow 2\,Al(OH)_3(s) + 3\,H_2S(g)$
b) $K(s) + H_2O(l) \rightarrow KOH(aq) + H_2(g)$
c) $SiO_2(s) + 4\,HF(aq) \rightarrow SiF_4(g) + 2\,H_2O(l)$
d) $PbO(s) + 2\,NH_3(g) \rightarrow Pb(s) + N_2(g) + H_2O(l)$

53. Human cells obtain energy from a reaction called respiration. Balance the skeletal equation for respiration.

$$C_6H_{12}O_6(aq) + O_2(g) \rightarrow CO_2(g) + H_2O(l)$$

54. Propane camping stoves produce heat by the combustion of gaseous propane (C_3H_8). Balance the skeletal equation for the combustion of propane.

$$C_3H_8(g) + O_2(g) \rightarrow CO_2(g) + H_2O(g)$$

55. Catalytic converters work to remove nitrogen oxides and carbon monoxide from exhaust. Balance the skeletal equation for one of the reactions that occurs in a catalytic converter.

$$NO(g) + CO(g) \rightarrow N_2(g) + CO_2(g)$$

56. Billions of pounds of urea are produced annually for use as a fertilizer. Balance the skeletal equation for the synthesis of urea.

$$NH_3(g) + CO_2(g) \rightarrow CO(NH_2)_2(s) + H_2O(l)$$

Solubility

57. Determine whether each of the following compounds is soluble or insoluble. For the soluble compounds, write the ions present in solution.
a) $NaNO_3$
b) $Pb(C_2H_3O_2)_2$
c) $CuCO_3$
d) $(NH_4)_2S$

58. Determine whether each of the following compounds is soluble or insoluble. For the soluble compounds, write the ions present in solution.
a) AgI
b) $AgNO_3$
c) $CoPO_4$
d) Na_3PO_4

59. Pair each cation on the left with an anion on the right that will form an *insoluble* compound and write a formula for the insoluble compound. Use each anion only once.

Ag^+	SO_4^{2-}
Ba^{2+}	Cl^-
Cu^{2+}	CO_3^{2-}
Fe^{3+}	S^{2-}

60. Pair each cation on the left with an anion on the right that will form a *soluble* compound and write a formula for the soluble compound. Use each anion only once.

Na^+	NO_3^-
Sr^{2+}	SO_4^{2-}
Co^{2+}	S^{2-}
Pb^{2+}	CO_3^{2-}

61. Determine whether each of the following compounds is in the correct column. If not, move it to the correct column.

Soluble	*Insoluble*
Hg_2I_2	Hg_2Cl_2
CaS	$Cu_3(PO_4)_2$
LiOH	Na_2S
$PbCl_2$	$PbSO_4$
K_2S	$CaSO_4$
NH_4Cl	SrS
K_2CO_3	AgCl

62. Determine whether each of the following compounds is in the correct column. If not, move it to the correct column.

Soluble	*Insoluble*
PbI_2	Na_2CO_3
$BaCl_2$	$Cu(OH)_2$
Li_2S	$Ca(C_2H_3O_2)_2$
Li_3PO_4	$SrSO_4$
CuI_2	Hg_2Br_2
$Pb(NO_3)_2$	$PbBr_2$
$CoCO_3$	BaS

Precipitation Reactions

63. Complete and balance each of the following equations. If no reaction occurs, write *NO REACTION*.
a) $NH_4Cl(aq) + AgNO_3(aq) \rightarrow$
b) $NaCl(aq) + CaS(aq) \rightarrow$
c) $CrCl_2(aq) + Li_2CO_3(aq) \rightarrow$
d) $KOH(aq) + FeCl_3(aq) \rightarrow$

64. Complete and balance each of the following equations. If no reaction occurs, write *NO REACTION*.
a) $KNO_3(aq) + NaCl(aq) \rightarrow$
b) $KCl(aq) + Hg_2(C_2H_3O_2)_2(aq) \rightarrow$
c) $(NH_4)_2SO_4(aq) + BaCl_2(aq) \rightarrow$
d) $LiI(aq) + SrS(aq) \rightarrow$

65. Write a molecular equation for the precipitation reaction that occurs (if any) when the following solutions are mixed. If no reaction occurs, write *NO REACTION*.
a) sodium carbonate and lead(II) nitrate
b) potassium sulfate and lead(II) acetate
c) copper(II) nitrate and barium sulfide
d) calcium nitrate and sodium iodide

66. Write a molecular equation for the precipitation reaction that occurs (if any) when the following solutions are mixed. If no reaction occurs, write *NO REACTION*.
a) potassium chloride and lead(II) acetate
b) lithium sulfate and strontium chloride
c) potassium bromide and calcium sulfide
d) chromium(III) nitrate and potassium phosphate

67. Determine if each of the following equations for precipitation reactions is correct. If not, write the correct equation. If no reaction occurs, write *NO REACTION*.
a) $Ba(NO_3)_2(aq) + (NH_4)_2SO_4(aq) \rightarrow BaSO_4(s) + 2\,NH_4NO_3(aq)$
b) $BaS(aq) + 2\,KCl(aq) \rightarrow BaCl_2(s) + K_2S(aq)$
c) $2\,KI(aq) + Pb(NO_3)_2(aq) \rightarrow PbI_2(s) + 2\,KNO_3(aq)$
d) $Pb(NO_3)_2(aq) + 2\,LiCl(aq) \rightarrow 2\,LiNO_3(s) + PbCl_2(aq)$

68. Determine if each of the following equations for precipitation reactions is correct. If not, write the correct equation. If no reaction occurs, write *NO REACTION*.
a) $AgNO_3(aq) + NaCl(aq) \rightarrow NaCl(s) + AgNO_3(aq)$
b) $K_2SO_4(aq) + Co(NO_3)_2(aq) \rightarrow CoSO_4(s) + 2\,KNO_3(aq)$
c) $Cu(NO_3)_2(aq) + (NH_4)_2S(aq) \rightarrow CuS(s) + 2\,NH_4NO_3(aq)$
d) $Hg_2(NO_3)_2(aq) + 2\,LiCl(aq) \rightarrow Hg_2Cl_2(s) + 2\,LiNO_3(aq)$

Ionic and Net Ionic Equations

69. Identify the spectator ions in the following complete ionic equation.

$$Pb^{2+}(aq) + 2\,C_2H_3O_2^-(aq) + 2\,K^+(aq) + 2\,Br^-(aq) \rightarrow PbBr_2(s) + 2\,C_2H_3O_2^-(aq) + 2\,K^+(aq)$$

70. Identify the spectator ions in the following complete ionic equation.

$$2\,Li^+(aq) + S^{2-}(aq) + Mg^{2+}(aq) + 2\,Cl^-(aq) \rightarrow MgS(s) + 2\,Li^+(aq) + 2\,Cl^-(aq)$$

71. Write balanced complete ionic and net ionic equations for each of the following reactions.
a) $AgNO_3(aq) + KCl(aq) \rightarrow AgCl(s) + KNO_3(aq)$
b) $CaS(aq) + CuCl_2(aq) \rightarrow CuS(s) + CaCl_2(aq)$
c) $NaOH(aq) + HNO_3(aq) \rightarrow H_2O(l) + NaNO_3(aq)$
d) $2\,K_3PO_4(aq) + 3\,NiCl_2(aq) \rightarrow Ni_3(PO_4)_2(s) + 6\,KCl(aq)$

72. Write balanced complete ionic and net ionic equations for each of the following reactions.
a) $HI(aq) + KOH(aq) \rightarrow H_2O(l) + KI(aq)$
b) $Na_2SO_4(aq) + CaI_2(aq) \rightarrow CaSO_4(s) + 2\,NaI(aq)$
c) $2\,HC_2H_3O_2(aq) + Na_2CO_3(aq) \rightarrow H_2O(l) + CO_2(g) + 2\,NaC_2H_3O_2(aq)$
d) $NH_4Cl(aq) + NaOH(aq) \rightarrow H_2O(l) + NH_3(g) + NaCl(aq)$

73. Mercury ions (Hg_2^{2+}) can be removed from solution by precipitation with Cl^-. Suppose a solution contains aqueous $Hg_2(NO_3)_2$. Write complete ionic and net ionic equations to show the reaction of aqueous $Hg_2(NO_3)_2$ with aqueous sodium chloride to form solid Hg_2Cl_2 and aqueous sodium nitrate.

74. Lead ions can be removed from solution by precipitation with sulfate ions. Suppose a solution contains lead(II) nitrate. Write a complete ionic and net ionic equation to show the reaction of aqueous lead(II) nitrate with aqueous potassium sulfate to form solid lead(II) sulfate and aqueous potassium nitrate.

75. Write complete ionic and net ionic equations for each of the reactions in Problem 65.

76. Write complete ionic and net ionic equations for each of the reactions in Problem 66.

Acid–Base and Gas Evolution Reactions

77. When a hydrochloric acid solution is combined with a potassium hydroxide solution, an acid–base reaction occurs. Write a balanced molecular equation and a net ionic equation for this reaction.

78. A beaker of nitric acid is neutralized with calcium hydroxide. Write a balanced molecular equation and a net ionic equation for this reaction.

79. Complete and balance each of the following equations for acid–base reactions.
a) $HCl(aq) + Ba(OH)_2(aq) \rightarrow$
b) $H_2SO_4(aq) + KOH(aq) \rightarrow$
c) $HClO_4(aq) + NaOH(aq) \rightarrow$

80. Complete and balance each of the following equations for acid–base reactions.
a) $HC_2H_3O_2(aq) + Ca(OH)_2(aq) \rightarrow$
b) $HBr(aq) + LiOH(aq) \rightarrow$
c) $H_2SO_4(aq) + Ba(OH)_2(aq) \rightarrow$

81. Complete and balance each of the following equations for gas evolution reactions.
a) $HBr(aq) + NaHCO_3(aq) \rightarrow$
b) $NH_4I(aq) + KOH(aq) \rightarrow$
c) $HNO_3(aq) + K_2SO_3(aq) \rightarrow$
d) $HI(aq) + Li_2S(aq) \rightarrow$

82. Complete and balance each of the following equations for gas evolution reactions.
a) $HClO_4(aq) + K_2CO_3(aq) \rightarrow$
b) $HC_2H_3O_2(aq) + LiHSO_3(aq) \rightarrow$
c) $(NH_4)_2SO_4(aq) + Ca(OH)_2(aq) \rightarrow$
d) $HCl(aq) + ZnS(s) \rightarrow$

Oxidation–Reduction and Combustion

83. Which of the following reactions are redox reactions?
a) $Ba(NO_3)_2(aq) + K_2SO_4(aq) \rightarrow BaSO_4(s) + 2\,KNO_3(aq)$
b) $Ca(s) + Cl_2(g) \rightarrow CaCl_2(s)$
c) $HCl(aq) + NaOH(aq) \rightarrow H_2O(l) + NaCl(aq)$
d) $Zn(s) + Fe^{2+}(aq) \rightarrow Zn^{2+}(aq) + Fe(s)$

84. Which of the following reactions are redox reactions?
a) $Al(s) + Ag^{+}(aq) \rightarrow Al^{3+}(aq) + Ag(s)$
b) $4\,K(s) + O_2(g) \rightarrow 2\,K_2O(s)$
c) $SO_3(g) + H_2O(l) \rightarrow H_2SO_4(aq)$
d) $Mg(s) + Br_2(l) \rightarrow MgBr_2(s)$

85. Complete and balance each of the following equations for combustion reactions.
a) $C_2H_6(g) + O_2(g) \rightarrow$
b) $Ca(s) + O_2(g) \rightarrow$
c) $C_3H_8O(l) + O_2(g) \rightarrow$
d) $C_4H_{10}S(l) + O_2(g) \rightarrow$

86. Complete and balance each of the following equations for combustion reactions.
a) $S(s) + O_2(g) \rightarrow$
b) $C_7H_{16}(l) + O_2(g) \rightarrow$
c) $C_4H_{10}O(l) + O_2(g) \rightarrow$
d) $CS_2(s) + O_2(g) \rightarrow$

Classifying Chemical Reactions by What Atoms Do

87. Classify each of the following chemical reactions as a synthesis, decomposition, single-displacement, or double-displacement reaction.
a) $K_2S(aq) + Co(NO_3)_2(aq) \rightarrow 2\,KNO_3(aq) + CoS(s)$
b) $3\,H_2(g) + N_2(g) \rightarrow 2\,NH_3(g)$
c) $Zn(s) + CoCl_2(aq) \rightarrow ZnCl_2(aq) + Co(s)$
d) $CH_3Br(g) \xrightarrow[\text{uv light}]{} CH_3(g) + Br(g)$

88. Classify each of the following chemical reactions as a synthesis, decomposition, single-displacement, or double-displacement reaction.
a) $CaSO_4(s) \xrightarrow[\text{heat}]{} CaO(s) + SO_3(g)$
b) $2\,Na(s) + O_2(g) \rightarrow Na_2O_2(s)$
c) $Pb(s) + 2\,AgNO_3(aq) \rightarrow Pb(NO_3)_2(aq) + 2\,Ag(s)$
d) $HI(aq) + NaOH(aq) \rightarrow H_2O(l) + NaI(aq)$

89. NO is a pollutant emitted by motor vehicles. It is formed by the following reaction.
a) $N_2(g) + O_2(g) \rightarrow 2\,NO(g)$
Once in the atmosphere, NO (through a series of reactions) adds one oxygen atom to form NO_2. NO_2 then interacts with UV light according to the following reaction.
b) $NO_2(g) \xrightarrow[\text{uv light}]{} NO(g) + O(g)$
These freshly formed oxygen atoms then react with O_2 in the air to form ozone (O_3), a main component of photochemical smog.
c) $O(g) + O_2(g) \rightarrow O_3(g)$
Classify each of the preceding reactions as a synthesis, decomposition, single-displacement, or double-displacement reaction.

90. A main source of sulfur oxide pollutants are smelters where sulfide ores are converted into metals. The first step in this process is the reaction of the sulfide ore with oxygen in reactions such as:
a) $2\,PbS(s) + 3\,O_2(g) \rightarrow 2\,PbO(s) + 2\,SO_2(g)$
Sulfur dioxide can then react with oxygen in air to form sulfur trioxide.
b) $2\,SO_2(g) + O_2(g) \rightarrow 2\,SO_3(g)$
Sulfur trioxide can then react with water from rain to form sulfuric acid that falls as acid rain.
c) $SO_3(g) + H_2O(l) \rightarrow H_2SO_4(aq)$
Classify each of the preceding reactions as a synthesis, decomposition, single-displacement, or double-displacement reaction.

Cumulative Problems

91. Predict the products of each of these reactions and write balanced complete ionic and net ionic equations for each. If no reaction occurs, write *NO REACTION*.
a) $NaI(aq) + Hg_2(NO_3)_2(aq) \rightarrow$
b) $HClO_4(aq) + Ba(OH)_2(aq) \rightarrow$
c) $Li_2CO_3(aq) + NaCl(aq) \rightarrow$
d) $HCl(aq) + Li_2CO_3(aq) + \rightarrow$

92. Predict the products of each of these reactions and write balanced complete ionic and net ionic equations for each. If no reaction occurs, write *NO REACTION*.
a) $LiCl(aq) + AgNO_3(aq) \rightarrow$
b) $H_2SO_4(aq) + Li_2SO_3(aq) + \rightarrow$
c) $HC_2H_3O_2(aq) + Ca(OH)_2(aq) \rightarrow$
d) $HCl(aq) + KBr(aq) \rightarrow$

93. Predict the products of each of these reactions and write balanced complete ionic and net ionic equations for each. If no reaction occurs, write *NO REACTION*.
a) $BaS(aq) + NH_4Cl(aq) \rightarrow$
b) $NaC_2H_3O_2(aq) + HCl(aq) \rightarrow$
c) $KHSO_3(aq) + HNO_3(aq) \rightarrow$
d) $MnCl_3(aq) + K_3PO_4(aq) \rightarrow$

94. Predict the products of each of these reactions and write balanced complete ionic and net ionic equations for each. If no reaction occurs, write *NO REACTION*.
a) $H_2SO_4(aq) + HNO_3(aq) \rightarrow$
b) $NaOH(aq) + LiOH(aq) \rightarrow$
c) $Cr(NO_3)_3(aq) + LiOH(aq) \rightarrow$
d) $HCl(aq) + Hg_2(NO_3)_2(aq) \rightarrow$

95. Predict the type of reaction (if any) that occurs between each of the following substances. Write balanced molecular equations for each. If no reaction occurs, write *NO REACTION*.
a) aqueous potassium hydroxide and aqueous acetic acid
b) aqueous hydrobromic acid and aqueous potassium carbonate
c) gaseous hydrogen and gaseous oxygen
d) aqueous ammonium chloride and aqueous lead(II) nitrate

96. Predict the type of reaction (if any) that occurs between each of the following substances. Write balanced molecular equations for each. If no reaction occurs, write *NO REACTION*.
a) aqueous hydrochloric acid and aqueous copper(II) nitrate
b) liquid pentanol ($C_5H_{12}O$) and gaseous oxygen
c) aqueous ammonium chloride and aqueous calcium hydroxide
d) aqueous strontium sulfide and aqueous copper(II) sulfate

97. Hard water often contains dissolved Ca^{2+} and Mg^{2+} ions. One way to soften water is to add phosphates. The phosphate ion forms insoluble precipitates with calcium and magnesium ions, removing them from solution. Suppose that a solution contains aqueous calcium chloride and aqueous magnesium nitrate. Write molecular, complete ionic, and net ionic equations showing how the addition of sodium phosphate precipitates the calcium and magnesium ions.

98. Lakes that have been acidified by acid rain (HNO_3 and H_2SO_4) can be neutralized by a process called *liming*, in which limestone ($CaCO_3$) is added to the acidified water. Write ionic and net ionic equations to show how limestone reacts with HNO_3 and H_2SO_4 to neutralize them. How would you be able to tell if the neutralization process was working?

99. A solution contains one or more of the following ions: Ag^+, Ca^{2+}, and Cu^{2+}. When sodium chloride is added to the solution, no precipitate occurs. When sodium sulfate is added to the solution, a white precipitate occurs. The precipitate is filtered off and sodium carbonate is added to the remaining solution producing a precipitate. Which ions were present in the original solution? Write net ionic equations for the formation of each of the precipitates observed.

100. A solution contains one or more of the following ions: Hg_2^{2+}, Ba^{2+}, and Fe^{2+}. When potassium chloride is added to the solution, a precipitate forms. The precipitate is filtered off and potassium sulfate is added to the remaining solution producing no precipitate. When potassium carbonate is added to the remaining solution, a precipitate occurs. Which ions were present in the original solution? Write net ionic equations for the formation of each of the precipitates observed.

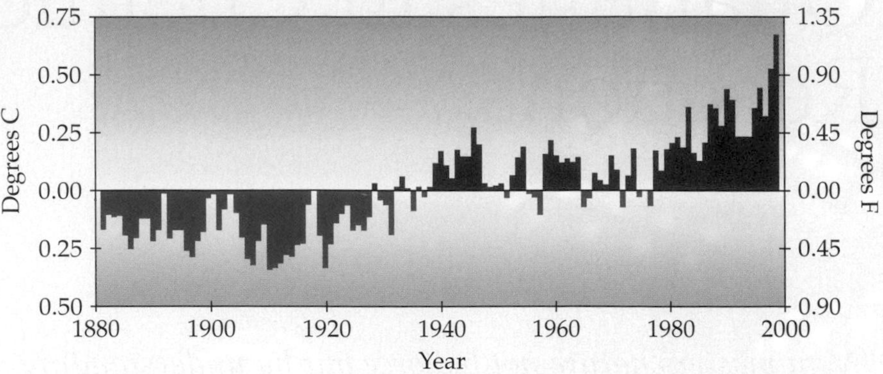

Figure 8.2 Yearly temperature differences from the 120-year average temperature. The earth's average temperature has increased by about 0.6°C since 1880.

The balanced chemical equation shows that 16 moles of CO_2 are produced for every 2 moles of octane burned. Since we know the world's annual fossil fuel consumption, we can estimate the world's annual CO_2 production. A simple calculation shows that the world's annual CO_2 production—from fossil fuel combustion—matches the measured annual atmospheric CO_2 increase, implying that fossil fuel combustion is indeed responsible for increased atmospheric CO_2 levels.

The numerical relationship between chemical quantities in a balanced chemical equation is called reaction **stoichiometry**. Stoichiometry allows us to predict the amounts of products that form in a chemical reaction based on the amounts of reactants. Stoichiometry also allows us to predict how much of the reactants is necessary to form a given amount of product. These calculations are central to chemistry, allowing the chemist to plan and carry out chemical reactions to obtain products in the desired quantities.

8.2 Making Pancakes: Relationships Among Ingredients

The concepts of stoichiometry are similar to the concepts in following a cooking recipe. Calculating the amount of carbon dioxide produced by the combustion of a given amount of a fossil fuel is similar to calculating the number of pancakes that can be made from a given number of eggs. For example, suppose you use the following pancake recipe.

$$1 \text{ cup flour} + 2 \text{ eggs} + \tfrac{1}{2} \text{ tsp baking powder} \rightarrow 5 \text{ pancakes}$$

A recipe gives numerical relationships between the ingredients and the number of pancakes.

The recipe shows the numerical relationships between the pancake ingredients. It says that if you have 2 eggs—and enough of everything else—you can make 5 pancakes. We can write this relationship as an equivalence.

$$2 \text{ eggs} \equiv 5 \text{ pancakes}$$

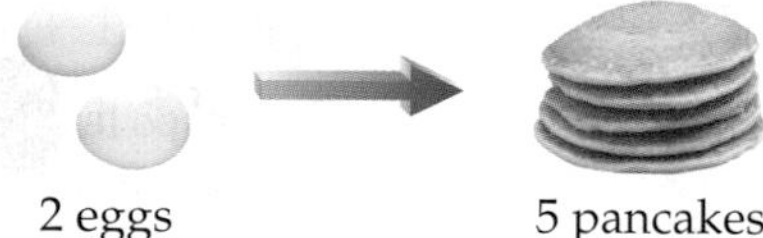

From our recipe, we know that 2 eggs are enough to make 5 pancakes; therefore 8 eggs are enough to make 20 pancakes.

The $\equiv$ sign, which we first encountered in chapter 6, means **equivalent** to. Therefore, 2 eggs are equivalent to 5 pancakes meaning that, for this recipe, they must occur in that ratio for the pancakes to turn out right. What if you have 8 eggs? Assuming that you have enough of everything else, how many pancakes can you make? Using the preceding equivalence as a conversion factor, 8 eggs are sufficient to make 20 pancakes.

$$8 \cancel{\text{eggs}} \times \frac{5 \text{ pancakes}}{2 \cancel{\text{eggs}}} = 20 \text{ pancakes}$$

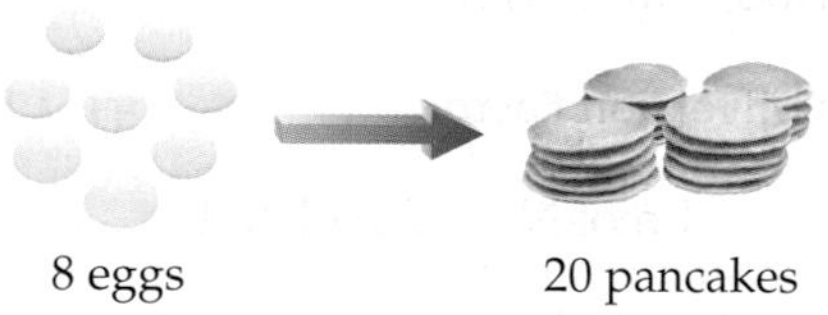

The pancake recipe contains numerical conversion factors between the pancake ingredients and the number of pancakes. Other conversion factors from this recipe include:

$$1 \text{ cup flour} \equiv 5 \text{ pancakes}$$

$$\tfrac{1}{2} \text{ tsp baking powder} \equiv 5 \text{ pancakes}$$

The recipe also gives us relationships among the ingredients themselves. For example, how much baking powder is required to go with 3 cups of flour? From the recipe:

$$1 \text{ cup flour} \equiv \tfrac{1}{2} \text{ tsp baking powder}$$

With this conversion factor, we can find the appropriate amount of baking powder.

$$3 \cancel{\text{cups flour}} \times \frac{\tfrac{1}{2} \text{ tsp baking powder}}{1 \cancel{\text{cup flour}}} = \tfrac{3}{2} \text{ tsp baking powder}$$

8.5 More Pancakes: Limiting Reactant, Theoretical Yield, and Percent Yield

Let's return to our pancake analogy to understand two more concepts important in reaction stoichiometry: limiting reactant and percent yield. Recall our pancake recipe:

$$1 \text{ cup flour} + 2 \text{ eggs} + \tfrac{1}{2} \text{ tsp baking powder} \rightarrow 5 \text{ pancakes}$$

Suppose we have 3 cups flour, 10 eggs, and 4 tsp baking powder. How many pancakes can we make?

We have enough flour to make:

$$3 \cancel{\text{cups flour}} \times \frac{5 \text{ pancakes}}{1 \cancel{\text{cup flour}}} = 15 \text{ pancakes}$$

We have enough eggs to make:

$$10 \cancel{\text{eggs}} \times \frac{5 \text{ pancakes}}{2 \cancel{\text{eggs}}} = 25 \text{ pancakes}$$

We have enough baking powder to make:

$$4 \cancel{\text{tsp baking powder}} \times \frac{5 \text{ pancakes}}{\frac{1}{2} \cancel{\text{tsp baking powder}}} = 40 \text{ pancakes}$$

We have enough flour for 15 pancakes, enough eggs for 25 pancakes, and enough baking powder for 40 pancakes. Consequently, unless we get more ingredients, we can only make 15 pancakes. The flour *limits* the number of pancakes we can make. If this were a chemical reaction, the flour would be the **limiting reactant**, the reactant that limits the amount of product in a chemical reaction. Notice that the limiting reactant is simply the reactant that makes *the least amount of product*. If this were a chemical reaction, 15 pancakes would be the **theoretical yield**, the amount of product that can be made in a chemical reaction based on the amount of limiting reactant.

The term limiting *reagent* is sometimes used in place of limiting *reactant*.

If this were a chemical reaction, the flour would be the limiting reactant and 15 pancakes would be the theoretical yeild.

Let us carry this analogy one step further. Suppose we go on to cook our pancakes. We accidentally burn three of them and one falls on the floor. So even though we had enough flour for 15 pancakes, we finished with only 11 pancakes. If this were a chemical reaction, the 11 pancakes would be our **actual yield**, the amount of product actually produced by a chemical reaction. Finally, our **percent yield**, the percentage of the theoretical yield that was actually attained, is:

$$\text{Percent yield} = \frac{11 \text{ pancakes}}{15 \text{ pancakes}} \times 100\% = 73\%$$

Since four of the pancakes were ruined, we only got 73% of our theoretical yield.

To summarize:

- ***Limiting reactant (or limiting reagent)*–the reactant that is completely consumed in a chemical reaction.**
- ***Theoretical yield*–the amount of product that can be made in a chemical reaction based on the amount of limiting reactant.**
- ***Actual yield*–the amount of product actually produced by a chemical reaction.**
- ***Percent yield*–** $\frac{\textbf{Actual yield}}{\textbf{Theoretical yield}} \times \mathbf{100\%}$

Now consider the following reaction.

$$Ti(s) + 2\,Cl_2(g) \rightarrow TiCl_4(s)$$

If we begin with 1.8 mol titanium and 3.2 mol chlorine, what is the limiting reactant and theoretical yield of $TiCl_4$ in moles? We can set up the problem according to our standard problem solving procedure.

Given: 1.8 mol Ti, 3.2 mol Cl_2

Find: limiting reactant, theoretical yield

Conversion Factors:

The conversion factors come from the balanced chemical equation and give the relationships between moles of each of the reactants and moles of product.

$$1 \text{ mol Ti} \equiv 1 \text{ mol } TiCl_4$$

$$2 \text{ mol } Cl_2 \equiv 1 \text{ mol } TiCl_4$$

Like our pancake analogy, we find the limiting reactant by calculating how much product can be made from each reactant. The reactant that makes the *least amount of product* is the limiting reactant.

Solution Map:

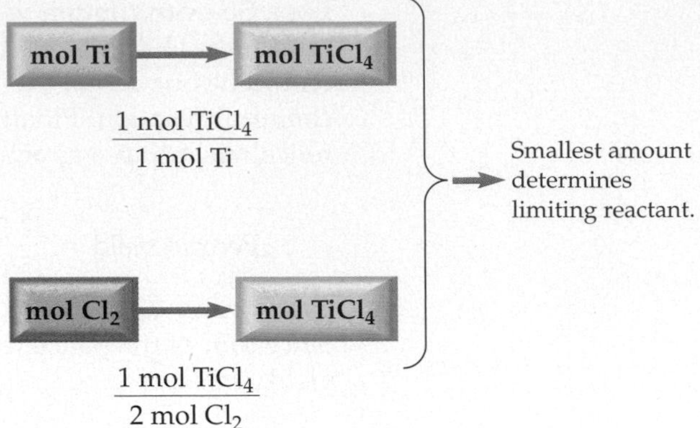

Solution:

$$1.8 \text{ mol Ti} \times \frac{1 \text{ mol TiCl}_4}{1 \text{ mol Ti}} = 1.8 \text{ mol TiCl}_4$$

$$3.2 \text{ mol Cl}_2 \times \frac{1 \text{ mol TiCl}_4}{2 \text{ mol Cl}_2} = 1.6 \text{ mol TiCl}_4$$

Limiting reactant **Least amount of product**

Since the 3.2 mol of Cl_2 make the least amount of $TiCl_4$, Cl_2 is the limiting reactant. Notice that we began with more moles of Cl_2 than Ti, but since the reaction requires 2 Cl_2 for each Ti, Cl_2 is still the limiting reagent. The theoretical yield is 1.6 mol of $TiCl_4$.

In many industrial applications, the more costly reactant or the reactant that is most difficult to remove from the product mixture is chosen to be the limiting reactant.

EXAMPLE 8.4 Limiting Reactant and Theoretical Yield From Initial Moles of Reactants

Consider the following reaction.

$$2\,Al(s) + 3\,Cl_2(g) \rightarrow 2\,AlCl_3(s)$$

If we begin with 0.552 mol aluminum and 0.887 mol chlorine, what is the limiting reactant and theoretical yield of $AlCl_3$ in moles? We set up the problem according to our problem solving procedure.

Given: 0.552 mol Al, 0.887 mol Cl_2

Find: limiting reagent, theoretical yield

Conversion Factors:

From the balanced chemical equation we have:

$$2 \text{ mol Al} \equiv 2 \text{ mol AlCl}_3$$
$$3 \text{ mol Cl}_2 \equiv 2 \text{ mol AlCl}_3$$

The solution map shows how to get from moles of each reactant to mol $AlCl_3$. The reactant that makes the *least amount of $AlCl_3$* is the limiting reactant.

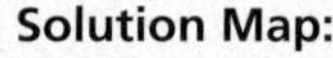

Solution Map:

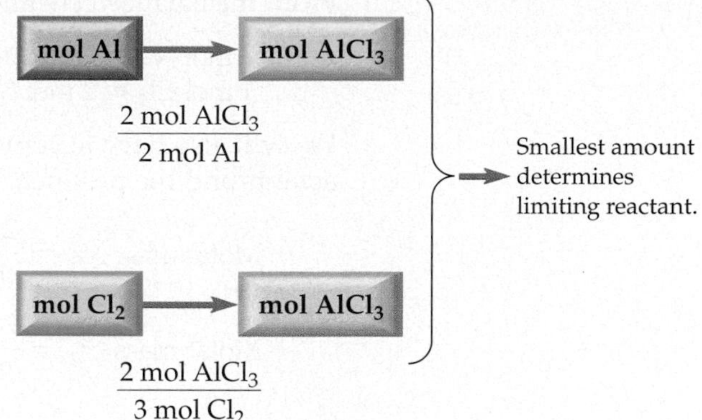

Solution:

$$0.552 \text{ mol Al} \times \frac{2 \text{ mol } AlCl_3}{2 \text{ mol Al}} = 0.552 \text{ mol } AlCl_3$$

Limiting reactant — Least amount of product

$$0.887 \text{ mol } Cl_2 \times \frac{2 \text{ mol } AlCl_3}{3 \text{ mol } Cl_2} = 0.591 \text{ mol } AlCl_3$$

Since the 0.552 mol of Al makes the least amount of $AlCl_3$, Al is the limiting reactant. The theoretical yield is 0.552 mol of $AlCl_3$.

SKILLBUILDER 8.4 **Limiting Reactant and Theoretical Yield From Initial Moles of Reactants**

Consider the following reaction.

$$2\,Na(s) + F_2(g) \rightarrow 2\,NaF(s)$$

If we begin with 4.8 mol sodium and 2.6 mol fluorine, what is the limiting reactant and theoretical yield of NaF in moles?

8.6 Limiting Reactant, Theoretical Yield, and Percent Yield From Initial Masses of Reactants

When working in the laboratory, we normally measure the initial amounts of reactants in grams. To find limiting reagents and theoretical yields from initial masses, we must add two steps to our calculations. Consider, for example, the following synthesis reaction.

$$2\,Na(s) + Cl_2(g) \rightarrow 2\,NaCl(s)$$

If we have 53.2 g Na and 65.8 g Cl_2, what is the limiting reactant and theoretical yield? We will approach this problem in our standard way.

Given: 53.2 g Na, 65.8 g Cl_2

Find: limiting reactant, theoretical yield

Conversion Factors:

From the balanced chemical equation, we get:

$$2 \text{ mol Na} \equiv 2 \text{ mol NaCl}$$
$$1 \text{ mol } Cl_2 \equiv 2 \text{ mol NaCl}$$

We will also have to convert between grams and moles for each of the reactants and the product, so we also need molar masses.

$$\text{Molar mass Na} = \frac{22.98 \text{ g Na}}{1 \text{ mol Na}}$$

$$\text{Molar mass } Cl_2 = \frac{70.91 \text{ g } Cl_2}{1 \text{ mol } Cl_2}$$

$$\text{Molar mass NaCl} = \frac{58.44 \text{ g NaCl}}{1 \text{ mol NaCl}}$$

We again find the limiting reactant by calculating how much product can be made from each reactant. Since we are given the initial amounts in grams, we must first convert to moles. After we convert to moles of product, we convert back to grams of product. The reactant that makes the *least amount of product* is the limiting reactant.

Solution Map:

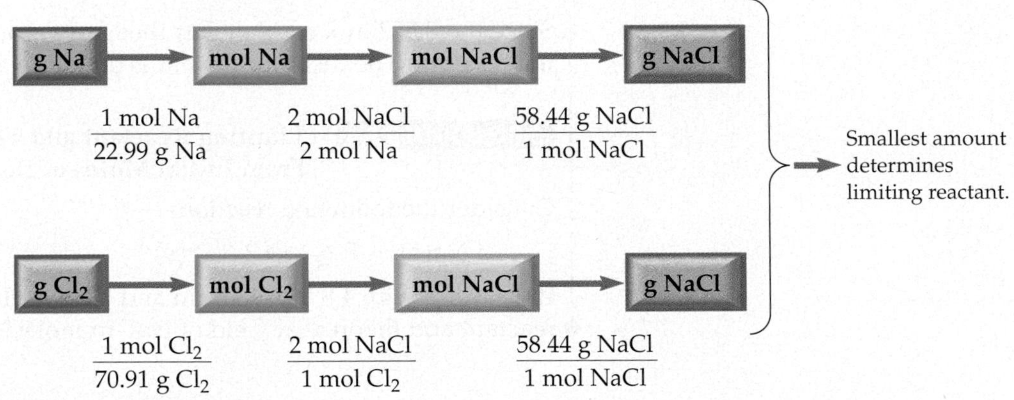

The limiting reactant can also be found by calculating the number of moles of NaCl (rather than grams) that can be made from each reactant. However, since theoretical yields are normally calculated in grams, we take the calculation all the way to grams to determine limiting reagent.

Solution:

Beginning with the actual amounts of each reactant, we follow the solution map to calculate how much product can be made from each.

$$53.2 \text{ g Na} \times \frac{1 \text{ mol Na}}{22.99 \text{ g Na}} \times \frac{2 \text{ mol NaCl}}{2 \text{ mol Na}} \times \frac{58.44 \text{ g NaCl}}{1 \text{ mol NaCl}} = 135 \text{ g NaCl}$$

$$65.8 \text{ g } Cl_2 \times \frac{1 \text{ mol } Cl_2}{70.91 \text{ g } Cl_2} \times \frac{2 \text{ mol NaCl}}{1 \text{ mol } Cl_2} \times \frac{58.44 \text{ g NaCl}}{1 \text{ mol NaCl}} = 108 \text{ g NaCl}$$

Limiting reactant (65.8 g Cl_2); Least amount of product (108 g NaCl)

The limiting reactant is NOT necessarily the reactant with the least mass.

Since Cl_2 makes the least amount of product (108 g NaCl is less than 135 g NaCl), it is the limiting reactant. Notice that the limiting reactant is not necessarily the reactant with least mass. In this case, we had fewer grams of Na than Cl_2, yet Cl_2 was the limiting reactant because it made less

NaCl. The theoretical yield is therefore 108 g of NaCl, the amount of product possible based on the limiting reactant.

Now suppose that when the synthesis was carried out, the actual yield of NaCl was 86.4 g. What is the percent yield? The percent yield is simply:

The actual yield is always less than the theoretical yield because at least a small amount of product is usually lost or does not form during a reaction.

$$\text{Percent yield} = \frac{\text{Actual yield}}{\text{Theoretical yield}} \times 100\% = \frac{86.4\text{ g}}{108\text{ g}} \times 100\% = 80.0\%$$

EXAMPLE 8.5 Finding Limiting Reactant and Theoretical Yield

Ammonia, NH_3, can be synthesized by the following reaction.

$$2\,NO(g) + 5\,H_2(g) \rightarrow 2\,NH_3(g) + 2\,H_2O(g)$$

What is the maximum amount of ammonia that can be synthesized from 45.8 g NO and 12.4 g H_2?

Although this problem does not specifically ask for the limiting reactant, it must be found to determine the theoretical yield, which is the maximum amount of ammonia that can be synthesized.

We set up the problem in the standard format.

Given: 45.8 g NO, 12.4 g H_2

Find: maximum amount of NH_3 (theoretical yield)

Conversion Factors:

$$2\text{ mol NO} \equiv 2\text{ mol NH}_3$$

$$5\text{ mol H}_2 \equiv 2\text{ mol NH}_3$$

$$\text{Molar mass NO} = \frac{30.01\text{ g NO}}{1\text{ mol NO}}$$

$$\text{Molar mass H}_2 = \frac{2.02\text{ g H}_2}{1\text{ mol H}_2}$$

$$\text{Molar mass NH}_3 = \frac{17.04\text{ g NH}_3}{1\text{ mol NH}_3}$$

We find the limiting reactant by calculating how much product can be made from each reactant. The reactant that makes the *least amount of product* is the limiting reactant.

Solution Map:

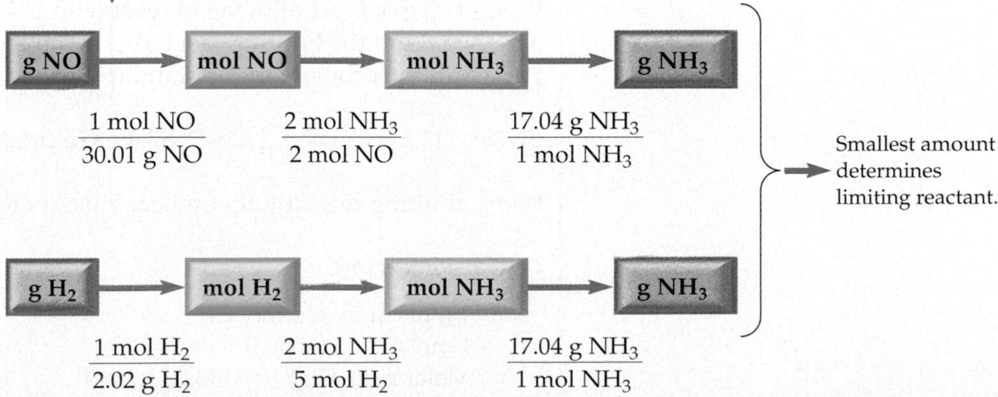

Solution:

We then follow the solution map, beginning with the actual amount of each reactant given, to calculate the amount of product that can be made from each reactant.

$$45.8 \text{ g NO} \times \frac{1 \text{ mol NO}}{30.01 \text{ g NO}} \times \frac{2 \text{ mol NH}_3}{2 \text{ mol NO}} \times \frac{17.04 \text{ g NH}_3}{1 \text{ mol NH}_3} = 26.0 \text{ g NH}_3$$

Limiting reactant — Least amount of product

$$12.4 \text{ g H}_2 \times \frac{1 \text{ mol H}_2}{2.02 \text{ g H}_2} \times \frac{2 \text{ mol NH}_3}{5 \text{ mol H}_2} \times \frac{17.04 \text{ g NH}_3}{1 \text{ mol NH}_3} = 41.8 \text{ g NH}_3$$

We have enough NO to make 26.0 g NH_3 and enough H_2 to make 41.8 g NH_3. Therefore, NO is the limiting reactant and the maximum amount of ammonia that we can possibly make is 26.0 g, the theoretical yield.

SKILLBUILDER 8.5 **Finding Limiting Reactant and Theoretical Yield**

Ammonia can also be synthesized by the following reaction.

$$3\, H_2(g) + N_2(g) \rightarrow 2\, NH_3(g)$$

What is the maximum amount of ammonia that can be synthesized from 25.2 g N_2 and 8.42 g H_2?

SKILLBUILDER PLUS

What is the maximum amount of ammonia, in kilograms, that can be synthesized from 5.22 kg of H_2 and 31.5 kg of N_2?

EXAMPLE 8.6 **Finding Limiting Reactant, Theoretical Yield, and Percent Yield**

Consider the following reaction.

$$Cu_2O(s) + C(s) \rightarrow 2\, Cu(s) + CO(g)$$

When 11.5 g of C are allowed to react with 114.5 g of Cu_2O, 87.4 g of Cu are obtained. Find the limiting reactant, theoretical yield, and percent yield. We set up the problem in the standard format.

Given: 11.5 g C; 114.5 g Cu_2O; 87.4 g Cu produced

Find: limiting reactant, theoretical yield, percent yield

Conversion Factors:

$1 \text{ mol } Cu_2O \equiv 2 \text{ mol Cu}$

$1 \text{ mol C} \equiv 2 \text{ mol Cu}$

Molar mass Cu_2O = 143.08 g/mol

Molar mass C = 12.01 g/mol
Molar mass Cu = 63.54 g/mol

The solution map shows how to find the mass of Cu formed by each of the initial masses of Cu_2O and C. The reactant that makes the least amount of product is the limiting reactant and determines the theoretical yield.

Solution Map:

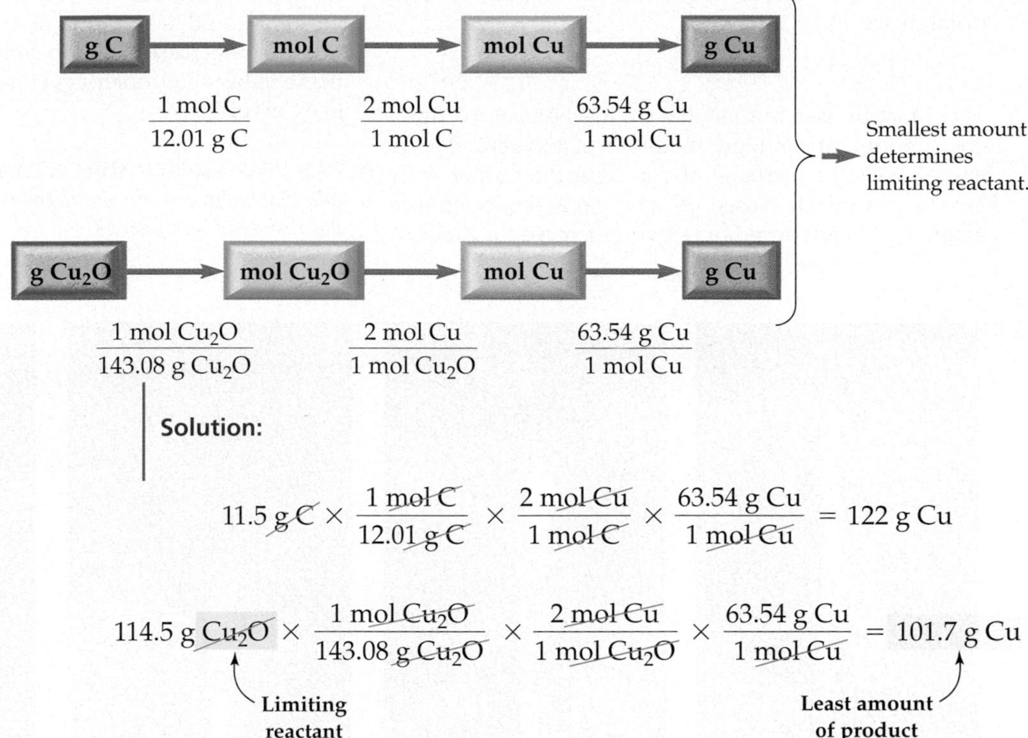

Solution:

$$11.5\ \text{g C} \times \frac{1\ \text{mol C}}{12.01\ \text{g C}} \times \frac{2\ \text{mol Cu}}{1\ \text{mol C}} \times \frac{63.54\ \text{g Cu}}{1\ \text{mol Cu}} = 122\ \text{g Cu}$$

$$114.5\ \text{g Cu}_2\text{O} \times \frac{1\ \text{mol Cu}_2\text{O}}{143.08\ \text{g Cu}_2\text{O}} \times \frac{2\ \text{mol Cu}}{1\ \text{mol Cu}_2\text{O}} \times \frac{63.54\ \text{g Cu}}{1\ \text{mol Cu}} = 101.7\ \text{g Cu}$$

Since Cu_2O makes the least amount of product, Cu_2O is the limiting reactant. The theoretical yield is simply the amount of product made by the limiting reactant.

Theoretical yield = 101.7 g Cu

The percent yield is:

$$\text{Percent yield} = \frac{\text{Actual yield}}{\text{Theoretical yield}} \times 100\%$$

$$= \frac{87.4\ \text{g}}{101.7\ \text{g}} \times 100\% = 85.9\%$$

SKILLBUILDER 8.6 **Finding Limiting Reactant, Theoretical Yield, and Percent Yield**

The following reaction is used to obtain iron from iron ore:

$$Fe_2O_3(s) + 3\,CO(g) \rightarrow 2\,Fe(s) + 3\,CO_2(g)$$

The reaction of 185 g Fe_2O_3 with 95.3 g CO produces 87.4 g Fe. Find the limiting reactant, theoretical yield, and percent yield.

EVERYDAY CHEMISTRY

Bunsen Burners

In the laboratory, we often use Bunsen burners as heat sources. These burners are normally fueled by methane. The balanced equation for methane (CH_4) combustion is:

$$CH_4(g) + 2\,O_2(g) \rightarrow CO_2(g) + 2\,H_2O(g)$$

Most Bunsen burners have a mechanism to adjust the amount of air (and therefore of oxygen) that is mixed with the methane. If you light the burner with the air completely closed off, you get a yellow, smoky flame that is not very hot. As you increase the amount of air going into the burner, the flame becomes bluer, less smoky, and hotter. When you reach the optimum adjustment, the flame has a sharp, inner blue triangle, no smoke, and is hot enough to easily melt glass. Continuing to increase the air beyond this point causes the flame to become cooler again and may cause the flame to extinguish.

CAN YOU ANSWER THIS? *Can you use the concepts from this chapter to explain the changes in the Bunsen burner as the air intake is adjusted?*

(a) No air

(b) Small amount of air

(c) Optimum

(d) Too much air

Bunsen burner at various stages of air intake adjustment.

CHAPTER IN REVIEW

Chemical Principles

Stoichiometry: A balanced chemical equation gives quantitative relationships between the amounts of reactants and products. For example, the reaction $2\,H_2 + O_2 \rightarrow 2\,H_2O$ says that 2 mol H_2 reacts with 1 mol O_2 to form 2 mol H_2O. These relationships can be used to calculate quantities such as the amount of product possible with a certain amount of reactant, or the amount of one reactant required to completely react with a certain amount of another reactant. The quantitative relationships between reactants and products in a chemical reaction are called reaction stoichiometry.

Relevance

Stoichiometry: Reaction stoichiometry is important because we often want to know numerical relationships between the reactants and products in a chemical reaction. For example, we might want to know how much carbon dioxide, a greenhouse gas, is formed upon burning a certain amount of a particular fossil fuel.

Limiting Reactant, Theoretical Yield, and Percent Yield: The limiting reactant in a chemical reaction is the reactant that, based on the amounts of each reactant used in a particular instance, limits the amount of product that can be made. The theoretical yield in a chemical reaction is the amount of product that can be made based on the amount of the limiting reactant. The actual yield in a chemical reaction is the amount of product actually produced. The percent yield in a chemical reaction is the actual yield divided by theoretical yield times 100%.

Limiting Reactant, Theoretical Yield, and Percent Yield: Calculations of limiting reactant, theoretical yield, and percent yield are central to chemistry because they bring quantitative understanding to chemical reactions. Just as you need to know relationships between ingredients to follow a recipe, so you must know relationships between reactants and products to carry out a chemical reaction. The percent yield in a chemical reaction is often used as a measure of the success of the reaction. Imagine following a recipe and only making 1% of the final product—your cooking would be a failure. Similarly, low percent yields in chemical reactions are usually considered poor, and high percent yields are considered good.

Chemical Skills

Examples

Mole-to-Mole Conversions (Section 8.3)

For mole-to-mole conversions, we follow our standard problem-solving procedure, writing down the given quantities, find quantities, and conversion factors. The conversion factor comes from the balanced chemical equation.

The solution map begins with moles of the given substance and then uses the conversion factor from the balanced chemical equation to get to moles of the substance you are trying to find.

To compute the answer, begin with the number of moles of the given substance and follow the solution map to get to the number of moles of the substance you are trying to find.

EXAMPLE 8.7 Mole-to-Mole Conversions

How many moles of sodium oxide can be synthesized from 4.8 mol sodium? Assume that there is more than enough oxygen present. The balanced equation is:

$$4\,\text{Na}(s) + \text{O}_2(g) \rightarrow 2\,\text{Na}_2\text{O}(s)$$

Given: 4.8 mol Na

Find: mol Na_2O

Conversion Factor:

$$4\text{ mol Na} \equiv 2\text{ mol Na}_2\text{O}$$

Solution Map:

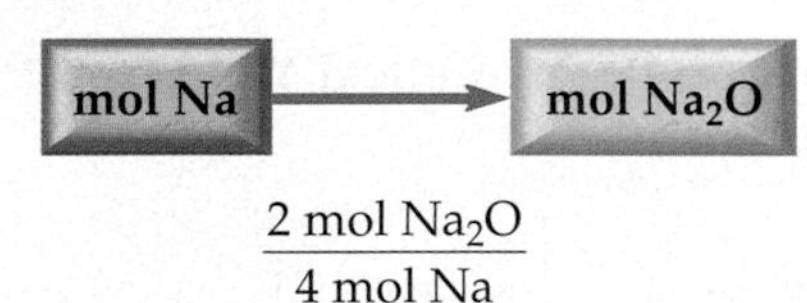

$$\frac{2\text{ mol Na}_2\text{O}}{4\text{ mol Na}}$$

Solution:

$$4.8\,\cancel{\text{mol Na}} \times \frac{2\text{ mol Na}_2\text{O}}{4\,\cancel{\text{mol Na}}} = 2.4\text{ mol Na}$$

Mass-to-Mass Conversions (Section 8.4)

EXAMPLE 8.8 Mass-to-Mass Conversions

How many grams of sodium oxide can be synthesized from 17.4 g sodium? Assume that there is more than enough oxygen present. The balanced equation is:

$$4\,\text{Na}(s) + \text{O}_2(g) \rightarrow 2\,\text{Na}_2\text{O}(s)$$

Given: 17.4 g Na

For mass-to-mass conversions, we also follow our standard problem-solving procedure, writing down the given quantities, find quantities, and conversion factors. The required conversion factors include those from the balanced chemical equation as well as the molar masses of the substances involved.

Find: g Na_2O

Conversion Factors:

4 mol Na = 2 mol Na_2O
Molar mass Na = 22.99 g/mol
Molar mass Na_2O = 61.98 g/mol

Draw the solution map by beginning with the mass of the given substance. Convert to moles using the molar mass and then convert to moles of the substance you are trying to find using the conversion factor obtained from the balanced chemical equation. Finally convert to mass of the substance you are trying to find using its molar mass.

Solution Map:

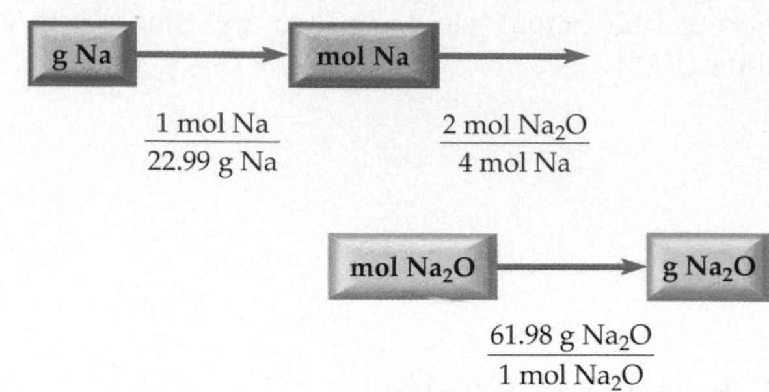

Compute the answer by beginning with the amount of the given substance and multiplying by the appropriate conversion factors to get to the mass of the substance you are trying to find.

Solution:

$$17.4 \text{ g Na} \times \frac{1 \text{ mol Na}}{22.99 \text{ g Na}} \times \frac{2 \text{ mol } Na_2O}{4 \text{ mol Na}} \times \frac{61.98 \text{ g } Na_2O}{1 \text{ mol } Na_2O} = 23.5 \text{ g } Na_2O$$

Limiting Reactant, Theoretical Yield, and Percent Yield (Sections 8.5, 8.6)

EXAMPLE 8.9 Limiting Reactant, Theoretical Yield, and Percent Yield

10.4 g Fe react with 11.8 g S to produce 14.2 g Fe_2S_3. Find the limiting reactant, theoretical yield, and percent yield for this reaction. The balanced chemical equation is:

$$2\,Fe(s) + 3\,S(l) \rightarrow Fe_2S_3(s)$$

For limiting reactant problems, we also follow our standard problem-solving procedure, writing down the given quantities, find quantities, and conversion factors. The required conversion factors include those from the balanced chemical equation as well as the molar masses of the substances involved.

Given: 10.4 g Fe; 11.8 g S; 14.2g Fe_2S_3

Find: limiting reactant, theoretical yield, percent yield

Conversion Factors:

2 mol Fe = 1 mol Fe_2S_3
3 mol S = 1 mol Fe_2S_3
Molar mass Fe = 55.85 g/mol
Molar mass S = 32.06 g/mol
Molar mass Fe_2S_3 = 207.9 g/mol

The solution map for limiting reactant problems shows how to convert from mass of each of the reactants to mass of the product for each reactant. These are mass-to-mass conversions involving the basic outline of mass → moles → moles → mass. The reactant that forms the least amount of product is the limiting reactant.

Solution Map:

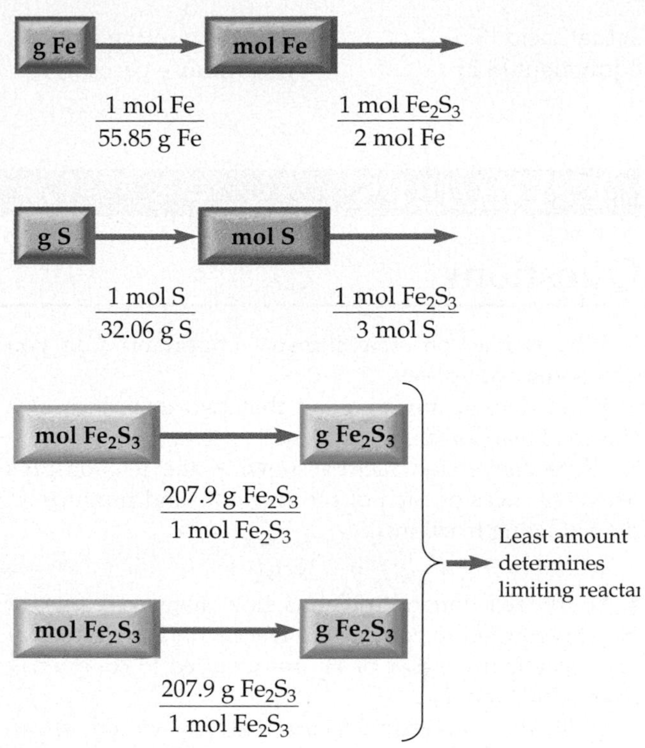

To compute the amount of product formed by each reactant, begin with the given amount of each reactant and multiply by the appropriate conversion factors, as shown in the solution map, to arrive at the mass of product for each reactant. The reactant that forms the least amount of product is the limiting reactant.

Solution:

$$10.4 \text{ g Fe} \times \frac{1 \text{ mol Fe}}{55.85 \text{ g Fe}} \times \frac{1 \text{ mol } Fe_2S_3}{2 \text{ mol Fe}} \times \frac{207.9 \text{ g } Fe_2S_3}{1 \text{ mol } Fe_2S_3} = 19.4 \text{ g } Fe_2S_3$$

Limiting reactant (Fe); **Least amount of product** (19.4 g Fe_2S_3)

$$11.8 \text{ g S} \times \frac{1 \text{ mol S}}{32.06 \text{ g S}} \times \frac{1 \text{ mol } Fe_2S_3}{3 \text{ mol S}} \times \frac{207.9 \text{ g } Fe_2S_3}{1 \text{ mol } Fe_2S_3} = 25.5 \text{ g } Fe_2S_3$$

The theoretical yield is the amount of product formed by the limiting reactant.

The percent yield is the actual yield divided by the theoretical yield times 100%.

The limiting reactant is Fe.

The theoretical yield is 19.4 g Fe_2S_3.

$$\text{Percent yield} = \frac{\text{Actual yield}}{\text{Theoretical yield}} \times 100\% = \frac{14.2 \text{ g}}{19.4 \text{ g}} \times 100\% = 73.2\%$$

KEY TERMS

actual yield [8.5]
equivalent [8.2]
global warming [8.1]
greenhouse gases [8.1]
limiting reactant [8.5]
percent yield [8.5]
stoichiometry [8.1]
theoretical yield [8.5]

EXERCISES

Questions

1. Why is reaction stoichiometry important? Can you give some examples?
2. What does it mean to say that two quantities in a chemical reaction are equivalent?
3. Write conversion factors showing the relationships between moles of each of the reactants and products in the following reaction.

$$N_2(g) + 3\,H_2(g) \rightarrow 2\,NH_3(g)$$

4. For the reaction in Problem 3, how many molecules of H_2 are required to completely react with 2 molecules of N_2? How many moles of H_2 are required to completely react with 2 mol N_2?
5. Write the conversion factor that you would use to convert *from* moles of Cl_2 *to* moles of NaCl in the following reaction.

$$2\,Na(s) + Cl_2(g) \rightarrow 2\,NaCl$$

6. What is wrong with this statement in reference to the reaction in Problem 5?
"Two grams of Na react with 1 g of Cl_2 to form 2 g NaCl." Correct the statement to make it true.
7. What is the general form of the solution map for problems in which you are given the mass of a reactant in a chemical reaction and asked to find the mass of the product that can be made from the given amount of reactant?
8. Consider the following recipe for making tomato and garlic pasta.

2 cups noodles + 12 tomatoes + 3 cloves garlic → 4 servings pasta

If you have 7 cups of noodles, 27 tomatoes, and 9 garlic cloves, how many servings of pasta can you make? Which ingredient limits the amount of pasta that is possible?
9. In a chemical reaction, what is the limiting reactant?
10. In a chemical reaction, what is the theoretical yield?
11. In a chemical reaction, what are the actual yield and percent yield?
12. If you are given a chemical equation and specific amounts of each reactant in grams, how would you determine how much product can possibly be made?
13. Consider the following generic chemical reaction.

$$A + 2\,B \rightarrow C + D$$

Suppose you have 12 g of A and 24 g of B. Which of the following statements are true?
a) A will definitely be the limiting reactant.
b) B will definitely be the limiting reactant.
c) A will be the limiting reactant if its molar mass is less than B.
d) A will be the limiting reactant if its molar mass is greater than B.
14. Consider the following generic chemical equation.

$$A + B \rightarrow C$$

Suppose 25 g of A were allowed to react with 8 g of B. Analysis of the final mixture showed that A was completely used up and 4 g of B remained. What was the limiting reactant?

Problems

Mole-to-Mole Conversions

15. Consider the following generic chemical reaction.

$$A + 2\,B \rightarrow C$$

How many moles of C are formed upon complete reaction of:
a) 1 mole A
b) 1 mole B
c) 2 mole A
d) 2 mole B

16. Consider the following generic chemical reaction.

$$2\,A + 3\,B \rightarrow 3\,C$$

How many moles of B are required to completely react with:
a) 2 moles A
b) 4 moles A
c) 5 moles A
d) 17 moles A

17. For the reaction shown, calculate how many moles of NO_2 form when each of the following completely reacts.

$$2\,N_2O_5(g) \rightarrow 4\,NO_2(g) + O_2(g)$$

a) 1.3 mol N_2O_5
b) 5.8 mol N_2O_5
c) 4.45×10^3 mol N_2O_5
d) 1.006×10^{-3} mol N_2O_5

18. For the reaction shown, calculate how many moles of NH_3 form when each of the following completely reacts.

$$3\,N_2H_4(l) \rightarrow 4\,NH_3(g) + N_2(g)$$

a) 5.3 mol N_2H_4
b) 2.28 mol N_2H_4
c) 5.8×10^{-2} mol N_2H_4
d) 9.76×10^7 mol N_2H_4

19. For each reaction, calculate how many moles of the product form when 1.48 mol of the reactant in italics completely reacts. Assume that there is more than enough of the other reactant.
a) $\mathit{H_2}(g) + Cl_2(g) \rightarrow 2\,HCl(g)$
b) $2\,H_2(g) + \mathit{O_2}(g) \rightarrow 2\,H_2O(l)$
c) $2\,\mathit{Na}(s) + O_2(g) \rightarrow Na_2O_2(s)$
d) $2\,S(s) + 3\,\mathit{O_2}(g) \rightarrow 2\,SO_3(g)$

20. For each reaction, calculate how many moles of the product form when 0.036 mol of the reactant in italics completely reacts. Assume that there is more than enough of the other reactant.
a) $2\,Ca(s) + \mathit{O_2}(g) \rightarrow 2\,CaO(s)$
b) $4\,Fe(s) + 3\,\mathit{O_2}(g) \rightarrow 2\,Fe_2O_3(s)$
c) $4\,\mathit{K}(s) + O_2(g) \rightarrow 2\,K_2O(s)$
d) $4\,Al(s) + 3\,\mathit{O_2}(g) \rightarrow 2\,Al_2O_3(s)$

21. For the reaction shown, calculate how many moles of each product form when the following amounts of each reactant completely react to form products. Assume that there is more than enough of the other reactant.

$$2\,PbS(s) + 3\,O_2(g) \rightarrow 2\,PbO(s) + 2\,SO_2(g)$$

a) 1.2 mol PbS
b) 1.2 mol O_2
c) 7.9 mol PbS
d) 7.9 mol O_2

22. For the reaction shown, calculate how many moles of each product form when the following amounts of each reactant completely react to form products. Assume that there is more than enough of the other reactant.

$$C_3H_8(g) + 5\,O_2(g) \rightarrow 3\,CO_2(g) + 4\,H_2O(g)$$

a) 8.7 mol C_3H_8
b) 8.7 mol O_2
c) 0.0918 mol C_3H_8
d) 0.0918 mol O_2

23. Consider the following balanced equation.

$$2\,N_2H_4(g) + N_2O_4(g) \rightarrow 3\,N_2(g) + 4\,H_2O(g)$$

Complete the following table showing the appropriate number of moles of reactants and products. If the number of moles of a reactant is provided, fill in the required amount of the other reactant, as well as the moles of each product formed. If the number of moles of a product is provided, fill in the required amount of each reactant to make that amount of product, as well as the amount of the other product that is made.

mol N_2H_4	*mol N_2O_4*	*mol N_2*	*mol H_2O*
4	____	____	____
____	3	____	____
____	____	____	6
5.8	____	____	____
____	6.7	____	____
____	____	22.5	____

24. Consider the following balanced equation.

$$SiO_2(s) + 3\,C(s) \rightarrow SiC(s) + 2\,CO(g)$$

Complete the following table showing the appropriate number of moles of reactants and products. If the number of moles of a reactant is provided, fill in the required amount of the other reactant, as well as the moles of each product formed. If the number of moles of a product is provided, fill in the required amount of each reactant to make that amount of product, as well as the amount of the other product that is made.

mol SiO_2	*mol C*	*mol SiC*	*mol CO*
2	____	____	____
____	9	____	____
____	____	____	8
5.1	____	____	____
____	12.5	____	____
____	____	17.4	____

25. Consider the following unbalanced equation for the combustion of butane.

$$C_4H_{10}(g) + O_2(g) \rightarrow CO_2(g) + H_2O(g)$$

Balance the equation and determine how many moles of O_2 are required to react completely with 4.9 mol C_4H_{10}.

26. Consider the following unbalanced equation for the neutralization of acetic acid.

$$HC_2H_3O_2(aq) + Ca(OH)_2(aq) \rightarrow H_2O(l) + Ca(C_2H_3O_2)_2(aq)$$

Balance the equation and determine how many moles of $Ca(OH)_2$ are required to completely neutralize 1.07 mol of $HC_2H_3O_2$.

27. Consider the following unbalanced equation for the reaction of solid lead with silver nitrate.

$$Pb(s) + AgNO_3(aq) \rightarrow Pb(NO_3)_2(aq) + Ag(s)$$

a) Balance the equation.
b) How many moles of silver nitrate are required to completely react with 9.3 mol lead?
c) How many moles of Ag(s) are formed by the complete reaction of 28.4 mol Pb?

28. Consider the following unbalanced equation for the reaction of aluminum with sulfuric acid.

$$Al(s) + H_2SO_4(aq) \rightarrow Al_2(SO_4)_3(aq) + H_2(g)$$

a) Balance the equation.
b) How many moles of H_2SO_4 are required to completely react with 8.3 mol Al?
c) How many moles of H_2 are formed by the complete reaction of 0.341 mol Al?

Mass-to-Mass Conversions

29. For the reaction shown, calculate how many grams of oxygen form when each of the following completely reacts.

$$2\,HgO(s) \rightarrow 2\,Hg(l) + O_2(g)$$

a) 2.88 g HgO
b) 5.21 g HgO
c) 1.3×10^3 g HgO
d) 4.6×10^{-3} g HgO

30. For the reaction shown, calculate how many grams of oxygen form when each of the following completely reacts.

$$2\,KClO_3(s) \rightarrow 2\,KCl(s) + 3\,O_2(g)$$

a) 8.91 g $KClO_3$
b) 0.549 g $KClO_3$
c) 8.4×10^6 g $KClO_3$
d) 3.3×10^{-4} g $KClO_3$

31. For each of the reactions shown, calculate how many grams of the product form when 1.8 g of the reactant in italics completely reacts. Assume that there is more than enough of the other reactant.
a) $2\,Na(s) + \mathit{Cl_2}(g) \rightarrow 2\,NaCl(s)$
b) $\mathit{CaO}(s) + CO_2(g) \rightarrow CaCO_3(s)$
c) $2\,\mathit{Mg}(s) + O_2(g) \rightarrow 2\,MgO(s)$
d) $\mathit{Na_2O}(s) + H_2O(l) \rightarrow 2\,NaOH(aq)$

32. For each of the reactions shown, calculate how many grams of the product form when 14.4 g of the reactant in italics completely reacts. Assume that there is more than enough of the other reactant.
a) $\mathit{Ca}(s) + Cl_2(g) \rightarrow CaCl_2(s)$
b) $2\,K(s) + \mathit{Br_2}(l) \rightarrow 2\,KBr(s)$
c) $4\,Cr(s) + 3\,O_2(g) \rightarrow 2\,Cr_2O_3(s)$
d) $2\,\mathit{Sr}(s) + O_2(g) \rightarrow 2\,SrO(s)$

33. For the reaction shown, calculate how many grams of each product forms when the following amounts of reactant completely react to form products. Assume that there is more than enough of the other reactant.

$$2\,Al(s) + Fe_2O_3(s) \rightarrow Al_2O_3(s) + 2\,Fe(l)$$

a) 4.7 g Al
b) 4.7 g Fe_2O_3

34. For the reaction shown, calculate how many grams of each product form when the following amounts of reactant completely react to form products. Assume that there is more than enough of the other reactant.

$$2\,HCl(aq) + Na_2CO_3(aq) \rightarrow 2\,NaCl(aq) + H_2O(l) + CO_2(g)$$

a) 10.8 g HCl
b) 10.8 g Na_2CO_3

35. For each of the following acid–base reactions, calculate how many grams of each acid are necessary to completely react with and neutralize 2.5 g of the base.
a) $HCl(aq) + NaOH(aq) \rightarrow H_2O(l) + NaCl(aq)$
b) $2\ HNO_3(aq) + Ca(OH)_2(aq) \rightarrow 2\ H_2O(l) + Ca(NO_3)_2(aq)$
c) $H_2SO_4(aq) + 2\ KOH(aq) \rightarrow 2\ H_2O(l) + K_2SO_4(aq)$

36. For each of the following precipitation reactions, calculate how many grams of the first reactant are necessary to completely react with 17.3 g of the second reactant.
a) $2\ KI(aq) + Pb(NO_3)_2(aq) \rightarrow PbI_2(s) + 2\ KNO_3(aq)$
b) $Na_2CO_3(aq) + CuCl_2(aq) \rightarrow CuCO_3(s) + 2\ NaCl(aq)$
c) $K_2SO_4(aq) + Sr(NO_3)_2(aq) \rightarrow SrSO_4(s) + 2\ KNO_3(aq)$

37. Sulfuric acid can dissolve aluminum metal according to the following reaction.

$$2\ Al(s) + 3\ H_2SO_4(aq) \rightarrow Al_2(SO_4)_3(aq) + 3\ H_2(g)$$

Suppose you wanted to dissolve an aluminum block with a mass of 22.5 g. What minimum amount of H_2SO_4 (in grams) would you need? How many grams of H_2 gas would be produced by the complete reaction of the aluminum block?

38. Hydrochloric acid can dissolve solid iron according to the following reaction.

$$Fe(s) + 2\ HCl(aq) \rightarrow FeCl_2(aq) + H_2(g)$$

How much HCl (in grams) would you need to dissolve a 2.8 g iron bar on a padlock? How much H_2 would be produced by the complete reaction of the iron bar?

Limiting Reactant, Theoretical Yield, and Percent Yield

39. Consider the following generic chemical equation.

$$2\ A + 4\ B \rightarrow 3\ C$$

What is the limiting reactant when each of the following amounts of A and B are allowed to react?
a) 2 mol A; 5 mol B
b) 1.8 mol A; 4 mol B
c) 3 mol A; 4 mol B
d) 22 mol A; 40 mol B

40. Consider the following generic chemical equation.

$$A + 3\ B \rightarrow C$$

What is the limiting reactant when each of the following amounts of A and B are allowed to react?
a) 1 mol A; 4 mol B
b) 2 mol A; 3 mol B
c) 0.5 mol A; 1.6 mol B
d) 24 mol A; 75 mol B

41. Determine the theoretical yield of C when each of the following amounts of A and B are allowed to react in the generic reaction:

$$A + 2\ B \rightarrow 3\ C$$

a) 1 mol A; 1 mol B
b) 2 mol A; 2 mol B
c) 1 mol A; 3 mol B
d) 32 mol A; 68 mol B

42. Determine the theoretical yield of C when each of the following amounts of A and B is allowed to react in the generic reaction:

$$2\ A + 3\ B \rightarrow 2\ C$$

a) 2 mol A; 4 mol B
b) 3 mol A; 3 mol B
c) 5 mol A; 6 mol B
d) 4 mol A; 5 mol B

43. For the reaction shown, find the limiting reactant for each of the following initial amounts of reactants.

$$2\ K(s) + Cl_2(g) \rightarrow 2\ KCl(s)$$

a) 1 mol K; 1 mol Cl_2
b) 1.8 mol K; 1 mol Cl_2
c) 2.2 mol K; 1 mol Cl_2
d) 14.6 mol K; 7.8 mol Cl_2

44. For the reaction shown, find the limiting reactant for each of the following initial amounts of reactants.

$$4\ Cr(s) + 3\ O_2(g) \rightarrow 2\ Cr_2O_3(s)$$

a) 1 mol Cr; 1 mol O_2
b) 4 mol Cr; 2.5 mol O_2
c) 12 mol Cr; 10 mol O_2
d) 14.8 mol Cr; 10.3 mol O_2

45. For the reaction shown, compute the theoretical yield of product (in moles) for each of the following initial amounts of reactants.

$$2\,Mn(s) + 3\,O_2(g) \rightarrow 2\,MnO_3(s)$$

a) 2 mol Mn; 2 mol O_2
b) 4.8 mol Mn; 8.5 mol O_2
c) 0.114 mol Mn; 0.161 mol O_2
d) 27.5 mol Mn; 43.8 mol O_2

46. For the reaction shown, compute the theoretical yield of the product (in moles) for each of the following initial amounts of reactants.

$$Ti(s) + 2\,Cl_2(g) \rightarrow TiCl_4(s)$$

a) 2 mol Ti; 2 mol Cl_2
b) 5 mol Ti; 9 mol Cl_2
c) 0.483 mol Ti; 0.911 mol Cl_2
d) 12.4 mol Ti; 15.8 mol Cl_2

47. For the reaction shown, find the limiting reactant for each of the following initial amounts of reactants.

$$2\,Li(s) + F_2(g) \rightarrow 2\,LiF(s)$$

a) 1.0 g Li; 1.0 g F_2
b) 10.5 g Li; 37.2 g F_2
c) 2.85×10^3 g Li; 6.79×10^3 g F_2

48. For the reaction shown, find the limiting reactant for each of the following initial amounts of reactants.

$$4\,Al(s) + 3\,O_2(g) \rightarrow 2\,Al_2O_3(s)$$

a) 1.0 g Al; 1.0 g O_2
b) 2.2 g Al; 1.8 g O_2
c) 0.353 g Al; 0.482 g O_2

49. For the reaction shown, compute the theoretical yield of the product (in grams) for each of the following initial amounts of reactants.

$$2\,Al(s) + 3\,Cl_2(g) \rightarrow 2\,AlCl_3(s)$$

a) 1.0 g Al; 1.0 g Cl_2
b) 5.5 g Al; 19.8 g Cl_2
c) 0.439 g Al; 2.29 g Cl_2

50. For the reaction shown, compute the theoretical yield of the product (in grams) for each of the following initial amounts of reactants.

$$Ti(s) + 2\,F_2(g) \rightarrow TiF_4(s)$$

a) 1.0 g Ti; 1.0 g F_2
b) 4.8 g Ti; 3.2 g F_2
c) 0.388 g Ti; 0.341 g F_2

51. If the theoretical yield of a reaction is 24.8 g and the actual yield is 18.5 g, what is the percent yield?

52. If the theoretical yield of a reaction is 0.118 g and the actual yield is 0.104 g, what is the percent yield?

53. Consider the following reaction.

$$CaO(s) + CO_2(g) \rightarrow CaCO_3(s)$$

A chemist allows 14.4 g CaO and 13.8 g CO_2 to react. When the reaction is finished, the chemist collects 19.4 g $CaCO_3$. Determine the limiting reactant, theoretical yield, and percent yield for the reaction.

54. Consider the following reaction.

$$SO_3(g) + H_2O(l) \rightarrow H_2SO_4(aq)$$

A chemist allows 61.5 g SO_3 and 11.2 g H_2O to react. When the reaction is finished, the chemist collects 54.9 g H_2SO_4. Determine the limiting reactant, theoretical yield, and percent yield for the reaction.

55. Consider the following reaction.

$$2\,NiS_2(s) + 5\,O_2(g) \rightarrow 2\,NiO(s) + 4\,SO_2(g)$$

When 11.2 g of NiS_2 are allowed to react with 5.43 g O_2, 4.86 g of NiO are obtained. Determine the limiting reactant, theoretical yield of NiO, and percent yield for the reaction.

56. Consider the following reaction.

$$4\,HCl(g) + O_2(g) \rightarrow 2\,H_2O(l) + 2\,Cl_2(g)$$

When 63.1 g HCl are allowed to react with 17.2 g O_2, 49.3 g Cl_2 are collected. Determine the limiting reactant, theoretical yield of Cl_2, and percent yield for the reaction.

57. Lead ions can be precipitated from solution with NaCl according to the following reaction.

$$Pb^{2+}(aq) + 2\,NaCl(aq) \rightarrow PbCl_2(s) + 2\,Na^+(aq)$$

When 135.8 g NaCl are added to a solution containing 195.7 g of Pb^{2+}, a $PbCl_2$ precipitate forms. The precipitate is filtered and dried and found to have a mass of 252.4 g. Determine the limiting reactant, theoretical yield of $PbCl_2$ and percent yield for the reaction.

58. Magnesium oxide can be made by heating magnesium metal in the presence of oxygen. The balanced equation for the reaction is:

$$2\,Mg(s) + O_2(g) \rightarrow 2\,MgO(s)$$

When 10.1 g of Mg are allowed to react with 10.5 g O_2, 11.9 g MgO are collected. Determine the limiting reactant, theoretical yield, and percent yield for the reaction.

Cumulative Problems

59. Sodium bicarbonate is often used as an antacid to neutralize excess hydrochloric acid in an upset stomach. How much hydrochloric acid (in grams) can be neutralized by 3.5 g of sodium bicarbonate? (Hint: Begin by writing a balanced equation for the reaction between aqueous sodium bicarbonate and aqueous hydrochloric acid.)

60. Toilet bowl cleaners often contain hydrochloric acid to dissolve the calcium carbonate deposits that accumulate within a toilet bowl. How much calcium carbonate (in grams) can be dissolved by 5.8 g of HCl? (Hint: Begin

61. The combustion of gasoline produces carbon dioxide and water. Assume gasoline to be pure octane (C_8H_{18}) and calculate how many kilograms of carbon dioxide are added to the atmosphere per 1.0 kg of octane burned. (Hint: Begin by writing a balanced equation for the combustion reaction.)

62. Many home barbeques are fueled with propane gas (C_3H_8). How much carbon dioxide (in kilograms) is produced upon the complete combustion of 18.9 L of propane (approximate contents of one 5-gal tank)? Assume that the density of the liquid propane in the tank is 0.621 g/mL. (Hint: Begin by writing a balanced equation for the combustion reaction.)

63. A hard water solution contains 4.8 g of calcium chloride. How much sodium phosphate (in grams) should be added to the solution to completely precipitate all of the calcium?

64. Magnesium ions can be precipitated from seawater by the addition of sodium hydroxide. How much sodium hydroxide (in grams) must be added to a sample of seawater to completely precipitate the 88.4 mg of magnesium present.

65. Hydrogen gas can be prepared in the laboratory by a single displacement reaction in which solid zinc reacts with hydrochloric acid. How much zinc (in grams) is required to make 14.5 g of hydrogen gas through this reaction?

66. Pure oxygen gas can be prepared in the laboratory by the decomposition of solid potassium chlorate to form solid potassium chloride and oxygen gas. How much oxygen gas (in grams) can be prepared from 45.8 g of potassium chlorate?

67. Aspirin can be made in the laboratory by reacting acetic anhydride ($C_4H_6O_3$) with salicylic acid ($C_7H_6O_3$) to form aspirin ($C_9H_8O_4$) and acetic acid ($C_2H_4O_2$). The balanced equation is:

$$C_4H_6O_3 + C_7H_6O_3 \rightarrow C_9H_8O_4 + C_2H_4O_2.$$

In a laboratory synthesis, a student begins with 5.00 mL of acetic anhydride (density = 1.08 g/mL) and 2.08 g of salicylic acid. Once the reaction is complete, the student collects 2.01 g of aspirin. Determine the limiting reactant, theoretical yield of aspirin, and percent yield for the reaction.

68. The combustion of liquid ethanol (C_2H_5OH) produces carbon dioxide and water. After 3.8 mL of ethanol (density = 0.789 g/mL) was allowed to burn in the presence of 12.5 g of oxygen gas, 3.10 mL of water (density = 1.00 g/mL) were collected. Determine the limiting reactant, theoretical yield of H_2O, and percent yield for the reaction. (Hint: Write a balanced equation for the combustion of ethanol.)

69. Urea (CH_4N_2O), a common fertilizer, can be synthesized by the reaction of ammonia (NH_3) with carbon dioxide:

$$2\,NH_3(aq) + CO_2(aq) \rightarrow CH_4N_2O(aq) + H_2O(l)$$

An industrial synthesis of urea obtains 87.5 kg of urea upon reaction of 68.2 kg of ammonia with 105 kg of carbon dioxide. Determine the limiting reactant, theoretical yield of urea, and percent yield for the reaction.

70. Silicon, which occurs in nature as SiO_2, is the material from which most computer chips are made. If SiO_2 is heated until it melts into a liquid, it will react with solid carbon to form liquid silicon and carbon monoxide gas. In an industrial preparation of silicon, 52.8 kg of SiO_2 reacted with 25.8 kg of carbon to produce 22.4 kg of silicon. Determine the limiting reactant, theoretical yield, and percent yield for the reaction.

Highlight Problems

71. A loud classroom demonstration involves igniting a hydrogen-filled balloon. The hydrogen within the balloon explosively reacts with oxygen in the air to form water.

$$2\,H_2(g) + O_2(g) \rightarrow 2\,H_2O(g)$$

If the balloon is filled with a mixture of hydrogen and oxygen, the explosion is even louder than if the balloon is filled with only hydrogen, and the intensity of the explosion depends on the relative amounts of oxygen and hydrogen within the balloon. Look at the following molecular views representing different amounts of hydrogen and oxygen in four different balloons. Based on the balanced chemical equation, which balloon will make the loudest explosion?

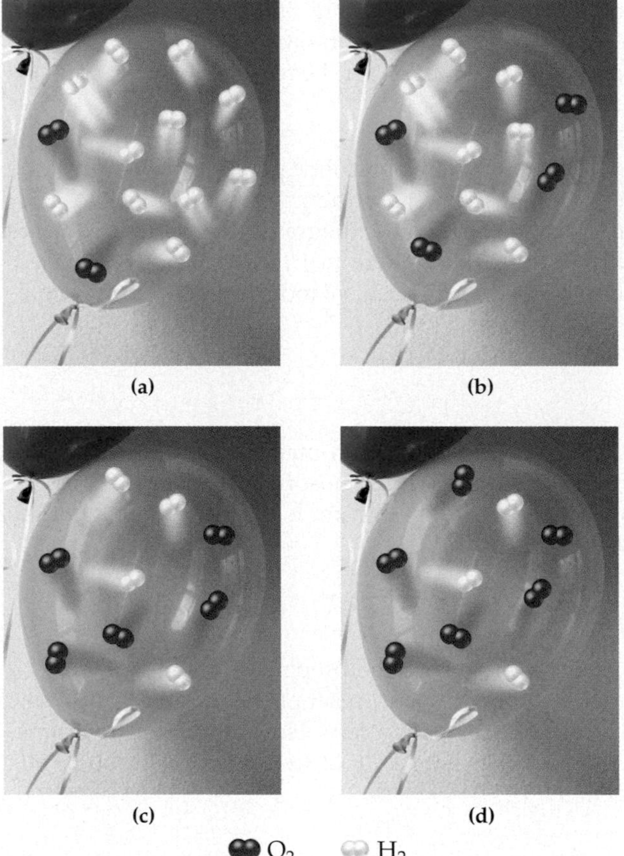

72. A hydrochloric acid solution will neutralize a sodium hydroxide solution. Look at the following molecular views showing one beaker of HCl and four beakers of NaOH. Which NaOH beaker will just neutralize the HCl beaker? Begin by writing a balanced chemical equation for the neutralization reaction.

73. As we have seen, scientists have grown increasingly worried about the potential for global warming caused by increasing atmospheric carbon dioxide levels. The world burns the fossil fuel equivalent of 7×10^{12} kg of petroleum per year. Assume that all of this petroleum is in the form of octane (C_8H_{18}) and calculate how much CO_2 (in kilograms) is produced by world fossil fuel combustion per year (Hint: Begin by writing a balanced equation for the combustion of octane). If the atmosphere currently contains approximately 3×10^{15} kg of CO_2, how long will it take for the world's fossil fuel combustion to double the amount of atmospheric carbon dioxide?

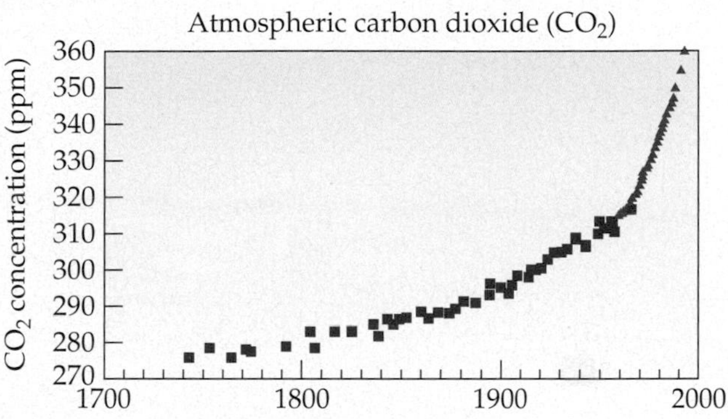

Atmospheric CO_2 levels 1860 to present.

74. Lakes that have been acidified by acid rain can be neutralized by the addition of limestone ($CaCO_3$). How much limestone (in kilograms) would be required to completely neutralize a 5.2×10^9 L lake containing 5.0×10^{-3}g H_2SO_4 per liter?

Answers to Skillbuilder Exercises

Skillbuilder 8.1 49.2 moles H_2O
Skillbuilder 8.2 6.89 g HCl
Skillbuilder 8.3 4.0×10^3 kg
Skillbuilder 8.4 Limiting reactant is Na; theoretical yield is 4.8 mol of NaF
Skillbuilder 8.5 30.7 g NH_3
Skillbuilder Plus, p. 262 29.4 kg
Skillbuilder 8.6 Limiting reactant is CO; theoretical yield = 127 g Fe; percent yield = 68.8%

TRO
Introductory Chemistry

Electrons in Atoms and the Periodic Table

9

"Anyone who is not shocked by quantum mechanics has not understood it."
Niels Bohr

9.1 Blimps, Balloons, and Models for the Atom

You have probably seen one of the Goodyear blimps floating in the sky. The Goodyear blimp is often present at championship sporting events such as the Rose Bowl, the Indy 500, and the U.S. Open Golf Tournament. It was present at the Statue of Liberty's 100th birthday party and has made appearances in countless movies and television shows. The blimp's inherent stability allows it to provide spectacular views of the world below for television and film.

The Goodyear blimp is similar to a large balloon. Unlike airplanes, which must be moving fast to stay in flight, a blimp or *airship* floats in air because it is filled with a gas that is lighter than air. The Goodyear blimp is filled with helium. Other airships in history, however, have used hydrogen for buoyancy. For example, the Hindenburg—the largest airship ever constructed—was filled with hydrogen, which turned out to be a poor choice. Hydrogen is a reactive and flammable gas. On May 6, 1937, while landing in New Jersey on its first transatlantic crossing, the Hindenburg burst into flames, destroying the airship and killing 36 of the 97 passengers. Apparently, as the Hindenburg was landing, a leak in the hydrogen gas ignited, resulting in an explosion that destroyed the ship.

The skin of the Hindenburg, which was constructed of a flammable material, may have also been partially to blame for its demise.

A similar accident cannot happen to the Goodyear blimp because it is filled with helium, an inert and therefore nonflammable gas. A spark or even a flame would actually be *extinguished* by helium.

◀ Modern blimps are filled with helium, an inert gas. In this chapter we learn models that explain the inertness of heliium and the reactivity of other elements.

Why is helium inert? What is it about helium *atoms* that makes helium *gas* inert? On the other hand, why is hydrogen reactive? Recall from

The Hindenburg was filled with hydrogen, a reactive and flammable gas. **Question:** What makes hydrogen reactive?

Chapter 5 that elemental hydrogen exists as a diatomic element. Hydrogen atoms are so reactive that they react with each other to form hydrogen molecules. What is it about hydrogen atoms that make them so reactive? What is the difference between hydrogen and helium that accounts for their different reactivities?

When we examine the properties of hydrogen and helium, we make observations about nature. Mendeleev's periodic law, first discussed in Chapter 4, summarizes the results of many similar observations on the properties of elements.

> When the elements are arranged in order of increasing atomic number, certain sets of properties recur periodically.

This is a modification of Mendeleev's original periodic law. Mendeleev originally listed elements in order of increasing *mass*; today we list them in order of increasing *atomic number*.

We know that hydrogen's reactivity recurs with lithium, sodium, and the other Group I metals. We also know that helium's inertness recurs with neon, argon, and the other noble gases. These are the first two steps in the scientific method: observing something about nature and formulating a law that summarizes a large number of observations. The next step in the scientific method is building a model or theory that gives the underlying reasons for the observations and laws.

Recall from Chapter 1 that the scientific method consists of observations, laws, theories and experiments. Laws summarize observations, while theories give the underlying reasons for them.

In this chapter, we examine two important models—the **Bohr model** and the **quantum-mechanical model**—that explain the underlying reasons for the inertness of helium, the reactivity of hydrogen, and the periodic law. These models explain how electrons exist in atoms and how those electrons affect the chemical and physical properties of elements. We have already learned much about the behavior of elements. We know, for example, that sodium tends to form +1 ions and that fluorine tends to form −1 ions. We know that some elements are metals and that others are nonmetals. We know that the noble gases are chemically inert and that the alkali metals are chemically reactive. But we do not know *why*. The models in this chapter explain why.

These models were developed in the early 1900s and, especially the quantum-mechanical model, caused a revolution in the physical sciences,

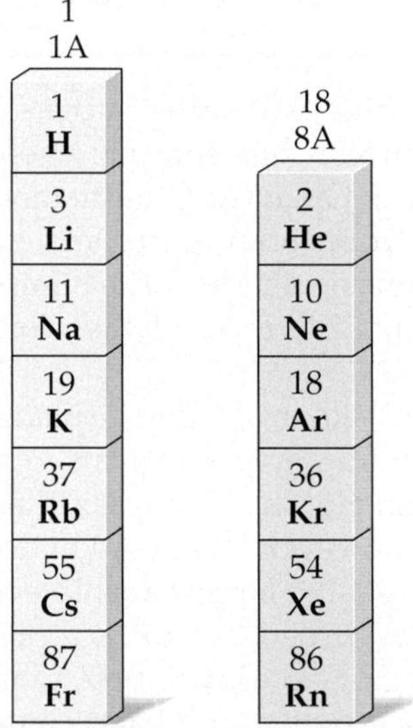

Alkali metals Noble gases

The noble gases are chemically inert and the alkali metals are chemically reactive. Why?

Neils Bohr (left) and Erwin Schrödinger (right), along with Albert Einstein, played a role in the discovery of quantum mechanics, yet they were bewildered by their discoveries.

changing our fundamental view of matter at its most basic level. The scientists who devised these models—including Niels Bohr, Erwin Schrödinger, and Albert Einstein—were bewildered by their discoveries. Bohr claims, "Anyone who is not shocked by quantum mechanics has not understood it." Schrödinger laments, "I don't like it, and I am sorry I ever had anything to do with it." Einstein disbelieved it, insisting that, "God does not play dice with the universe." However, the quantum-mechanical model has such explanatory power that it is rarely questioned today. It forms the basis of the modern periodic table and our understanding of chemical bonding. Its applications include lasers, computers, and semiconductors, and it has given us new ways to design drugs that cure disease. The quantum-mechanical model for the atom is, in many ways, the foundation of modern chemistry.

9.2 Light: Electromagnetic Radiation

Before we explore models for the atom, we must understand a few things about light, because the interaction of light with atoms helped to shape these models. Light is familiar to all of us—we see the world by it—but we may not know much about what light is. Unlike most of what we have encountered so far in this book, light is not matter—it has no mass. Light is a form of **electromagnetic radiation**, a type of energy that travels through space at a constant speed of 3.0×10^8 m/s (186,000 mi/s) and exhibits both wave-like and particle-like behaviors. At this speed, a flash of light generated at the equator would travel around the world in one-seventh of a second. This extremely fast speed is responsible for the delay between the time that you see a firework in the sky and the time that you hear the sound of its explosion. The light from the exploding firework reaches your eye almost instantaneously. The sound, traveling much slower, takes longer.

Figure 9.1 The wavelength of light (λ) is defined as the distance between adjacent crests in light waves.

The Greek letter *lambda*(λ) is pronounced "lam-duh."

Helpful Mneomic: ROY G BIV–Red, Orange, Yellow, Green, Blue, Indigo, Violet

nano = 10^{-9}

One of the unusual characteristics of light is its wave–particle duality. Certain properties of light are best described by thinking of it as a wave, while other properties are best described by thinking of it as a particle. A particle of light is called a **photon**, a packet of light energy. The wave nature of light is characterized by its **wavelength** ($\boldsymbol{\lambda}$), the distance between adjacent wave crests (Figure 9.1).

Wavelength determines the color of visible light. White light, as produced by the sun or by a light bulb, contains a spectrum of wavelengths and therefore a spectrum of color. We can see these colors—red, orange, yellow, green, blue, indigo, and violet—in a rainbow or when white light is passed through a prism (Figure 9.2). Red light, with a wavelength of 750 nm (nanometers), has the longest wavelength of visible light. Violet light, with a wavelength of 400 nm, has the shortest. The presence of color in white light is responsible for the colors we see. For example, a red shirt is red because it absorbs all colors except red, which it reflects (Figure 9.3). Our eyes see only the reflected light, making the shirt appear red.

The wavelength of light also determines the amount of energy carried in its photons—the shorter the wavelength, the greater the energy. Just as ocean waves carry more energy if their crests are closer together—think about surf pounding a beach—so light waves carry more energy if their crests are closer together. Therefore violet light (shorter wavelength) carries more energy per photon than red light (longer wavelength). Light is also often characterized by its **frequency** ($\boldsymbol{\nu}$), the number of cycles or

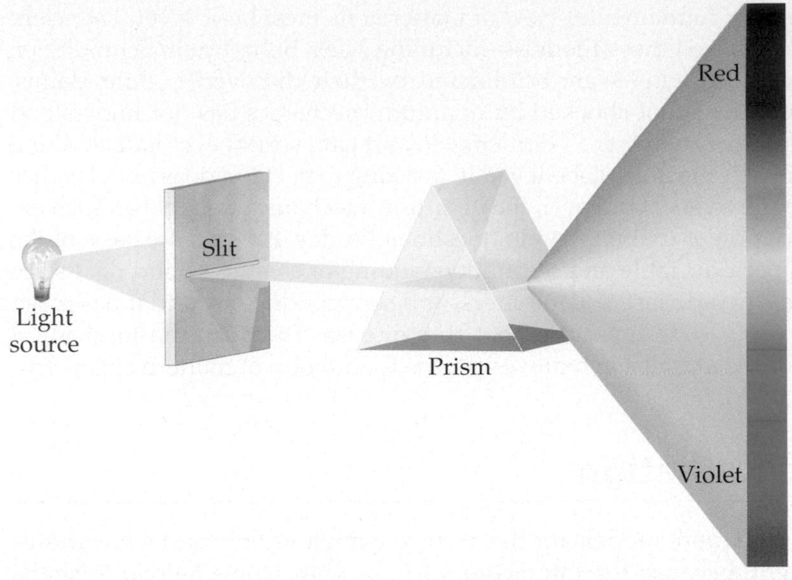

Figure 9.2 Light is separated into its constituent colors—red, orange, yellow, green, blue, indigo, and violet—when it is passed through a prism.

Figure 9.3 A red shirt appears red because it absorbs all colors except red, which it reflects.

The Greek letter *nu* (ν) is pronounced *noo*.

crests that pass through a stationary point in one second. Wavelength and frequency are inversely related—the higher the frequency, the shorter the wavelength.

To Summarize:

- ***Electromagnetic radiation is a form of energy that travels through space at a constant speed of 3.0×10^8 m/s and exhibits both wave-like and particle-like properties.***
- ***The wavelength of electromagnetic radiation determines the amount of energy carried by one of its photons. The longer the wavelength, the less the energy it carries.***
- ***The frequency of electromagnetic radiation is inversely related to its wavelength.***

9.3 The Electromagnetic Spectrum

Electromagnetic radiation ranges in wavelength from 10^{-16} m (gamma rays) to 10^6 m (radiowaves). Visible light only comprises a tiny portion of the **electromagnetic spectrum**, which includes all wavelengths of electromagnetic radiation. Figure 9.4 shows the entire electromagnetic spectrum, with short-wavelength, high-frequency radiation on the right, and long-wavelength, low-frequency radiation on the left. Visible light is only a small sliver in the middle.

Remember that the energy carried per photon is greater for short wavelengths than for long wavelengths. Therefore, the most energetic photons are those of **gamma rays**, the form of electromagnetic radiation with the shortest wavelength. Gamma rays are produced by the sun, by stars, and by certain unstable atomic nuclei on earth. Human exposure to

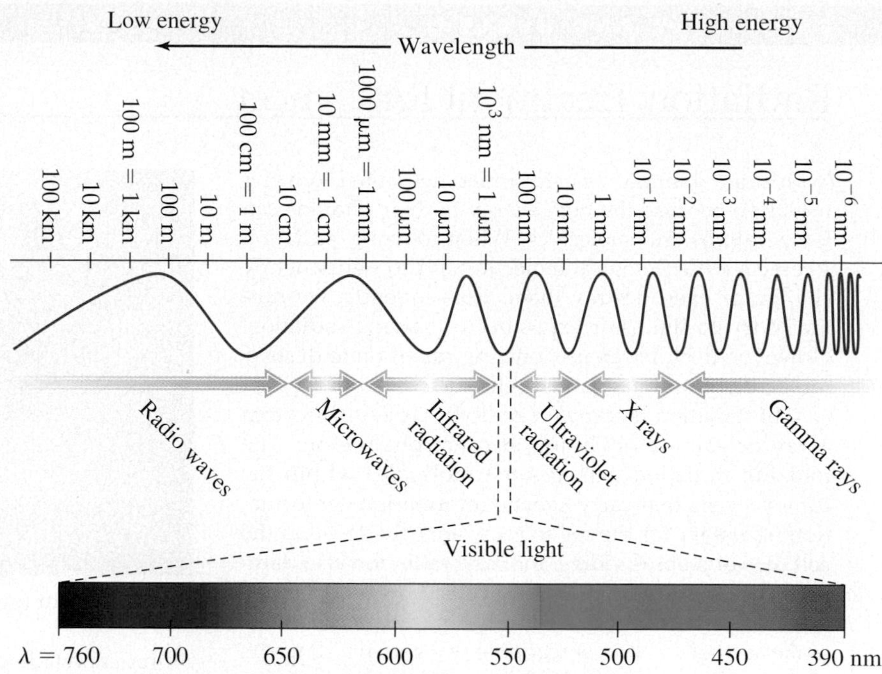

Figure 9.4 The electromagnetic spectrum

gamma rays is dangerous because the high energy of gamma-ray photons can damage biological molecules.

Next on the electromagnetic spectrum, at longer wavelengths than gamma rays, are **X-rays**, familiar to us from their medical use. X-rays pass through many substances that block visible light and are therefore used to image internal bones and organs. Like gamma-ray photons, X-ray photons carry enough energy to damage biological molecules. While several yearly exposures to X-rays are relatively harmless, excessive exposure to X-rays increases cancer risk.

Sandwiched between X-rays and visible light in the electromagnetic spectrum is **ultraviolet** or UV light, most familiar to us as the component of sunlight that produces a sunburn or suntan. While not as energetic as gamma ray or X-ray photons, ultraviolet photons still carry enough energy to damage biological molecules. Excessive exposure to ultraviolet light increases the risk of skin cancer and cataracts, and causes premature wrinkling of the skin. Next on the spectrum is **visible light** ranging from violet (shorter wavelength, higher energy) to red (longer wavelength, lower energy). Visible photons do not damage biological molecules. They do, however, cause molecules in our eyes to rearrange and send a signal to our brain that results in vision.

Beyond visible light lies **infrared light.** The heat you feel when you place your hand near a hot object is infrared light. All warm objects, including human bodies, emit infrared light. While infrared light is invisible to our eyes, infrared sensors can detect it and are often used in night vision technology to "see" in the dark. In the infrared region of the spectrum, warm objects—such as human bodies, for example—glow much as a light bulb glows in the visible region of the spectrum.

CHEMISTRY AND HEALTH

Radiation Treatment for Cancer

X-rays and gamma rays are sometimes called ionizing radiation because the high energy in their photons can ionize atoms and molecules. When ionizing radiation interacts with biological molecules, it can permanently change or even destroy them. Consequently, we normally try to limit our exposure to ionizing radiation. However, doctors can use ionizing radiation to destroy molecules within unwanted cells such as cancer cells.

In radiation therapy (or radiotherapy) doctors aim X-ray or gamma ray beams at cancerous tumors. The ionizing radiation damages the molecules within the tumor's cells that carry genetic information—information necessary for the cell to grow and divide—and the cell dies or stops dividing. Ionizing radiation also damages molecules within healthy cells; however, cancerous cells divide more quickly than healthy cells, making them more susceptible to genetic damage. Nonetheless, healthy cells are damaged during treatments resulting in side effects such as fatigue, skin lesions, and hair loss. Doctors try to minimize the exposure of healthy cells by appropriate shielding and by targeting the tumor from multiple directions, minimizing the exposure of healthy cells while maximizing the exposure of cancerous cells (Figure 9.5).

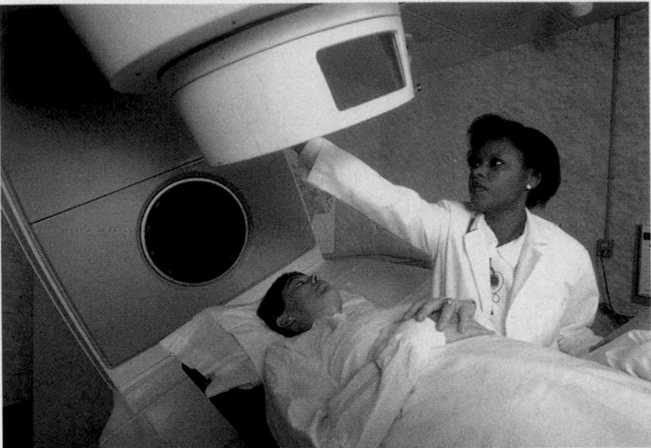

Cancer patient undergoing radiation therapy.

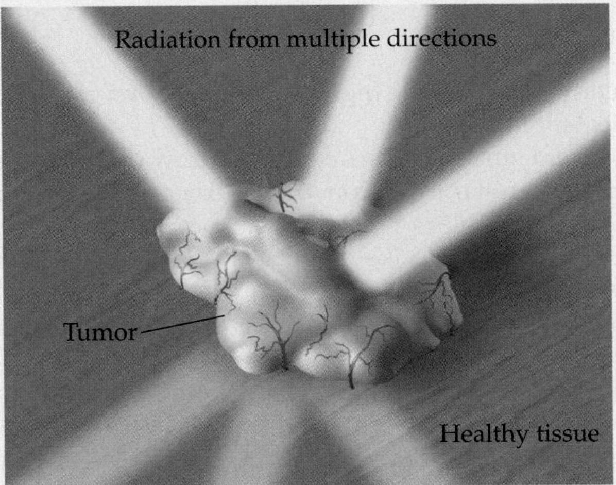

Figure 9.5 By targeting the tumor from various different directions, damage to healthy tissue is minimized.

Another side effect of exposing healthy cells to radiation is that they too may become cancerous. So a treatment for cancer may cause cancer. Why do we continue to use it? In radiation therapy, as in most other disease therapies, there is an associated risk. We take risks all the time, many for lesser reasons. For example, every time we drive a car, we risk injury or even death. Why? Because we perceive the benefit—such as getting to the grocery store to buy food—to be worth the risk. The situation is similar in cancer therapy or any other therapy for that matter. The benefit of cancer therapy (possibly curing a cancer that will certainly kill you) is worth the risk (a slight increase in the chance of developing a future cancer).

CAN YOU ANSWER THIS? *Why would visible light not work to destroy cancerous tumors?*

Beyond infrared light, at longer wavelengths still, are **microwaves,** used in microwave ovens. Although microwave light has longer wavelengths—and therefore lower energy per photon—than visible or infrared light, it is efficiently absorbed by water and can therefore heat substances that contain water. For this reason substances that contain water, such as food, are warmed in a microwave oven, but substances that do not contain water, such as a plate, are not.

Normal photograph | Infrared photograph

Warm objects, such as human or animal bodies, give off infrared light that is easily detected with an infrared camera.

The longest wavelengths are **radio waves,** which are used to transmit the signals responsible for AM and FM radio, cellular telephones, television, and other forms of communication.

EXAMPLE 9.1 Wavelength, Energy, and Frequency

Arrange the three types of electromagnetic radiation—visible light, X-rays, and microwaves—in order of increasing:

a) Wavelength
b) Frequency
c) Energy per photon

Solution:

a) Wavelength
We can examine Figure 9.4 to see that X-rays have the shortest wavelength, followed by visible light and then microwaves.

X-rays, Visible, Microwaves

b) Frequency
Since frequency and wavelength are inversely proportional—the longer the wavelength the shorter the frequency—the ordering with respect to frequency is exactly the reverse of the ordering with respect to wavelength.

Microwaves, Visible, X-rays

c) Energy per photon
Energy per photon decreases with increasing wavelength but increases with increasing frequency; therefore the ordering with respect to energy per photon is the same as frequency.

Microwaves, Visible, X-rays

SKILLBUILDER 9.1 Wavelength, Energy, and Frequency

Arrange the following colors of visible light—green, red, and blue—in order of increasing:

a) Wavelength
b) Frequency
c) Energy per photon

9.4 The Bohr Model: Atoms with Orbits

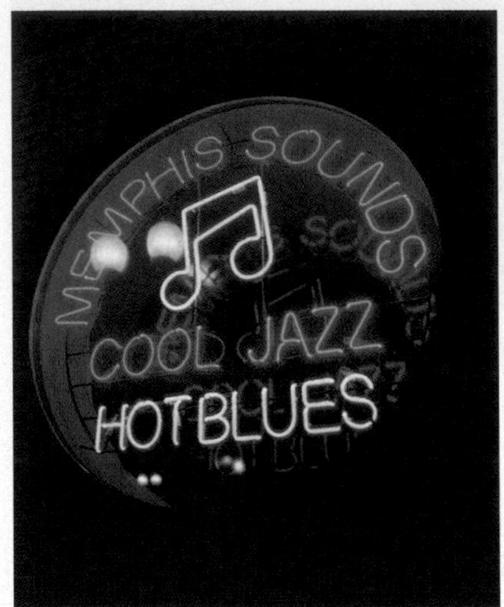

Figure 9.6 In a neon sign, neon atoms inside a glass tube absorb electrical energy and then re-emit the energy as light.

When an atom absorbs energy—in the form of heat, light, or electricity—it often re-emits that energy as light. For example, a neon sign is composed of one or more glass tubes filled with gaseous neon atoms. When an electrical current is passed through the tube, the neon atoms absorb some of the electrical energy and re-emit it as the familiar red light of a neon sign (Figure 9.6). If the atoms in the tube are different, the emitted light is a different color. In other words, atoms of a given element emit light of unique colors (or unique wavelengths). Hydrogen atoms, for example, emit light that appears pink and helium atoms emit light that appears yellow-orange (Figure 9.7).

Closer inspection of the light emitted by hydrogen, helium, and neon atoms reveals that each contains several distinct colors or wavelengths. Just as the white light from a light bulb can be separated into its constituent wavelengths by passing it through a prism, so the light emitted by glowing hydrogen, helium, or neon can also be separated into its constituent wavelengths (Figure 9.8) by passing it through a prism. The result is called an **emission spectrum**. Notice the differences between a white light spectrum and the emission spectra of hydrogen, helium and neon. The white light spectrum is *continuous,* meaning that the light intensity is uninterrupted or smooth across the entire spectrum. The emission spectra of hydrogen, helium, and neon, however, are not continuous. They consist of bright spots or lines at specific wavelengths with complete darkness in between. Since the emission of light in atoms is related to the motions of electrons within the atoms, a model for how electrons exist in atoms must account for these spectra.

Discrete means *noncontinuous.*

A major challenge in developing a model for electrons in atoms was the discrete or bright line nature of the emission spectra. Why did atoms, researchers wondered, emit light at discrete wavelengths when excited with energy? Why did they *not* emit a continuous spectrum? Niels Bohr developed a simple model to explain these results. In his model, now

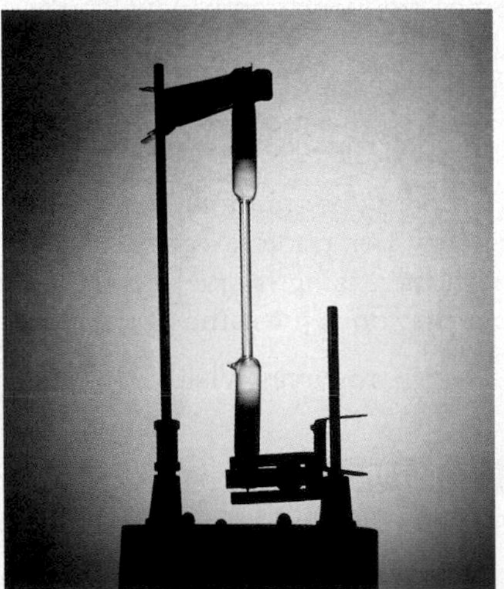

Hydrogen lamp

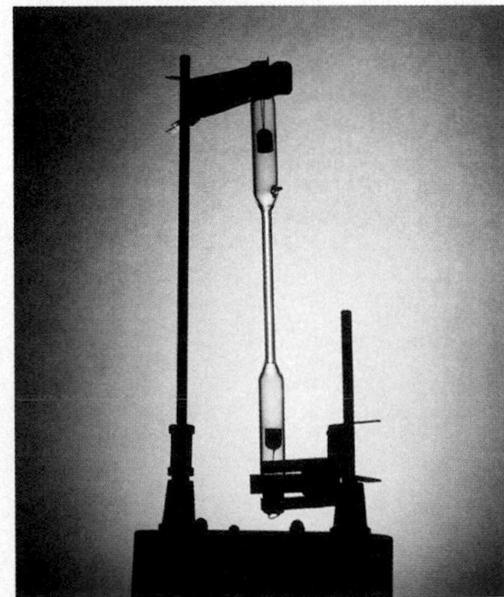

Helium lamp

Figure 9.7 Light emitted from a hydrogen lamp appears pink, and light emitted from a helium lamp appears yellow-orange.

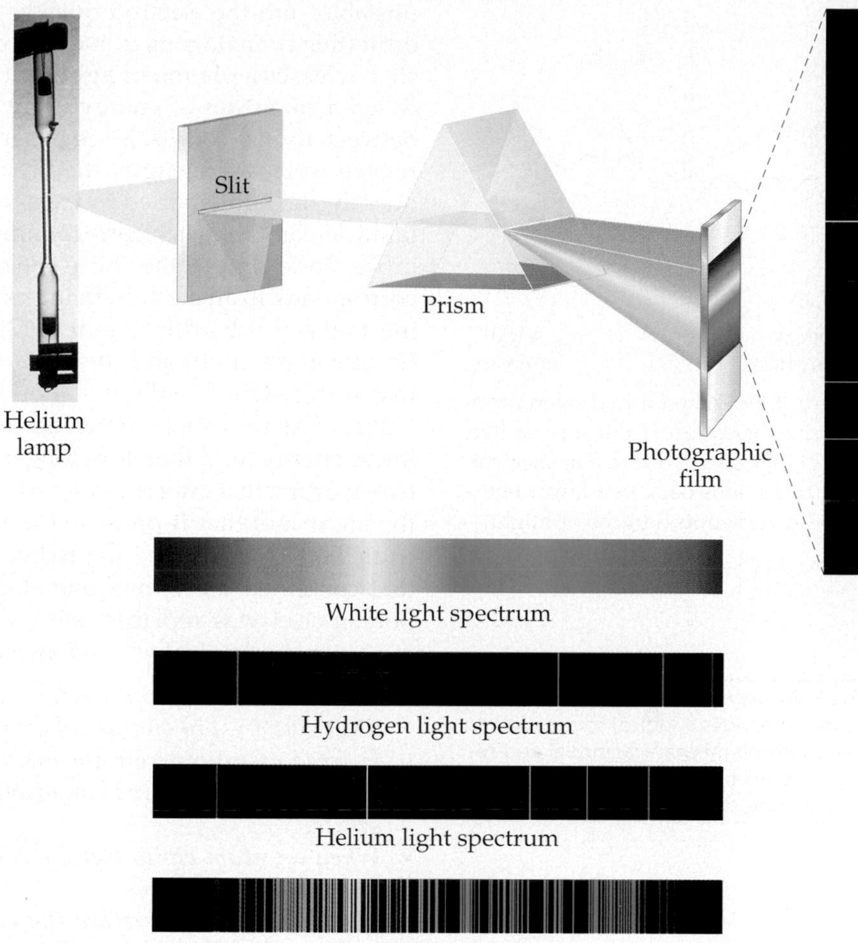

Figure 9.8 Each of the three elements has a different emission spectrum.

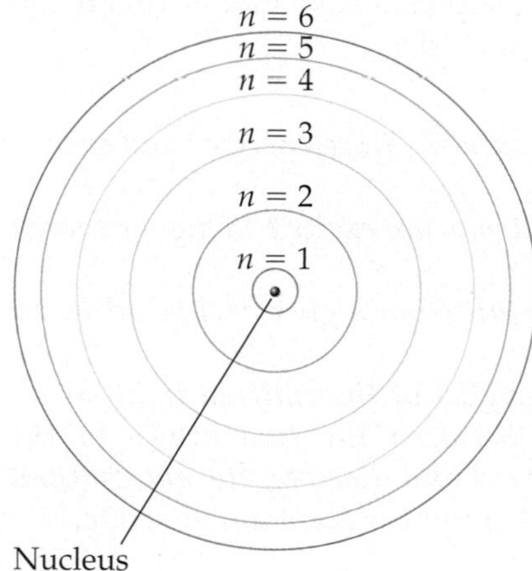

Figure 9.9 Bohr orbits

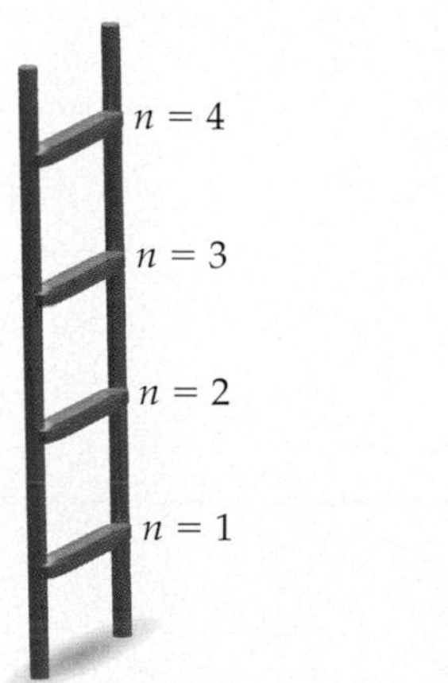

Figure 9.10 Bohr orbits are like steps in a ladder. It is possible to stand on one step or another, but it is impossible to stand between steps.

called the Bohr model, electrons travel around the nucleus in circular orbits that are similar to planetary orbits around the sun. However, unlike planets revolving around the sun—which can theoretically orbit at any distance whatsoever from the sun—electrons in the Bohr model can only orbit at *specific, fixed* distances from the nucleus (Figure 9.9).

The *energy* of each Bohr orbit, specified by **quantum numbers** $n = 1,2,3\ldots$, was also fixed, or **quantized**. Bohr orbits are like steps in a ladder (Figure 9.10), each at a specific distance from the nucleus and each at a specific energy. Just like it is impossible to stand *between steps* on a ladder, so it is impossible for an electron to exist *between orbits* in the Bohr model. An electron in an $n = 3$ orbit, for example, is farther from the nucleus and has more energy than an electron in an $n = 2$ orbit. However, an electron cannot exist at an intermediate distance or energy between the two orbits—the orbits are quantized. As long as an electron remains in a given orbit, it does not absorb or emit light, and its energy remains fixed and constant.

When an atom absorbs energy, an electron in one of these fixed orbits is *excited* or promoted to an orbit that is farther away from the nucleus (Figure 9.11) and therefore higher in energy (this is analogous to moving up a step on the ladder). However, in this new configuration, the atom is

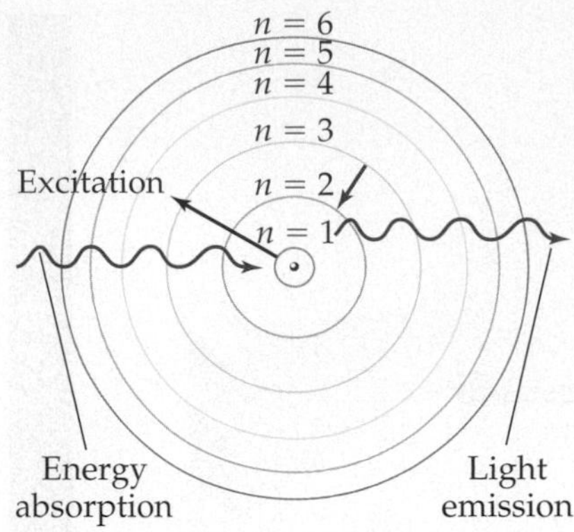

Figure 9.11 When a hydrogen atom absorbs energy, an electron is excited to a higher energy orbit. The electron then transitions back to a lower energy orbit and emits a photon of light.

unstable, and the electron quickly falls back or *relaxes* to a lower energy orbit (this is analogous to moving down a step on the ladder). As it does so, it releases a photon of light containing the precise amount of energy—called a **quantum** of energy—that corresponds to the energy difference between the two orbits. Since the amount of energy in a photon is directly related to its wavelength, the photon has a specific wavelength. Consequently, the light emitted by excited atoms consists of specific lines at certain wavelengths, each corresponding to a specific transition between two orbits. For example, the line at 486 nm in the hydrogen emission spectrum corresponds to an electron being excited to the $n = 4$ orbit and then relaxing to the $n = 2$ orbit (Figure 9.12). In the same way, the line at 657 nm (longer wavelength and therefore lower energy) corresponds to an electron being excited to the $n = 3$ orbit and then relaxing to the $n = 2$ orbit. Notice that transitions between orbits that are closer together produce lower energy (and therefore longer wavelength) light than transitions between orbits that are farther apart. The great success of the Bohr model of the atom was that it predicted the lines of the hydrogen emission spectrum. However, it failed to predict the emission spectra of other elements that contained more than one electron. For this, and other reasons, the Bohr model was replaced with a more sophisticated model called the quantum-mechanical or wave-mechanical model.

The Bohr model is still important because it provides a logical foundation to the quantum-mechanical model and reveals the historical development of scientific understanding.

To Summarize the Bohr model:

- ***Electrons exist in quantized orbits at specific, fixed energies and specific, fixed distances from the nucleus.***
- ***When energy is put into an atom, electrons are excited to higher energy orbits.***
- ***When an atom emits light, electrons fall from higher energy orbits to lower energy orbits.***
- ***The energy (and therefore the wavelength) of the emitted light corresponds to the difference in energy between the two orbits in the transition. Since these energies are fixed and discrete, the energy (and therefore the wavelength) of the emitted light is fixed and discrete.***

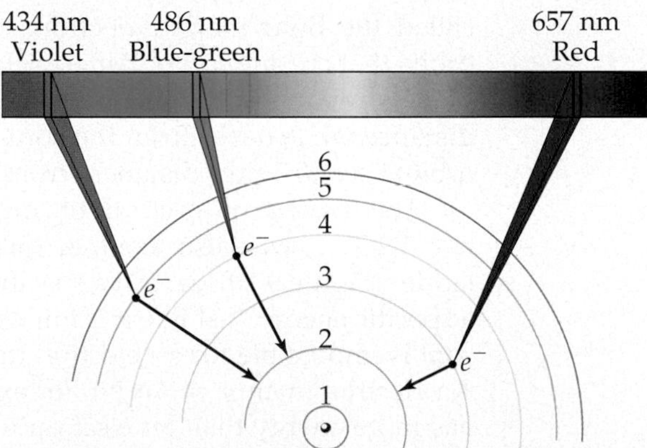

Figure 9.12 The 657 nm line of the hydrogen emission spectra corresponds to an electron relaxing from the $n = 3$ orbit to the $n = 2$ orbit. The 486 nm line corresponds to an electron relaxing from the $n = 4$ orbit to the $n = 2$ orbit, and the 434 nm line corresponds to an electron relaxing from $n = 5$ to $n = 2$.

9.5 The Quantum-Mechanical Model: Atoms with Orbitals

In the quantum-mechanical model for the atom, Bohr orbits are replaced with quantum-mechanical **orbitals.** Orbitals are different from orbits in that they represent, not specific paths that electrons follow, but probability maps that show a statistical distribution of where the electron is likely to be found. This is a non-intuitive property—one that is difficult to visualize—of electrons. A revolutionary concept in quantum mechanics is that electrons *do not* behave like particles flying through space. We cannot describe their exact paths. An orbital does not represent exactly how an electron moves—instead, it is a probability map showing where the electron is likely to be found.

Baseball Paths and Electron Probability Maps

To understand orbitals, let's contrast the behavior of a baseball with that of an electron. Imagine a baseball thrown from the pitcher's mound to a catcher at home plate (Figure 9.13). The baseball's path can easily be traced as it travels from the pitcher to the catcher. The catcher can watch the baseball as it travels through the air, and she can predict exactly where the baseball will cross over home plate. She can even place her

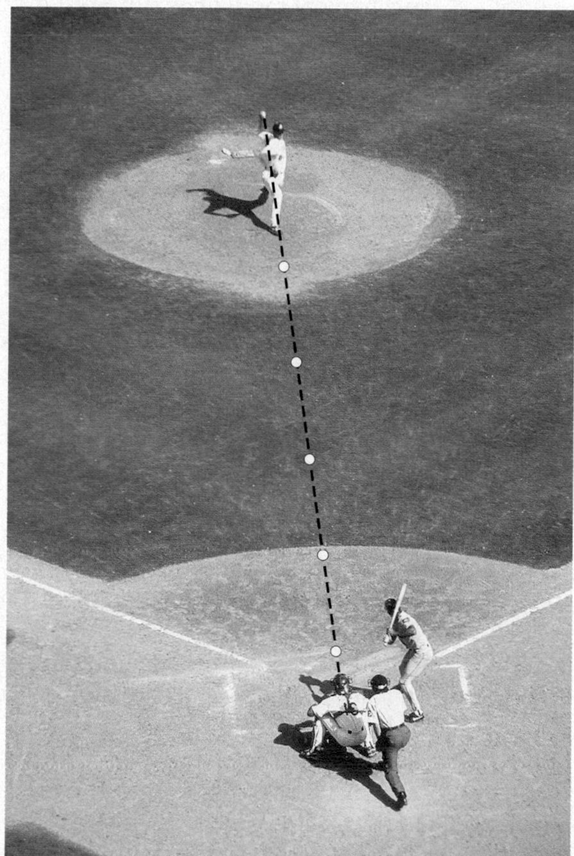

Figure 9.13 A baseball follows a well-defined path as it travels from the pitcher to the catcher.

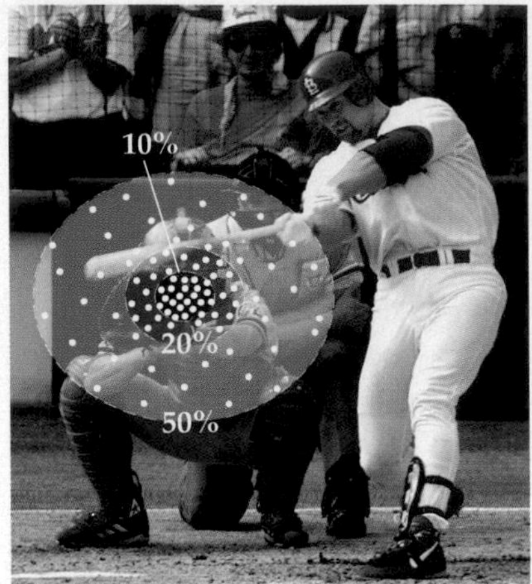

Figure 9.14 Probability map of where a "pitched" electron will cross home plate.

Electrons cannot really be thrown like a baseball can, but the analogy still holds in terms of its path.

mitt in the correct place to catch it. This would be impossible for an electron. Like photons, electrons have wave–particle duality. This duality leads to behavior that makes it impossible to trace an electron's path. If an electron were thrown from the pitcher's mound to home plate, it would land in a different place every time, *even if it were thrown in exactly the same way*. Baseballs have predictable paths—electrons do not.

In the quantum-mechanical world of the electron, the catcher could not know exactly where the electron will cross the plate for any given throw. She would have no way of putting her mitt in the right place to catch it. However, if the catcher kept track of hundreds of electron throws, she could observe a reproducible, statistical pattern of where the electron crosses the plate. She could even draw maps in the strike zone showing the probability of an electron crossing a certain area (Figure 9.14). These maps are called *probability maps*.

From Orbits to Orbitals

In the Bohr model, an *orbit* is a circular path—analogous to a baseball's path—that shows the electron's path around an atomic nucleus. In the quantum-mechanical model, an *orbital* is a probability map, analogous to the probability map drawn by our catcher, that shows the probability of where the electron will be found when the atom is probed. Just as the Bohr model has different orbits with different radii, the quantum-mechanical model has different orbitals with different shapes.

9.6 Quantum-Mechanical Orbitals

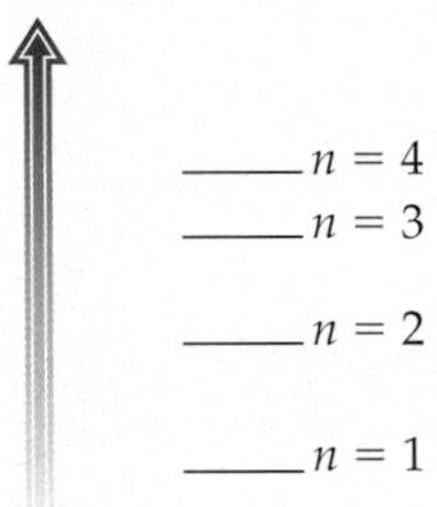

Figure 9.15 The principal quantum numbers ($n = 1$, $n = 2$, $n = 3$...) determine the energy of the hydrogen quantum-mechanical orbitals.

In the Bohr model for the atom, a single quantum number (n) specifies each orbit. In the quantum-mechanical model, a number and a letter are required to specify an orbital. For example, the lowest energy orbital in the quantum-mechanical model—analogous to the $n = 1$ orbit in the Bohr model—is called the *1s orbital*. It is specified by the number *1* and the letter *s*. The number is called the **principal quantum number** (n) and specifies the **principal shell** of the orbital. The higher the principal quantum number is, the higher the energy of the orbital. The possible principal quantum numbers are $n = 1, 2, 3 \ldots$. with energy increasing as n increases (Figure 9.15). Since the 1*s* orbital has the lowest possible principal quantum number, it is in the lowest energy shell and has the lowest possible energy. The letter indicates the **subshell** of the orbital and specifies its shape. The possible letters are *s*, *p*, *d*, or *f*, each with a different shape.

Orbitals within the *s* subshell have a spherical shape. Unlike the $n = 1$ Bohr orbit, which shows the electron's path, the 1*s* quantum mechanical orbital is a three-dimensional probability map. These probability maps are best represented by dots (Figure 9.16) where the dot density is proportional to the probability of finding the electron. We can understand these probability maps and dots better with another analogy. Imagine the electron moving randomly around the nucleus. Imagine also taking a photograph of the electron every second for ten or fifteen minutes. One second the electron is very close to the nucleus; the next second it is farther away and so on. Each photo shows a dot representing the electron's position relative to the nucleus at that time. If you took hundreds of photos and superimposed

This analogy is purely hypothetical. It is impossible to photograph electrons in this way.

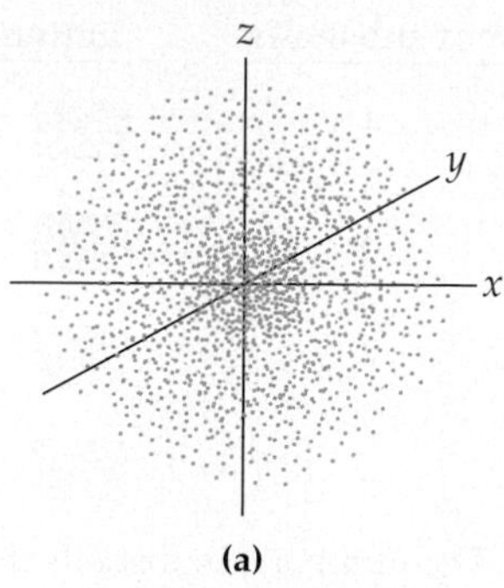

Figure 9.16 The 1*s* orbital. The dot density is proportional to the probability of finding the electron. The greater dot density near the middle represents a higher probability of finding the electron near the nucleus.

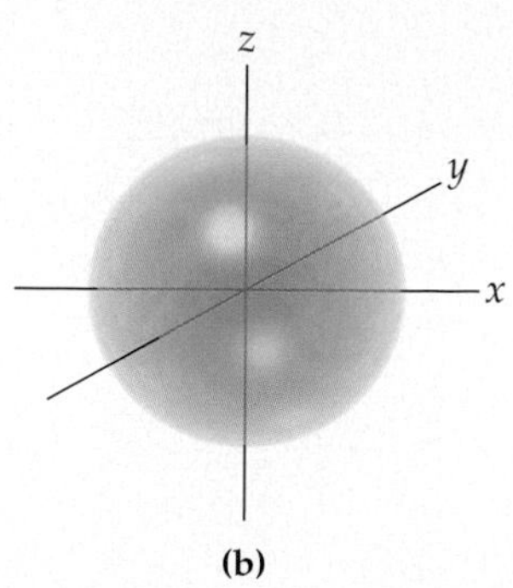

Figure 9.17 Shape representation of the 1*s* orbital. When the electron is in the 1*s* orbital, it is most likely found within this sphere.

all of them, you would have a probability map like Figure 9.16—a statistical representation of where the electron spends its time. Notice that the dot density for the 1*s* orbital is greatest near the nucleus and decreases farther away from the nucleus. This means that the electron is more likely to be found close to the nucleus than far away from it.

Orbitals can also be represented as geometric shapes that encompass most of the volume where the electron is likely to be found. For example, the 1*s* orbital can be represented as a sphere (Figure 9.17). If we superimpose the dot density representation of the 1*s* orbital on the shape representation (Figure 9.18), we can see that most of the dots are within the sphere, meaning that the electron is most likely to be found within the sphere when it is in the 1*s* orbital.

The single electron of an undisturbed hydrogen atom at room temperature is in the 1*s* orbital. This is called the **ground state,** or lowest energy state, of the hydrogen atom. However, like the Bohr model, the quantum-mechanical model allows transitions to higher-energy orbits upon the absorption of energy. What are these higher-energy orbitals? What do they look like?

The next orbitals are those with principal quantum number $n = 2$. Unlike the $n = 1$ principal shell, which contains only one subshell (specified by *s*), the $n = 2$ principal shell contains two subshells, specified by *s* and *p*.

The number of subshells in a given principal shell is equal to the value of *n*.

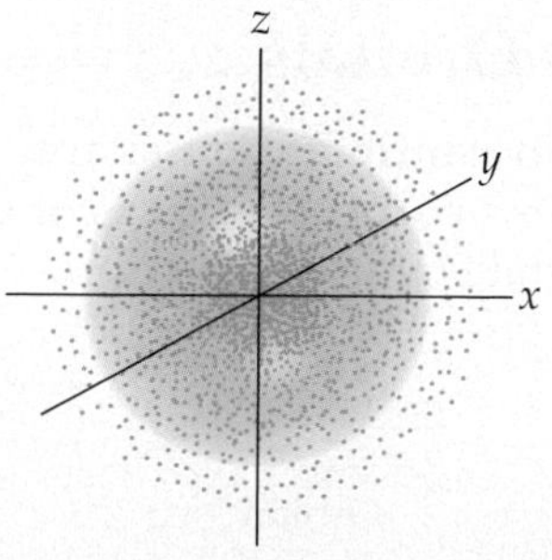

Figure 9.18 The shape representation of the 1*s* orbital superimposed on the dot density representation.

Shell	# of subshells	Letters specifying subshells			
$n = 4$	4	*s*	*p*	*d*	*f*
$n = 3$	3	*s*	*p*	*d*	
$n = 2$	2	*s*	*p*		
$n = 1$	1	*s*			

Figure 9.19 The number of subshells in a given principal shell is equal to the value of *n*.

Therefore the $n = 1$ principal shell has 1 subshell, the $n = 2$ principal shell has 2 subshells, etc. (Figure 9.19). The *s* subshell contains the 2*s* orbital, higher in energy than the 1*s* orbital and slightly larger (Figure 9.20), but otherwise similar in shape. The *p* subshell contains three 2*p* orbitals (Figure 9.21), all with the same shape but different orientations.

The next principal shell, $n = 3$, contains three subshells specified by *s*, *p*, and *d*. The *s* and *p* subshells contain the 3*s* and 3*p* orbitals, similar in shape to the 2*s* and 2*p* orbitals, but slightly larger and higher in energy. The *d* subshell contains the five *d*-orbitals shown in Figure 9.22. The next principal shell, $n = 4$, contains four subshells specified by *s*, *p*, *d*, and *f*. The *s*, *p*, and *d* subshells are similar to those in $n = 3$. The *f* subshell contains 7 orbitals (called the 4*f* orbitals) whose shape we do not consider in this book.

As we have already discussed, hydrogen's single electron is usually in the 1*s* orbital because electrons seek out the lowest energy orbital available. In hydrogen, the rest of the orbitals are normally empty. However, the absorption of energy by hydrogen atoms can cause electrons to transition from the 1*s* orbital to other orbitals. When the electron is in a higher energy orbital, the hydrogen atom is said to be in an excited state. Excited states are unstable, and the electron will usually relax back to a lower energy orbital, resulting in the emission of energy, often in the form of light. As in the Bohr model, the energy difference between the two orbitals involved in the transition determines the wavelength of the emitted light (the greater the energy difference, the shorter the wavelength). The quantum-mechanical model predicts the bright-line spectrum of hydrogen as well as the Bohr model. However, it can also predict the bright-line spectra of other elements as well.

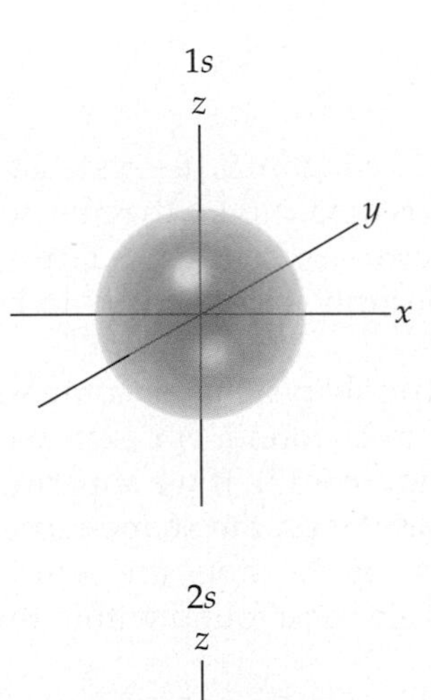

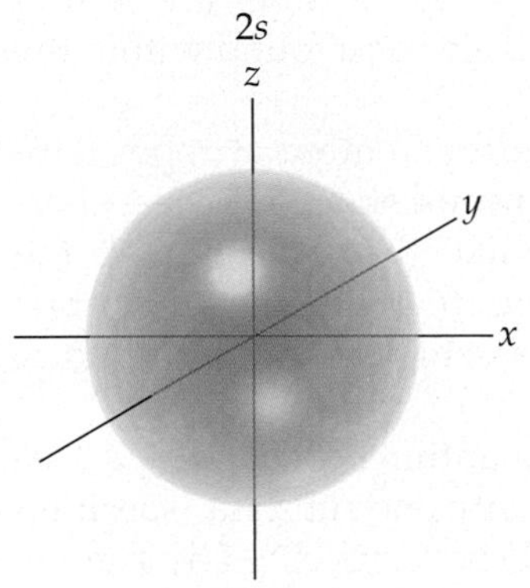

Figure 9.20 The 2*s* orbital is similar to the 1*s* orbital, but larger in size.

Electron Configurations: How Electrons Occupy Orbitals

An **electron configuration** simply shows the occupation of orbitals by electrons for a particular atom. For example, the electron configuration for a ground-state hydrogen atom is:

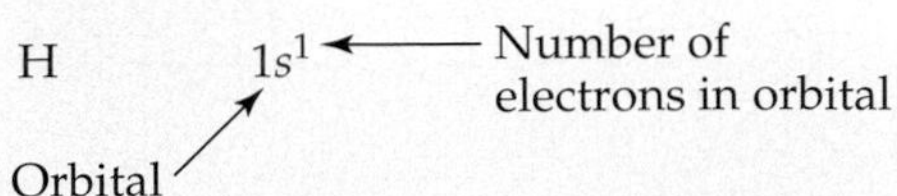

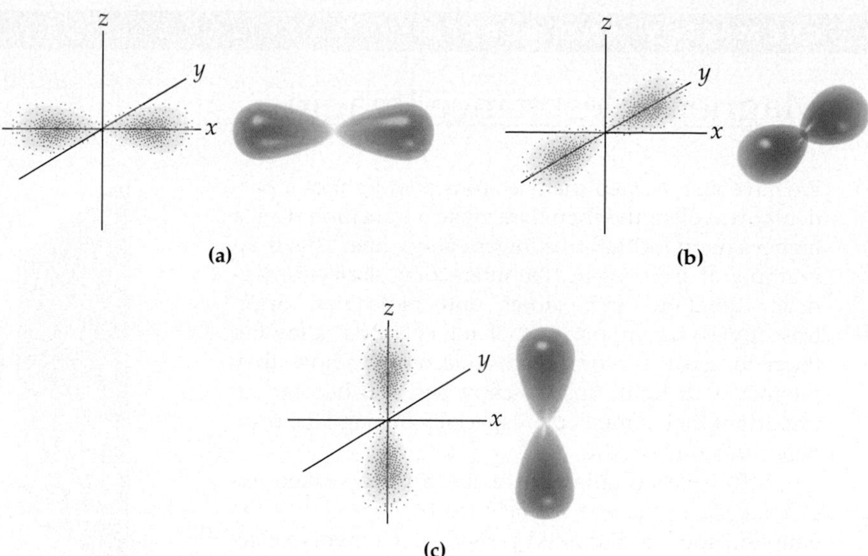

Figure 9.21 The 2p orbitals.

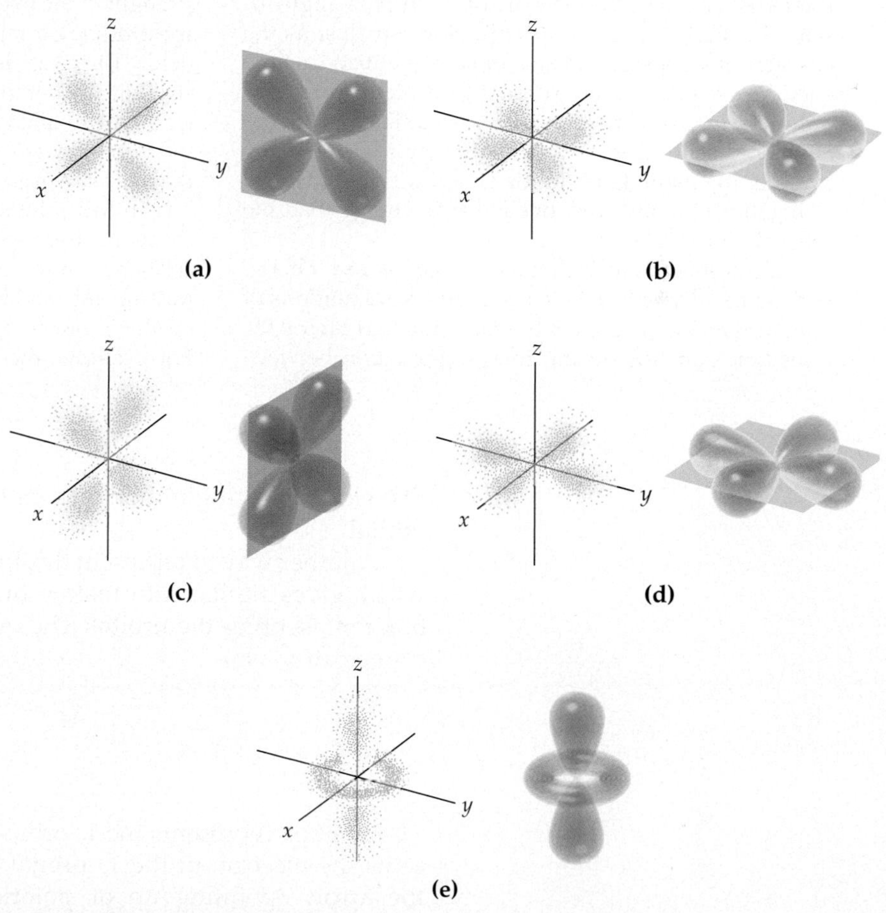

Figure 9.22 The 3d orbitals.

CHEMISTRY AND HEALTH

Magnetic Resonance Imaging

We have just learned that electrons emit light of a particular wavelength when they make a transition from a higher energy orbital to a lower energy one. This is an example of *spectroscopy*, the interaction of electromagnetic radiation with atoms and molecules. Spectroscopy is an important tool for chemists, allowing them to analyze atoms and molecules by how they interact with light. Spectroscopy has also become an important tool in medicine, especially in magnetic resonance imaging or *MRI*.

MRI is based on a type of spectroscopy called *nuclear magnetic resonance* or *NMR spectroscopy*. Unlike the emission spectra discussed earlier, which involve electrons making transitions from one energy level to another, NMR involves atomic nuclei making transitions from one energy level to another.

The best way to understand NMR is to think of atomic nuclei as tiny magnets (Figure 9.23). When these magnets are placed in an external magnetic field, they align themselves with this external field to minimize their energy (just as the needle of a compass aligns itself with Earth's magnetic field). However, just as the energies of an electron in an atom are quantized, so the energies of nuclei in an external field are quantized—only certain, fixed energies are allowed. These fixed energies correspond to fixed orientations relative to the external magnetic field. In the simplest case, only two orientations are allowed, one lower in energy than the other (Figure 9.24).

Electromagnetic radiation of the correct energy will cause a transition between the two orientations of the nucleus. The energy of the radiation that causes the transition depends on the energy separation between the two orientations, which in turn depends on the strength of the external magnetic field. In normal NMR spectroscopy, a sample is placed in a uniform magnetic field. The wavelength of electromagnetic radiation striking the sample is then varied to find the wavelength or frequency that causes the transition between the two allowed orientations. This frequency is called the *resonance frequency*.

In MRI, the sample is the patient. The nuclei are those of hydrogen atoms within water molecules contained in the patient's tissues. However, instead of putting the patient in a uniform magnetic field, he or she is put in a magnetic field that varies in space. For example, the magnetic field may be strongest on

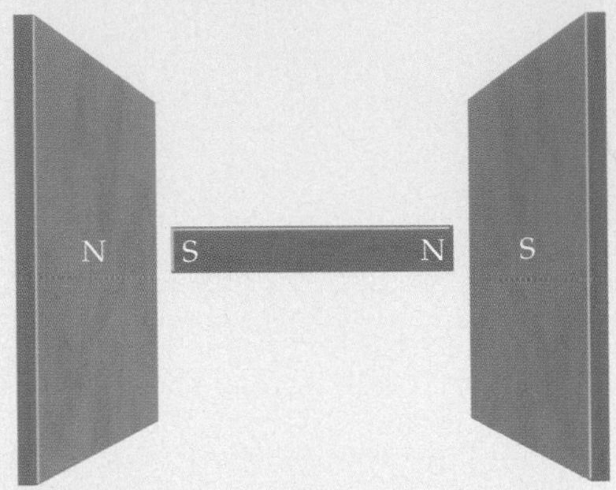

Figure 9.23 A nucleus is like a small magnet that aligns itself with an external magnetic field.

The electron configuration tells us that hydrogen's 1 electron is in the 1*s* orbital.

Another way to represent this information is with an **orbital diagram,** which gives similar information but shows the electrons as arrows in a box representing the orbital. The orbital diagram for a ground-state hydrogen atom is:

H [↑]
1*s*

The box represents the 1*s* orbital and the arrow within the box represents the electron in the 1*s* orbital. In orbital diagrams, the direction of the arrow (pointing up or pointing down) represents **electron spin,** a fundamental property of electrons. All electrons have spin. The **Pauli**

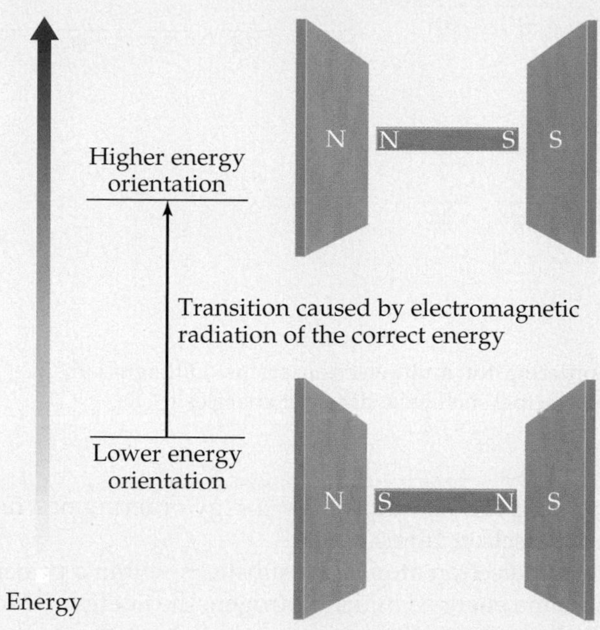

Figure 9.24 The energies (and therefore the orientations) of a nucleus in an external magnetic field are quantized. Light of the correct energy causes a transition from one orientation to the other.

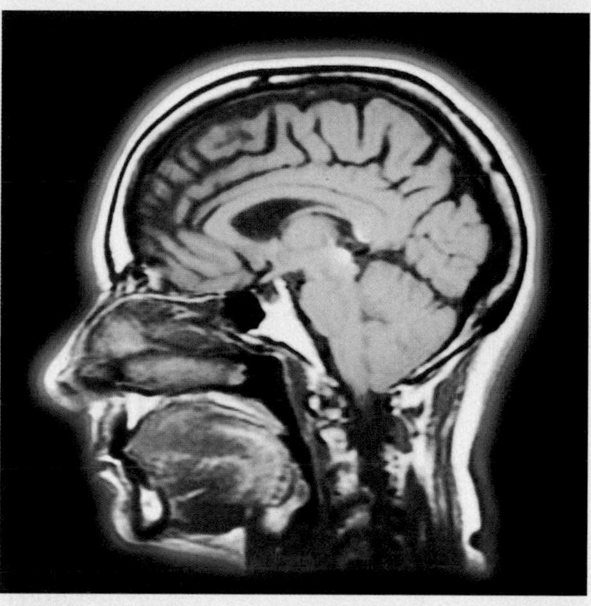

Figure 9.25 MRI produces remarkably clear images of a patient's internal tissues.

the left side of the patient and weaker on the right. Then those nuclei on the left side of the patient will have a resonance frequency of higher energy than those on the right side. By mapping these resonance frequencies—each resonance frequency corresponds to a different position in space—MRI can obtain a remarkably clear and detailed image of the patient's internal tissues (Figure 9.25).

CAN YOU ANSWER THIS? *Will the* **wavelength** *of electromagnetic radiation required to cause a transition in the preceding example (stronger magnetic field on the left side of the person than the right side) be longer or shorter on the left side versus the right? Explain.*

exclusion principle states that orbitals may hold no more than two electrons with opposing spins. We symbolize this as two arrows pointing in opposite directions ⇅. Helium atoms, for example, have two electrons. The electron configuration and orbital diagram for helium are:

	Electron configuration	Orbital diagram
He	$1s^2$	⇅ (1s)

Since we know that electrons occupy the lowest-energy orbitals available, and since we know that only two electrons (with opposing spins) are allowed in each orbital, we can continue to build ground state electron configurations for the rest of the elements as long as we know the energy

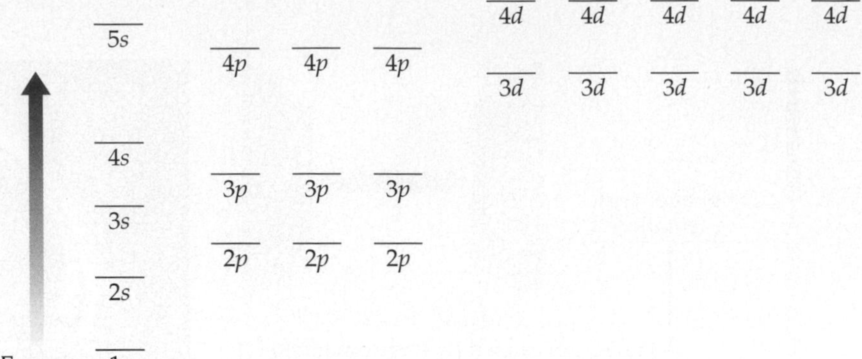

Figure 9.26 Energy ordering for multi-electron atoms. Different subshells within the same principal shell have different energies.

ordering of the orbitals. Figure 9.26 shows the energy ordering of a number of orbitals for multi-electron atoms.

The subshells within a principal shell do not have the same energy because of electron–electron interactions.

Notice that, for multi-electron atoms, the subshells within a principal shell **do not** have the same energy. Unlike hydrogen, the energy ordering is not determined by the principal quantum number alone. For example, the 4*s* subshell is lower in energy than the 3*d* subshell, even though its principal quantum number is higher. Using this relative energy ordering, we can write ground-state electron configurations and orbital diagrams for other elements. For lithium, which has three electrons, the electron configuration and orbital diagram are:

Remember that the number of electrons in an atom is equal to its atomic number.

	Electron configuration	**Orbital diagram**
Li	$1s^22s^1$	[⇅] [↑] (1s, 2s)

For carbon, which has six electrons, the electron configuration and orbital diagram are:

Electron configuration **Orbital diagram**

C $1s^22s^22p^2$

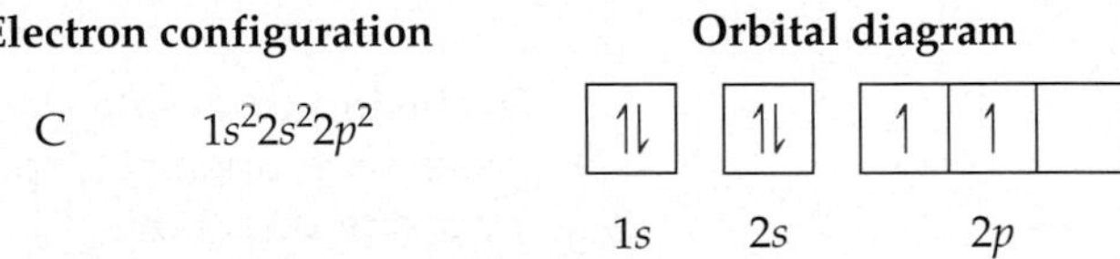

Notice that the 2*p* electrons occupy the *p* orbitals (of equal energy) singly, rather than pairing in one orbital. This is a result of **Hund's rule,** which states that when filling orbitals of equal energy, electrons fill them singly first, with parallel spins.

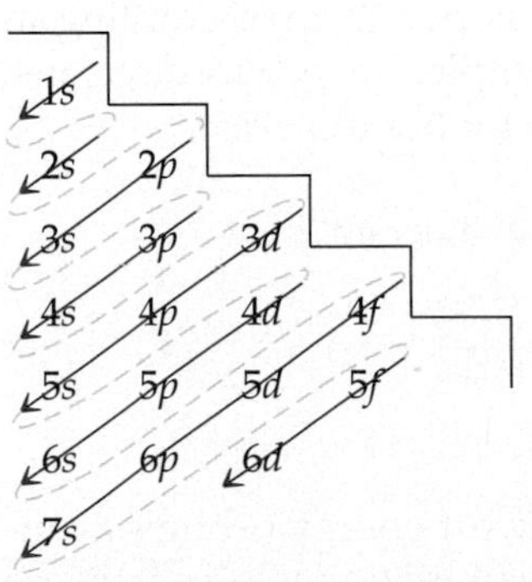

Figure 9.27 The arrows indicate the order in which orbitals fill.

Before we write electron configurations for other elements, let us summarize what we have learned so far:

- ***Electrons occupy orbitals so as to minimize the energy of the atom; therefore lower energy orbitals fill before higher energy orbitals. Orbitals fill in the following order:* 1*s* 2*s* 2*p* 3*s* 3*p* 4*s* 3*d* 4*p* 5*s* 4*d* 5*p* 6*s* *(Figure 9.27)***

- ***Orbitals can hold no more than two electrons each. When two electrons occupy the same orbital, they must have opposing spins. This is known as the Pauli exclusion principle.***
- ***When orbitals of identical energy are available, these are first occupied singly with parallel spins rather than in pairs. This is known as Hund's rule.***

Consider the electron configurations and orbital diagrams for elements with atomic numbers 3 through 10.

Symbol (#e^-)	Electron configuration	Orbital diagram 1s	2s	2p
Li (3)	$1s^22s^1$	⇅	↑	
Be (4)	$1s^22s^2$	⇅	⇅	
B (5)	$1s^22s^22p^1$	⇅	⇅	↑ | |
C (6)	$1s^22s^22p^2$	⇅	⇅	↑ | ↑ |
N (7)	$1s^22s^22p^3$	⇅	⇅	↑ | ↑ | ↑
O (8)	$1s^22s^22p^4$	⇅	⇅	⇅ | ↑ | ↑
F (9)	$1s^22s^22p^5$	⇅	⇅	⇅ | ⇅ | ↑
Ne (10)	$1s^22s^22p^6$	⇅	⇅	⇅ | ⇅ | ⇅

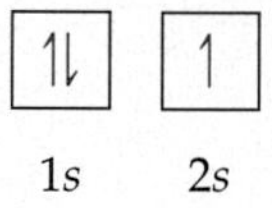

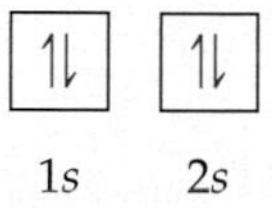

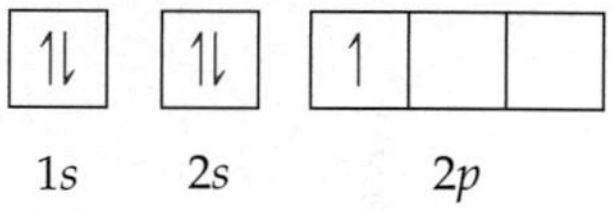

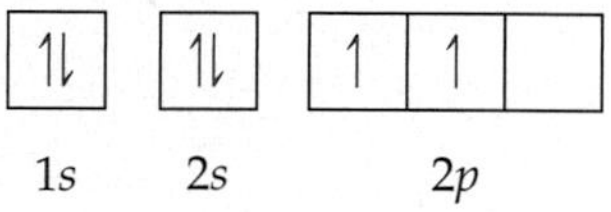

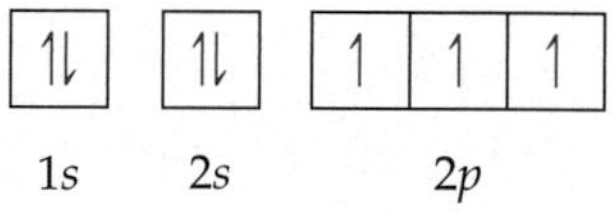

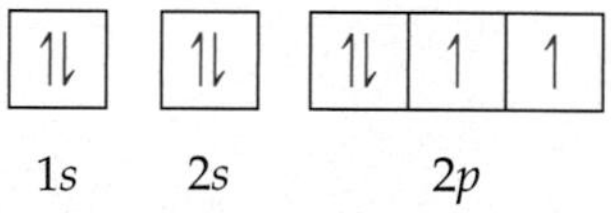

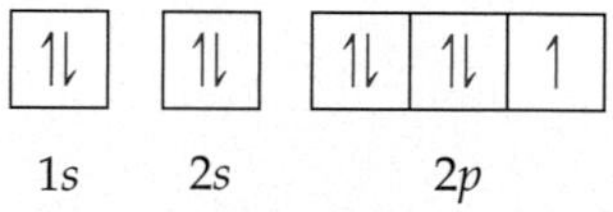

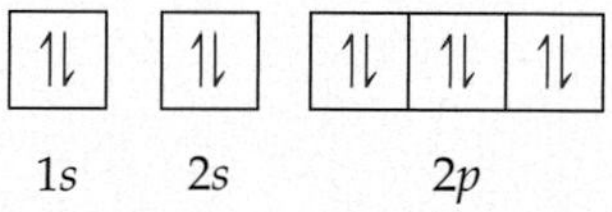

Notice how the p orbitals fill. As a result of Hund's rule, the p orbitals fill with single electrons before they fill with paired electrons. The electron configuration of neon represents the complete filling of the $n = 2$ principal shell. When writing electron configurations for elements beyond neon—or beyond any other noble gas—the electron configuration of the previous noble gas is often abbreviated by the symbol for the noble gas in brackets. For example, the electron configuration of sodium is:

Na $1s^22s^22p^63s^1$

This can also be written as:

Na $[Ne]3s^1$

where [Ne] represents $1s^22s^22p^6$, the electron configuration for neon.

To write an electron configuration for an element, first find its atomic number from the periodic table—this number equals the number of electrons. Then use the order of filling from Figure 9.26 or 9.27 to distribute the electrons in the appropriate orbitals. Remember that each orbital can hold a maximum of 2 electrons. Consequently:

- the *s* subshell has only 1 orbital and therefore can hold only 2 electrons.
- the *p* subshell has 3 orbitals and therefore can hold 6 electrons.
- the *d* subshell has 5 orbitals and therefore can hold 10 electrons.

EXAMPLE 9.2 Electron Configurations

Write electron configurations for each of the following elements.

a) Mg
b) S
c) Ga

Solution:

a) Magnesium has 12 electrons. We distribute two of these into the 1*s* orbital, two into the 2*s* orbital, six into the 2*p* orbitals, and two into the 3*s* orbital. The electron configuration is:

Mg $1s^22s^22p^63s^2$

We can write this more compactly as:

Mg $[Ne]3s^2$ where [Ne] represents $1s^22s^22p^6$

b) Sulfur has 16 electrons. We distribute two of these into the 1*s* orbital, two into the 2*s* orbital, six into the 2*p* orbitals, two into the 3*s* orbital, and four into the 3*p* orbitals. The electron configuration is:

S $1s^22s^22p^63s^23p^4$

We can write this more compactly as:

S $[Ne]3s^23p^4$ where [Ne] represents $1s^22s^22p^6$

c) Gallium has 31 electrons. We distribute two of these into the 1*s* orbital, two into the 2*s* orbital, six into the 2*p* orbitals, two into the 3*s* orbital, six into the 3*p* orbitals, two into the 4*s* orbital, ten into the 3*d* orbitals, and one into the 4*p* orbitals. The electron configuration is:

Ga $1s^22s^22p^63s^23p^64s^23d^{10}4p^1$

Notice that the *d* subshell has five orbitals and can therefore hold 10 electrons. We can write this more compactly as:

Ga $[Ar]4s^23d^{10}4p^1$ where [Ar] represents $1s^22s^22p^63s^23p^6$

SKILLBUILDER 9.2 Electron Configurations

Write electron configurations for each of the following elements.

a) Al
b) Br
c) Sr

SKILLBUILDER PLUS

Write electron configurations for each of the following ions.

a) Al^{3+}
b) Cl^-
c) O^{2-}

EXAMPLE 9.3 **Writing Orbital Diagrams**

Write an orbital diagram for silicon.

Solution:

Since silicon is atomic number 14, it has 14 electrons. Draw a box for each orbital, putting the lowest energy orbital (1*s*) on the far left and proceeding to orbitals of higher energy to the right.

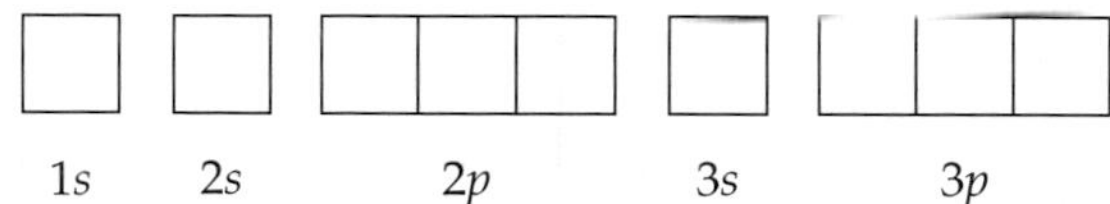

Distribute the 14 electrons into the orbitals, allowing a maximum of two electrons per orbital and remembering Hund's rule. The complete orbital diagram is:

	1s	2s	2p			3s	3p		
Si	⇅	⇅	⇅	⇅	⇅	⇅	↑	↑	

SKILLBUILDER 9.3 **Writing Orbital Diagrams**

Write an orbital diagram for argon.

9.7 Electron Configurations and the Periodic Table

Valence Electrons

Valence electrons are the electrons in the outermost principal shell (the principal shell with the highest principal quantum number, n). These electrons are important because, as we will see in the next chapter, they are involved in chemical bonding. Electrons that are *not* in the outermost principal shell are called **core electrons**. For example, silicon, with the electron configuration of $1s^22s^22p^63s^23p^2$ has 4 valence electrons (those in the $n = 3$ principal shell) and 10 core electrons.

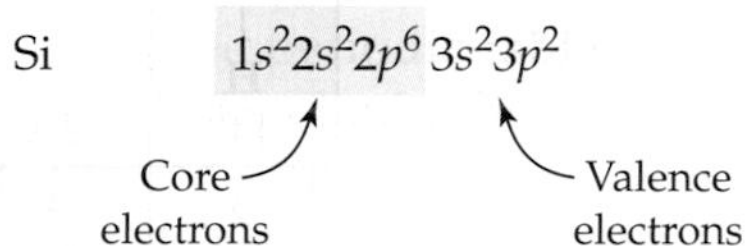

EXAMPLE 9.4 **Valence Electrons and Core Electrons**

Write an electron configuraton for selenium and identify the valence electrons and the core electrons.

Problems

Wavelength, Energy, and Frequency of Electromagnetic radiation

29. Which one of these types of electromagnetic radiation has the longest wavelength?
a) visible
b) ultraviolet
c) infrared
d) X-rays

30. Which one of these types of electromagnetic radiation has the shortest wavelength?
a) radiowaves
b) microwaves
c) infrared
d) ultraviolet

31. Write the following types of electromagnetic radiation in order of increasing energy per photon.
a) radiowaves
b) microwaves
c) infrared
d) ultraviolet

32. Write the following types of electromagnetic radiation in order of decreasing energy per photon.
a) gamma rays
b) radiowaves
c) microwaves
d) visible light

33. List two types of electromagnetic radiation with frequencies higher than visible light.

34. List two types of electromagnetic radiation with frequencies lower than infrared light.

35. Which of the following types of electromagnetic radiation—X-rays or microwaves—have the greatest:
a) energy per photon
b) frequency
c) wavelength

36. Which of the following types of electromagnetic radiation—visible or infrared—have the lowest:
a) energy per photon
b) frequency
c) wavelength

The Bohr Model

37. Bohr orbits have fixed _____ and fixed _____.

38. In the Bohr model, what happens when an electron makes transitions between orbits?

39. Two of the emission wavelengths in the hydrogen emission spectrum are 410 nm and 434 nm. One of these is due to the $n = 6$ to $n = 2$ transition and the other is due to the $n = 5$ to $n = 2$ transition. Which wavelength goes with which transition?

40. Two of the emission wavelengths in the hydrogen emission spectrum are 656 nm and 486 nm. One of these is due to the $n = 4$ to $n = 2$ transition and the other is due to the $n = 3$ to $n = 2$ transition. Which wavelength goes with which transition?

The Quantum-Mechanical Model

41. Make a sketch of the 1*s* and 2*p* orbitals. How would the 2*s* and 3*p* orbitals differ from the 1*s* and 2*p* orbitals?

42. Make a sketch of the 3*d* orbitals. How would the 4*d* orbitals differ from the 3*d* orbitals?

43. Which electron is, on average, closer to the nucleus: an electron in a 2*s* orbital or an electron in a 3*s* orbital?

44. Which electron is, on average, further from the nucleus: an electron in a 3*p* orbital or an electron in a 4*p* orbital?

45. According to the quantum-mechanical model for the hydrogen atom, which of the following electron transitions would produce light with longer wavelength: 2*p* to 1*s* or 3*p* to 1*s*?

46. According to the quantum-mechanical model for the hydrogen atom, which of the following transitions would produce light with longer wavelength: 3*p* to 2*s* or 4*p* to 2*s*?

Electron Configurations

47. Write full electron configurations for each of the following elements.
a) C
b) Na
c) Ar
d) Si

48. Write full electron configurations for each of the following elements.
a) N
b) S
c) Ne
d) K

49. Write full orbital diagrams for each of the following elements.
a) O
b) P
c) F
d) Mg

50. Write full orbital diagrams for each of the following elements.
a) S
b) Ca
c) Ne
d) C

51. Write electron configurations for each of the following elements. Use the symbol of the previous noble gas in brackets to represent the core electrons.
a) Ga
b) As
c) Rb
d) Sn

52. Write electron configurations for each of the following elements. Use the symbol of the previous noble gas in brackets to represent the core electrons.
a) Te
b) Br
c) I
d) Cs

53. Write electron configurations for each of the following transition metals.
a) Ti
b) V
c) Cr
d) Mn

54. Write electron configurations for each of the following transition metals.
a) Co
b) Ni
c) Cu
d) Zn

Valence Electrons and Core Electrons

55. Write full electron configurations for each of the following elements and indicate the valence electrons and the core electrons.
a) B
b) N
c) Sb
d) K

56. Write full electron configurations for each of the following elements and indicate the valence electrons and the core electrons.
a) Sr
b) Cl
c) Kr
d) Ge

57. How many valence electrons are in each of the following:
a) O
b) S
c) Br
d) Rb

58. How many valence electrons are in each of the following:
a) Ba
b) Al
c) Be
d) Se

Electron Configurations and the Periodic Table

59. Give the outer electron configuration for each of the following columns in the periodic table.
a) 1
b) 2
c) 5
d) 7

60. Give the outer electron configuration for each of the following columns in the periodic table.
a) 3
b) 4
c) 6
d) 8

61. Use the periodic table to write electron configurations for each of the following elements.
a) Al
b) Be
c) In
d) Zr

62. Use the periodic table to write electron configurations for each of the following elements.
a) Xe
b) Sc
c) Zr
d) Ba

63. Use the periodic table to write electron configurations for each of the following elements.
a) As
b) Ba
c) Ni
d) Bi

64. Use the periodic table to write electron configurations for each of the following elements.
a) Se
b) Sn
c) Pb
d) Cd

65. How many $2p$ electrons are in each of the following elements?
a) C
b) N
c) F
d) P

66. How many $3d$ electrons are in each of the following elements?
a) Fe
b) Zn
c) K
d) As

67. Give the number of elements in each of the following periods (rows) of the periodic table.
a) 2
b) 3
c) 4
d) 5

68. Which periods (rows) in the periodic table contain 32 elements?

69. Name an element in the third period (row) of the periodic table with:
a) 3 valence electrons
b) a total of four $3p$ electrons
c) six $3p$ electrons
d) two $3s$ electrons and no $3p$ electrons

70. Name an element in the fourth period of the periodic table with:
a) 5 valence electrons
b) a total of four $4p$ electrons
c) a total of three $3d$ electrons
d) a complete outer shell

71. Use the periodic table to identify the element with the following electron configuration.
a) $[Ne]3s^2 3p^5$
b) $[Ar]4s^2 3d^{10} 4p^1$
c) $[Ar]4s^2 3d^6$
d) $[Kr]5s^1$

72. Use the periodic table to identify the element with the following electron configuration.
a) $[Ne]3s^1$
b) $[Kr]5s^2 4d^{10}$
c) $[Xe]6s^2$
d) $[Kr]5s^2 4d^{10} 5p^3$

Periodic Trends

73. Choose the element with the highest ionization energy from each of the following pairs.
a) Na or Rb
b) Ga or Ge
c) P or I
d) P or Sn

74. Choose the element with the highest ionization energy from each of the following pairs.
a) As or At
b) Br or Bi
c) Si or Cl
d) P or Sb

75. Arrange the following elements in order of increasing ionization energy. Te, Pb, Cl, S, Sn

76. Arrange the following elements in order of increasing ionization energy. Ga, In, F, Si, N

77. Choose the element with the larger atoms from each of the following pairs.
a) Al or In
b) Si or N
c) P or Pb
d) C or F

78. Choose the element with the larger atoms from each of the following pairs.
a) Sn or Si
b) Br or Ga
c) Sn or Bi
d) Se or Sn

79. Arrange the following elements in order of increasing atomic size. Ca, Rb, S, Si, Ge, F

80. Arrange the following elements in order of increasing atomic size. Cs, Sb, S, Pb, Se

81. Choose the more metallic element from each of the following pairs.
a) Sr or Sb
b) As or Bi
c) Cl or O
d) S or As

82. Choose the more metallic element from each of the following pairs.
a) Sb or Pb
b) K or Ge
c) Ge or Sb
d) As or Sn

83. Arrange the following elements in order of increasing metallic character. Fr, Sb, In, S, Ba, Se

84. Arrange the following elements in order of increasing metallic character. Sr, N, Si, P, Ga, Al

Cumulative Problems

85. Use the electron configurations of the alkali metals to explain why they tend to form +1 ions.

86. Use the electron configurations of the halogens to explain why they tend to form −1 ions.

87. Write electron configurations for each of the following ions.
a) Ca^{2+}
b) K^{+}
c) S^{2-}
d) Br^{-}

88. Write electron configurations for each of the following ions.
a) F^{-}
b) P^{3}
c) Li^{+}
d) Al^{3+}

89. Examine Figure 4.12, which shows the division of the periodic table into metals, nonmetals, and metalloids. Use what you know about electron configurations to explain these divisions.

90. Examine Figure 4.14, which shows the elements that form predictable ions. Use what you know about electron configurations to explain these trends.

91. Explain what is wrong with each of the following electron configurations and write the correct configuration based on the number of electrons.
a) $1s^3 2s^3 2p^9$
b) $1s^2 2s^2 2p^6 2d^4$
c) $1s^2 1p^5$
d) $1s^2 2s^2 2p^8 3s^2 3p^1$

92. Explain what is wrong with each of the following electron configurations and write the correct configuration based on the number of electrons.
a) $1s^4 2s^4 2p^{12}$
b) $1s^2 2s^2 2p^6 3s^2 3p^6 3d^{10}$
c) $1s^2 2p^6 3s^2$
d) $1s^2 2s^2 2p^6 3s^2 3p^6 4s^2 4d^{10} 4p^3$

93. Bromine is a highly reactive liquid while krypton is an inert gas. Explain the difference based on their electron configurations.

94. Potassium is a highly reactive metal while argon is an inert gas. Explain the difference based on their electron configurations.

95. When an electron makes a transition from the $n = 3$ to the $n = 2$ hydrogen atom Bohr orbit, the energy difference between these two orbits (3.0×10^{-19} J) is given off in a photon of light. The relationship between the energy of a photon and its wavelength is given by $E = hc/\lambda$, where E is the energy of the photon in J, h is Planck's constant (6.626×10^{-34} J s) and c is the speed of light (3.00×10^8 m/s). Find the wavelength of light emitted by hydrogen atoms when an electron makes this transition.

96. When an electron makes a transition from the $n = 4$ to the $n = 2$ hydrogen atom Bohr orbit, the energy difference between these two orbits (4.1×10^{-19} J) is given off in a photon of light. The relationship between the energy of a photon and its wavelength is given by $E = hc/\lambda$, where E is the energy of the photon in J, h is Planck's constant (6.626×10^{-34} J s) and c is the speed of light (3.00×10^8 m/s). Find the wavelength of light emitted by hydrogen atoms when an electron makes this transition.

97. The distance from the sun to the earth is 1.496×10^8 km. How long does it take light to travel from the sun to earth?

98. The nearest star is Alpha Centauri, at a distance of 4.3 light years from earth. A light year is the distance that light travels in one year (365 days). How far away, in kilometers, is Alpha Centauri from earth?

99. In the beginning of this chapter, we learned that the quantum-mechanical model for the atom is the foundation for modern chemical understanding. Explain why this is so.

100. Niels Bohr said, "Anyone who is not shocked by quantum mechanics has not understood it." What did he mean by this?

Highlight Problems

101. Excessive exposure to sunlight increases the risk of skin cancer because some of the photons have enough energy to break chemical bonds in biological molecules. These bonds require approximately 250–800 kJ/mole of energy to break. The energy of a single photon is given by $E = hc/\lambda$, where E is the energy of the photon in J, h is Planck's constant (6.626×10^{-34} J s) and c is the speed of light (3.00×10^8 m/s). Determine which of the following kinds of light contain enough energy to break chemical bonds in biological molecules by calculating the total energy in 1 mol photons for light of each wavelength.
a) infrared light (1500 nm)
b) visible light (500 nm)
c) ultraviolet light (150 nm)

102. The quantum-mechanical model, besides revolutionizing chemistry, shook the philosophical world because of its implications regarding determinism. Determinism is the idea that the outcomes of future events are determined by preceding events. The trajectory of a baseball, for example, is deterministic; that is, its trajectory—and therefore its landing place—is determined by its position, speed, and direction of travel. Before quantum mechanics, most scientists thought that fundamental particles—such as electrons and protons—also behaved deterministically. The implication of this was that the entire universe must behave deterministically—its future must be determined by preceding events. Quantum mechanics challenged this reasoning because fundamental particles did not behave deterministically—their future paths were not determined by preceding events. Some scientists struggled with this idea. Einstein himself refused to believe it, stating, "God does not play dice with the universe." Explain what Einstein meant by this statement.

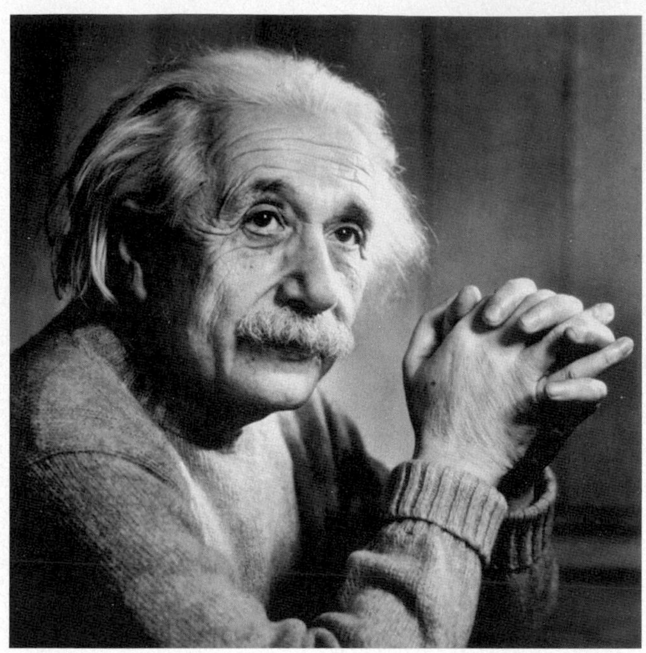

"God does not play dice with the universe."

Answers to Skillbuilder Exercises

Skillbuilder 9.1 **a)** blue, green, red **b)** red, green, blue **c)** red, green, blue

Skillbuilder 9.2

a) Al $1s^2 2s^2 2p^6 3s^2 3p^1$ or $[Ne]3s^2 3p^1$

b) Br $1s^2 2s^2 2p^6 3s^2 3p^6 4s^2 3d^{10} 4p^5$ or $[Ar]4s^2 3d^{10} 4p^5$

c) Sr $1s^2 2s^2 2p^6 3s^2 3p^6 4s^2 3d^{10} 4p^6 5s^2$ or $[Kr]5s^2$

Skillbuilder Plus, p. 297 Subtract one electron for each unit of positive charge. Add one electron for each unit of negative charge.

a) Al^{3+} $1s^2 2s^2 2p^6$

b) Cl^- $1s^2 2s^2 2p^6 3s^2 3p^6$

c) O^{2-} $1s^2 2s^2 2p^6$

Skillbuilder 9.3

	1s	2s	2p	3s	3p
Ar	⇅	⇅	⇅ ⇅ ⇅	⇅	⇅ ⇅ ⇅

Skillbuilder 9.4

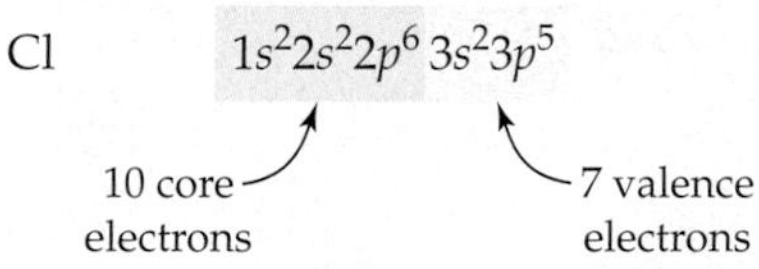

Skillbuilder 9.5 $[Kr]5s^2 4d^{10} 5p^2$

Skillbuilder 9.6 **a)** Mg **b)** Te **c)** cannot tell based on periodic properties **d)** F

Skillbuilder 9.7 **a)** Pb **b)** Rb **c)** cannot tell based on periodic properties **d)** Se

Skillbuilder 9.8 **a)** In **b)** cannot tell based on periodic properties **c)** Bi **d)** B

TATM 1540
29.25
TATM 1541 C20 MK1 902
9.772 1.00 29.96
N4 MK1 902
20.913 5.924 1.00 31.
Quade Paul '02

Chemical Bonding

10

"Never say, 'I tried it once and it did not work.'"
Ernest Rutherford

10.1 Bonding Models and AIDS Drugs

Proteins are discussed in more detail in Chapter 19.

In 1989, researchers discovered the structure of a molecule called HIV-protease. HIV-protease is a protein (a class of biological molecules) synthesized by the human immunodeficiency virus (HIV). HIV-protease is crucial to the virus's ability to replicate itself and cause acquired immune deficiency syndrome (AIDS). Without HIV-protease, HIV could not spread in the human body because the virus could not copy itself, and AIDS would not develop.

With knowledge of the HIV-protease structure, drug companies set out to design a molecule that would disable protease by sticking to the working part of the molecule (called the *active site*). To design such a molecule, researchers used **bonding theories**—models that predict how atoms bond together to form molecules—to simulate how potential drug molecules would interact with the protease molecule. By the early 1990s, these companies developed several drug molecules that seemed to work. Since these molecules inhibit the action of HIV-protease, they are called *protease inhibitors*. In human trials, protease inhibitors in combination with other drugs have decreased the viral count in HIV-infected individuals to undetectable levels. Many AIDS patients are still alive today because of the development of these drugs.

◀ The gold-colored structure on the computer screen is a representation of HIV-protease. The molecule shown in the center is Indinavir, a protease inhibitor.

Bonding theories are central to chemistry because they predict how atoms bond together to form molecules. They predict what combinations of atoms form molecules and what combinations do not. For example, bonding theories predict why salt is NaCl and not $NaCl_2$ and why water is

H_2O and not H_3O. Bonding theories also explain the shapes of molecules, which in turn determine many of their physical and chemical properties. The bonding theory you will learn in this chapter is called **Lewis theory,** named after the American chemist who developed it, **G.N. Lewis** (1875–1946). It consists of representing electrons as dots and drawing what are called **dot structures** or **Lewis structures** to represent molecules. These structures, which are fairly simple to draw, have tremendous predictive power. In just a few minutes, you can use Lewis theory to determine whether or not a particular set of atoms will form a stable molecule and what that molecule might look like. Although modern chemists also use more advanced bonding theories to better predict molecular properties, Lewis theory remains the simplest method for making quick, everyday predictions about molecules.

10.2 Representing Valence Electrons with Dots

Remember, the number of valence electrons for any main-group element (except helium, which has 2 valence electrons but is in Group 8A) is equal to the group number of the element.

In the previous chapter, we learned that valence electrons are those electrons in the outermost principal shell. Since valence electrons are most important in bonding, Lewis theory focuses on these. In Lewis theory, the valence electrons of an element are represented as dots surrounding the symbol of the element. The result is called a Lewis structure. For example, the electron configuration of O is:

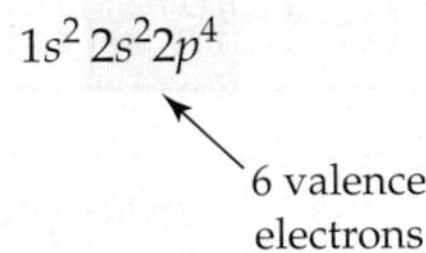

and the Lewis structure is:

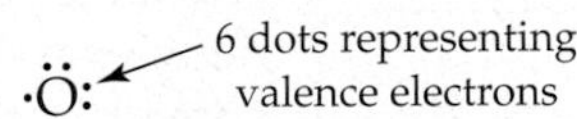

Number of valence electrons	Order of filling used in this book:
1	X·
2	X:
3	Ẋ:
4	·Ẋ:
5	·Ẋ: (one dot below)
6	·Ẍ: (one dot below)
7	:Ẍ: (one dot below)
8	:Ẍ: (two dots below)

Each dot represents a valence electron. The dots are placed around the element's symbol with a maximum of two dots per side. While the exact location of dots is not critical, the first two dots are usually placed on the right side of the atomic symbol. In this book, the rest of the dots are filled in singly first, in counter-clockwise order, and then paired.

The Lewis structures for all of the period 2 elements are:

Li· Be: Ḃ: ·Ċ: ·Ṅ: ·Ö: :F̤: :N̤e:

Lewis structures allow us to easily see the number of valence electrons in an atom. Notice that atoms with 8 valence electrons—which are particularly stable—are easily identified because they have eight dots, an **octet.**

Helium is somewhat of an exception. Its electron configuration and Lewis structure are:

$1s^2$ He:

The Lewis structure of helium contains only two dots (a **duet**). For helium, a duet represents a stable electron configuration.

In Lewis theory, a **chemical bond** is the sharing or transfer of electrons to attain stable electron configurations among the bonding atoms. If the electrons are transferred, the bond is an **ionic bond.** If the electrons are shared, the bond is a **covalent bond.** In either case, the bonding atoms get stable electron configurations; because this stable configuration is usually eight electrons, this is known as the **octet rule.**

EXAMPLE 10.1 **Writing Lewis Structures for Elements**

Write a Lewis structure for P.

Solution:

Since P is in Group 5A in the periodic table, it has five valence electrons. We represent these as 5 dots surrounding the symbol for phosphorus.

$$\cdot\dot{\text{P}}\text{:}$$

SKILLBUILDER 10.1 **Writing Lewis Structures for Elements**

Write a Lewis structure for Mg.

10.3 Lewis Structures for Ionic Compounds: Electrons Transferred

Recall from Chapter 5 that when metals bond with nonmetals, electrons are transferred from the metal to the nonmetal. The metal becomes a cation and the nonmetal becomes an anion. The attraction between the cation and the anion results in an ionic compound. In Lewis theory, we represent this by moving electron dots from the metal to the nonmetal. For example, potassium and chlorine have the following Lewis structures.

$$\text{K}\cdot \quad :\ddot{\underset{\cdot}{\text{Cl}}}:$$

When potassium and chlorine bond, potassium transfers its valence electron to chlorine.

Recall from Section 5.4 that ionic compounds do not exist as distinct molecules, but rather as part of a large lattice of alternating cations and anions.

$$\text{K}\cdot \quad :\ddot{\underset{\cdot}{\text{Cl}}}: \longrightarrow \text{K}^+ \; [:\ddot{\underset{\cdot\cdot}{\text{Cl}}}:]^-$$

The transfer of the electron gives chlorine an octet (shown as eight dots around chlorine), and leaves potassium without any valence electrons (which is effectively an octet in the previous principal shell). The potassium, because it lost an electron, becomes positively charged, while the chlorine, which gained an electron, becomes negatively charged. The Lewis structure of an anion is usually written within brackets with the charge in the upper right hand corner (outside the brackets). The positive and negative charges attract one another, which then results in the compound KCl.

Recall from Section 4.7 that atoms that lose electrons become positively charged and atoms that gain electrons become negatively charged.

EXAMPLE 10.2 Writing Ionic Lewis Structures

Write a Lewis structure for the compound MgO.

Solution:

The Lewis structures of magnesium and oxygen are:

Mg: ·Ö:

In MgO, magnesium loses its two valence electrons, forming a +2 charge, and oxygen gains two electrons, forming a −2 charge and acquiring an octet:

Mg^{2+} $[:\ddot{\underset{..}{O}}:]^{2-}$

SKILLBUILDER 10.2 Writing Ionic Lewis Structures

Write a Lewis structure for the compound NaBr.

Lewis theory predicts the correct chemical formulas for ionic compounds. For the compound that forms between K and Cl, for example, Lewis theory predicts one potassium cation to every chlorine anion, KCl. As another example, consider the ionic compound formed between sodium and sulfur. The Lewis structures for sodium and sulfur are:

Na· ·S̈:

Notice that sodium must lose its one valence electron to get an octet (in the previous principal shell), while sulfur must gain two electrons to get an octet. Consequently, the compound that forms between sodium and sulfur requires two sodium atoms to every one sulfur atom. The Lewis structure is:

Na^+ $[:\ddot{\underset{..}{S}}:]^{2-}$ Na^+

The two sodium atoms each lose their one valence electron, while the sulfur atom gains two electrons and gets an octet. The correct chemical formula is Na_2S.

EXAMPLE 10.3 Using Lewis Theory to Predict the Chemical Formula of an Ionic Compound

Use Lewis theory to predict the formula for the compound that forms between calcium and chlorine.

Solution:

The Lewis structures of calcium and chlorine are:

Ca: :C̈l:

Calcium must lose its two valence electrons (to effectively get an octet in its previous principal shell), while chlorine only needs to gain one electron to get an octet. Consequently, the compound that forms between

Ca and Cl must have two chlorine atoms to every one calcium atom. The Lewis structure is:

$$[:\ddot{\underset{..}{Cl}}:]^- \; Ca^{2+} \; [:\ddot{\underset{..}{Cl}}:]^-$$

The formula is therefore $CaCl_2$.

SKILLBUILDER 10.3 **Using Lewis Theory to Predict the Chemical Formula of an Ionic Compound**

Use Lewis theory to predict the formula for the compound that forms between magnesium and nitrogen.

10.4 Covalent Lewis Structures: Electrons Shared

Recall from Chapter 5 that when nonmetals bond with other nonmetals, a molecular compound results. Molecular compounds contain covalent bonds, in which electrons are shared between atoms rather than transferred. In Lewis theory, we represent covalent bonding by allowing neighboring atoms to share some of their valence electrons in order to attain octets (or duets for hydrogen). For example, hydrogen and oxygen have the following Lewis structures.

$$H\cdot \quad \cdot\ddot{\underset{\cdot}{O}}:$$

In water, hydrogen and oxygen share their electrons so that each hydrogen atom gets a duet and the oxygen atom gets an octet.

$$H:\ddot{\underset{..}{O}}:H$$

The shared electrons—those that appear in the space between the two atoms—count towards the octets (or duets) of *both of the atoms.*

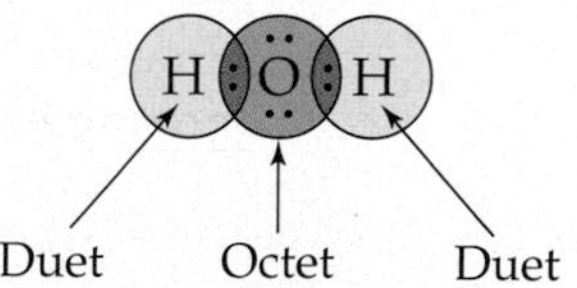

Sometimes lone pair electrons are also called nonbonding electrons.

Electrons that are shared between two atoms are called **bonding pair** electrons, while those that are only on one atom are called **lone pair** electrons.

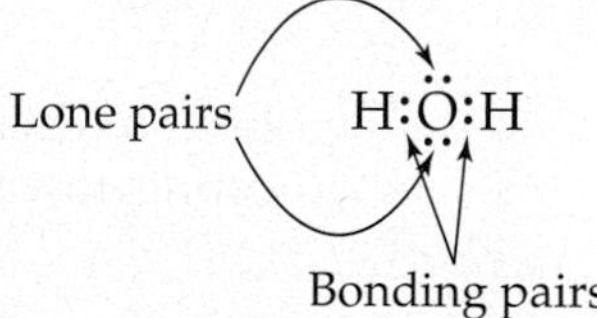

Bonding pair electrons are often represented by dashes to emphasize that they are a chemical bond.

$$H-\ddot{\underset{..}{O}}-H$$

Lewis theory also explains why the halogens form diatomic molecules. Consider the Lewis structure of chlorine.

$$:\ddot{\underset{\cdot}{Cl}}:$$

If two Cl atoms pair together, they can each get an octet.

:C̤̈l:C̤̈l: or :C̤̈l—C̤̈l:

When we examine elemental chlorine, it indeed exists as a diatomic molecule, just as Lewis theory predicts. The same is true for the other halogens.

Similarly, Lewis theory predicts that hydrogen, which has the following Lewis structure:

H·

should exist as H_2. When two hydrogen atoms share their valence electrons, they each get a duet, a stable configuration for hydrogen.

H:H or H—H

Again, Lewis theory is correct. In nature, elemental hydrogen exists as H_2 molecules.

Double and Triple Bonds

In Lewis theory, two atoms may share more than one electron pair to get octets. For example, we know from Chapter 5 that oxygen exists as the diatomic molecule, O_2. The Lewis structure of an oxygen atom is:

·Ö̤:

If we pair two oxygen atoms together and then try to write a Lewis structure, we do not have enough electrons to give each O atom an octet.

:Ö:Ö̤:

However, we can convert a lone pair into an additional bonding pair by moving it into the bonding region.

:Ö:Ö̤:

↓

:Ö::Ö: or :Ö=Ö:

Each oxygen atom now has an octet because the additional bonding pair counts towards the octet of both oxygen atoms.

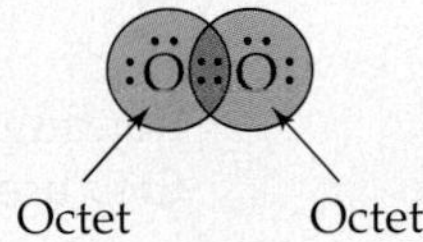

When two electron pairs are shared between two atoms, the resulting bond is a **double bond**. In general, double bonds are shorter and stronger than single bonds.

Atoms can also share three electron pairs. Consider the Lewis structure of N_2. Since each N atom has five valence electrons, the Lewis structure for N_2 has 10 electrons. A first attempt at writing the Lewis structure gives:

:N̈:N̈:

As with O_2, we do not have enough electrons to satisfy the octet rule for both N atoms. However, if we convert two additional lone pairs into bonding pairs, each nitrogen atom can get an octet.

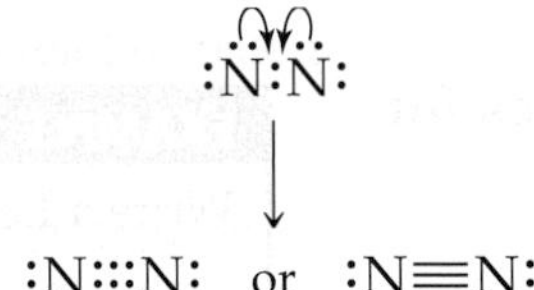

The resulting bond is called a **triple bond**. Triple bonds are even shorter and stronger than double bonds. When we examine nitrogen in nature, we find that it indeed exists as a diatomic molecule with a very strong bond between the two nitrogen atoms. The bond is so strong that it is difficult to break, making N_2 a relatively unreactive molecule.

10.5 Writing Lewis Structures for Covalent Compounds

To write a Lewis structure for a covalent compound, follow these steps.

1. ***Write the correct skeletal structure for the molecule.*** The Lewis structure of a molecule must have the atoms in the correct positions. For example, you could not write a Lewis structure for water if you started with the hydrogen atoms next to each other and the oxygen atom at the end (H H O). In nature, oxygen is the central atom and the hydrogens are **terminal atoms** (at the ends). The correct skeletal structure is H O H. The only way to absolutely know the correct skeletal structure for any molecule is by examining its structure in nature. However, we can write likely skeletal structures by remembering two guidelines. First, *hydrogen atoms will always be terminal.* Since hydrogen only requires a duet, it will never be a central atom because central atoms must form at least two bonds and hydrogen can only form one. Second, many *molecules tend to be symmetrical*, so when a molecule contains several atoms of the same type, these tend to be in terminal positions. This second guideline, however, has many exceptions. In cases where the skeletal structure is unclear, this text will provide you with the correct skeletal structure.

When guessing at skeletal structures, put the less metallic elements in terminal positions and the more metallic elements in central positions.

2. ***Calculate the total number of electrons for the Lewis structure by summing the valence electrons of each atom in the molecule.*** Remember that the number of valence electrons for any main-group element is equal to its group number in the periodic table. ***If you are writing a Lewis structure for a polyatomic ion, the charge of the ion must be considered when calculating the total number of electrons.*** Add one electron for each negative charge and subtract one electron for each positive charge.
3. ***Distribute the electrons among the atoms, giving octets (or duets for hydrogen) to as many atoms as possible.*** Begin by placing two electrons between each pair of atoms. These are the minimal number of bonding electrons. Then distribute the remaining electrons, first to terminal atoms, and then to the central atom.

As another example, consider the molecule H_2CO. Its Lewis structure is:

$$\begin{array}{c} :\!\ddot{O}\!: \\ \| \\ H-C-H \end{array}$$

This molecule has three electron groups around the central atom. These three electron groups get as far away from each other as possible, resulting in a bond angle of 120° and a **trigonal planar** geometry.

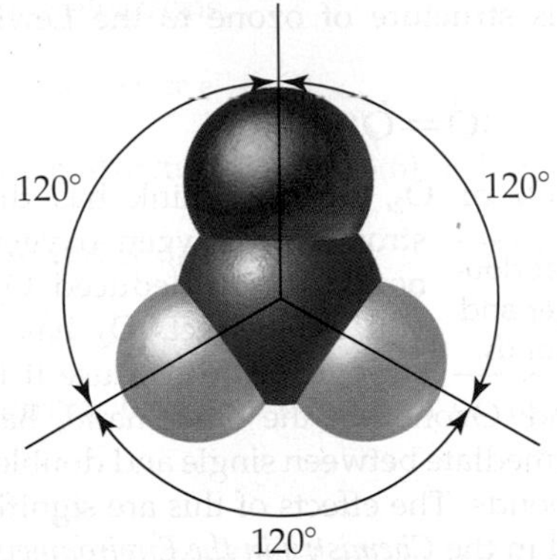

A tetrahedron is a geometrical shape with four faces.

If a molecule has four electron groups around the central atom, such as CH_4, it has a **tetrahedral** geometry with bond angles of 109.5°.

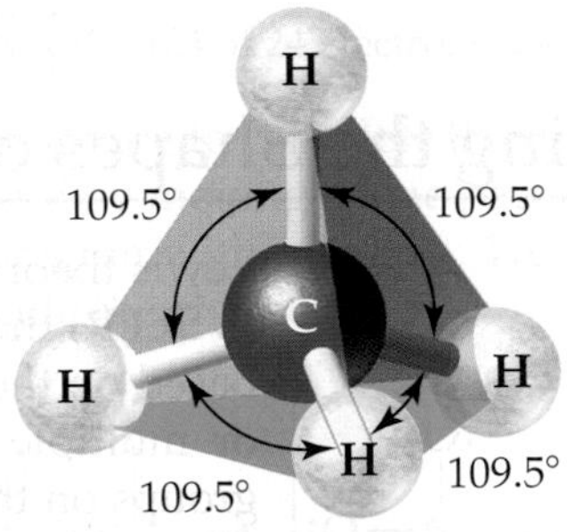

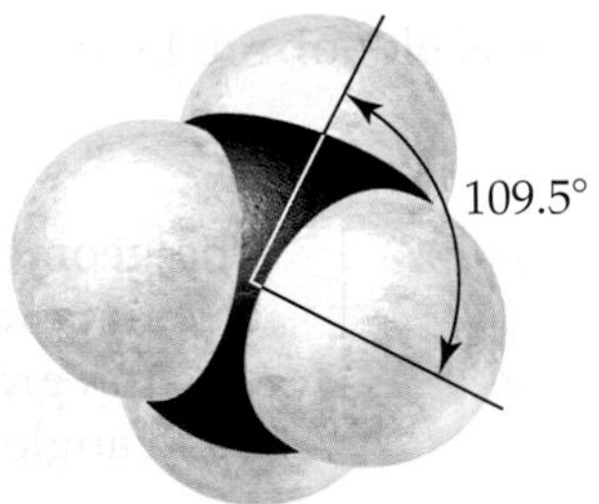

CH_4 is shown here with both a ball-and-stick model (above) and a space filling model (below). Although space-filling models more closely portray mlecules, ball-and-stick models are often used to clearly illustrate molecular geometries.

The electron groups repelling each other cause the tetrahedral shape—the tetrahedron allows the maximum separation among the four groups. When we write the structure of CH_4 on paper, it may seem that the molecule should be square planar, with bond angles of 90°. However, in three dimensions the electrons groups can get further away from each other by forming the tetrahedral geometry.

Each of the preceding examples only has bonding groups of electrons around the central atom. What happens in molecules with lone pairs around the central atom? These lone pairs also repel other electron groups. For example, consider NH_3.

H
|
H—N̤—H

The four electron groups (one lone pair and three bonding pairs) get as far away from each other possible. If we look only at the electrons, we find that the **electron geometry**—the geometrical arrangement of the electron groups—is tetrahedral.

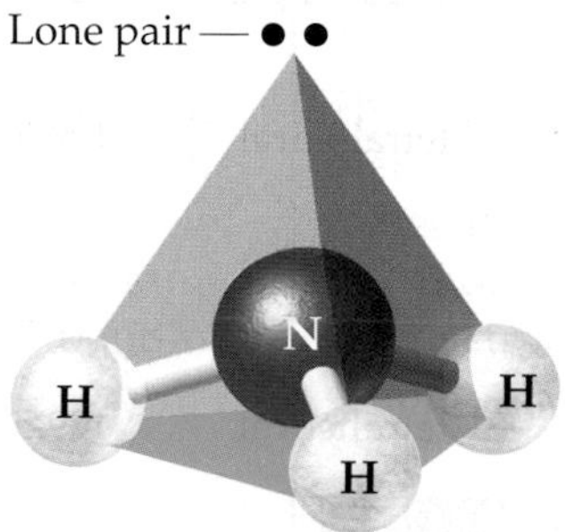

However, the **molecular geometry**—the geometrical arrangement of the atoms—is **trigonal pyramidal**.

Pyramidal structure

Notice that, although the electron geometry and the molecular geometry are different, the electron geometry is relevant to the molecular geometry. In other words, the lone pair exerts its influence on the bonding pairs.

Consider one last example, H_2O. Its Lewis structure is:

H—Ö̤—H

Since it has four electron groups, its electron geometry is also tetrahedral.

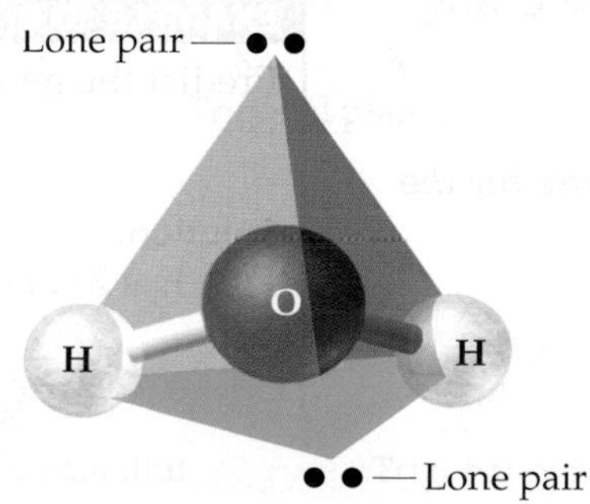

Writing Lewis Structures for Covalent Compounds (Sections 10.4, 10.5)

To write covalent Lewis structures, follow these steps:

1. ***Write the correct skeletal structure for the molecule.*** Hydrogen atoms will always be terminal and many molecules tend to be symmetrical
2. ***Calculate the total number of electrons for the Lewis structure by summing the valence electrons of each atom in the molecule.*** Remember that the number of valence electrons for any main-group element is equal to its group number in the periodic table. For polyatomic ions, add one electron for each negative charge and subtract one electron for each positive charge.
3. ***Distribute the electrons among the atoms, giving octets (or duets for hydrogen) to as many atoms as possible.*** Begin by placing two electrons between each pair of atoms. These are the bonding electrons. Then distribute the remaining electrons, first to terminal atoms and then to the central atom.
4. ***If any atoms lack an octet, form double or triple bonds as necessary to give them octets.*** Do this by moving lone electron pairs from terminal atoms into the bonding region with the central atom.

EXAMPLE 10.15 Writing Lewis Structures for Covalent Compounds

Write a Lewis structure for CS_2.

Solution:

S C S

$$\begin{aligned} \text{total \# e}^- &= 1 \times (\text{\# valence e}^- \text{ in C}) \\ &\quad + 2 \times (\text{\# valence e}^- \text{ in S}) \\ &= 4 + 2(6) \\ &= 16 \end{aligned}$$

$$\text{S:C:S} \quad (4 \text{ of } 16\ \text{e}^- \text{ used})$$

$$:\ddot{\underset{\cdot\cdot}{\text{S}}}:\text{C}:\ddot{\underset{\cdot\cdot}{\text{S}}}: \quad (16 \text{ of } 16\ \text{e}^- \text{ used})$$

$$:\ddot{\text{S}}::\text{C}::\ddot{\text{S}}: \quad \text{or} \quad :\ddot{\text{S}}{=}\text{C}{=}\ddot{\text{S}}:$$

Writing Resonance Structures (Section 10.6)

When two or more equivalent (or nearly equivalent) Lewis structures can be written for a molecule, the true structure is an average between these. Represent this by writing all of the correct structures (called resonance structures) with double-headed arrows between them.

EXAMPLE 10.16 Writing Resonance Structures

Write resonance structures for SeO_2.

Solution:

We can write a Lewis structure for SeO_2 by following the steps for writing covalent Lewis structures. We find that we can write two equally correct structures, so we draw them both as resonance structures.

$$:\ddot{\underset{\cdot\cdot}{\text{O}}}-\ddot{\text{Se}}{=}\ddot{\text{O}}: \longleftrightarrow :\ddot{\text{O}}{=}\ddot{\text{Se}}-\ddot{\underset{\cdot\cdot}{\text{O}}}:$$

Predicting the Shapes of Molecules (Section 10.7)

To determine the shape of a molecule, follow these steps:

1. ***Draw a Lewis structure for the molecule.***

EXAMPLE 10.17 Predicting the Shapes of Molecules

Predict the geometry of SeO_2.

Solution:

The Lewis structure for SeO_2 (as we saw in Example 10.16) is composed of the following two resonance structures.

$$:\ddot{\underset{\cdot\cdot}{\text{O}}}-\ddot{\text{Se}}{=}\ddot{\text{O}}: \longleftrightarrow :\ddot{\text{O}}{=}\ddot{\text{Se}}-\ddot{\underset{\cdot\cdot}{\text{O}}}:$$

2. *Determine the total number of electron groups around the central atom.* Lone pairs, single bonds, double bonds, or triple bonds, each count as one group.

Either of the resonance structures will give the same geometry.
Total number of electron groups = 3

3. *Determine the number of bonding groups and the number of lone pairs around the central atom.* These should sum to the result from Step 2. Bonding groups include single bonds, double bonds, or triple bonds.

Number of bonding groups = 2
Number of lone pairs = 1

4. *Use Table 10.1 to determine the electron geometry and molecular geometry.*

Electron geometry = Trigonal planar
Molecular geometry = Bent

Determining if a Molecule is Polar (Section 10.8)

1. *Determine if the molecule contains polar bonds.* A bond is polar if the two bonding atoms have different electronegativities. If there are no polar bonds, the molecule is nonpolar.
2. *Determine if the polar bonds add together to form a net dipole moment.* Use VSEPR to determine the geometry of the molecule. Then visualize each bond as a rope pulling on the central atom. Is the molecule highly symmetrical? Do the pulls of the ropes cancel? If so, there is no net dipole moment and the molecule is nonpolar. If the molecule is asymmetrical and the pulls of the rope do not cancel, the molecule is polar.

EXAMPLE 10.18 Determining if a Molecule is Polar

Determine if SeO_2 is polar.

Solution:

Se and O are nonmetals with different electronegativities (2.4 for Se and 3.5 for O). Therefore, the Se-O bonds are polar.

As we saw in Example 10.17, the geometry of SeO_2 is bent.

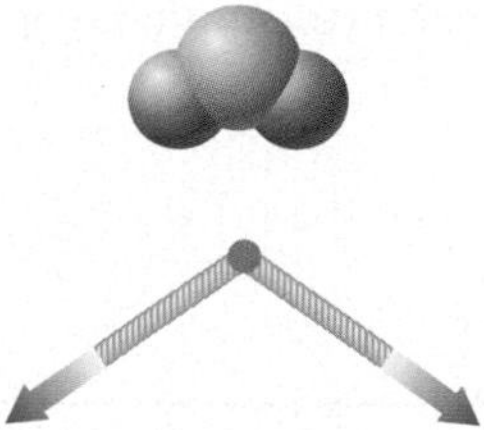

The polar bonds do not cancel but rather sum to give a net dipole moment. Therefore the molecule is polar.

KEY TERMS

bent [10.7]
bonding pair [10.4]
bonding theory [10.1]
chemical bond [10.2]
covalent bond [10.2]
dipole moment [10.8]
dot structure [10.1]
double bond [10.4]
duet [10.2]
electron geometry [10.7]
electron group [10.7]
electronegativity [10.8]
ionic bond [10.2]
G.N. Lewis [10.1]
Lewis structure [10.1]
Lewis theory [10.1]
linear [10.7]
lone pair [10.4]
molecular geometry [10.7]
nonpolar [10.8]
octet [10.2]
octet rule [10.2]
polar covalent bond [10.8]
polar molecule [10.8]
resonance structures [10.6]
terminal atom [10.5]
tetrahedral [10.7]
trigonal planar [10.7]
trigonal pyramidal [10.7]
triple bond [10.4]
valence shell electron pair repulsion theory (VSEPR) [10.7]

EXERCISES

Questions

1. Why are bonding theories important?
2. Give some examples of what bonding theories can predict.
3. Write the electron configurations for Ne and Ar. How many valence electrons do they each have?
4. In Lewis theory, what is an octet? What is a duet?
5. According to Lewis theory, what is a chemical bond?
6. What is the difference between ionic bonding and covalent bonding?
7. How can Lewis theory be used to determine the formula of ionic compounds? You may explain this with an example.
8. What is the difference between lone pair and bonding pair electrons?
9. How are double and triple bonds different from single bonds?
10. What is the procedure for writing a covalent Lewis structure?
11. How do you determine the number of electrons that go into the Lewis structure of a molecule?
12. How do you determine the number of electrons that go into the Lewis structure of a polyatomic ion?
13. Why does the octet rule have exceptions? Give some examples.
14. What are resonance structures? Why are they necessary?
15. Explain how VSEPR predicts the shapes of molecules.
16. If all of the electron groups around a central atom are bonding groups (i.e. there are no lone pairs), what is the molecular geometry for:
a) two electron groups
b) three electron groups
c) four electron groups
17. Give the bond angles for each of the geometries in Question 16.
18. What is the difference between electron geometry and molecular geometry in VSEPR?
19. What is electronegativity?
20. What is the most electronegative element on the periodic table?
21. What is a polar covalent bond?
22. What is a dipole moment?
23. Why is it important to know if a molecule is polar or nonpolar?
24. If a molecule has polar bonds, will the molecule itself be polar? Why or why not?

Problems

Writing Lewis Structures for Elements

25. Write the electron configuration for N. Then write a Lewis structure for N and show which electrons from the electron configuration are included in the Lewis structure.

26. Write the electron configuration for Ne. Then write a Lewis structure for Ne and show which electrons from the electron configuration are included in the Lewis structure.

27. Write Lewis structures for each of the following elements.
a) K
b) Al
c) P
d) Ar

28. Write Lewis structures for each of the following elements.
a) Mg
b) Si
c) S
d) Br

29. Write a generic Lewis structure for the halogens. Do the halogens tend to gain or lose electrons in chemical reactions? How many?

30. Write a generic Lewis structure for the alkali metals. Do the alkali metals tend to gain or lose electrons in chemical reactions? How many?

31. Write a generic Lewis structure for the alkaline earth metals. Do the alkaline earth metals tend to gain or lose electrons in chemical reactions. How many?

32. Write a generic Lewis structure for the elements in the oxygen family (Group 6). Do the elements in the oxygen family tend to gain or lose electrons in chemical reactions? How many?

33. Write a Lewis structure for each of the following ions.
a) Cl^-
b) Se^{2-}
c) Na^+
d) Mg^{2+}

34. Write a Lewis structure for each of the following ions.
a) N^{3-}
b) S^{2-}
c) Ca^{2+}
d) Al^{3+}

Lewis Structures for Ionic Compounds

35. Write a Lewis structure for each of the following ionic compounds.
a) NaF
b) CaO
c) $SrBr_2$
d) K_2O

36. Write a Lewis structure for each of the following ionic compounds.
a) SrO
b) Li_2S
c) CaI_2
d) RbF

37. Use Lewis theory to determine the formula for the compound that forms between:
a) Sr and Se
b) Ba and Cl
c) Na and S
d) Al and O

38. Use Lewis theory to determine the formula for the compound that forms between:
a) Ca and N
b) Mg and I
c) Ca and S
d) Cs and F

39. Draw the Lewis structure for the ionic compound that forms between Li and:
a) F
b) O
c) N

40. Draw the Lewis structure for the ionic compound that forms between Ba and:
a) F
b) O
c) N

41. Determine what is wrong with each of the following ionic Lewis structures and write the correct structure.

(a) $[\mathrm{Cs}\,:]^+\ [:\ddot{\mathrm{Cl}}:]^-$

(b) $\mathrm{Ba}^+\ [:\ddot{\underset{\cdot\cdot}{\mathrm{O}}}:]^-$

(c) $\mathrm{Ca}^{2+}\ [:\ddot{\underset{\cdot\cdot}{\mathrm{I}}}:]^-$

42. Determine what is wrong with each of the following ionic Lewis structures and write the correct structure.

(a) $[:\ddot{\underset{\cdot\cdot}{\mathrm{O}}}:]^{2-}\mathrm{Na}^+[:\ddot{\underset{\cdot\cdot}{\mathrm{O}}}:]^{2-}$

(b) $\mathrm{Mg}:\ddot{\underset{\cdot\cdot}{\mathrm{O}}}:$

(c) $[\mathrm{Li}:]^+[:\ddot{\underset{\cdot\cdot}{\mathrm{S}}}:]^-$

Lewis Structures for Covalent Compounds

43. Write a Lewis structure for each of the following molecules.
a) PH_3
b) SCl_2
c) F_2
d) HI

44. Write a Lewis structure for each of the following molecules.
a) CH_4
b) NF_3
c) OF_2
d) H_2O

45. Write a Lewis structure for each of the following molecules.
a) O_2
b) CO
c) HONO
d) SO_2

46. Write a Lewis structure for each of the following molecules.
a) N_2O (oxygen is terminal)
b) SiH_4
c) CI_4
d) Cl_2CO (carbon is central)

47. Write a Lewis structure for each of the following molecules.
a) C_2H_2
b) C_2H_4
c) N_2H_2
d) N_2H_4

48. Write a Lewis structure for each of the following molecules.
a) H_2CO (carbon is central)
b) H_3COH (carbon and oxygen are both central)
c) H_3COCH_3 (oxygen is between the two carbon atoms)
d) H_2O_2

49. Determine what is wrong with each of the following Lewis structures and write the correct structure.

(a) :N̈=N̈:

(b) :S̤̈—Si—S̤̈:

(c) H—H—Ö̤:

(d) :Ï̤—N—Ï̤:
with :Ï̤: bonded below N

50. Determine what is wrong with each of the following Lewis structures and write the correct structure.

(a) H—H—H—N̤̈:

(b) :C̤̈l=O=C̤̈l:

(c) H—C—Ö̤—H
with :Ö: bonded above C

(d) H=B̤̈r:

51. Write a Lewis structure for each of the following molecules or ions. Include resonance structures if necessary.
a) SeO_2
b) CO_3^{2-}
c) ClO^-
d) ClO_2^-

52. Write a Lewis structure for each of the following molecules or ions. Include resonance structures if necessary.
a) ClO_3^-
b) ClO_4^-
c) NO_3^-
d) SO_3

53. Write Lewis structures for each of the following ions. Include resonance structures if necessary.
a) PO_4^{3-}
b) CN^-
c) NO_2^-
d) SO_3^{2-}

54. Write Lewis structures for each of the following ions. Include resonance structures if necessary.
a) SO_4^{2-}
b) HSO_4^- (S is central; H attached to one of the O atoms)
c) NH_4^+
d) BrO_2^- (Br is central)

55. Write a Lewis structure for each of the following molecules that are exceptions to the octet rule.
a) BCl_3
b) NO_2
c) BH_3

56. Write a Lewis structure for each of the following molecules that are exceptions to the octet rule.
a) BBr_3
b) NO

Predicting the Shapes of Molecules

57. Determine the number of electron groups around the central atom for each of the following molecules.
a) PCl_3
b) SBr_2
c) CH_2Cl_2
d) CS_2

58. Determine the number of electron groups around the central atom for each of the following molecules.
a) CH_4
b) NF_3
c) OF_2
d) H_2S

59. Determine the number of bonding groups and the number of lone pairs for each of the molecules in Problem 57. The sum of these should equal your answer to Problem 57.

60. Determine the number of bonding groups and the number of lone pairs for each of the molecules in Problem 58. The sum of these should equal your answer to Problem 58.

61. Determine the molecular geometry of each of the following molecules.
a) CBr_4
b) H_2CO
c) CS_2
d) BH_3

62. Determine the molecular geometry of each of the following molecules.
a) SiO_2
b) BF_3
c) $CFCl_3$ (carbon is central)
d) H_2CS (carbon is central)

63. Determine the bond angles for each molecule in Problem 61.

64. Determine the bond angles for each molecule in Problem 62.

65. Determine the electron and molecular geometries of each of the following molecules.
a) N_2O (oxygen is terminal)
b) SO_2
c) H_2S
d) PF_3

66. Determine the electron and molecular geometries of each of the following molecules (Hint: Determine geometry around each of the two central atoms).
a) C_2H_2 (skeletal structure HCCH)
b) C_2H_4 (skeletal structure H_2CCH_2)
c) C_2H_6 (skeletal structure H_3CCH_3)

67. Determine the bond angles for each molecule in Problem 65.

68. Determine the bond angles for each molecule in Problem 66.

69. Determine the electron and molecular geometries of each of the following molecules. For those with two central atoms, indicate the geometry about each central atom.
a) N_2
b) N_2H_2 (skeletal structure HNNH)
c) N_2H_4 (skeletal structure H_2NNH_2)

70. Determine the electron and molecular geometries of each of the following molecules. For those with more than one central atom, indicate the geometry about each central atom.
a) CH_3OH (skeletal structure H_3COH)
b) H_3COCH_3 (skeletal structure H_3COCH_3)
c) H_2O_2 (skeletal structure HOOH)

71. Determine the molecular geometry of each of the following polyatomic ions.
a) CO_3^{2-}
b) ClO_2^-
c) NO_3^-
d) NH_4^+

72. Determine the molecular geometry of each of the following polyatomic ions.
a) ClO_4^-
b) BrO_2^-
c) NO_2^-
d) SO_4^{2-}

Electronegativity and Polarity

73. Use Figure 10.2 to determine the electronegativity of each of the following elements.
a) Mg
b) Si
c) Br

74. Use Figure 10.2 to determine the electronegativity of each of the following elements.
a) F
b) C
c) S

75. Use Figure 10.2 to find the electronegativity difference between each of the following pairs of elements and then use Table 10.2 to classify the bonds that occur between them as pure covalent, polar covalent, or ionic.
a) Mg and Br
b) Cr and F
c) Br and Br
d) Si and O

76. Use Figure 10.2 to find the electronegativity difference between each of the following pairs of elements and then use Table 10.2 to classify the bonds that occur between them as pure covalent, polar covalent, or ionic.
a) K and Cl
b) N and N
c) C and S
d) C and Cl

77. Classify each of the following diatomic molecules as polar or nonpolar.
a) CO
b) O_2
c) F_2
d) HBr

78. Classify each of the following diatomic molecules as polar or nonpolar.
a) I_2
b) NO
c) HCl
d) N_2

79. Classify each of the following molecules as polar or nonpolar.
a) CS_2
b) SO_2
c) CH_4
d) CH_3Cl

80. Classify each of the following molecules as polar or nonpolar.
a) H_2CO
b) CH_3OH
c) CH_2Cl_2
d) SiO_2

81. Classify each of the following molecules as polar or nonpolar.
a) BH_3
b) $CHCl_3$
c) C_2H_2
d) NH_3

82. Classify each of the following molecules as polar or nonpolar.
a) N_2H_2
b) H_2O_2
c) CF_4
d) NO_2

Cumulative Problems

83. Write electron configurations and Lewis structures for each of the following elements. Indicate which of the electrons in the electron configuration are shown in the Lewis structure.
a) Ca
b) Ga
c) As
d) I

84. Write electron configurations and Lewis structures for each of the following elements. Indicate which of the electrons in the electron configuration are shown in the Lewis structure.
a) Rb
b) Ge
c) Kr
d) Se

85. Determine if each of the following compounds is ionic or covalent and write an appropriate Lewis structure.
a) K_2S
b) CHFO (carbon is central)
c) MgSe
d) PBr_3

86. Determine if each of the following compounds is ionic or covalent and write an appropriate Lewis structure.
a) HCN
b) ClF
c) MgI_2
d) CaS

87. Write a Lewis structure for $OCCl_2$ (carbon is central) and determine if the molecule is polar. Draw a three-dimensional structure for the molecule.

88. Write a Lewis structure for CH_3COH and determine if the molecule is polar. Draw a three-dimensional structure for the molecule. The skeletal structure is:

```
  H O
H C C H
  H
```

89. Write a Lewis structure for acetic acid (a component of vinegar), CH_3COOH, and draw a three dimensional sketch of the molecule. The skeletal structure is:

```
  H O
H C C OH
  H
```

90. Write a Lewis structure for benzene, C_6H_6, and draw a three-dimensional sketch of the molecule. The skeletal structure is the ring shown below. (Hint: Include two resonance structures):

```
    H
    C
 HC   CH
 HC   CH
    C
    H
```

91. Consider the following neutralization reaction.

$$HCl(aq) + NaOH(aq) \rightarrow H_2O(l) + NaCl(aq)$$

Write the reaction showing the Lewis structures of each of the reactants and products.

92. Consider the following precipitation reaction.

$$Ba(NO_3)_2(aq) + 2\ LiCl(aq) \rightarrow BaCl_2(s) + 2\ LiNO_3(aq)$$

Write the reaction showing the Lewis structures of each of the reactants and products.

93. Each of the following compounds contains both ionic and covalent bonds. Write ionic Lewis structures for each of them including the covalent structure for the ion in brackets. Write resonance structures if necessary.
a) KOH
b) KNO_3
c) LiIO
d) $BaCO_3$

94. Each of the following compounds contains both ionic and covalent bonds. Write ionic Lewis structures for each of them including the covalent structure for the ion in brackets. Write resonance structures if necessary.
a) $RbIO_2$
b) $Ca(OH)_2$
c) NH_4Cl
d) $Sr(CN)_2$

Highlight Problems

95. Some theories on aging suggest that a class of molecules and ions called *free radicals* cause some diseases and aging. Free radicals are molecules or ions containing an unpaired electron. As you know from Lewis theory, such molecules are not chemically stable and will quickly react with other molecules. Free radicals may attack

molecules within the cell, such as DNA, changing them and causing cancer or other diseases. Free radicals may also attack molecules on the surfaces of cells, making them appear foreign to the body's immune system. The immune system then attacks the cell and destroys it, weakening the body. Draw Lewis structures for each of the following free radicals implicated in this theory of aging.

a) O_2^-
b) O^-
c) OH
d) CH_3OO (unpaired electron on terminal oxygen)

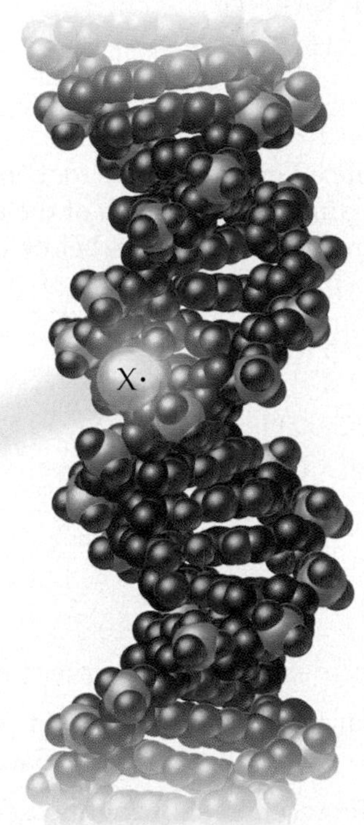

Free radicals, molecules containing unpaired electrons, may attack biological molecules.

96. Free radicals (see Problem 95) are also important in many environmentally significant reactions. For example, photochemical smog, which forms as a result of the action of sunlight on air pollutants, is formed in part by the following two steps.

$$NO_2 \xrightarrow{\text{UV-light}} NO + O$$
$$O + O_2 \longrightarrow O_3$$

The product of this reaction, ozone, is a pollutant in the lower atmosphere. Ozone is an eye and lung irritant and also accelerates the weathering of rubber products. Write Lewis structures for each of the reactants and products in the preceding reactions.

Ozone damages rubber products.

97. The following are formulas of several molecules along with a space filling model of their structure. Determine if the structure is correct and if not, make a sketch of the correct structure.

a) H_2Se

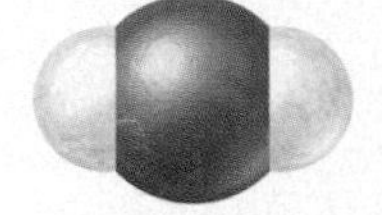

b) CSe_2

c) PCl_3

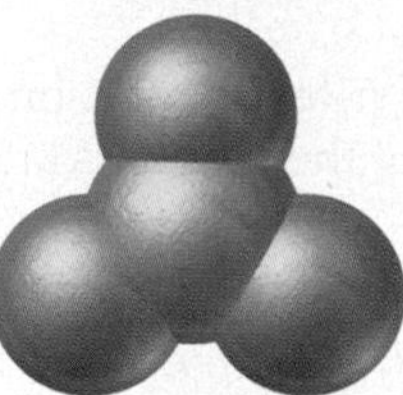

d) CF_2Cl_2

Answers to Skillbuilder Exercises

Skillbuilder 10.1 Mg:

Skillbuilder 10.2

$Na^+ [:\ddot{\underset{..}{Br}}:]^-$

Skillbuilder 10.3

Solution: Lewis structure: $Mg^{2+} [:\ddot{\underset{..}{N}}:]^{3-} Mg^{2+} [:\ddot{\underset{..}{N}}:]^{3-} Mg^{2+}$

Formula: Mg_3N_2

Skillbuilder 10.4

$:C \equiv O:$

Skillbuilder 10.5

$$\begin{array}{c} \ddot{O}: \\ \| \\ H-C-H \end{array}$$

Skillbuilder 10.6

$[:\ddot{\underset{..}{Cl}}:\ddot{\underset{..}{O}}:]^-$

Skillbuilder 10.7

$[:\ddot{O}=\ddot{N}-\ddot{\underset{..}{O}}:]^- \longleftrightarrow [:\ddot{\underset{..}{O}}-\ddot{N}=\ddot{O}:]^-$

Skillbuilder 10.8 bent
Skillbuilder 10.9 trigonal pyramidal
Skillbuilder 10.10 **a)** pure covalent **b)** ionic **c)** polar covalent
Skillbuilder 10.11 CH_4 is nonpolar.

11

Gases

"We live immersed at the bottom of a sea of elemental air."
Evangelista Torricelli

11.1 Extralong Straws

Like most kids, I grew up preferring fast-food restaurants to home cooking. My favorite stunt at the burger restaurant was drinking my orange soda from an extralong straw that I pieced together from several smaller straws. I would pinch the end of one straw and fit it into the end of another. By attaching several straws together, I could put my orange soda on the floor and drink it while standing on my chair (for some reason, my parents did not appreciate my scientific curiosity). I sometimes planned ahead and brought duct tape with me to form extra tight seals between adjacent straws. My brother and I would have contests to make the longest working straw. Since I was older, I usually won.

Sometimes I wondered how long the straw could be if I made perfect seals between the straws. Could I drink my orange soda from a cup on the ground while I sat in my tree house? Could I drink it from the top of a ten-story building? It seemed to me that I could. I was wrong. Even if the extended straw had perfect seals and rigid walls and I sucked hard enough to create a perfect vacuum (the absence of all air), I could never suck my orange soda from a straw longer than about 10.3 meters (34 feet). Why?

◀ When you drink from a straw, you remove some of the molecules from inside the straw. This creates a pressure difference between the inside of the straw and the outside of the straw that results in the liquid being pushed up the straw. The pushing is done by molecules in the atmosphere—primarily nitrogen and oxygen—as shown here.

Straws work because sucking creates a pressure difference between the inside of the straw and the outside. We define pressure more thoroughly later. For now think of pressure as the push (or force) exerted per unit area by gaseous molecules as they collide with the surfaces around them (Figure 11.1). Just as a ball exerts a force when it is bounced against

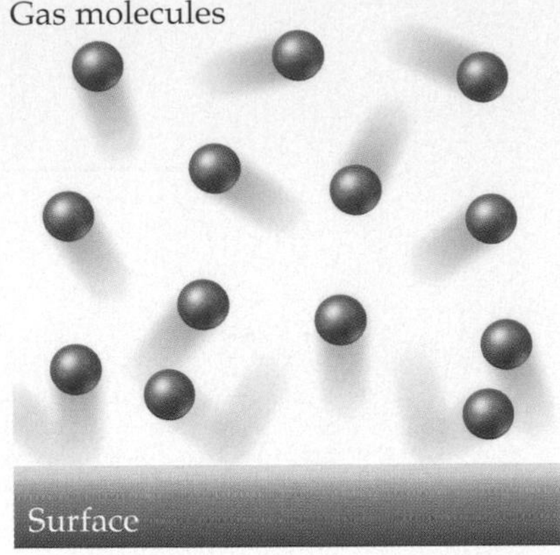

Figure 11.1 Pressure is the force exerted by gas molecules as they collide with the surfaces around them. **Question:** Do you think pressure would increase or decrease if you had more molecules per unit volume?

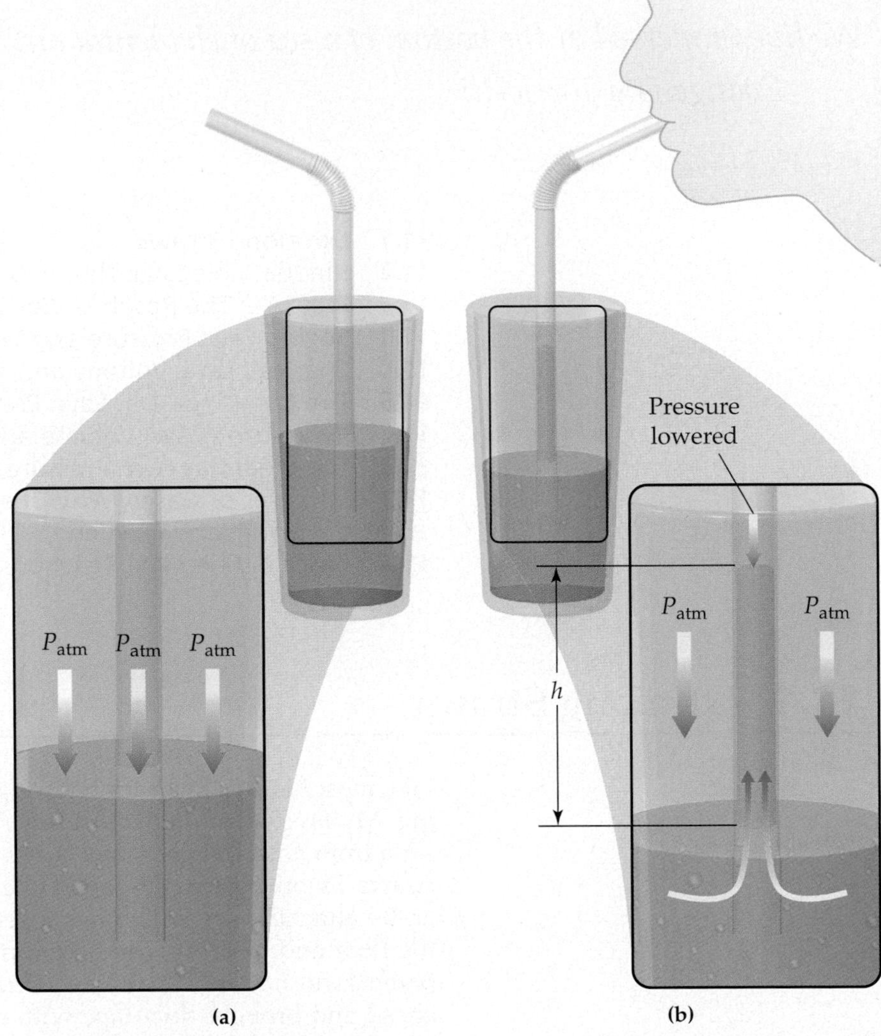

Figure 11.2 **a)** When a straw is put into a glass of orange soda, the pressure inside and outside of the straw are the same, so the liquid levels inside and outside of the straw are the same. **b)** When a person sucks on the straw, the pressure inside the straw is lowered. The greater pressure on the surface of the liquid outside of the straw pushes the liquid up the straw.

a wall, so a molecule exerts a force when it collides against a surface. The result of many of these collisions is pressure. The total amount of pressure exerted by a gas sample depends on several factors, including the concentration of gas molecules in the sample. On Earth at sea level, the gas molecules in our atmosphere exert an average pressure of 101,325 N/m^2 or, in English units, 14.7 lb/in^2.

The newton (N) is a metric unit of force. 1 lb = 4.45 N.

When you put a straw in orange soda, the pressure inside and outside of the straw are the same, so the soda does not rise within the straw (Figure 11.2a). When you suck on the straw, however, you remove some of

Air pressure pushes equally in all directions. The excess downward pressure created by sucking on the straw is converted to upward pressure by the liquid.

the air molecules, lowering the number of collisions that occur inside of the straw and therefore lowering the pressure (Figure 11.2b). However, the pressure outside of the straw remains the same. The result is a pressure differential—the pressure outside of the straw becomes greater than the pressure inside of the straw. This greater external pressure pushes the liquid up the straw and into your mouth.

How high can this greater external pressure push the liquid up the straw? If you formed a perfect vacuum within the straw, the pressure outside of the straw at sea level would be enough to push the orange soda (which is mostly water) to a total height of about 10.3 m (Figure 11.3). This is because a 10.3 m column of water exerts the same pressure—101,325 N/m^2 or 14.7 $lb/in.^2$—as the gas molecules in our atmosphere. In other words, the orange soda would rise up the straw until the pressure exerted by its weight equaled the pressure exerted by the molecules in our atmosphere.

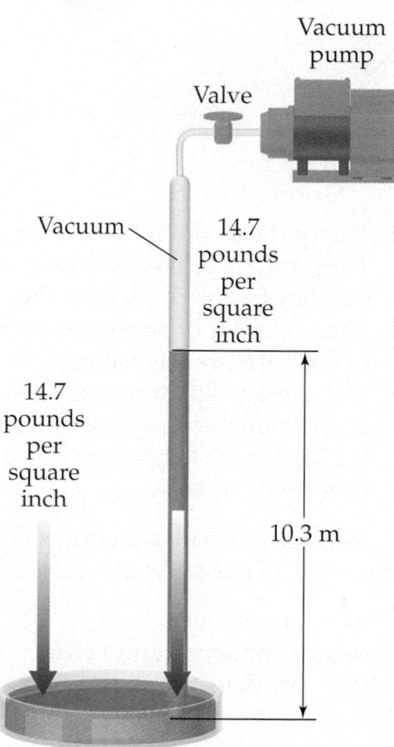

Figure 11.3 Even if you formed a perfect vacuum, atmospheric pressure could only push orange soda to a total height of about 10 m. This is because a column of water 10.3 m high exerts the same pressure (14.7 $lb/in.^2$) as the gas molecules in our atmosphere.

11.2 Kinetic Molecular Theory: A Model for Gases

In past chapters, we have learned the importance of models or theories in understanding nature. A simple model for gases is called the **kinetic molecular theory.** In this chapter, we examine this model and how it predicts the correct behavior for most gases under many conditions. Like other models, the kinetic molecular theory is not perfect and there are conditions under which it breaks down. In this book, however, we focus on conditions where it works well.

Kinetic molecular theory has the following assumptions (Figure 11.4).

1. A gas is a collection of particles (molecules or atoms) in constant, straight-line motion.

Kinetic molecular theory

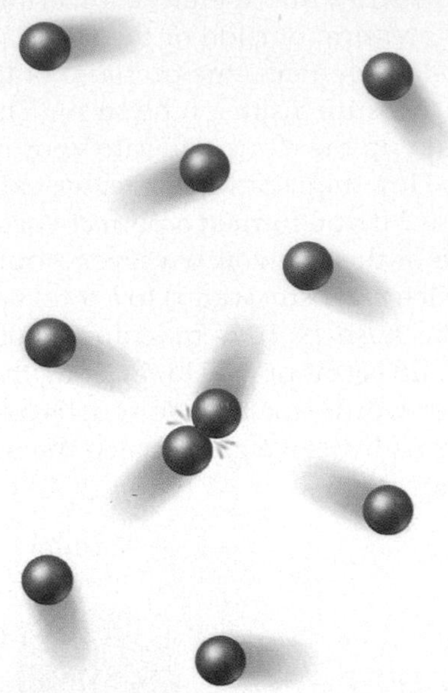

1. Collection of particles in constant motion
2. No attractions or repulsions between particles; collisions like billiard ball collisions
3. A lot of space between the particles compared to the size of the particles themselves
4. The speed that the particles move increases with increasing temperature

Figure 11.4

2. The particles do not attract or repel each other—they do not interact. The particles collide with each other and with the surfaces around them, but they bounce back from these collisions like perfect billiard balls.
3. There is a lot of space between the particles compared with the size of the particles themselves.
4. The average kinetic energy—energy due to motion—of the particles is proportional to the temperature of the gas in kelvin. This means that the higher the temperature, the more energy the particles have and the faster they move.

See Table 3.1.

Kinetic molecular theory is consistent with, and indeed predicts, the properties of gases. As we learned in Section 3.3 of Chapter 3, gases:

- are compressible;
- assume the shape and volume of their container;
- have low densities in comparison with liquids and solids.

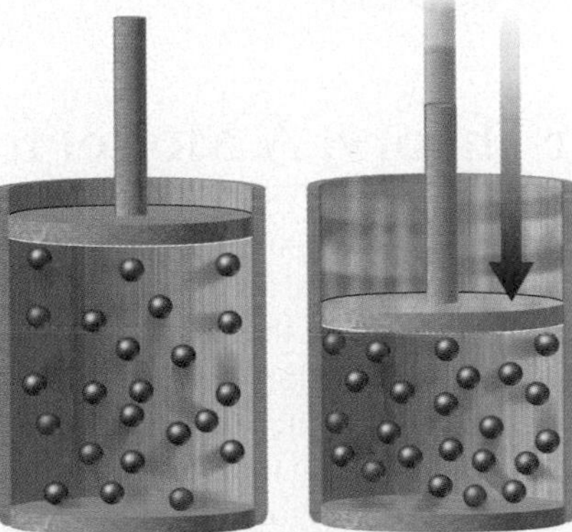

Figure 11.5 Gases are compressible because there is so much empty space between gas particles.

Figure 11.6 Liquids are not compressible because there is so little space between the liquid particles.

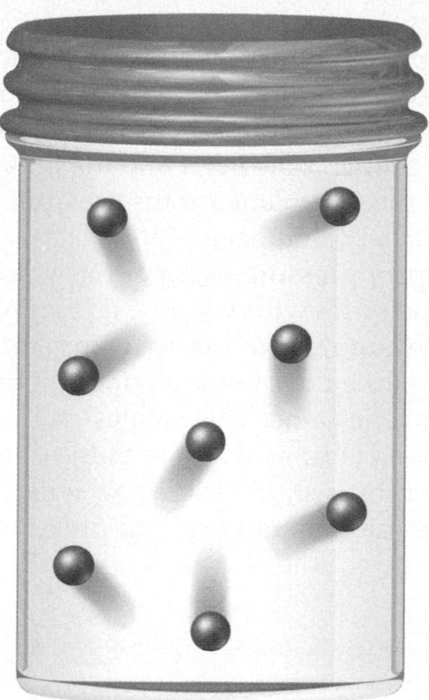

Figure 11.7 Molecules in a gas collectively assume the shape of their container.

Gases are compressible because the atoms or molecules that compose them have a lot of space between them. By applying external pressure to a gas sample, the atoms or molecules are forced closer together, compressing the gas. Liquids and solids, in contrast, are not compressible because the atoms or molecules composing them are already in close contact—they cannot be forced any closer together. The compressibility of a gas can be seen, for example, by pushing a piston into a cylinder containing a gas. The piston will go down (Figure 11.5) in response to the external pressure. If the cylinder were filled with a liquid or a solid, however, the piston would not move when pushed (Figure 11.6).

Gases assume the shape and volume of their container because gaseous atoms or molecules are in constant, straight-line motion. In contrast to a solid or liquid whose atoms or molecules interact with one another, the atoms or molecules in a gas do not interact with one another (or more correctly, their interactions are negligible.) They simply move in straight lines, colliding with each other and with the walls of their container. As a result, they fill the entire container, collectively assuming its shape (Figure 11.7).

Gases have a low density in comparison with solids and liquids because there is so much empty space between the atoms or molecules in a gas. For example, if the water in a 350 mL (12 oz) can of soda were converted to steam (gaseous water), the steam would occupy a volume of 595 L (the equivalent of 1700 soda cans). Gas particles are separated by a lot of empty space.

This calculation assumes normal atmospheric pressure and a temperature of 100 °C.

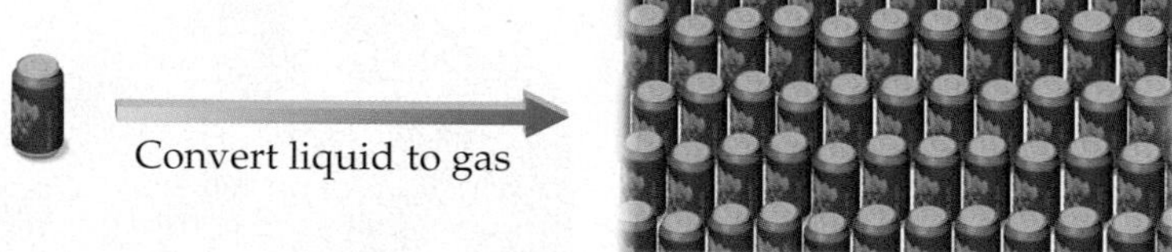

If all of the water in a 12 oz (350 mL) can of orange soda where converted to gaseous steam (at 1 atmosphere pressure and 100°C), the steam would occupy a volume equal to 1700 soda cans.

11.3 Pressure: The Result of Constant Molecular Collisions

A prediction of kinetic molecular theory—which we already encountered in explaining how straws work—is pressure. Pressure is the result of the constant collisions between the atoms or molecules in a gas and the surfaces around them. Because of pressure, we can drink from straws, inflate basketballs, and move air in and out of our lungs. Variation in pressure in the Earth's atmosphere creates wind and changes in pressure help predict weather. Pressure is all around us and even inside of us. The pressure exerted by a gas sample is defined as the force per unit area that results from the collisions of gas particles with surrounding surfaces.

$$\text{Pressure} = \frac{\text{Force}}{\text{Area}}$$

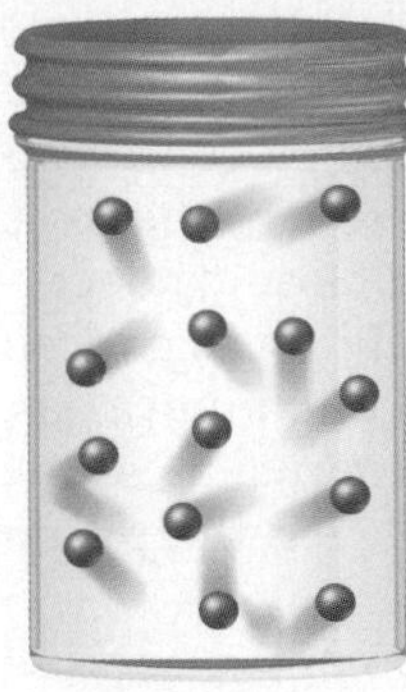

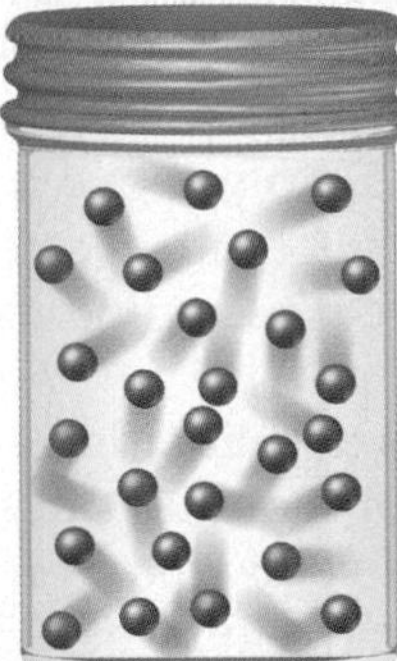

Figure 11.8 Since pressure is a result of collisions between gas particles and the surfaces around them, the amount of pressure increases when the number of particles in a given volume increases.

The pressure exerted by a gas depends on several factors. One of these factors is the number of gas particles in a given volume (Figure 11.8). The fewer the gas particles, the lower the pressure. The pressure decreases, for example, with increasing altitude. As we climb a mountain or ascend in an airplane, there are fewer molecules per unit volume in air and the pressure consequently drops. For this reason, most airplane cabins are artificially pressurized (see the *Everyday Chemistry* box on page 363.)

You can often feel the effect of a drop in pressure as a pain in your ears. This pain is caused by air-containing cavities within your ear (Figure 11.9). When you climb a mountain, for example, the external pressure (that pressure that surrounds you) drops while the pressure within your ear cavities (the internal pressure) remains the same. This creates an imbalance—the greater internal pressure causes your eardrum to bulge outward, causing pain. With time and a yawn or two, the excess air within your ears' cavities escapes, equalizing the internal and external pressure and relieving the pain.

Pressure Units

The simplest unit of pressure is the **atmosphere (atm),** defined as the average pressure at sea level.

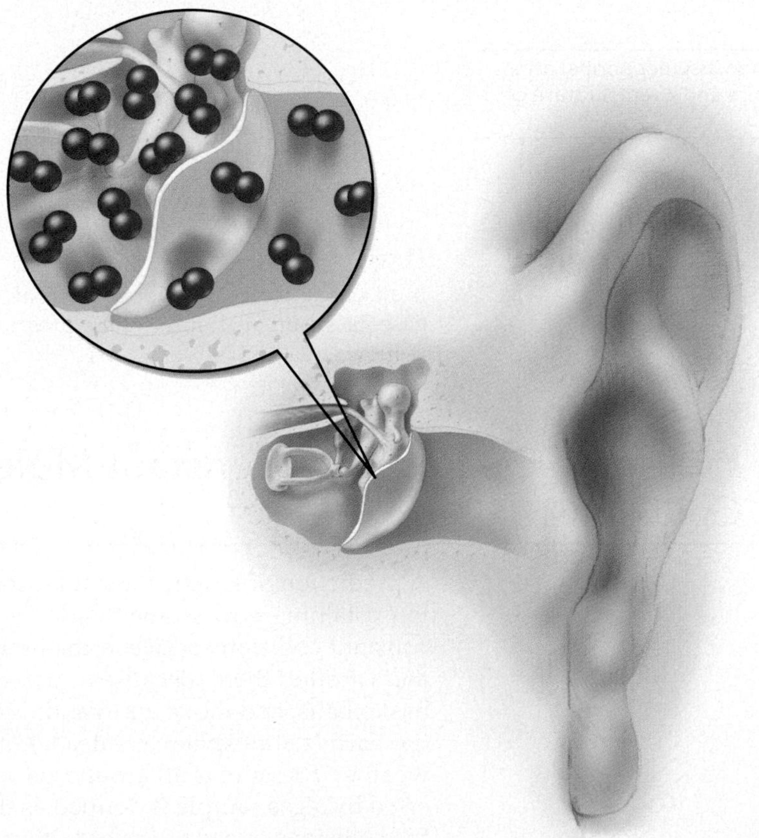

Figure 11.9 The pain you feel in your ears upon ascending a mountain or ascending in an airplane is caused by an imbalance of pressure between the cavities inside your ear and the outside air.

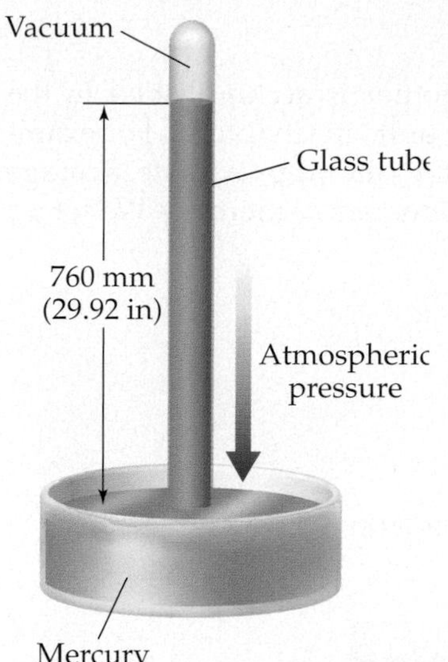

Figure 11.10 Mercury barometer. Average atmospheric pressure at sea level pushes a column of mercury to a height of 760 mm (29.92 in.). **Question:** What would happen to the height of the mercury column if the external pressure became lower? Higher?

Since mercury is 13.5 times as dense as water, it is pushed up 1/13.5 times as high as water by atmospheric pressure.

1 atm = Average pressure of air at sea level

For example, in atmospheres, a fully inflated mountain bike tire has a pressure of about 6 atm, and the pressure on top of Mt. Everest is about 0.311 atm.

The SI unit of pressure is the **pascal (Pa),** defined as 1 newton (N) per square meter.

$$1 \text{ Pa} = 1 \text{ N/m}^2$$

The pascal is a much smaller unit of pressure, with one atmosphere being equal to 101,325 Pa.

$$1 \text{ atm} = 101{,}325 \text{ Pa}$$

A third unit of pressure, **millimeter of mercury (mm Hg),** originates from how pressure is measured with a barometer (Figure 11.10). A barometer is an evacuated glass tube whose tip is submerged in a pool of mercury. As we learned in Section 11.1, a liquid is pushed up an evacuated tube by atmospheric gas pressure on the liquid's surface. We learned that water is pushed up to a height of 10.3 m by the average pressure at sea level. Mercury, however, with its higher density, is only pushed up to a height of 0.760 m or 760 mm by the average pressure at sea level. This shorter length—0.760 m instead of 10.3 m—makes a column of mercury a convenient way to measure pressure.

In a barometer, the mercury column rises or falls with changes in pressure. If the pressure increases, the level of mercury within the column rises. If the pressure decreases, the level of mercury within the column falls. Since one atmosphere of pressure pushes a column of mercury to a height of 760 mm, 1 atm and 760 mm Hg are equal.

$$1 \text{ atm} = 760 \text{ mm Hg}$$

A millimeter of mercury is also called a **torr** after Italian physicist **Evangelista Torricelli** (1608–1647) who invented the barometer.

$$1 \text{ mm Hg} = 1 \text{ torr}$$

Other common units of pressure include inches of mercury (in Hg) and pounds per square inch (psi).

$$1 \text{ atm} = 14.7 \text{ psi} \qquad 1 \text{ atm} = 29.92 \text{ in Hg}$$

These units are all summarized in Table 11.1.

TABLE 11.1

Common units of Pressure

Unit	*Average Air Pressure at Sea Level*
pascal (Pa)	101,325
atmosphere (atm)	1
millimeter of mercury (mm Hg)	760
torr (torr)	760
pounds per square inch (psi)	14.7
inches of mercury (in Hg)	29.92

Pressure Unit Conversion

See section 2.6.

The conversion of one pressure unit to another is accomplished in the same way that we converted between other units in Chapter 2. For example, suppose we want to convert 0.311 atm (the approximate average pressure at the top of Mount Everest) to millimeters of mercury. We set up the problem in the standard way.

Given: 0.311 atm
Find: mm Hg

Conversion Factor:

1 atm = 760 mm Hg (from Table 11.1)

The solution map shows how to convert from atm to mm Hg.

Solution Map:

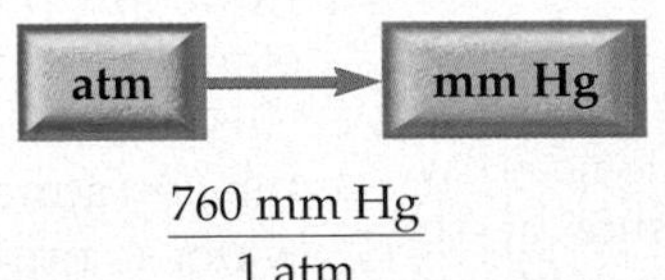

$$\frac{760 \text{ mm Hg}}{1 \text{ atm}}$$

Solution:
The solution begins with the given value (0.311 atm) and converts it to mm Hg.

$$0.311 \text{ atm} \times \frac{760 \text{ mm Hg}}{1 \text{ atm}} = 236 \text{ mm Hg}$$

EXAMPLE 11.1 Converting Between Pressure Units

A high performance road bicycle tire is inflated to a total pressure of 125 psi. What is this pressure in millimeters of mercury?

Given: 125 psi
Find: mm Hg

Conversion Factors:

1 atm = 14.7 psi (from Table 11.1)

760 mm Hg = 1 atm

Solution Map:

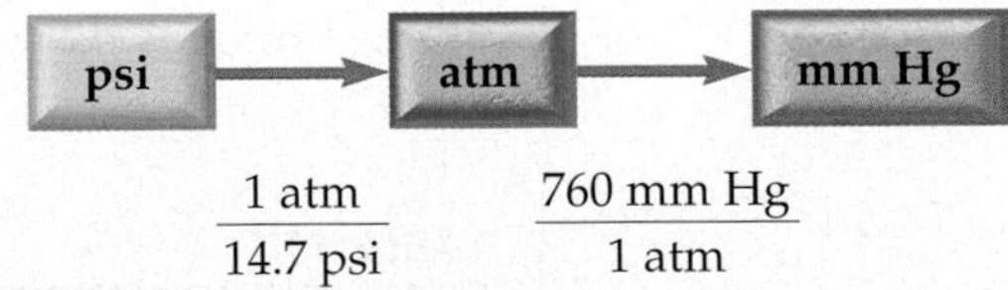

$$\frac{1 \text{ atm}}{14.7 \text{ psi}} \qquad \frac{760 \text{ mm Hg}}{1 \text{ atm}}$$

Solution:

$$125 \text{ psi} \times \frac{1 \text{ atm}}{14.7 \text{ psi}} \times \frac{760 \text{ mm Hg}}{1 \text{ atm}} = 6.46 \times 10^3 \text{ mm Hg}$$

SKILLBUILDER 11.1 Converting Between Pressure Units

Convert a pressure of 173 in Hg into pounds per square inch.

SKILLBUILDER PLUS Convert a pressure of 23.8 in Hg into kPa.

EVERYDAY CHEMISTRY

Airplane Cabin Pressurization

Most commercial airplanes fly at elevations between 20,000 and 30,000 ft. At these elevations, atmospheric pressure is below 0.50 atm, much less than the 1.0 atm of pressure to which our bodies are accustomed. The physiological effects of these lowered pressures—and the correspondingly lowered oxygen levels—include dizziness, headache, shortness of breath, and even unconsciousness. Consequently, commercial airplanes pressurize the air in their cabins. If, for some reason, an airplane cabin should lose its pressurization, passengers are directed to breathe oxygen through an oxygen mask.

Commercial airplane cabins must be pressurized to a pressure greater than the equivalent atmospheric pressure at an elevation of 8,000 ft.

Cabin air pressurization is accomplished as part of the cabin's overall air circulation system. As air flows into the plane's jet engines, the large turbines at the front of the engines compress it. Most of this compressed (or pressurized) air exits out the back of the engines, creating the thrust that drives the plane forward. However, some of the pressurized air is directed into the cabin, where it is cooled and mixed with existing cabin air. This air is then circulated through the cabin through the overhead vents. The air leaves the cabin through ducts that direct it into the lower portion of the airplane. About half of this exiting air is mixed with incoming, pressurized air to circulate again. The other half is vented out of the jet through an outflow valve. This valve is adjusted to maintain the desired cabin pressure. Federal regulations require that cabin pressures in commercial airliners be greater than the equivalent of outside air pressure at 8,000 ft.

CAN YOU ANSWER THIS? *Atmospheric pressure at elevations of 8,000 ft average about 0.72 atm. Convert this pressure to millimeters of mercury, inches of mercury, and pounds per square inch. Would a cabin pressurized at 500 mm Hg meet federal standards?*

11.4 Boyle's Law: Pressure and Volume

The pressure of a gas sample changes when its volume changes. If the temperature and the amount of gas are constant, the pressure of a gas sample *increases* for a *decrease* in volume and *decreases* for an *increase* in volume. A simple hand-pump, for example, works on this principle. A hand pump is basically a cylinder equipped with a moveable piston (Figure 11.11). The volume in the cylinder increases when you pull the handle up (the up-stroke) and decreases when you push the handle down (the down-stroke). On the up-stroke, the *increasing* volume causes a *decrease* in the internal pressure (the pressure within the pump's cylinder). This, in turn, draws air into the pump's cylinder through a one-way valve. On the down-stroke, the *decreasing* volume causes an *increase* in the internal pressure. This increase forces the air out of the pump, through a different one-way valve, and into the tire or whatever else is being inflated.

The relationships between gas properties—such as pressure and volume—are described by gas laws. These laws show how a change in one of these properties affects one or more of the others. The relationship between volume and pressure was discovered by **Robert Boyle** (1627–1691) and is called **Boyle's law**.

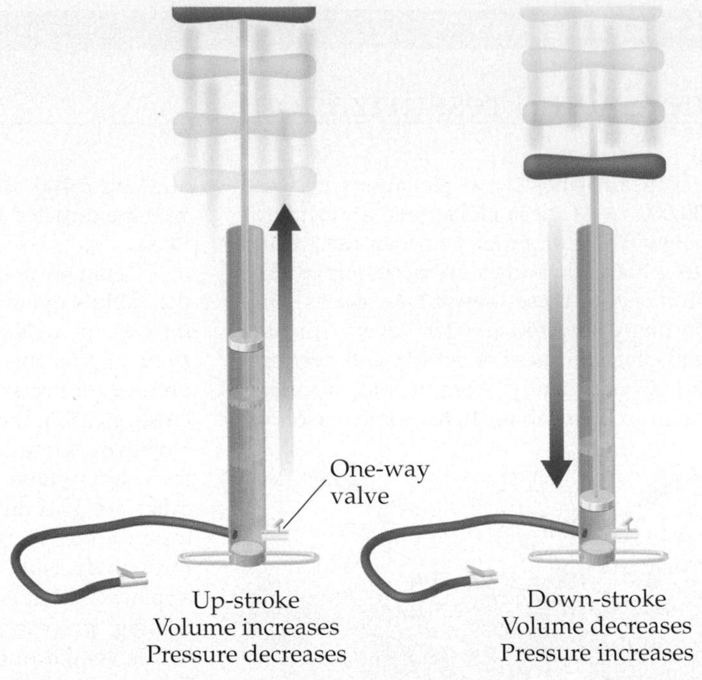

Figure 11.11

Boyle's law: The volume of a gas and its pressure are inversely proportional.

Boyle's law assumes constant temperature and constant number of gas particles.

$V \alpha \frac{1}{P}$ (α means "proportional to")

If two quantities are inversely proportional, then increasing one de-

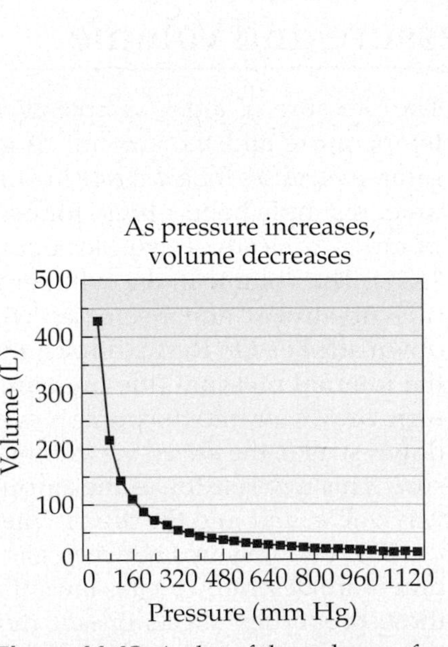

Figure 11.12 A plot of the volume of a gas as a function of pressure.

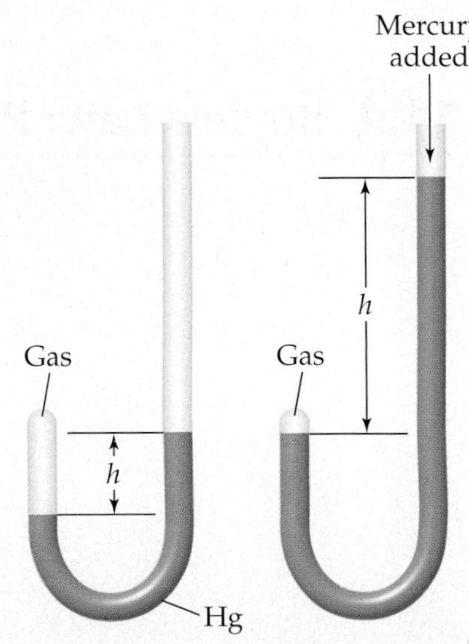

A J-tube, such as the one shown here, can be used to measure the volume of a gas at different pressures. By simply adding mercury to the J-tube, the pressure on the gas sample increases and its volume decreases.

creases the other (Figure 11.12). As we saw for our hand pump, when the volume of a sample of gas is decreased its pressure increases and vice versa. This follows from kinetic molecular theory. If the volume of a gas sample is decreased, the same number of gas particles is crowded into a smaller volume causing more collisions with the walls of the container and therefore increasing the pressure (Figure 11.13).

Scuba divers learn about Boyle's law during certification because it explains why ascending quickly toward the surface is dangerous. For every 10 m of depth that a diver descends in water, he experiences an additional 1 atm of pressure due to the weight of the water above him (Fig-

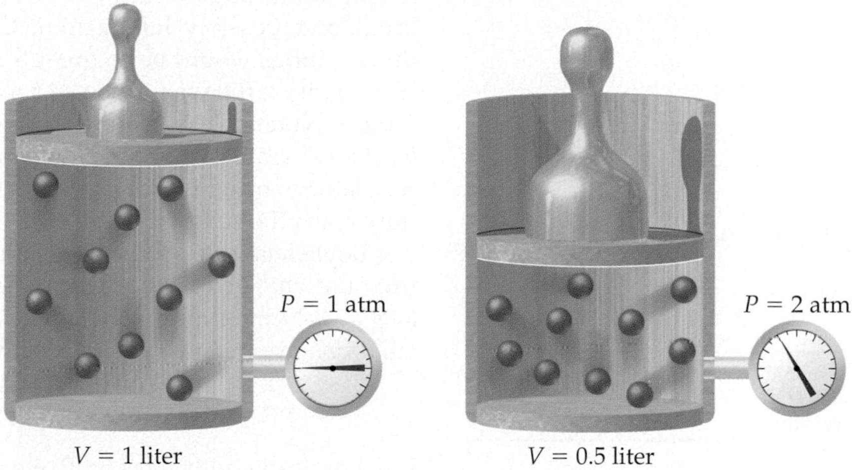

Figure 11.13 As the pressure of a sample of gas increases, its volume decreases.

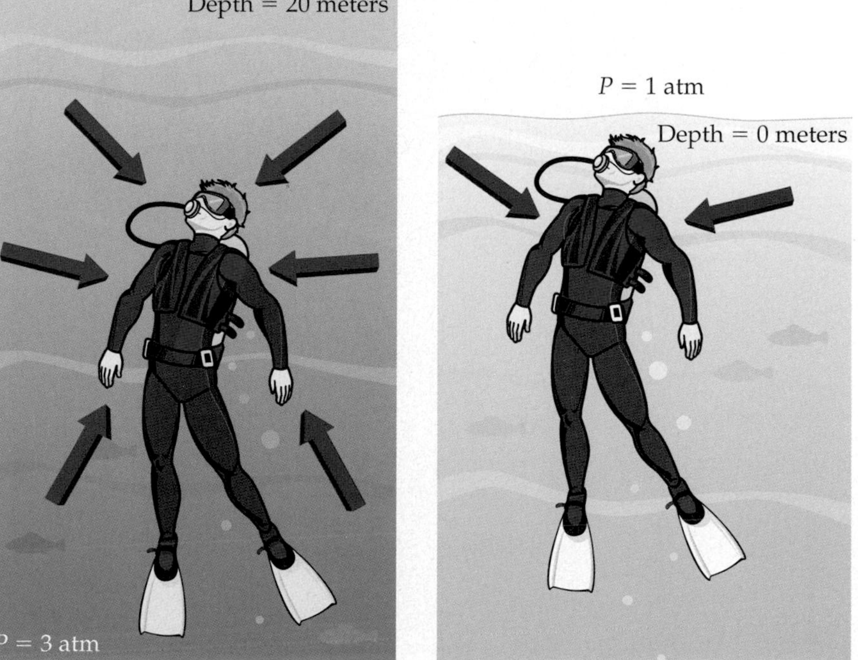

Figure 11.14 For every 10 m of depth, a diver experiences an additional 1 atm of pressure due to the weight of the water surrounding him. At 20 m, the diver experiences a total pressure of 3 atm (1 atm from atmospheric pressure plus an additional 2 atm from the weight of the water).

ure 11.14). The pressure regulator used in scuba diving delivers air at a pressure that matches the external pressure; otherwise the diver could not inhale the air (see the *Everyday Chemistry* box on page 368). For example, when a diver is at 20 m of depth, the regulator delivers air at a pressure of 3 atm to match the 3 atm of pressure around the diver (1 atm due to normal atmospheric pressure and 2 additional atmospheres due to the weight of the water at 20 m). Suppose that a diver inhaled a lungful of 3 atm air and swam quickly to the surface (where the pressure drops to 1 atm) while holding his breath. What would happen to the volume of air in his lungs? Since the pressure decreases by a factor of 3, the volume of the air in his lungs would increase by a factor of 3, severely damaging his lungs and possibly killing him. Of course, the volume increase in the diver's lungs would be so great that the diver would not be able to hold his breath all the way to the surface—the air would force itself out of his mouth. Nonetheless, the most important rule in diving is *never hold your breath.* Divers must ascend slowly and breathe continuously, allowing the regulator to bring the air pressure in their lungs back to 1 atm by the time they reach the surface.

Boyle's law can be used to compute the volume of a gas following a pressure change or the pressure of a gas following a volume change *as long as the temperature and the amount of gas remain constant.* For these calculations, we must write Boyle's law in a slightly different way.

If two quantities are proportional, then one is equal to the other multiplied by a constant.

$$\text{Since } V \alpha \frac{1}{P}, \text{ then, } V = \frac{\text{Constant}}{P}$$

If we multiply both sides by P, we get:

$$PV = \text{constant}$$

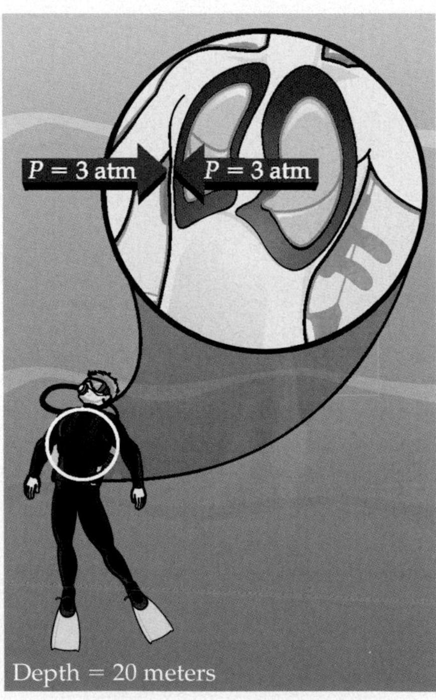

(a)

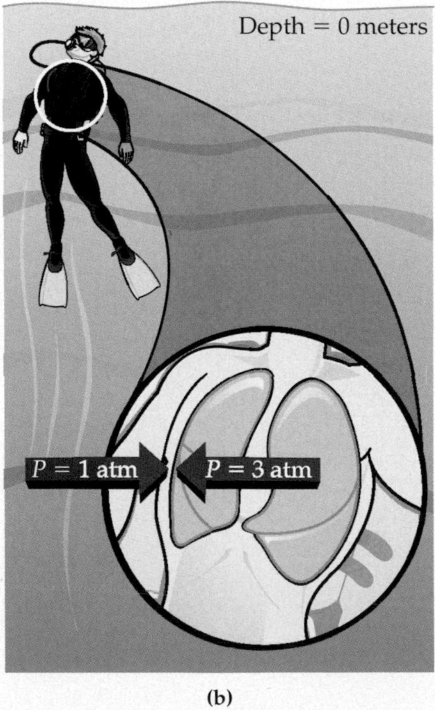

(b)

a) A diver at 20 m experiences an external pressure of 3 atm and breathes air pressurized at 3 atm. **b)** If the diver shoots toward the surface with lungs full of 3 atm air, his lungs will expand as the external pressure drops to 1 atm.

This relationship is true because if the pressure increases, the volume decreases, but the product $P \times V$ is always equal to the same constant. For two different sets of conditions, we can say that

$$P_1V_1 = \text{Constant} = P_2V_2, \text{ or}$$
$$\boldsymbol{P_1V_1 = P_2V_2}$$

where P_1 and V_1 are the initial pressure and volume of the gas, and P_2 and V_2 are the final volume and pressure. For example, suppose we want to calculate the pressure of a gas, which was initially at 765 mm Hg and 1.78 L and later compressed to 1.25 L. We set up the problem as follows:

Based on Boyle's law, and before doing any calculations, do you expect P_2 to be greater than or less than P_1?

Given: $P_1 = 765$ mm Hg
$V_1 = 1.78$ L
$V_2 = 1.25$ L

Find: P_2

Equation:

Since this problem involves an equation, we write the appropriate equation here.

$$P_1V_1 = P_2V_2$$

Solution Map:

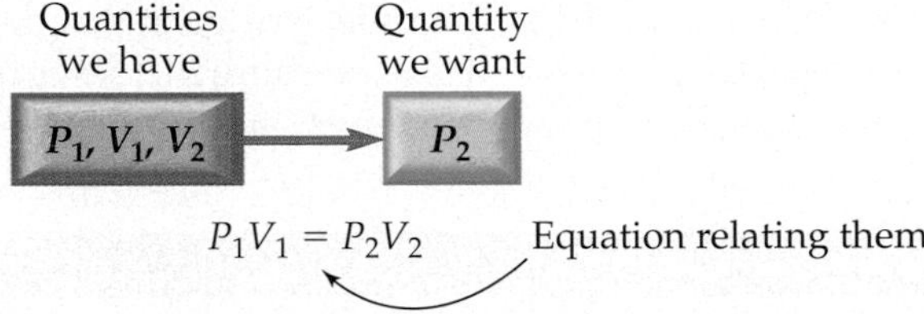

The solution map shows how the equation takes us from the given quantities (what we have) to the find quantity (what we want).

Solution:
We then solve the equation for the quantity we are trying to find (P_2).

$$P_1V_1 = P_2V_2$$
$$P_2 = \frac{V_1}{V_2}P_1$$

Lastly, we substitute the numerical values into the equation and compute the answer.

$$P_2 = \frac{V_1}{V_2}P_1$$
$$= \frac{1.78\ \cancel{\text{L}}}{1.25\ \cancel{\text{L}}}\, 765 \text{ mm Hg}$$
$$= 1.09 \times 10^3 \text{ mm Hg}$$

EXAMPLE 11.2 Boyle's Law

A cylinder equipped with a moveable piston has an applied pressure of 4.0 atm and a volume of 6.0 L. What is the volume of the cylinder if the applied pressure is decreased to 1.0 atm?

Given: $P_1 = 4.0\ \text{atm}$
$V_1 = 6.0\ \text{L}$
$P_2 = 1.0\ \text{atm}$

Find: V_2

Equation: $P_1V_1 = P_2V_2$

Solution Map:

$P_1, V_1, P_2 \rightarrow V_2$

$P_1V_1 = P_2V_2$

Solution:

$$P_1V_1 = P_2V_2$$

$$V_2 = \frac{P_1}{P_2}V_1$$

$$= \frac{4.0\ \cancel{\text{atm}}}{1.0\ \cancel{\text{atm}}}\ 6.0\ \text{L}$$

$$= 24\ \text{L}$$

SKILLBUILDER 11.2 **Boyle's Law**

A snorkeler takes a syringe filled with 16 mL of air from the surface, where the pressure is 1.0 atm, to an unknown depth. The volume of the air in the syringe at this depth is 7.5 mL. What is the pressure at this depth? If the pressure increases by an additional 1 atm for every 10 m of depth, how deep is the snorkeler?

EVERYDAY CHEMISTRY

Extralong Snorkels

Several episodes of *The Flintstones* cartoon featured Fred Flintstone and Barney Rubble snorkeling. Their snorkels, however, were not the modern kind, but long reeds that stretched from the surface of the water down to many meters of depth. Fred and Barney swam around in deep water while breathing air provided to them by these extralong snorkels. Would this work? Why do people bother with scuba diving equipment if they could simply use 10-m snorkels the way that Fred and Barney did?

When we breathe, we expand the volume of our lungs, lowering the pressure within them (Boyle's law). Air from outside of our lungs then flows into them. Extralong snorkels, such as those used by Fred and Barney, do not work because of the pressure caused by water at depth. A diver at 10 m experiences a pressure

Fred and Barney used reeds to breathe air from the surface, even when they were at depth. This would not work because the pressure at depth would push air out of their lungs, preventing them from breathing.

11.5 Charles's Law: Volume and Temperature

Recall from Section 2.9 that density = mass/volume. If the volume increases, and the mass remains constant, the density must decrease.

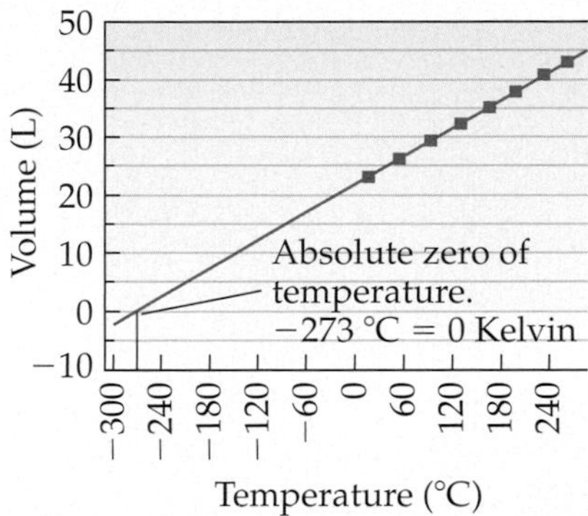

Figure 11.16 The volume of a gas increases linearly with increasing temperature. **Question:** How does this graph demonstrate that −273 °C is the coldest possible temperature?

Have you ever noticed that hot air rises? You may have walked upstairs in your house and noticed it getting warmer. Or you may have witnessed a hot air balloon take flight. The air that fills a hot air balloon is warmed with a burner, which then causes the balloon to rise in the cooler air around it. Why does hot air rise?

Hot air rises because the volume of a gas sample at constant pressure increases with increasing temperature. As long as the amount of gas (and therefore its mass) remains constant, warming it decreases its density because density is mass divided by volume. A lower density gas floats in a higher density gas just as lower density wood floats in higher density water.

Suppose you keep the pressure of a gas sample constant and measure its volume at a number of different temperatures. The results of a number of such measurements are shown in Figure 11.16. From the plot we can see the relationship between volume and temperature: the volume of a gas increases with increasing temperature. Looking at the plot, however, reveals more; temperature and volume are *linearly related*. If two variables are linearly related, then plotting one against the other produces a straight line.

Another interesting feature arises if we extend the line on our plot backwards from the lowest measured point—a process called *extrapolation*. Our extrapolated line shows that the gas should have a zero volume

of 2 atm that compresses the air in his lungs to a pressure of 2 atm. If the diver had a snorkel that went to the surface—where the air pressure is 1 atm—air would flow out of his lungs, not into them. It would be impossible to breathe.

CAN YOU ANSWER THIS? *Suppose a diver takes a balloon with a volume of 2.5 L from the surface, where the pressure is 1.0 atm, to a depth of 20 m, where the pressure is 3.0 atm. What would happen to the volume of the balloon? What if the end of the balloon was on a long tube that went to the surface and was attached to another balloon (Figure 11.15)? Which way would air flow as the diver descends?*

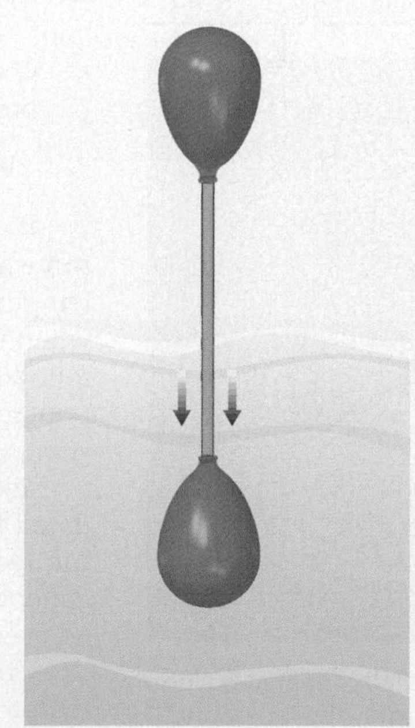

Figure 11.15 Question: If one end of a long tube with balloons tied on both ends were submerged in water, in which direction would air flow?

Hot air balloons float because they are filled with hot air. Since hot air occupies more volume than colder air, it is less dense, causing the balloon to rise above the ground.

The extrapolated line could not be measured experimentally because all gases would condense into liquids before −273 °C is reached.

Section 3.9 summarizes the three different temperature scales.

Charles's law assumes constant pressure and constant amount of gas.

at −273 °C. Recall from Chapter 3 that −273 °C corresponds to 0 K, the coldest possible temperature. Our extrapolated line shows that below −273 °C, our gas would have a negative volume, which is physically impossible. For this reason, we refer to 0 K as **absolute zero**—colder temperatures do not exist.

The first person to carefully quantify the relationship between the volume of a gas and its temperature was **J.A.C. Charles** (1746–1823), a French mathematician and physicist. Charles was interested in gases and was among the first people to ascend in a hydrogen-filled balloon. The law he formulated is called Charles's law.

Charles's law: The volume (V) of a gas and its temperature (T) expressed in kelvin are directly proportional.

$$V \alpha T$$

If two variables are directly proportional, then increasing one by some factor increases the other by the same factor. For example, when the temperature (in kelvin) of a gas sample is doubled, its volume doubles, when the temperature is tripled, its volume triples, and so on. This also follows from kinetic molecular theory. If the temperature of a gas sample is increased, the gas particles move faster and, if the pressure remains constant, will collectively occupy more space (Figure 11.17).

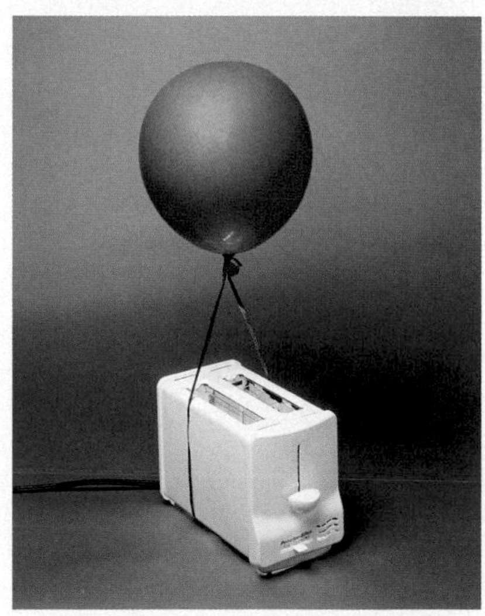

If you hold a partially inflated balloon over a warm toaster, the balloon will expand as the air within the balloon warms.

You can experience Charles's law directly by holding a partially inflated balloon over a warm toaster. As the air in the balloon warms, you can feel the balloon expanding. Alternatively, you can a put an inflated balloon in the freezer or take it outside on very cold day (below freezing) and see that it becomes smaller as it cools.

Charles's law can be used to compute the volume of a gas following a temperature change or the temperature of a gas following a volume change *as long as the pressure and the amount of gas are constant.* For these calculations, we express Charles's law in a different way as follows:

Figure 11.17 If a balloon is moved from an ice water bath into a boiling water bath, its volume increases because, as the molecules move faster due to increased temperature, they collectively occupy more volume.

Since $V \alpha T$, then, $V = \text{Constant} \times T$

If we divide both sides by T, we get:

$V/T = \text{Constant}$

If the temperature increases, the volume increases in direct proportion so that the quotient, V/T, is always equal to the same constant. So, for two different measurements, we can say that

$V_1/T_1 = \text{Constant} = V_2/T_2$, or

$$\frac{\boldsymbol{V_1}}{\boldsymbol{T_1}} = \frac{\boldsymbol{V_2}}{\boldsymbol{T_2}}$$

where V_1 and T_1 are the initial volume and temperature of the gas and V_2 and T_2 are the final volume and temperature. *All temperatures must be expressed in kelvin.*

For example, suppose we want to calculate the volume of a gas, which was initially at 298 K and 2.37 L and later heated to 354 K. We set up the problem in the usual way.

Given:

$$T_1 = 298 \text{ K}$$
$$V_1 = 2.37 \text{ L}$$
$$T_2 = 354 \text{ K}$$

Based on Charles's law, and before doing any calculations, do you expect V_2 to be greater than or less than V_1?

Find: V_2

Equation:

$$\frac{V_1}{T_1} = \frac{V_2}{T_2}$$

The solution map shows how the equation takes us from the given quantities to the find quantity.

Solution Map:

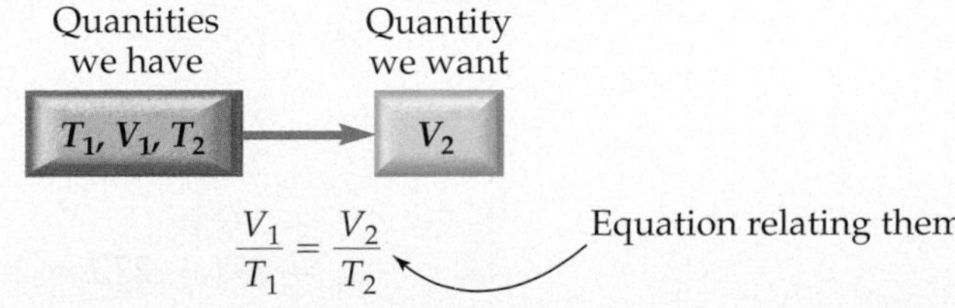

Solution:

We then solve the equation for the quantity we are trying to find (V_2).

$$\frac{V_1}{T_1} = \frac{V_2}{T_2}$$

$$V_2 = \frac{V_1}{T_1} T_2$$

Lastly, we substitute the numerical values into the equation and compute the answer.

$$V_2 = \frac{V_1}{T_1} T_2$$

$$= \frac{2.37 \text{ L}}{298 \text{ K}} 354 \text{ K}$$

$$= 2.82 \text{ L}$$

EXAMPLE 11.3 **Charles's Law**

A sample of gas has a volume of 2.80 L at an unknown temperature. When the sample is submerged into ice water at $t = 0\ °\text{C}$, its volume decreases to 2.57 L. What was its initial temperature (in kelvin and in Celsius)?

To differentiate between temperatures in Celsius and in kelvin, we use lowercase t to represent temperature in Celsius, and capital T to represent temperature in kelvin.

Given:

$V_1 = 2.80 \text{ L}$
$V_2 = 2.57$
$t_2 = 0\ °\text{C}$

Find: T_1 and t_1

Equation:

$$\frac{V_1}{T_1} = \frac{V_2}{T_2}$$

Solution Map:

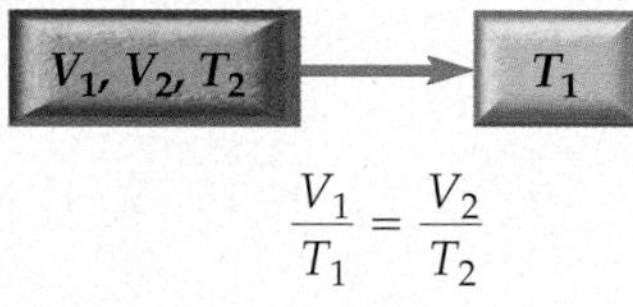

$$\frac{V_1}{T_1} = \frac{V_2}{T_2}$$

Solution:

$$\frac{V_1}{T_1} = \frac{V_2}{T_2}$$

$$T_1 = \frac{V_1}{V_2} T_2$$

Before we substitute in the numerical values, we must convert the temperature to kelvin (K). **Remember, gas law problems must always be worked in kelvin.**

$$T_2 = 0 + 273 = 273 \text{ K}$$

$$T_1 = \frac{V_1}{V_2} T_2$$

$$= \frac{2.80 \text{ L}}{2.57 \text{ L}} 273 \text{ K}$$

$$= 297 \text{ K}$$

$$t_1 = 297 - 273 = 24\ °\text{C}$$

SKILLBUILDER 11.3 **Charles's Law**

A gas in a cylinder with a moveable piston with an initial volume of 88.2 mL is heated from 35 °C to 155 °C. What is the final volume of the gas (in milliliters)?

11.6 The Combined Gas Law: Pressure, Volume, and Temperature

Boyle's law shows how *P* and *V* are related, and Charles's law shows how *V* and *T* are related. But what if two of these variables change at once? For example, what happens to the volume of a gas if both its temperature and its pressure are changed? For these calculations, we need the **combined gas law:**

The combined gas law assumes that the amount of gas is constant. All temperatures in the combined gas law must be expressed in kelvin.

$$\frac{P_1V_1}{T_1} = \frac{P_2V_2}{T_2}$$

Suppose you carry a cylinder with a moveable piston that has an initial volume of 3.65 L up a mountain. The pressure at the bottom of the mountain is 755 mm Hg and the temperature is 302 K. The pressure at the top of the mountain is 687 mm Hg and the temperature is 291 K. What is the volume of the cylinder at the top of the mountain? We set up the problem as follows:

Given:

$$P_1 = 755 \text{ mm Hg}$$
$$T_1 = 302 \text{ K}$$
$$V_1 = 3.65 \text{ L}$$
$$P_2 = 687 \text{ mm Hg}$$
$$T_2 = 291 \text{ K}$$

Find: V_2

Equation:

$$\frac{P_1V_1}{T_1} = \frac{P_2V_2}{T_2}$$

The solution map shows how the equation takes us from the given quantities to the find quantity.

Solution Map:

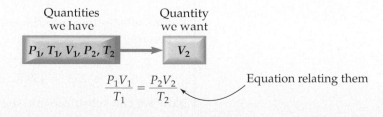

Quantities we have | Quantity we want

P_1, T_1, V_1, P_2, T_2 → V_2

$$\frac{P_1V_1}{T_1} = \frac{P_2V_2}{T_2}$$

Equation relating them

Solution:

We then solve the equation for the quantity we are trying to find (V_2).

$$\frac{P_1V_1}{T_1} = \frac{P_2V_2}{T_2}$$

$$V_2 = \frac{P_1V_1T_2}{T_1P_2}$$

Lastly, we substitute in the appropriate values and compute the answer.

$$V_2 = \frac{P_1V_1T_2}{T_1P_2}$$

$$= \frac{755 \text{ }\cancel{\text{mm Hg}} \times 3.65 \text{ L} \times 291 \text{ }\cancel{\text{K}}}{302 \text{ }\cancel{\text{K}} \times 687 \text{ }\cancel{\text{mm Hg}}}$$

$$= 3.87 \text{ L}$$

EXAMPLE 11.4 The Combined Gas Law

A sample of gas has an initial volume of 158 mL at a pressure of 735 mm Hg and a temperature of 34 °C. If the gas is compressed to a volume of 108 mL and heated to a temperature of 85 °C, what is its final pressure (in millimeters of mercury)?

Given:

$P_1 = 735$ mm Hg
$t_1 = 34$ °C $t_2 = 85$ °C
$V_1 = 158$ mL $V_2 = 108$ mL

Find: P_2

Equation:

$$\frac{P_1V_1}{T_1} = \frac{P_2V_2}{T_2}$$

Solution Map:

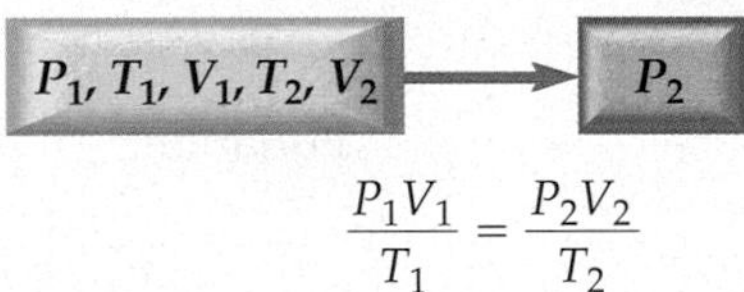

$$\frac{P_1V_1}{T_1} = \frac{P_2V_2}{T_2}$$

Solution:

$$\frac{P_1V_1}{T_1} = \frac{P_2V_2}{T_2}$$

$$P_2 = \frac{P_1V_1T_2}{T_1V_2}$$

Before substituting into the equation, we must convert the temperatures to kelvin.

$$T_1 = 34 + 273 = 307 \text{ K}$$
$$T_2 = 85 + 273 = 358 \text{ K}$$

Now we can substitute into the equation and compute P_2.

$$P_2 = \frac{735 \text{ mm Hg} \times 158 \text{ mL} \times 358 \text{ K}}{307 \text{ K} \times 108 \text{ mL}}$$

$$= 1.25 \times 10^3 \text{ mm Hg}$$

SKILLBUILDER 11.4 **The Combined Gas Law**

A balloon has a volume of 3.7 L at a pressure of 1.1 atm and a temperature of 30 °C. If the balloon is submerged in water to a depth where the pressure is 4.7 atm and the temperature is 15 °C, what will its volume be (assume that any changes in pressure caused by the skin of the balloon are negligible)?

11.7 Avogadro's Law: Volume and Moles

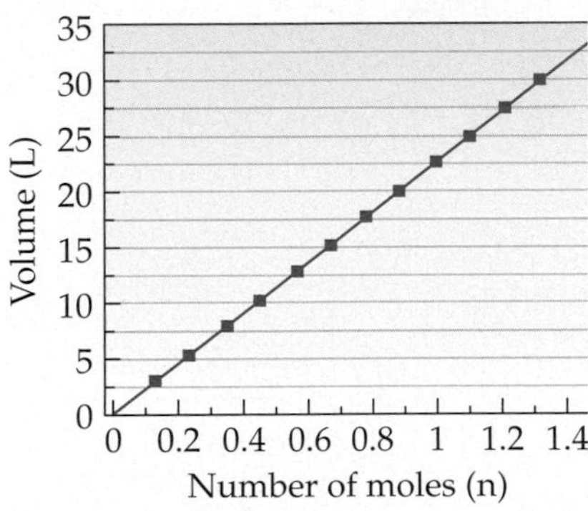

Figure 11.18 The volume of a gas sample increases linearly with the number of moles in the sample.

So far, we have learned how V, P, and T are interrelated, but we have only considered a constant amount of a gas. What happens when the amount of gas changes? If we make several measurements on the volume of a gas sample (at constant temperature and pressure) while varying the number of moles in the sample, the results look like Figure 11.18. We can see that the relationship between volume and number of moles is linear. An extrapolation to zero moles shows a zero volume, as we might expect. This relationship was first stated formally by Amadeo Avogadro (1776–1856), and is called **Avogadro's law.**

Avogadro's law: The volume of a gas and the amount of the gas in moles *(n)* are directly proportional.

$$V \alpha n$$

Avogadro's law assumes constant temperature and pressure.

When the amount of gas in a sample is increased, its volume increases in direct proportion. This also follows from kinetic molecular theory. If the number of gas particles increases at constant pressure and temperature, the particles occupy more volume.

You experience Avogadro's law when you inflate a balloon, for example. With each exhaled breath, you add more gas particles to the inside of the balloon, increasing its volume (Figure 11.19). Avogadro's law can be used to compute the volume of a gas following a change in the amount of the gas *as long as the pressure and temperature of the gas are constant.* For these calculations, Avogadro's law is expressed as

Since $V \alpha n$, then $\frac{V}{n}$ = constant. If the number of moles increases, then the volume increases in direct proportion so that the quotient, $\frac{V}{n'}$, is always equal to the same constant. So, for two different measurements, we can say that $\frac{V_1}{n_1} = \text{constant} = \frac{V_2}{n_2}$ or $\frac{V_1}{n_1} = \frac{V_2}{n_2}$.

$$\frac{V_1}{n_1} = \frac{V_2}{n_2}$$

where V_1 and n_1 are the initial volume and number of moles of the gas and V_2 and n_2 are the final volume and number of moles. In calculations, Avogadro's law is used in a manner similar to the other gas laws, as shown in the following example.

Figure 11.19 As you exhale into a balloon, you add gas molecules to the inside of the balloon, increasing its volume.

EXAMPLE 11.5 Avogadro's Law

A 4.8 L sample of helium gas contains 0.22 mol helium. How many additional moles of helium gas must be added to the sample to obtain a volume of 6.4 L? Assume constant temperature and pressure.

Given:

$$V_1 = 4.8\text{ L}$$
$$n_1 = 0.22\text{ mol}$$
$$V_2 = 6.4\text{ L}$$

Find: n_2

Equation:

$$\frac{V_1}{n_1} = \frac{V_2}{n_2}$$

Solution Map:

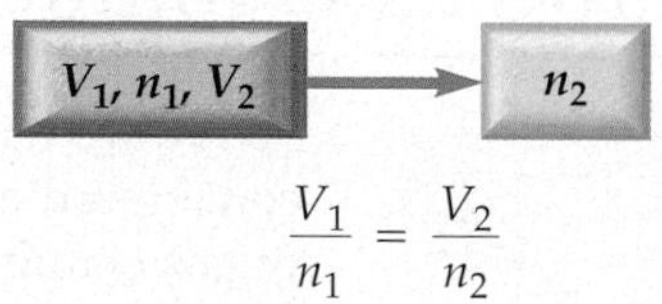

Solution:

$$\frac{V_1}{n_1} = \frac{V_2}{n_2}$$

$$n_2 = \frac{V_2}{V_1} n_1$$

$$= \frac{6.4\text{ L}}{4.8\text{ L}} 0.22\text{ mol}$$

$$= 0.29\text{ mol}$$

Since the balloon already contains 0.22 mol, 0.07 additional mol helium must be added to bring the total to 0.29 mol.

SKILLBUILDER 11.5 Avogadro's Law

A chemical reaction occurring in a cylinder equipped with a moveable piston produces 0.58 mol of a gaseous product. If the cylinder contained 0.11 mol of gas before the reaction and had an initial volume of 2.1 L, what was its volume after the reaction?

11.8 The Ideal Gas Law: Pressure, Volume, Temperature and Moles

The relationships that we have learned so far can be combined into a single law that encompasses all of them. So far, we know that:

$$V \alpha \frac{1}{P} \quad \text{(Boyle's law)}$$

$$V \alpha T \quad \text{(Charles's law)}$$

$$V \alpha n \quad \text{(Avogadro's law)}$$

Combining these three expressions we get:

$$V \alpha \frac{nT}{P}$$

The volume of a gas is directly proportional to the number of moles of gas and the temperature of the gas and is inversely proportional to the pressure of the gas. We can replace the proportional sign with an equal sign by adding R, a proportionality constant called the **ideal gas constant.**

$$V = \frac{RnT}{P}$$

Rearranging, we get:

$$PV = nRT$$

The preceding equation is called the **ideal gas law.** The value of R, the **ideal gas constant** is:

$$R = 0.0821 \frac{\text{L} \cdot \text{atm}}{\text{mol} \cdot \text{K}}$$

The ideal gas law contains within it the simple gas laws we have learned. For example, recall that Boyle's law states that $V\alpha\frac{1}{P}$ when the amount of gas (n) and the temperature of the gas (T) are kept constant. We can rearrange the ideal gas law as follows:

$$PV = nRT$$

First, divide both sides by P.

$$V = \frac{nRT}{P}$$

Then put the variables that are constant in parentheses.

$$V = (nRT)\frac{1}{P}$$

Since n and T are constant in this case and since R is always a constant

$$V = (\text{Constant}) \times \frac{1}{P}$$

which gives us Boyle's law $\left(V\alpha\frac{1}{P}\right)$.

The ideal gas law also shows how other pairs of variables are related. For example, from Charles's law we know that volume is proportional to temperature at constant pressure and constant number of moles. But what if we heat a sample of gas at constant *volume* and constant number of moles? This question applies to the warning labels on aerosol cans such as hair spray or deodorants. These labels warn the user against excessive heating or incineration of the can, even after the contents are used up.

Why? A seemingly empty aerosol can is not really empty but contains a fixed amount of gas trapped in a fixed volume. What would happen if you heated the can? Let's rearrange the ideal gas law to clearly see the relationship between pressure and temperature at constant volume and constant number of moles.

$$PV = nRT$$

Divide both sides by V.

$$P = \frac{nRT}{V}$$

$$P = \left(\frac{nR}{V}\right)T$$

Since n and V are constant and since R is always a constant:

$$P = \text{Constant} \times T$$

The relationship between pressure and temperature is also known as Gay-Lussac's law.

As the temperature of a fixed amount of gas in a fixed volume increases, the pressure increases. In an aerosol can, this pressure increase can cause the can to explode, which is why aerosol cans should not be heated or incinerated.

The ideal gas law can also be used to determine the value of any one of the four variables (P, V, n, or T) given the other three. However, each of the quantities in the ideal gas law *must be expressed* in the units within R.

- Pressure (P) must be expressed in atmospheres.
- Volume (V) must be expressed in liters.
- Moles (n) must be expressed in moles.
- Temperature (T) must be expressed in kelvin.

For example, suppose we want to know the pressure of 0.18 mol of a gas in a 1.2 L flask at 298 K.

Given:

$$n = 0.18 \text{ mol}$$
$$V = 1.2 \text{ L}$$
$$T = 298 \text{ K}$$

Find: P

Equation:

$$PV = nRT$$

The equation that relates these is the ideal gas law.

Solution Map:

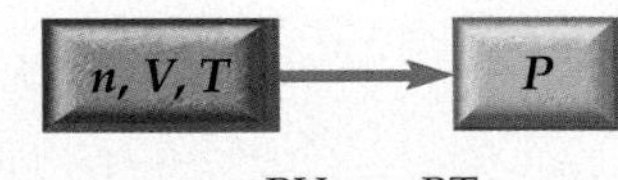

$PV = nRT$

The solution map shows how the ideal gas law takes us from the given quantities to the find quantity.

Solution:

We then solve the equation for the quantity we are trying to find (in this case, P).

$$PV = nRT$$

$$P = \frac{nRT}{V}$$

We then substitute in the numerical values and compute the answer.

$$P = \frac{0.18 \text{ mol} \times 0.0821 \dfrac{\text{L} \cdot \text{atm}}{\text{mol} \cdot \text{K}} \times 298 \text{ K}}{1.2 \text{ L}}$$

$$= 3.7 \text{ atm}$$

Notice that all units cancel except the units of the quantity we need (atm).

EXAMPLE 11.6 The Ideal Gas Law

Calculate the volume occupied by 0.845 mol nitrogen gas at a pressure of 1.37 atm and a temperature of 315 K.

Given:

$n = 0.845$ mol
$P = 1.37$ atm
$T = 315$ K

Find: V

Equation:

$$PV = nRT$$

Solution Map:

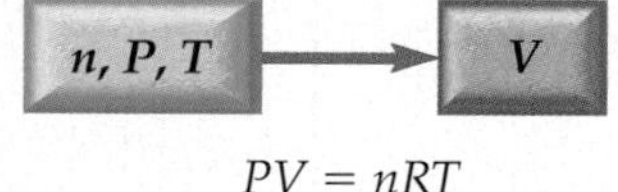

$PV = nRT$

Solution:

$$PV = nRT$$

$$V = \frac{nRT}{P}$$

$$V = \frac{0.845 \text{ mol} \times 0.0821 \dfrac{\text{L} \cdot \text{atm}}{\text{mol} \cdot \text{K}} \times 315 \text{ K}}{1.37 \text{ atm}}$$

$$= 16.0 \text{ L}$$

SKILLBUILDER 11.6 The Ideal Gas Law

An 8.5 L tire is filled with 0.55 mol of gas at a temperature of 305 K. What is the pressure of the gas in the tire?

If the units given in an ideal gas law problem are different from those of the ideal gas constant (atm, L, mol, and K), you must convert to the correct units before you substitute into the ideal gas equation as demonstrated in the following example.

EXAMPLE 11.7 The Ideal Gas Law Requiring Units Conversions

Calculate the number of moles of gas in a basketball inflated to a total pressure of 24.3 psi with a volume of 3.2 L at 25 °C.

The total pressure is not the same as the gauge pressure. The gauge pressure (the pressure read on a pressure gauge) is the difference between the total pressure and atmospheric pressure. In this case, if atmospheric pressure is 14.7 psi, the gauge pressure would be 9.6 psi. However, for calculations involving the ideal gas law, you must use the total pressure of 24.3 psi.

Given:

$$P = 24.3 \text{ psi}$$
$$V = 3.2 \text{ L}$$
$$t = 25 \text{ °C}$$

Find: n

Equation: $PV = nRT$

Solution Map:

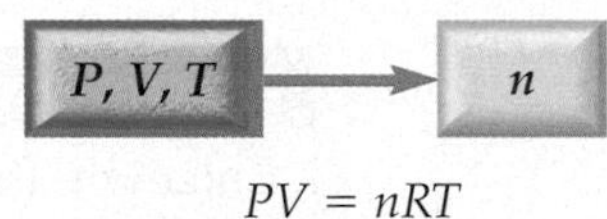

Solution:

$$PV = nRT$$
$$n = \frac{PV}{RT}$$

Before substituting into the equation, we must convert P and t into the correct units.

Since this is an intermediate answer, we mark the least significant digit but don't round until the end. See Chapter 2.

$$P = 24.2 \text{ psi} \times \frac{1 \text{ atm}}{14.7 \text{ psi}} = 1.6\underline{4}62 \text{ atm}$$

$$T = 25 + 273 = 298 \text{ K}$$

Now we can substitute into the equation and compute n.

$$n = \frac{1.6\underline{4}62 \text{ atm} \times 3.2 \text{ L}}{0.0821 \dfrac{\text{L} \cdot \text{atm}}{\text{mol} \cdot \text{K}} \times 298 \text{ K}}$$
$$= 0.22 \text{ mol}$$

SKILLBUILDER 11.7 **The Ideal Gas Law Requiring Unit Conversion**

How much volume does 0.556 mol of gas occupy when its pressure is 715 mm Hg and its temperature is 58 °C?

SKILLBUILDER PLUS

Find the pressure in millimeters of mercury of a 0.133 g sample of helium gas at 32°C and contained in a 648 mL container.

Molar Mass of a Gas from the Ideal Gas Law

The ideal gas law can be used in combination with mass measurements to calculate the molar mass of a gas. For example, a sample of gas has a mass of 0.136 g. Its volume is 0.112 L at a temperature of 298 K and a pressure of 1.06 atm. Find its molar mass.

We first set up the problem.

Given:

$m = 0.136 \text{ g}$
$V = 0.112 \text{ L}$
$T = 298 \text{ K}$
$P = 1.06 \text{ atm}$

Find: molar mass (g/mol)

Equations:

We need two equations to solve this problem, the ideal gas law and the definition of molar mass.

$$PV = nRT$$

$$\text{Molar mass} = \frac{\text{Mass}}{\text{Moles}}$$

Solution Map:

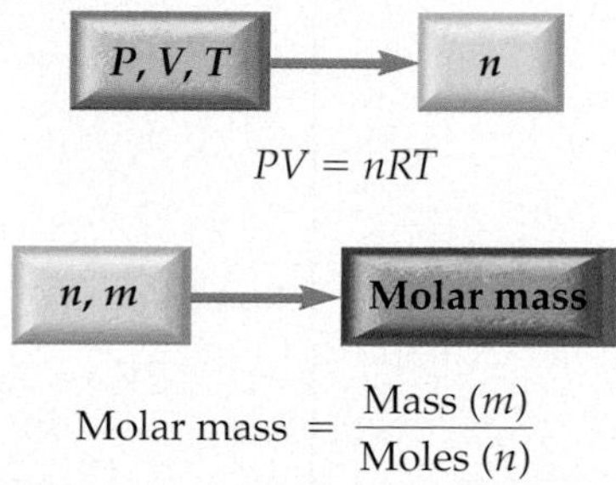

The solution map has two parts. In the first part, we use P, V, and T to find the number of moles of gas. In the second part, we use the number of moles of gas and the given mass to find the molar mass.

Solution:

$$PV = nRT$$

$$n = \frac{PV}{RT}$$

$$= \frac{1.06 \text{ atm} \times 0.112 \text{ L}}{0.0821 \frac{\text{L} \cdot \text{atm}}{\text{mol} \cdot \text{K}} \times 298 \text{ K}}$$

$$= 4.8525 \times 10^{-3} \text{ mol}$$

$$\text{Molar mass} = \frac{\text{Mass (m)}}{\text{Moles (n)}}$$

$$= \frac{0.136 \text{ g}}{4.8525 \times 10^{-3} \text{ mol}}$$

$$= 28.0 \text{ g/mol}$$

EXAMPLE 11.8 Molar Mass Using the Ideal Gas Law and Mass Measurement

A sample of gas has a mass of 0.311 g. Its volume is 0.225 L at a temperature of 55 °C and a pressure of 886 mm Hg. Find its molar mass.
We first set up the problem.

Given:

$$m = 0.311\ \text{g}$$
$$V = 0.225\ \text{L}$$
$$t = 55\ °\text{C}$$
$$P = 886\ \text{mm Hg}$$

Find: molar mass (g/mol)

Equations:

$$PV = nRT$$

$$\text{Molar mass} = \frac{\text{Mass (m)}}{\text{Moles (n)}}$$

Solution Map:

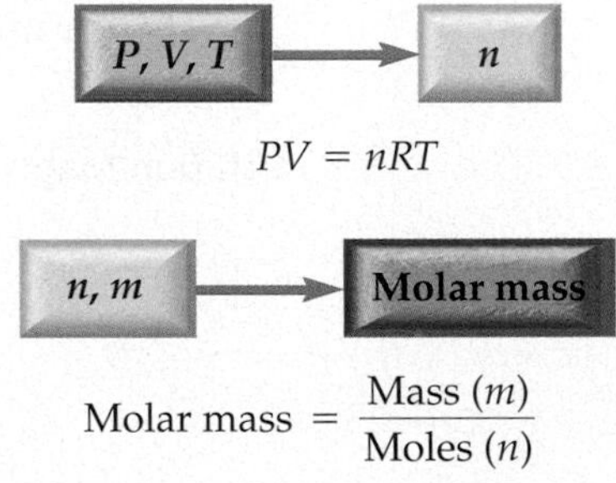

Solution:

$$PV = nRT$$

$$n = \frac{PV}{RT}$$

Before we substitute into the equation, we must convert the pressure to atm and temperature to K.

$$P = 886\ \cancel{\text{mm Hg}} \times \frac{1\ \text{atm}}{760\ \cancel{\text{mm Hg}}} = 1.1\underline{6}58\ \text{atm}$$

$$T = 55\ °\text{C} + 273 = 328\ \text{K}$$

Now, we can substitute into the equation and compute n, the number of moles.

$$n = \frac{1.1\underline{6}58\ \cancel{\text{atm}} \times 0.225\ \cancel{\text{L}}}{0.0821 \frac{\cancel{\text{L}} \cdot \cancel{\text{atm}}}{\text{mol} \cdot \cancel{\text{K}}} \times 328\ \cancel{\text{K}}}$$

$$= 9.7\underline{4}06 \times 10^{-3}\ \text{mol}$$

Finally, we use the number of moles and the given mass (m) to find the molar mass.

$$\text{Molar mass} = \frac{\text{Mass (m)}}{\text{Moles (n)}}$$

$$= \frac{0.311\ \text{g}}{9.7\underline{4}06 \times 10^{-3}\ \text{mol}}$$

$$= 31.9\ \text{g/mol}$$

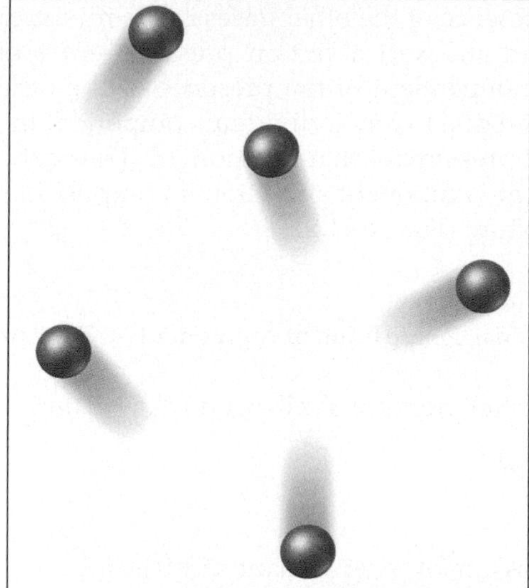

Ideal gas conditions
•High temperature
•Low pressure
•Particle size small compared to space between particles.
•Interactions between particles are insignificant.

Figure 11.20

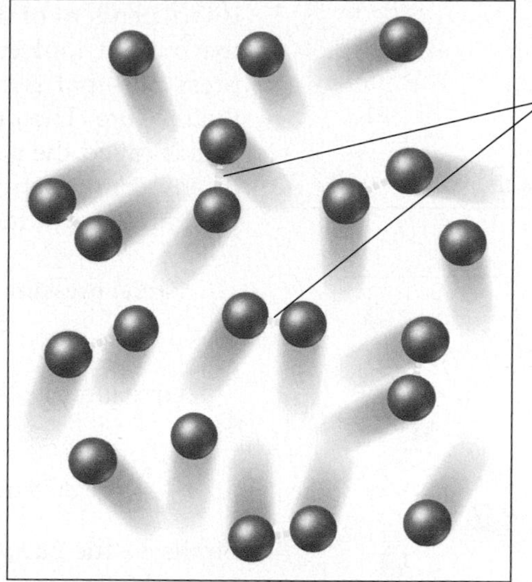

Non-ideal gas conditions
•Low temperature
•High pressure
Intermolecular Interactions
•Particle size significant compared to space between particles.
•Interactions between particles are significant.

Figure 11.21

SKILLBUILDER 11.8 **Molar Mass Using the Ideal Gas Law and Mass Measurement**

A sample of gas has a mass of 827 mg. Its volume is 0.270 L at a temperature of 88 °C and a pressure of 975 mm Hg. Find its molar mass.

Although a complete derivation is beyond the scope of this text, the ideal gas law follows directly from the kinetic molecular theory of gases. Consequently, the ideal gas law only holds under conditions where the kinetic molecular theory holds. The ideal gas law only works exactly for gases that are acting ideally (Figure 11.20), which means that a) the volume of the gas particles is small compared to the space between them and b) the forces between the gas particles are not significant. These assumptions break down (Figure 11.21) under conditions of high pressure (when the space between the gas particles becomes smaller) or low temperatures (when the gas particles move so slowly that their interactions become significant). For all of the problems encountered in this book, you may assume ideal gas behavior.

11.9 Mixtures of Gases: Why Deep Sea Divers Breathe a Mixture of Helium and Oxygen

Many gas samples are not pure but consist of mixtures of gases. The air we breathe, for example, is a mixture containing 78% nitrogen, 21% oxygen, 0.9% argon, 0.03% carbon dioxide, (Table 11.2) and a few other gases in smaller amounts.

TABLE 11.2

Composition of Dry Air

Gas	*Percent by Volume (%)*
nitrogen (N_2)	78
oxygen (O_2)	21
argon (Ar)	0.9
carbon dioxide (CO_2)	0.03

According to the kinetic molecular theory, each of the components in a gas mixture acts independently of the others. For example, the nitrogen molecules in air exert a certain pressure—78% of the total pressure—that is independent of the presence of the other gases in the mixture. Likewise, the oxygen molecules in air exert a certain pressure—21% of the total pressure—that is also independent of the presence of the other gases in the mixture. The pressure due to any individual component in a gas mixture is called the **partial pressure** of that component. The partial pressure of any component is that component's fractional composition times the total pressure of the mixture (Figure 11.22).

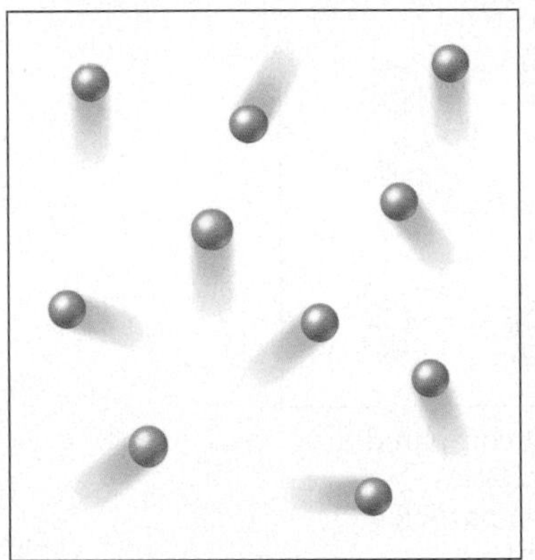

Gas mixture (80% He, 20% Ne)
$P_{tot} = 1.0$ atm
$P_{He} = 0.80$ atm
$P_{Ne} = 0.20$ atm

Figure 11.22 A gas mixture at a total pressure of 1.0 atm consisting of 80% helium and 20% neon will have a helium partial pressure of 0.80 atm and a neon partial pressure of 0.20 atm.

Partial pressure of component =
Fractional composition of component × Total pressure

The fractional composition is the percent composition divided by one-hundred.

For example, the partial pressure of nitrogen (P_{N_2}) in air at 1.0 atm, is:

$$P_{N_2} = 0.78 \times 1 \text{ atm} = 0.78 \text{ atm}$$

Similarly, the partial pressure of oxygen in air at 1.0 atm is:

$$P_{O_2} = 0.21 \times 1 \text{ atm} = 0.21 \text{ atm}$$

The sum of the partial pressures of each of the components in a gas mixture must equal the total pressure

$$P_{tot} = P_a + P_b + P_c + \ldots..$$

where P_{tot} is the total pressure and $P_a, P_b, P_c \ldots$ are the partial pressures of the components. This is known as **Dalton's law of partial pressures.** For 1 atm air:

$$P_{tot} = P_{N_2} + P_{O_2} + P_{Ar}$$
$$P_{tot} = 0.78 \text{ atm} + 0.21 \text{ atm} + 0.01 \text{ atm} = 1.0 \text{ atm}$$

We can ignore the contribution of the CO_2 and other trace gases because they are so small.

EXAMPLE 11.9 Total Pressure and Partial Pressures

A mixture of helium, neon, and argon has a total pressure of 558 mm Hg. If the partial pressure of helium is 341 mm Hg and the partial pressure of neon is 112 mm Hg, what is the partial pressure of argon?

Given:

$P_{tot} = 558$ mm Hg
$P_{He} = 341$ mm Hg
$P_{Ne} = 112$ mm Hg

Find: P_{Ar}

Equation:

$$P_{tot} = P_a + P_b + P_c + \ldots..$$

Solution:

$$\begin{aligned} P_{tot} &= P_{He} + P_{Ne} + P_{Ar} \\ P_{Ar} &= P_{tot} - P_{He} - P_{Ne} \\ &= 558 \text{ mm Hg} - 341 \text{ mm Hg} - 112 \text{ mm Hg} \\ &= 105 \text{ mm Hg} \end{aligned}$$

SKILLBUILDER 11.9 **Total Pressure and Partial Pressures**

A sample of hydrogen gas is mixed with water vapor. The mixture has a total pressure of 745 torr and the water vapor has a partial pressure of 24 torr. What is the partial pressure of the hydrogen gas?

Deep Sea Diving and Partial Pressure

Our lungs have evolved to breathe oxygen at a partial pressure of $P_{O_2} = 0.21$ atm. If the total pressure decreases—when climbing a mountain, for example—the partial pressure of oxygen also decreases. For example, on top of Mt. Everest, where the total pressure is only 0.311 atm, the partial pressure of oxygen is only 0.065 atm. As we learned earlier, low oxygen levels have physiological effects, a condition called *hypoxia* or oxygen starvation. Mild hypoxia causes dizziness, headache, and shortness of breath. Severe hypoxia, which occurs when P_{O_2} drops below 0.1 atm, may cause unconsciousness or even death. For this reason, climbers hoping to make the summit of Mt. Everest usually carry oxygen to breathe.

High oxygen levels can also have physiological effects. Scuba divers, as we have learned, breathe pressurized air. At 30 m, a scuba diver breathes air at a total pressure of 4.0 atm, making P_{O_2} about 0.84 atm. This increased partial pressure of oxygen results in a higher density of oxygen molecules in the lungs (Figure 11.23), which results in a higher concentration of oxygen

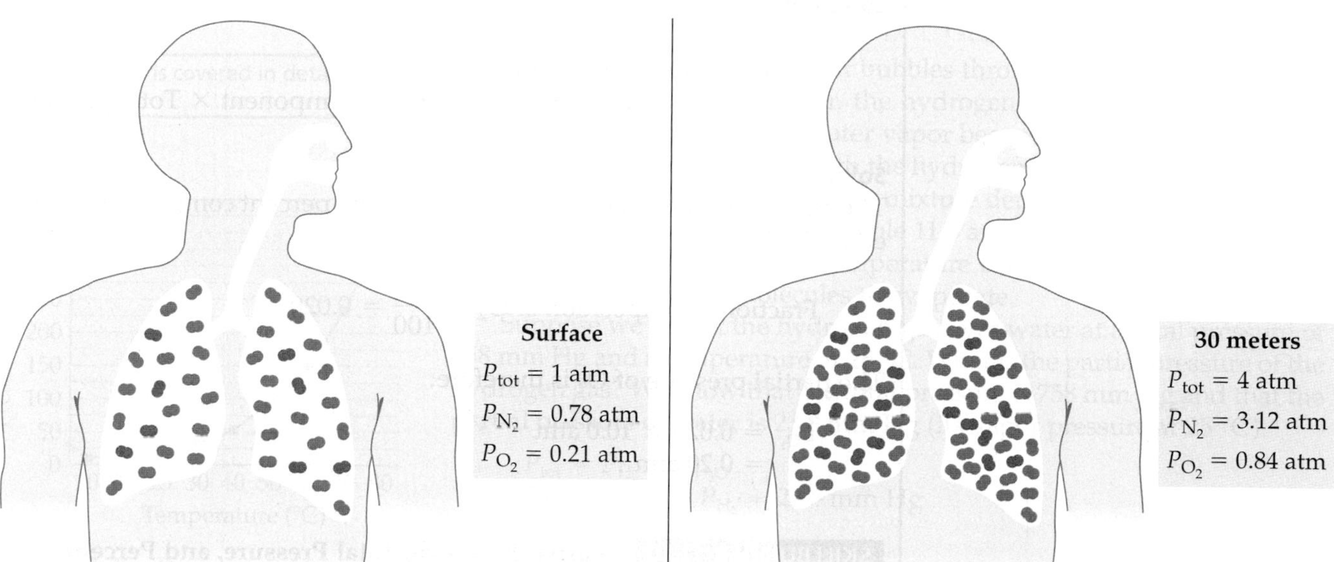

Figure 11.23 When breathing compressed air, there is a larger partial pressure of oxygen in the lungs. A large oxygen partial pressure in the lungs results in a larger amount of oxygen in bodily tissues. When the oxygen partial pressure increases beyond 1.4 atm, oxygen toxicity results (In this figure, the red molecules are oxygen and the blue ones are nitrogen.).

45. A balloon contains 0.128 mol of gas and has a volume of 2.76 L. If an additional 0.073 mol of gas are added to the balloon, what will its final volume be?

46. A cylinder with a moveable piston contains 0.87 mol of gas and has a volume of 334 mL. What will its volume be if an additional 0.22 mol of gas are added to the cylinder?

Combined Gas Law

47. A sample of gas with an initial volume of 32.5 L at a pressure of 755 mm Hg and a temperature of 315 K is compressed to a volume of 15.8 L and warmed to a temperature of 395 K. What is the final pressure of the gas?

48. A cylinder with a moveable piston contains 188 mL of nitrogen gas at a pressure of 1.12 atm and a temperature of 298 K. What must the final volume be for the pressure of the gas to be 1.42 atm at a temperature of 345 K?

49. A scuba diver takes a 2.8 L balloon from the surface, where the pressure is 1.0 atm and the temperature is 34 °C, to a depth of 25 m, where the pressure is 3.5 atm and the temperature is 18 °C. What is the volume of the balloon at this depth?

50. A bag of potato chips contains 585 mL of air at 25 °C and a pressure of 765 mm Hg. Assuming the bag does not break, what will be its volume at the top of a mountain where the pressure is 442 mm Hg and the temperature is 5.0 °C?

51. A gas sample with a volume of 5.3 L has a pressure of 735 mm Hg at 28 °C. What is the pressure of the sample if the volume remains at 5.3 L but the temperature rises to 86 °C?

52. The total pressure in a 11.7 L automobile tire is 44 psi at 11 °C. By how much does the pressure in the tire rise if it warms to a temperature of 37 °C and the volume remains at 11.7 L?

Ideal Gas Law

53. What is the volume occupied by 0.118 moles of helium gas at a pressure of 0.97 atm and a temperature of 305 K?

54. What is the pressure in a 10.0 L cylinder filled with 0.448 moles of nitrogen gas at a temperature of 315 K?

55. A cylinder contains 28.5 L of oxygen gas at a pressure of 1.8 atm and a temperature of 298 K. How many moles of gas are in the cylinder?

56. What is the temperature of 0.52 moles of gas at a pressure of 1.3 atm and a volume of 11.8 L?

57. A cylinder contains 11.8 L of air at a total pressure of 43.2 psi and a temperature of 25 °C. How many moles of gas does the cylinder contain?

58. What is the pressure (in millimeters of mercury) of 0.0115 mol of helium gas with a volume of 214 mL at a temperature of 45 °C?

59. How many moles of gas must be forced into a 3.5 L ball to give it a gauge pressure of 9.4 psi at 25 °C? The gauge pressure is relative to atmospheric pressure. Assume that atmospheric pressure is 14.7 psi so that the total pressure in the ball is 24.1 psi.

60. How many moles of gas must be forced into a 4.8 L tire to give it a gauge pressure of 32.4 psi at 25 °C? The gauge pressure is relative to atmospheric pressure. Assume that atmospheric pressure is 14.7 psi so that the total pressure in the tire is 47.1 psi.

61. An experiment shows that a 248 mL gas sample has a mass of 0.433 g at a pressure of 745 mm Hg and a temperature of 28 °C. What is the molar mass of the gas?

62. An experiment shows that a 113 mL gas sample has a mass of 0.171 g at a pressure of 721 mm Hg and a temperature of 32 °C. What is the molar mass of the gas?

63. A sample of gas has a mass of 38.8 mg. Its volume is 224 mL at a temperature of 55 °C and a pressure of 886 torr. Find the molar mass of the gas.

64. A sample of gas has a mass of 0.555 g. Its volume is 117 mL at a temperature of 85 °C and a pressure of 753 mm Hg. Find the molar mass of the gas.

Partial Pressure

65. A gas mixture contains each of the following gases at the indicated partial pressure.

N_2	355 torr
O_2	128 torr
He	229 torr

What is the total pressure of the mixture?

66. A gas mixture contains each of the following gases at the indicated partial pressure.

CO_2	455 mm Hg
Ar	124 mm Hg
O_2	167 mm Hg
H_2	92 mm Hg

What is the total pressure of the mixture?

67. A heliox deep-sea diving mixture delivers an oxygen partial pressure of 0.30 atm when the total pressure is 11.0 atm. What is the partial pressure of helium in this mixture?

68. A mixture of helium, nitrogen, and oxygen has a total pressure of 752 mm Hg. The partial pressures of helium and nitrogen are 234 mm Hg and 197 mm Hg, respectively. What is the partial pressure of oxygen in the mixture?

69. The hydrogen gas formed in a chemical reaction is collected over water at 30 °C at a total pressure of 732 mm Hg. What is the partial pressure of the hydrogen gas collected in this way?

70. The oxygen gas emitted from an aquatic plant during photosynthesis is collected over water at a temperature of 25 °C and a total pressure of 753 torr. What is the partial pressure of the oxygen gas?

71. A gas mixture contains 78% nitrogen and 22% oxygen. If the total pressure is 1.12 atm, what are the partial pressures of each component?

72. An air sample contains 0.038% CO_2. If the total pressure is 758 mm Hg, what is the partial pressure of CO_2?

73. A heliox deep-sea diving mixture contains 4.0% oxygen and 96.0% helium. What is the partial pressure of oxygen when this mixture is delivered at a total pressure of 8.5 atm?

74. A scuba diver breathing normal air descends to 100 m of depth where the total pressure is 11 atm. What is the partial pressure of oxygen that the diver experiences at this depth? Is the diver in danger of experiencing oxygen toxicity?

Molar Volume

75. Calculate the volume of each of the following gas samples at STP.
a) 2.5 mol helium
b) 5.9 mol nitrogen
c) 32.7 mol Cl_2
d) 41 mol CH_4

76. Calculate the volume of each of the following gas samples at STP.
a) 7.8 mol C_2H_6
b) 0.435 mol CO
c) 0.298 mol CO_2
d) 38.9 mol N_2O

77. Calculate the volume of each of the following gas samples at STP.
a) 73.9 g of N_2
b) 42.9 g of O_2
c) 148 g of NO_2
d) 245 mg of CO_2

78. Calculate the volume of each of the following gas samples at STP.
a) 48.9 g He
b) 45.2 g Xe
c) 48.2 mg of Cl_2
d) 3.83 kg of SO_2

79. What is the mass of 178 mL of CO_2 at STP?

80. What is the mass of 5.82 L of NO at STP?

Gases in Chemical Reactions

81. Consider the following chemical reaction.

$$C(s) + H_2O(g) \rightarrow CO(g) + H_2(g)$$

How many liters of hydrogen gas are formed from the complete reaction of 1.45 mol C? Assume that the hydrogen gas is collected at a pressure of 1.0 atm and temperature of 355 K.

82. Consider the following chemical reaction.

$$2\,H_2O(l) \rightarrow 2\,H_2(g) + O_2(g)$$

How many moles of H_2O are required to form 1.8 L of O_2 at a temperature of 315 K and a pressure of 0.957 atm?

83. CH_3OH can be synthesized by the following reaction.

$$CO(g) + 2\,H_2(g) \rightarrow CH_3OH(g)$$

How many liters of H_2 gas, measured at 748 mm Hg and 86°C, are required to synthesize 0.55 mol CH_3OH? How many liters of CO gas, measured under the same conditions, are required?

84. Oxygen gas reacts with powdered aluminum according to the following reaction.

$$4\,Al(s) + 3\,O_2(g) \rightarrow 2\,Al_2O_3(s)$$

How many liters of O_2 gas, measured at 782 mm Hg and 25°C, are required to completely react with 2.4 mol Al?

85. Nitrogen reacts with powdered aluminum according to the following reaction.

$$2\,Al(s) + N_2(g) \rightarrow 2\,AlN(s)$$

How many liters of N_2 gas, measured at 892 torr and 95°C, are required to completely react with 18.5 g of Al?

86. Sodium reacts with chlorine gas according to the following reaction.

$$2\,Na(s) + Cl_2(g) \rightarrow 2\,NaCl(s)$$

What volume of Cl_2 gas, measured at 687 torr and 35°C, is required to form 28 g of NaCl?

87. How many grams of NH_3 form when 24.8 L of $H_2(g)$ (measured at STP) reacts with N_2 to form NH_3 according to the following reaction?

$$N_2(g) + 3\,H_2(g) \rightarrow 2\,NH_3(g)$$

88. Lithium reacts with nitrogen gas according to the following reaction.

$$6\,Li(s) + N_2(g) \rightarrow 2\,Li_3N(s)$$

How many grams of lithium are required to completely react with 58.5 mL of N_2 gas at STP?

89. How many grams of calcium are consumed when 156.8 mL of oxygen gas, measured at STP, reacts with calcium according to the following reaction?

$$2\,Ca(s) + O_2(g) \rightarrow 2\,CaO(s)$$

90. How many grams of magnesium oxide are formed when 14.8 L of oxygen gas, measured at STP, completely reacts with magnesium metal according to the following reaction?

$$2\,Mg(s) + O_2(g) \rightarrow 2\,MgO(s)$$

Cumulative Problems

91. Use the ideal gas law to show that the molar volume of a gas at STP is 22.4 L.

92. Use the ideal gas law to show that 28.0 g of nitrogen gas and 4.00 g of helium gas occupy the same volume at any temperature and pressure.

93. The mass of an evacuated 255 mL flask is 143.187 g. The mass of the flask filled with 267 torr of an unknown gas at 25 °C is 143.289 g. Calculate the molar mass of the unknown gas.

94. A 118 mL flask is evacuated and massed at 97.129 g. When the flask is filled with 768 torr of helium gas at 35 °C, it is found to have a mass of 97.171 g. Was the gas pure helium?

95. A gaseous hydrogen-and-carbon-containing compound is decomposed and found to contain 82.66% carbon and 17.34% hydrogen by mass. The mass of 158 mL of the gas, measured at 556 mm Hg and 25 °C, was found to be 0.275 g. What is the molecular formula of the compound?

96. A gaseous hydrogen-and-carbon-containing compound is decomposed and found to contain 85.63% C and 14.37% H by mass. The mass of 258 mL of the gas, measured at STP, was found to be 0.646 g. What is the molecular formula of the compound?

97. The following reaction is carried out as a source of hydrogen gas in the laboratory.

$$Zn(s) + 2\,HCl(aq) \rightarrow ZnCl_2(aq) + H_2(g)$$

If 325 mL of hydrogen gas is collected over water at 25 °C at a total pressure of 748 mm Hg, how many grams of Zn reacted?

98. Consider the following reaction.

$$2\,NiO(s) \rightarrow 2\,Ni(s) + O_2(g)$$

If O_2 is collected over water at 40 °C and a total pressure of 745 mm Hg, what volume of gas will be collected for the complete reaction of 24.78 g of NiO?

99. How many grams of hydrogen are collected in a reaction where 1.78 L of hydrogen gas are collected over water at a temperature of 40 °C and a total pressure of 748 torr?

100. How many grams of oxygen are collected in a reaction where 235 mL of oxygen gas are collected over water at a temperature of 25 °C and a total pressure of 697 torr?

101. The following reaction forms 15.8 g of Ag(s).

$$2\,Ag_2O(s) \rightarrow 4\,Ag(s) + O_2(g)$$

What total volume of gas forms if it is collected over water at a temperature of 25 °C and a total pressure of 752 mm Hg?

102. The following reaction consumes 2.45 kg of CO(g).

$$CO(g) + H_2O(g) \rightarrow CO_2(g) + H_2(g)$$

How many total liters of gas are formed if the products are collected at STP?

103. Consider the following reaction.

$$2\,SO_2(g) + O_2(g) \rightarrow 2\,SO_3(g)$$

a) If 285.5 mL of SO_2 are allowed to react with 158.9 mL of O_2 (both measured at STP), what is the limiting reactant and the theoretical yield of SO_3?
b) If 187.2 mL of SO_3 are collected (measured at STP), what is the percent yield for the reaction?

104. Consider the following reaction.

$$P_4(s) + 6\,H_2(g) \rightarrow 4\,PH_3(g)$$

a) If 88.6 liters of $H_2(g)$, measured at STP, are allowed to react with 158.3 g of P_4, what is the limiting reactant?
b) If 48.3 L of PH_3, measured at STP, forms, what is the percent yield?

105. Consider the following equation for the synthesis of nitric acid.

$$3\,NO_2(g) + H_2O(l) \rightarrow 2\,HNO_3(aq) + NO(g)$$

a) If 12.8 L of $NO_2(g)$, measured at STP, is allowed to react with 14.9 g of water, find the limiting reagent and the theoretical yield of HNO_3 (in grams).
b) If 14.8 g of HNO_3 form, what is the percent yield?

106. Consider the following equation for the production of NO_2 from NO.

$$2\,NO(g) + O_2(g) \rightarrow 2\,NO_2(g)$$

a) If 84.8 L of $O_2(g)$, measured at 35 °C and 632 mm Hg, is allowed to react with 158.2 g of NO, find the limiting reagent.
b) If 97.3 L of NO_2 form, measured at 35 °C and 632 mm Hg, what is the percent yield?

Highlight Problems

107. Which of the following gas samples, all at the same temperature, will have the greatest pressure? Explain.

a)

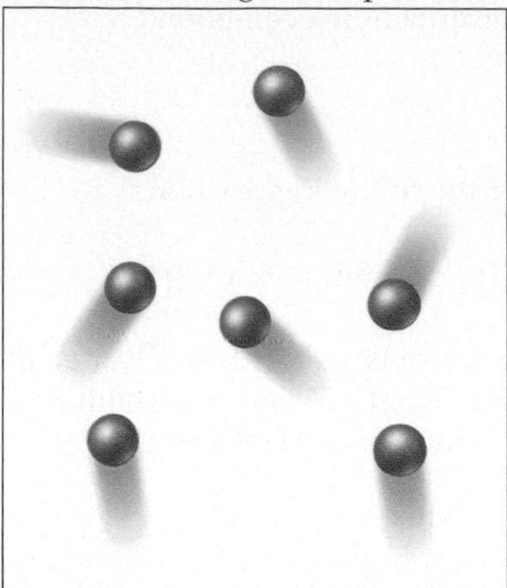

b)

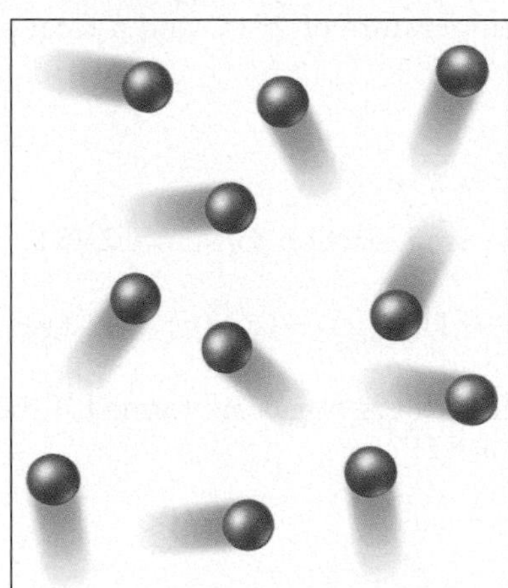

c)

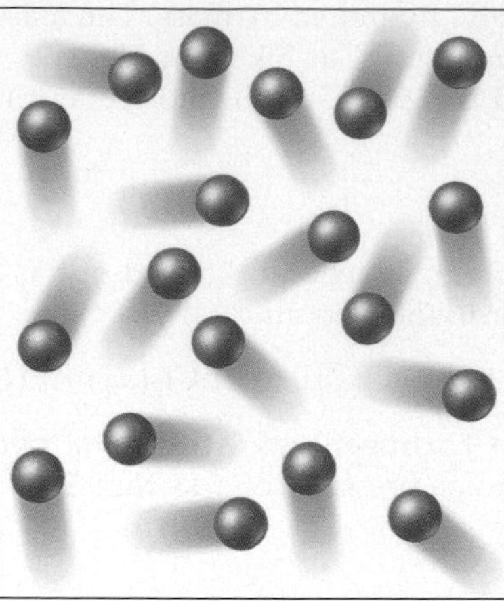

108. The following picture represents a sample of gas at a pressure of 1 atm, a volume of 1 L, and a temperature of 25 °C. Draw a similar picture showing what happens if the volume were reduced to 0.5 L and the temperature increased to 250 °C. What happens to the pressure?

$V = 1.0$ L
$T = 25$°C
$P = 1.0$ atm

109. Automobile air bags inflate following a serious impact. The impact triggers the following chemical reaction.

$$2\,NaN_3(s) \rightarrow 2\,Na(s) + 3\,N_2(g)$$

If an automobile air bag has a volume of 11.8 L, how much NaN_3 (in grams) is required to fully inflate the air bag upon impact? Assume STP conditions.

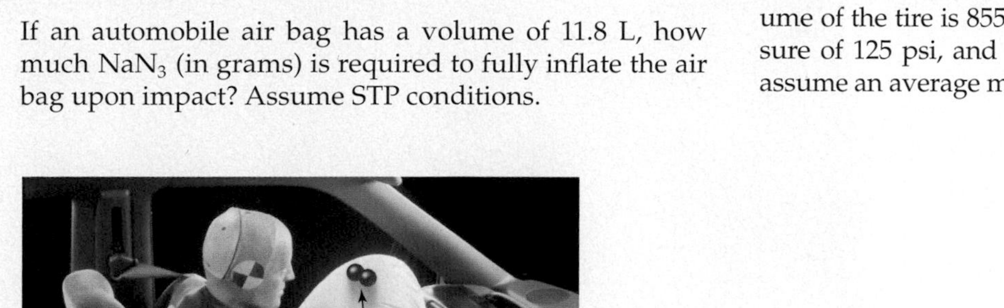

110. Olympic cyclists fill their tires with helium to make them lighter. Calculate the mass of air in an air-filled tire and the mass of helium in a helium-filled tire. What is the mass difference between the two? Assume that the volume of the tire is 855 mL, that it is filled with a total pressure of 125 psi, and that the temperature is 25 °C. Also, assume an average molar mass for air of 28.8 g/mole.

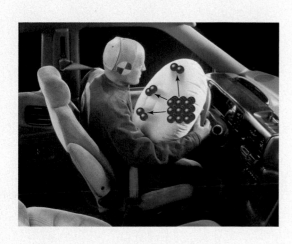

Answers to Skillbuilder Exercises

Skillbuilder 11.1 85.0 psi
Skillbuilder Plus, p. 362 80.6 kPa
Skillbuilder 11.2 $P_2 = 2.1$ atm; depth is approximately 11 m
Skillbuilder 11.3 123 mL
Skillbuilder 11.4 0.82 L
Skillbuilder 11.5 11 L
Skillbuilder 11.6 1.6 atm
Skillbuilder 11.7 16.1 L
Skillbuilder Plus, p. 380 1.28 atm
Skillbuilder 11.8 70.8 g/mol
Skillbuilder 11.9 721 torr
Skillbuilder 11.10 $P_{tot} = 4.2$ atm
Skillbuilder 11.11 82.3 g
Skillbuilder 11.12 6.53 L O_2

The interactions between bitter molecules in coffee and molecular receptors on the tongue are highly specific. However, less specific intermolecular forces exist between all molecules and atoms. These intermolecular forces are responsible for the existence of liquids and solids. The state of a sample of matter—solid, liquid, or gas—depends on the magnitude of intermolecular forces relative to the amount of thermal energy in the sample. Recall from Chapter 3 (see Section 3.9), that the molecules and atoms that compose matter are in constant random motion that increases with increasing temperature. The energy associated with this motion is called *thermal energy*. The weaker the intermolecular forces relative to thermal energy, the more likely the sample will be gaseous. The stronger the intermolecular forces relative to thermal energy, the more likely the sample will be liquid or solid.

12.2 Properties of Solids and Liquids

We are all familiar with solids and liquids. Water, gasoline, rubbing alcohol, and nail polish remover are all common liquids that you have probably encountered. Ice, dry ice, and diamond are familiar solids. In contrast to gases—in which molecules or atoms are separated by large distances—the molecules or atoms that compose liquids and solids are in close contact with one another (Figure 12.1).

The difference between solids and liquids is in the freedom of movement of the constituent molecules or atoms. In liquids, even though the atoms or molecules are in close contact, they are still free to move around each other. In solids, the atoms or molecules are fixed in their positions, although thermal energy causes them to vibrate about a fixed point. These molecular properties of solids and liquids result in the following macroscopic properties.

Properties of Liquids

- High densities in comparison to gases.
- Indefinite shape; they assume the shape of their container.
- Definite volume; they are not easily compressed.

Figure 12.1 Gas, liquid, and solid states.

TABLE 12.1

Properties of the Phases of Matter

Phase	*Density*	*Shape*	*Volume*	*Strength of Intermolecular Forces**	*Example*
gas	low	indefinite	indefinite	weak	carbon dioxide gas (CO_2)
liquid	high	indefinite	definite	moderate	liquid water (H_2O)
solid	high	definite	definite	strong	sugar ($C_{12}H_{22}H_{11}$)

*relative to thermal energy

Properties of Solids

- High densities in comparison to gases.
- Definite shape; they do not assume the shape of their container.
- Definite volume; they are not easily compressed.
- May be crystalline (ordered) or amorphous (disordered)

These properties, as well as the properties of gases for comparison, are summarized in Table 12.1.

In comparison to gases, liquids have high densities because the atoms or molecules that compose liquids are much closer together. The density of liquid water, for example, is 1.0 g/cm^3 (at 25 °C), while the density of gaseous water at 100 °C and 1 atm is 0.59 g/L (or 5.9×10^{-4} g/cm^3). Liquids assume the shape of their containers because the atoms or molecules that compose them are free to flow. When you pour water into a flask, the water flows and assumes the shape of the flask (Figure 12.2). Liquids are not easily compressed because the molecules or atoms that compose them are in close contact—they cannot be pushed closer together.

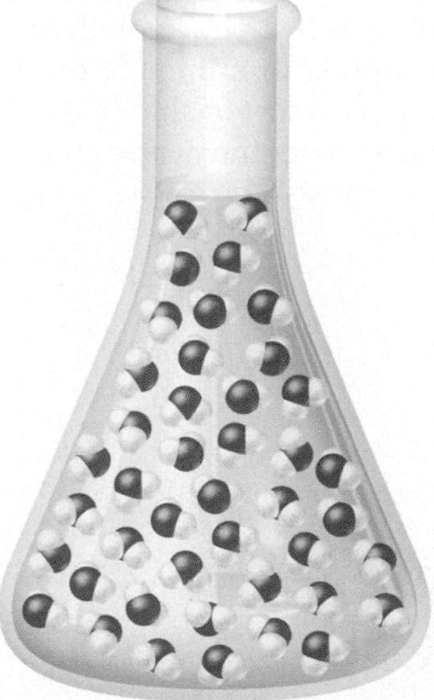

Figure 12.2 Since water molecules in liquid water are free to move around each other, they flow to assume the shape of their container.

Figure 12.3 In a solid such as ice, the molecules are fixed in place. However, the molecules vibrate about a fixed point.

Like liquids, solids have high densities in comparison to gases because the atoms or molecules that compose solids are also close together. The densities of solids are usually just slightly greater than their corresponding liquid. A major exception, however, is water, whose solid (ice) is slightly less dense than liquid water. Solids have a definite shape because, in contrast to liquids or gases, the molecules or atoms that compose solids are fixed in place—each molecule or atom only vibrates about a fixed point (Figure 12.3). Like liquids, solids have a definite volume and cannot be compressed because the molecules or atoms composing them are in close contact. Solids may be crystalline, in which case the atoms or molecules that compose them arrange themselves in a well-ordered, three-dimensional array, or they may be amorphous, in which case the atoms or molecules that compose them have no long-range order.

As we will see in Section 12.8, ice is less dense than liquid water because water expands when it freezes due to its unique crystalline structure.

12.3 Intermolecular Forces in Action: Surface Tension and Viscosity

The most important manifestation of intermolecular forces is the existence of liquids and solids. Without intermolecular forces, solids or liquids would not exist. In liquids, we also observe several other manifestations of intermolecular forces.

Surface Tension

A fly fisherman delicately casts a small metal hook (with a few feathers and strings attached to make it look like a fly) onto the surface of a moving stream. The hook floats on the surface of the water and attracts trout (Figure 12.4). Why? The hook floats because of **surface tension,** the tendency of liquids to minimize their surface area. This tendency causes liquids to have a sort of skin that resists penetration. For the fisherman's hook to sink into the water, the water's surface area must increase

Figure 12.4 Fly fishing lures float on the surface of water because of surface tension.

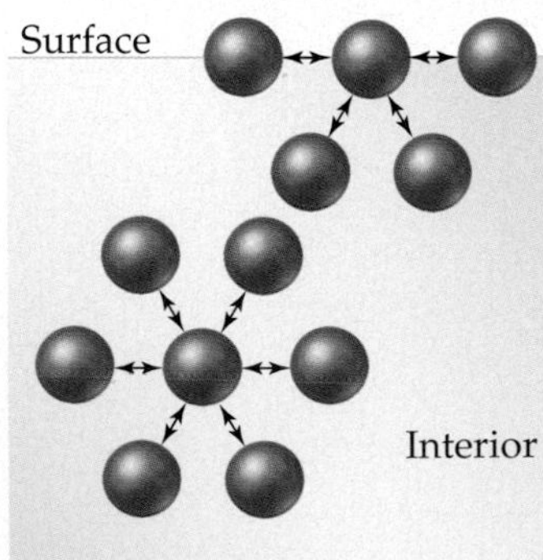

Figure 12.5 Molecules at the surface of a liquid have fewer neighbors with which to interact.

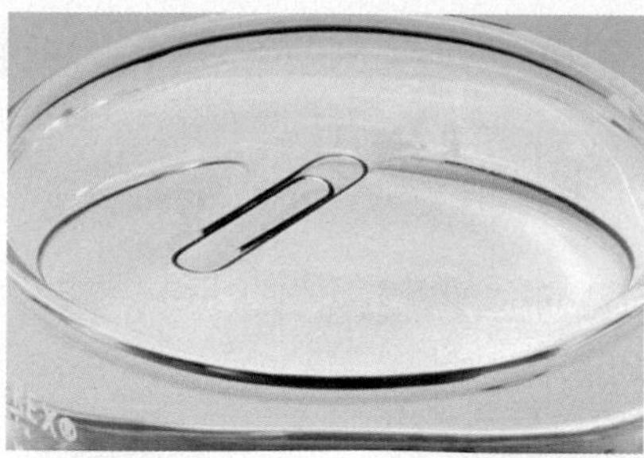

Figure 12.6 A paper clip will float on water if it is carefully placed on the surface of the water. It is held up by surface tension.

Figure 12.7 Maple syrup is viscous.

slightly. This is resisted because molecules at the surface have fewer neighbors with which to interact (Figure 12.5). You can observe surface tension by carefully placing a paper clip on the surface of water (Figure 12.6). The paper clip, even though it is denser than water, will float on the surface of the water. A slight tap on the clip will overcome surface tension and cause the clip to sink. Surface tension increases with increasing intermolecular forces. If you try to float a paper clip on gasoline, for example, you can't, because the intermolecular forces among the molecules composing gasoline are weaker than the intermolecular forces among water molecules.

Viscosity

Another manifestation of intermolecular forces is **viscosity,** the resistance of a liquid to flow. Motor oil, for example, is more viscous than gasoline, and maple syrup is more viscous than water (Figure 12.7). Viscosity is greater in substances with stronger intermolecular forces because molecules cannot move around each other as freely, hindering flow.

12.4 Evaporation and Condensation

Leave a glass of water in the open for several days and the water level within the glass slowly drops. Why? Water molecules at the surface of the water—which experience fewer attractions to neighboring molecules and are therefore held less tightly—break away from the rest of the liquid resulting in **evaporation** or **vaporization,** a physical change in which a substance is converted from its liquid form to its gaseous form (Figure 12.11). If you take the same water and spill it on the table, it evaporates faster, probably within a few hours. The surface area of the spilled water is greater, leaving more molecules susceptible to evaporation. If you warm the water, it also evaporates faster, because the greater thermal energy causes more molecules at the surface to break away. If you fill the glass with rubbing alcohol instead of water, the liquid again evaporates faster because the intermolecular forces between the alcohol molecules are

EVERYDAY CHEMISTRY

Why Are Water Drops Spherical?

Figure 12.8 If a water droplet is small enough, it will be free of the distorting effects of gravity and be perfectly spherical.

Figure 12.10 These magnetic marbles tend to arrange themselves in a spherical shape.

Have you ever seen a close-up photograph of tiny water droplets (Figure 12.8) or carefully watched water in free fall? In both cases, the distorting effects of gravity are diminished, and the water forms nearly perfect spheres. On the space shuttle, the complete absence of gravity results in floating spheres of water (Figure 12.9). Why? Water drops are spherical because of the surface tension caused by the attractive forces between water molecules. Just as gravity pulls matter in a planet or star into a sphere, so intermolecular forces pull a water drop into a sphere. The sphere minimizes the surface-area-to-volume ratio, thereby minimizing the number of molecules at the surface. A collection of magnetic marbles provides a good physical model of a water drop. Each magnetic marble is like a water molecule, attracted to the marbles around it. If you agitate these marbles slightly, so that they can find their preferred configuration, they tend towards a spherical shape (Figure 12.10) because the attractions between the marbles cause them to minimize the number of marbles at the surface.

Figure 12.9 In the absence of gravity, as in this picture taken on the space shuttle, water will assume the shape of a sphere.

CAN YOU ANSWER THIS? *How would the tendency of a liquid to form spherical drops depend on the strength of intermolecular forces? Would liquids with weaker intermolecular forces have a higher or lower tendency to form spherical drops?*

weaker than the intermolecular forces between water molecules. In general, the rate of vaporization increases with:

- Increasing surface area
- Increasing temperature
- Decreasing strength of intermolecular forces

Figure 12.11 Since molecules on the surface of a liquid are held less tightly than those in the interior, they can break away into the gas phase. This is called *evaporation*.

In evaporation or vaporization, a substance is converted from its liquid form into its gaseous form.

We call this *dynamic* equilibrium, because condensation and evaporation of individual molecules are still occurring, albeit at the same rate.

Liquids that evaporate easily are termed **volatile,** while those that do not vaporize easily are termed **nonvolatile.** Rubbing alcohol, for example, is more volatile than water. Motor oil is nonvolatile.

If you leave water in a closed container, it will not evaporate away because the molecules that leave the liquid are trapped in the air space above the water. These gaseous molecules bounce off of the walls of the container and eventually hit the surface of the water again and recondense. **Condensation** is a physical change in which a substance is converted from its gaseous form to its liquid form. Evaporation and condensation are opposites: evaporation is a liquid turning into a gas, and condensation is a gas turning into liquid. When liquid water is first put into a closed container, more evaporation happens than condensation, because there are so few gaseous water molecules in the space above the water initially (Figure 12.12a). However, as the number of gaseous water molecules increases, the rate of condensation also increases (Figure 12.12b). At the point where the rates of condensation and evaporation become equal (Figure 12.12c), **dynamic equilibrium** is reached and the number of gaseous water molecules above the liquid remains constant. Vapor pressure is the partial pressure of a gas in dynamic equilibrium with its liquid. For water at 25 °C, the vapor pressure is 23.8 mm Hg. Vapor pressure increases with:

- Increasing temperature
- Decreasing strength of intermolecular forces

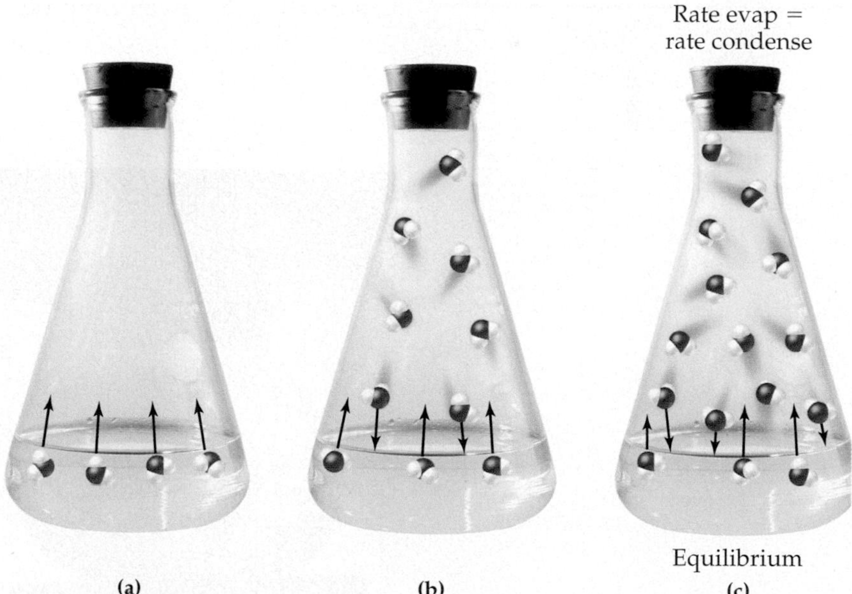

Figure 12.12 **a)** When water is first put into a closed container, water molecules begin to evaporate. **b)** As the number of gaseous molecules increases, some of the molecules begin to recondense into liquid. **c)** When the rate of evaporation equals the rate of condensation, dynamic equilibrium occurs, and the number of gaseous molecules remains constant.

Boiling

As you increase the temperature of water in an open container, the thermal energy causes molecules to leave the surface and vaporize at a faster and faster rate. At the **boiling point**—the temperature at which the vapor pressure of a liquid is equal to the pressure above it—the thermal energy is enough for molecules within the interior of the water (not just those at the surface) to break free into the gas phase (Figure 12.13). Water's **normal boiling point**—its boiling point at a pressure of 1 atmosphere—is 100 °C. When a sample of water reaches 100 °C, you see bubbles form within the water. These bubbles are pockets of gaseous water that have formed within the liquid water. The bubbles float to the surface and leave as gaseous water or steam. Once the boiling point of a liquid is reached, additional heating only causes more rapid boiling; it does not raise the temperature of the liquid above its boiling point (Figure 12.14). Therefore, boiling water at 1 atm will always have a temperature of 100 °C. After all the water has been converted to steam, the temperature of the steam can continue to rise beyond 100 °C.

Sometimes you see bubbles begin to form in hot water below 100 °C. These bubbles are dissolved air—not gaseous water—leaving the liquid. Dissolved air comes out of water as you heat it because the solubility of a gas in a liquid decreases with increasing temperature. See Section 13.4.

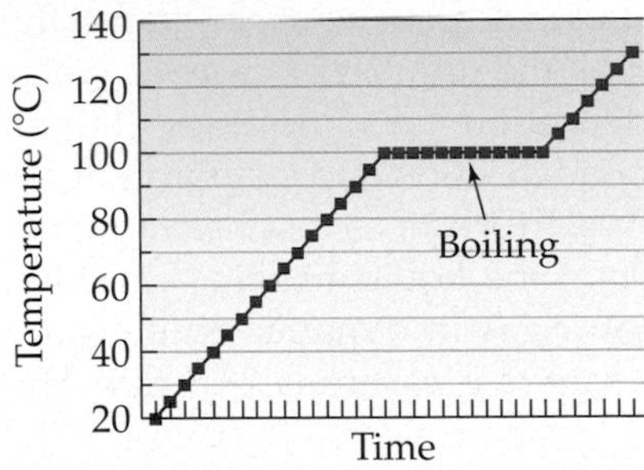

Figure 12.14 The temperature of water as it is heated from room temperature through boiling. During boiling, the temperature remains at 100 °C until all the liquid is evaporated.

Energetics of Evaporation and Condensation

Evaporation is **endothermic**—heat is absorbed when a liquid is converted into a gas.

In an endothermic process, heat is absorbed. In an exothermic process, heat is emitted.

If you turn off the heat beneath a boiling pot of water, it will quickly stop boiling as the heat lost due to vaporization causes the water to cool below its boiling point. Our bodies use evaporation as a cooling mechanism. When we get overheated, we sweat, causing our skin to be covered with liquid water. As this water evaporates, it absorbs heat from our bodies, cooling our skin. A fan makes us feel cooler because it blows newly vaporized water away from our skin, allowing more sweat to vaporize

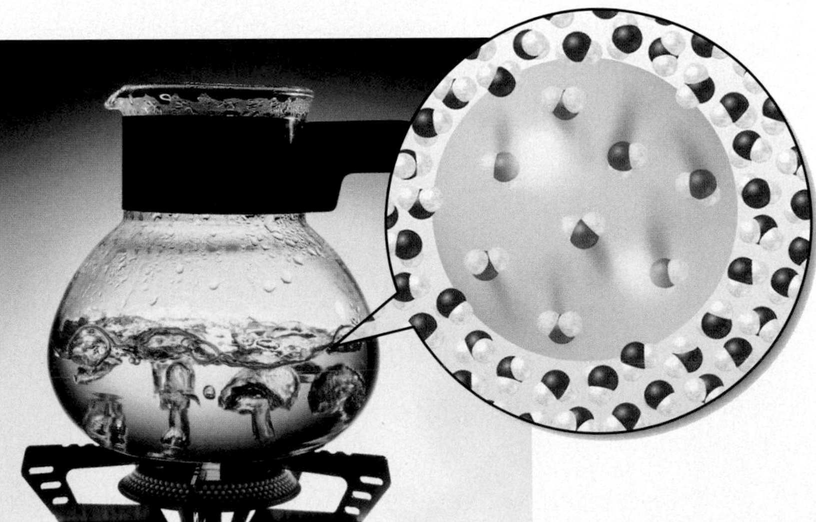

Figure 12.13 During boiling, thermal energy is enough to cause water molecules in the interior of the liquid to become gaseous, forming bubbles containing gaseous water molecules.

and causing even more cooling. High humidity, on the other hand, slows down evaporation, preventing cooling. When the air already contains high amounts of water vapor, sweat does not evaporate easily, making our cooling system less efficient.

Condensation, the opposite of evaporation, is **exothermic**—heat is released when a gas condenses to a liquid. If you have ever accidentally put your hand above a steaming kettle, you may have experienced a *steam burn*. As the steam condenses to a liquid on your skin, it releases heat, causing a severe burn. The condensation of water vapor is also the reason that winter overnight temperatures in coastal cities, which tend to have water vapor in the air, do not get as low as in deserts, which tend to have dry air. As the air temperature in a coastal city drops, water condenses out of the air, releasing heat and preventing the temperature from dropping further. In deserts, there is little moisture in the air to condense, so the temperature drop is greater.

Heat of Vaporization

The amount of heat required to vaporize one mole of liquid is called the **heat of vaporization** (ΔH_{vap}). The heat of vaporization of water at its normal boiling point (100 °C) is 40.6 kJ/mole.

By writing the 40.6 kJ as a reactant in the chemical equation, we show that the reaction absorbs 40.6 kJ of heat for every 1 mol of water that is vaporized.

$$40.6\text{ kJ} + H_2O(l) \rightarrow H_2O(g) \qquad (\text{at } 100\ °\text{C})$$

The same amount of heat is involved when 1 mol of gas condenses, but the heat is emitted rather than absorbed.

By writing the 40.6 kJ as a product in the chemical equation, we show that the reaction emits 40.6 kJ of heat for every one mole of water that condenses.

$$H_2O(g) \rightarrow H_2O(l) + 40.6\text{ kJ} \qquad (\text{at } 100\ °\text{C})$$

Different liquids have different heats of vaporization (Table 12.2). Heats of vaporization are also temperature dependent. The higher the temperature, the easier it is to vaporize a given liquid and therefore the lower the heat of vaporization.

The heat of vaporization of a liquid can be used to calculate the amount of heat energy required to vaporize a given amount of the liquid. The heat of vaporization can be viewed as a conversion factor between moles of a liquid and the amount of heat required to vaporize it. For

TABLE 12.2

Heats of Vaporization of Several Liquids at Their Boiling Points and at 25 °C

Liquid	*Chemical Formula*	*Normal Boiling Point (°C)*	*Heat of Vaporization (kJ/mole) at Boiling Point*	*(kJ/mole) at 25 °C*
water	H_2O	100	40.6	44.0
rubbing alcohol (isopropyl alcohol)	C_3H_8O	82.3	39.9	45.4
acetone	C_3H_6O	56.1	29.1	31.0
diethyl ether	$C_4H_{10}O$	34.5	26.5	27.1

example, suppose we want to calculate the amount of heat required to vaporize 25.0 g of water at its boiling point. We set up the problem in the standard way.

Given: 25.0 g H_2O

Find: heat (kJ)

Conversion Factors:

$$\Delta H_{vap} = 40.6 \text{ kJ/mole (at 100 °C)}$$
$$1 \text{ mol } H_2O = 18.02 \text{ g } H_2O$$

Solution Map:

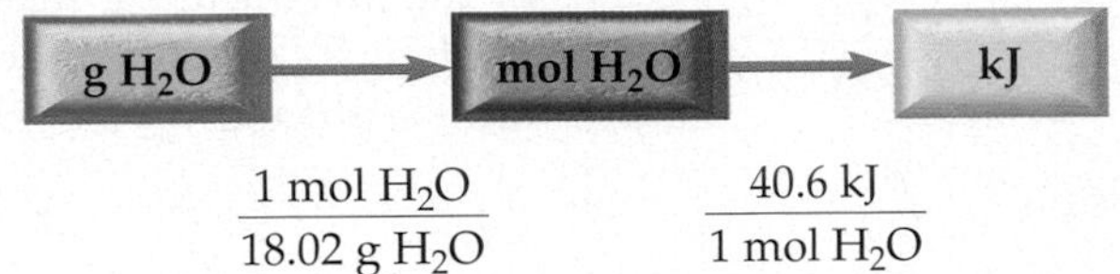

$$\frac{1 \text{ mol } H_2O}{18.02 \text{ g } H_2O} \qquad \frac{40.6 \text{ kJ}}{1 \text{ mol } H_2O}$$

The solution map begins with g of water, shows the conversion to mol of water using the molar mass of water, and then the conversion to kJ using ΔH_{vap}.

Solution:

$$25.0 \text{ g } H_2O \times \frac{1 \text{ mol } H_2O}{18.02 \text{ g } H_2O} \times \frac{40.6 \text{ kJ}}{1 \text{ mol } H_2O} = 56.3 \text{ kJ}$$

EXAMPLE 12.1 Using the Heat of Vaporization in Calculations

Calculate the amount of water (in grams) that can be vaporized at its boiling point with 155 kJ of heat.
Set up the problem in standard way.

Given: 155 kJ

Find: g H_2O

Conversion Factors:

$$\Delta H_{vap} = 40.6 \text{ kJ/mole (at 100 °C)}$$
$$18.02 \text{ g } H_2O = 1 \text{ mol } H_2O$$

Solution Map:

155 kJ → mol H_2O → g H_2O

$$\frac{1 \text{ mol } H_2O}{40.6 \text{ kJ}} \qquad \frac{18.02 \text{ g } H_2O}{1 \text{ mol } H_2O}$$

Solution:

$$155 \text{ kJ} \times \frac{1 \text{ mol } H_2O}{40.6 \text{ kJ}} \times \frac{18.02 \text{ g}}{1 \text{ mol } H_2O} = 68.8 \text{ g}$$

SKILLBUILDER 12.1 Using the Heat of Vaporization in Calculations

Calculate the amount of heat (in kilojoules) required to vaporize 2.58 kg of water at its boiling point.

SKILLBUILDER PLUS

A drop of water weighing 0.48 g condenses on the surface of a 55 g block of aluminum that is initially at 25 °C. If the heat released during condensation goes only toward heating the metal, what is the final temperature (in Celsius) of the metal block? (The specific heat capacity of aluminum is 0.903 J/g °C.)

12.5 Melting, Freezing, and Sublimation

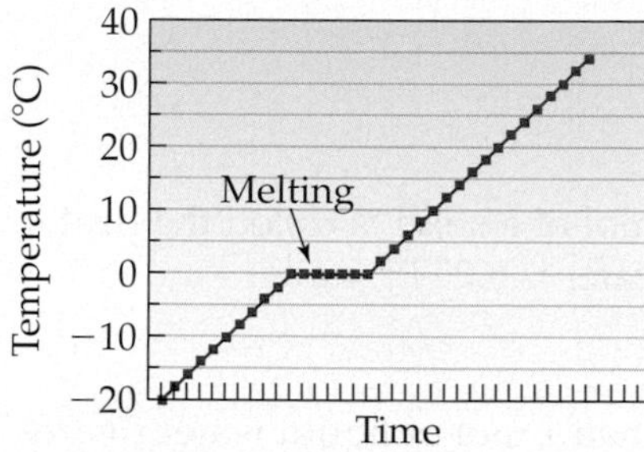

Figure 12.15 A graph of the temperature of ice as it is heated from −20 °C to 35 °C. During melting, the temperature remains at 0 °C until the entire solid is melted.

As you increase the temperature of a solid, thermal energy causes the molecules and atoms composing the solid to vibrate faster. At the **melting point,** atoms and molecules have enough thermal energy to overcome the intermolecular forces that hold them at their stationary points, and the solid turns into a liquid. The melting point of ice, for example, is 0 °C. Once the melting point of a solid is reached, additional heating only causes more rapid melting; it does not raise the temperature of the solid above its melting point (Figure 12.15). Only after all of the ice has melted will additional heating raise the temperature of the liquid water past 0 °C. A mixture of water *and* ice will always have a temperature of 0 °C (at 1 atm pressure).

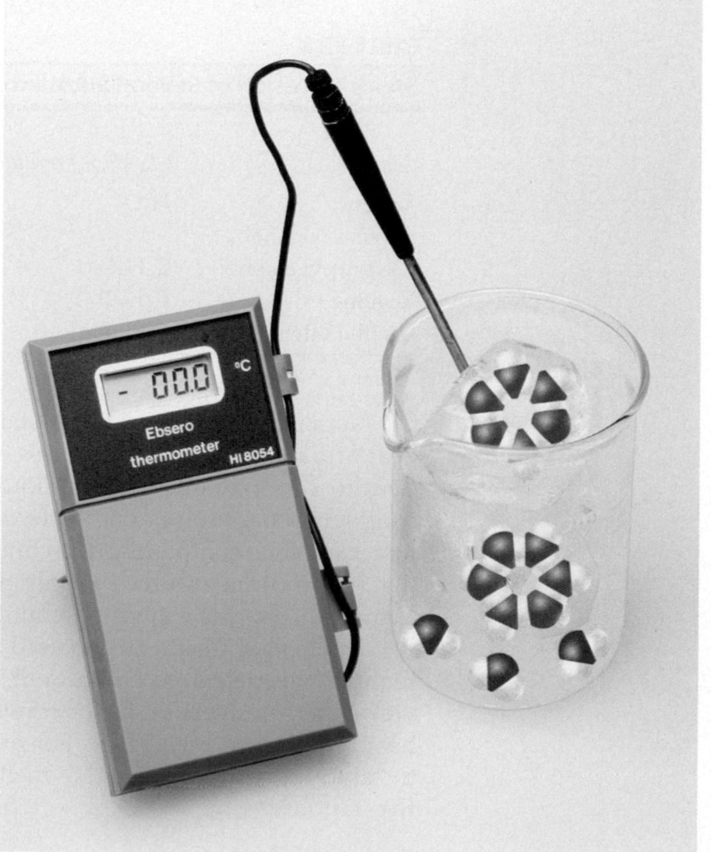

When ice melts, water molecules break free from the solid structure and become a liquid. As long as ice and water are both present, the temperature will be 0.0 °C.

Energetics of Melting and Freezing

The most common way to cool down a drink is to drop several ice cubes into it. As the ice melts, the drink cools because melting is endothermic—heat is absorbed when a solid is converted into a liquid. The melting ice absorbs heat from the liquid in the drink and cools the liquid.

Freezing, the opposite of melting, is exothermic—heat is released when a liquid freezes into a solid. For example, as water in your freezer turns into ice, it releases heat, which must be removed by the refrigeration system of the freezer. If the refrigeration system did not remove the heat, the water would not completely freeze into ice. The heat released as it began to freeze would warm the freezer, preventing further freezing.

Heat of Fusion

The amount of heat required to melt 1 mol of a solid is called the **heat of fusion** (ΔH_{fus}). The heat of fusion for water is 6.02 kJ/mole.

$$6.02 \text{ kJ} + H_2O(s) \rightarrow H_2O(l)$$

The same amount of heat is involved when 1 mol of liquid water freezes, but the heat is emitted rather than absorbed.

$$H_2O(l) \rightarrow H_2O(s) + 6.02 \text{ kJ}$$

Different substances have different heats of fusion (Table 12.3).

TABLE 12.3

Heats of Fusion of Several Substances

Liquid	*Chemical Formula*	*Melting Point (°C)*	*Heat of Fusion (kJ/mole)*
water	H_2O	0.00	6.02
rubbing alcohol (isopropyl alcohol)	C_3H_8O	−89.5	5.37
acetone	C_3H_6O	−94.8	5.69
diethyl ether	$C_4H_{10}O$	−116.3	7.27

Notice that, in general, the heat of fusion is significantly less than the heat of vaporization. It takes less energy to melt one mole of ice than it does to vaporize one mole of liquid water. Why? Vaporization requires complete separation of one molecule from another, so the intermolecular forces must be completely overcome. Melting, on the other hand, requires that intermolecular forces be only partially overcome, allowing molecules to move around one another while still remaining in contact.

The heat of fusion can be used to calculate the amount of heat energy required to melt a given amount of a solid. The heat of fusion can be viewed as a conversion factor between moles of a solid and the amount of heat required to melt them. For example, suppose we want to calculate the amount of heat required to melt 25.0 g of ice. We set up the problem in the standard way.

Given: 25.0 g H_2O

Find: heat (kJ)

Conversion Factor:

$$\Delta H_{fus} = 6.02 \text{ kJ/mole}$$

Solution Map:

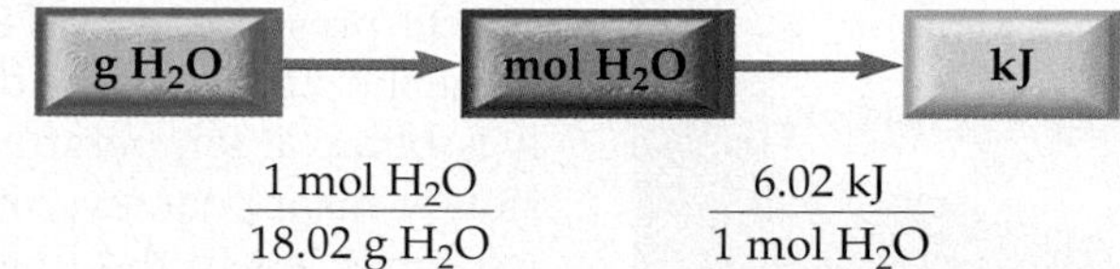

$$\frac{1 \text{ mol } H_2O}{18.02 \text{ g } H_2O} \qquad \frac{6.02 \text{ kJ}}{1 \text{ mol } H_2O}$$

The solution map begins with grams of ice, shows the conversion to moles of ice using the molar mass of water, and then makes the conversion to kJ using ΔH_{fus}.

Solution:

$$25.0 \text{ g } H_2O \times \frac{1 \text{ mol } H_2O}{18.02 \text{ g } H_2O} \times \frac{6.02 \text{ kJ}}{1 \text{ mol } H_2O} = 8.35 \text{ kJ}$$

EXAMPLE 12.2 Using the Heat of Fusion in Calculations

Calculate the amount of ice (in grams) that, upon melting, absorbs 237 kJ. We set up the problem in the standard way.

Given: 237 kJ

Find: g H_2O (ice)

Conversion Factors:

$$\Delta H_{fus} = 6.02 \text{ kJ/mole}$$
$$1 \text{ mol } H_2O = 18.02 \text{ g } H_2O$$

Solution Map:

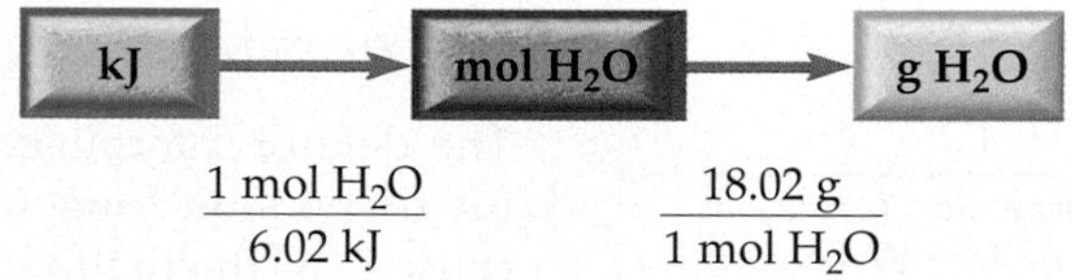

$$\frac{1 \text{ mol } H_2O}{6.02 \text{ kJ}} \qquad \frac{18.02 \text{ g}}{1 \text{ mol } H_2O}$$

Solution:

$$237 \text{ kJ} \times \frac{1 \text{ mol } H_2O}{6.02 \text{ kJ}} \times \frac{18.02 \text{ g}}{1 \text{ mol } H_2O} = 709 \text{ g}$$

SKILLBUILDER 12.2 Using the Heat of Fusion in Calculations

Calculate the amount of heat absorbed when a 15.5 g ice cube melts.

SKILLBUILDER PLUS

A 5.6 g ice cube is placed into 195 g of water initially at room temperature. If the heat absorbed for melting the ice comes only from the 195 g of water, what is the temperature change of the 195 g of water?

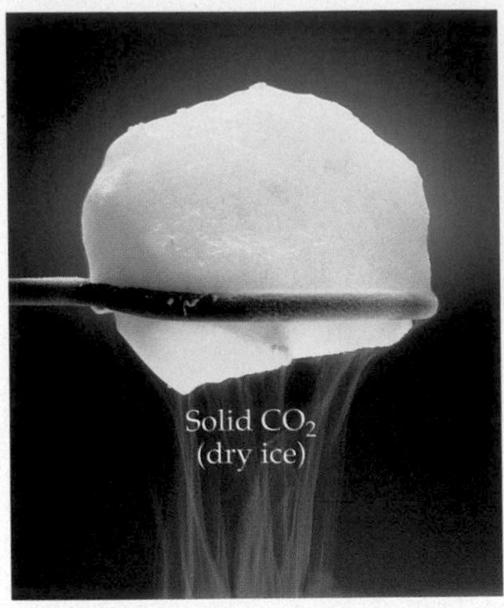

Dry ice is solid carbon dioxide. The solid does not melt but rather sublimes. It goes directly from solid carbon dioxide to gaseous carbon dioxide.

Sublimation

Sublimation is a physical change in which a substance is converted from its solid form directly into its gaseous form. When a substance sublimes, molecules leave the surface of the solid and become gaseous. For example, dry ice, which is solid carbon dioxide, does not melt under atmospheric pressure. At −78 °C the CO_2 molecules have enough energy to leave the surface of the dry ice and become gaseous. Regular ice will slowly sublime at temperatures below 0 °C. In cold climates, ice and snow will slowly disappear, even if the temperature remains below 0 °C. Similarly, ice cubes left in the freezer for a long time slowly become smaller, even though the freezer is always below 0 °C. In both cases, the ice is subliming, turning directly into water vapor.

Ice also sublimes out of frozen foods. This can be seen, for example, when food is frozen in an airtight plastic bag for a long time. The ice crystals that form in the bag are water that has sublimed out of the food and redeposited on the surface of the bag. For this reason, food that remains frozen for too long becomes dried out. This can be avoided to some degree by freezing foods to colder temperatures, a process called deep-freezing. The colder temperature lowers the rate of sublimation and preserves the food longer.

12.6 Types of Intermolecular Forces: Dispersion, Dipole–Dipole, and Hydrogen Bonding

The strength of the intermolecular forces between the molecules or atoms that compose a substance determines the state—solid, liquid, or gas—of the substance at room temperature. Strong intermolecular forces tend to result in liquids and solids (high melting and boiling points). Weak intermolecular forces tend to result in gases (low melting and boiling points). There are three different types of intermolecular forces. In order of increasing strength, they are the dispersion force, the dipole–dipole force, and the hydrogen bond.

Dispersion Force

The nature of dispersion forces was first recognized by Fritz W. London (1900–1954), a German American physicist.

The default intermolecular force, present in all molecules and atoms, is the **dispersion force** (also called the *London force*). Dispersion forces are caused by fluctuations in the electron distribution within molecules or atoms. Since all atoms and molecules have electrons, they all have dispersion forces. The electrons in an atom or molecule may, at any one instant, be unevenly distributed. For example, imagine a frame-by-frame movie of a helium atom in which each "frame" captures the position of the helium atom's two electrons (Figure 12.16). In any one frame, the electrons are not symmetrically arranged around the nucleus. In frame 3, for example, helium's two electrons are on the left side of the helium atom. The left side then acquires a slightly negative charge (δ^-). The right side of the atom, which is void of electrons, acquires a slightly positive charge (δ^+). This fleeting charge separation is called an **instantaneous dipole** or a **temporary dipole**. An instantaneous dipole on one helium atom induces an instantaneous dipole on its neighboring atoms (Figure 12.17) because the positive end of the instantaneous dipole attracts electrons in the neighboring atoms. The neighboring atoms then attract one another—the positive

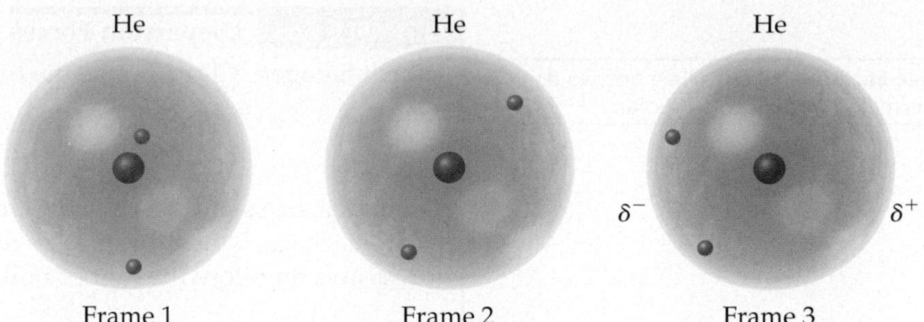

Figure 12.16 Random fluctuations in the electron distribution of a helium atom cause instantaneous dipoles to form.

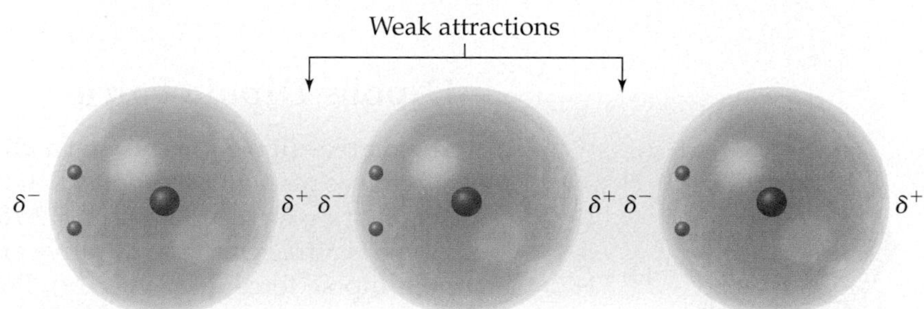

Figure 12.17 An instantaneous dipole on any one helium atom induces instantaneous dipoles on neighboring atoms. The neighboring atoms then attract one another. This attraction is called the *dispersion force*.

end of one instantaneous dipole attracts the negative end of another. This attraction is the dispersion force.

To *polarize* means to form a dipole moment.

The magnitude of the dispersion force depends on how easily the electrons in the atom or molecule can move or *polarize* in response to an instantaneous dipole, which in turn depends on the size of the electron cloud. A larger electron cloud results in a greater dispersion force because the electrons are held less tightly by the nucleus and therefore can polarize more easily. If all other variables are constant, the dispersion force increases with increasing molar mass. For example, consider the boiling points of the noble gases displayed in Table 12.4.

TABLE 12.4

Noble Gas Boiling Points

Noble Gas	*Molar Mass (g/mole)*	*Boiling Point (°C)*
He	4.00	−269
Ne	20.18	−246
Ar	39.95	−186
Kr	83.80	−153
Xe	131.30	−108

As the molar mass of the noble gas increases, its boiling point increases. While molar mass alone does not determine the magnitude of the dispersion force, it can be used as a guide when comparing dispersion forces within a family of similar elements or compounds.

See Section 10.8 to review how to determine if a molecule is polar.

EXAMPLE 12.3 Dispersion Forces

Which halogen, Cl_2 or I_2, has the higher boiling point?

Solution:

The molar mass of Cl_2 is 70.90 g/mole and the molar mass of I_2 is 253.81 g/mole. Since I_2 has the higher molar mass, it has stronger dispersion forces and therefore the higher boiling point.

SKILLBUILDER 12.3 Dispersion Forces

Which hydrocarbon, CH_4 or C_2H_6, has the higher boiling point?

Dipole–Dipole Force

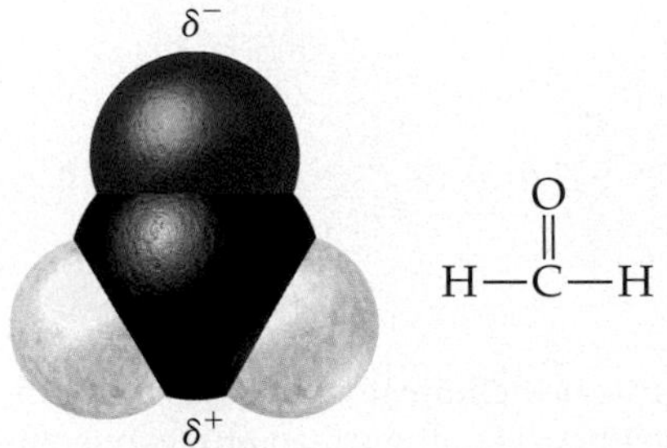

Figure 12.18 Molecules such as formaldehyde are polar and therefore have a permanent dipole.

The dipole–dipole force exists in all molecules that are polar. Polar molecules have **permanent dipoles** that interact with the permanent dipoles of neighboring molecules (Figure 12.18). The positive end of one permanent dipole is attracted to the negative end of another; this attraction is the dipole–dipole force (Figure 12.19). Polar molecules, therefore, have higher melting and boiling points than nonpolar molecules of similar molar mass. Remember that all molecules (including polar ones) have dispersion forces. In addition, polar molecules have dipole–dipole forces. This additional attractive force raises their melting and boiling points relative to nonpolar molecules of similar molar mass. For example, consider the following two compounds.

Name	Formula	Molar mass (g/mol)	Structure	bp (°C)	mp (°C)
Formaldehyde	CH_2O	30.0	O ‖ H—C—H	−19.5	−92
Ethane	C_2H_6	30.0	H H | | H—C—C—H | | H H	−88	−172

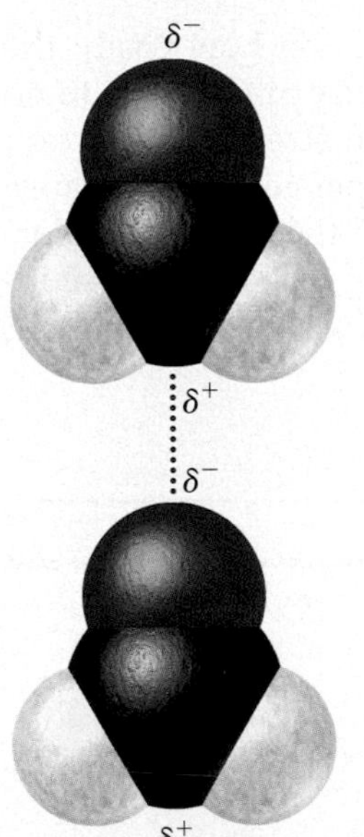

Figure 12.19 The positive end of a polar molecule is attracted to the negative end of its neighbor. This attraction is called a dipole–dipole attraction.

Formaldehyde is polar, and therefore has a higher melting point and boiling point than nonpolar ethane, even though the two compounds have the same molar mass.

The polarity of molecules composing liquids is also important in determining the **miscibility**—the ability to mix without separating into two phases—of liquids. In general, polar liquids are miscible with other polar liquids but are not miscible with nonpolar liquids. For example, water, a polar liquid, is not miscible with pentane (C_5H_{12}), a nonpolar liquid (Figure 12.20). Similarly, water and oil (also nonpolar) do not mix. Consequently, oily hands or oily stains on clothes cannot be washed with plain water.

Figure 12.20 Pentane, a nonpolar compound, does not mix with water, a polar compound.

EXAMPLE 12.4 Dipole–Dipole Forces

Which of the following molecules have dipole–dipole forces?

a) CO_2
b) CH_2Cl_2
c) CH_4

Solution:

A molecule will have dipole–dipole forces if it is polar. To determine if a molecule is polar we must:

1. *determine if the molecule contains polar* bonds and
2. *determine if the polar bonds add together to form a net dipole moment (Section 10.8)*

a) Since the electronegativities of carbon and oxygen are 2.5 and 3.5, respectively (Figure 10.2), CO_2 has polar bonds. The geometry of CO_2 is linear.

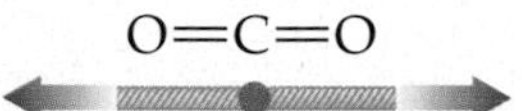

Consequently, the polar bonds cancel; the molecule is not polar and does not have dipole–dipole forces.

b) The electronegativities of C, H, and Cl are 2.5, 2.1, and 3.5, respectively. Consequently, CH_2Cl_2 has two polar bonds (C—Cl) and two bonds that are nearly nonpolar (C—H). The geometry of CH_2Cl_2 is tetrahedral.

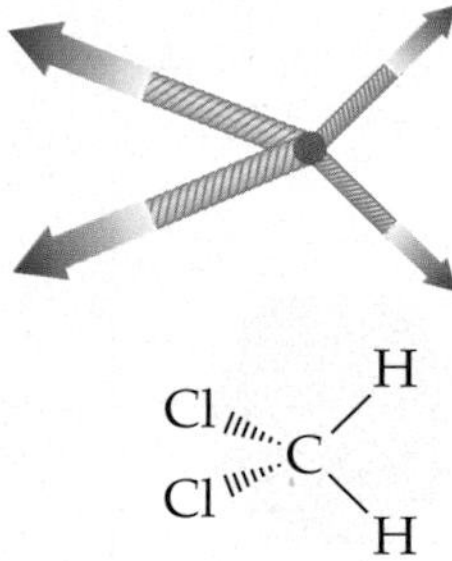

Since the the C—Cl bonds and the C—H bonds are different, they do not cancel, but sum to a net dipole moment. Therefore the molecule is polar and has dipole–dipole forces.

c) Since the electronegativities of C and H are 2.5 and 2.1, respectively, the C—H bonds are nearly nonpolar. In addition, since the geometry of the molecule is tetrahedral, any slight polarities that the bonds might have will cancel.

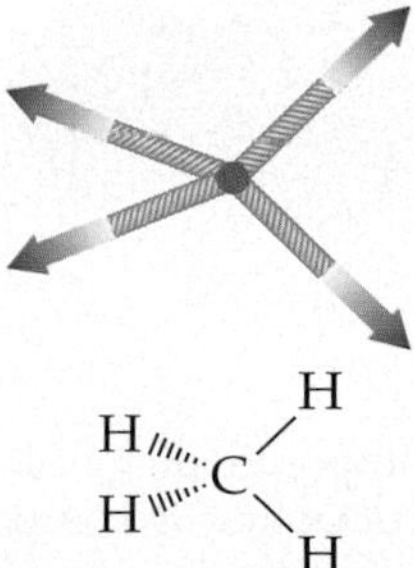

CH_4 is therefore nonpolar and does not have dipole–dipole forces.

SKILLBUILDER 12.4 **Dipole–Dipole Forces**

Which of the following molecules have dipole–dipole forces?

a) CI_4
b) CH_3Cl
c) HCl

CHEMISTRY AND HEALTH

Hydrogen Bonding in DNA

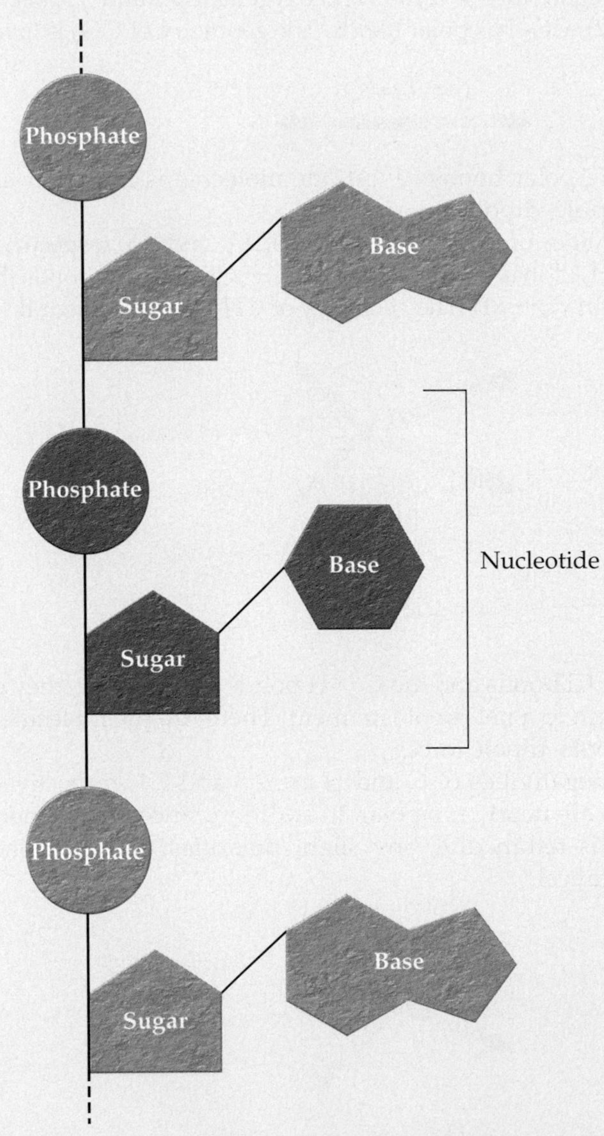

Figure 12.21 DNA is composed of repeating units called nucleotides. Each nucleotide is composed of a sugar, a phosphate, and a base.

DNA is a long chain-like molecule that acts as a blueprint for living organisms. Copies of DNA are passed from parent to offspring, which is why we inherit traits from our parents. A DNA molecule is composed of thousands of repeating units called *nucleotides* (Figure 12.21). Each nucleotide contains one of four different bases: adenine, thymine, cytosine, and guanine (abbreviated *A, T, C,* and *G*). The order of these bases along DNA contains the code that specifies how proteins—the workhorse molecules in living organisms—are made. Proteins determine many human characteristics including how we look, how we fight infections, and even how we behave. Consequently, human DNA is a blueprint for how humans are made.

Each human cell actually contains two complete and complementary copies of DNA, both necessary for replication. The replicating mechanism is related to the structure of DNA, discovered in 1953 by James Watson and Francis Crick. DNA consists of two complementary strands wrapped around each other in the now famous double helix. Each strand is held to the other by hydrogen bonds that occur between the bases on each strand. DNA replicates because each base (A, T, C, and G) has a complementary partner with which it hydrogen bonds (Figure 12.22). Adenine (A) hydrogen bonds with thymine (T) and cytosine (C) hydrogen bonds with guanine (G). The hydrogen bonds are so specific that each base will pair only with its complementary partner. When a cell is going to divide, the DNA unzips across the hydrogen bonds that run along its length. Then new bases, complementary to the bases in each half, add along each of the halves, forming hydrogen bonds with their complement. The result is two identical copies of the original DNA (See Chapter 19).

CAN YOU ANSWER THIS? *Why would dispersion forces not work as a way to hold the two halves of DNA together? Why would covalent bonds not work?*

Hydrogen Bonding

Polar molecules containing hydrogen atoms bonded directly to fluorine, oxygen, or nitrogen have an additional intermolecular force called a **hydrogen bond.** HF, NH_3 and H_2O, for example, all undergo hydrogen bonding. The hydrogen bond is a sort of *super* dipole–dipole force. The large electronegativity difference between hydrogen and these electronegative

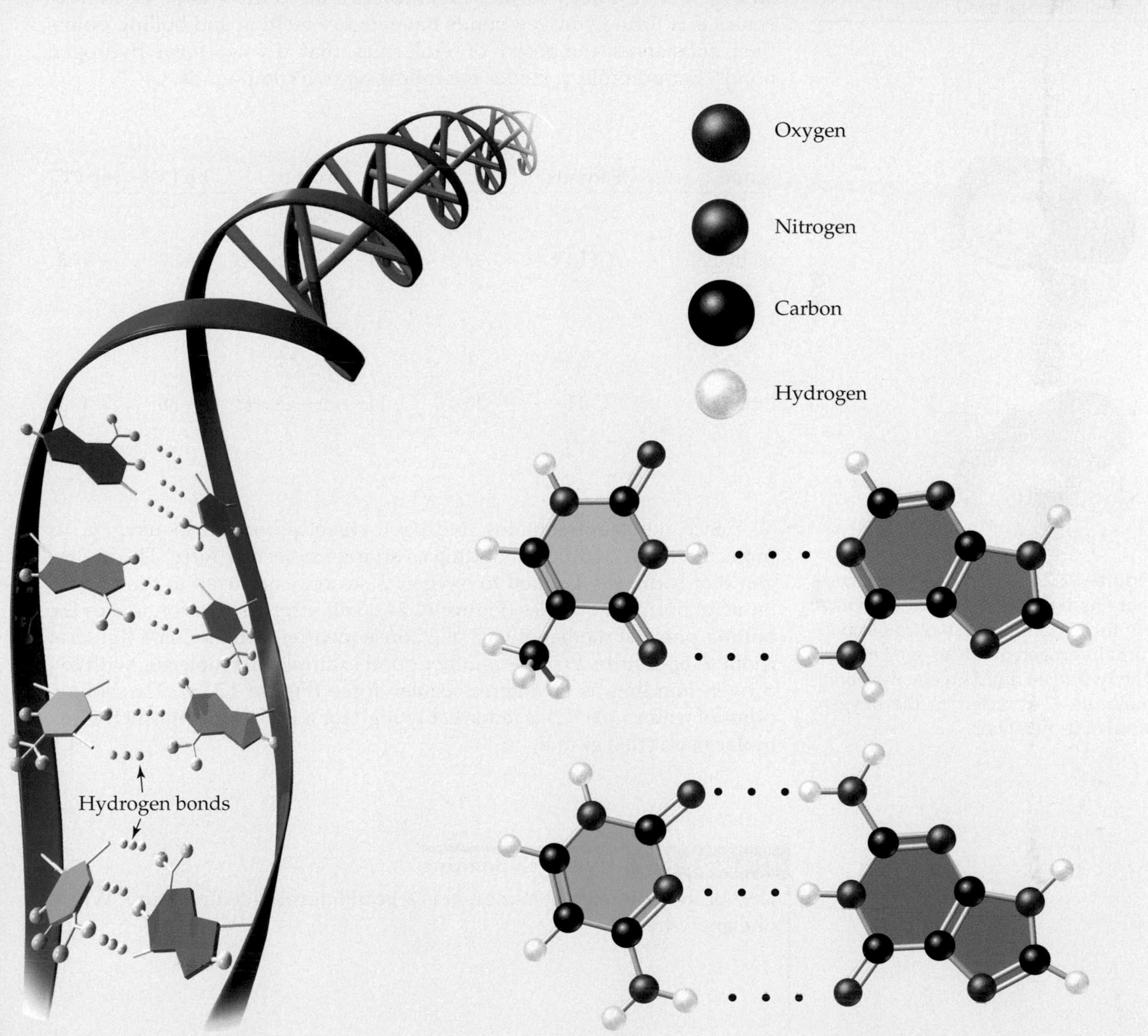

Figure 12.22 The two halves of the DNA double helix are held together by hydrogen bonds.

H—F ······ H—F ······ H—F

Figure 12.23 In HF the hydrogen on one molecule is strongly attracted to the fluorine on its neighbors. This attraction is called a *hydrogen bond*.

The strength of a hydrogen bond is only 2 to 5% of the strength of a typical covalent chemical bond.

elements, as well as the small size of these atoms, allows for a strong attraction between the hydrogen in each of these molecules and the F, O, or N on its neighbors. This attraction is the hydrogen bond. For example, in HF the hydrogen is strongly attracted to the fluorine on neighboring molecules (Figure 12.23).

Hydrogen bonds should not be confused with chemical bonds. Chemical bonds occur between individual atoms within a molecule and are generally much stronger than hydrogen bonds. Hydrogen bonds—like dispersion forces and dipole–dipole forces—are intermolecular forces that occur between molecules. Hydrogen bonds are, however, the strongest of the three intermolecular forces. Substances composed of molecules that form hydrogen bonds have higher melting and boiling points than substances composed of molecules that do not form hydrogen bonds. For example, consider the following two compounds.

Name	Formula	Molar mass (g/mol)	Structure	bp (°C)	mp (°C)
Methanol	CH_4O	32.0	$H_3C{-}O{-}H$ (H—C—O—H with two H on C)	64.7	−97.8
Ethane	C_2H_6	30.0	$H_3C{-}CH_3$ (H—C—C—H with two H on each C)	−88	−172

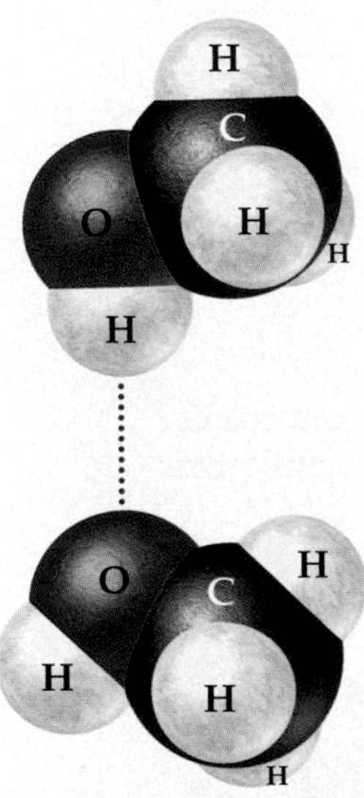

Figure 12.24 Since methanol contains hydrogen atoms directly bonded to oxygen, methanol molecules form hydrogen bonds to one another. The hydrogen atom on one methanol molecule is attracted to the oxygen atom of its neighbor.

Since methanol contains hydrogen directly bonded to oxygen, its molecules have hydrogen bonding as an intermolecular force. The hydrogen that is directly bonded to oxygen is strongly attracted to the oxygen on neighboring molecules (Figure 12.24). This strong attraction makes the boiling point of methanol 64.7 °C. Consequently, methanol is a liquid at room temperature. Water is another good example of a molecule with hydrogen bonding as an intermolecular force (Figure 12.25). The boiling point of water (100 °C) is remarkably high for a molecule with such a low molar mass (18.0 g/mol)

EXAMPLE 12.5 Hydrogen Bonding

One of the following compounds is a liquid at room temperature. Which one and why?

Formaldehyde	Fluoromethane	Hydrogen peroxide
H—C(=O)—H	H—C—F (with H above and below C)	H—O—O—H

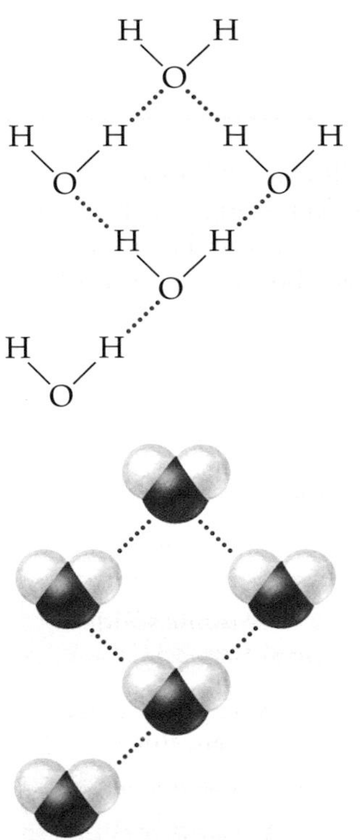

Figure 12.25 Water molecules form strong hydrogen bonds with one another.

Solution:
The three compounds have similar molar masses.

formaldehyde	30.03 g/mole
fluoromethane	34.03 g/mole
hydrogen peroxide	34.02 g/mole

Therefore, the strengths of their dispersion forces are similar. All three compounds are also polar, so they have dipole–dipole forces. Hydrogen peroxide, however, is the only compound to also contain H bonded directly to F, O, or N. Therefore it also has hydrogen bonding and is most likely to have the highest boiling point of the three. Since the example stated that only one of the compounds was a liquid, we can safely assume that hydrogen peroxide is the liquid. Note that, although fluoromethane *contains* both H and F, H is not *directly bonded* to F, so fluoromethane does not have hydrogen bonding as an intermolecular force. Similarly, although formaldehyde *contains* both H and O, H is not *directly bonded* to O, so formaldehyde does not have hydrogen bonding either.

SKILLBUILDER 12.5 **Hydrogen Bonding**

Which has the higher boiling point, HF or HCl? Why?

The different types of intermolecular forces are summarized in Table 12.5. Remember that dispersion forces, the weakest kind of intermolecular force, are present in all molecules and atoms and increase with increasing molar mass. These forces are always weak in small molecules, but they become substantial in molecules with high molar masses. Dipole–dipole forces are present in polar molecules. Hydrogen bonds, the strongest kind of intermolecular force, are present in molecules containing hydrogen bonded directly to fluorine, oxygen, or nitrogen.

TABLE 12.5

Types of Intermolecular Forces

Type Of Force	*Relative Strength*	*Present in What Kind of Molecules*	*Example*
dispersion force (or London force)	weak but increases with increasing molar mass	present in all atoms and molecules	H_2
dipole–dipole force	moderate	present only in polar molecules	HCl
hydrogen bond	strong	present in molecules containing H bonded directly to F, O, or N.	HF

13

Solutions

"I have no doubt that in reality the future will be vastly more surprising than anything I can imagine. Now my own suspicion is that the universe is not only queerer than we suppose, but queerer than we can suppose."

John Burdon Sanderson Haldane

13.1 Tragedy in Cameroon

Most people living near Lake Nyos in Cameroon, West Africa, had an ordinary day on August 22, 1986. Unfortunately, the day ended in tragedy. On that evening, a large cloud of carbon dioxide gas, burped up from the depths of Lake Nyos, killed over 1700 people and about 3000 head of cattle. Survivors tell of smelling rotten eggs, feeling a warm sensation, and then losing consciousness. Two years before that, a similar tragedy had occurred in Lake Monoun, just 60 miles away, killing 37 people. Today, scientists are taking steps to prevent these lakes, both of which are in danger of burping again, from accumulating the carbon dioxide that caused the disaster.

The concentration is the amount of carbon dioxide in a given amount of water.

◀ Carbon dioxide bubbled out of Lake Nyos and flowed into the adjacent valley. The carbon dioxide came from the bottom of the Lake Nyos, where it was held in solution by the pressure of the water above it. When the layers in the lake were disturbed, the carbon dioxide came out of solution due to the decrease in pressure.

Lake Nyos is a water-filled volcanic crater. Some 50 miles beneath the surface of the lake, molten volcanic rock (magma) produces carbon dioxide gas that seeps into the lake through the volcano's plumbing system. The carbon dioxide mixes with the lake water, and the high pressure at the bottom of the deep lake allows the mixture to become highly concentrated in carbon dioxide (just as the pressure in a beer can allows beer to be highly concentrated in carbon dioxide). Over time, the carbon dioxide and water mixture at the bottom of the lake became so concentrated that—either because of the high concentration itself or because of some other natural trigger such as a landslide—some gaseous carbon dioxide

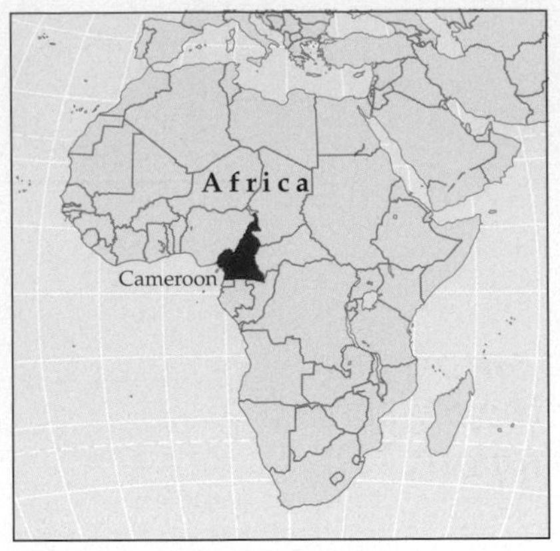

Cameroon is in West Africa.

escaped. The rising bubbles disrupted the stratified layers of lake water, causing the highly concentrated carbon dioxide and water mixture at the bottom of the lake to rise, lowering the pressure on it. The drop in pressure on the mixture released more carbon dioxide bubbles just as the drop in pressure upon opening a beer can releases carbon dioxide bubbles. This in turn caused more churning and more carbon dioxide release. Since carbon dioxide is heavier than air, it traveled down the sides of the volcano and into the nearby valley, displacing air and asphyxiating many of the local residents.

In efforts to prevent the tragedy from occurring again—by the year 2001 carbon dioxide concentrations had already returned to dangerously high levels—scientists are building a piping system that slowly vents carbon dioxide from the lake bottom. This system slowly releases the carbon dioxide into the atmosphere, preventing a repeat of the tragedy.

13.2 Solutions: Homogenous Mixtures

A solution is a homogeneous mixture of two or more substances.

The carbon dioxide and water mixture at the bottom of the Lake Nyos is an example of a **solution,** a homogenous mixture of two or more substances. A solution may be composed of a gas and a liquid (such as carbon dioxide and water), but it may also be composed of a solid and a liquid, a liquid and another liquid, a solid and a gas, and other combinations (see Table 13.1).

Aqueous comes from the latin *aqua*, meaning water.

The most common solutions, however, are those containing a solid, a liquid, or a gas and water. These are called aqueous solutions, and they are the main focus of this chapter. For example, sugar or salt readily dissolves in water to form solutions of solids and water. Similarly, ethyl alcohol—the alcohol in alcoholic beverages—readily mixes with water to form a solution of a liquid with water, and we have already seen an example of a gas and water solution in Lake Nyos.

Polar and nonpolar are defined in section 10.8.

A solution has at least two components. The majority component is called the **solvent,** and the minority component is called the **solute.** In our carbon dioxide-and-water solution, carbon dioxide was the solute and water was the solvent. In a salt-and-water solution, salt is the solute and water is the solvent. Because water is so abundant on earth, it is a common solvent. However, other solvents are often used in the laboratory and even in the home, especially to form solutions with nonpolar solutes. For

TABLE 13.1

Common Types of Solutions

Solution Phase	*Solute Phase*	*Solvent Phase*	*Example*
gaseous solutions	gas	gas	air (mainly oxygen and nitrogen)
	liquid	gas	humid air (water and air)
liquid solutions	gas	liquid	soda water (CO_2 and water)
	liquid	liquid	vodka (ethanol and water)
	solid	liquid	seawater (salt and water)
solid solutions	solid	solid	brass (copper and zinc) and other alloys

example, you may use paint thinner, a nonpolar solvent, to remove grease from a dirty bicycle chain or from ball bearings. The paint thinner dissolves (or forms a solution with) the grease, removing it from the metal. In general, polar solvents dissolve polar or ionic solutes and nonpolar solvents dissolve nonpolar solutes. This trend is described in the rule *like dissolves like.* Similar kinds of solvents dissolve similar kinds of solutes. Table 13.2 lists some common polar and nonpolar laboratory solvents.

Like dissolves like—polar solvents dissolve polar solutes and nonpolar solvents dissolve nonpolar solutes.

TABLE 13.2

Common Laboratory Solvents

Common Polar Solvents	*Common Nonpolar Solvents*
water (H_2O)	hexane (C_6H_6)
acetone (CH_3COCH_3)	ethyl ether ($CH_3CH_2OCH_2CH_3$)
methyl alcohol (CH_3OH)	toluene (C_7H_8)

13.3 Solutions of Solids Dissolved in Water: How to Make Rock Candy

When most people think of a solution, they think of a solid dissolved in water. The ocean, for example, is a solution of salt and other solids dissolved in water. A sweetened cup of coffee is a solution of sugar and other solids dissolved in water. Our blood is a solution of several solids (and some gases) dissolved in water. Not all solids, however, dissolve in water. We already know that nonpolar solids—such as lard or shortening, for example—do not dissolve in water. However, solids such as calcium carbonate or sand do not dissolve either.

When a solid is put into water, there is a competition between the attractive forces that hold the solid together (the solute–solute interactions) and the attractive forces occurring between the water molecules and the particles that compose the solid (the solvent–solute interactions). For example, when sodium chloride is put into water, there is a competition between the attractions of Na^+ cations to Cl^- anions and the attraction of water molecules to Na^+ and Cl^- (Figure 13.1). In the case of NaCl, the attraction to water wins, and sodium chloride dissolves (Figure 13.2). In the case of calcium carbonate ($CaCO_3$) the attractions between Ca^{2+} ions and CO_3^{2-} ions win and calcium carbonate does not dissolve in water.

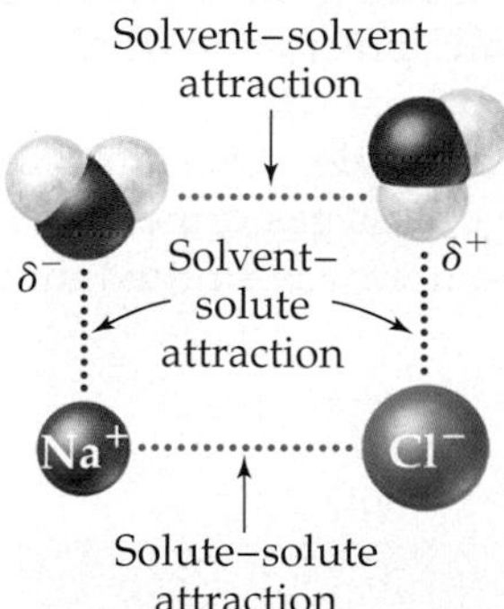

Figure 13.1 When NaCl is put into water, the attraction between water molecules and Na^+ and Cl^- ions (solvent–solute attraction) overcomes the attraction between Na^+ and Cl^- (solute–solute attraction).

Solubility and Saturation

The **solubility** of a compound is defined as the amount of the compound, usually in grams, that will dissolve in a certain amount of liquid. For example, the solubility of sodium chloride at 25 °C is 36 g NaCl per 100 g water, while the solubility of calcium carbonate is close to zero. A solution that has 36 g of NaCl per 100 g water is called a saturated sodium chloride solution. A **saturated solution** is one that holds the maximum amount of solute under the solution conditions (as described by the solubility of the solute). If additional solute is added to a saturated solution, it will not dissolve. An **unsaturated solution** is one holding less than the maximum

Figure 13.2 In an NaCl solution, the Na^+ ions and the Cl^- ions are dispersed in the water.

amount of solute. If additional solute is added to an unsaturated solution, it will dissolve. A **supersaturated solution** is one holding more than the maximum amount of solute. The solute will normally precipitate from (or come out of) a supersaturated solution. As the carbon dioxide and water solution rose from the bottom of Lake Nyos, for example, it became supersaturated because of the drop in pressure. The excess gas came out of the solution and rose to the surface of the lake where it was emitted into the surrounding air.

Supersaturated solutions can form under special circumstances, such as the sudden release in pressure that occurs in a soda can when it is opened.

A supersaturated solution is holding more than the maximum amount of solute. In some cases, such as the sodium acetate solution pictured here, a supersaturated solution may be temporarily stable. Any disturbance however, such as dropping in a small piece of solid sodium acetate **(a)** will cause the solid to come out of solution **(b)** and **(c)**.

Table 7.2 lists the solubility rules for ionic compounds.

In Chapter 7 we learned the solubility rules, which give us a qualitative description of the solubility of ionic solids. Molecular solids may also be soluble in water depending on whether or not the solid is polar. Table sugar ($C_{12}H_{22}O_{11}$), for example, is polar and soluble in water. Nonpolar solids, such as lard or vegetable shortening, are usually insoluble in water.

Electrolyte Solutions: Dissolved Ionic Solids

There is an important difference, however, between a sugar solution (containing a molecular solid) and a salt solution (containing an ionic solid) (Figure 13.3). In a salt solution the dissolved particles are ions, while in a sugar solution the dissolved particles are molecules. The ions in the salt solution are mobile charged particles and can therefore conduct electricity. As we learned in section 7.6, a solution containing a solute that dissociates into ions is called a **strong electrolyte solution.** The sugar solution contains dissolved sugar molecules and cannot conduct electricity; it is called a **nonelectrolyte solution.** In general, soluble ionic solids form strong electrolyte solutions, while soluble molecular solids form nonelectrolyte solutions.

How Solubility Varies with Temperature

The solubility of solids in water can be highly dependent on temperature. Have you ever noticed the difference between dissolving sugar into hot tea versus cold tea? It is much easier to dissolve the sugar into the *hot* tea. In general, the solubility of *solids* in water increases with increasing temperature (Figure 13.4). For example, the solubility of potassium nitrate (KNO_3) at room temperature is about 37 g KNO_3 per 100 g of water. However, at 50 °C, the solubility rises to 88 g KNO_3 per 100 g of water. A common way to purify a solid is a technique called **recrystallization.** In this technique, the solid is put into water (or some other solvent) at an elevated temperature. Enough solid is added to the solvent to create a

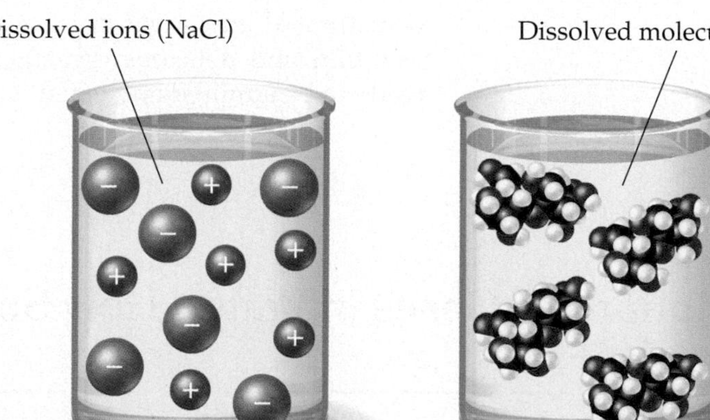

Figure 13.3 Electrolyte solutions contain dissolved ions (charged particles) and therefore conduct electricity. Nonelectrolyte solutions contain dissolved molecules (neutral particles) and do not conduct electricity.

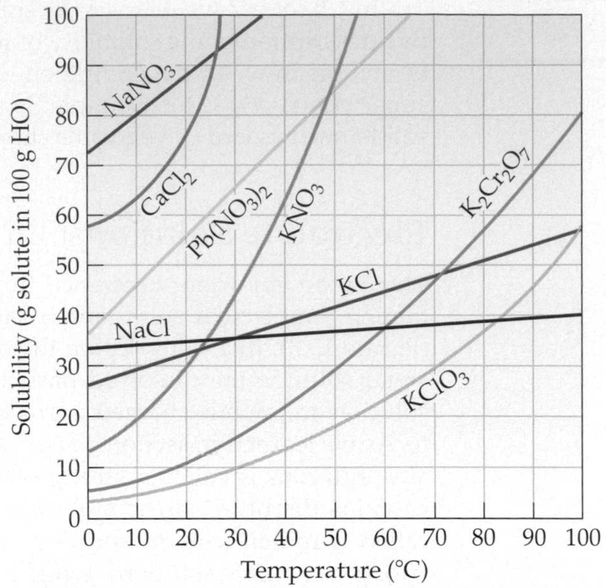

Figure 13.4 The solubility of several ionic solids as a function of temperature.

Rock candy is composed of sugar crystals that have been grown through recrystallization.

saturated solution at the elevated temperature. As the solution cools, it becomes supersaturated, and the excess solid begins to come out of solution. If the solution cools slowly, the solid will form crystals as it comes out of solution. The crystalline structure tends to reject impurities, resulting in a purer solid.

Rock Candy

A similar effect can be seen if you make rock candy. To make rock candy, a saturated sucrose (table sugar) solution is prepared at an elevated temperature. A string is left to dangle in the solution, and the solution is allowed to cool and stand for several days. As the solution cools, it becomes supersaturated and sugar crystals grow on the string. After several days, beautiful and delicious crystals, or "rocks," of sugar cover the string, ready to be admired and eaten.

13.4 Solutions of Gases in Water: How Soda Pop Gets Its Fizz

The water at the bottom of Lake Nyos and a can of soda pop are both examples of a gas dissolved in a liquid. They are solutions of carbon dioxide and water. Most liquids exposed to air contain some dissolved gases. Fish, for example, breathe dissolved oxygen in lake or sea water. Our blood contains dissolved nitrogen, oxygen, and carbon dioxide. Even tap water contains dissolved nitrogen and oxygen.

You can see the dissolved gases in ordinary tap water by heating it on a stove. Before the water reaches its boiling point, you can see small bubbles develop in the water. These bubbles are dissolved air (mostly nitrogen and oxygen) coming out of solution. Once the water boils, the bubbling becomes more vigorous—these larger bubbles are composed of water vapor. The dissolved air comes out of solution because—unlike solids, whose solubility *increases* with increasing temperature—the solubility of gases in water *decreases* with increasing temperature. As the temperature of the water rises, the solubility of the dissolved nitrogen and oxygen decreases and these gases come out of solution, forming small bubbles around the bottom of the pot. The decrease in the solubility of gases with increasing temperature is the reason that warm soda pop bubbles more than cold soda pop and also the reason that warm beer goes flat faster than cold beer. The carbon dioxide comes out of solution faster (bubbles more) at room temperature than at lower temperature because it is less soluble at room temperature.

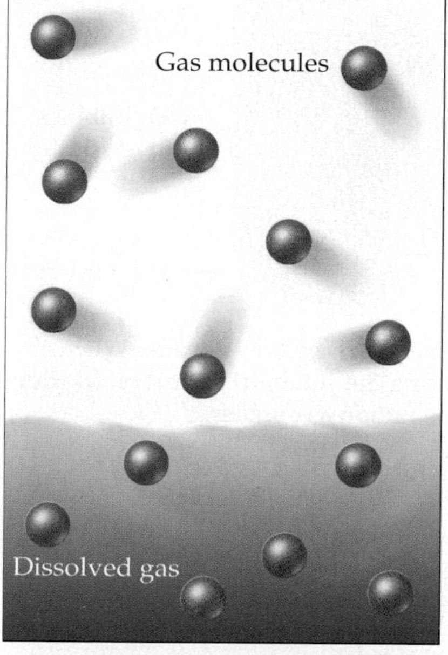

Gas at low pressure over a liquid

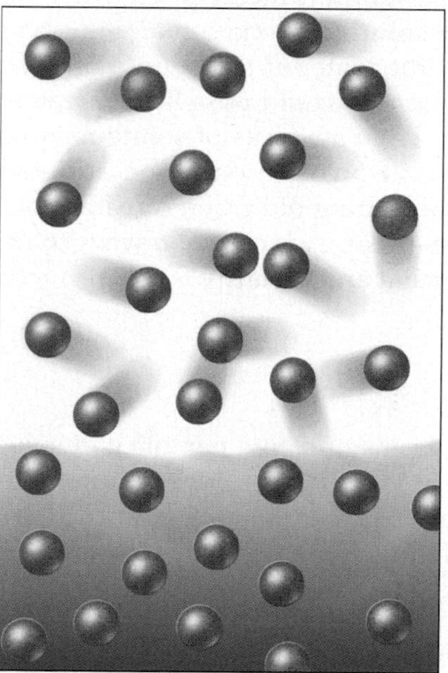

Gas at high pressure over a liquid

Figure 13.5 The higher the pressure above a liquid, the more soluble the gas is in the liquid.

Warm soda pop fizzes more than cold soda pop because the solubility of the dissolved carbon dioxide decreases with increasing temperature.

The solubility of gases also depends on pressure. The higher the pressure above a liquid, the more soluble the gas is in the liquid (Figure 13.5). In a can of soda pop and in Lake Nyos, carbon dioxide is maintained in solution by high pressure. In soda pop, the pressure is provided by a high amount of carbon dioxide gas that is pumped into the can before sealing it. When the can is opened, the pressure is released and the solubility of carbon dioxide decreases, resulting in bubbling (Figure 13.6). In Lake Nyos, the pressure is provided by the mass of the lake water itself pushing down on the carbon-dioxide-rich water at the bottom of the lake. When the stratification (or layering) of the lake is disturbed, the pressure on the carbon dioxide solution is lowered and the solubility of carbon dioxide decreases, resulting in the release of excess carbon dioxide gas.

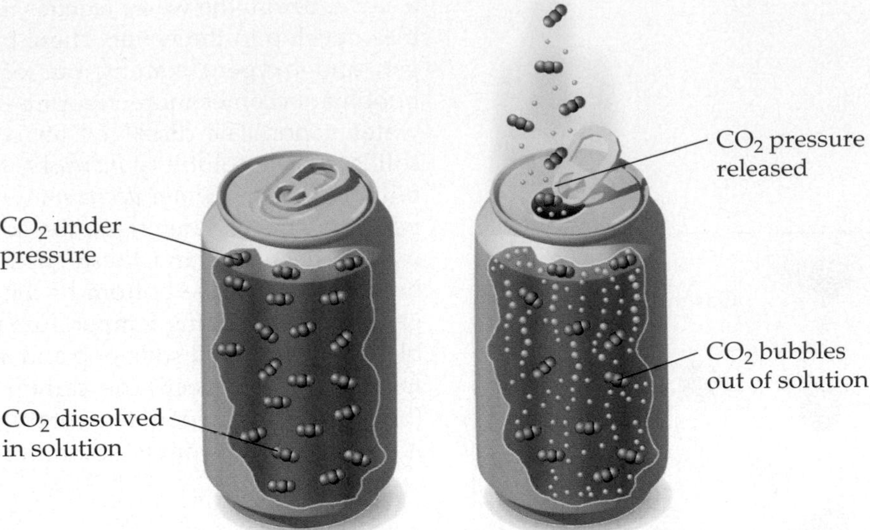

Figure 13.6 A can of soda pop is pressurized with carbon dioxide. When the can is opened the pressure is released, lowering the solubility of carbon dioxide in the solution and causing it to come out of solution as bubbles.

13.5 Specifying Solution Concentration: Mass Percent

As we have seen, the amount of solute in a solution is an important property of the solution. For example, the amount of carbon dioxide in the water at the bottom of Lake Nyos is an important predictor of when the deadly event may repeat itself. A **dilute solution** is one containing small amounts of solute relative to solvent. If the water at the bottom of Lake Nyos were a dilute carbon dioxide solution, it would pose little threat. A **concentrated solution** is one containing large amounts of solute relative to solvent. If the water at the bottom of Lake Nyos becomes concentrated in carbon dioxide (through the continual feeding of carbon dioxide from magma into the lake), it becomes a large threat. Two common ways to report solution concentration are mass percent and molarity.

Mass Percent

Mass percent is simply the number of grams of solute per 100 g of solution. So a solution with a concentration of 14% by mass, for example, contains 14 g of solute per 100 g of solution.

Also in common use are *parts per million (ppm)*, the number of grams of solute per 1 million g of solution, and *parts per billion (ppb)*, the number of grams of solute per 1 billion g of solution.

To calculate mass percent, simply divide the mass of the solute by the mass of the solution (solute *and* solvent) and multiply by 100%.

$$\text{Mass percent} = \frac{\text{Mass solute}}{\text{Mass solution}} \times 100\%$$

Note that the denominator is the mass of *solution*, not the mass of solvent.

For example, suppose you wanted to calculate the mass percent of NaCl in a solution containing 15.3 g of NaCl and 155.0 g of water. Set up the problem as you normally would.

Given: 15.3 g NaCl
155.0 g H_2O

Find: mass percent

Equation:

This problem requires the use of the equation that defines mass percent.

$$\text{Mass percent} = \frac{\text{Mass solute}}{\text{Mass solution}} \times 100\%$$

Solution:
The mass of solution is simply the mass of NaCl plus the mass of H_2O.

$$\begin{aligned}\text{Mass solution} &= \text{Mass NaCl} + \text{Mass } H_2O \\ &= 15.3\text{ g} + 155.0\text{ g} \\ &= 170.3\text{ g}\end{aligned}$$

We then substitute the correct quantities into the equation.

$$\begin{aligned}\text{Mass percent} &= \frac{\text{Mass solute}}{\text{Mass solution}} \times 100\% \\ &= \frac{15.3\text{ g}}{170.3\text{ g}} \times 100\% \\ &= 8.98\%\end{aligned}$$

The solution is 8.98% NaCl by mass.

EXAMPLE 13.1 Calculating Mass Percent

Calculate the mass percent of a solution containing 27.5 g of ethanol (C_2H_6O) and 175 mL of H_2O. Assume that the density of water is 1.00 g/mL.
We begin by setting up the problem.

Given: 27.5 g C_2H_6O
175 mL H_2O

Find: mass percent

Equation:

$$\text{Mass percent} = \frac{\text{Mass solute}}{\text{Mass solution}} \times 100\%$$

Conversion Factor:

This problem requires converting mL to g using the density of water as the conversion factor.

$$d(H_2O) = \frac{1.00\text{ g}}{\text{mL}}$$

Solution:
To find the mass percent, we simply substitute into the equation for mass percent. We need the mass of the solution, which is simply the mass of ethanol plus the mass of water. The mass of water is obtained from the volume of water by using the density as a conversion factor.

As an example of using mass percent as a conversion factor, consider a water sample from the bottom of Lake Nyos containing 8.5% carbon dioxide by mass. How much carbon dioxide (in grams) is contained in 28.5 L of the water solution? (Assume that the density of the solution is 1.03 g/mL.) We set up the problem in the standard way.

Given: 8.5% CO_2 by mass
28.6 L solution

Find: g CO_2

Conversion Factors:

The main conversion factor is the mass percent of the solution. We write it as g CO_2 per 100 g solution.

$$\frac{8.5 \text{ g } CO_2}{100 \text{ g solution}}$$

We will also need the density of the solution and the conversion factor between L and mL.

$$\frac{1.03 \text{ g}}{\text{mL}}$$

$$1000 \text{ mL} = 1 \text{ L}$$

Solution Map:

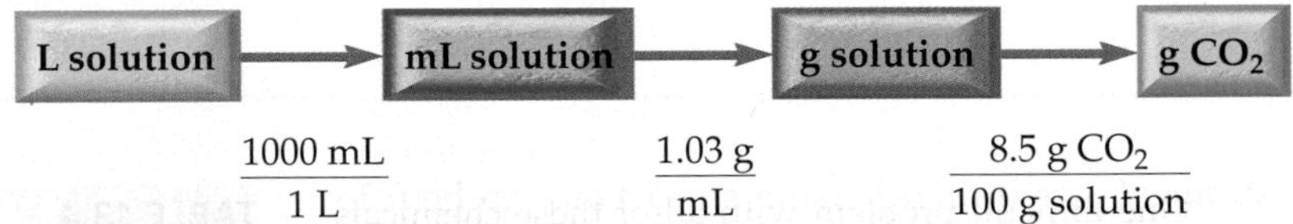

The solution map begins with L solution and shows the conversion to mL solution and then to g solution using the density. The map then proceeds from g solution to g CO_2, using the mass percent (expressed as a fraction) as a conversion factor.

Solution:

Finally, we follow the solution map to compute the answer.

$$28.6 \text{ } \cancel{\text{L solution}} \times \frac{1000 \text{ } \cancel{\text{mL}}}{\cancel{\text{L}}} \times \frac{1.03 \text{ g}}{\cancel{\text{mL}}} \times \frac{8.5 \text{ g } CO_2}{100 \text{ } \cancel{\text{g solution}}} = 2.5 \times 10^3 \text{ g } CO_2$$

In this example, we used mass percent to convert from a given amount of *solution* to the amount of *solute* present in the solution. In Example 13.2, we use mass percent to convert from a given amount of *solute* to the amount of *solution* containing that solute.

EXAMPLE 13.2 Using Mass Percent in Calculations

A soft drink contains 11.5% sucrose ($C_{12}H_{22}O_{11}$) by mass. What volume of the soft drink solution (in mL) contains 85.2 g of sucrose? (Assume a density of 1.00 g/mL).

Given: 11.5% $C_{12}H_{22}O_{11}$ by mass
85.2 g $C_{12}H_{22}O_{11}$

Find: mL solution (soft drink)

Conversion Factors:

$$\frac{11.5 \text{ g } C_{12}H_{22}O_{11}}{100 \text{ g solution}}$$

$$d = \frac{1.00 \text{ g}}{\text{mL}}$$

Solution Map:

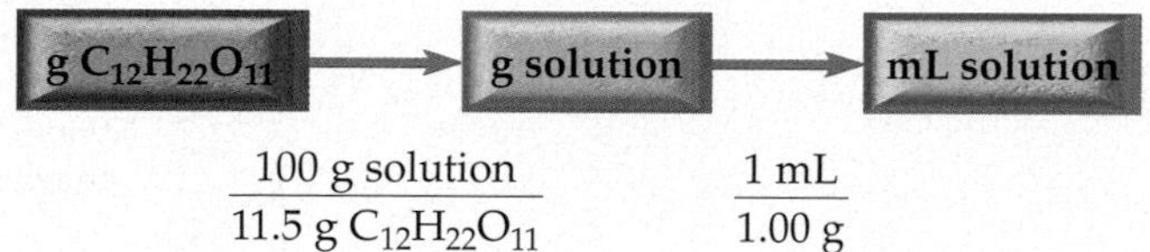

$$\frac{100 \text{ g solution}}{11.5 \text{ g } C_{12}H_{22}O_{11}} \qquad \frac{1 \text{ mL}}{1.00 \text{ g}}$$

In this case, we convert from g solute ($C_{12}H_{22}O_{11}$) to g solution using the mass percent in fractional form as the conversion factor. We then convert to mL using the density.

Solution:

$$85.2 \text{ g } C_{12}H_{22}O_{11} \times \frac{100 \text{ g solution}}{11.5 \text{ g } C_{12}H_{22}O_{11}} \times \frac{1 \text{ mL solution}}{1.00 \text{ g}} = 741 \text{ mL solution}$$

SKILLBUILDER 13.2 **Using Mass Percent in Calculations**

How much sucrose ($C_{12}H_{22}O_{11}$), in grams, is contained in 355 mL (12 oz) of the soft drink in Example 13.2?

13.6 Specifying Solution Concentration: Molarity

A second way to express solution concentration is **molarity (M)**, moles of solute divided by liters of solution.

$$\text{Molarity (M)} = \frac{\text{Moles solute}}{\text{Liters solution}}$$

Note that molarity is moles of solute per liter of *solution*, not per liter of solvent. To make a solution of a specified molarity, you usually put the solute into a flask and then add water to the desired volume of solution. For example, to make 1.0 L of a 1 M NaCl solution, you add 1 mol of NaCl to a flask and then add water to make 1 L of solution {Figure 13.7}. You *do not* combine 1 mol of NaCl with 1 L of water because that would result in a total volume exceeding 1 L and therefore a molarity of less than 1 M. Molarity is moles of solute per liter of *solution*.

To calculate molarity, simply divide the moles of the solute by the volume of the solution (solute *and* solvent) in liters. For example, calculate the molarity of a sucrose ($C_{12}H_{22}O_{11}$) solution made with 1.58 mol of sucrose diluted to a total volume of 5.0 L of solution. Set up the problem in the standard way.

How to prepare a 1 molar NaCl solution.
1 mole NaCl
(58.44 g)
Water
Add water until
solid is dissolved.
Then add additional
water until the 1 liter
mark is reached.
Mix
First add 1 mole of NaCl.
A 1 molar NaCl solution

Figure 13.7 To make 1.0 L of a 1.0 M NaCl solution, you add 1.0 mol (58.44 g) of sodium chloride to a flask and then dilute to 1.0 L of total volume. **Question:** What would happen if you added 1 L of water to 1 mol of sodium chloride? Would the resulting solution be 1 M?

Given: 1.58 mol $C_{12}H_{22}O_{11}$
5.0 L solution

Find: molarity (M)

Equation:

This problem requires the use of the equation that defines molarity.

$$\text{Molarity (M)} = \frac{\text{Moles solute}}{\text{Liters solution}}$$

Solution:
Simply substitute the correct values into the equation and compute the answer.

$$\begin{aligned}\text{Molarity (M)} &= \frac{\text{Moles solute}}{\text{Liters solution}}\\ &= \frac{1.58 \text{ mol } C_{12}H_{22}O_{11}}{5.0 \text{ L solution}}\\ &= 0.32 \text{ M}\end{aligned}$$

EXAMPLE 13.3 Calculating Molarity

Calculate the molarity of a solution made by adding 15.5 g NaCl into a beaker and adding water to make 1.50 L of NaCl solution.
Set up the problem in the standard way.

Given: 15.5 g NaCl
1.50 L solution

Find: molarity (M)

Equation and Conversion Factor:

$$\text{Molarity (M)} = \frac{\text{Moles solute}}{\text{Liters solution}} \qquad \text{Molar mass of NaCl} = \frac{58.44\text{ g}}{1\text{ mol}}$$

Solution:

To calculate molarity, we simply substitute the correct values into the equation and compute the answer. However, we must first convert the amount of NaCl from g to mol using the molar mass of NaCl (58.44 g/mol).

$$\text{Mol NaCl} = 15.5\ \cancel{\text{g NaCl}} \times \frac{1\text{ mol NaCl}}{58.44\ \cancel{\text{g NaCl}}} = 0.26\underline{5}2\text{ mol NaCl}$$

$$\begin{aligned}\text{Molarity (M)} &= \frac{\text{Moles solute}}{\text{Liters solution}} \\ &= \frac{0.26\underline{5}2\text{ mol NaCl}}{1.50\text{ L solution}} \\ &= 0.177\text{ M}\end{aligned}$$

SKILLBUILDER 13.3 **Calculating Molarity**

Calculate the molarity of a solution made by adding 55.8 g of $NaNO_3$ to a beaker and diluting to 2.50 L.

Using Molarity in Calculations

The molarity of a solution can be used as a conversion factor between moles of the solute and liters of the solution. For example, a 0.500 M NaCl solution contains 0.500 mol NaCl for every liter of solution.

$$\frac{0.500\text{ mol NaCl}}{\text{Liters solution}} \qquad \text{converts L solution} \rightarrow \text{mol NaCl}$$

This conversion factor converts from L solution to mol NaCl. If you want to go the other way, simply invert the conversion factor.

$$\frac{\text{L solution}}{0.500\text{ mol NaCl}} \qquad \text{converts mol NaCl} \rightarrow \text{L solution}$$

For example, how many grams of sucrose ($C_{12}H_{22}O_{11}$) are contained in 1.72 L of 0.758 M sucrose solution? We set up the problem in the standard way.

Given: 0.758 M $C_{12}H_{22}O_{11}$
1.72 L solution

Find: g $C_{12}H_{22}O_{11}$

Conversion Factors:

The main conversion factor is the molarity of the solution. We write it as mol $C_{12}H_{22}O_{11}$ per L solution.

$$\frac{0.758 \text{ mol } C_{12}H_{22}O_{11}}{\text{L solution}}$$

We will also need the molar mass of sucrose.

$$\text{Molar mass of } C_{12}H_{22}O_{11} = \frac{342.34 \text{ g}}{\text{mol}}$$

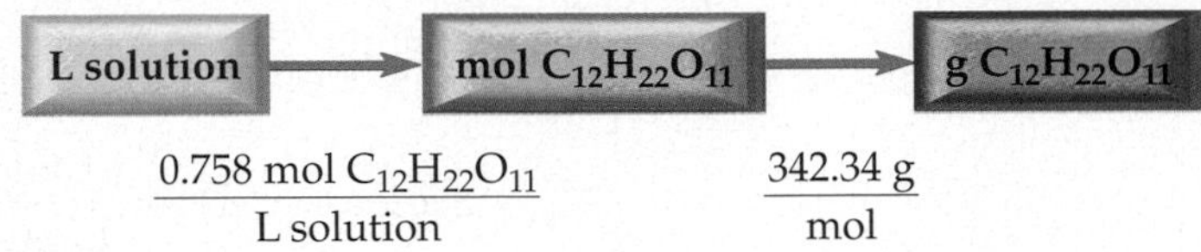

$$\frac{0.758 \text{ mol } C_{12}H_{22}O_{11}}{\text{L solution}} \qquad \frac{342.34 \text{ g}}{\text{mol}}$$

Solution Map:

The solution map begins with L solution and shows the conversion to moles of $C_{12}H_{22}O_{11}$ using the molarity and then the conversion to g using the molar mass.

Solution:

We then follow the solution map to compute the answer.

$$1.72 \text{ L solution} \times \frac{0.758 \text{ mol } C_{12}H_{22}O_{11}}{\text{L solution}} \times \frac{342.34 \text{ g } C_{12}H_{22}O_{11}}{\text{mol } C_{12}H_{22}O_{11}} = 446 \text{ g } C_{12}H_{22}O_{11}$$

In this example, we used molarity to convert from a given amount of *solution* to the amount of *solute* in that solution. In the example that follows, we use molarity to convert from a given amount of *solute* to the amount of *solution* containing that solute.

Using Molarity in Calculations

How many liters of a 0.114 M NaOH solution contains 1.24 mol of NaOH?

Given: 0.114 M NaOH
1.24 mol NaOH

Find: L solution

Conversion Factor:

$$\frac{0.114 \text{ mol NaOH}}{\text{L solution}}$$

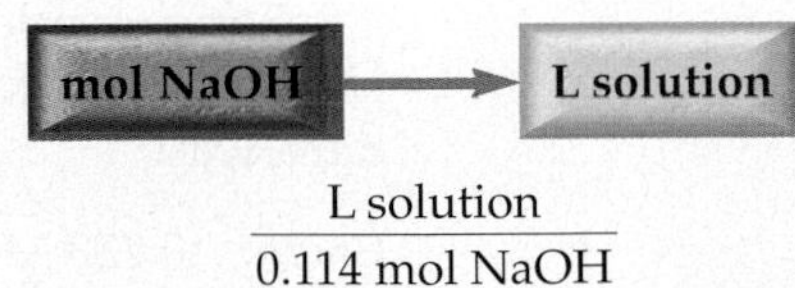

$$\frac{\text{L solution}}{0.114 \text{ mol NaOH}}$$

Solution Map:

The solution map begins with mol NaOH and shows the conversion to L of solution using the molarity.

Solution:

$$1.24 \text{ mol NaOH} \times \frac{\text{L solution}}{0.114 \text{ mol NaOH}} = 10.9 \text{ L solution}$$

SKILLBUILDER 13.4 **Using Molarity in Calculations**

13.7 Solution Dilution

When diluting acids, always add the concentrated acid to the water. Never add water to concentrated acid solutions.

To save space in laboratory storerooms, solutions are often stored in concentrated forms called **stock solutions**. For example, hydrochloric acid is often stored as a 12-M stock solution. However, many lab procedures call for much less-concentrated hydrochloric acid solutions, so chemists must dilute the stock solution to the required concentration. This is normally done by diluting a certain amount of the stock solution with water. How do we know how much of the stock solution to use? The easiest way to solve these problems is to use the following dilution equation

This equation works because the molarity multiplied by the volume gives the number of moles of solute ($M \times V = \text{mol}$), which is the same in both solutions.

$$M_1V_1 = M_2V_2$$

where M_1 and V_1 are the molarity and volume of the initial concentrated solution and M_2 and V_2 are the molarity and volume of the final diluted solution. For example, suppose a laboratory procedure calls for 5.00 L of a 1.50 M KCl solution. How should you prepare this solution from a 12.0 M stock solution? Set up the problem in the standard way.

Given: $M_1 = 12.0 \text{ M}$
$M_2 = 1.50 \text{ M}$
$V_2 = 5.00 \text{ L}$

Find: V_1

Equation:

$$M_1V_1 = M_2V_2$$

Solution:

We solve the equation for V_1, the volume of the stock solution required for the dilution, and then substitute in the correct values to compute it.

$$\begin{aligned} M_1V_1 &= M_2V_2 \\ V_1 &= \frac{M_2V_2}{M_1} \\ &= \frac{1.50 \frac{\text{mol}}{\text{L}} \times 5.00 \text{ L}}{12.0 \frac{\text{mol}}{\text{L}}} \\ &= 0.625 \text{ L} \end{aligned}$$

Consequently, we make the solution by diluting 0.625 L of the stock solution to a total volume of 5.00 L (V_2). The resulting solution will be 1.50 M in KCl (Figure 13.8).

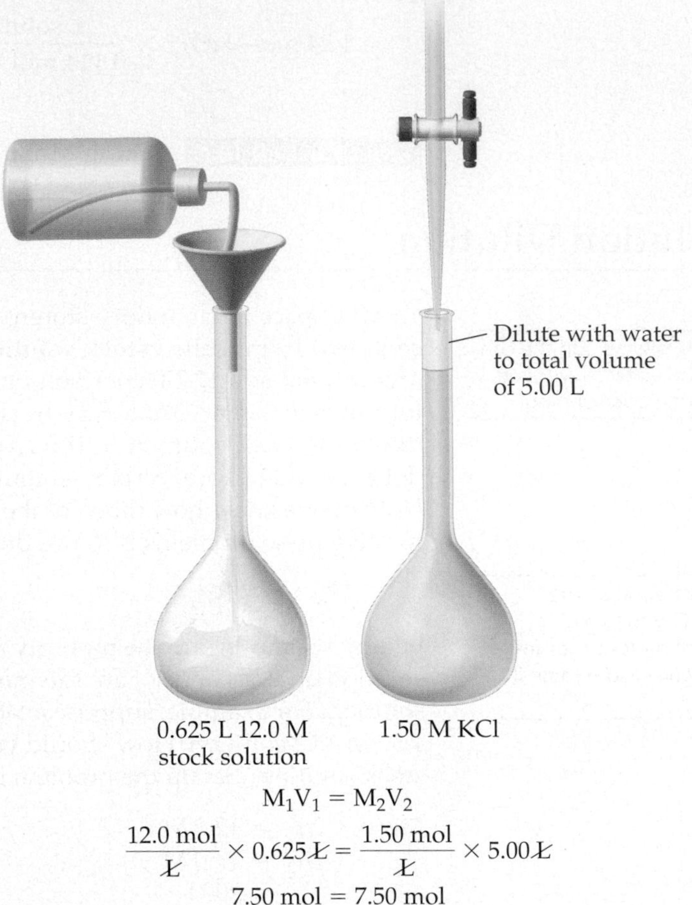
How to make 5.00 L of a 1.50 M KCl
from a 12.0 M stock solution.
Dilute with water
to total volume
of 5.00 L
0.625 L 12.0 M
stock solution
1.50 M KCl

$$M_1V_1 = M_2V_2$$

$$\frac{12.0 \text{ mol}}{\cancel{L}} \times 0.625 \cancel{L} = \frac{1.50 \text{ mol}}{\cancel{L}} \times 5.00 \cancel{L}$$

$$7.50 \text{ mol} = 7.50 \text{ mol}$$

Figure 13.8 Making a solution by dilution of a more concentrated solution.

EXAMPLE 13.5 Solution Dilution

To what volume should you dilute 0.100 L of a 15.0 M NaOH solution to obtain a 1.0 M NaOH solution?

Set up the problem in the standard way.

Given: $M_1 = 15$ M

$M_2 = 1.0$ M

$V_1 = 0.100$ L

Find: V_2

Equation:

$$M_1V_1 = M_2V_2$$

Solution:

In this case, we solve the equation for V_2, the volume of the final solution.

$$M_1V_1 = M_2V_2$$
$$V_2 = \frac{M_1V_1}{M_2}$$
$$= \frac{15\,\frac{\cancel{\text{mol}}}{\cancel{\text{L}}} \times 0.100\text{ L}}{1.0\,\frac{\cancel{\text{mol}}}{\cancel{\text{L}}}}$$
$$= 1.5\text{ L}$$

Consequently, we make the solution by diluting 0.100 L of the stock solution to a total volume of 1.5 L (V_2). The resulting solution will have a concentration of 1.0 M.

SKILLBUILDER 13.5 Solution Dilution

How much of 6.0 M $NaNO_3$ solution should be used to make 0.585 L of a 1.2 M $NaNO_3$ solution?

13.8 Solution Stoichiometry

See Sections 8.2–8.4 for a review of reaction stoichiometry.

As we discussed in Chapter 7, many chemical reactions take place in aqueous solutions. Precipitation reactions, neutralization reactions, and gas evolution reactions, for example, all occur in aqueous solutions. In Chapter 8, we learned how the coefficients in chemical equations are used as conversion factors between moles of reactants and moles of products in stoichiometric calculations. These conversion factors are often used to determine, for example, the amount of product obtained in a chemical reaction based on a given amount of reactant or the amount of one reactant needed to completely react with a given amount of another reactant. The general solution map for these kinds of calculations is

where A and B are two different substances involved in the reaction and the conversion factor between them comes from the stoichiometric coefficients in the balanced chemical equation.

In reactions involving aqueous reactant and products, it is often convenient to specify the amount of reactants or products in terms of their volume and concentration. We can then use the volume and concentration to calculate moles of reactants or products and then use the stoichiometric coefficients to convert to other quantities in the reaction. The general solution map for these kinds of calculation is

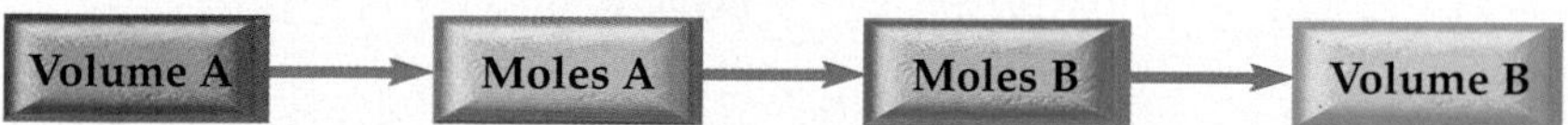

where the conversions between volume and moles are achieved using the molarities of the solutions. For example, consider the following reaction for the neutralization of sulfuric acid.

$$H_2SO_4(aq) + 2\,NaOH(aq) \rightarrow Na_2SO_4(aq) + 2\,H_2O(l)$$

How much 0.125 M NaOH solution is required to completely neutralize 0.225 L of 0.175 M H_2SO_4 solution?

We set up the problem in the standard way.

Given: 0.225 L H_2SO_4 solution
0.175 M H_2SO_4
0.125 M NaOH

Find: L NaOH solution

Conversion Factors:

The conversion factors for this problem are the molarities of the two solutions expressed in mol of solute per L of solution and the stoichiometric relationship (from the balanced equation) between mol of H_2SO_4 and mol of NaOH.

$$\text{M}\ (H_2SO_4) = \frac{0.175\ \text{mol}\ H_2SO_4}{\text{L}\ H_2SO_4\ \text{solution}}$$

$$\text{M}\ (\text{NaOH}) = \frac{0.125\ \text{mol NaOH}}{\text{L NaOH solution}}$$

$$1\ \text{mol}\ H_2SO_4 \equiv 2\ \text{mol NaOH}$$

Solution Map:

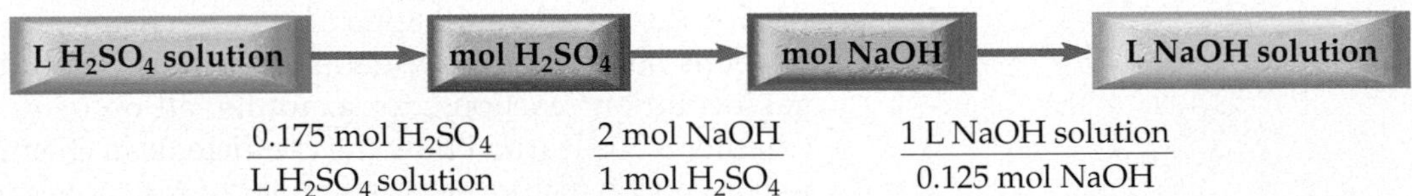

The solution map for this problem is similar to the solution maps for other stoichiometric problems. We first use the volume and molarity of H_2SO_4 solution to get mol of H_2SO_4. Then we use the stoichiometric coefficients from the equation to convert mol of H_2SO_4 to mol of NaOH. Finally, we use the molarity of NaOH to get to L of NaOH solution.

Solution:
To solve the problem, we follow the solution map and compute the answer.

$$0.225\ \text{L}\ H_2SO_4\ \text{solution} \times \frac{0.175\ \text{mol}\ H_2SO_4}{\text{L}\ H_2SO_4\ \text{solution}} \times \frac{2\ \text{mol NaOH}}{1\ \text{mol}\ H_2SO_4} \times \frac{1\ \text{L NaOH solution}}{0.125\ \text{mol NaOH}}$$

$$= 0.630\ \text{L NaOH solution}$$

It will take 0.630 L of the NaOH solution to completely neutralize the H_2SO_4.

EXAMPLE 13.6 Solution Stoichiometry

Consider the following precipitation reaction:

$$2\ \text{KI}(aq) + \text{Pb}(NO_3)_2(aq) \rightarrow \text{PbI}_2(s) + 2\ \text{KNO}_3(aq)$$

How much 0.115M KI solution (in L) is required to completely precipitate the Pb^{2+} in 0.104 L of 0.225 M $Pb(NO_3)_2$ solution?

We set up the problem in the standard way.

Given: 0.104 L $Pb(NO_3)_2$ solution
0.115 M KI
0.225 M $Pb(NO_3)_2$

Find: L KI solution

Conversion Factors:

$$M\ (KI) = \frac{0.115\ \text{mol KI}}{\text{L KI solution}}$$

$$M\ [Pb(NO_3)_2] = \frac{0.225\ \text{mol } Pb(NO_3)_2}{\text{L } Pb(NO_3)_2\ \text{solution}}$$

$$2\ \text{mol KI} \equiv 1\ \text{mol } Pb(NO_3)_2$$

Solution Map:

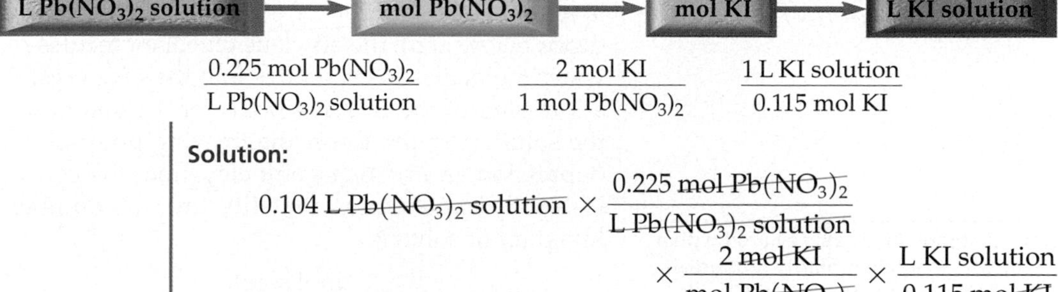

Solution:

$$0.104\ \text{L } \cancel{Pb(NO_3)_2\ \text{solution}} \times \frac{0.225\ \cancel{\text{mol } Pb(NO_3)_2}}{\cancel{\text{L } Pb(NO_3)_2\ \text{solution}}} \times \frac{2\ \cancel{\text{mol KI}}}{\cancel{\text{mol } Pb(NO_3)_2}} \times \frac{\text{L KI solution}}{0.115\ \cancel{\text{mol KI}}}$$

$$= 0.407\ \text{L KI solution}$$

SKILLBUILDER 13.6 **Solution Stoichiometry**

How many milliliters of 0.112 M Na_2CO_3 are necessary to completely react with 27.2 mL of 0.135 M HNO_3 according to the following reaction?

$$2\ HNO_3(aq) + Na_2CO_3(aq) \rightarrow H_2O(l) + CO_2(g) + 2\ NaNO_3(aq)$$

SKILLBUILDER PLUS

A 25.0 mL sample of HNO_3 solution requires 35.7 mL of 0.108 M Na_2CO_3 to completely react with all of the HNO_3 in the solution. What was the concentration of the HNO_3 solution?

13.9 Freezing Point Depression and Boiling Point Elevation: Making Water Freeze Colder and Boil Hotter

Have you ever wondered why salt is added to ice in an ice-cream maker? Or why salt is often scattered on icy roads in cold climates? Salt actually lowers the melting point of ice. A salt and water solution will remain a liquid even below 0 °C. By adding salt to ice in the ice-cream maker, you form a mixture of ice, salt, and water that reaches a temperature of about

Adding salt to icy roads lowers the melting point of ice, allowing it to melt even if the temperature is below 0 °C.

−10 °C, causing the cream to freeze. On the road, the salt allows the ice to melt, even if the ambient temperature is below freezing.

Adding a nonvolatile solute to a liquid extends the temperature range over which the liquid remains a liquid. The solution has a lower melting point and a higher boiling point than the pure solvent. These effects are called **freezing point depression** and **boiling point elevation**. Freezing point depression and boiling point elevation depend only on the number of solute particles in solution, not on the type of solute particles. Properties such as these—which depend on the amount of solute and not the type of solute—are called **colligative properties**.

Freezing Point Depression

The freezing point of a solution containing a nonvolatile solute is lower than the freezing point of the pure solvent because the solute molecules interfere with the freezing of the solvent molecules. For example, antifreeze, used to prevent the freezing of engine blocks in cold climates, is an aqueous solution of ethylene glycol ($C_2H_6O_2$). If the temperature drops below zero, the ethylene glycol molecules make it more difficult for water molecules to crystallize into their ice crystal structure. The result is a lowering of the freezing point for the solution. The more concentrated the solution is, the lower the freezing point becomes. For freezing point depression and boiling point elevation, the concentration of the solution is usually expressed in **molality (m)**, the number of moles of solute per kilogram of solvent.

Note that molality is abbreviated with a lower case *m* while molarity is abbreviated with a capital *M*.

$$\text{Molality (m)} = \frac{\text{mol solute}}{\text{kg solvent}}$$

Notice that molality is defined with respect to kilograms of *solvent* not kilograms of *solution*.

EXAMPLE 13.7 Calculating Molality

Calculate the molality of a solution containing 17.2 g of ethylene glycol ($C_2H_6O_2$) dissolved in 0.500 kg of water.
Set up the problem in the standard way.

Given: 17.2 g $C_2H_6O_2$
0.500 kg H_2O

Find: molality (m)

Equation and Conversion Factor:

$$\text{Molality (m)} = \frac{\text{Moles solute}}{\text{Kilograms solvent}}$$

$$\text{Molar mass of } C_2H_6O_2 = \frac{62.08 \text{ g}}{\text{mol}}$$

Solution:

To calculate molality, we simply substitute the correct values into the equation and compute the answer. However, we must first convert the amount of $C_2H_6O_2$ from g to mol using the molar mass of $C_2H_6O_2$ (62.07 g/mol).

$$\text{mol } C_2H_6O_2 = 17.2 \text{ g } C_2H_6O \times \frac{1 \text{ mol } C_2H_6O}{62.08 \text{ g } C_2H_6O} = 0.2771 \text{ mol } C_2H_6O$$

$$\text{Molality (m)} = \frac{\text{Moles solute}}{\text{Kilograms solvent}}$$

$$= \frac{0.2771 \text{ mol } C_2H_6O}{0.500 \text{ kg } H_2O}$$

$$= 0.554 \text{ m}$$

SKILLBUILDER 13.7 Calculating Molality

Calculate the molality (m) of a sucrose ($C_{12}H_{22}O_{11}$) solution containing 50.4 g sucrose and 0.332 kg of water.

The amount that the freezing point is lowered for solutions is given by the following equation

$$\Delta T_f = m \times K_f$$

where

- ΔT_f is the change in temperature of the freezing point in °C (from the freezing point of the pure solvent).
- m is the molality of the solution in $\frac{\text{mol solute}}{\text{kg solvent}}$
- K_f is the freezing point depression constant for the solvent.

For water:

$$K_f = 1.86 \frac{°\text{C kg solvent}}{\text{mol solute}}$$

Different solvents have different values of K_f.

Calculating the freezing point of a solution involves substituting into the preceding equation as the following example demonstrates.

EXAMPLE 13.8 Freezing Point Depression

Calculate the freezing point of a 1.7 m ethylene glycol solution. Set up the problem in the standard way.

Given: 1.7 m solution

Find: ΔT_f

Equation:

$$\Delta T_f = m \times K_f$$

Solution:

To solve this problem, simply substitute the values into the equation for freezing point depression and calculate ΔT_f.

$$\Delta T_f = m \times K_f$$

$$= 1.7 \frac{\text{mol solute}}{\text{kg solvent}} \times 1.86 \frac{°\text{C kg solvent}}{\text{mol solute}}$$

$$= 3.2 \text{ °C}$$

The freezing point will be the freezing point of pure water (0.00 °C) − ΔT_f.

$$\text{Freezing point} = 0.00 \text{ °C} - 3.2 \text{ °C}$$

$$= -3.2 \text{ °C}$$

SKILLBUILDER 13.8 **Freezing Point Depression**

Calculate the freezing point of a 2.6 m sucrose solution.

Boiling Point Elevation

The boiling point of a solution containing a nonvolatile solute is higher than the boiling point of the pure solvent because the solute molecules interfere with the vaporization of the solvent molecules. In automobiles, antifreeze not only prevents the freezing of water within engine blocks in cold climates, it also prevents the boiling of water within engine blocks in hot climates. The amount that the boiling point is raised for solutions is given by the following equation

$$\Delta T_b = m \times K_b$$

EVERYDAY CHEMISTRY

Antifreeze in Frogs

On the outside, wood frogs *(rana sylvatica)* look like most other frogs They are only a few inches long and have characteristic greenish-brown skin. However, wood frogs survive cold winters in a remarkable way—they partially freeze. In this state, the wood frog has no heartbeat, no blood circulation, no breathing, and no brain activity. Within 1–2 hours of thawing, however, these vital functions return, and the wood frog hops off to find food. How does the wood frog do this?

Most cold-blooded animals cannot survive freezing temperatures because the water within their cells freezes. As we learned in Section 12.8, when water freezes, it expands, irreversibly damaging cells. When the wood frog hibernates for the winter, however, it produces large amounts of glucose that circulate into the blood and fill the interior of its cells. When the temperature drops below freezing, extracellular bodily fluids, such as those in the abdominal cavity, freeze solid. Fluids within cells, however, remain liquid because the high glucose concentration lowers their freezing point. In other words, the concentrated glucose solution within the frog's cells acts as antifreeze, preventing the water within the cells from freezing and allowing the frog to survive.

The wood frog survives cold winters by partially freezing. The fluids within the frog's cells, however, remain liquid to temperatures as low as −8 °C. This is accomplished by flooding these fluids with glucose, which acts as antifreeze, lowering the freezing point of intracellular fluids.

CAN YOU ANSWER THIS? *The tree frog can survive at body temperatures a low as −8.0 °C. Calculate the molality of a glucose solution* ($C_6H_{12}O_6$) *required to lower the freezing point of water to −8.0 °C.*

where

- ΔT_b is change in temperature of the boiling point in °C (from the boiling point of the pure solvent).
- m is the molality of the solution in $\frac{\text{mol solute}}{\text{kg solvent}}$
- K_b is the boiling point elevation constant for the solvent.

For water:

$$K_b = 0.512 \frac{\text{°C kg solvent}}{\text{mol solute}}$$

Different solvents have different values of K_b.

The boiling point of solutions is calculated by simply substituting into the preceding equation as the following example demonstrates.

EXAMPLE 13.9 Boiling Point Elevation

Calculate the boiling point of a 1.7 m ethylene glycol solution.
Set up the problem in the standard way.

Given: 1.7 m solution

Find: boiling point

Equation:

$$\Delta T_b = m \times K_b$$

Solution:
To solve this problem, simply substitute the values into the equation for boiling point elevation and calculate ΔT_b.

$$\begin{aligned}\Delta T_b &= m \times K_b \\ &= 1.7 \frac{\cancel{\text{mol solute}}}{\cancel{\text{kg solvent}}} \times 0.512 \frac{\text{°C } \cancel{\text{kg solvent}}}{\cancel{\text{mol solute}}} \\ &= 0.87 \text{ °C}\end{aligned}$$

The boiling point will be the boiling point of pure water (100.00 °C) + ΔT_b.

$$\begin{aligned}\text{Boiling point} &= 100.00 \text{ °C} + 0.87 \text{ °C} \\ &= 100.87 \text{ °C}\end{aligned}$$

SKILLBUILDER 13.9 Boiling Point Elevation

Calculate the boiling point of 3.5 m glucose solution.

13.10 Osmosis: Why Drinking Saltwater Causes Dehydration

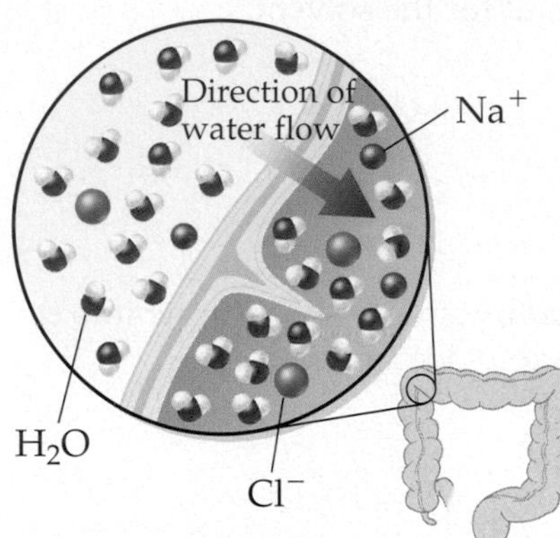

Figure 13.9 Drinking seawater promotes dehydration because seawater is a *thirsty solution*. As the seawater flows through the stomach and intestine it draws water *out of* bodily tissues.

Humans adrift at sea are surrounded by water, yet drinking that water would only accelerate their dehydration. Why? Saltwater causes dehydration because of **osmosis**, the flow of solvent from a lower-concentration solution to a higher-concentration solution. Solutions containing a high concentration of solutes draw solvent from solutions containing a lower concentration of solutes. In other words, aqueous solutions with high concentrations of solutes, such as seawater, are actually *thirsty solutions*—they draw water away from other less-concentrated solutions, including the human body (Figure 13.9).

Figure 13.10 shows an osmosis cell. The left side of the cell contains a concentrated saltwater solution, and the right side of the cell contains pure water. A **semipermeable membrane**—a membrane that selectively allows some substances to pass through but not others—separates the two halves of the cell. Through osmosis, water flows from the pure-water side of the cell, through the semipermeable membrane, and into the saltwater side. Over time, the water level on the left side of the cell rises, while the water level on the right side of the cell falls. The pressure required to stop the osmotic flow is called **osmotic pressure.**

The membranes of living cells act as semipermeable membranes. Consequently, if you put a living cell into seawater, it loses water through osmosis and becomes dehydrated. Similarly, if you drink seawater, the seawater actually draws water out of your body as it passes through your stomach and intestines. All of that extra water in your intestine promotes dehydration of bodily tissues and diarrhea. Consequently, seawater should never be consumed as drinking water.

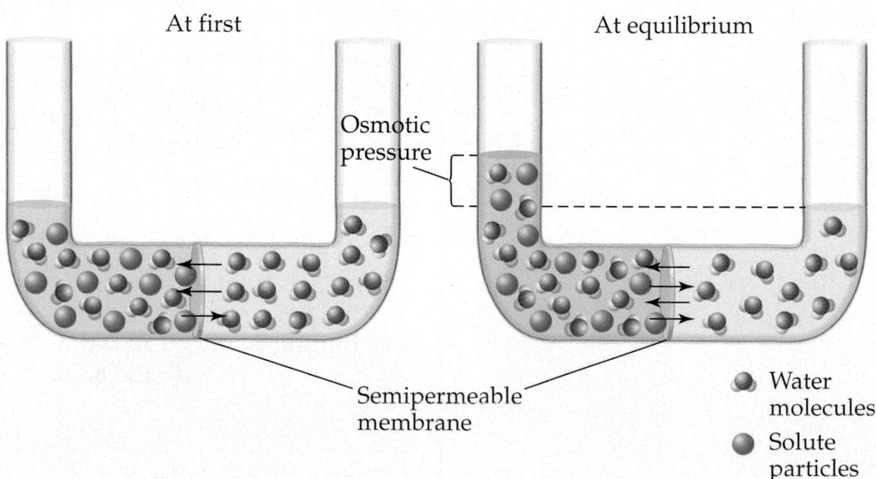

Figure 13.10 Osmosis cell. In an osmosis cell, water flows towards the more concentrated solution.

CHEMISTRY AND HEALTH

Solutions in Medicine

Doctors and others working in health fields must often administer solutions to patients. The osmotic pressure of these solutions is controlled for the desired effect on the patient. Solutions having osmotic pressures greater than bodily fluids are called *hyperosmotic*. These solutions tend to take water out of cells and tissues. When a human cell is placed in a hyperosmotic solution, it tends to shrivel as it loses water to the surrounding solution. Solutions having osmotic pressures less than bodily fluids are called *hypoosmotic*. These solutions tend to pump water into cells. When a human cell is placed in a hypoosmotic solution—such as pure water, for example—water enters the cell, sometimes causing it to burst.

Intravenous solutions—those that are administered directly into a patient's veins—must have osmotic pressures equal to bodily fluids. These solutions are called **isoosmotic**. When a patient is given an I.V. in a hospital, the majority of the fluid is usually an isoosmotic saline solution—a solution containing 0.9 g NaCl per 100 mL of solution. In medicine and in other health-related fields, solution concentrations are often reported in units that indicate the mass of the solute per a given volume of solution. Also common is *percent-mass-to-volume*—which is simply the mass of the solute in grams divided by volume of the solution in mL times 100%. In these units, the concentration of an isoosmotic saline solution is 0.9% mass/volume.

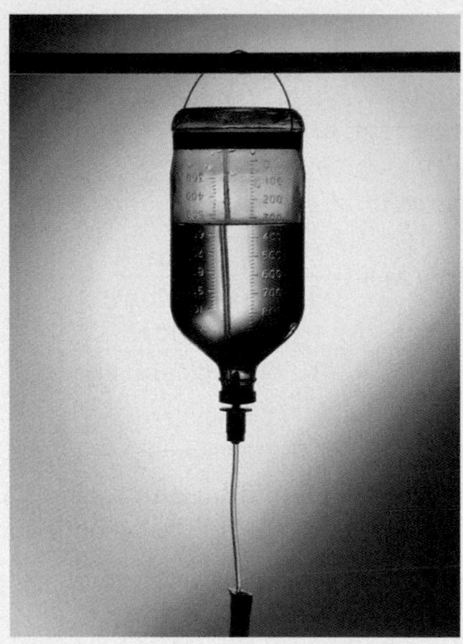

Intravenous fluids must always be saline solutions that have an osmotic pressure equal to bodily fluids. **Question:** Why would it be dangerous to administer intravenous fluids that do hot have an osmotic pressure equal to bodily fluids?

CAN YOU ANSWER THIS? *An isoosmotic sucrose ($C_{12}H_{22}O_{11}$) solution has a concentration of 0.30 M. Calculate its concentration in percent mass to volume.*

CHAPTER IN REVIEW

Chemical Principles

Solutions: A solution is a homogeneous mixture with two or more components. The solvent is the majority component and the solute is the minority component. Aqueous solutions are those with water as the solvent.

Solid and Water Solutions: The solubility—the amount of solute that dissolves in a certain amount of solvent—of solids in liquids increases with increasing temperature. Recrystallization involves dissolving a solid into hot solvent to saturation and then allowing it to cool. As the solution cools, it becomes supersaturated and the solid crystallizes.

Relevance

Solutions: Common solutions include seawater (solid and liquid), soda pop (gas and liquid), and alcoholic spirits such as vodka (liquid and liquid).

Solids and Water Solutions: When most people think of a solution, they think of a solid dissolved in water. Common solid/water solutions include seawater, coffee, and sugar water. Recrystallization is used extensively in the laboratory to purify solids. It is also used in making rock candy.

Gas and Water Solutions: The solubility of gases in liquids decreases with increasing temperature but increases with increasing pressure.

Gas and Water Solutions: The temperature and pressure dependence of gas solubility is the reason that soda pop fizzes when opened and also the reason that warm beer goes flat.

Solution Concentration:

Three common ways to express solution concentration are mass percent, molarity, and molality.

$$\text{Mass percent} = \frac{\text{Mass solute}}{\text{Mass solution}} \times 100\%$$

$$\text{Molarity (M)} = \frac{\text{Mol solute}}{\text{Liters solution}}$$

$$\text{Molality (m)} = \frac{\text{Mol solute}}{\text{Kilograms solvent}}$$

Solution Concentration: Solution concentration is used to specify how much of the solute is present in a given amount of solvent. It is also useful in converting between amounts of solute and solution. Mass percent and molarity are the most common concentration units. Molality is useful in quantifying colligative properties such as freezing point depression and boiling point elevation.

Solution Dilution: Solution dilution problems are most conveniently solved using the following equation.

$$M_1V_1 = M_2V_2$$

Solution Dilution: Since many solutions are stored in concentrated form, it is often necessary to dilute them to a desired concentration.

Freezing Point Depression and Boiling Point Elevation: A nonvolatile solute will extend the liquid temperature range of a solution relative to the pure solvent. The freezing point of a solution is lower than the freezing point of the pure solvent and the boiling point of a solution is higher than the boiling point of the pure solvent. These relationships are quantified by the following equations.

Freezing point depression:

$$\Delta T_f = m \times K_f$$

Boiling point elevation:

$$\Delta T_b = m \times K_b$$

Freezing Point Depression and Boiling Point Elevation: Salt is often added to ice in making ice cream and to roads in icy weather. The salt lowers the freezing point of water, allowing the cream within the ice cream maker to freeze and the ice on icy roads to melt.

Osmosis: Osmosis is the flow of water from a high concentration solution to a low concentration solution through a semipermeable membrane.

Osmosis: Osmosis explains why drinking seawater causes dehydration. As seawater goes through the stomach and intestines, it draws water away from the body through osmosis, resulting in diarrhea and dehydration.

Chemical Skills

Examples

Calculating Mass Percent (Section 13.5)

EXAMPLE 13.10 Calculating Mass Percent

Find the mass percent concentration of a solution containing 19 g of solute and 158 g of solvent.

Begin by setting up the problem in the standard way.

Given: 19 g solute
158 g solvent

Find: mass percent

To calculate mass percent concentration, divide the mass of the solute by the mass of the solution (solute *and* solvent) and multiply by 100%.

Equation:

$$\text{Mass percent} = \frac{\text{Mass solute}}{\text{Mass solution}} \times 100\%$$

Solution:

$$\text{Mass solution} = 19\text{ g} + 158\text{ g} = 177\text{ g}$$

$$\text{Mass percent} = \frac{19\text{ g}}{177\text{ g}} \times 100\%$$
$$= 11\%$$

Using Mass Percent in Calculations (Section 13.5)

When using mass percent in calculations, set up the problem in the standard way. When writing the conversion factors, express the mass percent as:

$$\frac{\text{g solute}}{100\text{ g solution}}$$

This will be your key conversion factor in converting from the amount of solution to the amount of solute. If you are asked to convert from the amount of solute to the amount of solution, invert the conversion factor.

When writing the solution map, begin with the given quantity and convert to grams of solution (through other units as necessary). Then convert to grams of solute. If you are asked to convert from the amount of solute to the amount of solution, first convert to the amount of solute in grams, then use the concentration (inverted) to convert to the amount of solution in grams, and then to whatever units the problem asks. Finally, follow your solution map to compute the answer.

EXAMPLE 13.11 Using Mass Percent in Calculations

How much KCl, in g, is in 0.337 L of a 5.8% mass percent KCl solution? (Assume that the density of the solution is 1.05 g/mL.)

Given: 5.8% KCl by mass
0.337 L solution

Find: g KCl

Conversion Factors:

$$\frac{5.8\text{ g KCl}}{100\text{ g solution}}$$

$$\frac{1.05\text{ g}}{\text{mL}}$$

$$1000\text{ mL} = 1\text{ L}$$

Solution Map:

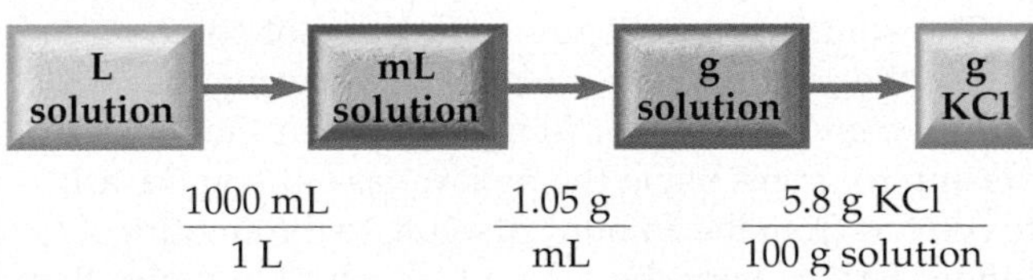

Solution:

$$0.337\ \cancel{\text{L solution}} \times \frac{1000\ \cancel{\text{mL}}}{\cancel{\text{L}}} \times \frac{1.05\ \cancel{\text{g}}}{\cancel{\text{mL}}} \times \frac{5.8\text{ g KCl}}{100\ \cancel{\text{g solution}}}$$
$$= 20.5\text{ g KCl}$$

Calculating Molarity (Section 13.6)

Begin by setting up the problem in the standard way.

To calculate molarity, divide the moles of solute by the volume of the solution.

EXAMPLE 13.12 Calculating Molarity

Calculate the molarity of a KCl solution containing 0.22 mol of KCl in 0.455 L of solution.

Given: 0.22 mol KCl
0.455 L solution

Find: molarity (M)

Equation:

$$\text{Molarity (M)} = \frac{\text{Moles solute}}{\text{Liters solution}}$$

Solution:

$$\text{Molarity (M)} = \frac{0.22 \text{ mol KCl}}{0.455 \text{ L solution}} = 0.48 \text{ M}$$

Using Molarity in Calculations (Section 13.5)

When using molarity in calculations, set up the problem in the standard way.

When writing the conversion factors, express the molarity as:

$$\frac{\text{Moles solute}}{\text{Liters solution}}$$

This will be your key conversion factor in converting from the amount of solution to the amount of solute. If you are asked to convert from the amount of solute to the amount of solution, invert the conversion factor. You may also need molar mass of the solute as a conversion factor.

The solution map begins with liters of solution and shows the conversion to moles of solute using the molarity as a conversion factor. You should then show the conversion to grams using the molar mass. If you are asked to convert from the amount of solute to the amount of solution, first convert the amount of solute to moles then use the molarity (inverted) to convert to the amount of solution in liters.

Finally, follow your solution map to compute the answer.

EXAMPLE 13.13 Using Molarity in Calculations

How much KCl, in grams, is contained in 0.488 L of 1.25 M KCl solution?

Given: 1.25 M KCl
0.488 L solution

Find: g KCl

Conversion Factors:

$$\frac{1.25 \text{ mol KCl}}{\text{L solution}}$$

$$\text{KCl molar mass} = \frac{74.55 \text{ g KCl}}{\text{mol KCl}}$$

Solution Map:

L KCl solution → mol KCl → g KCl

$$\frac{1.25 \text{ mol KCl}}{\text{L solution}} \qquad \frac{74.55 \text{ g KCl}}{\text{mol KCl}}$$

Solution:

$$0.488 \text{ }\cancel{\text{L solution}} \times \frac{1.25 \text{ }\cancel{\text{mol KCl}}}{\cancel{\text{L solution}}} \times \frac{74.55 \text{ g KCl}}{\cancel{\text{mol KCl}}} = 45.5 \text{ g KCl}$$

Solution Dilution (Section 13.7)

EXAMPLE 13.14 Solution Dilution

How much of an 8.0 M HCl solution should be used to make 0.400 L of a 2.7 M HCl solution?

Begin by setting up the problem in the normal way.

Given: $M_1 = 8.0\text{ M}$
$M_2 = 2.7\text{ M}$
$V_2 = 0.400\text{ L}$

Find: V_1

Most solution dilution problems will use equation $M_1V_1 = M_2V_2$.

Equation:

$$M_1V_1 = M_2V_2$$

Solve the equation for the quantity you are trying to find (in this case V_1) and then substitute in the correct values to compute it.

Solution:

$$M_1V_1 = M_2V_2$$

$$V_1 = \frac{M_2V_2}{M_1} = \frac{2.7\frac{\text{mol}}{\text{L}} \times 0.400\text{ L}}{8.0\frac{\text{mol}}{\text{L}}} = 0.14\text{ L}$$

Solution Stoichiometry (Section 13.8)

EXAMPLE 13.15 Solution Stoichiometry

Consider the following reaction:

$$\text{Hcl}(aq) + \text{NaOH}(aq) \rightarrow \text{NaCl}(aq) + \text{H}_2\text{O}(l)$$

How much 0.113 M NaOH solution is required to completely neutralize 1.25 L of 0.228 M HCl solution?

Begin by setting up the problem in the standard way.

Given: 1.25 L HCl solution
0.228 M HCl
0.113 M NaOH

Find: L NaOH solution

The conversion factors for these kinds of problems are the molarities of the solutions expressed in moles of solute per liter of solution and the stoichiometric relationships (from the balanced equation) between the reactants and products of interest.

Conversion Factors:

$$\text{M (HCl)} = \frac{0.228\text{ mol HCl}}{\text{L HCl solution}}$$

$$\text{M (NaOH)} = \frac{0.113\text{ mol NaOH}}{\text{L NaOH solution}}$$

$$1\text{ mol HCl} \equiv 1\text{ mol NaOH}$$

In the solution map, use the volume and molarity of A (in this case HCl) to get to moles of A. Then use the stoichiometric coefficients to convert to moles of other reactants or products (in this case NaOH). Finally, convert back to volume of B (in this case NaOH) using the molarity of B.

Solution Map:

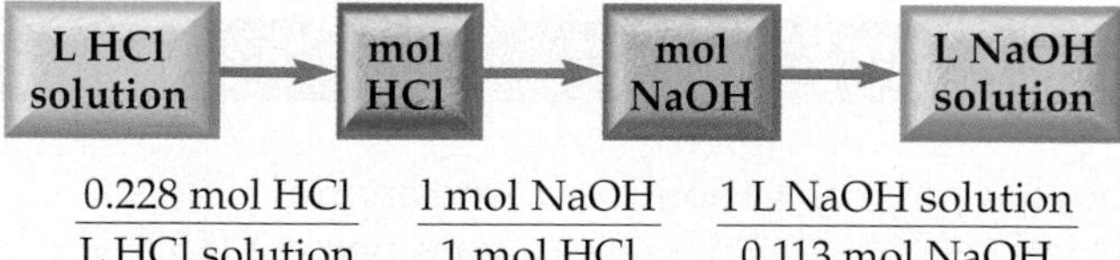

$$\frac{0.228\text{ mol HCl}}{\text{L HCl solution}} \quad \frac{1\text{ mol NaOH}}{1\text{ mol HCl}} \quad \frac{1\text{ L NaOH solution}}{0.113\text{ mol NaOH}}$$

Follow the solution map to compute the answer.

Solution:

$$1.25\text{ L }\cancel{\text{HCl solution}} \times \frac{0.228\ \cancel{\text{mol HCl}}}{\text{L }\cancel{\text{HCl solution}}} \times \frac{\cancel{\text{mol NaOH}}}{\cancel{\text{mol HCl}}} \times \frac{\text{L NaOH solution}}{0.113\ \cancel{\text{mol NaOH}}} = 2.52\text{ L NaOH solution}$$

Calculating Molality (Section 13.9)

EXAMPLE 13.16 Calculating Molarity

Calculate the molality of a solution containing 0.183 mol of sucrose dissolved in 1.10 kg of water.

Begin by setting up the problem in the standard way.

Given: 0.183 mol of sucrose
1.10 kg H_2O

Find: molality (m)

To calculate molality, you need the definition of molality which is moles of solute per kilogram of solvent.

Equation:

$$\text{Molality (m)} = \frac{\text{Moles solute}}{\text{Kilograms solvent}}$$

Substitute the correct values into the definition of molality and compute the answer. If any of the quantities are not in the correct units, convert them into the correct units before substituting into the equation.

Solution:

$$\text{Molality (m)} = \frac{0.183 \text{ mol sucrose}}{1.10 \text{ kg } H_2O} = 0.166 \text{ m}$$

Freezing Point Depression and Boiling Point Elevation (Section 13.9)

EXAMPLE 13.17 Freezing Point Depression and Boiling Point Elevation

Calculate the freezing point of a 2.5 m sucrose solution.

Begin by setting up the problem in the standard way.

Given: 2.5 m solution

Find: ΔT_f

For freezing point depression problems use $\Delta T_f = m \times K_f$. For boiling point elevation problems use $\Delta T_b = m \times K_b$.

Equation:

$$\Delta T_f = m \times K_f$$

To compute ΔT_f or ΔT_b simply substitute the values into the equation and compute.

The freezing point will be the freezing point of pure water $(0.00\ ^\circ\text{C}) - \Delta T_f$.

The boiling point will be the boiling point of pure water $(100.00\ ^\circ\text{C}) + \Delta T_b$.

Solution:

$$\Delta T_f = m \times K_f$$

$$= 2.5 \frac{\text{mol solute}}{\text{kg solvent}} \times 1.86 \frac{^\circ\text{C kgsolvent}}{\text{mol solute}}$$

$$= 4.7\ ^\circ\text{C}$$

$$\text{Freezing point} = 0.00\ ^\circ\text{C} - 4.7\ ^\circ\text{C} = -4.7\ ^\circ\text{C}$$

KEY TERMS

boiling point elevation [13.9]
colligative properties [13.9]
concentrated solution [13.5]
dilute solution [13.5]
freezing point depression [13.9]
isoosmotic
mass percent [13.5]
molality (m) [13.9]
molarity (M) [13.6]
nonelectrolyte solution [13.3]
osmosis [13.10]
osmotic pressure [13.10]
recrystallization [13.3]
saturated solution [13.3]
semipermeable membrane [13.10]
solubility [13.3]
solute [13.2]
solution [13.2]
solvent [13.2]
stock solution [13.7]
strong electrolyte solution [13.3]
supersaturated solution [13.3]
unsaturated solution [13.3]

EXERCISES

Questions

1. What is a solution? Give some examples.
2. What is an aqueous solution?
3. In a solution, what is the solvent? What is the solute? Give some examples.
4. Explain what "Like dissolves like" means.
5. What is solubility?
6. Describe what happens when additional solute is added to:
a) a saturated solution
b) an unsaturated solution
c) a supersaturated solution
7. Explain the difference between a strong electrolyte solution and a nonelectrolyte solution. What kinds of solutes form strong electrolyte solutions?
8. How does the solubility of gases depend on temperature?
9. Explain recrystallization.
10. How is rock candy made?
11. When you heat water on a stove, bubbles form on the bottom of the pot *before* the water boils. What are these bubbles? Why do they form?
12. Explain why warm soda pop goes flat faster than cold soda pop.
13. How does the solubility of gases depend on pressure?
14. Explain why a can of soda pop fizzes when opened.
15. What is the difference between a dilute solution and a concentrated solution?
16. What is mass percent?
17. What is molarity?
18. What is a stock solution?
19. How does the presence of a nonvolatile solute affect the boiling point and melting point of a solution relative to the boiling point and melting point of the pure solvent?
20. What are colligative properties?
21. Define molality.
22. What is osmosis?
23. Two shipwreck survivors were rescued from a life raft. One had drunk seawater while the other had not. The one who had drunk the seawater was more severely dehydrated than the one who did not. Explain.
24. Why are intravenous fluids always isoosmotic saline solutions? What would happen if pure water were administered intravenously?

Problems

Solutions

25. Which of the following are solutions?
a) sand and water mixture
b) oil and water mixture
c) salt and water mixture
d) sterling silver cup

26. Which of the following are solutions?
a) air
b) carbon dioxide and water mixture
c) a blueberry muffin
d) a brass buckle

27. Identify the solute and solvent in each of the following solutions.
a) salt water
b) sugar water
c) soda water

28. Identify the solute and solvent in each of the following solutions.
a) 80-proof vodka (40% ethyl alcohol)
b) oxygenated water
c) antifreeze (ethylene glycol and water)

29. Pick an appropriate solvent from Table 13.2 to dissolve:
a) motor oil (nonpolar)
b) sugar (polar)
c) lard (nonpolar)

30. Pick an appropriate solvent from Table 13.2 to dissolve:
a) glucose (polar)
b) salt (ionic)
c) vegetable oil (nonpolar)

Solids Dissolved in Water

31. What are the dissolved particles in a solution containing an ionic solute? What is the name for this kind of solution?

32. What are the dissolved particles in a solution containing a molecular solute? What is the name for this kind of solution?

33. A solution contains 25 g of NaCl per 100 g of water at 25 °C. Is the solution unsaturated, saturated, or supersaturated? (Use Figure 13.4)

34. A solution contains 32 g of KNO_3 per 100 g of water at 25 °C. Is the solution unsaturated, saturated, or supersaturated? (Use Figure 13.4)

35. A KNO_3 solution containing 45 g of KNO_3 per 100 g of water is cooled from 40 °C to 0 °C. What will happen during cooling? (Use Figure 13.4)

36. A KCl solution containing 42 g of KCl per 100 g of water is cooled from 60 °C to 0 °C. What will happen during cooling? (Use Figure 13.4)

Gases Dissolved in Water

37. Some laboratory procedures involving oxygen-sensitive reactants or products call for using preboiled water. Explain why this is so.

38. A person preparing a fish tank uses preboiled water to fill it. When the fish is put into the tank, it dies. Explain.

39. Scuba divers breathing air at increased pressure can suffer from nitrogen narcosis—a condition resembling drunkenness—when the partial pressure of nitrogen exceeds about 4 atm. What property of gas/water solutions causes this to happen? How could the diver reverse this effect?

40. Scuba divers breathing air at increased pressure can suffer from oxygen toxicity—too much oxygen in their bloodstream—when the partial pressure of oxygen exceeds about 1.4 atm. What happens to the amount of oxygen in their bloodstream when they breathe oxygen at elevated pressures? How can this be reversed?

Mass Percent

41. Calculate the concentration of each of the following solutions in mass percent.
a) 12.8 g NaCl in 145 g H_2O
b) 55.1 g $C_{12}H_{22}O_{11}$ in 478 g H_2O
c) 355 mg $C_6H_{12}O_6$ in 5.22 g H_2O

42. Calculate the concentration of each of the following solutions in mass percent.
a) 11.3 g C_2H_6O in 67.4 g H_2O
b) 104 g KCl in 558 g H_2O
c) 38.2 mg KNO_3 in 2.58 g H_2O

43. A soft drink contains 45 g of sugar in 309 g of H_2O. What is the concentration of sugar in the soft drink in mass percent?

44. A soft drink contains 35 mg of sodium in 315 g of H_2O. What is the concentration of sodium in the soft drink in mass percent?

45. Ocean water contains 3.5% NaCl by mass. How much salt can be obtained from 274 g of seawater?

46. A saline solution contains 1.1% NaCl by mass. How much NaCl is present in 87.2 g of this solution?

47. Determine the amount of sucrose in each of the following solutions.
a) 55 g of a solution containing 4.9% sucrose by mass
b) 122 mg of a solution containing 11.2% sucrose by mass
c) 5.8 kg of a solution containing 17.1% sucrose by mass

48. Determine the amount of potassium chloride in each of the following solutions.
a) 22.9 g of a solution containing 1.33% KCl by mass
b) 38.2 kg of a solution containing 22.5% KCl by mass
c) 52 mg of a solution containing 14% KCl by mass

49. Determine how much (in grams) of each of the following NaCl solutions contains 1.5 g NaCl.
a) 0.045% NaCl by mass
b) 1.94% NaCl by mass
c) 9.82% NaCl by mass

50. Determine how much (in grams) of each of the following sucrose solutions contains 12 g sucrose.
a) 5.8% sucrose by mass
b) 2.2% sucrose by mass
c) 14.9% sucrose by mass

51. $AgNO_3$ solutions are often used to plate silver onto other metals. What is the maximum amount of silver (in grams) that can be plated out of 4.8 L of an $AgNO_3$ solution containing 3.4% Ag by mass? Assume that the density of the solution is 1.01 g/mL.

52. A dioxin-contaminated water source contains 0.085% dioxin by mass. How much dioxin is present in 2.5 L of this water? Assume a density of 1.00 g/mL.

53. Ocean water contains 3.5% NaCl by mass. How much ocean water (in grams) contains 45.8 g of NaCl?

54. A hard water sample contains 0.0085% Ca by mass (in the form of Ca^{2+} ions). How much water (in grams) contains 1.2 g of Ca? (1.2 g of Ca is the recommended daily allowance of calcium for 19–24 year olds.)

55. Lead is a toxic metal that affects the central nervous system. A Pb-contaminated water sample contains 0.0011% Pb by mass. How much of the water (in mL) contains 150 mg of Pb? (Assume a density of 1.0 g/mL.)

56. Benzene is a carcinogenic (cancer-causing) compound. A benzene-contaminated water sample contains 0.000037% benzene by mass. How much of the water (in L) contains 125 mg of benzene? (Assume a density of 1.0 g/mL.)

Molarity

57. Calculate the molarity of each of the following solutions.
a) 1.3 mol of KCl in 2.5 L solution
b) 0.225 mol of KNO_3 in 0.855 L of solution
c) 0.117 mol of sucrose in 588 mL of solution

58. Calculate the molarity of each of the following solutions.
a) 0.11 mol of $LiNO_3$ in 5.2 L of solution
b) 1.77 mol of LiCl in 33.2 L of solution
c) 0.0441 mol of glucose in 84.2 mL of solution

59. Calculate the molarity of each of the following solutions.
a) 22.6 g $C_{12}H_{22}O_{11}$ in 0.442 L of solution
b) 42.6 g NaCl in 1.58 L of solution
c) 315 mg $C_6H_{12}O_6$ in 58.2 mL of solution

60. Calculate the molarity of each of the following solutions.
a) 33.2 g KCl in 0.895 L of solution
b) 61.3 g C_2H_6O in 3.4 L of solution
c) 38.2 mg KI in 112 mL of solution

61. A 205-mL sample of ocean water is found to contain 7.2 g of NaCl. What is the molarity of the solution with respect to NaCl?

62. A 355-mL can of soda pop is found to contain 45 g of sucrose ($C_{12}H_{22}O_{11}$). What is the molarity of the solution with respect to sucrose?

63. How many moles of NaCl are contained in each of the following?
a) 1.5 L of a 1.2 M NaCl solution
b) 0.448 L of a 0.85 M NaCl solution
c) 144 mL of a 1.65 M NaCl solution

64. How many moles of sucrose are contained in each of the following?
a) 3.4 L of a 0.100 M sucrose solution
b) 0.952 L of a 1.88 M sucrose solution
c) 21.5 mL of a 0.528 M sucrose solution

65. What volume of each of the following solutions contains 0.10 mol of KCl?
a) 0.255 M KCl
b) 1.8 M KCl
c) 0.995 M KCl

66. What volume of each of the following solutions contains 0.225 mol of NaI?
a) 0.152 M NaI
b) 0.982 M NaI
c) 1.76 M NaI

67. Calculate the mass of NaCl in a 55-mL sample of a 1.5 M NaCl solution.

68. Calculate the mass of glucose ($C_6H_{12}O_6$) in a 125-mL sample of a 1.15 M glucose solution.

69. A chemist wants to make 2.5 L of a 0.100 M KCl solution. How much KCl (in grams) should the chemist use?

70. A laboratory procedure calls for making 500.0 mL of a 1.4 M KNO_3 solution. How much KNO_3 (in grams) is needed?

71. How many liters of a 0.500 M sucrose ($C_{12}H_{22}O_{11}$) solution contain 1.5 kg of sucrose?

72. What volume of a 0.35 M $Mg(NO_3)_2$ solution contains 87 g of $Mg(NO_3)_2$?

Solution Dilution

73. A 158-mL sample of a 1.2 M sucrose solution is diluted to 500.0 mL. What is the molarity of the diluted solution?

74. A 2.5-L sample of a 5.8 M NaCl solution is diluted to 55 L. What is the molarity of the diluted solution?

75. Describe how you would make 2.5 L of a 0.100 M KCl solution from a 5.5 M stock KCl solution.

76. Describe how you would make 500.0 mL of a 0.200 M NaOH solution from a 15.0 M stock NaOH solution.

77. To what volume should you dilute 25 mL of a 12 M stock HCl solution to obtain a 0.500 M HCl solution?

78. To what volume should you dilute 75 mL of a 10.0 M H_2SO_4 solution to obtain a 1.75 M H_2SO_4 solution?

79. How much of a 12.0 M HNO_3 solution should you use to make 850.0 mL of a 0.250 M HNO_3 solution?

80. How much of a 5.0 M sucrose solution should be used to make 85.0 mL of a 0.040 M solution?

Solution Stoichiometry

81. Determine the volume of 0.150 M NaOH solution required to neutralize each of the following samples of hydrochloric acid. The neutralization reaction is:

$$NaOH(aq) + HCl(aq) \rightarrow H_2O(l) + NaCl\ (aq)$$

a) 25 mL of a 0.150 M HCl solution
b) 55 mL of 0.055 M HCl solution
c) 175 mL of a 0.885 M HCl solution

82. Determine the volume of 0.225 M KOH solution required to neutralize each of the following samples of sulfuric acid. The neutralization reaction is:

$$H_2SO_4(aq) + 2\ KOH(aq) \rightarrow K_2SO_4(aq) + 2\ H_2O(l)$$

a) 45 mL of 0.225 M H_2SO_4
b) 185 mL of 0.125 M H_2SO_4
c) 75 mL of 0.100 M H_2SO_4

83. Consider the following reaction.

$$2\ K_3PO_4(aq) + 3\ NiCl_2(aq) \rightarrow Ni_3(PO_4)_2(s) + 6\ KCl(aq)$$

What volume of 0.225 M K_3PO_4 solution is necessary to completely react with 134 mL of 0.0112 M $NiCl_2$?

84. Consider the following reaction.

$$K_2S(aq) + Co(NO_3)_2(aq) \rightarrow 2\ KNO_3(aq) + CoS(s)$$

What volume of 0.225 M K_2S solution is required to completely react with 175 mL of 0.115 M $Co(NO_3)_2$?

c)

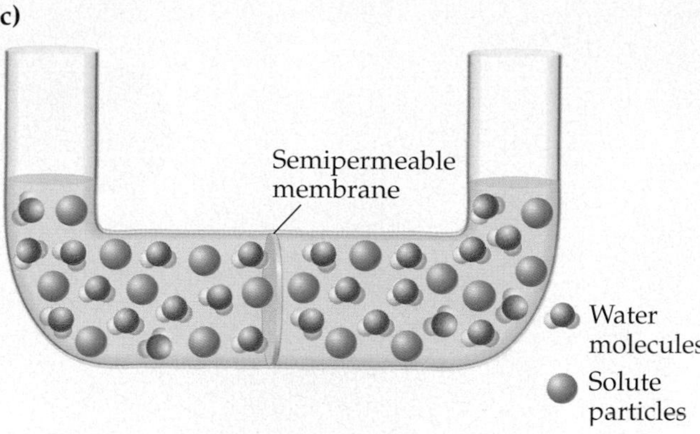

112. What is wrong with the following molecular view of a sodium chloride solution? What would make the picture correct?

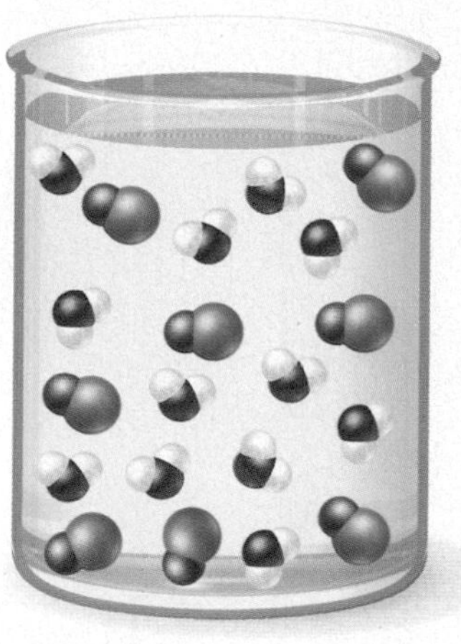

113. The Safe Drinking Water Act (SDWA) sets a limit for mercury—a toxin to the central nervous system—at 0.002 mg/L. Water suppliers must periodically test their water to ensure that mercury levels do not exceed 0.002 mg/L. Suppose water becomes contaminated with mercury at twice the legal limit (0.004 mg/L). How much of this water would have to be consumed to ingest 0.100 g of mercury?

114. Water softeners often replace calcium ions in hard water with sodium ions. Since sodium compounds are soluble, the presence of sodium ions in water does not cause the white, scaly residues caused by calcium ions. However, calcium is more beneficial to human health than sodium. Calcium is a necessary part of the human diet, while high levels of sodium intake are linked to increases in blood pressure. The Food and Drug Administration (FDA) recommends that adults ingest less than 2.4 g of sodium per day. How many liters of softened water, containing a sodium concentration of 0.050% sodium by mass, have to be consumed to exceed the FDA recommendation? (Assume a water density of 1.0 g/mL.)

Answers to Skillbuilder Exercises

Skillbuilder 13.1 2.67%
Skillbuilder 13.2 40.8 g
Skillbuilder 13.3 0.263 M
Skillbuilder 13.4 3.33 L
Skillbuilder 13.5 0.12 L
Skillbuilder 13.6 16.4 mL
Skillbuilder Plus, p. 463 0.308 M
Skillbuilder 13.7 0.443 m
Skillbuilder 13.8 −4.8 °C
Skillbuilder 13.9 101.8 °C

REALEMON
LEMON JUICE
BEST YET

14

Acids and Bases

"Science is all those things which are confirmed to such a degree that it would be unreasonable to withhold one's provisional consent."

Stephen Jay Gould

14.1 Sour Patch Kids and International Spy Movies

Gummy candies, those chewy, sweet little shapes that children and adults both love. From the introduction of the gummy bear to the gummy worm to just about any shape you can imagine, these candies have been incredibly popular. A variation is the sour gummy candy, whose best-known incarnation is the sour patch kid. Sour patch kids are gummy candies shaped like kids and coated with a white powder. When you first put a sour patch kid in your mouth, it tastes incredibly sour. The sour taste is caused by the white powder coating, a mixture of citric acid and tartaric acid. Like all acids, citric and tartaric acid have a sour taste.

◀ Acids are found in many common foods. The molecules shown here are as follows: citric acid (upper left), the acid found in lemon and limes; acetic acid (upper right), the acid present in vinegar; and tartaric acid (lower left), one of the acids used to coat sour gummy candies.

A number of other foods contain acids as well. The sour taste in lemons and limes, the bite of sourdough bread, and the tang of a ripe tomato are all caused by acids. Acids are substances that—by one definition that we will elaborate on later—produce H^+ ions in solution. When the citric and tartaric acids from a sour patch kid combine with saliva in your mouth, they produce H^+ ions. Those H^+ ions then react with protein molecules on your tongue. The protein molecules change shape, sending an electrical signal to your brain that you experience as a sour taste (Figure 14.1).

Acids have also been made famous by their use in spy movies. James Bond, for example, often carries an acid-filled gold pen. When Bond is captured and put in jail—as inevitably happens at least one time in each movie—he squirts some acid out of his pen and onto the iron bars of the jail cell. The acid quickly dissolves the metal, providing Bond with a way of escape. While acids do not dissolve iron bars with the ease depicted in the movies, they do dissolve metals. A small piece of aluminum placed in hydrochloric acid, for example, dissolves away in about ten minutes (Figure 14.2). With enough acid, it would be possible to dissolve the iron bars of a prison cell, but it would take more than the amount that fits in a pen.

Bond's pen is made of gold because gold is one of the few metals that is not dissolved by most acids (see Section 16.5).

When we say that acids *dissolve* metals, we mean that acids react with metals in a way that causes them to go into solution as metal cations.

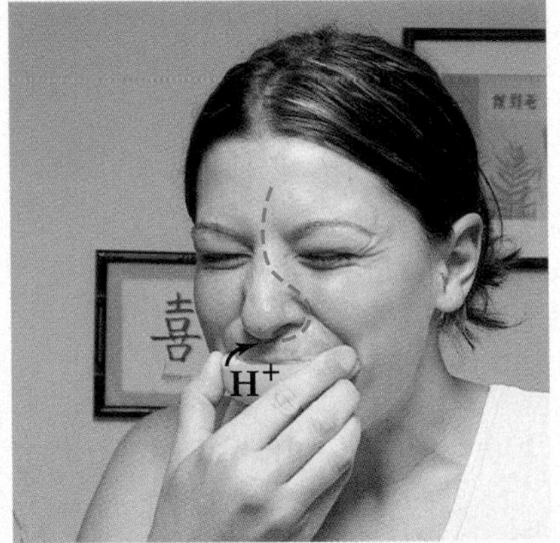

Figure 14.1 When a person eats a sour food, H^+ ions from the acid in the food react with protein molecules on the tongue. The protein molecules change shape, sending an electrical signal to the brain that the person experiences as a sour taste.

Figure 14.2 When aluminum is put into hydrochloric acid, the aluminum dissolves. **Question:** What happens to the aluminum atoms? Where do they go?

14.2 Acids: Properties and Examples

Acids have the following properties.

- Acids have a sour taste.
- Acids dissolve many metals.
- Acids turn litmus paper red.

Never taste or touch laboratory chemicals.

TABLE 14.1

Some Common Acids

Name	*Uses*
hydrochloric acid (HCl)	metal cleaning; food preparation; ore refining, main component of stomach acid
sulfuric acid (H_2SO_4)	fertilizer and explosive manufacturing; dye and glue production; automobile batteries
nitric acid (HNO_3)	fertilizer and explosive manufacturing; dye and glue production
acetic acid ($HC_2H_3O_2$)	plastic and rubber manufacturing; food preservative; active component of vinegar
carbonic acid (H_2CO_3)	found in carbonated beverages due to the reaction of carbon dioxide with water
hydrofluoric acid (HF)	metal cleaning; glass frosting and etching

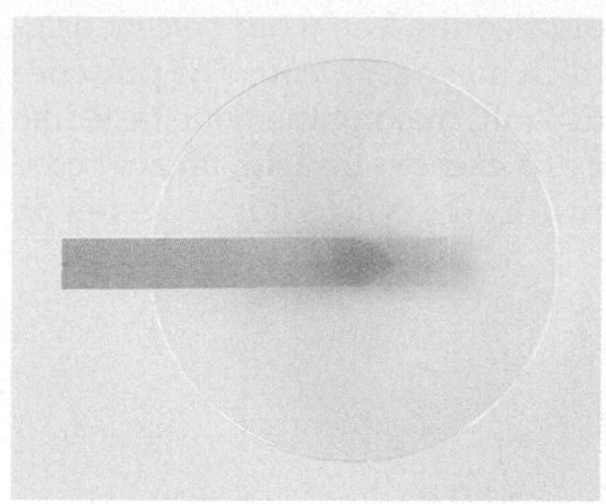

Figure 14.3 Litmus paper turns red in acid.

We have just seen examples of the sour taste of acids (sour patch kids) and their ability to dissolve metals (spy movies). Acids also turn litmus paper red. Litmus paper contains a dye that turns red in acidic solutions (Figure 14.3). In the laboratory, litmus paper is used routinely to test the acidity of solutions.

Some common acids are listed in Table 14.1. Hydrochloric acid is

Hydrochloric acid

found in most chemistry laboratories. It is used in industry to clean metals, to prepare and process some foods, and to refine metal ores. Hydrochloric acid is also the main component of stomach acid. In the stomach, hydrochloric acid helps break down food and kills harmful bacteria that might enter the body through food. The sour taste sometimes associated with indigestion is caused by the stomach's hydrochloric acid refluxing up into the esophagus (the tube that joins the stomach and the mouth) and throat.

For a review of naming acids, see Section 5.9.

Sulfuric acid and nitric acid are also commonly used in the laboratory.

Sulfuric acid

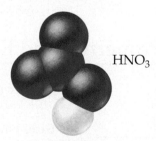

HNO_3

Nitric acid

In addition, they are used in the manufacture of fertilizers, explosives, dyes and glue. Sulfuric acid is contained in most automobile batteries. Acetic acid is found in most people's homes as the active component of

Acetic acid

Acetic acid is the active component in vinegar.

vinegar. It is also found in improperly stored wines. The word *vinegar* originates from the French words *vin aigre,* which mean *sour wine*. The presence of vinegar in wines is considered a serious fault, making the wine taste like salad dressing. Acetic acid is an example of a **carboxylic acid,** an acid containing the following grouping of atoms. Carboxylic acids, covered in

Carboxylic acid group

more detail in Chapter 18 (Section 18.15), are often found in substances derived from living organisms. Other carboxylic acids include citric acid, the main acid in lemons and limes, and malic acid, an acid found in apples, grapes, and wine.

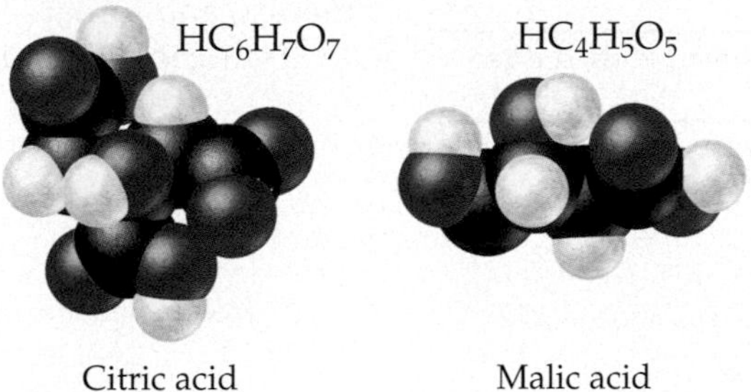

Citric acid   Malic acid

14.3 Bases: Properties and Examples

Bases have the following properties.

- Bases have a bitter taste.
- Bases have a slippery feel.
- Bases turn litmus paper blue.

Never taste or touch laboratory chemicals.

Bases are less common in foods than acids because of their bitter taste. A sour patch kid coated with a base would never sell. Our repulsion to the taste of bases is probably an adaptation against **alkaloids,** organic bases found in plants. Alkaloids are often poisonous—the active component of hemlock, for example, is the alkaloid coniine—so their bitter taste keeps us from eating them. Nonetheless, some foods, such as coffee, contain small amounts of base. Many people enjoy the bitter taste, but only after acquiring the taste over time.

Coffee is acidic overall but, but bases present in coffee—such as caffeine, for example—impart a bitter flavor.

Bases are also slippery. Soap, for example, is basic and its slippery feel is characteristic of bases. Some household cleaning solutions, such as ammonia, are also basic and have the characteristic slippery feel of a base.

Bases feel slippery because they react with oils on your skin to form soap-like substances.

TABLE 14.2

Common Bases

Name	*Uses*
sodium hydroxide (NaOH)	petroleum processing; soap and plastic manufacturing
potassium hydroxide (KOH)	cotton processing; electroplating; soap production
sodium bicarbonate ($NaHCO_3$)*	antacid; ingredient of baking soda; source of CO_2
ammonia (NH_3)	detergent; fertilizer and explosive manufacturing; synthetic fiber production

*Sodium bicarbonate is a salt whose anion (HCO_3^-) is the conjugate base of a weak acid (see Section 14.4) and acts as a base.

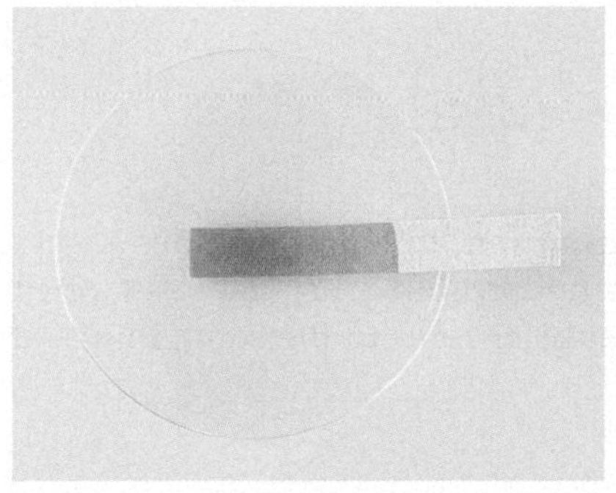

Figure 14.4 Base turns litmus paper blue.

Bases turn litmus paper blue (Figure 14.4). In the laboratory, litmus paper is routinely used to test the basicity of solutions.

Some common bases are listed in Table 14.2. Sodium hydroxide and potassium hydroxide are found in most chemistry laboratories. They are also used in processing petroleum and cotton and in soap and plastic manufacturing. Sodium hydroxide is the active ingredient in products such as Drano that work to unclog drains. Sodium bicarbonate can be found in most homes as baking soda and is also an active ingredient in many antacids. When taken as an antacid, sodium bicarbonate neutralizes stomach acid (see Section 14.5), relieving heartburn and sour stomach.

These consumer products all contain bases.

14.4 Molecular Definitions of Acids and Bases

Arrhenius Definition

Arrhenius definition of acids and bases.

In the 1880s, Swedish chemist **Svante Arrhenius** proposed the following molecular definitions of acids and bases.

Acid–An acid produces H^+ ions in aqueous solution.
Base–A base produces OH^- ions in aqueous solution.

For example, under the **Arrhenius definition,** HCl is an acid because it produces H^+ ions in solution (Figure 14.5).

$$HCl(aq) \rightarrow H^+(aq) + Cl^-(aq)$$

Ionization is the forming of ions.

HCl is a covalent compound and does not contain ions. However, in water it **ionizes** to form H^+ (aq) ions and Cl^- (aq) ions. The H^+ ions are highly reactive. In aqueous solution, they bond to water molecules according to the following reaction.

$$H^+ + :\ddot{\underset{\cdot\cdot}{O}}\!:\!H \ (\text{with H above O}) \longrightarrow \left[H\!:\!\ddot{\underset{\cdot\cdot}{O}}\!:\!H\ (\text{with H above O})\right]^+$$

The H_3O^+ ion is called the **hydronium ion.** In water, H^+ ions *always* associate with H_2O molecules to form hydronium ions. Chemists often use $H^+(aq)$ and $H_3O^+(aq)$ interchangeably, to refer to the same thing—a hydronium ion.

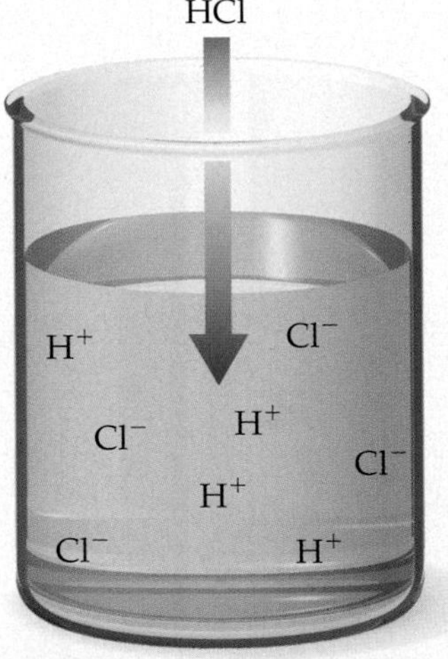

$HCl(aq) \longrightarrow H^+(aq) + Cl^-(aq)$

Figure 14.5 The Arrhenius definition states that an acid is a substance that produces H^+ ions in solution. These H^+ ions associate with H_2O to form H_3O^+ ions.

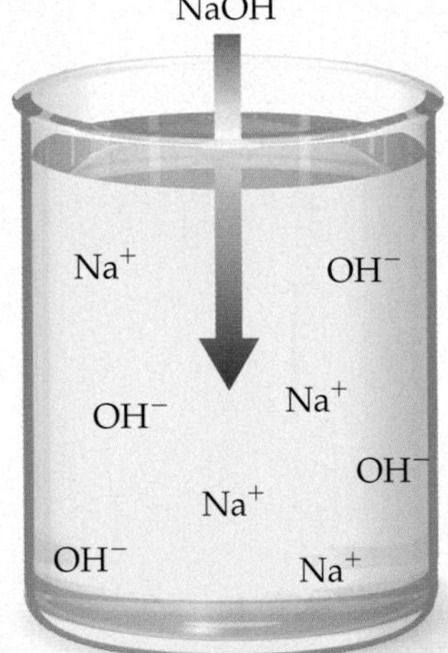

$NaOH(aq) \longrightarrow Na^+(aq) + OH^-(aq)$

Figure 14.6 The Arrhenius definition states that a base is a substance that produces OH^- ions in solution.

NaOH is an Arrhenius base because it produces OH^- ions in solution (Figure 14.6).

$$NaOH(aq) \rightarrow Na^+(aq) + OH^-(aq)$$

NaOH is an ionic compound and therefore contains Na^+ and OH^- ions. When NaOH is added to water, it **dissociates** or breaks apart into its component ions.

Ionic compounds such as NaOH are composed of positive and negative ions. In solution, soluble ionic compounds dissociate into their component ions.

Under the Arrhenius definition, acids and bases naturally combine to form water, neutralizing each other in the process.

$$H^+(aq) + OH^-(aq) \rightarrow H_2O(l)$$

The Brønsted-Lowry Definition

A second definition of acids and bases, called the **Brønsted-Lowry definition,** was introduced in 1923 and applies to a larger range of acid–base phenomena. This definition focuses on the transfer of H^+ ions in an acid–base reaction. Since an H^+ ion is a proton—a hydrogen atom with its electron taken away—this definition uses the idea of a proton donor and a proton acceptor.

Acid–An acid is a proton (H^+ ion) *donor*.
Base–A base is a proton (H^+ ion) *acceptor*.

According to this definition, HCl is an acid because, in solution, it donates a proton to water.

$$HCl(aq) + H_2O(l) \rightarrow H_3O^+(aq) + Cl^-(aq)$$

This definition more clearly shows what happens to the H^+ ion from an acid—it associates with a water molecule to form H_3O^+ (a hydronium ion). The Brønsted-Lowry definition also works well with bases (such as NH_3) that do not inherently contain OH^- ions but that still produce OH^- ions in solution. In the Brønsted-Lowry definition, NH_3 is a base because it accepts a proton from water.

The double arrows in this equation indicate that the reaction does not go to completion. We discuss this concept in more detail in Section 14.7.

$$NH_3(aq) + H_2O(l) \rightleftharpoons NH_4^+(aq) + OH^-(aq)$$

In the Brønsted-Lowry definition, acids (proton donors) and bases (proton acceptors) always occur together. In the reaction between HCl and H_2O, HCl is the proton donor (acid) and H_2O is the proton acceptor (base).

$$\underset{\substack{\text{Acid}\\\text{(Proton donor)}}}{HCl(aq)} + \underset{\substack{\text{Base}\\\text{(Proton acceptor)}}}{H_2O(l)} \rightarrow H_3O^+(aq) + Cl^-(aq)$$

In the reaction between NH_3 and H_2O, H_2O is the proton donor (acid) and NH_3 is the proton acceptor (base).

$$\underset{\substack{\text{Base}\\\text{(Proton acceptor)}}}{NH_3(aq)} + \underset{\substack{\text{Acid}\\\text{(Proton donor)}}}{H_2O(l)} \rightleftharpoons NH_4^+(aq) + OH^-(aq)$$

Notice that under the Brønsted-Lowry definition, some substances—such as water in the previous two equations—can act as acids *or* bases. Substances that can act as acids or bases are termed **amphoteric.** Notice also what happens when an equation representing Brønsted-Lowry acid–base behavior is reversed.

$$\underset{\substack{\text{Acid} \\ \text{(Proton donor)}}}{NH_4^+(aq)} + \underset{\substack{\text{Base} \\ \text{(Proton acceptor)}}}{OH^-(aq)} \rightleftharpoons NH_3(aq) + H_2O(l)$$

In this reaction, NH_4^+ is the proton donor (acid) and OH^- is the proton acceptor (base). What was the base (NH_3) has become the acid (NH_4^+) and vice versa. NH_4^+ and NH_3 are often referred to as a **conjugate acid–base pair,** two substances related to each other by the transfer of a proton (Figure 14.7). Going back to the original forward reaction, we can identify the conjugate acid-base pairs as follows:

$$\underset{\text{Base}}{NH_3(aq)} + \underset{\text{Acid}}{H_2O(l)} \rightleftharpoons \underset{\substack{\text{Conjugate} \\ \text{acid}}}{NH_4^+(aq)} + \underset{\substack{\text{Conjugate} \\ \text{base}}}{OH^-(aq)}$$

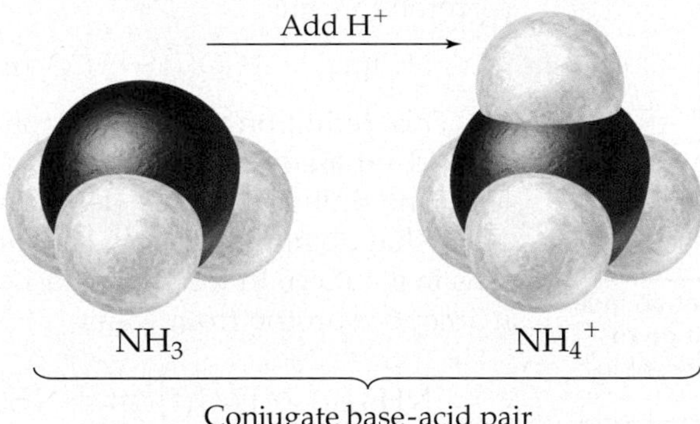

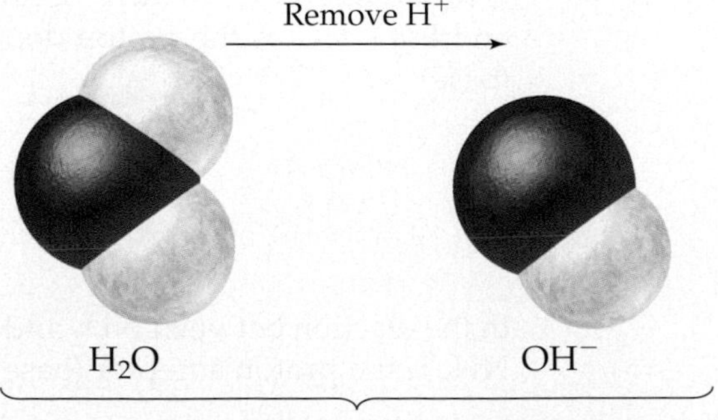

Figure 14.7 A conjugate acid–base pair is any two substances related to each other by the transfer of a proton.

In an acid–base reaction, a base accepts a proton and becomes a conjugate acid. An acid donates a proton and becomes a conjugate base.

Identifying Brønsted-Lowry Acids and Bases and Their Conjugates

In each of the following reactions, identify the Brønsted-Lowry acid, the Brønsted-Lowry base, the conjugate acid, and the conjugate base.

a) $H_2SO_4(aq) + H_2O(l) \rightarrow H_3O^+(aq) + HSO_4^-(aq)$
b) $HCO_3^-(aq) + H_2O(l) \rightleftharpoons H_2CO_3(aq) + OH^-(aq)$

Solution:

a) Since H_2SO_4 donates a proton to H_2O in this reaction, it is the acid

$$\underset{\text{Acid}}{H_2SO_4(aq)} + \underset{\text{Base}}{H_2O(l)} \longrightarrow \underset{\text{Conjugate base}}{HSO_4^-(aq)} + \underset{\text{Conjugate acid}}{H_3O^+(aq)}$$

(proton donor). After H_2SO_4 donates the proton, it becomes HSO_4^-, the conjugate base. Since H_2O accepts a proton, it is the base (proton acceptor). After H_2O accepts the proton it becomes H_3O^+, the conjugate acid.

b) Since H_2O donates a proton to HCO_3^- in this reaction, it is the acid

$$\underset{\text{Base}}{HCO_3^-(aq)} + \underset{\text{Acid}}{H_2O(l)} \longrightarrow \underset{\text{Conjugate acid}}{H_2CO_3(aq)} + \underset{\text{Conjugate base}}{OH^-(aq)}$$

(proton donor). After H_2O donates the proton, it becomes OH^-, the conjugate base. Since HCO_3^- accepts a proton, it is the base (proton acceptor). After HCO_3^- accepts the proton it becomes H_2CO_3, the conjugate acid.

SKILLBUILDER 14.1 **Identifying Brønsted-Lowry Acids and Bases and Their Conjugates**

14.5 Reactions of Acids and Bases

Neutralization Reactions

Neutralization reactions, also known as acid–base reactions, are covered in Section 7.8.

One of the most important acid–base reactions is **neutralization,** first introduced in Chapter 7. When an acid and a base are mixed, the $H^+(aq)$ from the acid combines with the $OH^-(aq)$ from the base to form $H_2O(l)$. For example, consider the reaction between hydrochloric acid and potassium hydroxide.

$$HCl(aq) + KOH(aq) \rightarrow H_2O(l) + KCl(aq)$$

Acid Base Water Salt

Acid–base reactions generally form water and a **salt**—an ionic compound—that usually remains dissolved in the solution. The salt contains the cation from the base and the anion from the acid.

Net ionic equations are defined in Section 7.7.

$$\text{Acid} + \text{Base} \longrightarrow \text{Water} + \text{Salt}$$

Salt: Ionic compound that contains the cation from the base and the anion from the acid

The net ionic equation for many neutralization reactions is:

$$H^+(aq) + OH^-(aq) \rightarrow H_2O(l)$$

A slightly different but common type of neutralization reaction involves an acid reacting with carbonates or bicarbonates (compounds containing CO_3^{2-} or HCO_3^-.) This type of neutralization reaction produces water, gaseous carbon dioxide, and a salt. As an example, consider the reaction of hydrochloric acid and sodium bicarbonate.

$$HCl(aq) + NaHCO_3(aq) \rightarrow H_2O(l) + CO_2(g) + NaCl(aq)$$

Since this reaction produces gaseous CO_2, it is also called a *gas evolution reaction.*

The reaction of carbonates or bicarbonates with acids produces water, gaseous carbon dioxide, and a salt.

See Section 7.8 for more on gas evolution reactions.

EXAMPLE 14.2 Writing Equations for Neutralization Reactions

Write a molecular equation for the reaction between aqueous HCl and aqueous $Ca(OH)_2$.

Solution:

We must identify the acid and the base and know that they react to form water and a salt. Notice that $Ca(OH)_2$ contains 2 mol of OH^- for every one mol of $Ca(OH)_2$ and will therefore require 2 moles of H^+ to neutralize it. We first write the skeletal reaction.

$$HCl(aq) + Ca(OH)_2(aq) \rightarrow H_2O(l) + CaCl_2(aq)$$

We then balance the equation.

$$2\,HCl(aq) + Ca(OH)_2(aq) \rightarrow 2\,H_2O(l) + CaCl_2(aq)$$

SKILLBUILDER 14.2 Writing Equations for Neutralization Reactions

Write a molecular equation for the reaction that occurs between aqueous H_3PO_4 and aqueous NaOH. (Hint: H_3PO_4 is a triprotic acid, meaning that 1 mol H_3PO_4 requires 3 mol OH^- to completely react with it.)

$$2\,HCl(aq) + Mg(s) \longrightarrow H_2(g) + MgCl_2(aq)$$

The reaction between an acid and a metal usually produces hydrogen gas and a dissolved salt containing the metal ion.

Acid Reactions

In Section 14.1, we saw that acids dissolve metals or more precisely, that acids react with metals in a way that causes them to go into solution. The reaction between an acid and a metal usually produces hydrogen gas and a dissolved salt containing the metal ion as the cation. For example, hydrochloric acid reacts with magnesium metal to form hydrogen gas and magnesium chloride.

$$\underset{\text{Acid}}{2\,HCl(aq)} + \underset{\text{Metal}}{Mg(s)} \rightarrow \underset{\text{Hydrogen gas}}{H_2(g)} + \underset{\text{Salt}}{MgCl_2(aq)}$$

Similarly, sulfuric acid reacts with zinc to form hydrogen gas and zinc sulfate.

$$\underset{\text{Acid}}{H_2SO_4(aq)} + \underset{\text{Metal}}{Zn(s)} \rightarrow \underset{\text{Hydrogen gas}}{H_2(g)} + \underset{\text{Salt}}{ZnSO_4(aq)}$$

It is through reactions such as these that the acid in James Bond's pen from our opening example dissolves the metal bars that imprison him. For example, if the bars were made of iron and the acid in the pen were hydrochloric acid, the reaction would be:

$$\underset{\text{Acid}}{2\,HCl\,(aq)} + \underset{\text{Metal}}{Fe(s)} \rightarrow \underset{\text{Hydrogen gas}}{H_2(g)} + \underset{\text{Salt}}{FeCl_2(aq)}$$

We should note, however, that some metals do not react with acids. If the bars that imprisoned James Bond were made of gold, for example, a pen filled with hydrochloric acid would not dissolve the bars.

The way to determine if a particular metal dissolves in an acid is presented in section 16.5.

Acids also react with metal oxides to produce water and a dissolved salt. For example, hydrochloric acid reacts with potassium oxide to form water and potassium chloride.

$$\underset{\text{Acid}}{2\,HCl(aq)} + \underset{\text{Metal oxide}}{K_2O(s)} \rightarrow \underset{\text{Water}}{H_2O(l)} + \underset{\text{Salt}}{2\,KCl(aq)}$$

Similarly, hydrobromic acid reacts with magnesium oxide to form water and magnesium bromide.

$$\underset{\text{Acid}}{2\,HBr(aq)} + \underset{\text{Metal oxide}}{MgO(s)} \rightarrow \underset{\text{Water}}{H_2O(l)} + \underset{\text{Salt}}{MgBr_2(aq)}$$

EXAMPLE 14.3 Writing Equations for Acid Reactions

Write an equation for each of the following:

a) The reaction of hydroiodic acid with potassium metal
b) The reaction of hydrobromic acid with sodium oxide

Solution:

a) The reaction of hydroiodic acid with potassium metal forms hydrogen gas and a salt. The salt contains the ionized form of the metal (K^+) as the cation and the anion of the acid (I^-) as the anion. We first write the skeletal equation.

$$HI(aq) + K(s) \rightarrow H_2(g) + KI(aq)$$

We then balance the equation.

$$2\,HI(aq) + 2\,K(s) \rightarrow H_2(g) + 2\,KI(aq)$$

b) The reaction of hydrobromic acid with sodium oxide forms water and a salt. The salt contains the cation from the metal oxide (Na^+) and the anion of the acid (Br^-). We first write the skeletal equation.

$$HBr(aq) + Na_2O(s) \rightarrow H_2O(l) + NaBr(aq)$$

We then balance the equation.

$$2\,HBr(aq) + Na_2O(s) \rightarrow H_2O(l) + 2\,NaBr(aq)$$

SKILLBUILDER 14.3 **Writing Equations for Acid Reactions**

Write an equation for each of the following:

a) The reaction of hydrochloric acid with strontium metal

b) The reaction of hydroiodic acid with barium oxide

Base Reactions

The most important base reactions are those in which a base neutralizes an acid (see the beginning of this section). The only other kind of base reaction that we cover in this book is the reaction of sodium hydroxide with aluminum and water.

$$2\,NaOH(aq) + 2\,Al(s) + 6\,H_2O(l) \rightarrow 2\,NaAl(OH)_4(aq) + 3\,H_2(g)$$

Aluminum is one of the few metals that dissolves in a base. Consequently, it is safe to use NaOH (the main ingredient in many drain-opening products) to unclog your drain as long as your pipes are not made of aluminum, which is against most building codes.

EVERYDAY CHEMISTRY

What Is in My Antacid?

Heartburn, a burning sensation in the lower throat and above the stomach, is caused by the reflux or backflow of stomach acid into the esophagus (the tube that joins the stomach to the throat). In healthy individuals, this occurs only occasionally, especially after large meals. Physical activity—such as bending, stooping, or lifting—after meals also aggravates heartburn. Antacids such as Mylanta or Phillips' milk of magnesia contain bases that neutralize the refluxed stomach acid, alleviating heartburn. In some people, the flap between the esophagus and the stomach that normally prevents acid reflux becomes damaged, in which case heartburn becomes a regular occurrence.

CAN YOU ANSWER THIS? *Look at the labels of Mylanta (left) and Phillips' milk of magnesia (right). Can you identify the bases responsible for the antacid action? Write chemical equations showing the reactions of these bases with stomach acid (HCl).*

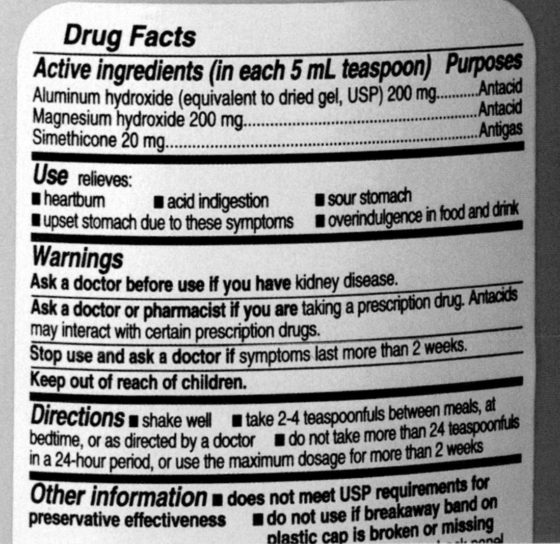

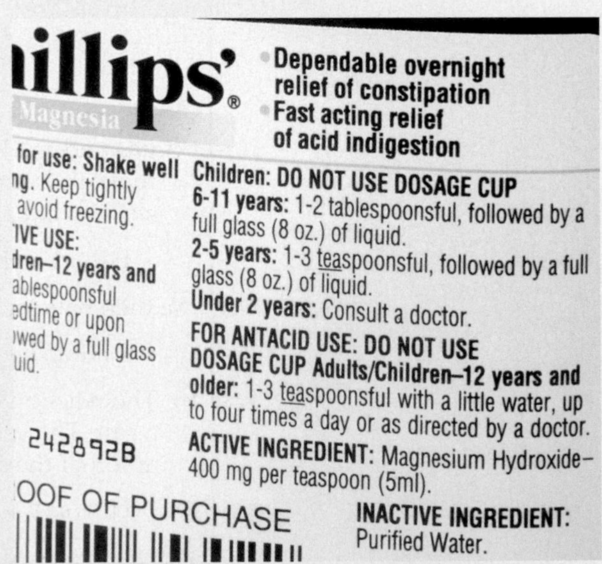

14.6 Acid–Base Titration: A Way to Quantify the Amount of Acid or Base in a Solution

The principles we learned in Chapter 13 (Section 13.8) on solution stoichiometry can be applied to a common laboratory procedure called a **titration.** In a titration, a reactant in a solution of known concentration is reacted with another reactant in a solution of unknown concentration. By precisely measuring the volume of each solution required to reach the **endpoint** of the reaction—the point at which the reactants are in exact stoichiometric proportions—the concentration of the unknown solution can be determined.

In an acid–base titration, the reactants are an acid and a base, and the reaction is a neutralization reaction. The endpoint occurs when the number of moles of H^+ from the acid equals the number of moles of OH^- from the base. The endpoint is usually detected with an **indicator,** a substance that changes color with acidity level.

(a)

(b)

(c)

In this titration, NaOH is added to an HCl solution. When the NaOH and HCl reach stoichiometric proportions (1 mol OH^- for every 1 mol of H^+), the indicator (phenolphthalein) changes to pink, signaling the endpoint of the titration.

Phenolphthalein is an indicator that is colorless in acidic solution and pink in basic solution.

EXAMPLE 14.4 Acid-Base Titration

The titration of 10.00 mL of an HCl solution of unknown concentration requires 12.54 mL of a 0.100 M NaOH solution to reach the endpoint. What is the concentration of the unknown HCl solution?

Set up the problem in the standard way, including an equation for the neutralization reaction of HCl and NaOH.

Given: 10.00 mL HCl solution

12.54 mL of a 0.100 M NaOH solution

Find: concentration of HCl solution (mol/L)

Equations:

$$HCl(aq) + NaOH(aq) \rightarrow H_2O(l) + NaCl(aq)$$

$$\text{Molarity (M)} = \frac{\text{mol solute}}{\text{L solution}}$$

Solution Maps:

In the first solution map, we use the volume of NaOH required to reach the endpoint to calculate the moles of HCl in the solution. The final conversion factor comes from the balanced neutralization equation.

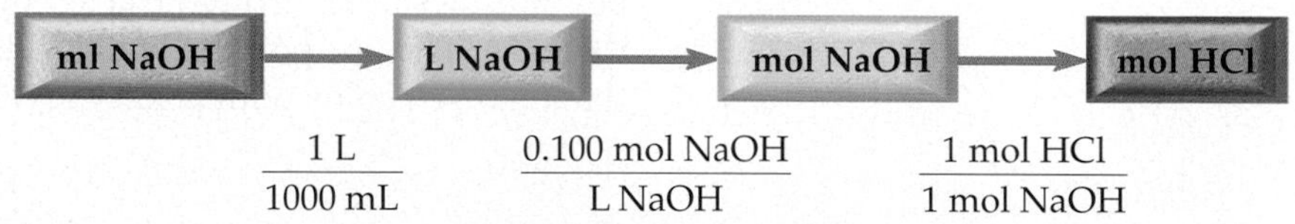

$$\frac{1\ \text{L}}{1000\ \text{mL}} \qquad \frac{0.100\ \text{mol NaOH}}{\text{L NaOH}} \qquad \frac{1\ \text{mol HCl}}{1\ \text{mol NaOH}}$$

In the second solution map, we use the moles of HCl and the volume of HCl solution to determine the molarity of the HCl solution.

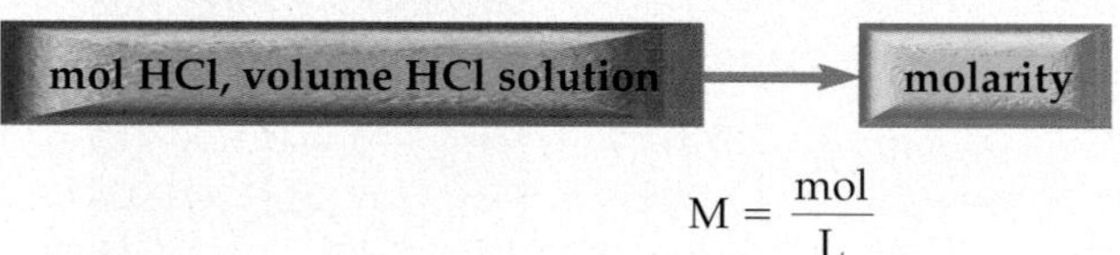

$$M = \frac{\text{mol}}{\text{L}}$$

Solution:

$$12.54\ \text{mL NaOH} \times \frac{1\ \text{L}}{1000\ \text{mL}} \times \frac{0.100\ \text{mol NaOH}}{\text{L NaOH}} \times \frac{1\ \text{mol HCl}}{1\ \text{mol NaOH}} = 1.25 \times 10^{-3}\ \text{mol HCl}$$

This is the number of moles of HCl in the unknown solution. To get the concentration of the solution, we divide the number of moles of HCl by the volume of the HCl solution in L. A volume of 10.00 mL is equivalent to 0.01000 L.

$$\text{Molarity} = \frac{1.25 \times 10^{-3}\ \text{mol HCl}}{0.01000\ \text{L}} = 0.125\ \text{M}$$

The unknown HCl solution has a concentration of 0.125 M.

SKILLBUILDER 14.4 **Acid-Base Titration**

The titration of a 20.0 mL sample of an H_2SO_4 solution of unknown concentration requires 22.87 mL of a 0.158 M KOH solution to reach the endpoint. What is the concentration of the unknown H_2SO_4 solution?

14.7 Strong and Weak Acids and Bases

Strong Acids

Hydrochloric acid (HCl) and hydrofluoric acid (HF) appear similar, but there is an important difference between these two acids. HCl is an example of a **strong acid,** one that completely ionizes in solution.

Single arrow indicates complete ionization

$$HCl(aq) + H_2O(l) \longrightarrow H_3O^+(aq) + Cl^-(aq)$$

We show the *complete* ionization of HCl with a single arrow pointing to the right in the equation. An HCl solution contains no intact HCl; the HCl has all reacted with water to form $H_3O^+(aq)$ and $Cl^-(aq)$ (Figure 14.8). A 1.0 M HCl solution will have an H_3O^+ concentration of 1.0 M. The concentration of H_3O^+ is often abbreviated as $[H_3O^+]$. Using this notation, a 1 M HCl solution has $[H_3O^+] = 1.0$ M.

Strong electrolyte solutions were first defined in Section 7.6.

A strong acid is an example of a **strong electrolyte,** a substance whose aqueous solutions are good conductors of electricity (Figure 14.9). Aqueous solutions require the presence of charged particles to conduct electricity. Pure water is not a good conductor of electricity because it has relatively few charged particles. The danger of using electrical devices—such as a hair dryer, for example—while sitting in water is that water is seldom pure and often contains dissolved ions. If the device were to come in contact with the water, dangerously high levels of electricity could flow through it. Strong acid solutions are also strong electrolyte solutions because each acid molecule ionizes into positive and negative ions. These mobile ions are good conductors of electricity.

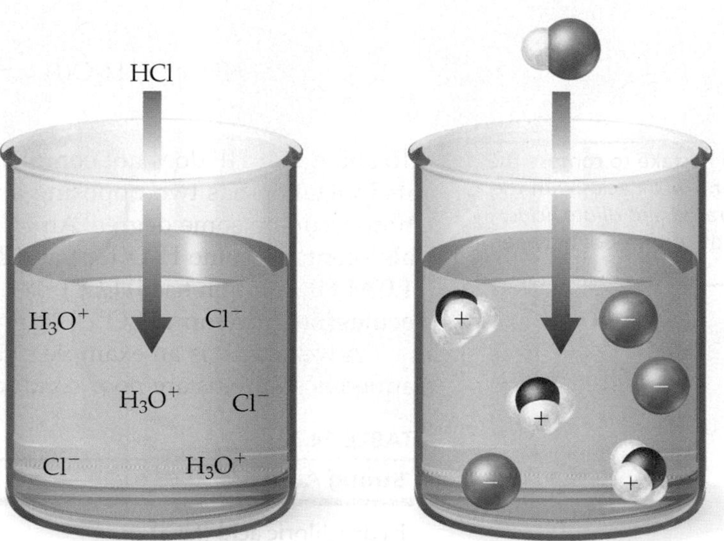

Figure 14.8 When HCl dissolves in water, it completely ionizes into H_3O^+ and Cl^- ions. The solution contains no intact HCl.

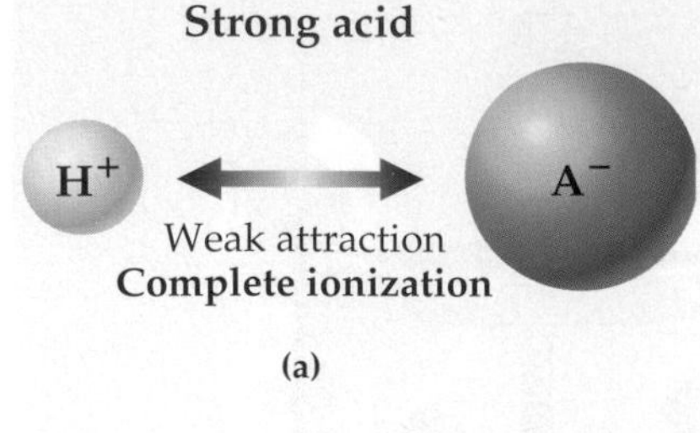

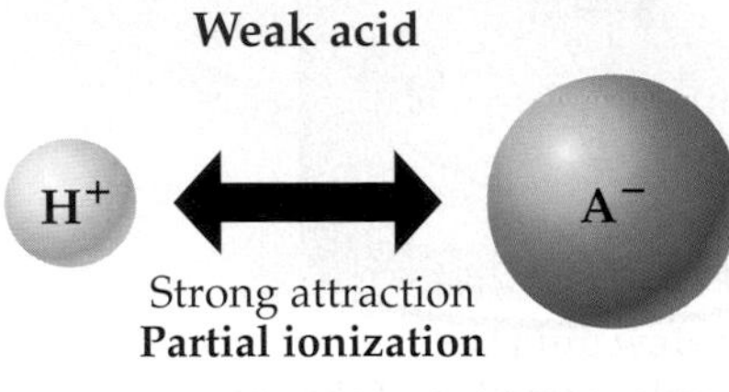

Figure 14.12 **a)** In a strong acid, the attraction between H^+ and A^- is low, resulting in complete ionization. **b)** In a weak acid, the attraction between H^+ and A^- is high, resulting in partial ionization.

Notice that the strength of a conjugate base is related to its attraction to H^+ in solution.

If the attraction between H^+ and A^- is *weak,* then the reaction favors the forward direction and the acid is *strong* (Figure 14.12a). If the attraction between H^+ and A^- is *strong,* then the reaction favors the reverse direction and the acid is *weak* (Figure 14.12b).

For example, in HCl, the conjugate base (Cl^-) has a relatively weak attraction to H^+, meaning that the reverse reaction does not occur to any significant extent. In HF, on the other hand, the conjugate base (F^-) has a greater attraction to H^+, meaning that the reverse reaction occurs to a significant degree. *In general, the stronger the acid, the weaker the conjugate base and vice versa.* This means that if the forward reaction (that of the acid) has a high tendency to occur, then the reverse reaction (that of the conjugate base) has a low tendency to occur. Table 14.4 lists some common weak acids.

Notice that two of the weak acids in Table 14.4 are diprotic, (meaning they have two ionizable protons, and one is triprotic, meaning that it has three ionizable protons. Let us return to sulfuric acid for a moment. Sulfuric acid is a diprotic acid that is strong in its first ionizable proton

$$H_2SO_4(aq) + H_2O(l) \rightarrow H_3O^+(aq) + HSO_4^-(aq)$$

but weak in its second ionizable proton.

$$HSO_4^-(aq) + H_2O(l) \rightleftharpoons H_3SO^+(aq) + O_4^{2-}(aq)$$

Sulfurous acid and carbonic acid are weak in both of their ionizable protons, and phosphoric acid is weak in all three of its ionizable protons.

TABLE 14.4

Weak Acids	
hydrofluoric acid (HF)	sulfurous acid (H_2SO_3) *(diprotic)*
acetic acid ($HC_2H_3O_2$)	carbonic acid (H_2CO_3) *(diprotic)*
formic acid ($HCHO_2$)	phosphoric acid (H_3PO_4) *(triprotic)*

EXAMPLE 14.5 **Determining $[H_3O^+]$ in Acid Solutions**

What is the H_3O^+ concentration in each of the following solutions?

a) 1.5 M HCl
b) 3.0 M $HC_2H_3O_2$
c) 2.5 M HNO_3

Solution:

a) Since HCl is a strong acid, it completely ionizes. The concentration of H_3O^+ will be 1.5 M.

$$[H_3O^+] = 1.5\ M$$

b) Since $HC_2H_3O_2$ is a weak acid, it partially ionizes. The calculation of the exact concentration of H_3O^+ is beyond the scope of this text, but we know it will be less than 3.0 M.

$$[H_3O^+] < 3.0\ M$$

c) Since HNO_3 is a strong acid, it completely ionizes. The concentration of H_3O^+ will be 2.5 M.

$$[H_3O^+] = 2.5\ M$$

SKILLBUILDER 14.5 Determining $[H_3O^+]$ in Acid Solutions

What is the H_3O^+ concentration in each of the following solutions?

a) 0.50 M $HCHO_2$
b) 1.25 M HI
c) 0.75 M HF

Strong Bases

In analogy to the definition of a strong acid, a **strong base** is a base that completely dissociates in solution. NaOH, for example, is a strong base.

$$NaOH(aq) \rightarrow Na^+(aq) + OH^-(aq)$$

An NaOH solution contains no intact NaOH—it has all dissociated to form $Na^+(aq)$ and $OH^-(aq)$ (Figure 14.13). In other words, a 1.0 M NaOH solution will have $[OH^-] = 1.0$ M and $[Na^+] = 1.0$ M. Some common strong bases are listed in Table 14.5.

Some of these strong bases, such as $Sr(OH)_2$, contain 2 OH^- ions. These bases completely dissociate, producing two moles of OH^- per mole of the base. For example, $Sr(OH)_2$ dissociates as follows:

$$Sr(OH)_2(aq) \rightarrow Sr^{2+}(aq) + 2\,OH^-(aq)$$

Unlike diprotic acids, which ionize in two steps, bases containing 2 OH^- ions dissociate in one step.

Figure 14.13 When NaOH dissolves in water, it completely dissociates into Na^+ and OH^-. **Question:** The solution contains no intact NaOH. Would NaOH be a strong or weak electrolyte?

TABLE 14.5

Strong Bases	
lithium hydroxide (LiOH)	strontium hydroxide ($Sr(OH)_2$)
sodium hydroxide (NaOH)	calcium hydroxide ($Ca(OH)_2$)
potassium hydroxide (KOH)	barium hydroxide ($Ba(OH)_2$)

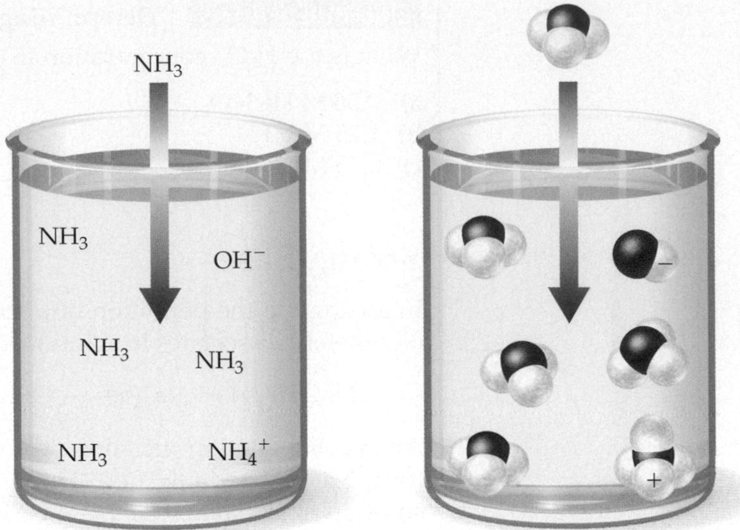

Figure 14.14 When NH_3 dissolves in water, it partially ionizes to form NH_4^+ and OH^-. However, only a fraction of the molecules ionize. Most NH_3 molecules remain as NH_3. **Question:** Would NH_3 be a strong or weak electrolyte?

Weak Bases

A **weak base** is analogous to a weak acid. Unlike strong bases that contain OH^- and dissociate in water, the most common weak bases produce OH^- by accepting a proton from water, ionizing water to form OH^-.

$$B(aq) + H_2O(l) \rightleftharpoons BH^+(aq) + OH^-(aq)$$

In this equation; *B* is simply generic for a weak base. Ammonia, for example, ionizes water according to the following reaction.

$$NH_3(aq) + H_2O(l) \rightleftharpoons NH_4^+(aq) + OH^-(aq)$$

The double arrow indicates that the ionization is not complete. An NH_3 solution contains NH_3, NH_4^+, and OH^- (Figure 14.14). A 1.0 M NH_3 solution will have $[OH^-] < 1.0$ M. Table 14.6 lists some common weak bases.

Calculating exact $[OH^-]$ for weak bases is beyond the scope of this text.

TABLE 14.6

Weak Bases	Ionization Reaction
ammonia (NH_3)	$NH_3(aq) + H_2O(l) \rightleftharpoons NH_4^+(aq) + OH^-(aq)$
pyridine (C_5H_5N)	$C_5H_5N(aq) + H_2O(l) \rightleftharpoons C_5H_5NH^+(aq) + OH^-(aq)$
methyl amine (CH_3NH_2)	$CH_3NH_2(aq) + H_2O(l) \rightleftharpoons CH_3NH_3^+(aq) + OH^-(aq)$
ethyl amine ($C_2H_5NH_2$)	$C_2H_5NH_2(aq) + H_2O(l) \rightleftharpoons C_2H_5NH_3^+(aq) + OH^-(aq)$
bicarbonate ion (HCO_3^-)*	$HCO_3^-(aq) + H_2O(l) \rightleftharpoons H_2CO_3(aq) + OH^-(aq)$

*The bicarbonate ion must occur with a positively charged ion such as Na^+ that serves to balance the charge, but does not have any part in the ionization reaction. It is the bicarbonate ion that makes sodium bicarbonate ($NaHCO_3$) basic.

EXAMPLE 14.6 Determining [OH^-] in Base Solutions

What is the OH^- concentration in each of the following solutions?

a) 2.25 M KOH
b) 0.35 M CH_3NH_2
c) 0.025 M $Sr(OH)_2$

Solution:

a) Since KOH is a strong base, it completely dissociates into K^+ and OH^- in solution. The concentration of OH^- will be 2.25 M.

$$[OH^-] = 2.25 \text{ M}$$

b) Since CH_3NH_2 is a weak base, it only partially ionizes water. We cannot calculate the exact concentration of OH^-, but we know it will be less than 0.35 M.

$$[OH^-] < 0.35 \text{ M}$$

c) Since $Sr(OH)_2$ is a strong base, it completely dissociates into $Sr^{2+}(aq)$ and $2\,OH^-(aq)$. $Sr(OH)_2$ forms 2 mol of OH^- for every 1 mol of $Sr(OH)_2$. Consequently, the concentration of OH^- will be twice the concentration of $Sr(OH)_2$.

$$[OH^-] = 2(0.025\text{M}) = 0.050 \text{ M}$$

SKILLBUILDER 14.6 Determining [OH^-] in Base Solutions

What is the OH^- concentration in each of the following solutions?

a) 0.055 M $Ba(OH)_2$
b) 1.05 M C_5H_5N
c) 0.45 M NaOH

14.8 Water: Acid and Base in One

We saw earlier that water acts as a base when it reacts with HCl and as an acid when it reacts with NH_3.

Water acting as a base

$$\underset{\substack{\text{Acid} \\ \text{(Proton donor)}}}{HCl(aq)} + \underset{\substack{\text{Base} \\ \text{(Proton acceptor)}}}{H_2O(l)} \longrightarrow H_3O^+(aq) + Cl^-(aq)$$

Water acting as an acid

$$\underset{\substack{\text{Base} \\ \text{(Proton acceptor)}}}{NH_3(aq)} + \underset{\substack{\text{Acid} \\ \text{(Proton donor)}}}{H_2O(l)} \rightleftharpoons NH_4^+(aq) + OH^-(aq)$$

Water is *amphoteric*; it can act as either an acid or base. Even in pure water, water acts as an acid and a base with itself, a process called self-ionization.

Water acting as both an acid and a base

$$\underset{\substack{\text{Acid} \\ \text{(Proton donor)}}}{H_2O(l)} + \underset{\substack{\text{Base} \\ \text{(Proton acceptor)}}}{H_2O(l)} \rightleftharpoons H_3O^+(aq) + OH^-(aq)$$

The units of K_w are normally dropped.

We can quantify the degree of water self-ionization with the **ion product constant for water (K_w)**, the product of the H_3O^+ ion concentration and the OH^- ion concentration in an aqueous solution. At room temperature, $K_w = 1.0 \times 10^{-14}$.

$$[H_3O^+][OH^-] = K_w = 1.0 \times 10^{-14}$$

where $[H_3O^+]$ = the concentration of H_3O^+ in M

and $[OH^-]$ = the concentration of OH^- in M

The preceding equation holds true for all aqueous solutions at room temperature. The concentration of H_3O^+ times the concentration of OH^- will always be 1.0×10^{-14} at room temperature. In pure water, since H_2O is the only source of these ions, there is one H_3O^+ ion for every OH^- ion. Consequently, the concentrations of H_3O^+ and OH^- are equal. Such a solution is a **neutral solution**.

$$[H_3O^+] = [OH^-] = \sqrt{K_w} = 1.0 \times 10^{-7}\,M \quad \text{(in pure water)}$$

As you can see, in pure water, the concentrations of H_3O^+ and OH^- are *very small* (1.0×10^{-7} M).

An **acidic solution** contains an acid that creates additional H_3O^+ ions, causing $[H_3O^+]$ to increase. However, the *ion product constant still applies.*

$$[H_3O^+][OH^-] = K_w = 1.0 \times 10^{-14}$$

If $[H_3O^+]$ increases, then $[OH^-]$ must decrease for the ion product to remain 1.0×10^{-14}. For example, suppose $[H_3O^+] = 1.0 \times 10^{-3}$ M, then $[OH^-]$ can be found by solving the ion product expression for $[OH^-]$.

$$[1.0 \times 10^{-3}][OH^-] = 1.0 \times 10^{-14}$$

$$[OH^-] = \frac{1.0 \times 10^{-14}}{1.0 \times 10^{-3}} = 1.0 \times 10^{-11}\,M$$

In an acidic solution $[H_3O^+] > 1.0 \times 10^{-7}$ M and $[OH^-] < 1.0 \times 10^{-7}$ M.

A **basic solution** contains a base that creates additional OH^- ions causing the $[OH^-]$ to increase and the $[H_3O^+]$ to decrease. For example, suppose $[OH^-] = 1.0 \times 10^{-2}$ M, then $[H_3O^+]$ can be found by solving the ion product expression for $[H_3O^+]$.

$$[H_3O^+][1.0 \times 10^{-2}] = 1.0 \times 10^{-14}$$

$$[H_3O^+] = \frac{1.0 \times 10^{-14}}{1.0 \times 10^{-2}} = 1.0 \times 10^{-12}\,M$$

In a basic solution $[OH^-] > 1.0 \times 10^{-7}$ M and $[H_3O^+] < 1.0 \times 10^{-7}$ M.

To summarize:

- ***A neutral solution contains $[H_3O^+] = [OH^-] = 1.0 \times 10^{-7}$ M***
- ***An acidic solution contains***
  $[H_3O^+] > 1.0 \times 10^{-7}$ M $[OH^-] < 1.0 \times 10^{-7}$ M
- ***A basic solution contains***
  $[OH^-] > 1.0 \times 10^{-7}$ M $[H_3O^+] < 1.0 \times 10^{-7}$
- ***In all aqueous solutions at 25 °C $[H_3O^+][OH^-] = K_w = 1.0 \times 10^{-14}$***

EXAMPLE 14.7 Using K_w in Calculations

Calculate $[OH^-]$ in each of the following solutions and determine if the solution is acidic, basic or neutral.

a) $[H_3O^+] = 7.5 \times 10^{-5}$ M
b) $[H_3O^+] = 1.5 \times 10^{-9}$ M
c) $[H_3O^+] = 1.0 \times 10^{-7}$ M

Solution:

a) To find $[OH^-]$ we use the ion product constant.

$$[H_3O^+][OH^-] = K_w = 1.0 \times 10^{-14}$$

We substitute the given value for $[H_3O^+]$ and solve the equation for $[OH^-]$.

$$[7.5 \times 10^{-5}][OH^-] = 1.0 \times 10^{-14}$$

$$[OH^-] = \frac{1.0 \times 10^{-14}}{7.5 \times 10^{-5}} = 1.3 \times 10^{-10}\ \text{M}$$

Since $[H_3O^+] > 1.0 \times 10^{-7}$ M and $[OH^-] < 1.0 \times 10^{-7}$ M, the solution is acidic.

b) We again substitute the given value for $[H_3O^+]$ and solve the ion product equation for $[OH^-]$.

$$[1.5 \times 10^{-9}][OH^-] = 1.0 \times 10^{-14}$$

$$[OH^-] = \frac{1.0 \times 10^{-14}}{1.5 \times 10^{-9}} = 6.7 \times 10^{-6}\ \text{M}$$

Since $[H_3O^+] < 1.0 \times 10^{-7}$ M and $[OH^-] > 1.0 \times 10^{-7}$ M, the solution is basic.

c) We again substitute the given value for $[H_3O^+]$ and solve the ion product equation for $[OH^-]$.

$$[1.0 \times 10^{-7}][OH^-] = 1.0 \times 10^{-14}$$

$$[OH^-] = \frac{1.0 \times 10^{-14}}{1.0 \times 10^{-7}} = 1.0 \times 10^{-7}\ \text{M}$$

Since $[H_3O^+] = 1.0 \times 10^{-7}$ M and $[OH^-] = 1.0 \times 10^{-7}$ M, the solution is neutral.

SKILLBUILDER 14.7 Using K_w in Calculations

Calculate $[H_3O^+]$ in each of the following solutions and determine if the solution is acidic, basic, or neutral.

a) $[OH^-] = 1.5 \times 10^{-2}$ M
b) $[OH^-] = 1.0 \times 10^{-7}$ M
c) $[OH^-] = 8.2 \times 10^{-10}$ M

14.9 The pH Scale: A Way to Express Acidity and Basicity

TABLE 14.7

The pH of Some Common Substances

Substance	*pH*
gastric (human stomach) acid	1.0–3.0
limes	1.8–2.0
lemons	2.2–2.4
soft drinks	2.0–4.0
plums	2.8–3.0
wine	2.8–3.8
apples	2.9–3.3
peaches	3.4–3.6
cherries	3.2–4.0
beer	4.0–5.0
rainwater (unpolluted)	5.6
human blood	7.3–7.4
egg whites	7.6–8.0
milk of magnesia	10.5
household ammonia	10.5–11.5
4% NaOH solution	14

Notice that an *increase* of 1 in pH corresponds to a *decrease* of a power of 10 in $[H_3O^+]$.

$\log 10^1 = 1$; $\log 10^2 = 2$; $\log 10^3 = 3$; $\log 10^{-1} = -1$; $\log 10^{-2} = -2$; $\log 10^{-3} = -3$
The log of a number is the exponent to which 10 must be raised to obtain that number.

When you take the log of a quantity, the result should have the same number of decimal places as the number of significant figures in the original quantity.

The **pH** scale is a compact way to specify the acidity of a solution. In general, if:

- pH $<$ 7 the solution is *acidic.*
- pH $>$ 7 the solution is *basic.*
- pH $=$ 7 the solution is *neutral.*

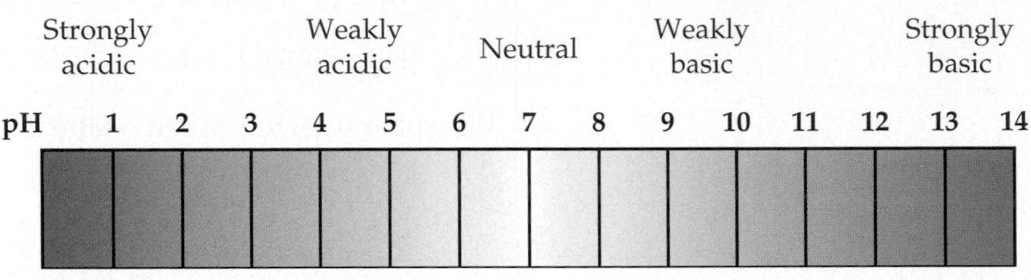

The pH scale.

Table 14.7 lists the pH of some common substances. Notice that, as we discussed in Section 14.3, many foods, especially fruits, are acidic and therefore have low pH values. The foods with the lowest pH values are limes and lemons, and they are among the sourest. Relatively few foods, however, are basic.

The pH scale is a **logarithmic scale,** therefore a change of 1 pH unit corresponds to a tenfold change in H_3O^+ concentration. For example, a lime with a pH of 2.0 is 10 times more acidic than a plum with a pH of 3.0 and 100 times more acidic than a cherry with a pH of 4.0. Each change of 1 in pH scale corresponds to a change of 10 in $[H_3O^+]$ (Figure 14.15).

Calculating pH from $[H_3O^+]$

The pH of a solution is defined as follows:

$$\text{pH} = -\log\,[H_3O^+]$$

To calculate pH, you must be able to find logarithms. In the following example, we need to compute the log of 1.5×10^{-7}.
A solution having an $[H_3O^+] = 1.5 \times 10^{-7}$ M (acidic) has a pH of:

$$\begin{aligned}\text{pH} &= -\log\,[H_3O^+]\\ &= -\log 1.5 \times 10^{-7}\\ &= -(-6.82391)\\ &= 6.82\end{aligned}$$

Notice that the pH is reported to two decimal places here. This is because only the numbers to the right of the decimal place are significant in a log. Since our original value for the concentration had two significant figures, the log of that number has two decimal places.

$$\log 1.0 \times 10^{-3} = 3.00$$

(2 decimal places)

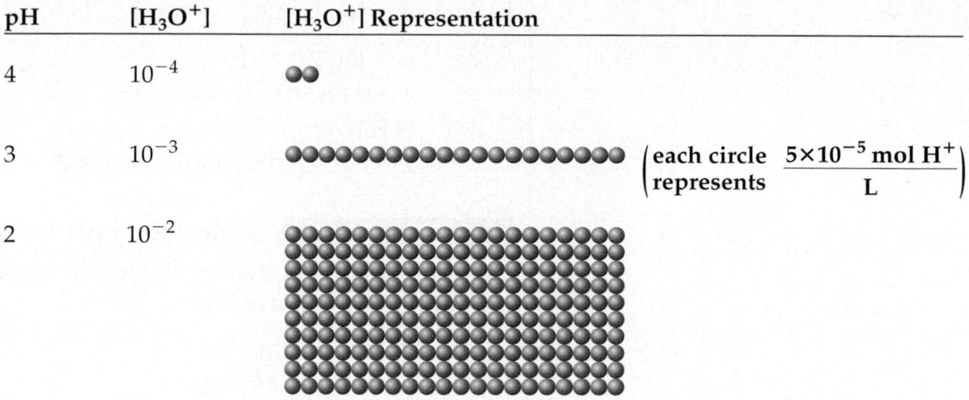

pH	$[H_3O^+]$	$[H_3O^+]$ Representation
4	10^{-4}	
3	10^{-3}	(each circle represents $\frac{5\times10^{-5}\text{ mol H}^+}{\text{L}}$)
2	10^{-2}	

Figure 14.15 The pH scale is a logarithmic scale. A *decrease* of 1 unit on the pH scale corresponds to an *increase* of 10 in H_3O^+ concentration. **Question:** How much of an increase in H_3O^+ concentration corresponds to a decrease of 2 pH units?

If the original number had three significant digits, the log would be reported to three decimal places:

$$-\log 1.00 \times 10^{-3} = 3.000 \quad \text{(3 decimal places)}$$

A solution having $[H_3O^+] = 1.0 \times 10^{-7}$ M (neutral) has a pH of:

$$\begin{aligned} \text{pH} &= -\log [H_3O^+] \\ &= -\log 1.0 \times 10^{-7} \\ &= -(-7.00) \\ &= 7.00 \end{aligned}$$

EXAMPLE 14.8 Calculating pH from $[H_3O^+]$

Calculate the pH of each of the following solutions and indicate whether the solution is acidic or basic.

a) $[H_3O^+] = 1.8 \times 10^{-4}$ M
b) $[H_3O^+] = 7.2 \times 10^{-9}$ M

Solution:

To calculate pH, simply substitute the given $[H_3O^+]$ into the pH equation.

a) $$\begin{aligned} \text{pH} &= -\log [H_3O^+] \\ &= -\log 1.8 \times 10^{-4} \\ &= -(-3.74) \\ &= 3.74 \end{aligned}$$

Since the pH < 7, this solution is acidic.

b) $pH = -\log [H_3O^+]$
$= -\log 7.2 \times 10^{-9}$
$= -(-8.14)$
$= 8.14$

Since the $pH > 7$, this solution is basic.

SKILLBUILDER 14.8 **Calculating pH from $[H_3O^+]$**

Calculate the pH of each of the following solutions and indicate whether the solution is acidic or basic.

a) $[H_3O^+] = 9.5 \times 10^{-9}$ M
b) $[H_3O^+] = 6.1 \times 10^{-3}$ M

SKILLBUILDER PLUS

Calculate the pH of a solution with $[OH^-] = 1.3 \times 10^{-2}$ M and indicate whether the solution is acidic or basic. (Hint: Begin by using K_w to find $[H_3O^+]$.

Calculating $[H_3O^+]$ from pH

The inverse log is sometimes called the *antilog*.

To calculate $[H_3O^+]$ from a pH value, you must *undo* the log. The log can be undone using the inverse log function *(Method 1)* on most calculators or it can be undone using the 10^x key *(Method 2)*. Both of these do the same thing; which one you use depends on your calculator.

The invlog "undoes" log. invlog(log x) = x.

Ten raised to the log of a number is equal to that number. $10^{\log x} = x$

Method 1: Inverse Log Function

$pH = -\log [H_3O^+]$
$-pH = \log [H_3O^+]$
$\text{invlog}(-pH) = \text{invlog}(\log [H_3O^+])$
$\text{invlog}(-pH) = [H_3O^+]$

Method 2: 10^x Function

$pH = -\log [H_3O^+]$
$-pH = \log [H_3O^+]$
$10^{-pH} = 10^{\log [H_3O^+]}$
$10^{-pH} = [H_3O^+]$

So, to calculate $[H_3O^+]$ from a pH value, simply take the inverse log of negative the pH value *(Method 1)* or raise ten to the negative of the pH value *(Method 2)*.

EXAMPLE 14.9 **Calculating $[H_3O^+]$ from pH**

Calculate the H_3O^+ concentration for a solution with a pH of 4.80.

Solution:

To find the $[H_3O^+]$ from pH, we must undo the log function. Use either Method 1 or Method 2 shown below.

Method 1: Inverse Log Function

$pH = -\log [H_3O^+]$
$4.80 = -\log [H_3O^+]$
$-4.80 = \log [H_3O^+]$
$\text{invlog}(-4.80) = \text{invlog}(\log [H_3O^+])$
$\text{invlog}(-4.80) = [H_3O^+]$
$[H_3O^+] = 1.6 \times 10^{-5}$ M

Method 2: 10^x function

$pH = -\log [H_3O^+]$
$4.80 = -\log [H_3O^+]$
$-4.80 = \log [H_3O^+]$
$10^{-4.80} = 10^{\log [H_3O^+]}$
$10^{-4.80} = [H_3O^+]$
$[H_3O^+] = 1.6 \times 10^{-5}$ M

SKILLBUILDER 14.9 **Calculating $[H_3O^+]$ from pH**

Calculate the H_3O^+ concentration for a solution with a pH of 8.37.

SKILLBUILDER PLUS

Calculate the OH^- concentration for a solution with a pH of 3.66.

CHEMISTRY AND HEALTH

Ulcers

An ulcer is a lesion that forms on the wall of the stomach or small intestine (Figure 14.16). Under normal circumstances, a thick layer of mucous lines the stomach wall and protects it from the hydrochloric acid and other gastric juices in the stomach. When that mucous layer is damaged, however, gastric juices come in direct contact with the stomach wall and begin to digest it, resulting in an ulcer. The main symptom of an ulcer is a burning or gnawing pain in the stomach.

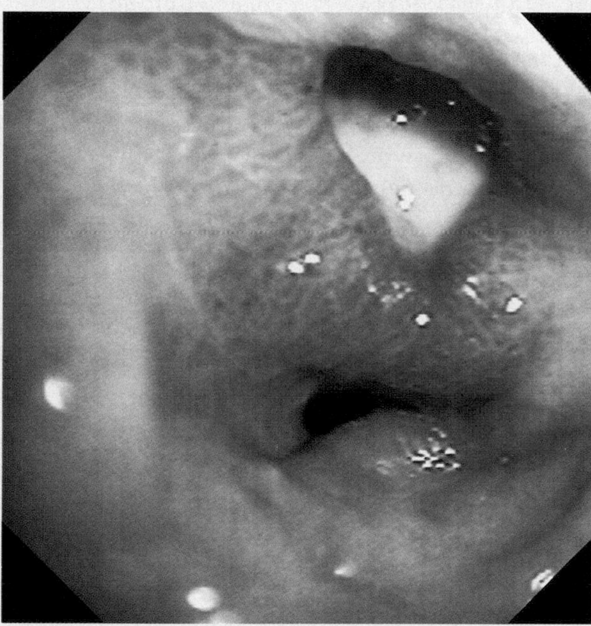

Figure 14.16 An ulcer is a lesion on the stomach wall caused by the breakdown of the stomach's mucus lining. A similar lesion is a cold sore. **Question:** How would that feel in your stomach?

Acidic drugs, such as aspirin, and acidic foods, such as citrus fruits and pickling fluids, irritate ulcers. When consumed, these substances increase the acidity of the stomach, worsening the irritation to the stomach wall. On the other hand, antacids—which contain bases—relieve ulcers. Common antacids include Tums and milk of magnesia.

The cause of ulcers remains unclear. For many years, a stressful lifestyle and a rich diet were blamed. However, recent theories propose that a bacterial infection of the stomach lining may be responsible. Long-term use of some over-the-counter pain relievers, such as aspirin, is also believed to cause ulcers.

CAN YOU ANSWER THIS? *Which of the following are likely to irritate an ulcer? Soothe an ulcer?*

Food	*pH*
limes	1.8–2.0
vinegar	2.0
wine	2.8–3.8
apples	2.9–3.3
egg whites	7.6–8.0
milk of magnesia	10.5

14.10 Buffers: Solutions That Resist pH Change

Most solutions will rapidly become more acidic (lower pH) upon addition of an acid or more basic (higher pH) upon addition of a base. A **buffer,** however, resists pH change by neutralizing added acid or added base. Human blood, for example, is a buffer. Any acid or base added to blood is neutralized by components within blood, resulting in a nearly constant pH. In healthy individuals, blood pH is between 7.36 and 7.40. If blood pH were to drop below 7.0 or increase beyond 7.8, death would result.

How does blood maintain such a narrow pH range? Like all buffers, blood contains *significant* amounts of *both a weak acid and its conjugate base.* When additional base is added to blood, the weak acid reacts with the base, neutralizing it. When additional acid is added to blood, the conjugate base reacts with the acid, neutralizing it. In this way, blood can maintain a constant pH.

A simple buffer can be made by mixing both acetic acid ($HC_2H_3O_2$) and its conjugate base, sodium acetate ($NaC_2H_3O_2$), into water (Figure 14.17). (The sodium in sodium acetate is just a spectator ion and does not contribute to buffer action.) Since $HC_2H_3O_2$ is a weak acid and since $C_2H_3O_2^-$ is its conjugate base a solution containing both of these is a buffer. Notice that a weak acid by itself, even though it partially ionizes to form some of its conjugate base, does not contain sufficient base to be a buffer. A buffer must contain *significant* amounts of *both* a weak acid and its conjugate base. Suppose that additional base, in the form of NaOH, were added to the buffer solution containing acetic acid and sodium acetate. The acetic acid would neutralize the base according to the following reaction.

$$NaOH(aq) + HC_2H_3O_2(aq) \rightarrow H_2O(l) + NaC_2H_3O_2(aq)$$

As long as the amount of NaOH added was less than the amount of $HC_2H_3O_2$ in solution, the solution would neutralize the NaOH, and the resulting pH change would be small. Suppose, on the other hand, that additional acid, in the form of HCl, were added to the solution. Then the conjugate base, $NaC_2H_3O_2$, would neutralize the added HCl according to the following reaction.

$$HCl(aq) + NaC_2H_3O_2(aq) \rightarrow HC_2H_3O_2(aq) + NaCl(aq)$$

As long as the amount of HCl added was less than the amount of $NaC_2H_3O_2$ in solution, the solution would neutralize the HCl and the resulting pH change would be small.

To summarize:

- ***Buffers resist pH change.***
- ***Buffers contain significant amounts of both a weak acid and its conjugate base.***
- ***The weak acid neutralizes added base.***
- ***The conjugate base neutralizes added acid.***

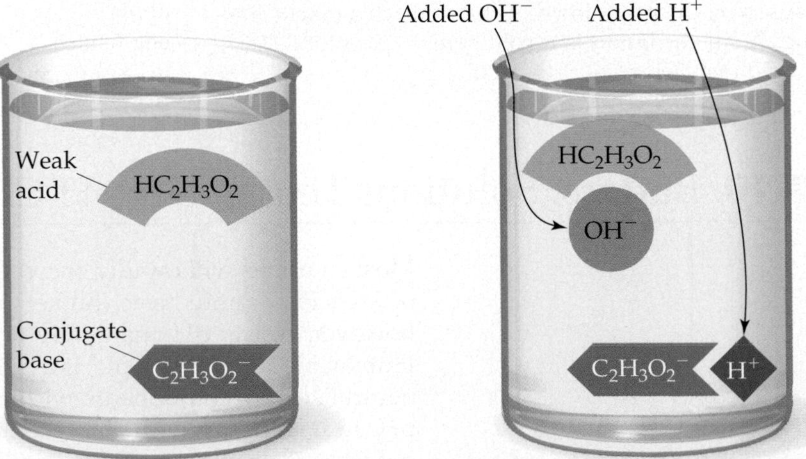

Figure 14.17 A buffer contains significant amounts of a weak acid and its conjugate base. The acid consumes any added base, and the base consumes any added acid. In this way, a buffer resists pH change.

CHEMISTRY AND HEALTH

The Danger of Antifreeze

Most types of antifreeze used in cars are solutions of ethylene glycol. Every year, thousands of dogs and cats die from ethylene glycol poisoning because they consume improperly stored antifreeze or antifreeze that has leaked out of a radiator. The antifreeze has somewhat of a sweet taste, which attracts a curious dog or cat. Young children are also at risk for ethylene glycol poisoning.

The first stage of ethylene glycol poisoning is a drunken state. Ethylene glycol is an alcohol, and it affects the brain of a dog or cat much as an alcoholic beverage would. Once ethylene glycol begins to metabolize, however, the second and more deadly stage begins. Ethylene glycol is metabolized in the liver into glycolic acid ($HC_2H_3O_3$), which enters the bloodstream. If the original quantities of consumed antifreeze are significant, the glycolic acid overwhelms the blood's natural buffering system, causing blood pH to drop to dangerously low levels. At this point, the cat or dog may begin hyperventilating in an effort to overcome the acidic blood's reduced ability to carry oxygen. If no treatment is administered, the animal will eventually go into a coma and die.

One treatment for ethylene glycol poisoning is the administration of ethyl alcohol (the alcohol found in alcoholic beverages). The liver enzyme that metabolizes ethylene glycol is the same one that metabolizes ethyl alcohol, but it has higher affinity for ethyl alcohol than for ethylene glycol. Consequently, the enzyme preferentially metabolizes ethyl alcohol, allowing the unmetabolized ethylene glycol to escape through the urine. If administered early, this treatment can save the life of a dog or cat that has consumed ethylene glycol.

CAN YOU ANSWER THIS? *One of the main buffering systems found in blood consists of carbonic acid (H_2CO_3) and bicarbonate ion (HCO_3^-). Write an equation showing how this buffering system would neutralize glycolic acid ($HC_2H_3O_3$) that might enter the blood from ethylene glycol poisoning. Suppose a cat has 0.15 mole of HCO_3^- and 0.15 mole of H_2CO_3 in its bloodstream. How many grams of $HC_2H_3O_2$ could be neutralized before the buffering system in the blood is overwhelmed?*

14.11 Acid Rain: An Environmental Problem Related to Fossil Fuel Combustion

About 90% of U.S. energy comes from fossil fuel combustion. Fossil fuels include petroleum, natural gas, and coal. Some fossil fuels, especially coal, contain small amounts of sulfur impurities. During combustion, these impurities react with oxygen to form SO_2. In addition, during combustion of any fossil fuel, nitrogen from the air reacts with oxygen to form NO_2. SO_2 and NO_2 emitted from fossil fuel combustion then react with water in the atmosphere to form sulfuric acid and nitric acid.

These equations are simplified versions of those that actually occur.

$$2\,SO_2 + O_2 + 2\,H_2O \rightarrow 2\,H_2SO_4$$
$$4\,NO_2 + O_2 + 2\,H_2O \rightarrow 4\,HNO_3$$

These acids combine with rain to form **acid rain.** The problem is greatest in the northeastern portion of the United States because many midwestern power plants burn coal. The sulfur and nitrogen oxides produced from coal combustion in the Midwest are carried toward the Northeast by natural air currents, making rain in that portion of the country significantly acidic.

Rain is naturally somewhat acidic because of atmospheric carbon dioxide. Carbon dioxide combines with rainwater to form carbonic acid.

$$CO_2 + H_2O \rightarrow H_2CO_3$$

However, carbonic acid is a relatively weak acid. Even rain that is saturated with CO_2 only has a pH of about 5.6, which is mildly acidic. However, when nitric acid and sulfuric acid mix with rain, the pH of the rain can fall as low 4.3 (Figure 14.18). Remember that, because of the logarithmic nature of the pH scale, rain with a pH of 4.3 has an $[H_3O^+]$ that is 20 times greater than rain with a pH of 5.6. Rain that is this acidic has negative consequences on the environment.

Acid Rain Damage

See Section 7.8 for the reaction between acids and carbonates.

Since acids dissolve metals, acid rain damages metal structures. Bridges, railroads, and even automobiles can be damaged by acid rain. Since acids react with carbonates, acid rain also damages building materials that contain carbonates (CO_3^{2-}), including marble, cement, and limestone. Statues, buildings, and pathways in the Northeast show significant signs of acid rain damage (Figure 14.19).

Acid rain can also accumulate in lakes and rivers and affect aquatic life. In the northeastern U.S., over 2000 lakes and streams have increased acidity levels due to acid rain. Aquatic plants, frogs, salamanders, and some species of fish, are sensitive to acid levels and cannot live in the acidified lakes. Trees can also be affected by acid rain because the acid removes nutrients from the soil, making it more difficult for trees to survive.

Acid Rain Legislation

Acid rain has been targeted by legislation. In 1990, the U.S. Congress passed amendments to the Clean Air Act that target acid rain. These amendments force electrical utilities—which are the most significant

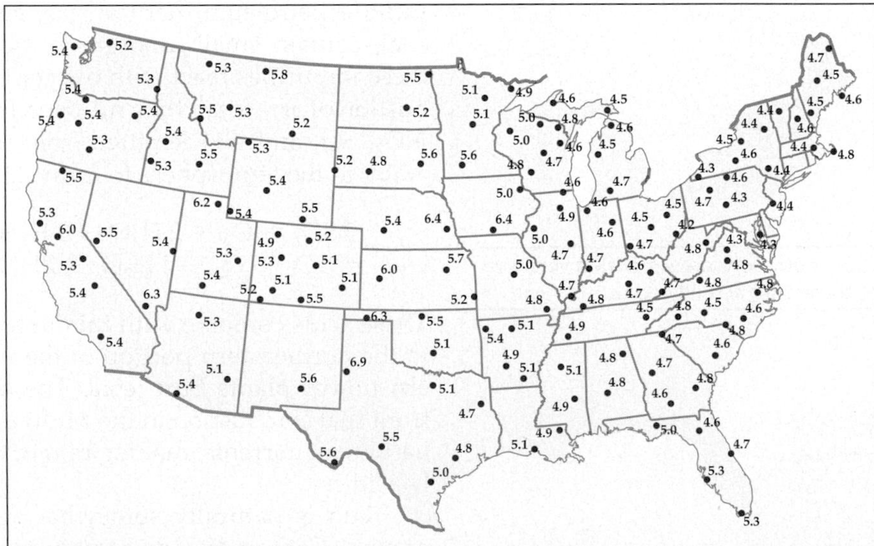

Figure 14.18 Average pH of precipitation in the United States for the period December 25, 2000, to January 1, 2001.

Figure 14.19 Many national monuments, such as the Lincoln Memorial, show signs of acid rain damage.

source of SO_2—to lower their SO_2 emissions gradually over time (Figure 14.20). The acidity of rain in the Northeast has already stabilized and should decrease in the coming years. Scientists expect most lakes, streams, and forests to recover once the pH of the rain returns to normal levels.

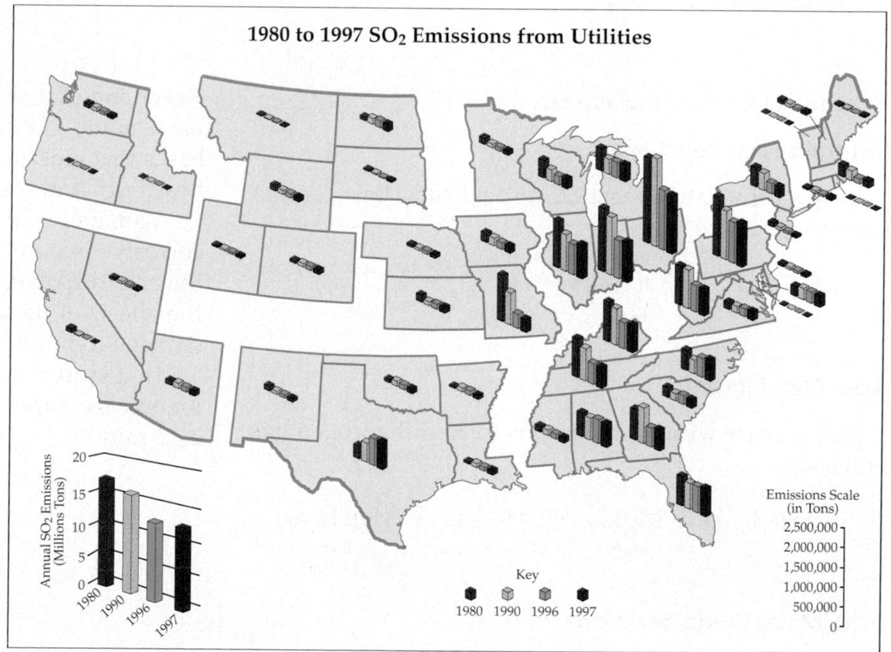

Figure 14.20 SO_2 emissions from utilities between the years 1980 and 1997. The height of each bar represents annual SO_2 emissions for that region in the year noted. Under the clean air act and its amendments SO_2 emissions are decreasing in most parts of the county.

CHAPTER IN REVIEW

Chemical Principles

Acid Properties:

- Acids have a sour taste.
- Acids dissolve many metals.
- Acids turn litmus paper red.

Relevance

Acid Properties: Acids are responsible for the sour taste in many foods such as lemons, limes, and vinegar. They also find widespread use in the laboratory and in industry.

Base Properties:

- Bases have a bitter taste.
- Bases have a slippery feel.
- Bases turn litmus paper blue.

Base Properties: Bases are less common in foods, but their presence in some foods—such as coffee and beer—is enjoyed by many as an acquired taste. Bases also have widespread use in the laboratory and in industry.

Molecular Definitions of Acids and Bases:

Arrhenius definition

Acid–substance that produces H^+ ions in solution
Base–substance that produces OH^- ions in solution

Brønsted-Lowry definition

Acid–proton donor
Base–proton acceptor

Molecular Definitions of Acids and Bases: The Arrhenius definition is the more straightforward and easier to use. It also shows how an acid and a base neutralize each other to form water ($H^+ + OH^- \rightarrow H_2O$). The Brønsted-Lowry definition helps us to see that in water, H^+ ions usually associate with water molecules to form H_3O^+. It also easily shows how bases that do not contain OH^- ions can still act as bases by accepting a proton from water.

Reactions of Acids and Bases:

Neutralization Reactions

In a neutralization reaction, an acid and a base react to form water and a salt.

$$\underset{\text{Acid}}{HCl(aq)} + \underset{\text{Base}}{KOH(aq)} \rightarrow \underset{\text{Water}}{H_2O(l)} + \underset{\text{Salt}}{KCl(aq)}$$

Acid–Metal Reactions

Acids react with many metals to form hydrogen gas and a salt.

$$\underset{\text{Acid}}{2\,HCl(aq)} + \underset{\text{Metal}}{Mg(s)} \rightarrow \underset{\text{Hydrogen gas}}{H_2(g)} + \underset{\text{Salt}}{MgCl_2(aq)}$$

Acid–Metal Oxide Reactions

Acids react with many metal oxides to form water and a salt.

$$\underset{\text{Acid}}{2\,HCl(aq)} + \underset{\text{Metal oxide}}{K_2O(s)} \rightarrow \underset{\text{Water}}{H_2O(l)} + \underset{\text{Salt}}{2\,KCl(aq)}$$

Reactions of Acids and Bases: Neutralization reactions have many common uses. Antacids, for example, are bases that react with acids from the stomach to alleviate heartburn and sour stomach.

Acid–metal and acid–metal oxide reactions show the corrosive nature of acids. In both of these reactions, the acid dissolves the metal or the metal oxide. Some of the effects of these kinds of reactions can be seen by the damage to building materials caused by acid rain. Since acids dissolve metals and metal oxides, any building materials composed of these substances are damaged by acid rain.

Acid–Base Titration: In an acid–base titration, an acid (or base) of known concentration is added to a base (or acid) of unknown concentration. The two reactants are combined until they are in exact stoichiometric proportions (moles H^+ = moles OH^-) which marks the endpoint of the titration. Since you know the moles H^+ (or OH^-) that you added, you can determine the moles of OH^- (or H^+) in the unknown solution.

Acid–Base Titration: An acid–base titration is a laboratory procedure often used to determine the unknown concentration of an acid or base.

Strong and Weak Acids and Bases: Strong acids completely ionize and strong bases completely dissociate in aqueous solutions. For example:

$$HCl(aq) + H_2O(l) \rightarrow H_3O^+(aq) + Cl^-(aq)$$
$$NaOH(aq) \rightarrow Na^+(aq) + OH^-(aq)$$

A 1 M HCl solution has $[H_3O^+] = 1$ M and a 1 M NaOH solution has $[OH^-] = 1$ M

Weak acids only partially ionize in solution. Most weak bases partially ionize water in solution. For example:

$$HF(aq) + H_2O(l) \rightleftharpoons H_3O^+(aq) + F^-(aq)$$
$$NH_3(aq) + H_2O(l) \rightleftharpoons NH_4^+(aq) + OH^-$$

A 1 M HF solution has $[H_3O^+] < 1$ M and a 1 M NH_3 solution has $[OH^-] < 1$ M

Strong and Weak Acids and Bases: Whether an acid is strong or weak depends on the conjugate base: the stronger the conjugate base, the weaker the acid. Since the acidity or basicity of a solution depends on $[H_3O^+]$ and $[OH^-]$, we must know if an acid is strong or weak to know the degree of acidity or basicity.

Self-Ionization of Water: Water can act as both an acid and a base with itself.

$$\underset{\text{Acid}}{H_2O(l)} + \underset{\text{Base}}{H_2O(l)} \rightleftharpoons H_3O^+(aq) + OH^-(aq)$$

The product of $[H_3O^+]$ and $[OH^-]$ in aqueous solutions will always be equal to the ion product constant, $K_w(10^{-14})$.

$$[H_3O^+][OH^-] = K_w = 1.0 \times 10^{-14}$$

Self-Ionization of Water: The self-ionization of water shows us that aqueous solutions always contain some H_3O^+ and some OH^-. In a neutral solution, the concentration of these is equal (1.0×10^{-7} M) When an acid is added to water, $[H_3O^+]$ increases and $[OH^-]$ decreases. When a base is added to water, the opposite happens. The ion product constant, however, still equals 1.0×10^{-14}, allowing us to calculate $[H_3O^+]$ given $[OH^-]$ and vice versa.

pH Scale:

$$pH = -\log [H_3O^+]$$
pH > 7 basic
pH < 7 acidic
pH = 7 neutral

pH Scale: pH is a convenient way to specify acidity or basicity. Since the pH scale is logarithmic, a change of one on the pH scale corresponds to a change of ten in the $[H_3O^+]$.

Buffers: Buffers are solutions containing significant amounts of both a weak acid and its conjugate base. Buffers resist pH change by neutralizing added acid or base.

Buffers: Buffers are important in blood chemistry, for example. Blood must maintain a narrow pH range in order to carry oxygen.

Acid Rain: Acid rain is the result of sulfur oxides and nitrogen oxides emitted by fossil fuel combustion. These oxides react with water to form sulfuric acid and nitric acid which then fall as acid rain.

Acid Rain: Since acids are corrosive, acid rain damages building materials. Since many aquatic plants and animals cannot survive in acidic water, acid rain also affects lakes and rivers, making them too acidic for the survival of some species.

Chemical Skills

Examples

Identifying Brønsted-Lowry Acids and Bases and Their Conjugates (Section 14.4)

The substance that donates the proton is the acid (proton donor) and becomes the conjugate base (as a product). The substance that accepts the proton (proton acceptor) is the base and becomes the conjugate acid (as a product).

EXAMPLE 14.10 Identifying Brønsted-Lowry Acids and Bases and Their Conjugates

Identify the Brønsted-Lowry acid, the Brønsted-Lowry base, the conjugate acid, and the conjugate base in the following reaction.

$$HNO_3(aq) + H_2O(l) \rightarrow H_3O^+(aq) + NO_3^-(aq)$$

Solution:

$$\underset{\text{Acid}}{HNO_3(aq)} + \underset{\text{Base}}{H_2O(l)} \rightarrow \underset{\text{Conjugate acid}}{H_3O^+(aq)} + \underset{\text{Conjugate base}}{NO_3^-(aq)}$$

Writing Equations for Neutralization Reactions (Section 14.5)

In a neutralization reaction, an acid and a base usually react to form water and a salt (ionic compound).

Acid + Base → Water + Salt

Write the skeletal equation first, making sure to write the formula of the salt so that it is charge neutral. Then balance the equation.

EXAMPLE 14.11 Writing Equations for Neutralization Reactions

Write a molecular equation for the reaction between aqueous HBr and aqueous $Ca(OH)_2$.

Solution:

Skeletal equation

$$HBr(aq) + Ca(OH)_2(aq) \rightarrow H_2O(l) + CaBr_2(aq)$$

Balanced equation

$$2\,HBr(aq) + Ca(OH)_2(aq) \rightarrow 2\,H_2O(l) + CaBr_2(aq)$$

Writing Equations for the Reactions of Acids with Metals and with Metal Oxides (Section 14.5)

Acids react with many metals to form hydrogen gas and a salt.

Acid + Metal → Hydrogen gas + Salt

Write the skeletal equation first, making sure to write the formula of the salt so that it is charge neutral. Then balance the equation.

Acids react with many metal oxides to form water and a salt.

Acid + Metal oxide → Water + Salt

Write the skeletal equation first, making sure to write the formula of the salt so that it is charge neutral. Then balance the equation.

EXAMPLE 14.12 Writing Equations for the Reactions of Acids with Metals and with Metal Oxides

Write equations for the reaction of hydrobromic acid with calcium metal and for the reaction of hydrobromic acid with calcium oxide.

Solution:

Skeletal equation

$$HBr(aq) + Ca(s) \rightarrow H_2(g) + CaBr_2(aq)$$

Balanced equation

$$2\,HBr(aq) + Ca(s) \rightarrow H_2(g) + CaBr_2(aq)$$

Skeletal equation

$$HBr(aq) + CaO(s) \rightarrow H_2O(l) + CaBr_2(aq)$$

Balanced equation

$$2\,HBr(aq) + CaO(s) \rightarrow H_2O(l) + CaBr_2(aq)$$

Acid–Base Titrations (Section 14.6)

EXAMPLE 14.13 Acid–Base Titrations

A 15.00 mL sample of a NaOH solution of unknown concentration requires 17.88 mL of a 0.1053 M H_2SO_4 solution to reach the endpoint in a titration. What is the concentration of the NaOH solution?

Set up the problem in the standard way.

Given: 15.00 mL NaOH

17.88 mL of a 0.1053 M H_2SO_4 solution

Find: concentration of NaOH solution (moles/L)

One of the equations should be the balanced equation for the reaction of the acid and the base that is given in the problem. The other equation is simply the definition of molarity (if the problem is asking for molarity).

Equations:

$$H_2SO_4(aq) + 2\,NaOH(aq) \rightarrow 2\,H_2O(l) + Na_2SO_4(aq)$$

$$\text{Molarity(M)} = \frac{\text{mol solute}}{\text{L solution}}$$

Use the volume and concentration of the known reactant to determine moles of the known reactant. (You may have to convert from milliliters to liters first.) Then use the stoichiometric ratio from the balanced equation to get moles of the unknown reactant.

If the problem is asking for the molarity of the unknown solution, add a second part to the solution map showing how moles and volume can be used to determine molarity.

Solution Map:

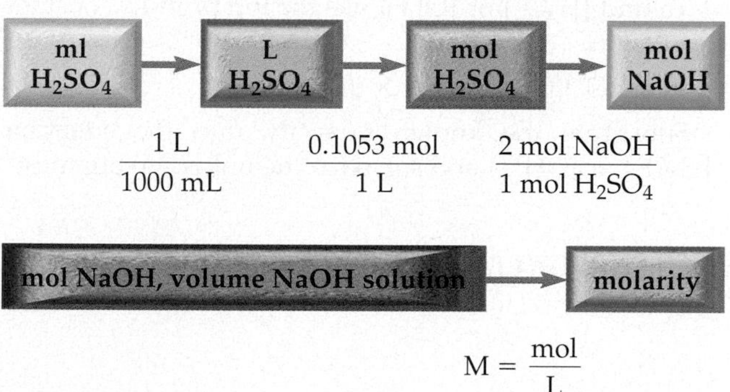

Solution:

Follow the solution map to solve the problem. The first part of the solution gives you moles of the unknown reactant.

$$17.88\text{ mL }H_2SO_4 \times \frac{1\text{ L}}{1000\text{ mL}} \times \frac{0.1053\text{ mol }H_2SO_4}{\text{L }H_2SO_4} \times \frac{2\text{ mol NaOH}}{1\text{ mol }H_2SO_4} = 3.76\underline{5}5 \times 10^{-3}\text{ mol NaOH}$$

In the second part of the solution, divide the moles from the first part by the volume to obtain molarity.

$$M = \frac{\text{mol}}{\text{L}} = \frac{3.76\underline{5}5 \times 10^{-3}\text{ mol}}{0.01500\text{ L}} = 0.2510\text{ M}$$

The unknown NaOH solution has a concentration of 0.1255 M.

Determining $[H_3O^+]$ in Acid Solutions (Section 14.7)

EXAMPLE 14.14 Determining $[H_3O^+]$ in Acid Solutions

In a strong acid, $[H_3O^+]$ will be equal to the concentration of the acid. In a weak acid, $[H_3O^+]$ will be less than the concentration of the acid.

What is the H_3O^+ concentration in a 0.25 M HCl solution and in a 0.25 M HF solution?

Solution:

In the 0.25 M HCl solution (strong acid), $[H_3O^+] = 0.25\text{ M}$. In the 0.25 M HF solution, $[H_3O^+] < 0.25\text{ M}$.

Determining $[OH^-]$ in Base Solutions (Section 14.7)

In a strong base, $[OH^-]$ is equal to the concentration of the base times the number of hydroxide ions in the base. In a weak base, $[OH^-]$ is less than the concentration of the base.

EXAMPLE 14.15 Determining $[OH^-]$ in Base Solutions

What is the OH^- concentration in a 0.25 M NaOH solution, in a 0.25 M $Sr(OH)_2$ solution, and in a 0.25 M NH_3 solution?

Solution:
In the 0.25 M NaOH solution (strong base), the $[OH^-] = 0.25$ M. In the 0.25 M $Sr(OH)_2$ solution (strong base), the $[OH^-] = 0.50$ M. In the 0.25 M NH_3 solution (weak base), the $[OH^-] < 0.25$ M.

Finding the Concentration of $[H_3O^+]$ or $[OH^-]$ from K_w (Section 14.8)

To find $[H_3O^+]$ or $[OH^-]$ use the ion product constant expression.

$$[H_3O^+][OH^-] = 1.0 \times 10^{-14}$$

Substitute the known quantity into the equation ($[H_3O^+]$ or $[OH^-]$) and solve for the unknown quantity.

EXAMPLE 14.16 Finding the Concentration of $[H_3O^+]$ or $[OH^-]$ from K_w

Calculate $[OH^-]$ in each a solution with $[H_3O^+] = 1.5 \times 10^{-4}$ M.

Solution:

$$[H_3O^+][OH^-] = 1.0 \times 10^{-14}$$
$$[1.5 \times 10^{-4}][OH^-] = 1.0 \times 10^{-14}$$
$$[OH^-] = \frac{1.0 \times 10^{-14}}{1.5 \times 10^{-4}} = 6.7 \times 10^{-11} M$$

Calculating pH from $[H_3O^+]$ (Section 14.9)

To calculate the pH of a solution from the $[H_3O^+]$, simply take the log of the $[H_3O^+]$, and change the sign of the result.

$$pH = -\log [H_3O^+]$$

EXAMPLE 14.17 Calculating pH from $[H_3O^+]$

Calculate the pH of a solution with $[H_3O^+] = 2.4 \times 10^{-5}$ M.

Solution:

$$\begin{aligned} pH &= -\log [H_3O^+] \\ &= -\log 2.4 \times 10^{-5} \\ &= -(-4.62) \\ &= 4.62 \end{aligned}$$

Calculating $[H_3O^+]$ from pH (Section 14.9)

You can calculate $[H_3O^+]$ from pH by taking the inverse log of the negative of the pH value (Method 1):

$$[H_3O^+] = \text{INVLOG}(-pH)$$

You can also calculate $[H_3O^+]$ from pH by raising 10 to the negative of the pH (Method 2):

$$[H_3O^+] = 10^{-pH}$$

EXAMPLE 14.18 Calculating $[H_3O^+]$ from pH

Calculate the $[H_3O^+]$ concentration for a solution with a pH of 6.22.

Solution:

Method 1: Inverse Log Function

$$\begin{aligned} [H_3O^+] &= \text{INVLOG}(-pH) \\ &= \text{INVLOG}(-6.22) \\ &= 6.0 \times 10^{-7} \end{aligned}$$

Method 2: 10^x Function

$$\begin{aligned} [H_3O^+] &= 10^{-pH} \\ &= 10^{-6.22} \\ &= 6.0 \times 10^{-7} \end{aligned}$$

KEY TERMS

acid
acid rain [14.11]
acidic solution [14.8]
alkaloids [14.3]
amphoteric [14.4]
Svante Arrhenius [14.4]
Arrhenius acid [14.4]
Arrhenius base [14.4]
Arrhenius definition [14.4]
base
basic solution [14.8]
Brønsted-Lowry acid [14.4]
Brønsted-Lowry base [14.4]
Brønsted-Lowry definition [14.4]
buffer [14.10]
carboxylic acid [14.2]
conjugate acid–base pair [14.4]
diprotic acid [14.7]
dissociation [14.4]
endpoint [14.6]
hydronium ion [14.4]
indicator [14.6]
ion product constant (K_w) [14.8]
ionization [14.4]
logarithmic scale [14.9]
monoprotic acid [14.7]
neutral solution [14.8]
neutralization [14.5]
organic acid
pH [14.9]
salt [14.5]
strong acid [14.7]
strong base [14.7]
strong electrolyte [14.7]
titration [14.6]
weak acid [14.7]
weak base [14.7]
weak electrolyte [14.7]

EXERCISES

Questions

1. What makes sour gummy candies, such as sour patch kids, sour?
2. Give some examples of foods that contain acids.
3. What are the properties of acids?
4. What is the main component of stomach acid? Why do we have stomach acid?
5. What are organic acids? Give two examples of organic acids.
6. What are the properties of bases?
7. What are alkaloids?
8. Give some examples of common substances that contain bases.
9. Give the Arrhenius definition of an acid and demonstrate the definition with a chemical equation.
10. Give the Arrhenius definition of a base and demonstrate the definition with a chemical equation.
11. Give the Brønsted-Lowry definitions of acids and bases and demonstrate each with a chemical equation.
12. According to the Brønsted-Lowry definition of acids and bases, what is a conjugate acid–base pair? Give an example.
13. What is an acid–base neutralization reaction? Give an example.
14. Give an example of a reaction between an acid and a metal.
15. Give an example of a reaction between an acid and a metal oxide.
16. Name a metal that dissolves in a base and write an equation for the reaction.
17. What is a titration? What is the endpoint?
18. If a solution contains 0.85 moles of OH^-, how many moles of H^+ would be required to reach the endpoint in a titration?
19. What is the difference between a strong acid and a weak acid?
20. How is the strength of an acid related to the strength of its conjugate base?
21. What are monoprotic and diprotic acids?
22. What is the difference between a strong base and a weak base?
23. Does pure water contain any H_3O^+ ions? Explain.
24. What happens to $[OH^-]$ in an aqueous solution when $[H_3O^+]$ increases?
25. Give a possible value of $[OH^-]$ and $[H_3O^+]$ in a solution that is:
a) acidic
b) basic
c) neutral
26. How is pH defined? A change of one pH unit corresponds to how much of a change in $[H_3O^+]$?
27. What is a buffer?
28. What are the main components in a buffer?
29. What is the cause of acid rain?
30. Write equations for the chemical reactions by which acid rain forms in the atmosphere.
31. What are the effects of acid rain?
32. How is the problem of acid rain being addressed in the U.S.?

Problems

Acid and Base Definitions

33. Identify each of the following as an acid or a base and write a chemical equation showing how it is an acid or a base according to the Arrhenius definition.
a) $HNO_3(aq)$
b) $KOH(aq)$
c) $HC_2H_3O_2(aq)$

34. Identify each of the following as an acid or a base and write a chemical equation showing how it is an acid or a base according to the Arrhenius definition.
a) $NaOH(aq)$
b) $HBr(aq)$
c) $Sr(OH)_2(aq)$

35. For each of the following, identify the Brønsted-Lowry, the Brønsted-Lowry base, the conjugate acid, and the conjugate base.
a) $HBr(aq) + H_2O(l) \rightarrow H_3O^+(aq) + Br^-(aq)$
b) $NH_3(aq) + H_2O(l) \rightleftharpoons NH_4^+(aq) + OH^-(aq)$
c) $HNO_3(aq) + H_2O(l) \rightarrow H_3O^+(aq) + NO_3^-(aq)$
d) $C_5H_5N(aq) + H_2O(l) \rightleftharpoons C_5H_5NH^+(aq) + OH^-(aq)$

36. For each of the following, identify the Brønsted-Lowry acid, the Brønsted-Lowry base, the conjugate acid, and the conjugate base.
a) $HI(aq) + H_2O(l) \rightarrow H_3O^+(aq) + I^-(aq)$
b) $CH_3NH_2(aq) + H_2O(l) \rightleftharpoons CH_3NH_3^+(aq) + OH^-(aq)$
c) $CO_3^{2-}(aq) + H_2O(l) \rightleftharpoons HCO_3^{2-}(aq) + OH^-(aq)$
d) $H_2CO_3(aq) + H_2O(l) \rightleftharpoons H_3O^+(aq) + HCO_3^-(aq)$

37. Which of the following are conjugate acid–base pairs?
a) NH_3, NH_4^+
b) HCl, HBr
c) $C_2H_3O_2^-$, $HC_2H_3O_2$
d) HCO_3^-, NO_3^-

38. Which of the following are conjugate acid-base pairs?
a) HI, I^-
b) $HCHO_2$, SO_4^{2-}
c) PO_4^{3-}, HPO_4^{2-}
d) CO_3^{2-}, HCl

39. Write the formula for the conjugate base of each of the following acids.
a) HCl
b) H_2SO_3
c) $HCHO_2$
d) HF

40. Write the formula for the conjugate base of each of the following acids.
a) HBr
b) H_2CO_3
c) $HClO_4$
d) $HC_2H_3O_2$

41. Write the formula for the conjugate acid of each of the following bases.
a) NH_3
b) ClO_4^-
c) HSO_4^-
d) CO_3^{2-}

42. Write the formula for the conjugate acid of each of the following bases.
a) CH_3NH_2
b) C_5H_5N
c) Cl^-
d) F^-

Acid–Base Reactions

43. Write neutralization reactions for each of the following acids and bases.
a) $HI(aq)$ and $NaOH(aq)$
b) $HBr(aq)$ and $KOH(aq)$
c) $HNO_3(aq)$ and $Ba(OH)_2(aq)$
d) $HClO_4(aq)$ and $Sr(OH)_2(aq)$

44. Write neutralization reactions for each of the following acids and bases.
a) $HF(aq)$ and $Ba(OH)_2(aq)$
b) $HClO_4(aq)$ and $NaOH(aq)$
c) $HBr(aq)$ and $Ca(OH)_2(aq)$
d) $HCl(aq)$ and $KOH(aq)$

45. Write a balanced chemical equation showing how each of the following metals is dissolved by HBr.
a) K
b) Ca
c) Sr
d) Rb

46. Write a balanced chemical equation showing how each of the following metals is dissolved by HCl.
a) Al
b) Mg
c) Cs
d) Ba

47. Write a balanced chemical equation showing how each of the following metal oxides is dissolved by HI.
a) MgO
b) K_2O
c) Rb_2O
d) CaO

48. Write a balanced chemical equation showing how each of the following metal oxides is dissolved by HCl.
a) SrO
b) Na_2O
c) Li_2O
d) BaO

Acid–Base Titrations

49. Four solutions of unknown HCl concentration are titrated with solutions of NaOH. The following table lists the volume of each unknown HCl solution, the volume of NaOH solution required to reach the endpoint, and the concentration of each NaOH solution. Calculate the concentration of the unknown HCl solution in each case.

HCl Volume (mL)	NaOH Volume (mL)	[NaOH] (M)
a) 25.00 mL	28.44 mL	0.1231 M
b) 15.00 mL	21.22 mL	0.0972 M
c) 20.00 mL	14.88 mL	0.1178 M
d) 5.00 mL	6.88 mL	0.1325 M

50. Four solutions of unknown NaOH concentration are titrated with solutions of HCl. The following table lists the volume of each unknown NaOH solution, the volume of HCl solution required to reach the endpoint, and the concentration of each HCl solution. Calculate the concentration of the unknown NaOH solution in each case.

NaOH Volume (mL)	HCl Volume (mL)	[HCl] (M)
a) 5.00 mL	9.77 mL	0.1599 M
b) 15.00 mL	11.34 mL	0.1311 M
c) 10.00 mL	10.55 mL	0.0889 M
d) 30.00 mL	36.18 mL	0.1021 M

51. A 25.00 mL sample of an H_2SO_4 solution of unknown concentration is titrated with a 0.1328 M KOH solution. A volume of 38.33 mL of KOH was required to reach the endpoint. What is the concentration of the unknown H_2SO_4 solution?

52. A 5.00 mL sample of an H_3PO_4 solution of unknown concentration is titrated with a 0.1221 M NaOH solution. A volume of 5.99 mL of the NaOH solution was required to reach the endpoint. What is the concentration of the unknown H_3PO_4 solution?

Strong and Weak Acids and Bases

53. Classify each of the following acids as strong or weak.
a) HCl
b) HF
c) HBr
d) H_2SO_3

54. Classify each of the following acids as strong or weak.
a) $HCHO_2$
b) H_2SO_4
c) HNO_3
d) H_2CO_3

55. Determine $[H_3O^+]$ in each of the following acid solutions. For weak acids, give what $[H_3O^+]$ is less than.
a) 2.5 M HI
b) 1.2 M $HClO_2$
c) 0.25 M H_2CO_3
d) 2.25 M $HCHO_2$

56. Determine $[H_3O^+]$ in each of the following acid solutions. For weak acids, give what $[H_3O^+]$ is less than.
a) 0.100 M HNO_3
b) 0.75 M H_3PO_4
c) 3.8 M HI
d) 0.85 M H_2SO_3

57. Classify each of the following bases as strong or weak.
a) $LiOH$
b) NH_4OH
c) $Ca(OH)_2$
d) NH_3

58. Classify each of the following bases as strong or weak.
a) C_5H_5N
b) $NaOH$
c) $Ba(OH)_2$
d) KOH

59. Determine $[OH^-]$ in each of the following base solutions. For weak bases, give what $[OH^-]$ is less than.
a) 0.88 M $NaOH$
b) 0.88 M NH_3
c) 0.88 M $Sr(OH)_2$
d) 1.55 M KOH

60. Determine $[OH^-]$ in each of the following base solutions. For weak bases, give what $[OH^-]$ is less than.
a) 3.7 M KOH
b) 1.10 M NH_3
c) 0.110 M $Ba(OH)_2$
d) 2.2 M C_5H_5N

Acidity, Basicity, and K_w

61. Determine if each of the following solutions is acidic, basic, or neutral.
a) $[H_3O^+] = 1 \times 10^{-5}$ M; $[OH^-] = 1 \times 10^{-9}$ M
b) $[H_3O^+] = 1 \times 10^{-6}$ M; $[OH^-] = 1 \times 10^{-8}$ M
c) $[H_3O^+] = 1 \times 10^{-7}$ M; $[OH^-] = 1 \times 10^{-7}$ M
d) $[H_3O^+] = 1 \times 10^{-8}$ M; $[OH^-] = 1 \times 10^{-6}$ M

62. Determine if each of the following solutions is acidic, basic, or neutral.
a) $[H_3O^+] = 1 \times 10^{-9}$ M; $[OH^-] = 1 \times 10^{-5}$ M
b) $[H_3O^+] = 1 \times 10^{-10}$ M; $[OH^-] = 1 \times 10^{-4}$ M
c) $[H_3O^+] = 1 \times 10^{-2}$ M; $[OH^-] = 1 \times 10^{-12}$ M
d) $[H_3O^+] = 1 \times 10^{-13}$ M; $[OH^-] = 1 \times 10^{-1}$ M

63. Calculate $[OH^-]$ given $[H_3O^+]$ in each of the following aqueous solutions and classify the solution as acidic or basic.
a) $[H_3O^+] = 1.5 \times 10^{-9}$ M
b) $[H_3O^+] = 9.3 \times 10^{-9}$ M
c) $[H_3O^+] = 2.2 \times 10^{-6}$ M
d) $[H_3O^+] = 7.4 \times 10^{-4}$ M

64. Calculate $[OH^-]$ given $[H_3O^+]$ in each of the following aqueous solutions and classify the solution as acidic or basic.
a) $[H_3O^+] = 1.3 \times 10^{-3}$ M
b) $[H_3O^+] = 9.1 \times 10^{-12}$ M
c) $[H_3O^+] = 5.2 \times 10^{-4}$ M
d) $[H_3O^+] = 6.1 \times 10^{-9}$ M

65. Calculate $[H_3O^+]$ given $[OH^-]$ in each of the following aqueous solutions and classify each solution as acidic or basic.
a) $[OH^-] = 2.7 \times 10^{-12}$ M
b) $[OH^-] = 2.5 \times 10^{-2}$ M
c) $[OH^-] = 1.1 \times 10^{-10}$ M
d) $[OH^-] = 3.3 \times 10^{-4}$ M

66. Calculate $[H_3O^+]$ given $[OH^-]$ in each of the following aqueous solutions and classify each solution as acidic or basic.
a) $[OH^-] = 2.1 \times 10^{-11}$ M
b) $[OH^-] = 7.5 \times 10^{-9}$ M
c) $[OH^-] = 2.1 \times 10^{-4}$ M
d) $[OH^-] = 1.0 \times 10^{-2}$ M

pH

67. Classify each of the following solutions as acidic, basic, or neutral based on the pH value.
a) pH = 9.0
b) pH = 7.0
c) pH = 2.0
d) pH = 6.0

68. Classify each of the following solutions as acidic, basic, or neutral based on the pH value.
a) pH = 5.0
b) pH = 4.0
c) pH = 14.0
d) pH = 0.5

69. Calculate the pH of each of the following solutions.
a) $[H_3O^+] = 1.7 \times 10^{-8}$ M
b) $[H_3O^+] = 1.0 \times 10^{-7}$ M
c) $[H_3O^+] = 2.2 \times 10^{-6}$ M
d) $[H_3O^+] = 7.4 \times 10^{-4}$ M

70. Calculate the pH of each of the following solutions.
a) $[H_3O^+] = 2.4 \times 10^{-10}$ M
b) $[H_3O^+] = 7.6 \times 10^{-2}$ M
c) $[H_3O^+] = 9.2 \times 10^{-13}$ M
d) $[H_3O^+] = 3.4 \times 10^{-5}$ M

71. Calculate $[H_3O^+]$ for each of the following solutions.
a) pH = 8.55
b) pH = 11.23
c) pH = 2.87
d) pH = 1.22

72. Calculate $[H_3O^+]$ for each of the following solutions.
a) pH = 1.76
b) pH = 3.88
c) pH = 8.43
d) pH = 12.32

73. Calculate the pH of each of the following solutions.
a) $[OH^-] = 1.9 \times 10^{-7}$ M
b) $[OH^-] = 2.6 \times 10^{-8}$ M
c) $[OH^-] = 7.2 \times 10^{-11}$ M
d) $[OH^-] = 9.5 \times 10^{-2}$ M

74. Calculate the pH of each of the following solutions.
a) $[OH^-] = 2.8 \times 10^{-11}$ M
b) $[OH^-] = 9.6 \times 10^{-3}$ M
c) $[OH^-] = 3.8 \times 10^{-12}$ M
d) $[OH^-] = 6.4 \times 10^{-4}$ M

75. Calculate $[OH^-]$ for each of the following solutions.
a) pH = 4.25
b) pH = 12.53
c) pH = 1.50
d) pH = 8.25

76. Calculate $[OH^-]$ for each of the following solutions.
a) pH = 1.82
b) pH = 13.28
c) pH = 8.29
d) pH = 2.32

Buffers and Acid Rain

77. Locate where you live on the map in Figure 14.8. What is the pH of rain where you live? What is the $[H_3O^+]$?

78. Locate the area of the United States with the most acidic rainfall on the map in Figure 14.8. What is the pH of the rain? What is the $[H_3O^+]$?

79. Which of the following mixtures are buffers?
a) HCl and HF
b) NaOH and NH_3
c) HF and NaF
d) $HC_2H_3O_2$ and $KC_2H_3O_2$

80. Which of the following mixtures are buffers?
a) HBr and NaCl
b) $HCHO_2$ and $NaCHO_2$
c) HCl and HBr
d) KOH and NH_3

81. Write reactions showing how each of the buffers in Problem 79 would neutralize added HCl.

82. Write reactions showing how each of the buffers in Problem 80 would neutralize added NaOH.

Cumulative Exercises

83. How much 0.100 M HCl is required to completely neutralize 20.0 mL of 0.250 M NaOH?

84. How much 0.200 M KOH is required to completely neutralize 25.0 mL of 0.150 M $HClO_4$?

85. What is the minimum volume of 5.0 M HCl required to completely dissolve 10.0 g of magnesium metal?

86. What is the minimum volume of 3.0 M HBr required to completely dissolve 15.0 g of potassium metal?

87. When 18.5 g of K_2O is completely dissolved by HI, how many grams of KI(*aq*) are formed in solution?

88. When 5.88 g of CaO is completely dissolved by HBr, how many grams of $CaBr_2(aq)$ are formed in solution?

89. A 0.125 g sample of a monoprotic acid of unknown molar mass is dissolved in water and titrated with 0.1003 M NaOH. The endpoint is reached after adding 20.77 mL of the base. What is the molar mass of the unknown acid?

90. A 0.105 g sample of a diprotic acid of unknown molar mass is dissolved in water and titrated with 0.1288 M NaOH. The endpoint is reached after adding 15.2 mL of base. What is the molar mass of the unknown acid?

91. For each of the following $[H_3O^+]$, determine the pH and state whether the solution is acidic or basic.
a) $[H_3O^+] = 0.0025$ M
b) $[H_3O^+] = 1.8 \times 10^{-12}$ M
c) $[H_3O^+] = 9.6 \times 10^{-9}$ M
d) $[H_3O^+] = 0.0195$ M

92. For each of the following $[OH^-]$, determine the the pH and state whether the solution is acidic or basic.
a) $[OH^-] = 1.8 \times 10^{-5}$ M
b) $[OH^-] = 8.9 \times 10^{-12}$ M
c) $[OH^-] = 3.1 \times 10^{-2}$ M
d) $[OH^-] = 1.96 \times 10^{-9}$ M

93. Complete the following table. (The first one is completed for you.)

$[H_3O^+]$	$[OH^-]$	*pH*	*Acidic or Basic*
1.0×10^{-4}	*1.0×10^{-10}*	*4.00*	*acidic*
5.5×10^{-3}	___	___	___
___	3.2×10^{-6}	___	___
4.8×10^{-9}	___	___	___
___	___	7.55	___

94. Complete the following table. (The first one is completed for you.)

$[H_3O^+]$	$[OH^-]$	*pH*	*Acidic or Basic*
1.0×10^{-8}	*1.0×10^{-6}*	*8.00*	*basic*
___	___	3.55	___
1.7×10^{-9}	___	___	___
___	___	13.5	___
___	8.6×10^{-11}	___	___

95. For each of the following strong acid solutions, determine $[H_3O^+]$, $[OH^-]$, and pH.
a) 0.0150 M HCl
b) 1.5×10^{-3} M HBr
c) 9.77×10^{-4} M HI
d) 0.0878 M HNO_3

96. For each of the following strong acid solutions, determine $[H_3O^+]$, $[OH^-]$, and pH.
a) 0.1150 M $HClO_4$
b) 4.5×10^{-5} M HCl
c) 0.0226 M HBr
d) 1.7×10^{-3} M HNO_3

97. For each of the following strong base solutions, determine $[OH^-]$, $[H_3O^+]$, and pH.
a) 0.15 M NaOH
b) 1.5×10^{-3} M $Ca(OH)_2$
c) 4.8×10^{-4} M $Sr(OH)_2$
d) 8.7×10^{-5} M KOH

98. For each of the following strong base solutions, determine $[OH^-]$, $[H_3O^+]$, and pH.
a) 8.77×10^{-3} M LiOH
b) 0.0112 M $Ba(OH)_2$
c) 1.9×10^{-4} M KOH
d) 5.0×10^{-4} M $Ca(OH)_2$

Highlight Problems

99. Consider the following molecular views of acid solutions. Based on the molecular view, determine if the acid is weak or strong.

a)

b)

c)

d)

100. Lakes that have been acidified by acid rain can be neutralized by liming, the addition of limestone ($CaCO_3$). How much limestone (in kilograms) is required to completely neutralize a 3.8×10^9 L lake with a pH of 5.5?

101. Acid rain over the Great Lakes has a pH of about 4.5. Calculate the $[H_3O^+]$ of this rain and compare that value to the $[H_3O^+]$ of rain over the West Coast that has a pH of 5.4. How many times more concentrated is the acid in rain over the Great Lakes?

Answers to Skillbuilder Exercises

Skillbuilder 14.1

a) $\underset{\text{base}}{C_5H_5N(aq)} + \underset{\text{acid}}{H_2O(l)} \rightleftharpoons \underset{\text{conjugate acid}}{C_5H_5NH^+(aq)} + \underset{\text{conjugate base}}{OH^-(aq)}$

b) $\underset{\text{acid}}{HNO_3(aq)} + \underset{\text{base}}{H_2O(l)} \rightarrow \underset{\text{conjugate acid}}{H_3O^+(aq)} + \underset{\text{conjugate base}}{NO_3^-(aq)}$

Skillbuilder 14.2
$H_3PO_4(aq) + 3\,NaOH(aq) \rightarrow 3\,H_2O(l) + Na_3PO_4(aq)$

Skillbuilder 14.3
a) $2\,HCl(aq) + Sr(s) \rightarrow H_2(g) + SrCl_2(aq)$
b) $2\,HI(aq) + BaO(s) \rightarrow H_2O(l) + BaI_2(aq)$

Skillbuilder 14.4
9.03×10^{-2} M H_2SO_4

Skillbuilder 14.5
a) $[H_3O^+] < 0.50$ M
b) $[H_3O^+] = 1.25$ M c) $[H_3O^+] < 0.75$ M

Skillbuilder 14.6 **a)** $[OH^-] = 0.11$ M
b) $[OH^-] < 1.05$ M **c)** $[OH^-] = 0.45$ M

Skillbuilder 14.7 **a)** $[H_3O^+] = 6.7 \times 10^{-13}$ M; basic
b) $[H_3O^+] = 1.0 \times 10^{-7}$ M; neutral
c) $[H_3O^+] = 1.2 \times 10^{-5}$ M; acidic

Skillbuilder 14.8 **a)** pH = 8.02; basic
b) pH = 2.21; acidic

Skillbuilder Plus, p. 508 pH = 12.11; basic

Skillbuilder 14.9 4.3×10^{-9} M

Skillbuilder Plus, p. 509 4.6×10^{-11} M

Chemical Equilibrium

15

"Old chemists never die, they just reach equilibrium."
Anonymous

15.1 Life: Controlled Disequilibrium

◀ Dynamic equilibrium involves two opposing processes occurring at the same rate. This image draws an analogy between a chemical equilibrium ($N_2O_4 \rightleftharpoons 2\ N_2O$) in which the two opposing reactions occur at the same rate and a freeway with traffic moving in opposing directions at the same rate.

Have you ever tried to define life? If you have, you know that defining life is difficult. How are living things different from nonliving things? You may try to define living things as those things that can move. But of course many living things do not move—many plants, for example, do not move very much—and some nonliving things, such as glaciers and the earth itself, do move. So motion is neither unique nor definitive to life. You may try to define living things as those things that can reproduce. But again, many living things, such as mules or sterile humans, cannot reproduce; yet they are alive. In addition, some nonliving things, such as crystals for example, reproduce (in some sense). So what is unique about living things?

One definition of life uses the concept of equilibrium. We will define *chemical* equilibrium more carefully soon. For now, we can think more generally of equilibrium as *sameness and changelessness*. When an object is in equilibrium with its surroundings, some property of the object has reached sameness with the surroundings and is no longer changing. For example, a cup of hot water is not in equilibrium with its surroundings with respect to temperature. If left alone, the cup of hot water will slowly cool until it reaches equilibrium with its surroundings. At that point, the temperature of the water is the *same as* the surroundings (sameness) and it *no longer changes* (changelessness).

So equilibrium involves sameness and changelessness. Part of a definition for living things, then, is that living things *are not* in equilibrium with their surroundings. Our body temperature, for example, is not the same as the temperature of our surroundings. When we jump into a swimming pool, the pH of our blood does not become the same as the pH of the surrounding water. Living things, even the simplest ones, maintain some measure of *disequilibrium* with their environment.

We must add one more concept, however, to complete our definition of life with respect to equilibrium. Our cup of hot water is in disequilibrium with its environment, yet it is not alive. However, the cup of hot water has no control over its disequilibrium and will slowly come to equilibrium with its environment. In contrast, living things—as long as they are alive—maintain and *control* their disequilibrium. Your body temperature, for example, is not *only* in disequilibrium with your surroundings—it is in *controlled* disequilibrium. Your body maintains your temperature within a specific range that is not in equilibrium with the surrounding temperature.

So one definition for life is that living things are in *controlled disequilibrium* with their environment. A living thing comes into equilibrium with its surroundings only after it dies. In this chapter, we will examine the concept of equilibrium, especially chemical equilibrium—the state that involves sameness and changelessness.

15.2 The Rate of a Chemical Reaction

Reaction rates are related to chemical equilibrium because, as we will see in Section 15.3, a chemical system is at equilibrium when the rate of the forward reaction equals the rate of the reverse reaction.

A reaction rate can also be defined as the amount of a product that forms in a given period of time.

Before we probe more deeply into the concept of chemical equilibrium, we must first understand something about the rates of chemical reactions. The **rate of a chemical reaction** is the amount of reactant that changes to product in a given period of time. A reaction with a fast rate proceeds quickly, with a large amount of reactant going to product in a certain period of time (Figure 15.1). A reaction with a slow rate proceeds slowly, with only a small amount of reactant converting to product in the same period of time (Figure 15.2).

Chemists seek to control reaction rates for many chemical reactions. For example, the space shuttle is propelled by the reaction of hydrogen and oxygen to form water. If the reaction proceeds too slowly, the shuttle will not lift off the ground. If, however, the reaction proceeds too quickly, the shuttle can explode. Reaction rates can be controlled if we understand the factors that influence them.

Collision Theory

According to **collision theory**, most chemical reactions occur through collisions between molecules or atoms. For example, consider the following gas phase chemical reaction between $H_2(g)$ and $I_2(g)$ to form $HI(g)$.

$$H_2(g) + I_2(g) \rightarrow 2\,HI(g)$$

The reaction begins when an H_2 molecule collides with an I_2 molecule. If the collision occurs with enough energy—that is, if the colliding molecules are moving fast enough—the product molecules (HI) form. If the collision occurs without enough energy, the reactant molecules (H_2 and I_2) simply

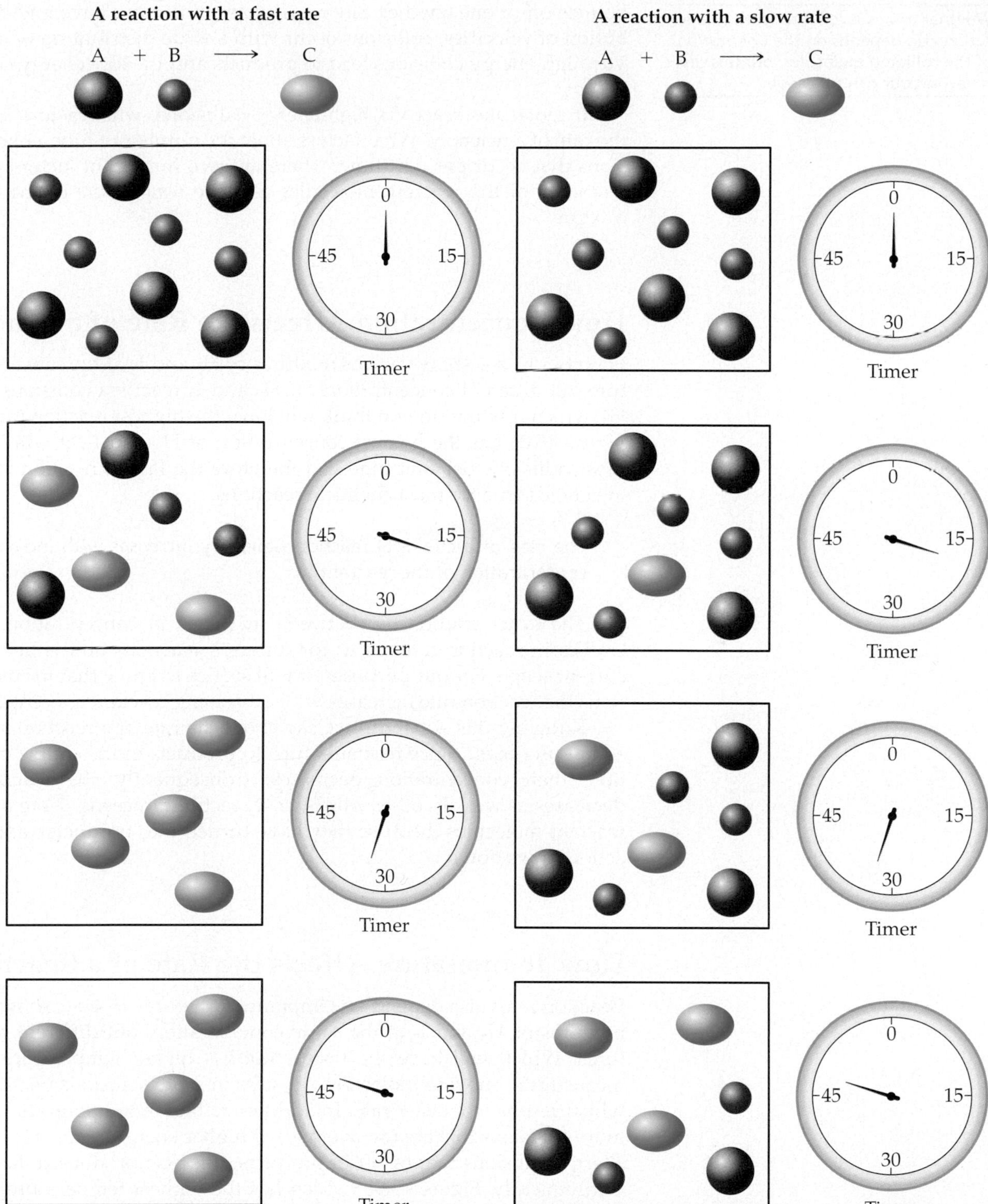

Figure 15.1 In a reaction with a fast rate, the reactants react to form products in a short period of time.

Figure 15.2 In a reaction with a slow rate, the reactants react to form products over a long period of time.

Whether or not a collision leads to a reaction also depends on the *orientation* of the colliding molecules, but this topic is beyond our current scope.

bounce off of one another. Since gas-phase molecules have a wide distribution of velocities, collisions occur with a wide distribution of energies. The high-energy collisions lead to products and the low energy collisions do not.

If molecules react via high-energy collisions, what factors influence the rate of a reaction? What factors affect the number of high-energy collisions that occur per unit time? There are two important factors: the *concentration* of the reacting molecules and the *temperature* of the reaction mixture.

How Concentration Affects the Rate of a Reaction

Figures 15.3 a–c show various mixtures of H_2 and I_2 at the same temperature but different concentrations. If H_2 and I_2 react via collisions to form HI, which mixture do you think will have the highest reaction rate? Since Figure 15.3c has the highest concentration of H_2 and I_2, it will have the most collisions per unit time and therefore the fastest reaction rate. This idea holds true for most chemical reactions.

The rate of a chemical reaction generally increases with increasing concentration of the reactants.

The exact relationship between increases in concentration and increases in reaction rate varies for different reactions and is beyond our current scope. For our purposes, it will suffice to know that for most reactions the reaction rate increases with increasing reactant concentration.

Knowing this, what can we say about the rate of a reaction as the reaction proceeds? Since reactants turn to products in the course of a reaction, their concentration decreases. Consequently, the reaction rate decreases as well. In other words, as a reaction proceeds, there are fewer reactant molecules (because they have turned into products), and the reaction slows down.

How Temperature Affects the Rate of a Reaction

Reaction rates also depend on temperature. Figures 15.4 a–c show various mixtures of H_2 and I_2 at the same concentration, but different temperatures. Which will have the fastest rate? A higher temperature—which means faster moving molecules—results in more collisions per unit time, which results in a faster rate. In addition, a higher temperature results in more collisions that are (on average) of higher energy. Since it is the high-energy collisions that result in products, this also produces a faster rate. Consequently, Figure 15.4c (which has the highest temperature) has the fastest reaction rate. This idea also holds true for most chemical reactions.

The rate of a chemical reaction generally increases with increasing temperature of the reaction mixture.

The reason that higher energy collisions are more likely to lead to products is related to a concept called the **activation energy** of a reaction.

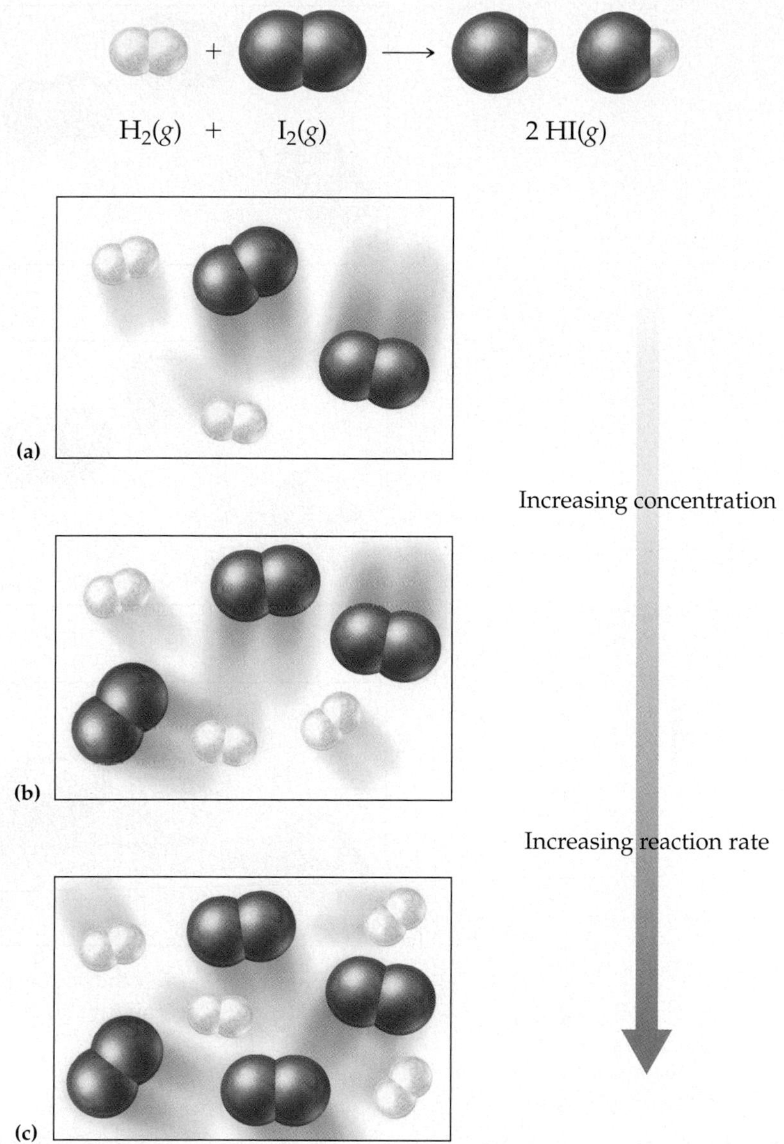

Figure 15.3 **Question:** Which reaction mixture will have the fastest initial rate? The mixture in (c) is fastest because it has the highest concentration of reactants and therefore the highest rate of collisions.

The activation energy for chemical reactions is discussed in more detail in Section 15.12. For now, we can think of the activation energy as a barrier that must be overcome for the reaction to proceed. For example, in the case of H_2 reacting with I_2 to form HI, the product (HI) can only begin to form after the H-H bond and the I-I bond each begin to break. The activation energy is the energy required to begin to break these bonds.

To summarize:

- ***Reaction rates generally increase with increasing reactant concentration.***
- ***Reaction rates generally increase with increasing temperature.***
- ***Reaction rates generally decrease as a reaction proceeds.***

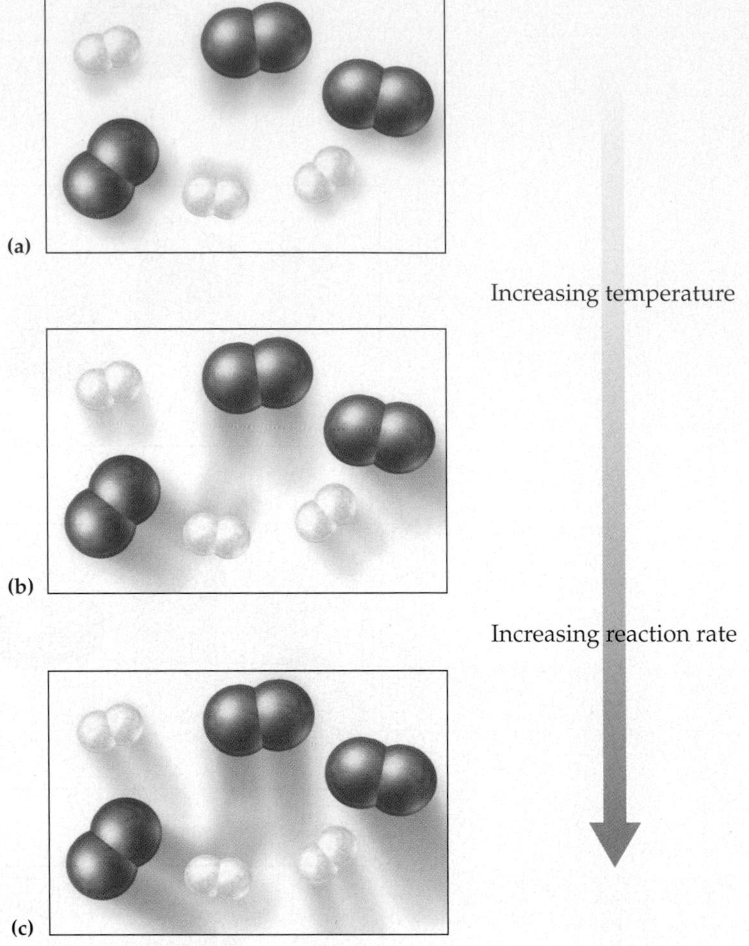

Figure 15.4 Question: Which reaction mixture will have the fastest initial rate? The mixture in (c) is fastest because it has the highest temperature.

15.3 The Idea of Dynamic Chemical Equilibrium

What would happen if our reaction between H_2 and I_2 to form HI were able to proceed in both the forward and reverse directions?

$$H_2(g) + I_2(g) \rightleftharpoons 2\,HI(g)$$

Now, H_2 and I_2 can collide and react to form 2 HI molecules, but the 2 HI molecules can also collide and react to reform H_2 and I_2. A reaction that can proceed in both the forward and reverse directions is said to be **reversible**.

Suppose we begin with only H_2 and I_2 in a container (Figure 15.5a). What happens initially? H_2 and I_2 begin to react to form HI (Figure 15.5b). However, as H_2 and I_2 react their concentration decreases, which in turn decreases the rate of the forward reaction. At the same time, HI begins to form. As the concentration of HI increases, the reverse reaction begins to occur at an increasingly faster rate because there are more HI collisions. Eventually the rate of the reverse reaction (which is increasing) equals the

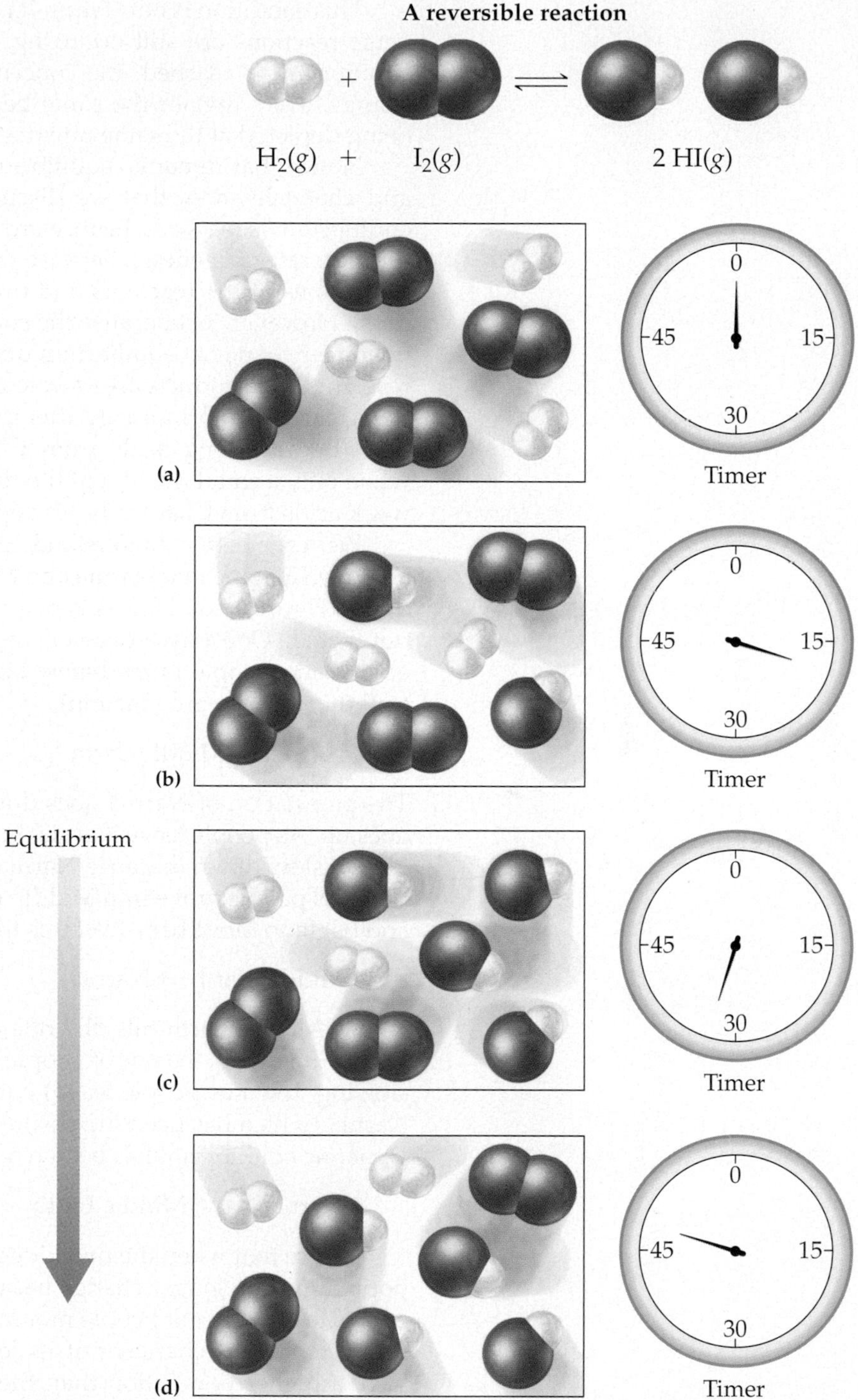

Figure 15.5 When the concentrations of the reactants and products no longer change, equilibrium has been reached.

rate of the forward reaction (which is decreasing). At that point, **dynamic equilibrium** is reached (Figure 15.5c and 15.5d).

Dynamic equilibrium – In a chemical reaction, the condition in which the rate of the forward reaction equals the rate of the reverse reaction.

This condition is not static—it is dynamic because the forward and reverse reactions are still occurring, but at constant rate. When dynamic equilibrium is reached, the concentrations of H_2, I_2, and HI no longer change. They remain the same because the reactants and products are being depleted at the same rate that they are being formed.

Notice that dynamic equilibrium includes the concepts of sameness and changelessness that we discussed in Section 15.1. When dynamic equilibrium is reached, the forward reaction rate is the same as the reverse reaction rate (sameness). Because the reaction rates are the same, the concentrations of the reactants and products no longer change (changelessness). However, just because the concentrations of reactants and products no longer change at equilibrium does *not* imply that the concentrations of reactants and products are *equal* to one another at equilibrium. Some reactions reach equilibrium only after most of the reactants have formed products (Recall strong acids from Chapter 14). Others reach equilibrium when only a small fraction of the reactants have formed products. (Recall weak acids from Chapter 14.) It depends on the reaction.

We can better understand dynamic equilibrium with a simple analogy. Imagine that Narnia and Middle Earth are two neighboring kingdoms (Figure 15.6). Narnia is overpopulated and Middle Earth is underpopulated. One day, however, the border between the two kingdoms opens, and people immediately begin to leave Narnia for Middle Earth (call this the forward reaction).

$$\text{Narnia} \rightarrow \text{Middle Earth} \qquad \text{(forward reaction)}$$

The population of Narnia goes down as the population of Middle Earth goes up. As people leave Narnia, however, the *rate* at which they leave begins to slow down (because Narnia becomes less crowded). On the other hand, as people move into Middle Earth, some decide it was not for them and begin to move back (call this the reverse reaction).

$$\text{Middle Earth} \rightarrow \text{Narnia} \qquad \text{(reverse reaction)}$$

As Middle Earth fills, the rate of people moving back to Narnia gets faster. Eventually the *rate* of people moving out of Narnia (which has been slowing down as people leave) equals the *rate* of people moving back to Narnia (which has been increasing as Middle Earth gets more crowded). Dynamic equilibrium has been reached.

$$\text{Narnia} \rightleftharpoons \text{Middle Earth}$$

Notice that when the two kingdoms reach dynamic equilibrium, their populations no longer change because the number of people moving out equals the number of people moving in. However one kingdom—because of its charm, the character of its leader, or whatever other reason—may have a higher population than the other kingdom, even when dynamic equilibrium is reached.

Similarly, when a chemical reaction reaches dynamic equilibrium, the rate of the forward reaction (analogous to people moving out of Narnia) equals the rate of the reverse reaction (analogous to people moving back into Narnia), and the relative concentrations of reactants and products (analogous to the relative populations of the two kingdoms) become constant. Also, like our two kingdoms, the concentrations of reactants and products will not necessarily be equal at equilibrium, just as the populations of the two kingdoms are not equal at equilibrium.

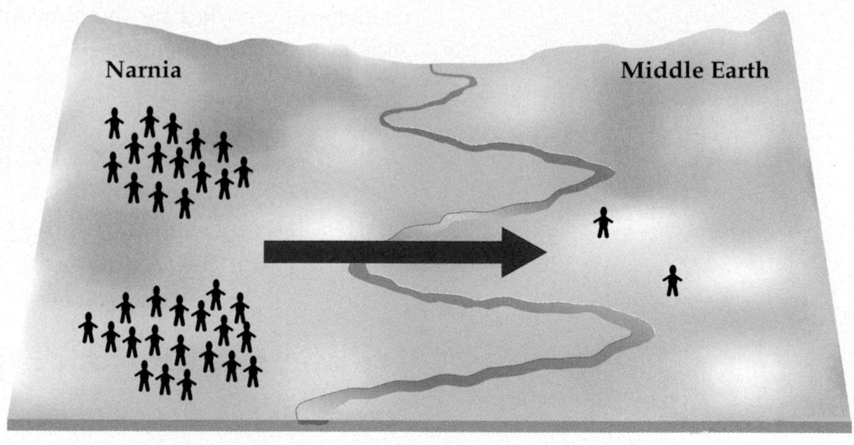

Initial

Represents population

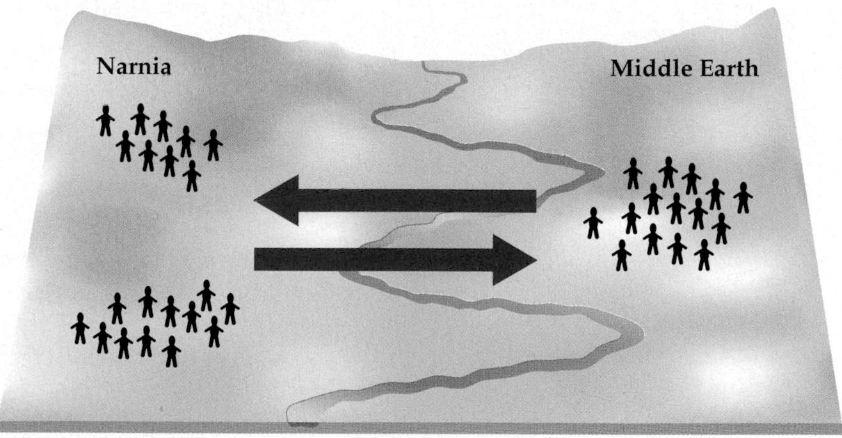

Equilibrium

Figure 15.6 Population analogy of a chemical reaction proceeding to equilibrium.

15.4 The Equilibrium Constant: A Measure of How Far a Reaction Goes

We have just learned that the *concentrations* of reactants and products are not equal at equilibrium—it is the *rates* of the forward and reverse reactions that are equal. But what about the concentrations? What can we know about them? The equilibrium constant (K_{eq}) is a way to quantify the concentrations of the reactants and products at equilibrium. Consider the following generic chemical reaction

$$aA + bB \rightleftharpoons cC + dD$$

where A and B are reactants, C and D are products, and a, b, c, and d are the respective stoichiometric coefficients in the chemical equation. The **equilibrium constant (K_{eq})** for the reaction is defined as the ratio—at equilibrium—of the concentrations of the products raised to their stoichiometric coefficients divided by the concentrations of the reactants raised to their stoichiometric coefficients.

$$K_{eq} = \frac{[C]^c\,[D]^d}{[A]^a\,[B]^b}$$

The equilibrium constant quantifies the relative concentrations of reactants and products at equilibrium.

Writing Equilibrium Expressions for Chemical Reactions

To write an equilibrium expression for a chemical reaction, simply examine the chemical equation and follow the preceding definition. For example, suppose we want to write an equilibrium expression for the following reaction.

$$2\,N_2O_5(g) \rightleftharpoons 4\,NO_2(g) + O_2(g)$$

The equilibrium constant is $[NO_2]$ raised to the fourth power multiplied by $[O_2]$ raised to the first power divided by $[N_2O_5]$ raised to the second power.

$$K_{eq} = \frac{[NO_2]^4\,[O_2]}{[N_2O_5]^2}$$

Notice that the *coefficients* in the chemical equation become the *exponents* in the equilibrium expression.

$$2\,N_2O_5(g) \rightleftharpoons 4\,NO_2(g) + O_2(g)$$

$$K_{eq} = \frac{[NO_2]^4\,[O_2]}{[N_2O_5]^2}$$

EXAMPLE 15.1 Writing Equilibrium Expressions for Chemical Reactions

Write an equilibrium expression for the following chemical equation.

$$CO(g) + 2\,H_2(g) \rightleftharpoons CH_3OH(g)$$

Solution:

The equilibrium expression is the concentrations of the products raised to their stoichiometric coefficients divided by the concentrations of the reactants raised to their stoichiometric coefficients.

$$K_{eq} = \frac{[CH_3OH]}{[CO][H_2]^2}$$

Notice that the expression is a ratio of products over reactants. Notice also that the coefficients in the chemical equation are the exponents in the equilibrium expression.

SKILLBUILDER 15.1 **Writing Equilibrium Expressions for Chemical Reactions**

Write an equilibrium expression for the following chemical equation.

$$H_2(g) + F_2(g) \rightleftharpoons 2\,HF(g)$$

The Significance of the Equilibrium Constant

Given this definition of an equilibrium constant, what does it mean? What, for example, does a large equilibrium constant ($K_{eq} \gg 1$) imply about a reaction? It means that the forward reaction is largely favored and that there will be more products than reactants when equilibrium is reached. For example, consider the following reaction.

$$H_2(g) + Br_2(g) \rightleftharpoons 2\,HBr(g) \qquad K_{eq} = 1.9 \times 10^{19} \text{ at } 25\ ^\circ\text{C}$$

The equilibrium constant is large, meaning that at equilibrium the reaction lies far to the right—high concentrations of products, low concentrations of reactants (Figure 15.7).

Conversely, what does a *small* equilibrium constant ($K_{eq} \ll 1$) mean? It means that the reverse reaction is favored and that there will be more reactants than products when equilibrium is reached. For example, consider the following reaction.

$$N_2(g) + O_2(g) \rightleftharpoons 2\,NO(g) \qquad K_{eq} = 4.1 \times 10^{-31} \text{ at } 25\ ^\circ\text{C}$$

The equilibrium constant is very small, meaning that at equilibrium the reaction lies far to the left—high concentrations of reactants, low concentrations of products (Figure 15.8). This is fortunate because N_2 and O_2 are the main components of air. If this equilibrium constant were large, much of the N_2 and O_2 in air would react to form NO, a toxic gas.

$$H_2(g) + Br_2(g) \rightleftharpoons 2\,HBr(g)$$

$$K = \frac{[HBr]^2}{[H_2][Br_2]} = \textbf{Large Number}$$

Figure 15.7 A large equilibrium constant means that there will be a high concentration of products and a low concentration of reactants at equilibrium.

$$N_2(g) + O_2(g) \rightleftharpoons 2\,NO(g)$$

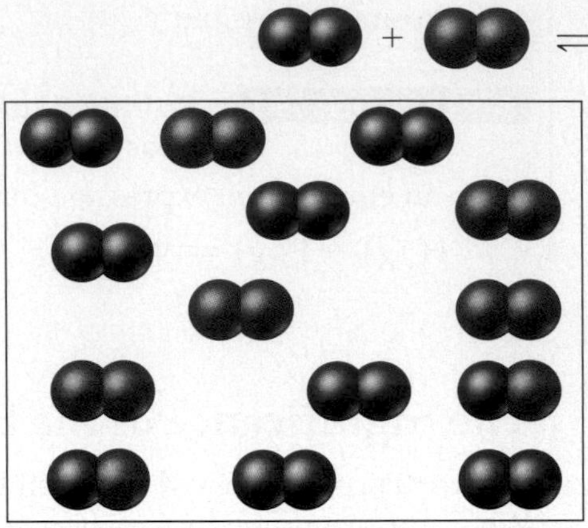

$$K = \frac{[NO]^2}{[N_2][O_2]} = \textbf{Small Number}$$

Figure 15.8 A small equilibrium constant means that there will be a high concentration of reactants and a low concentration of products at equilibrium.

The symbol $\approx$ means "approximately equal to."

To summarize:

- **$K_{eq} \ll 1$** ***Reverse reaction is favored; forward reaction does not proceed very far.***
- **$K_{eq} \approx 1$** ***Neither direction is favored; forward reaction proceeds about halfway.***
- **$K_{eq} \gg 1$** ***Forward reaction is favored; forward reaction proceeds to completion.***

15.5 Heterogeneous Equilibria: The Equilibrium Expression for Reactions Involving a Solid or a Liquid

Consider the following chemical reaction.

$$2\,CO(g) \rightleftharpoons CO_2(g) + C(s)$$

We might expect the expression for the equilibrium constant to be:

$$K_{eq} = \frac{[CO_2][C]}{[CO]^2} \qquad \text{(incorrect)}$$

However, since carbon is a solid, its concentration is constant—it does not change. Adding more or less carbon to the reaction mixture does not change the concentration of C. Consequently, pure solids—those reactants or products labeled in the chemical equation with an (*s*)—are not included in the equilibrium expression. The correct equilibrium expression is therefore:

$$K_{eq} = \frac{[CO_2]}{[CO]^2} \qquad \text{(correct)}$$

The concentrations of pure solids or pure liquids are excluded from equilibrium expressions because they are constant. Consequently, they simply become incorporated into the value of the equilibrium constant.

Similarly, the concentration of a pure liquid does not change. Consequently, pure liquids—those reactants or products labeled in the chemical equation with an (l)—are also excluded from the equilibrium expression. For example, what is the equilibrium expression for the following reaction?

$$CO_2(g) + H_2O(l) \rightleftharpoons H^+(aq) + HCO_3^-(aq)$$

Since $H_2O(1)$ is pure liquid, it is omitted from the equilibrium expression.

$$K_{eq} = \frac{[H^+][HCO_3^-]}{[CO_2]}$$

EXAMPLE 15.2 **Writing Equilibrium Expressions for Reactions Involving a Solid or a Liquid**

Write an equilibrium expression for the following chemical equation.

$$CaCO_3(s) \rightleftharpoons CaO(s) + CO_2(g)$$

Solution:

Since $CaCO_3(s)$ and $CaO(s)$ are both solids, they are omitted from the equilibrium expression.

$$K_{eq} = [CO_2]$$

SKILLBUILDER 15.2 **Writing Equilibrium Expressions for Reactions Involving a Solid or a Liquid**

Write an equilibrium expression for the following chemical equation.

$$4\,HCl(g) + O_2(g) \rightleftharpoons 2\,H_2O(l) + 2\,Cl_2(g)$$

15.6 Calculating and Using Equilibrium Constants

Calculating Equilibrium Constants

The most direct way to get a value for the equilibrium constant of a reaction is to measure the concentrations of the reactants and products in a reaction mixture at equilibrium. For example, consider the following reaction.

$$H_2(g) + I_2(g) \rightleftharpoons 2\,HI(g)$$

Suppose a mixture of H_2 and I_2 is allowed to come to equilibrium at 445 °C. The measured equilibrium concentrations are $[H_2] = 0.11$ M, $[I_2] = 0.11$ M, and $[HI] = 0.78$ M. What is the value of the equilibrium constant? We begin by setting up the problem in the standard format.

Equilibrium constants depend on temperature, so temperatures will often be included with equilibrium data. However, the temperature is not a part of the equilibrium expression.

Given: $[H_2] = 0.11$ M
$[I_2] = 0.11$ M
$[HI] = 0.78$ M

Find: K_{eq}

Solution:

The expression for K_{eq} can be written from the balanced equation.

$$K_{eq} = \frac{[HI]^2}{[H_2][I_2]}$$

TABLE 15.1

Initial and Equilibrium Concentrations for the Reaction $H_2(g) + I_2(g) \rightleftharpoons 2\,HI(g)$

Initial			*Equilibrium*			*Equilibrium Constant*
$[H_2]$	$[I_2]$	$[HI]$	$[H_2]$	$[I_2]$	$[HI]$	$K_{eq} = \frac{[HI]^2}{[H_2][I_2]}$
0.50	0.50	0.0	0.11	0.11	0.78	$\frac{[0.78]^2}{[0.11][0.11]} = 50$
0.0	0.0	0.50	0.055	0.055	0.39	$\frac{[0.39]^2}{[0.055][0.055]} = 50$
0.50	0.50	0.50	0.16$\underline{5}$	0.16$\underline{5}$	1.17	$\frac{[1.17]^2}{[0.16\underline{5}][0.16\underline{5}]} = 50$
1.0	0.5	0.0	0.53	0.033	0.93$\underline{4}$	$\frac{[0.93\underline{4}]^2}{[0.53][0.033]} = 50$

To calculate the value of K_{eq}, simply substitute the correct equilibrium concentrations into the expression for K_{eq}.

$$K_{eq} = \frac{[HI]^2}{[H_2][I_2]}$$

$$= \frac{[0.78]^2}{[0.11][0.11]}$$

$$= 5.0 \times 10^1$$

The concentrations in an equilibrium expression should always be in units of molarity (M), but the units themselves are normally dropped.

The concentrations within K_{eq} should always be written in moles per liter (M); however, the units are normally dropped in expressing the equilibrium constant so that K_{eq} is unitless.

The particular concentrations of reactants and products for a reaction at equilibrium will not always be the same for a given reaction—they will depend on the initial concentrations. However, the *equilibrium constant* will always be the same at a given temperature, regardless of the initial concentrations. For example, Table 15.1 shows several different equilibrium concentrations of H_2, I_2, and HI, each from a different set of initial concentrations. Notice that the equilibrium constant is always the same, regardless of the initial concentrations. In other words, no matter what the initial concentrations are, the reaction will always go in a direction so that the equilibrium concentrations—when substituted into the equilibrium expression—give the same constant, K_{eq}.

EXAMPLE 15.3 Calculating Equilibrium Constants

Consider the following reaction.

$$2\,CH_4(g) \rightleftharpoons C_2H_2(g) + 3\,H_2(g)$$

A mixture of CH_4, C_2H_2, and H_2 is allowed to come to equilibrium at 1700 °C. The measured equilibrium concentrations are $[CH_4] = 0.0203$ M, $[C_2H_2] = 0.0451$ M, and $[H_2] = 0.112$ M. What is the value of the equilibrium constant at this temperature?

Based on the relative amounts of reactants and products at equilibrium, do you expect the equilibrium in this example constant to be large, small, or intermediate?

Begin by setting up the problem in the standard format.

Given: $[CH_4] = 0.0203\ M$
$[C_2H_2] = 0.0451\ M$
$[H_2] = 0.112\ M.$

Find: K_{eq}

Solution:
The expression for K_{eq} is written from the balanced equation.

$$K_{eq} = \frac{[C_2H_2][H_2]^3}{[CH_4]^2}$$

To calculate the value of K_{eq}, simply substitute the correct equilibrium concentrations into the expression for K_{eq}.

$$K_{eq} = \frac{[0.0451][0.112]^3}{[0.0203]^2}$$

$$= 0.154$$

SKILLBUILDER 15.3 **Calculating Equilibrium Constants**

Consider the following reaction.

$$CO(g) + 2\,H_2(g) \rightleftharpoons CH_3OH(g)$$

A mixture of CO, H_2, and CH_3OH is allowed to come to equilibrium at 225 °C. The measured equilibrium concentrations are $[CO] = 0.489\ M$, $[H_2] = 0.146\ M$, and $[CH_3OH] = 0.151\ M$. What is the value of the equilibrium constant at this temperature?

SKILLBUILDER PLUS

Suppose the preceding reaction is carried out at a different temperature and that the initial concentrations of the reactants are $[CO] = 0.500\ M$ and $[H_2] = 1.00\ M$. Assuming that there is no product at the beginning of the reaction, and that at equilibrium $[CO] = 0.15\ M$, find the equilibrium constant at this new temperature. (Hint: Use the stoichiometric relationships from the balanced equation to find the equilibrium concentrations of H_2 and CH_3OH.)

Using Equilibrium Constants in Calculations

The equilibrium constant can also be used to calculate the equilibrium concentration of one of the reactants or products given the equilibrium concentrations of the others. For example, consider the following reaction.

$$2\,COF_2(g) \rightleftharpoons CO_2(g) + CF_4(g) \qquad K_{eq} = 2.00 \text{ at } 1000\ °C$$

In an equilibrium mixture, the concentration of COF_2 is 0.255 M and the concentration of CF_4 is 0.118 M. What is the equilibrium concentration of CO_2? Again, we set up the problem in the standard way.

Given: $[COF_2] = 0.255\ M$
$[CF_4] = 0.118\ M$
$K_{eq} = 2.00$

Find: $[CO_2]$

Solution Map:

$[COF_2], [CF_4], K_{eq}$ → $[CO_2]$

$$K_{eq} = \frac{[CO_2][CF_4]}{[COF_2]}$$

In this problem, we are given K_{eq} and the concentrations of one reactant and one product. We are asked to find the concentration of the other product. We can calculate this by using the expression for K_{eq}.

Solution:

We first write the equilibrium expression for the reaction, and then solve it for the quantity we are trying to find ($[CO_2]$).

$$K_{eq} = \frac{[CO_2][CF_4]}{[COF_2]^2}$$

$$[CO_2] = K_{eq}\frac{[COF_2]^2}{[CF_4]}$$

Now simply substitute the appropriate values and compute $[CO_2]$.

$$[CO_2] = 2.00\frac{[0.255]^2}{[0.118]}$$
$$= 1.10\ M$$

EXAMPLE 15.4 Using Equilibrium Constants in Calculations

Consider the following reaction.

$$H_2(g) + I_2(g) \rightleftharpoons 2\,HI(g) \qquad K_{eq} = 69 \text{ at } 340\ °C$$

In an equilibrium mixture, the concentrations of H_2 and I_2 are both 0.020 M. What is the equilibrium concentration of HI?

Begin by setting up the problem in the standard way.

Given: $[H_2] = [I_2] = 0.020\ M$

$K_{eq} = 69$

Find: [HI]

Solution Map:

$[H_2], [I_2], K_{eq}$ → $[HI]$

$$K_{eq} = \frac{[HI]^2}{[H_2][I_2]}$$

Solution:

We first solve the equilibrium expression for [HI] and then substitute in the appropriate values to compute it.

$$K_{eq} = \frac{[HI]^2}{[H_2][I_2]}$$

$$[HI]^2 = K_{eq}[H_2][I_2]$$
$$[HI] = \sqrt{K_{eq}[H_2][I_2]}$$
$$= \sqrt{69[0.020][0.020]}$$
$$= 0.17\ M$$

SKILLBUILDER 15.4 **Using Equilibrium Constants in Calculations**

Diatomic iodine (I_2) decomposes at high temperature to form I atoms according to the following reaction.

$$I_2(g) \rightleftharpoons 2\,I(g) \qquad K_{eq} = 0.011 \text{ at } 1200\ °C$$

In an equilibrium mixture, the concentration of I_2 is 0.10 M. What is the equilibrium concentration of I?

15.7 Disturbing a Reaction at Equilibrium: Le Châtelier's Principle

We have seen that a chemical system not in equilibrium tends to go towards equilibrium and that the concentrations of the reactants and products at equilibrium correspond to the equilibrium constant, K_{eq}. What happens, however, when a chemical system already at equilibrium is disturbed? **Le Châtelier's Principle** states that the chemical system will respond to minimize the disturbance.

> **Le Châtelier's Principle**–When a chemical system at equilibrium is disturbed, the system shifts in a direction that minimizes the disturbance.

In other words, a system at equilibrium tries to maintain that equilibrium—it fights back when disturbed.

We can understand Le Châtelier's Principle by returning to our Narnia and Middle Earth analogy. Suppose the populations of Narnia and Middle Earth are at equilibrium. Then, the rate of people moving out of Narnia and into Middle Earth is equal to the rate of people moving into Narnia and out of Middle Earth, and the populations of the two kingdoms are stable. Now imagine disturbing that balance (Figure 15.9). Suppose we add extra people to Middle Earth. What happens? Since Middle Earth suddenly becomes more crowded, the rate of people leaving Middle Earth increases. The net flow of people is out of Middle Earth and into Narnia. Notice what happened. We disturbed the equilibrium by adding more people to Middle Earth. The system responded by moving people out of Middle Earth—it shifted in the direction that minimized the disturbance.

On the other hand, what happens if we add extra people to Narnia? Since Narnia suddenly gets more crowded, the rate of people leaving Narnia goes up. The net flow of people is out of Narnia and into Middle

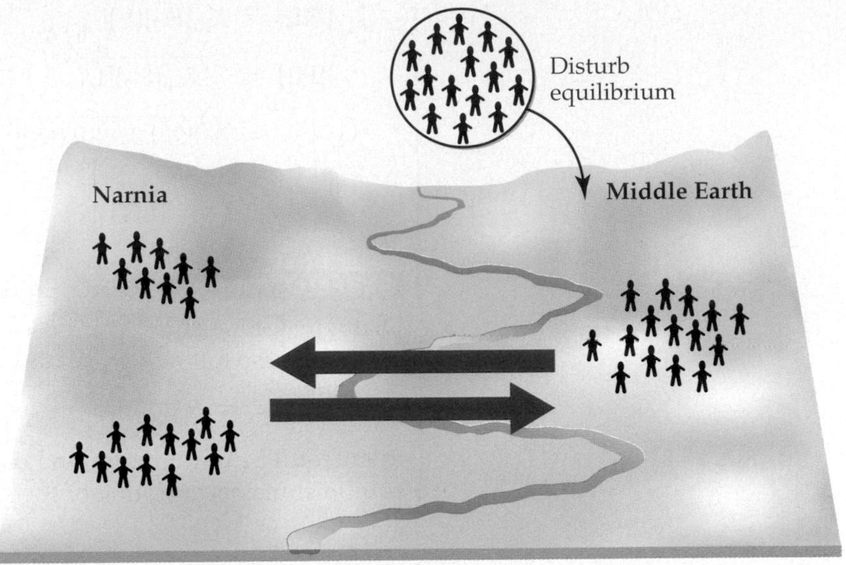

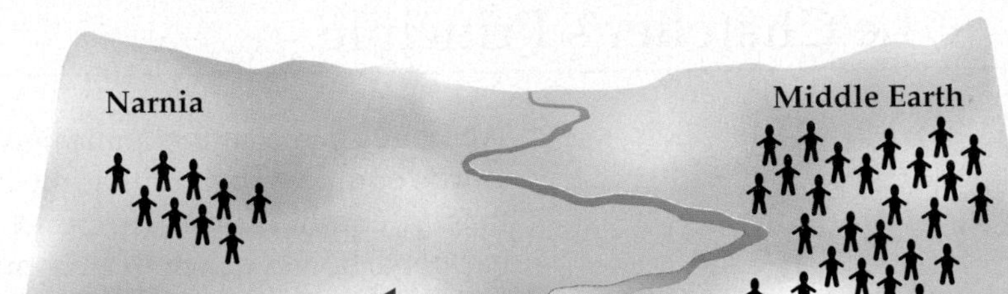

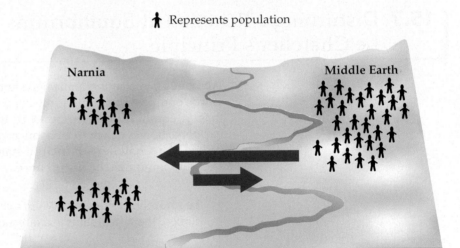

Figure 15.9 When a system at equilibrium is disturbed, it shifts to minimize the disturbance. In this case, adding population to Middle Earth (the disturbance) causes population to move out of Middle Earth (minimizing the disturbance.) **Question:** What would happen if you disturbed the equilibrium by taking population out of Middle Earth? In which direction would the population move to minimize the disturbance?

Earth. We added people to Narnia and the system responded by moving people out of Narnia. When systems at equilibrium are disturbed, they react to counter the disturbance. Chemical systems behave similarly. There are several ways to disturb a system in a chemical equilibrium. We consider each of these separately.

15.8 The Effect of a Concentration Change on Equilibrium

Consider the following reaction in chemical equilibrium.

$$N_2O_4(g) \rightleftharpoons 2\,NO_2(g)$$

Suppose we disturb the equilibrium by adding NO_2 to the equilibrium mixture (Figure 15.10). In other words, we increase the concentration of NO_2. What happens? According to Le Châtelier's Principle, the system shifts in a direction to minimize the disturbance. The reaction goes to the left (it proceeds in the reverse direction), consuming some of the added NO_2 and bringing its concentration back down.

When we say that a reaction *shifts to the left* we mean that it proceeds in the reverse direction, consuming products and forming reactants.

$$N_2O_4(g) \rightleftharpoons 2\,NO_2(g)$$

Shift to left ← Add NO_2

On the other hand, what happens if we add extra N_2O_4, increasing its concentration? In this case, the reaction shifts to the right, consuming some of the added N_2O_4, and bringing *its* concentration back down (Figure 15.11).

$$N_2O_4(g) \rightleftharpoons 2\,NO_2(g)$$

Add N_2O_4 Shift to right →

When we say that a reaction *shifts to the right* we mean that it proceeds in the forward direction, consuming reactants and forming products.

Figure 15.10 When a system at equilibrium is disturbed, it shifts to minimize the disturbance. In this case, adding NO_2 (the disturbance) causes the reaction to shift left, consuming NO_2 and minimizing the disturbance.

15.10 The Effect of a Temperature Change on Equilibrium

According to Le Châtelier's Principle, if the temperature of a system at equilibrium is changed, the system should shift in a direction to counter that change. So, if the temperature is increased, the reaction should shift in the direction that attempts to decrease the temperature and vice versa. Recall from Chapter 3 that energy changes are often associated with chemical reactions (see Section 3.8). If we want to predict the direction that a reaction will shift upon a temperature change, we must understand how a shift in the reaction affects the temperature.

We can classify chemical reactions according to whether they absorb or emit heat energy in the course of the reaction. An **exothermic reaction** emits heat.

$$\text{Exothermic reaction} \qquad A + B \rightleftharpoons C + D + \text{Heat}$$

In an exothermic reaction, you can think of heat as a product. Consequently, raising the temperature of an exothermic reaction—think of this as adding heat—causes the reaction to shift left. For example, the reaction of nitrogen with hydrogen to form ammonia is exothermic.

$$N_2(g) + 3\,H_2(g) \rightleftharpoons 2\,NH_3 + \text{Heat}$$

Shift to left ← ; Add heat ↑ (Heat)

Raising the temperature of an equilibrium mixture of these three gases causes the reaction to shift left, absorbing some of the added heat. Conversely, lowering the temperature of an equilibrium mixture of these three gases causes the reaction to shift right, releasing heat.

$$N_2(g) + 3\,H_2(g) \rightleftharpoons 2\,NH_3 + \text{Heat}$$

Shift to right → ; Remove heat ↓ (Heat)

In contrast, an **endothermic reaction** absorbs heat.

$$\text{Endothermic reaction:} \qquad A + B + \text{Heat} \rightleftharpoons C + D$$

In an endothermic reaction, you can think of heat as a reactant. Consequently, raising the temperature (or adding heat) causes an endothermic reaction to shift right. For example, the following reaction is endothermic.

$$\underset{\text{Colorless}}{N_2O_4(g)} + \text{Heat} \rightleftharpoons \underset{\text{Brown}}{2\,NO_2}$$

Add heat ↑ (Heat) ; Shift to right →

Raising the temperature of an equilibrium mixture of these two gases causes the reaction to shift right, absorbing some of the added heat. Since N_2O_4 is colorless and NO_2 is brown, the effects of changing the temperature of this reaction are easily seen (Figure 15.14). On the other hand,

Figure 15.14 $N_2O_4(g) \rightleftharpoons 2\,NO_2(g)$ equilibrium as a function of temperature. Since the reaction is endothermic, cool temperatures cause a shift to the left to colorless N_2O_4. Warm temperatures cause a shift to the right to brown NO_2.

lowering the temperature of a reaction mixture of these two gases causes the reaction to shift left, releasing heat.

$$\underset{\text{Colorless}}{N_2O_4(g)} + \text{Heat} \rightleftharpoons \underset{\text{Brown}}{2\,NO_2}$$

Remove heat ↓ ← Shift to left

To summarize:

In an exothermic chemical reaction, heat is a product and:

- **Increasing** *the temperature causes the reaction to shift left (in the direction of the reactants).*
- **Decreasing** *the temperature causes the reaction to shift right (in the direction of the products).*

In an endothermic chemical reaction heat is a reactant and:

- **Increasing** *the temperature causes the reaction to shift right (in the direction of the products).*
- **Decreasing** *the temperature causes the reaction to shift left (in the direction of the reactants).*

EXAMPLE 15.7 The Effect of a Temperature Change on Equilibrium

The following reaction is endothermic.

$$CaCO_3(s) \rightleftharpoons CaO(s) + CO_2(g)$$

What is the effect of increasing the temperature of the reaction mixture? Decreasing the temperature?

Solution:

Since the reaction is endothermic, we can think of heat as a reactant.

$$\text{Heat} + CaCO_3(s) \rightleftharpoons CaO(s) + CO_2(g)$$

Raising the temperature is adding heat, causing the reaction to shift to the right. Lowering the temperature is removing heat, causing the reaction to shift to the left.

SKILLBUILDER 15.7 The Effect of a Temperature Change on Equilibrium

The following reaction is exothermic.

$$2\,SO_2(g) + O_2(g) \rightleftharpoons 2\,SO_3(g)$$

What is the effect of increasing the temperature of the reaction mixture? Decreasing the temperature?

15.11 Solubility-Product Constant

Recall from Chapter 7 (see Section 7.7) that a compound is considered soluble if it dissolves in water and insoluble if it does not. Recall also that, through the *solubility rules*, we classified ionic compounds as soluble or insoluble. We can better understand the solubility of an ionic compound with the concept of equilibrium. The process by which an ionic compound dissolves is an equilibrium process. For example, we can represent the dissolving of calcium fluoride in water with the following chemical equation.

$$CaF_2(s) \rightleftharpoons Ca^{2+}(aq) + 2\,F^-(aq)$$

The equilibrium expression for a chemical equation that represents the dissolving of an ionic compound is called the **solubility-product constant (K_{sp})**. For CaF_2, the solubility-product constant is:

$$K_{sp} = [Ca^{2+}][F^-]^2$$

Notice that, as we discussed in Section 15.5, solids are omitted from the equilibrium expression.

The K_{sp} value is a measure of the solubility of a compound. A large K_{sp} (forward reaction favored) means that the compound is very soluble. A small K_{sp} (reverse reaction favored) means that the compound is not very soluble. Table 15.2 lists the value of K_{sp} for a number of ionic compounds.

EXAMPLE 15.8 Writing Expressions for K_{sp}

Write expressions for K_{sp} for each of the following ionic compounds.

a) $BaSO_4$
b) $Mn(OH)_2$
c) Ag_2CrO_4

Solution:

To write the expression for K_{sp}, first write the chemical reaction showing the solid compound in equilibrium with its dissolved aqueous ions. Then write the equilibrium expression based on this equation.

a) $BaSO_4(s) \rightleftharpoons Ba^{2+}(aq) + SO_4^{2-}(aq)$

$$K_{sp} = [Ba^{2+}][SO_4^{2-}]$$

b) $Mn(OH)_2(s) \rightleftharpoons Mn^{2+}(aq) + 2\,OH^-(aq)$

$K_{sp} = [Mn^{2+}][OH^-]^2$

c) $Ag_2CrO_4(s) \rightleftharpoons 2\,Ag^+(aq) + CrO_4^{2-}(aq)$

$K_{sp} = [Ag^+]^2[CrO_4^{2-}]$

SKILLBUILDER 15.8 **Writing Expressions for K_{sp}**

Write expressions for K_{sp} for each of the following ionic compounds.

a) AgI

b) $Ca(OH)_2$

Using K_{sp} to Determine Molar Solubility

Recall from Section 13.3 that the solubility of a compound is the amount of the compound that dissolves in a certain amount of liquid. The **molar solubility** is simply the solubility in units of moles per liter. The molar solubility of a compound can be computed directly from K_{sp}. For example, consider silver chloride.

$$AgCl(s) \rightleftharpoons Ag^+(aq) + Cl^-(aq) \quad K_{sp} = 1.77 \times 10^{-10}$$

How can we find the molar solubility of AgCl from K_{sp}? First, notice that K_{sp} is *not* the molar solubility; it is the solubility-product constant. Second, notice that the concentration of either Ag^+ or Cl^- at equilibrium will be equal to the amount of AgCl that dissolved. We know this from the relationship of the stoichiometric coefficients in the balanced equation.

$$1 \text{ mol AgCl} \equiv 1 \text{ mol } Ag^+ \equiv 1 \text{ mol } Cl^-$$

TABLE 15.2

Selected Solubility-Product Constants (K_{sp})

Compound	*Formula*	K_{sp}
barium sulfate	$BaSO_4$	1.07×10^{-10}
calcium carbonate	$CaCO_3$	4.96×10^{-9}
calcium fluoride	CaF_2	1.46×10^{-10}
calcium hydroxide	$Ca(OH)_2$	4.68×10^{-6}
calcium sulfate	$CaSO_4$	7.10×10^{-5}
copper(II) sulfide	CuS	1.27×10^{-36}
iron(II) carbonate	$FeCO_3$	3.07×10^{-11}
iron(II) hydroxide	$Fe(OH)_2$	4.87×10^{-17}
lead(II) chloride	$PbCl_2$	1.17×10^{-5}
lead(II) sulfate	$PbSO_4$	1.82×10^{-8}
lead(II) sulfide	PbS	9.04×10^{-29}
magnesium carbonate	$MgCO_3$	6.82×10^{-6}
magnesium hydroxide	$Mg(OH)_2$	2.06×10^{-13}
silver chloride	AgCl	1.77×10^{-10}
silver chromate	Ag_2CrO_4	1.12×10^{-12}
silver iodide	AgI	8.51×10^{-17}

Solubility-Product Constant, K_{sp}: The solubility-product constant of an ionic compound is the equilibrium constant for the chemical equation that describes the dissolving of the compound.

Solubility-Product Constant, K_{sp}: The solubility-product constant reflects the solubility of a compound. The greater the solubility-product constant, the greater the solubility of the compound.

Reaction Paths and Catalysts: Most chemical reactions must overcome an energy hump, called the *activation energy,* as they proceed from reactants to products. Increasing the temperature of a reaction mixture increases the fraction of reactant molecules that make it over the energy hump, therefore increasing the rate. A catalyst—a substance that increases the rate of the reaction but is not consumed by it—lowers the activation energy so that it is easier to get over the energy hump.

Reaction Paths and Catalysts: Catalysts are used in many chemical reactions to increase the rates. Without catalysts, many reactions occur too slowly to be of any value. The thousands of reactions that occur in living organisms are controlled by biological catalysts called *enzymes*.

Chemical Skills

Examples

Writing Equilibrium Expressions for Chemical Reactions (Section 15.5)

Examine the definition of the equilibrium constant in the Chemical Principles section (page 561). To write the equilibrium expression for a reaction, write the concentrations of the products raised to their stoichiometric coefficients divided by the concentrations of the reactants raised to their stoichiometric coefficients. Remember that if the reaction contains reactants or products that are liquids or solids, these are omitted from the equilibrium expression.

EXAMPLE 15.10 Writing Equilibrium Expressions for Chemical Reactions

Write an equilibrium expression for the following chemical equation.

$$2\ NO(g) + Br_2(g) \rightleftharpoons 2\ NOBr(g)$$

Solution:

$$K_{eq} = \frac{[NOBr]^2}{[NO]^2[Br_2]}$$

Calculating Equilibrium Constants (Section 15.6)

EXAMPLE 15.11 Calculating Equilibrium Constants

An equilibrium mixture of the following reaction had [I] = 0.075 M and $[I_2]$ = 0.88 M. What is the value of the equilibrium constant?

$$I_2(g) \rightleftharpoons 2\ I(g)$$

Begin by setting up the problem in the standard format.

Given: [I] = 0.075 M

$[I_2]$ = 0.88 M

Find: K_{eq}

Then write the expression for K_{eq} from the balanced equation.

To calculate the value of K_{eq}, simply substitute the correct equilibrium concentrations into the expression for K_{eq}. The concentrations within K_{eq} should always be written in moles per liter, M. Units are normally dropped in expressing the equilibrium constant so that K_{eq} is unitless.

Solution:

$$K_{eq} = \frac{[I]^2}{[I_2]} = \frac{[0.075]^2}{[0.88]} = 0.0064$$

Using the Equilibrium Constant to Find the Concentration of a Reactant or Product at Equilibrium (Section 15.6)

EXAMPLE 15.12 Using the Equilibrium Constant to Find the Concentration of a Reactant or Product at Equilibrium

Consider the following reaction.

$$N_2(g) + 3\,H_2(g) \rightleftharpoons 2\,NH_3(g)$$
$$K_{eq} = 152 \text{ at } 225\ ^\circ C$$

In an equilibrium mixture, $[N_2] = 0.110$ M and $[H_2] = 0.0935$ M. What is the equilibrium concentration of NH_3?

Begin by setting up the problem in the standard way.

Given: $[N_2] = 0.110$ M
$[H_2] = 0.0935$ M
$K_{eq} = 152$

Find: $[NH_3]$

Write a solution map that shows how you can use the given concentrations and the equilibrium constant to get to the unknown concentration.

Solution Map:

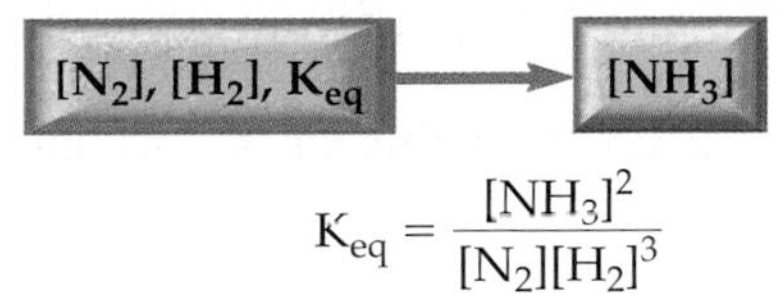

$$K_{eq} = \frac{[NH_3]^2}{[N_2][H_2]^3}$$

Next, solve the equilibrium expression for the quantity you are trying to find and then substitute in the appropriate values to compute the unknown quantity.

Solution:

$$K_{eq} = \frac{[NH_3]^2}{[N_2][H_2]^3}$$

$$[NH_3]^2 = K_{eq}[N_2][H_2]^3$$

$$[NH_3] = \sqrt{K_{eq}[N_2][H_2]^3}$$

$$= \sqrt{(152)[0.110][0.0935]^3}$$

$$= 0.117 \text{ M}$$

Using LeChâtelier's Principle (Sections 15.8, 15.9, 15.10)

EXAMPLE 15.13 Using LeChâtelier's Principle

Consider the following *endothermic* chemical reaction.

$$C(s) + H_2O(g) \rightleftharpoons CO(g) + H_2(g)$$

Predict the effect of:

a) Increasing [CO]
b) Increasing $[H_2O]$
c) Increasing the reaction volume
d) Increasing the temperature

To apply LeChâtelier's Principle, review the effects of concentration, volume, and temperature in the Chemical Principles section (page 561). For each disturbance, predict the direction that the reaction can go to counter the disturbance.

Solution:

a) Shift left
b) Shift right
c) Shift right (more moles of gas on right)
d) Shift right (heat is a reactant)

cases, sodium atoms lose electrons to become positive ions—sodium is oxidized.

$$Na \rightarrow Na^+ + e^-$$

In redox reactions between a metal and a nonmetal, the metal is oxidized and the nonmetal is reduced.

The electrons lost by sodium are gained by the nonmetals, which become negative ions—the nonmetals are reduced.

$$O_2 + 4\,e^- \rightarrow 2\,O_2^-$$
$$Cl_2 + 2\,e^- \rightarrow 2\,Cl^-$$

A more fundamental definition of **oxidation,** then, is simply the *loss of electrons,* and a more fundamental definition of **reduction** is the *gain of electrons.*

Oxidation is the loss of the electrons. Reduction is the gain of electrons.

Notice that oxidation and reduction must occur together. If one substance loses electrons (oxidation), then another substance must gain electrons (reduction) (Figure 16.4). The substance that is oxidized is called the **reducing agent** because it causes the reduction of the other substance. Similarly, the substance that is reduced is called the **oxidizing agent** because it causes the oxidation of the other substance. For example, consider our hydrogen–oxgyen fuel-cell reaction.

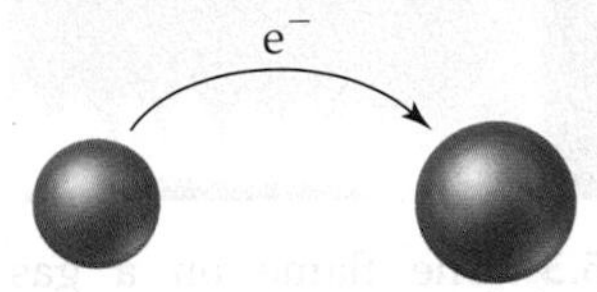
e⁻
The substance losing electrons is oxidized.
The substance gaining electrons is reduced.

Figure 16.4 In a redox reaction, one substance loses electrons, and another substance gains electrons.

$2\,H_2(g)$	+	$O_2(g)$	→	$2\,H_2O(g)$
reducing agent		oxidizing agent		

In this reaction, hydrogen is oxidized, making it the reducing agent. Oxygen is reduced, making it the oxidizing agent. Substances such as oxygen, which have a strong tendency to attract electrons, are good oxidizing agents—they tend to cause the oxidation of other substances. Substances such as hydrogen, which have a strong tendency to give up electrons, are good reducing agents—they tend to cause the reduction of other substances.

To summarize:

Oxidation–the loss of electrons
Reduction–the gain of electrons
Oxidizing agent–the substance being reduced
Reducing agent–the substance being oxidized

Helpful mnemonics: OIL RIG–**O**xidation **I**s **L**oss; **R**eduction **I**s **G**ain.
LEO GER–**L**ose **E**lectrons **O**xidation; **G**ain **E**lectrons **R**eduction

EXAMPLE 16.1 Identifying Oxidation and Reduction

For each of the following reactions, identify the substance being oxidized and the substance being reduced.

a) $2\,Mg(s) + O_2(g) \rightarrow 2\,MgO(s)$
b) $Fe(s) + Cl_2(g) \rightarrow FeCl_2(s)$
c) $Zn(s) + Fe^{2+}(aq) \rightarrow Zn^{2+}(aq) + Fe(s)$

Solution:

a) $2\,Mg(s) + O_2(g) \rightarrow 2\,MgO(s)$

In this reaction, magnesium is gaining oxygen and losing electrons to oxygen. Mg is therefore oxidized and O_2 is reduced.

b) $Fe(s) + Cl_2(g) \rightarrow FeCl_2(s)$

In this reaction, a metal (Fe) is reacting with an electronegative nonmetal (Cl_2). Fe loses electrons and is therefore oxidized, while Cl_2 gains electrons and is therefore reduced.

c) $Zn(s) + Fe^{2+}(aq) \rightarrow Zn^{2+}(aq) + Fe(s)$

In this reaction, electrons are transferred from the Zn to the Fe^{2+}. Zn loses electrons and is oxidized. Fe^{2+} gains electrons and is reduced.

SKILLBUILDER 16.1 **Identifying Oxidation and Reduction**

For each of the following reactions, identify the substance being oxidized and the substance being reduced.

a) $2\,K(s) + Cl_2(g) \rightarrow 2\,KCl(s)$
b) $2\,Al(s) + 3\,Sn^{2+}(aq) \rightarrow 2\,Al^{3+}(aq) + 3\,Sn(s)$
c) $C(s) + O_2(g) \rightarrow CO_2(g)$

EXAMPLE 16.2 **Identifying Oxidizing and Reducing Agents**

For each of the following reactions, identify the oxidizing agent and the reducing agent.

a) $2\,Mg(s) + O_2(g) \rightarrow 2\,MgO(s)$
b) $Fe(s) + Cl_2(g) \rightarrow FeCl_2(s)$
c) $Zn(s) + Fe^{2+}(aq) \rightarrow Zn^{2+}(aq) + Fe(s)$

Solution:

In the previous example, we identified the substance being oxidized and reduced for these reactions. The substance being oxidized is the reducing agent and the substance being reduced is the oxidizing agent.

a) Mg is oxidized and is therefore the reducing agent; O is reduced and is therefore the oxidizing agent.
b) Fe is oxidized and is therefore the reducing agent; Cl_2 is reduced and is therefore the oxidizing agent.
c) Zn is oxidized and is therefore the reducing agent; Fe^{2+} is reduced and is therefore the oxidizing agent.

SKILLBUILDER 16.2 **Identifying Oxidizing and Reducing Agents**

For each of the following reactions, identify the oxidizing agent and the reducing agent.

a) $2\,K(s) + Cl_2(g) \rightarrow 2\,KCl(s)$
b) $2\,Al(s) + 3\,Sn^{2+}(aq) \rightarrow 2\,Al^{3+}(aq) + 3\,Sn(s)$
c) $C(s) + O_2(g) \rightarrow CO_2(g)$

16.3 Oxidation States: Electron Bookkeeping

For many redox reactions, such as those involving oxygen or other highly electronegative elements, it is easy to identify (by inspection) the substances being oxidized and reduced. For other redox reactions, it is more

97. Silver can be electroplated at the cathode of an electrolysis cell by the following half-reaction.

$$Ag^{+}(aq) + e^{-} \rightarrow Ag(s)$$

How many moles of electrons are required to electroplate 5.8 g of Ag?

98. Gold can be electroplated at the cathode of an electrolysis cell by the following half-reaction.

$$Au^{3+}(aq) + 3\,e^{-} \rightarrow Au(s)$$

How many moles of electrons are required to electroplate 1.40 g of Au?

99. Determine if HI can dissolve each of the following metal samples. If so, write a balanced chemical reaction showing how the metal dissolves in HI and determine the minimum amount of 3.5 M HI required to completely dissolve the sample.

a) 5.95 g Cr
b) 2.15 g Al
c) 4.85 g Cu
d) 2.42 g Au

100. Determine if HCl can dissolve each of the following metal samples. If so, write a balanced chemical reaction showing how the metal dissolves in HCl and determine the minimum amount of 6.0 M HCl required to completely dissolve the sample.

a) 5.90 g Ag
b) 2.55 g Pb
c) 4.83 g Sn
d) 1.25 g Mg

Highlight Problems

101. Consider the following molecular views of an Al strip and Cu^{2+} solution. Draw a similar sketch showing what happens to the atoms and ions if the Al strip is submerged into the solution for a few minutes.

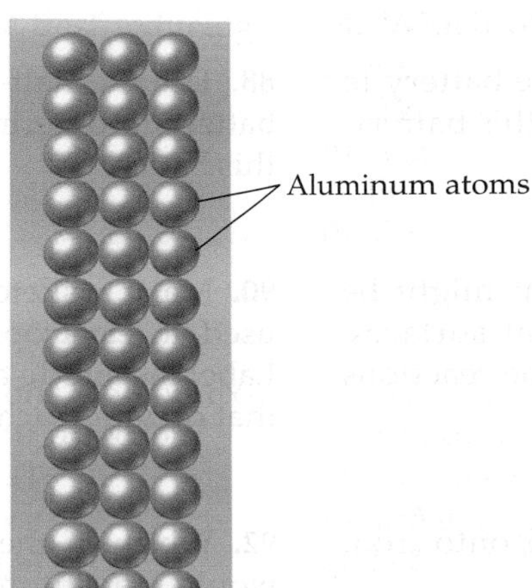

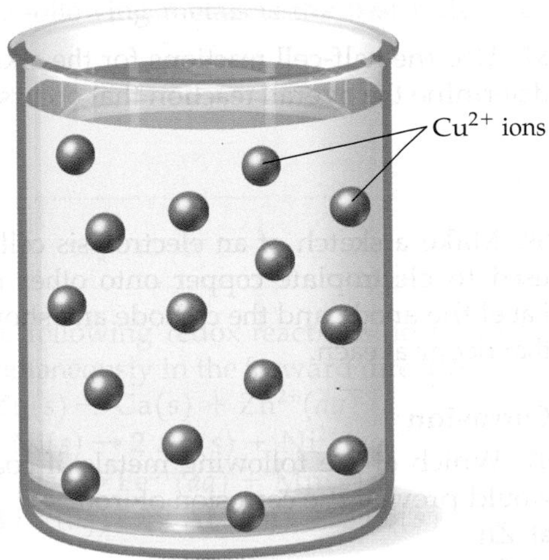

102. Suppose a fuel-cell generator was used to produce electricity for a house. If each H_2 molecule produces $2\,e^{-}$, how many kilograms of hydrogen would be required to generate the electricity needed for a typical house? Assume the home uses about 850 kilowatt-hours of electricity per month, which corresponds to approximately 2.65×10^{4} mol of electrons at the voltage of a fuel cell.

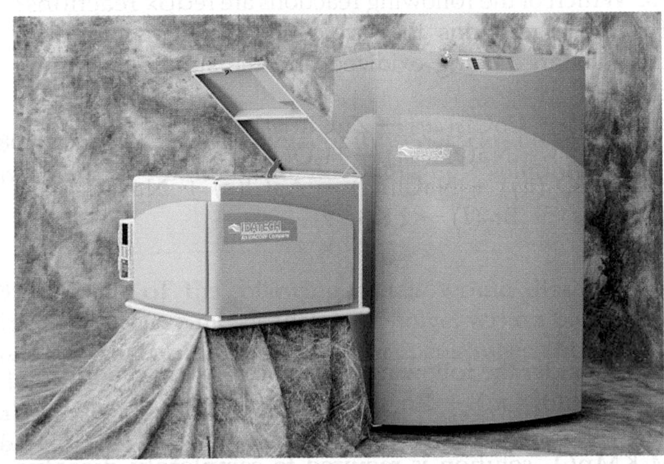

A 5-kilowatt home-fuel-cell power plant.

103. Consider the following molecular view of an electrochemical cell involving the following overall reaction.

$Zn(s) + Ni^{2+}(aq) \rightarrow Zn^{2+}(aq) + Ni(s)$

Draw a similar sketch showing how the cell might appear after it has generated a substantial amount of electrical current.

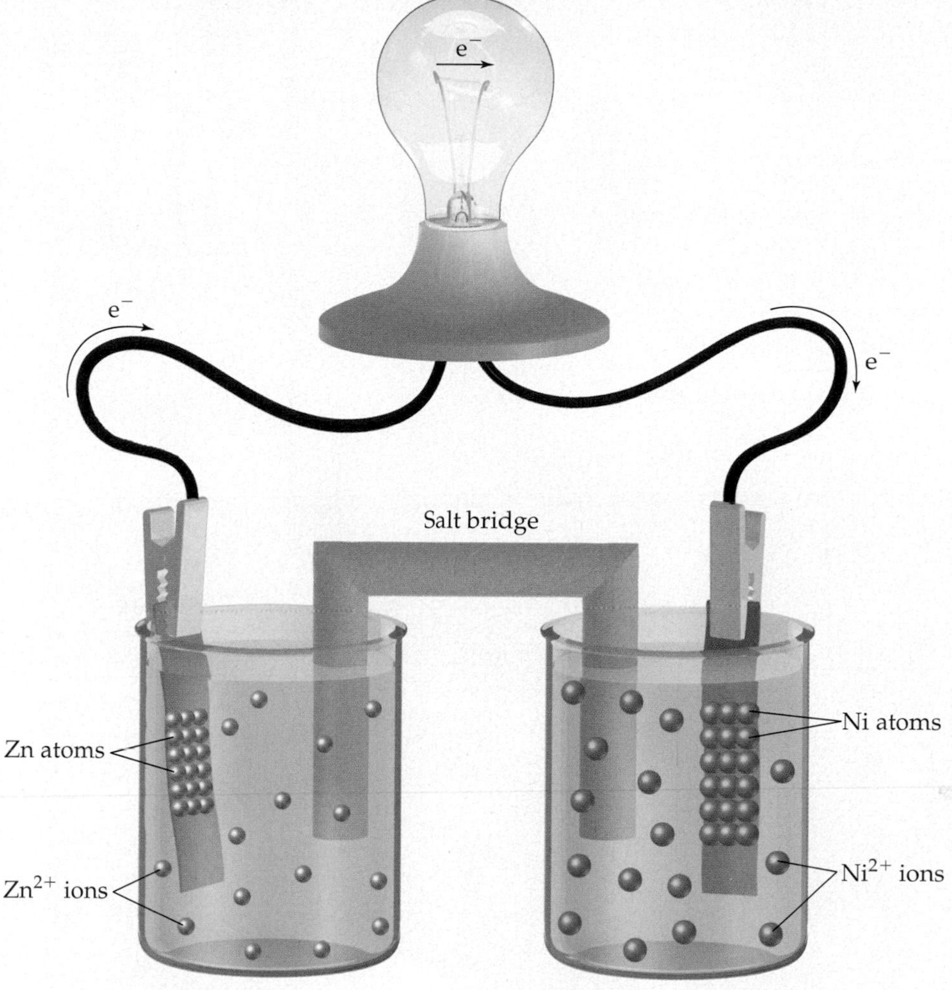

Anode: $Zn(s) \longrightarrow Zn^{2+}(aq) + 2e^-$ **Cathode:** $Ni^{2+}(aq) + 2e^- \longrightarrow Ni(s)$

Answers to Skillbuilder Exercises

Skillbuilder 16.1
a) K is oxidized, Cl_2 is reduced
b) Al is oxidized, Sn^{2+} is reduced
c) C is oxidized, O_2 is reduced

Skillbuilder 16.2
a) K is the reducing agent; Cl_2 is the oxidizing agent
b) Al is the reducing agent; Sn^{2+} is the oxidizing agent
c) C is the reducing agent; O_2 is the oxidizing agent.

Skillbuilder 16.3

a) $\underset{0}{Zn}$

b) $\underset{+2}{Cu^{2+}}$

c) $\underset{+2}{Ca}\ \underset{-1}{Cl_2}$

d) $\underset{+4}{C}\ \underset{-1}{F_4}$

e) $\underset{+3}{N}\ \underset{-2}{O_2^-}$

f) $\underset{+6}{S}\ \underset{-2}{O_3}$

Skillbuilder 16.4 Sn oxidized $(0 \rightarrow +4)$; N reduced $(+5 \rightarrow +4)$

Skillbuilder 16.5
$6\,H^+(aq) + 2\,Cr(s) \rightarrow 3\,H_2(g) + 2\,Cr^{3+}(aq)$

Skillbuilder 16.6
$Cu(s) + 4\,H^+(aq) + 2\,NO_3^- \rightarrow Cu^{2+}(aq) + 2\,NO_2(g) + 2\,H_2O(l)$

Skillbuilder 16.7
$5\,Sn(s) + 16\,H^+(aq) + 2\,MnO_4^-(aq) \rightarrow 5\,Sn^{2+}(aq) + 2\,Mn^{2+}(aq) + 8\,H_2O(l)$

Skillbuilder 16.8 **a)** spontaneous **b)** nonspontaneous

Skillbuilder 16.9 No

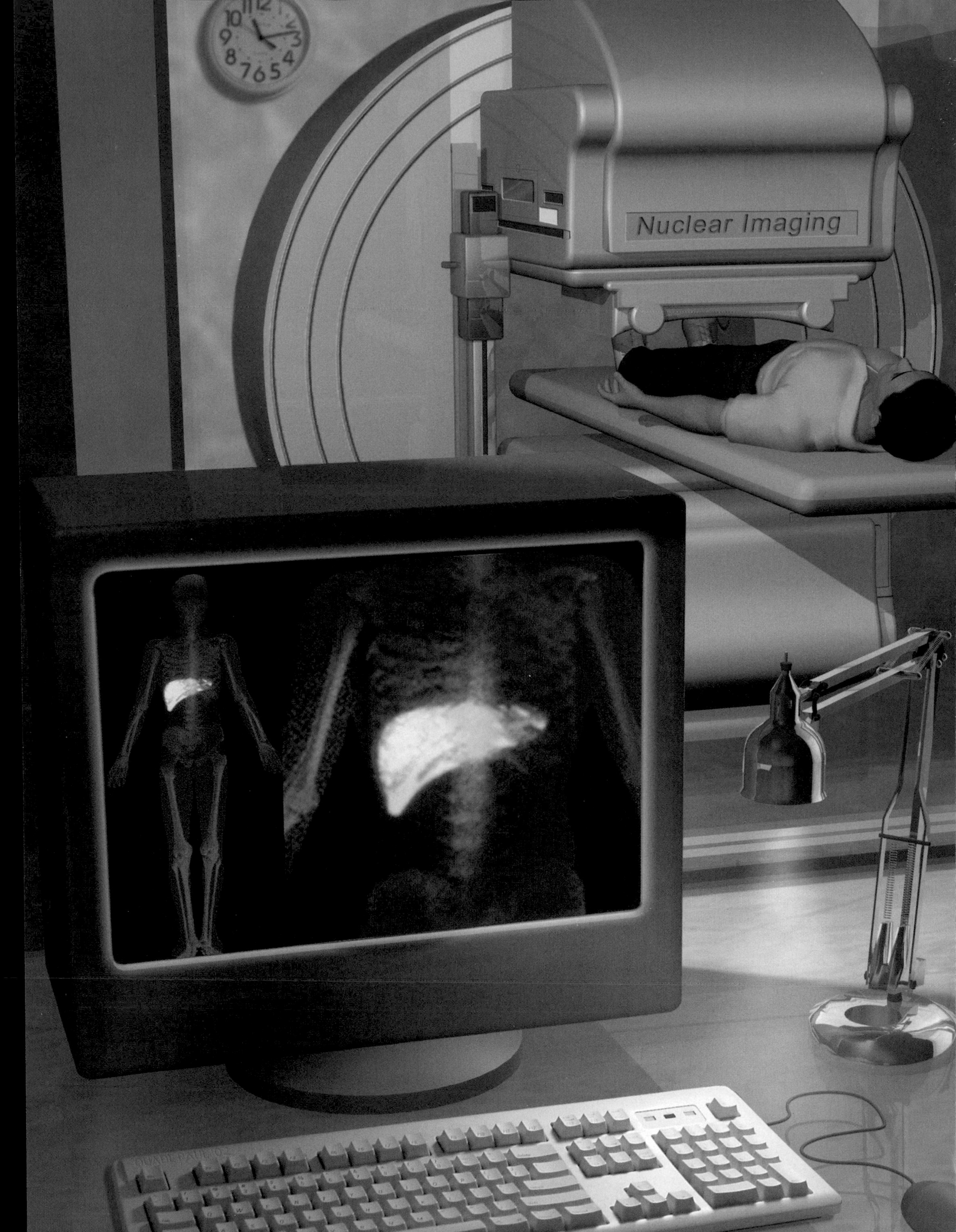
Nuclear Imaging

Radioactivity and Nuclear Chemistry 17

"Nuclear energy is incomparably greater than the molecular energy which we use today . . . What is lacking is the match to set the bonfire alight . . . The scientists are looking for this."

Winston Spencer Churchill (in 1934)

17.1 Diagnosing Appendicitis

Several years ago I awoke with a dull pain on the lower right side of my belly. The pain worsened over several hours, so I went to the hospital emergency room for evaluation. I was examined by a doctor who said I might have appendicitis, the inflammation of the appendix. The appendix, which has no known function, is a small intestinal pouch that extends from the right side of the large intestine. Occasionally, it becomes diseased and requires surgical removal.

Patients with appendicitis usually have a high number of white blood cells, so the hospital performed a blood test to determine my white blood cell count. The test was negative—I had a normal white blood cell count. Although my symptoms were consistent with appendicitis, the negative blood test clouded the diagnosis. The doctor gave me the choice of either having my appendix removed (with the chance of it being healthy) or performing an additional test to confirm appendicitis. I chose the additional test.

◀ In nuclear medicine, radioactivity is used to obtain clear images of internal organs.

The additional test involved nuclear medicine, an area of medical practice that uses radioactivity to diagnose and treat disease. **Radioactivity** is the emission of tiny, invisible particles by the nuclei of certain atoms (Figure 17.1). Many of these particles can pass right through matter. Atoms that emit these particles are known as being **radioactive.**

To perform the test, antibodies—naturally occurring molecules that fight infection—tagged with radioactive atoms were injected into my bloodstream. Since antibodies attack infection, they congregate in areas of the body that contain infection. If my appendix was infected, the antibodies would accumulate there. After waiting about an hour, I was taken to a room and laid on a table. A photographic film was inserted in a panel above me. Although radioactivity is invisible to the eye, it does expose photographic film. If my appendix were indeed infected, it would now contain the radioactively tagged antibodies, and the film would show a bright spot emanating from my appendix (Figure 17.2). In this test, I—or my appendix—was the source that would expose film. The test, however, was negative. No radioactivity was emanating from my appendix. It was healthy. After several hours, the pain in my belly subsided and I went home, appendix and all. I never did find out what caused the pain.

This example from medicine is just one of the many applications of radioactivity. Radioactivity is also used to diagnose and treat many other diseases including cancer, thyroid disease, abnormal kidney and bladder function, and heart disease. Naturally occurring radioactivity allows us to estimate the age of fossils and rocks. Radioactivity also led to the discovery of nuclear fission, used for electricity generation and nuclear weapons. In this chapter, you will learn about radioactivity—how it was discovered, what it is, and how it is used.

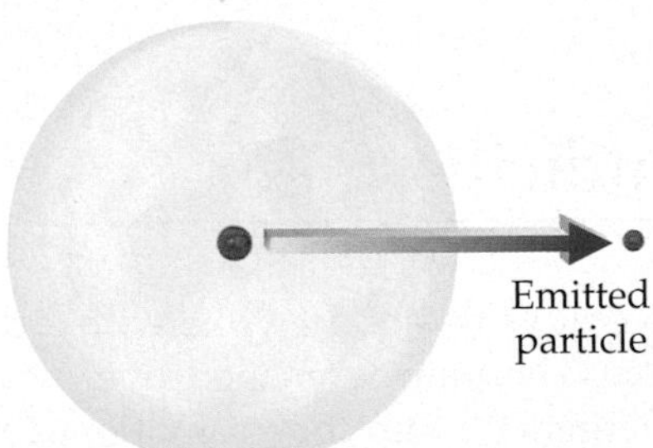

Figure 17.1 Radioactivity is the emission of tiny, energetic, invisible particles by the nuclei of certain atoms.

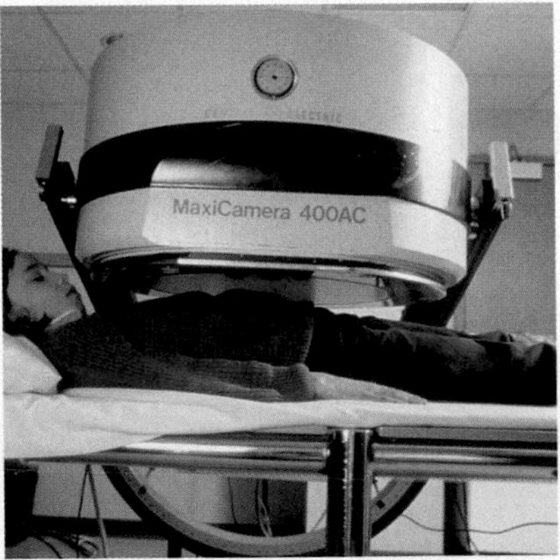

Figure 17.2 In a nuclear medicine test for appendicitis, radioactively tagged antibodies are given to the patient. If the patient has an infection in the appendix, the antibodies accumulate there and the emitted radiation can be detected on photographic film.

17.2 The Discovery of Radioactivity

In 1896, a French scientist named **Antoine-Henri Becquerel** (1852–1908) discovered radioactivity. Becquerel was interested in the newly discovered X-rays (see Section 9.3), which were the hot topic of physics research in his time. He hypothesized that X-rays were emitted in conjunction with **phosphorescence**. Phosphorescence is the long-lived *emission* of light that sometimes follows the *absorption* of light by some atoms and molecules. Phosphorescence is probably most familiar to you as the *glow* in glow-in-the-dark toys. After one of these toys is exposed to light, it reemits some of that light, usually at slightly longer wavelengths. If you turn off the room lights or put the toy in the dark, you can see the greenish glow of the emitted light. Becquerel hypothesized that the visible greenish glow was associated with the emission of X-rays (which are invisible).

To test his hypothesis, Becquerel placed crystals—composed of potassium uranyl sulfate, a compound known to phosphoresce—on top of a photographic plate wrapped in black cloth (Figure 17.3). He then placed the wrapped plate and the crystals outside to expose them to sunlight. He knew the crystals phosphoresced because he could see the emitted light when he brought them back into the dark. If the crystals also emitted X-rays, the X-rays would pass through the black cloth and expose the underlying photographic plate. Becquerel performed the experiment several times and always got the same result—the photographic plate showed a bright exposure spot where the crystals had been. Becquerel believed his hypothesis was correct and presented the results—that phosphorescence and X-rays were linked—to the French Academy of Sciences.

Marie Curie

Becquerel later retracted his results, however, when he discovered that a photographic plate with the same crystals showed a bright exposure spot even when the plate and the crystals were stored in a dark drawer and not exposed to sunlight. Becquerel realized that the crystals themselves were constantly emitting something that exposed the photographic plate, independent of whether or not they phosphoresced. Becquerel concluded that it was the uranium within the crystals that was the source of the emissions, and he called the emissions *uranic rays.*

Soon after Becquerel's discovery, a young graduate student named **Marie Sklodowska Curie** (1867–1934) (one of the first women in France

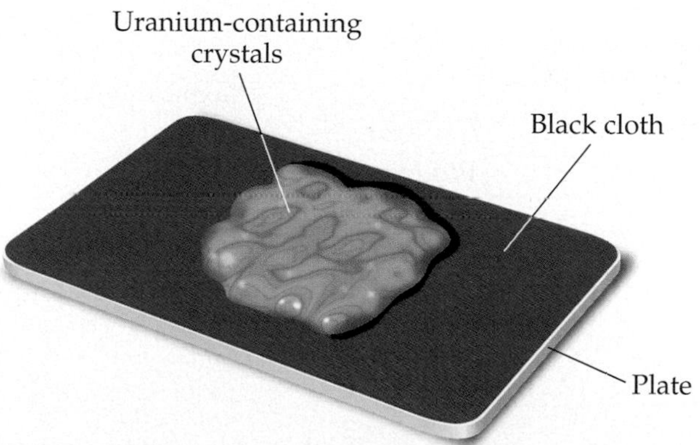

Figure 17.3 Becquerel's experiment.

to attempt doctoral work) decided to pursue the study of uranic rays for her doctoral thesis. Her first task was to determine if any other substances besides uranium (the heaviest known element at the time) emitted these rays. In her search, Curie discovered two new elements, both of which also emitted uranic rays. Curie named one of her newly discovered elements *polonium* after her home country of Poland. The other element she named *radium* because of the very high amount of radioactivity that it produced. Radium was so radioactive that it gently glowed in the dark and emitted significant amounts of heat. Since it was now clear that these rays were not unique to uranium, Curie changed the name of uranic rays to *radioactivity*. In 1903, Curie received the Nobel Prize in physics—which she shared with Becquerel and her husband, Pierre Curie—for the discovery of radioactivity. In 1911, Curie was also awarded a second Nobel Prize, this time in chemistry, for her discovery of the two new elements.

Element 96 (curium) is named in honor of Marie Curie and her contributions to our understanding of radioactivity.

In the past, glowing radium was added to some paints that were used on watch dials. The radium made the dial glow.

17.3 Types of Radioactivity: Alpha, Beta, and Gamma Decay

While Curie focused her work on discovering the different kinds of radioactive elements, **Ernest Rutherford** and others focused on characterizing the radioactivity itself. These scientists found that the emitted particles were produced by the nuclei of radioactive atoms. These nuclei were unstable and would emit small pieces of themselves to gain stability. These *emitted pieces* were the radioactivity that Becquerel and Curie detected. There are several different types of radioactivity including: alpha (α) rays, beta (β) rays, gamma (γ) rays, and positrons.

In order to understand these different types of radioactivity, we must briefly review the notation to symbolize isotopes that was first introduced in Section 4.8. Recall that any isotope can be represented with the following notation.

Remember that the atomic number equals the number of protons and that the mass number equals the number of protons and neutrons.

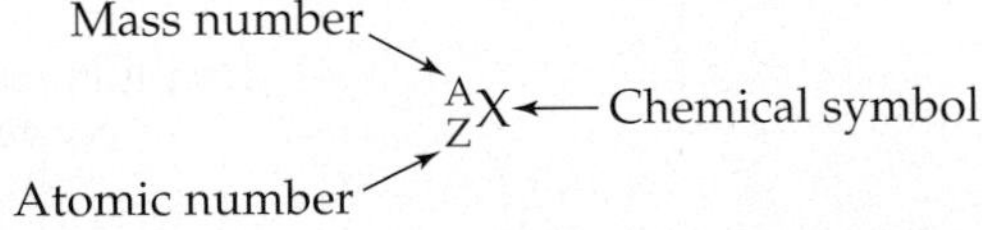

For example, the symbol

$$^{21}_{10}\text{Ne}$$

represents the neon isotope containing 10 protons and 11 neutrons. The symbol

$$^{20}_{10}\text{Ne}$$

represents the neon isotope containing 10 protons and 10 neutrons. Remember that any element can have several different isotopes.

The main subatomic particles—protons, neutrons, and electrons—can all be represented with similar notation.

Proton symbol $^{1}_{1}\text{p}$ Neutron symbol $^{1}_{0}\text{n}$ Electron symbol $^{0}_{-1}\text{e}$

Alpha (α) Radiation

Nuclei are unstable when they are too large or when they contain an unbalanced ratio of neutrons to protons. Small nuclei need about 1 neutron to every proton to be stable while larger nuclei need about 1.5 neutrons to every proton.

In nuclear chemistry, we are primarily interested in changes within the nucleus; therefore, the 2+ charge that we would normally write for a helium nucleus is omitted for an alpha particle.

Nuclear equations represent the changes that occur during radioactivity and other nuclear processes.

The term *nuclide* is used in nuclear chemistry to mean a specific isotope.

Alpha (α) radiation occurs when an unstable nucleus emits a small piece of itself composed of 2 protons and 2 neutrons (Figure 17.4).

Since 2 protons and 2 neutrons are identical to a helium-4 nucleus, the symbol for an **alpha (α) particle** is identical to the symbol for helium-4.

alpha (α) $^{4}_{2}\text{He}$

An α particle

When an atom emits an alpha particle, it becomes a lighter atom. We represent this with a **nuclear equation,** an equation that represents nuclear processes such as radioactivity. For example, the nuclear equation for the alpha decay of uranium-238 is:

Parent nuclide → $^{238}_{92}\text{U}$; Daughter nuclides → $^{234}_{90}\text{Th}$, $^{4}_{2}\text{He}$

$$^{238}_{92}\text{U} \longrightarrow {}^{234}_{90}\text{Th} + {}^{4}_{2}\text{He}$$

The original atom is called the **parent nuclide** and the products are called the **daughter nuclides.** Notice that when an element emits an alpha particle, the number of protons in its nucleus changes, transforming it into a different element. In this case, uranium-238 becomes thorium-234. Unlike a chemical reaction, in which elements retain their identity, a nuclear reaction often results in elements changing their identities. Like a chemical equation, however, nuclear equations must be balanced.

The sum of the atomic numbers on both sides of a nuclear equation must be equal, and the sum of the mass numbers on both sides must also be equal.

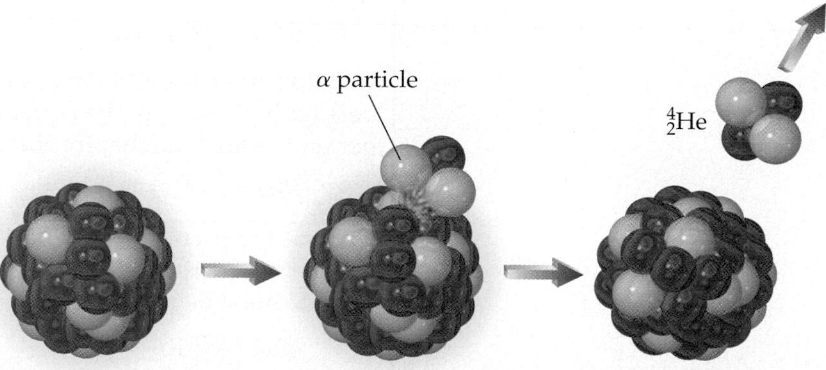

Figure 17.4 Alpha radiation occurs when an unstable nucleus emits a particle composed of 2 protons and 2 neutrons. **Question:** What happens to the atomic number of an element upon emission of an alpha particle?

$$^{238}_{92}\text{U} \longrightarrow {}^{234}_{90}\text{Th} + {}^{4}_{2}\text{He}$$

Left Side	*Right Side*
Sum of mass numbers = 238	Sum of mass numbers = 234 + 4 = 238
Sum of atomic numbers = 92	Sum of atomic numbers = 90 + 2 = 92

The identity and symbol of the daughter nuclide of any alpha decay can be deduced from the mass and atomic numbers of the parent nuclide. During alpha decay, the mass number decreases by 4 and the atomic number decreases by 2. For example, to write a nuclear equation for the alpha decay of Th-232, we begin with the symbol for Th-232 on the left side of the equation and the symbol for an alpha particle on the right side.

$$^{232}_{90}\text{Th} \longrightarrow {}^{x}_{y}? + {}^{4}_{2}\text{He}$$

We can then deduce the mass number and atomic number of the unknown daughter nuclide because the equation must be balanced.

$$^{232}_{90}\text{Th} \longrightarrow \overbrace{\underbrace{{}^{x}_{y}? + {}^{4}_{2}\text{He}}_{y + 2 = 90;\ y = 88}}^{x + 4 = 232;\ x = 228}$$

Therefore,

$$^{232}_{90}\text{Th} \longrightarrow {}^{228}_{88}? + {}^{4}_{2}\text{He}$$

The atomic number must be 88 and the mass number must be 228. Finally, we can deduce the identity of the daughter nuclide and its symbol from its atomic number. Since the atomic number is 88, the unknown daughter nuclide is radium (Ra).

$$^{232}_{90}\text{Th} \longrightarrow {}^{228}_{88}\text{Ra} + {}^{4}_{2}\text{He}$$

EXAMPLE 17.1 Writing Nuclear Equations for Alpha (α) Decay

Write a nuclear equation for the alpha decay of Ra-224.

Solution:

Begin with the symbol for Ra-224 on the left side of the equation and the symbol for an alpha particle on the right side.

$$^{224}_{88}\text{Ra} \longrightarrow {}^{x}_{y}? + {}^{4}_{2}\text{He}$$

Equalize the sum of the mass numbers and the sum of the atomic numbers on both sides of the equation by writing the appropriate mass number and atomic number for the unknown daughter nuclide.

$$^{224}_{88}\text{Ra} \longrightarrow {}^{220}_{86}? + {}^{4}_{2}\text{He}$$

Deduce the identity of the unknown daughter nuclide from the atomic number and write its symbol. Since the atomic number is 86, the daughter nuclide must be radon (Rn).

$$^{224}_{88}\text{Ra} \longrightarrow {}^{220}_{86}\text{Rn} + {}^{4}_{2}\text{He}$$

SKILLBUILDER 17.1 Writing Nuclear Equations for Alpha (α) Decay

Write a nuclear equation for the alpha decay of Po-216.

Alpha radiation is the semi-truck of radioactivity. The alpha particle is by far the most massive of all particles emitted by radioactive nuclei. Consequently, alpha radiation has the most potential to interact with and damage other molecules, including biological ones. Radiation interacts with other molecules and atoms by ionizing them. If radiation ionizes the molecules within the cells of living organisms, those molecules become damaged and the cell can die or begin to reproduce abnormally. The ability of radiation to ionize other molecules and atoms is called its **ionizing power**. Of all types of radioactivity, alpha radiation has the highest ionizing power.

To ionize means *to create ions (charged particles).*

However, because of its large size, alpha radiation also has the lowest **penetrating power**—the ability to penetrate matter. (Imagine a semi-truck trying to get through a traffic jam.) In order for radiation to damage important molecules within living cells, it must penetrate into the cell. Alpha radiation does not easily penetrate into cells because it can be stopped by a sheet of paper, by clothing, or even by air. Consequently, a low-level alpha emitter kept outside of the body is relatively safe. However, if an alpha emitter is ingested, it becomes very dangerous because the alpha particles then have direct access to the molecules that compose organs and tissues.

To summarize:

- ***Alpha particles are composed of 2 protons and 2 neutrons.***
- ***Alpha particles have the symbol $^{4}_{2}$He.***
- ***Alpha particles have a high ionizing power.***
- ***Alpha particles have a low penetrating power.***

Beta (β) Radiation

This kind of beta radiation is also called beta minus (β^-) radiation due to its negative charge.

Beta radiation occurs when an unstable nucleus emits an electron (Figure 17.5). How does a nucleus, which contains only protons and neutrons, emit an electron? Where do the electrons come from? They come from neutrons as they convert to protons. In other words, in some unstable nuclei, a neutron will change into a proton and emit an electron in the process.

Beta decay $\quad$ Neutron $\rightarrow$ Proton + Electron

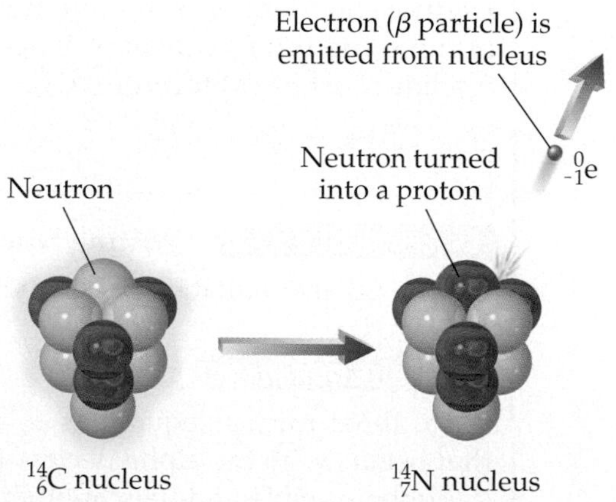

Figure 17.5 Beta radiation occurs when an unstable nucleus emits an electron. As the emission occurs, a neutron turns into a proton. **Question:** What happens to the atomic number of an element upon emission of a beta particle?

The symbol for a **beta (β) particle** in a nuclear equation is:

Beta (β) particle $^{0}_{-1}e$

Remember that the mass number is defined as the sum of the number of protons and neutrons. Since electrons have no protons or neutrons, their mass number is zero.

The zero in the upper left corner reflects the mass number of the electron. The -1 in the lower left corner reflects the charge of the electron which is equivalent to an atomic number of -1 in a nuclear equation. When an atom emits a beta particle, its atomic number increases by one because it now has an additional proton. For example, the nuclear equation for the beta decay of radium-228 is:

$$^{228}_{88}\text{Ra} \longrightarrow {}^{228}_{89}\text{Ac} + {}^{0}_{-1}\text{e}$$

Notice that the nuclear equation is still balanced—the sum of the mass numbers on both sides is equal and the sum of the atomic numbers on both sides is equal.

$$^{228}_{88}\text{Ra} \longrightarrow {}^{228}_{89}\text{Ac} + {}^{0}_{-1}\text{e}$$

Left side	*Right side*
Sum of mass numbers = 228	Sum of mass numbers = 228 + 0 = 228
Sum of atomic numbers = 88	Sum of atomic numbers = 89 − 1 = 88

The identity and symbol of the daughter nuclide of any beta decay can be determined in a manner similar to alpha decay as shown in the following example.

EXAMPLE 17.2 Writing Nuclear Equations for Beta (β) Decay

Write a nuclear equation for the beta decay of Bk-249.

Solution:

Begin with the symbol for Bk-249 on the left side of the equation and the symbol for a beta particle on the right side.

$$^{249}_{97}\text{Bk} \longrightarrow {}^{x}_{y}? + {}^{0}_{-1}\text{e}$$

Equalize the sum of the mass numbers and the sum of the atomic numbers on both sides of the equation by writing the appropriate mass number and atomic number for the unknown daughter nuclide.

$$^{249}_{97}\text{Bk} \longrightarrow {}^{249}_{98}? + {}^{0}_{-1}\text{e}$$

Deduce the identity of the unknown daughter nuclide from the atomic number and write its symbol. Since the atomic number is 98, the daughter nuclide must be californium (Cf).

$$^{249}_{97}\text{Bk} \longrightarrow {}^{249}_{98}\text{Cf} + {}^{0}_{-1}\text{e}$$

SKILLBUILDER 17.2 Writing Nuclear Equations for Beta (β) Decay

Write a nuclear equation for the beta decay of Ac-228.

SKILLBUILDER PLUS

Write three nuclear equations to represent the nuclear decay sequence that begins with the alpha decay of U-235 followed by a beta decay of the daughter nuclide and then another alpha decay.

Beta radiation is the mid-sized car of radioactivity. Beta particles are much less massive than alpha particles and consequently have a lower ionizing power. However, because of their smaller size, beta particles have a higher penetrating power and require a sheet of metal or a thick piece of wood to stop them. Consequently, a low-level beta emitter outside of the body poses a higher risk than an alpha emitter. Inside the body, however, the beta emitter does less damage than an alpha emitter.

To summarize:

- ***Beta particles are electrons emitted from atomic nuclei when a neutron changes into a proton.***
- ***Beta particles have the symbol*** ${}^{0}_{-1}\text{e}$.
- ***Beta particles have intermediate ionizing power.***
- ***Beta particles have intermediate penetrating power.***

Gamma (γ) Radiation

Gamma (γ) radiation is significantly different than alpha or beta radiation. Gamma radiation is *electromagnetic* radiation. Gamma rays are high energy (short wavelength) photons and therefore have zero mass. The symbol for a gamma ray is:

See Section 9.3 for a review of electromagnetic radiation.

Gamma (γ) ray ${}^{0}_{0}\gamma$

A gamma ray has no charge and no mass. When a gamma ray is emitted from a radioactive atom, it does not change the mass number or the atomic number of the element. Gamma rays, however, are usually emitted in conjunction with other types of radiation. For example, the alpha emission of U-238 (discussed previously) is also accompanied by the emission of a gamma ray.

$${}^{238}_{92}\text{U} \rightarrow {}^{234}_{90}\text{Th} + {}^{4}_{2}\text{He} + {}^{0}_{0}\gamma$$

Gamma rays are the motorbikes of radioactivity. They have the lowest ionizing power, but the highest penetrating power. (Imagine a motorbike zipping through a traffic jam.) Stopping gamma rays requires several inches of lead shielding or thick slabs of concrete.

To summarize:

- ***Gamma rays are electromagnetic radiation—high energy, short wavelength photons.***
- ***Gamma rays have the symbol*** ${}^{0}_{0}\gamma$.
- ***Gamma rays have low ionizing power.***
- ***Gamma rays have high penetrating power.***

Positron Emission

Positron emission occurs when an unstable nucleus emits a positron (Figure 17.6). A **positron** has the mass of an electron but carries a +1 charge. A positron is emitted from a proton as it converts into a neutron. In some unstable nuclei, a proton will change into a neutron and emit a positron in the process.

Positron emission can be thought of as a branch of beta emission. It sometimes referred to as beta-plus emission (β^{+}).

Positron emission Proton $\rightarrow$ Neutron + Positron

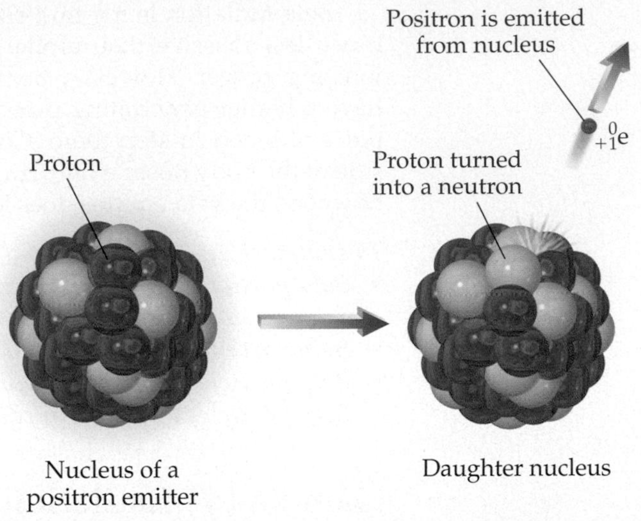

Figure 17.6 Positron emission occurs when an unstable nucleus emits a positron. As the emission occurs, a proton turns into a neutron. **Question:** What happens to the atomic number of an element upon positron emission?

The symbol for a positron in a nuclear equation is:

Positron $^{0}_{+1}e$ ●

The zero in the upper left corner reflects that a positron has a mass number of zero. The +1 in the lower left corner reflects the charge of the positron, which is equivalent to an atomic number of +1 in a nuclear equation. When an atom emits a positron, its atomic number decreases by one because it now has one less proton. For example, the nuclear equation for the positron emission of phosphorus-30 is:

$$^{30}_{15}P \rightarrow ^{30}_{14}Si + ^{0}_{+1}e$$

The identity and symbol of the daughter nuclide of any positron emission can be determined in a manner similar to alpha and beta decay as shown in the following example. Positron emission is similar to beta emission in its ionizing and penetrating power.

EXAMPLE 17.3 Writing Nuclear Equations for Positron Emission

Write a nuclear equation for the positron emission of potassium-40.

Solution:

Begin with the symbol for K-40 on the left side of the equation and the symbol for a positron on the right side.

$$^{40}_{19}K \rightarrow ^{x}_{y}? + ^{0}_{+1}e$$

Equalize the sum of the mass numbers and the sum of the atomic numbers on both sides of the equation by writing the appropriate mass number and atomic number for the unknown daughter nuclide.

$$^{40}_{19}K \rightarrow ^{40}_{18}? + ^{0}_{+1}e$$

Deduce the identity of the unknown daughter nuclide from the atomic number and write its symbol. Since the atomic number is 18, the daughter nuclide must be argon (Ar).

$$^{40}_{19}\text{K} \rightarrow ^{40}_{18}\text{Ar} + ^{0}_{+1}\text{e}$$

SKILLBUILDER 17.3 **Writing Nuclear Equations for Positron Emission**

Write a nuclear equation for the positron emission of sodium-22.

17.4 Detecting Radioactivity

The particles emitted from radioactive nuclei contain a large amount of energy and can therefore be detected with high sensitivity. Radiation detectors detect radioactive particles through the interactions of those particles with atoms or molecules. The simplest radiation detectors are pieces of photographic film that become exposed when radiation passes through them. **Film-badge dosimeters**—which consist of photographic film held in a small case that is pinned to clothing—are standard for most people working with or near radioactive substances. These badges are collected and processed (or developed) regularly as a way to monitor a person's exposure. The more exposed the film has become in a given period of time, the more the person has been exposed to radioactivity.

Radioactivity can be instantly detected with devices such as a **Geiger-Müller counter**. In a Geiger-Müller counter, the radioactive particles pass through an argon-filled chamber. The energetic particles create a trail of ionized argon atoms as they pass through the chamber. If the applied voltage is high enough, these newly formed ions produce an electrical signal that can be detected on a meter or turned into an audible click. Each click corresponds to a radioactive particle passing through the argon gas chamber. This clicking is the stereotypical sound most people associate with a radiation detector.

A second type of device commonly used to detect radiation instantly is called a **scintillation counter**. In a scintillation counter, the radioactive particles pass through a material (such as NaI or CsI) that emits ultraviolet or visible light in response to excitation by radioactive particles. This light is then detected and turned into an electrical signal that can be read on a meter.

17.5 Natural Radioactivity and Half-Life

Radioactivity is a natural component of our environment. The ground beneath you most likely contains radioactive atoms that emit radiation into the air around you. The food you eat contains a residual amount of radioactive atoms that enter into your body fluids and tissues. Small amounts of radiation from space make it through our atmosphere and constantly bombard the earth. Humans and other living organisms have evolved in this environment and have adapted to survive in it.

One reason for the radioactivity in our environment is the instability of all atomic nuclei beyond atomic number 83 (bismuth). In addition, some isotopes of elements with fewer than 83 protons are also unstable and radioactive.

Different radioactive nuclides decay into their daughter nuclides at different rates. Some nuclides decay quickly while others decay slowly. The time it takes for one-half of the parent nuclides in a radioactive sample to decay to the daughter nuclides is called the **half-life**. Nuclides that decay quickly have short half-lives and are very active (many decay events per unit time), while those that decay slowly have long half-lives and are less active (fewer decay events per unit time). For example, Th-232 is an alpha emitter that decays according to the following nuclear reaction.

A *decay event* is the emission of radiation by a single radioactive nuclide.

$$^{232}_{90}\text{Th} \rightarrow {}^{228}_{88}\text{Ra} + {}^{4}_{2}\text{He}$$

Th-232 has a half-life of 1.4×10^{10} years or 14 billion years. If we start with a sample of Th-232 containing one million atoms, the sample would decay to half a million atoms in 14 billion years and then to a quarter of a million in another 14 billion years and so on (Figure 17.7).

$$\begin{matrix}\text{1 million} \\ \text{Th-232 atoms}\end{matrix} \xrightarrow{\text{14 billion years}} \begin{matrix}\tfrac{1}{2}\text{ million} \\ \text{Th-232 atoms}\end{matrix} \xrightarrow{\text{14 billion years}} \begin{matrix}\tfrac{1}{4}\text{ million} \\ \text{Th-232 atoms}\end{matrix}$$

Notice that a radioactive sample **does not** decay to zero atoms in two half lives—*you can't add two half-lives together to get a "whole" life*. The amount that remains after one half-life is always half of what was present at the start. The amount that remains after two half-lives is a quarter of what was present at the start and so on.

Some nuclides have short half-lives. Radon-220, for example, has a half-life of approximately one minute. If we had a sample of radon-220 that contained one million atoms, it would be diminished to $\frac{1}{4}$ million radon-220 atoms in just 2 minutes.

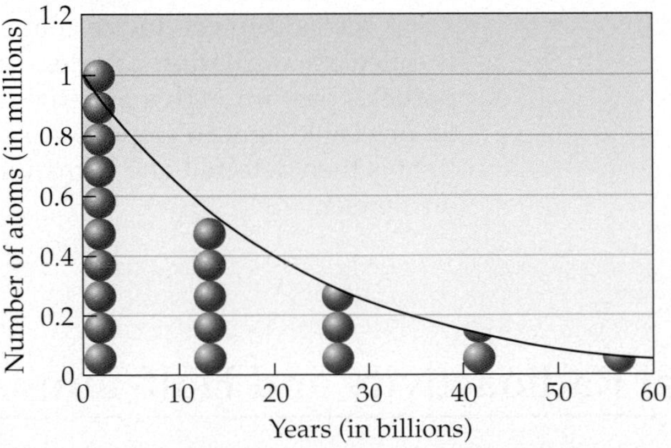

Figure 17.7 Number of Th-232 atoms in a sample initially containing 1 million atoms as a function of time. Th-232 has a half-life of 14 billion years.

$$\begin{array}{ccccc} \text{1 million} & & \frac{1}{2}\text{ million} & & \frac{1}{4}\text{ million} \\ \text{Rn-220 atoms} & \xrightarrow{\text{1 minute}} & \text{Rn-220 atoms} & \xrightarrow{\text{1 minute}} & \text{Rn-220 atoms} \end{array}$$

Rn-220 is therefore much more active than Th-232 because it undergoes many more decay events in a given period of time.

EXAMPLE 17.4 Half-Life

How long does it take for a 1.80-mol sample of Th-228 (which has a half-life of 1.9 years) to decay to 0.225 moles?

Solution:

It is easiest to draw a table showing the amount of Th-228 as a function of number of half-lives. For each half-life, simply divide the amount of Th-228 by two.

Amount of Th-228	*Number of Half-Lives*	*Time in Years*
1.80 mol	0	0
0.900 mol	1	1.9
0.450 mol	2	3.8
0.225 mol	3	5.7

Therefore, it takes three half-lives or 5.7 years for the sample to decay to 0.225 mol.

SKILLBUILDER 17.4 Half-Life

A radium-226 sample initially contains 0.112 mol. How much radium-226 is left in the sample after 6,400 years? The half-life of radium-226 is 1600 years.

A Natural Radioactive Decay Series

The radioactive elements in our environment are all undergoing radioactive decay. The reason that they are in our environment at all is because either they have very long half-lives (billions of years) or they are continuously being formed by some other process in the environment. In many cases, the daughter nuclide of a radioactive decay is itself radioactive and in turn produces another daughter that is radioactive and so on, resulting in a radioactive decay series. For example, uranium (atomic number 92) is the heaviest naturally occurring element. It is an alpha emitter that decays to Th-234 with a half-life of 4.47 billion years.

$$^{238}_{92}\text{U} \longrightarrow {}^{234}_{90}\text{Th} + {}^{4}_{2}\text{He}$$

The daughter nuclide, Th-234, is itself radioactive—it is a beta emitter that decays to Pa-234 with a half-life of 24.1 days.

$$^{234}_{90}\text{Th} \longrightarrow {}^{234}_{91}\text{Pa} + {}^{0}_{-1}\text{e}$$

Pa-234 is also radioactive, decaying to U-234 via beta emission with a half-life of 244,500 years. This process continues until it reaches Pb-206, which is stable. The entire uranium-238 decay series is shown in Figure 17.8. In other words, all of the uranium-238 in the environment is slowly decaying away to lead. Since the half-life for the first step in the series is so long,

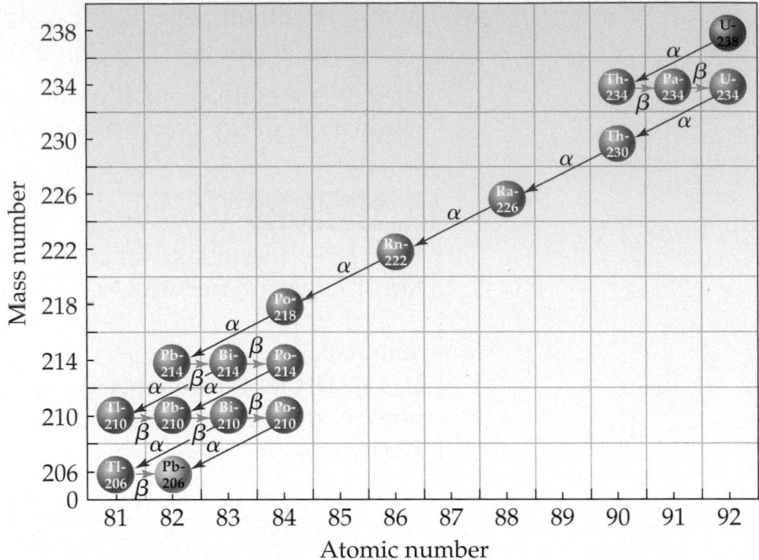

Figure 17.8 Uranium-238 decay series.

however, there is still plenty of uranium-238 in the environment. All of the other nuclides in the decay series are also present in the environment in varying amounts, depending on their half-lives.

CHEMISTRY AND HEALTH

Environmental Radon

Radon—a radioactive gas—is one of the products of the radioactive decay series of uranium. Wherever there is uranium in the ground, there is likely to be radon seeping up into the air. Because radon is a gas, it—and its daughter nuclides that attach to dust particles—can be inhaled into the lungs where they decay and increase lung cancer risk. The radioactive decay of radon is by far the single greatest source of human radiation exposure.

Homes built in areas with significant uranium deposits in the ground pose the greatest risk. These homes can accumulate radon levels that are above what the Environmental Protection Agency (EPA) considers safe. Simple test kits are available to test indoor air and determine radon levels. The higher the level is, the greater the risk. The risk is even higher for smokers who live in these houses. Excessively high indoor radon levels require the installation of a ventilation system to purge radon from the house. Lower levels can be ventilated by keeping windows and doors open.

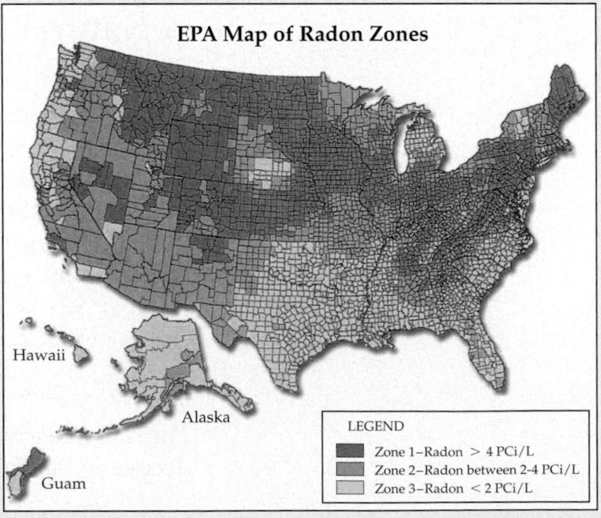

Map of U.S. showing radon levels. Zone 1 counties have the highest levels and zone 3 counties have the lowest.

CAN YOU ANSWER THIS? *Suppose a house contains* 1.80×10^{-3} *moles of radon-222 (which has a half-life of 3.8 days). If no new radon entered the house, how long would it take for the radon to decay to* 4.50×10^{-4} *moles?*

17.6 Radiocarbon Dating: Using Radioactivity to Measure the Age of Fossils and Other Artifacts

Archeologists, geologists, anthropologists, and other scientists take advantage of the presence of natural radioactivity in our environment to estimate the ages of fossils and artifacts through a technique called **radiocarbon dating**. For example, in 1947, young shepherds searching for a stray goat near the Dead Sea (east of Jerusalem) entered a cave and discovered ancient scrolls stuffed into jars. These scrolls—now named the Dead Sea Scrolls—are 2000-year-old biblical manuscripts, predating other existing manuscripts by almost one thousand years.

The Dead Sea Scrolls—like other ancient artifacts—contain a radioactive signature that reveals their age. This signature results from the presence of carbon-14—which is radioactive—in the environment. Carbon-14 is constantly formed in the upper atmosphere by the neutron bombardment of nitrogen.

$$^{14}_{7}\text{N} + ^{1}_{0}\text{n} \rightarrow ^{14}_{6}\text{C} + ^{1}_{1}\text{H}$$

Carbon-14 then decays back to nitrogen by beta emission with a half-life of 5,730 years.

$$^{14}_{6}\text{C} \rightarrow ^{14}_{7}\text{N} + ^{0}_{-1}\text{e}$$

The continuous formation of carbon-14 in the atmosphere and its continuous decay back to nitrogen-14 produces a nearly constant equilibrium concentration of atmospheric carbon-14. That carbon-14 is oxidized to carbon dioxide and then incorporated into plants by photosynthesis. It is also incorporated into animals since animals ultimately depend on plants for food. Consequently, all living organisms contain residual amounts of carbon-14. When a living organism dies, it stops incorporating new carbon-14 into its tissues. The carbon-14 present at the time of death decays with a half-life of 5,730 years. Since many artifacts, such as the Dead Sea Scrolls, are made from materials that were once living—such as papyrus, wood, or other plant and animal derivatives—the amount of carbon-14 in these artifacts indicates their age.

The concentration of carbon-14 in all *living* organisms is the same.

For example, suppose an ancient artifact has a carbon-14 concentration that is 50% of that found in living organisms. How old is the artifact? Since it contains half as much carbon-14 as a living organism, it must be one half-life or 5,730 years old. If the artifact has a carbon-14 concentration that is 25% of that found in living organisms, its age is two half-lives or 11,460 years old. Table 17.1 shows the age of an object based on its

TABLE 17.1

Age of Object Based on Its Carbon-14 Content

Concentration of C-14 (% Relative to Living Organisms)	*Age of Object in Years*
100.0	0
50.0	5,730
25.00	11,460
12.50	17,190
6.250	22,920
3.125	28,650
1.563	34,380

CHEMISTRY IN THE MEDIA

The Shroud of Turin

The shroud of Turin—kept in the Cathedral of Turin in Italy—is an old linen cloth that bears the image of a man who appears to have been crucified (Figure 17.9). The image becomes clearer if the shroud is photographed and viewed as a negative. Many believe that the shroud is the original burial cloth of Jesus Christ, miraculously imprinted with his image. In 1988, three independent laboratories were chosen by the Roman Catholic Church to perform radiocarbon dating on the shroud. The laboratories took samples from the shroud and measured the carbon-14 content. They all got similar results—the shroud was made from linen originating in about 1325 A.D. Although some have disputed the results, and although no scientific test is 100% reliable, newspapers around the world quickly announced that the shroud was a fake.

CAN YOU ANSWER THIS? *Suppose an artifact is claimed to be from 3000 B.C. Examination of the C-14 content of the artifacts reveals that the concentration of C-14 is 55% of that found in living organisms. Could the artifact be authentic?*

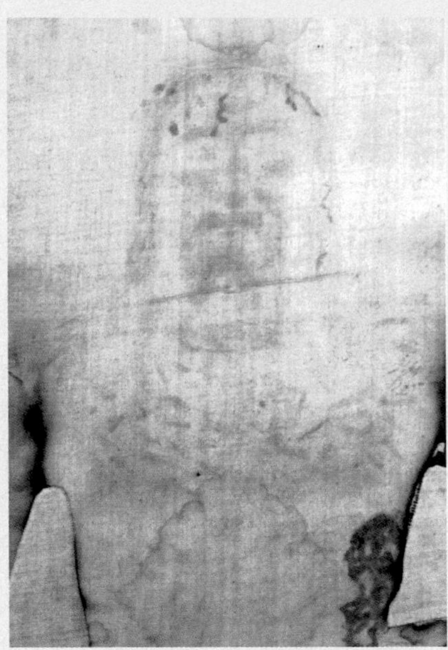

Figure 17.9 The shroud of Turin.

carbon-14 content. Ages of less than one half-life, or intermediate between a whole number of half-lives, can also be estimated, but the method for doing so is beyond the scope of this book.

We know that carbon-14 dating is accurate because it can be checked against objects whose ages are known from other methods. For example, old trees can be dated by counting the tree rings within their trunks and by carbon-14 dating. Except in cases where the carbon-14 content has been modified by some other source, carbon-14 dating has proven reliable. However, this method is not dependable when attempting to date objects that are more than 50,000 years old: the amount of carbon-14 becomes too low to measure.

EXAMPLE 17.5 Radiocarbon Dating

A skull believed to belong to an early human being is found to have a carbon-14 content of 3.125% of that found in living organisms. How old is the skull?

Solution:

We can examine Table 17.1 and see that a carbon-14 content of 3.125% of that found in living organisms corresponds to an age of 28,650 years.

SKILLBUILDER 17.5 Radiocarbon Dating

An ancient scroll is claimed to have originated from Greek scholars in about 500 B.C. A measure of its carbon-14 content reveals it to contain 100.0% of that found in living organisms. Is the scroll authentic?

17.7 The Discovery of Fission and the Atomic Bomb

The element with atomic number 100 is named *fermium*, in honor of Enrico Fermi.

In the mid-1930s **Enrico Fermi** (1901–1954), an Italian physicist, tried to synthesize a new element by bombarding uranium—the heaviest known element at that time—with neutrons. Fermi hypothesized that if a neutron were incorporated into the nucleus of a uranium atom, the nucleus might undergo beta decay, converting a neutron into a proton. If that happened, a new element, with atomic number 93, would be synthesized for the first time. The nuclear equation for the process is:

$$^{238}_{92}\text{U} + \underbrace{^{1}_{0}\text{n}}_{\text{Neutron}} \longrightarrow {}^{239}_{92}\text{U} \longrightarrow \underbrace{^{239}_{93}\text{X}}_{\text{Newly synthesized element}} + {}^{0}_{-1}\text{e}$$

Fermi performed the experiment and detected the emission of beta particles. However, his results were inconclusive. Had he synthesized a new element? Fermi never chemically examined the products to determine their composition and therefore could not say with certainty whether he had.

The element with atomic number 109 is named *meitnerium*, in honor of Lise Meitner.

Three researchers in Germany—**Lise Meitner** (1878–1968), **Fritz Strassmann** (1902–1980), and **Otto Hahn** (1879–1968)—repeated Fermi's experiments but then performed careful chemical analysis of the products. What they found in the products—several elements *lighter* than uranium—would change the world forever. On January 6, 1939, Meitner, Strassmann, and Hahn reported that the neutron bombardment of uranium resulted in **nuclear fission**—the splitting of the atom. The nucleus of the neutron-bombarded uranium atom had fallen apart into barium, krypton, and other smaller products. They also realized that the process emitted enormous amounts of energy. The following is a nuclear equation for a fission reaction, showing how uranium breaks apart into the daughter nuclides.

Lise Meitner

$$^{235}_{92}\text{U} + {}^{1}_{0}\text{n} \rightarrow {}^{142}_{56}\text{Ba} + {}^{91}_{36}\text{Kr} + 3\,^{1}_{0}\text{n} + \text{Energy}$$

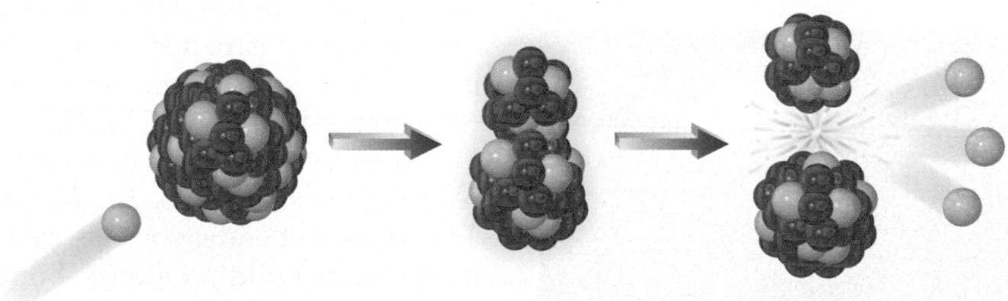

Notice that the initial uranium atom is the U-235 isotope, which comprises less than 1% of all naturally occurring uranium. U-238, the most abundant uranium isotope, does not undergo fission. Notice also that the process produces three neutrons, which have the potential to initiate fission in three other U-235 atoms.

U.S. scientists quickly realized that a sample rich in U-235 could undergo a **chain reaction** in which neutrons produced by the fission of one uranium nucleus would induce fission in other uranium nuclei (Figure 17.10).

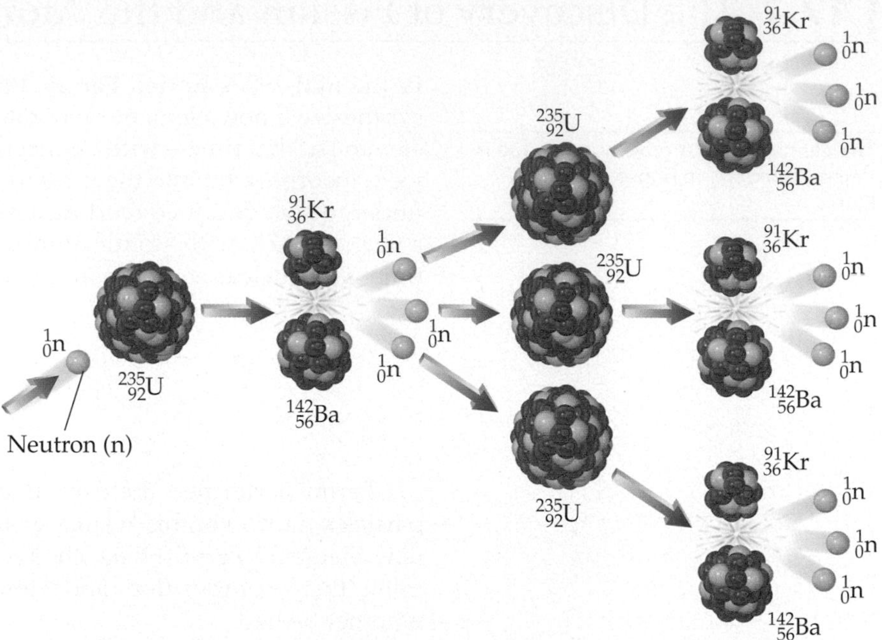

Figure 17.10 Fission chain reaction. The neutrons produced by the fission of one uranium nucleus induce fission in other uranium nuclei to produce a self-amplifying reaction. **Question:** Why is it necessary that each fission event produce more than one neutron for the chain reaction to amplify?

The result would be a self-amplifying reaction capable of producing an enormous amount of energy—an atomic bomb. However, to make a bomb, a **critical mass** of U-235—enough U-235 to produce a self-sustaining reaction—would be necessary. Fearing that Nazi Germany would develop such a bomb, several U.S. scientists persuaded Albert Einstein, the most well-known scientist of the time, to write a letter to President Franklin Roosevelt warning of this possibility. Einstein wrote, "... and it is conceivable—though much less certain—that extremely powerful bombs of a new type may thus be constructed. A single bomb of this type, carried by boat and exploded in a port, might very well destroy the whole port together with some of the surrounding territory."

The testing of the world's first nuclear bomb at Alamogordo, New Mexico, in 1945.

Roosevelt was convinced by Einstein's letter, and in 1941 he assembled the resources to begin the costliest scientific project ever attempted. The top-secret endeavor was called the **Manhattan Project,** and its main goal was to build an atomic bomb before the Germans did. The project was led by physicist **J.R. Oppenheimer** (1904–1967) at a high-security research facility in Los Alamos, New Mexico. Four years later, on July 16, 1945, the world's first nuclear weapon was successfully detonated at a test site in New Mexico. The first atomic bomb exploded with a force equivalent to 18,000 tons of dynamite. Ironically, the Germans—who had *not* made a successful nuclear bomb—had already been defeated by this time. Instead, the atomic bomb was used on Japan. One bomb was dropped on Hiroshima, and a second bomb was dropped on Nagasaki. Together, the bombs killed approximately 200,000 people and forced Japan to surrender. World War II was over. The atomic age had begun.

17.8 Nuclear Power: Using Fission to Generate Electricity

Nuclear reactions, such as fission, generate enormous amounts of energy. In a nuclear bomb, the energy is released all at once. However, the energy can also be released more slowly and used for other purposes such as electricity generation. In the United States, about 20% of electricity is generated by nuclear fission. In some other countries, as much as 70% of electricity is generated by nuclear fission. To get an idea of the amount of energy released during fission, imagine a hypothetical nuclear-powered car. Suppose the fuel for such a car was a uranium cylinder about the size of a pencil. How often would you have to refuel the car? The energy content of the uranium cylinder would be equivalent to about 1000 twenty-gallon tanks of gasoline. If you refuel your gasoline-powered car once a week, your nuclear-powered car could go 1000 weeks—almost 20 years—before refueling. Imagine a pencil-sized fuel rod lasting for twenty years of driving!

A nuclear-powered car really is hypothetical because the amount of uranium-235 needed for a critical mass is beyond the energy needs of a car.

Similarly, a nuclear-powered electrical plant can produce a lot of electricity with a small amount of fuel. Nuclear power plants generate electricity by using fission to generate heat (Figure 17.11). The heat is used to boil water and create steam, which then turns the turbine on a generator to produce electricity. The fission reaction itself occurs in the nuclear core of the power

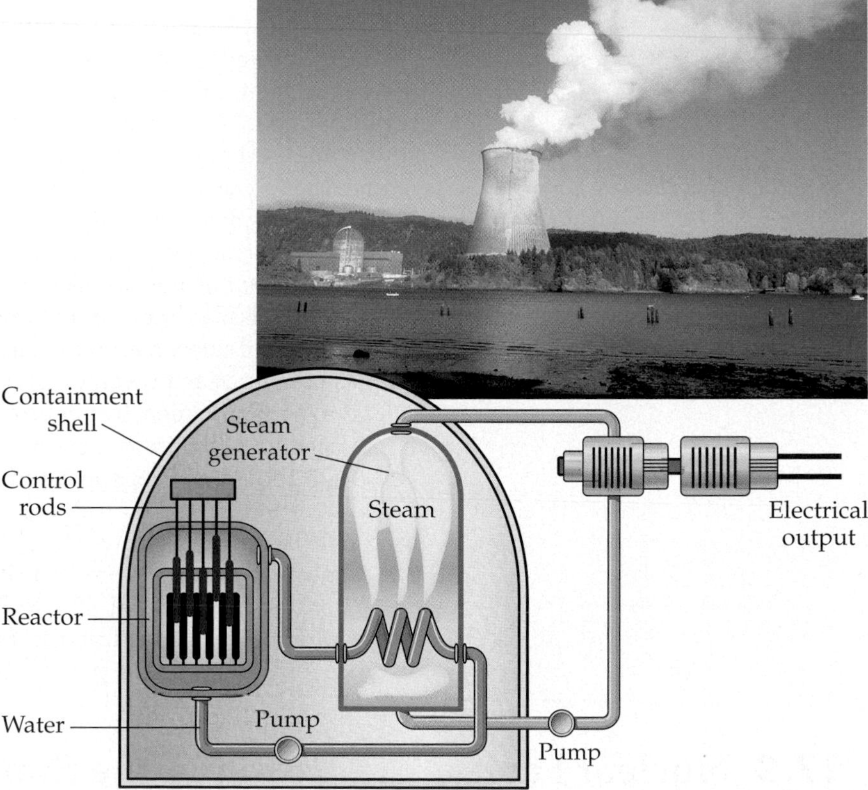

Figure 17.11 In a nuclear power plant, fission generates heat that is used to boil water and create steam. The steam then turns a turbine on a generator to produce electricity. Note that the water carrying heat from the reactor core is contained within separate pipes and does not come into direct contact with the steam that drives the turbines.

plant. The core consists of uranium fuel rods—enriched to about 3.5% U-235—interspersed between retractable neutron-absorbing control rods. When the control rods are fully retracted from the fuel rod assembly, the chain reaction can occur. However, when the control rods are fully inserted into the fuel assembly, they absorb the neutrons that would otherwise induce fission, shutting down the chain reaction. By inserting or retracting the control rods, the operator can control the rate of fission. If more heat is needed, the control rods are retracted slightly. If the fission reaction begins to get too hot, the control rods are inserted a little more. In this way, the fission reaction is controlled to produce the right amount of heat needed for electricity generation. In case of a power failure, the fuel rods automatically drop into the fuel rod assembly, shutting down the fission reaction.

A typical nuclear power plant generates enough electricity for a city of about 1 million people and uses about 50 kg of fuel per day. In contrast, a coal-burning power plant uses about 2,000,000 kg of fuel to generate the same amount of electricity. Furthermore, a nuclear power plant generates no air pollution and no greenhouse gases. A coal-burning power plant, on the other hand, emits pollutants such as carbon monoxide, nitrogen oxides, and sulfur oxides. Coal-burning power plants also emit carbon dioxide, a greenhouse gas.

Nuclear power generation, however, is not without problems. Foremost among them is the danger of nuclear accidents. In spite of safety precautions, the fission reaction occurring in nuclear power plants can overheat. The most famous example of this occurred in Chernobyl in the former Soviet Union on April 26, 1986. Operators of the plant were performing an experiment designed to reduce maintenance costs. In order to perform the experiment, however, many of the safety features of the reactor core were disabled. The experiment failed, with disastrous results. The nuclear core, composed partly of graphite, overheated and began to burn. The accident caused 31 deaths and produced a fire that scattered radioactive debris into the atmosphere. It is important to understand, however, that a nuclear power plant *cannot* become a nuclear bomb. The uranium fuel used in electricity generation is not sufficiently enriched in U-235 to produce a nuclear detonation. It is also important to understand that U.S. nuclear power plants have additional safety features designed to prevent similar accidents. For example, U.S. nuclear power plants have large containment structures designed to contain radioactive debris in the event of an accident.

Reactor cores in the U.S. are not made of graphite and could not burn in the way that the Chernobyl core did.

A second problem associated with nuclear power is waste disposal. Although the amount of fuel used in electricity generation is small compared to other fuels, the products of the reaction are radioactive and have long half-lives. What do we do with this waste? Currently, in the United States, nuclear waste is stored on site at the nuclear power plants. However, a permanent disposal site is being developed in Yucca Mountain, Nevada. The site is scheduled to be operational in 2010, but political resistance may alter current plans.

17.9 Nuclear Fusion: The Power of the Sun

As we have learned, nuclear fission is the splitting of a heavy nucleus to form two or more lighter ones. **Nuclear fusion,** in contrast, is the combination of two light nuclei to form a heavier one. Both fusion and fission emit large amounts of energy. Nuclear fusion is the energy source of stars, including our sun. In stars, hydrogen atoms fuse together to form helium atoms, emitting energy in the process.

Nuclear fusion is also the basis of modern nuclear weapons called hydrogen bombs. A modern hydrogen bomb has up to 1000 times the explosive force of the first atomic bombs. These bombs employ the following fusion reaction.

$$ {}^{2}_{1}\text{H} + {}^{3}_{1}\text{H} \rightarrow {}^{4}_{2}\text{He} + {}^{1}_{0}\text{n} $$

In this reaction, deuterium (the isotope of hydrogen with one neutron) and tritium (the isotope of hydrogen with two neutrons) combine to form helium-4 and a neutron. Because fusion reactions require two positively charged nuclei (which repel each other) to fuse together, extremely high temperatures are required. In a hydrogen bomb, a small fission bomb is detonated first, providing temperatures high enough for fusion to proceed.

Nuclear fusion has been intensely investigated as a way to produce electricity. Because of the higher energy production—fusion provides about ten times more energy per gram of fuel than fission—and because the products of the reaction are less problematic than fission, fusion holds promise as a future energy source. However, in spite of intense efforts, fusion electricity generation remains elusive. One of the main problems is the high temperature required for fusion to occur—no material can withstand these temperatures. After years of pouring billions of dollars into fusion research, the U.S. Congress has reduced funding for these projects. Whether fusion will ever be a viable energy source remains to be seen.

17.10 The Effects of Radiation on Life

As we have seen, the energy in radiation can ionize molecules. When radiation ionizes important molecules in living cells, problems can develop. The ingestion of radioactive materials—especially alpha and beta emitters—is particularly dangerous because the radioactivity is then inside the body and can do more damage. The effects of radiation can be divided into three different types: acute radiation damage, increased cancer risk, and genetic effects.

Acute Radiation Damage

Acute radiation damage results from exposures to large amounts of radiation in a short period of time. The main sources of this kind of exposure are nuclear bombs or exposed nuclear reactor cores. These high levels of radiation kill large numbers of cells. Rapidly dividing cells, such as those in the immune system and the intestinal lining, are most susceptible. Consequently, people exposed to high levels of radiation have weakened immune systems and a lowered ability to absorb nutrients from food. In milder cases, recovery is possible with time. In more extreme cases death results, often from unchecked infection.

TABLE 17.2

Effects of Radiation Exposure

Dose (rem)	*Probable Outcome*
20–100	decreased white blood cell count; possible increase in cancer risk
100–400 rem	radiation sickness; skin lesions, increase in cancer risk
500 rem	death

Increased Cancer Risk

Lower doses of radiation over extended periods of time can increase cancer risk. Radiation increases cancer risk because it can damage DNA, the molecules in cells that carry instructions for cell growth and replication. When the DNA within a cell is damaged, the cell normally dies. Occasionally, however, changes in DNA cause cells to grow abnormally and to become cancerous. These cancerous cells grow into tumors that can spread and, in some cases, cause death. Cancer risk increases with increased radiation exposure. However, cancer is so prevalent and has so many convoluted causes that it is difficult to determine an exact threshold for increased cancer risk from radiation exposure.

DNA and its function in the body are explained in more detail in Chapter 19.

Genetic Defects

Another possible effect of radiation exposure is genetic defects in offspring. If radiation damages the DNA of reproductive cells—such as eggs or sperm—then the offspring that develop from those cells may have genetic abnormalities. Genetic defects of this type have been observed in laboratory animals exposed to high levels of radiation. However, such genetic defects—with a clear causal connection to radiation exposure—have yet to be observed in humans, even in studies of Hiroshima survivors.

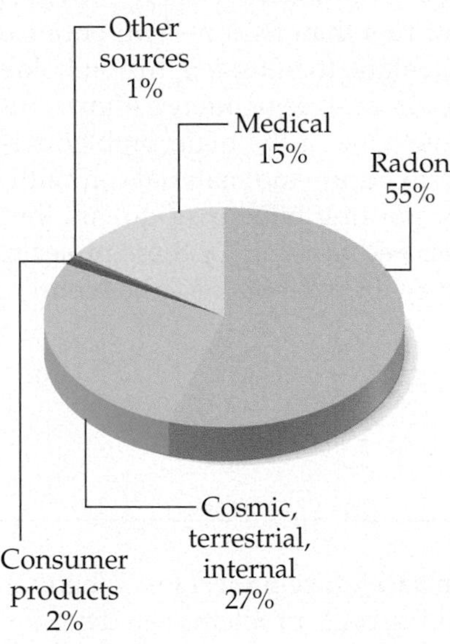

Figure 17.12 Radiation exposure by source.

Measuring Radiation Exposure

Radiation exposure is often reported in a unit called the **rem**. The rem, which stands for *roentgen equivalent man,* is a weighted measure of radiation exposure that accounts for the ionizing power of the different types of radiation. On average, each of us is exposed to approximately one-third of a rem of radiation per year. This radiation comes primarily from natural sources, especially radon, one of the products in the uranium decay series (Figure 17.12). It takes much more radiation than the natural amount to produce measurable health effects in humans. The first measurable effects, a decreased white blood cell count, occur at instantaneous exposures of approximately 20 rem (Table 17.2). Exposures of 100 rem show a definite increase in cancer risk, and exposures of over 500 rem often result in death.

Other radiation units include the *curie,* defined as 3.7×10^{10} decay events per second, and the *roentgen,* defined as the amount of radiation that produces 2.58×10^{-4}C of charge per kilogram of air.

17.11 Radioactivity in Medicine

Radioactivity is often perceived as dangerous; however, it is also incredibly useful to physicians in the diagnosis and treatment of disease. The use of radioactivity in medicine can be broadly divided into **isotope scanning** and **radiotherapy.**

Isotope Scanning

An example of isotope scanning was presented in the opening section of this chapter. In isotope scanning, a radioactive isotope is introduced into the body. Then the radiation emitted by the isotope is detected using a photographic film or a scintillation counter. Since different isotopes are taken up by different organs or tissues, isotope scanning has a variety of uses. For example, the radioactive isotope phosphorus-32 is preferentially taken up by cancerous tissue. A cancer patient can be given this isotope to find and identify cancerous tumors. Other isotopes commonly used in medicine include iodine-131, used to diagnose thyroid disorders, and technetium-99, which can produce images of several different internal organs.

Radiotherapy

Because radiation kills cells, and because it is particularly effective at killing rapidly dividing cells, it is often used as a therapy for cancer (cancer cells are rapidly dividing). Gamma rays are focused on internal tumors to kill them (Figure 17.13). The gamma ray beam is usually moved in a circular path around the tumor maximizing the exposure of the tumor while minimizing the exposure of the healthy tissue around the tumor. Nonetheless, cancer patients usually develop the symptoms of radiation sickness, which include vomiting, skin burns, and hair loss.

Some people wonder how radiation—which is known to cause cancer—can be used to treat cancer. The answer lies in risk analysis. A cancer patient is normally exposed to radiation doses of about 100 rem. Such a dose increases cancer risk by about 1%. However, if the patient has a 100% chance of dying from the cancer that he already has, such a risk becomes acceptable, especially since there is a significant chance of curing the cancer.

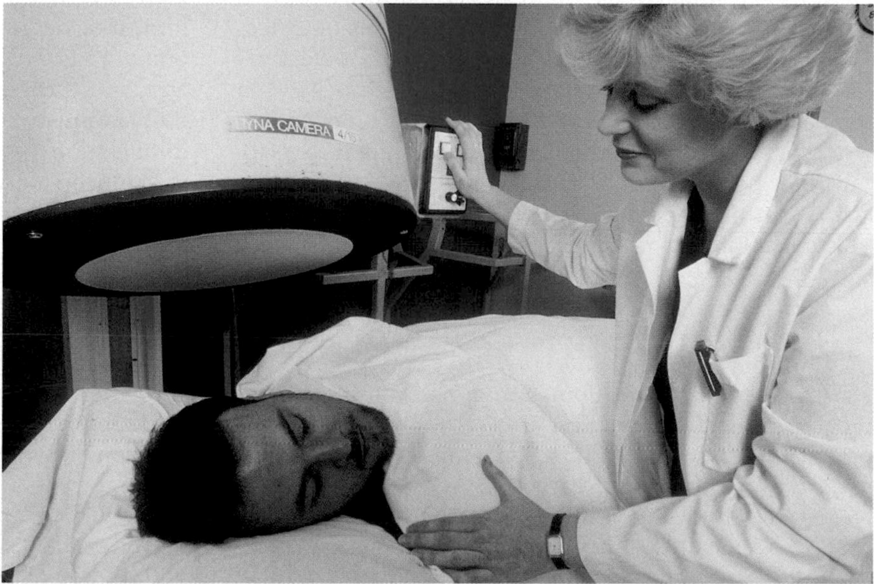

Figure 17.13 Radiotherapy for cancer treatment involves exposing the tumor to gamma rays. The beam is moved in a circular pattern around the tumor to maximize exposure of the tumor while minimizing exposure of healthy cells.

CHAPTER IN REVIEW

Chemical Principles

The Nature and Discovery of Radioactivity: Radioactivity is the emission of particles from unstable atomic nuclei. It was discovered in 1896 and described in the years that followed. Radioactivity can be divided into four types.

- **Alpha particles**: composed of 2 protons and 2 neutrons and given the symbol ${}_{2}^{4}\text{He}$. Alpha particles have a high ionizing power but a low penetrating power.
- **Beta particles**: electrons emitted from atomic nuclei when a neutron changes into a proton. Beta particles have the symbol ${}_{-1}^{0}\text{e}$ and have intermediate ionizing power and intermediate penetrating power.
- **Gamma rays**: high energy, short wavelength photons. Gamma rays have the symbol ${}_{0}^{0}\gamma$ and have low ionizing power and high penetrating power.
- **Positrons**: emitted from atomic nuclei when protons change into neutrons. Positrons have the symbol ${}_{+1}^{0}\text{e}$ and have intermediate ionizing power and intermediate penetrating power.

Relevance

The Nature and Discovery of Radioactivity: Radioactivity is a fundamental part of the behavior of some atoms, and it also has many applications. For example, radioactivity is used to diagnose and treat disease including cancer, thyroid diseases, abnormal kidney and bladder function, and heart disease. Natural radioactivity is part of our environment and can be used to date ancient objects. The discovery of radioactivity led to the discovery of fission, which in turn led to the development of nuclear bombs and nuclear energy.

Detecting Radioactivity: Radioactive particles carry a large amount of energy and are therefore easily detected. The most common and inexpensive way to detect radioactivity is with photographic film, which is used in film badge dosimeters to monitor exposure in people working with or near radioactive sources. Radioactivity can also be detected with devices such as a Geiger-Müller counter or a scintillation counter, both of which give instantaneous readings of radiation levels.

Detecting Radioactivity: Since radioactivity is invisible, it must be detected using film or instruments. The detection of radioactivity is important as both a scientific tool and a practical one. Our understanding of what radioactivity is and our continuing research to understand it and its effects on living organisms require the ability to detect it. Our need for safety in areas where radioactive particles are used also requires our ability to detect radiation.

Half-Life and Radiocarbon Dating: The half-life of a radioactive nuclide is the time it takes for one-half of the parent nuclides in a radioactive sample to decay. The presence of radioactive carbon-14 (with a half-life of 5,730 years) in the environment provides a natural clock by which to estimate the age of many artifacts and fossils. All living things contain carbon-14. When they die, the carbon-14 decays with its characteristic half-life. A measurement of the amount of carbon-14 in a fossil or artifact can therefore reveal its age.

Half-Life and Radiocarbon Dating: The half-life of radioactive nuclide tells the activity of the nuclide and how long it will be radioactive. Nuclides with short half-lives are very active (many decay events per unit time) but will not be radioactive for long. Nuclides with long half-lives are less active (fewer decays per unit time) but will be radioactive for a long time.

Fission, the Atomic Bomb, and Nuclear Power: Fission—the splitting of the atom into smaller fragments—was discovered in 1939. Fission occurs when a neutron is absorbed by a U-235 nucleus. The nucleus becomes unstable, falling apart and producing barium, krypton, neutrons, and a lot of energy.

Fission, the Atomic Bomb, and Nuclear Power: The discovery of fission in 1939 changed the world. Within six years, the U.S. developed and tested fission nuclear bombs, which ended World War II. Fission can also be used to generate electricity. The fission reaction heats water to create steam, which then turns the turbine on an electrical generator. Nuclear electricity generates about 20% of the electricity for the United States. Some countries generate up to 70% of their electricity from nuclear power.

Nuclear Fusion: Nuclear fusion is the combination of two light nuclei to form a heavier one. Nuclear fusion is the energy source of stars, including our sun.

Nuclear Fusion: Modern nuclear weapons are fusion bombs with 1000 times the explosive force of the first fission bombs. Nuclear fusion is still being explored as a way to generate electricity but has not proven successful to date.

The Effects of Radiation on Life and Nuclear Medicine: Radiation can damage important molecules within living cells. This can have detrimental effects but can also be used for positive purposes. Instantaneous high exposure to radiation can lead to radiation sickness and even death. Long-term lower exposure levels can increase cancer risk. However, the same radiation, when targeted at cancerous tumors, can be used to treat cancer. Radiation can also be used to image internal organs through isotope scanning.

The Effects of Radiation on Life and Nuclear Medicine: Radiation can be used for both good and harm. We have seen how the destructive effects of radiation can be employed in a nuclear bomb. However, we have also seen how radiation can be a precise tool in the physician's arsenal against disease.

Chemical Skills

Writing Nuclear Equations for Alpha Decay (Section 17.3)

Begin with the symbol for the isotope undergoing decay on the left side of the equation and the symbol for an alpha particle on the right side. Leave a space or a question mark for the unknown daughter nuclide.

Equalize the sum of the mass numbers and the sum of the atomic numbers on both sides of the equation by writing the appropriate mass number and atomic number for the unknown daughter nuclide.

Deduce the identity of the unknown daughter nuclide from the atomic number and write its symbol.

Examples

EXAMPLE 17.6 Writing Nuclear Equations for Alpha Decay

Write a nuclear equation for the alpha decay of Po-214.

Solution:

$${}^{214}_{84}\text{Po} \rightarrow {}^{?}_{?}? + {}^{4}_{2}\text{He}$$

$${}^{214}_{84}\text{Po} \rightarrow {}^{210}_{82}? + {}^{4}_{2}\text{He}$$

$${}^{214}_{84}\text{Po} \rightarrow {}^{210}_{82}\text{Pb} + {}^{4}_{2}\text{He}$$

Writing Nuclear Equations for Beta Decay (Section 17.3)

Begin with the symbol for the isotope undergoing decay on the left side of the equation and the symbol for a beta particle on the right side.

Equalize the sum of the mass numbers and the sum of the atomic numbers on both sides of the equation by writing the appropriate mass number and atomic number for the unknown daughter nuclide.

Deduce the identity of the unknown daughter nuclide from the atomic number and write its symbol.

EXAMPLE 17.7 Writing Nuclear Equations for Beta Decay

Write a nuclear equation for the beta decay of Bi-214.

Solution:

$${}^{214}_{83}\text{Bi} \rightarrow {}^{?}_{?}? + {}^{0}_{-1}\text{e}$$

$${}^{214}_{83}\text{Bi} \rightarrow {}^{214}_{84}? + {}^{0}_{-1}\text{e}$$

$${}^{214}_{83}\text{Bi} \rightarrow {}^{214}_{84}\text{Po} + {}^{0}_{-1}\text{e}$$

Writing Nuclear Equations for Positron Decay (Section 17.3)

Begin with the symbol for the isotope undergoing decay on the left side of the equation and the symbol for a positron on the right side.

Equalize the sum of the mass numbers and the sum of the atomic numbers on both sides of the equation by writing the appropriate mass number and atomic number for the unknown daughter nuclide.

Deduce the identity of the unknown daughter nuclide from the atomic number and write its symbol.

EXAMPLE 17.8 Writing Nuclear Equations for Positron Decay

Write a nuclear equation for the beta decay of C-11.

Solution:

$${}^{11}_{6}\text{C} \rightarrow {}^{?}_{?}? + {}^{0}_{+1}\text{e}$$

$${}^{11}_{6}\text{C} \rightarrow {}^{11}_{5}? + {}^{0}_{+1}\text{e}$$

$${}^{11}_{6}\text{C} \rightarrow {}^{11}_{5}\text{B} + {}^{0}_{+1}\text{e}$$

Using Half-Life (Section 17.5)

To use half-life to determine the time it takes for a sample to decay to a specified amount or the amount of a sample left after a specified time, it is easiest to draw a table showing the amount of the nuclide as a function of number of half-lives. For each half-life, simply divide the amount of parent nuclide by two.

EXAMPLE 17.9 Using Half-Life

Po-210 is an alpha emitter with a half-life of 138 days. How many grams of Po-210 remain after 552 days if the sample initially contained 5.80 g of Po-210?

Solution:

Po-210 (g)	*Number of Half-Lives*	*Time in Days*
5.80	0	0
2.90	1	138
1.45	2	276
0.725	3	414
0.363	4	552

Therefore, the amount of Po-210 left after 552 days is 0.363 g.

Using Carbon-14 Content to Determine the Age of Fossils or Artifacts (Section 17.6)

To determine the age of an artifact or fossil based on its carbon-14 content, you can either examine Table 17.1, or you can build your own table beginning with 100% carbon-14 (relative to living organisms) and reducing the amount by a factor of one-half for each half-life.

EXAMPLE 17.10 Using Carbon-14 Content to Determine the Age of Fossils or Artifacts

Some wood ashes from a fire pit in the ruins of an ancient village have a carbon-14 content that is 25% of the amount found in living organisms. How old are the ashes and, by implication, the village?

Solution:

*C-14 (%)**	*Number of Half-Lives*	*Time in Years*
100	0	0
50.0	1	5,730
25.0	2	11,460

*Percent relative to living organisms

The ashes are from wood that was living 11,460 years ago.

KEY TERMS

alpha (α) radiation [17.3]
alpha particle [17.3]
Antoine-Henri Becquerel [17.2]
beta (β) radiation [17.3]
beta particle [17.3]
chain reaction [17.7]
critical mass [17.7]
Marie Sklodowska Curie [17.2]
daughter nuclide [17.3]
Enrico Fermi [17.7]
film-badge dosimeter [17.4]
gamma (γ) radiation [17.3]
gamma ray [17.3]
Geiger-Müller counter [17.4]
Otto Hahn [17.7]
half-life [17.5]
ionizing power [17.3]
isotope scanning [17.11]
Manhattan Project [17.7]
Lise Meitner [17.7]
nuclear equation [17.3]
nuclear fission [17.7]
nuclear fusion [17.9]
J.R. Oppenheimer [17.7]
parent nuclide [17.3]
penetrating power [17.3]
phosphorescence [17.2]
positron [17.3]
positron emission [17.3]
radioactive [17.1]
radioactivity [17.1]
radiocarbon dating [17.6]
radiotherapy [17.11]
rem [17.10]
Ernest Rutherford [17.3]
scintillation counter [17.4]
Fritz Strassmann [17.7]

EXERCISES

Questions

1. What is radioactivity? What does it mean for an atom to be radioactive?
2. How can radioactivity be used to diagnose appendicitis?
3. How was radioactivity first discovered? By whom?
4. What are uranic rays?
5. What role did Marie Sklodowska Curie play in the discovery of radioactivity?
6. How was Marie Sklodowska Curie acknowledged for her work in radioactivity?
7. Explain what is represented by each symbol in the following notation.

$$^{A}_{Z}X$$

8. Radioactivity originates from the _____ of radioactive atoms.
9. What is alpha radiation? What is the symbol for an alpha particle?
10. What happens to an atom when it emits an alpha particle?
11. How do the ionizing power and penetrating power of alpha particles compare to other types of radiation?
12. What is beta radiation? What is the symbol for a beta particle?
13. What happens to an atom when it emits a beta particle?
14. How do the ionizing power and penetrating power of beta particles compare to other types of radiation?
15. What is gamma radiation? What is the symbol for a gamma ray?

16. What happens to an atom when it emits a gamma ray?
17. How do the ionizing power and penetrating power of gamma particles compare to other types of radiation?
18. What is positron emission? What is the symbol for a positron?
19. What happens to an atom when it emits a positron?
20. How do the ionizing power and penetrating power of positrons compare to other types of radiation?
21. What is a nuclear equation? What does it mean for a nuclear equation to be balanced?
22. Identify the parent nuclides and daughter nuclides in the following nuclear equation. What kind of radioactive decay is involved?

$$^{231}_{91}\text{Pa} \rightarrow ^{227}_{89}\text{Ac} + ^{4}_{2}\text{He}$$

23. What is a film-badge dosimeter and how does it work?
24. How does a Geiger-Müller counter detect radioactivity?
25. Explain how a scintillation counter works.
26. What are some sources of natural radioactivity?
27. Explain the concept of half-life.
28. What is a radioactive decay series?
29. What is the source of radon in our environment? Why is radon problematic?
30. What is the source of carbon-14 in our environment?
31. Why do all living organisms contain a uniform amount of carbon-14?
32. What happens to the carbon-14 in a living organism when it dies? How can this be used to establish how long ago the organism died?
33. How do we know that carbon-14 (or radiocarbon) dating is accurate? What is the age limit for which carbon-14 dating is useful?
34. Explain Fermi's experiment in which he bombarded uranium with neutrons. Include a nuclear equation in your answer.
35. What is nuclear fission? How and by whom was it discovered?
36. Why can nuclear fission be used in a bomb? Include the concept of a chain reaction in your explanation.
37. What is a critical mass?
38. What was the main goal of the Manhattan Project? Who was the project leader?
39. How can nuclear fission be used to generate electricity?
40. Explain the purpose of the control rods in a nuclear reactor core. How do they work?
41. What are the main advantages of nuclear electricity generation?
42. What are the main problems associated with nuclear electricity generation?
43. Can a nuclear reactor detonate the way a nuclear bomb can? Why or why not?
44. What is nuclear fusion?
45. Do modern nuclear weapons use fission or fusion or both? Explain.
46. Can nuclear fusion be used to generate electricity? What are the advantages of the fusion over fission for electricity generation? What are the problems with fusion?
47. How does radiation affect the molecules within living organisms?
48. What is acute radiation damage to living organisms?
49. Explain how radiation can increase cancer risk.
50. Explain how radiation can cause genetic defects. Has this ever been observed in laboratory animals? In humans?
51. What is the main unit of radiation exposure? How much radiation is the average American exposed to per year?
52. Describe the outcomes of radiation exposure at different doses (in rem).
53. Explain the medical use of radioactivity termed isotope scanning.
54. How is radioactivity used to treat cancer?

Problems

Isotopic and Nuclear Particle Symbols

55. Provide a symbol for the isotope of lead that contains 125 neutrons.

56. Provide a symbol for the isotope of bismuth that contains 128 neutrons.

57. How many protons and neutrons are in the following nuclide?

$$^{207}_{81}\text{Tl}$$

58. How many protons and neutrons are in the following nuclide?

$$^{219}_{86}\text{Rn}$$

59. Identify each of the following as an alpha particle, a beta particle, a gamma ray, a positron, a neutron, or a proton.
a) $^{0}_{-1}\text{e}$
b) $^{1}_{0}\text{n}$
c) $^{0}_{0}\gamma$

60. Identify each of the following as an alpha particle, a beta particle, a gamma ray, a positron, a neutron, or a proton.
a) $^{1}_{1}\text{p}$
b) $^{4}_{2}\text{He}$
c) $^{0}_{+1}\text{e}$

Radioactive Decay

61. Write a nuclear equation for the alpha decay of each of the following nuclides.
a) U-234
b) Th-230
c) Ra-226
d) Rn-222

62. Write a nuclear equation for the alpha decay of each of the following nuclides.
a) Po-218
b) Po-214
c) Po-210
d) Th-227

63. Write a nuclear equation for the beta decay of each of the following nuclides.
a) Pb-214
b) Bi-214
c) Th-231
d) Ac-227

64. Write a nuclear equation for the beta decay of each of the following nuclides.
a) Pb-211
b) Tl-207
c) Th-234
d) Pa-234

65. Write a nuclear equation for positron emission by each of the following nuclides.
a) C-11
b) N-13
c) O-15

66. Write a nuclear equation for positron emission by each of the following nuclides.
a) Co-55
b) Na-22
c) F-18

67. Fill in the blanks in the following partial decay series.

$$^{241}_{94}\text{Pu} \rightarrow {}^{241}_{95}\text{Am} + \_\_$$
$$^{241}_{95}\text{Am} \rightarrow {}^{237}_{93}\text{Np} + \_\_$$
$$^{237}_{93}\text{Np} \rightarrow \_\_ + {}^{4}_{2}\text{He}$$
$$\_\_ \rightarrow {}^{233}_{92}\text{U} + {}^{0}_{-1}\text{e}$$

68. Fill in the blanks in the following partial decay series.

$$^{225}_{88}\text{Ra} \rightarrow {}^{225}_{89}\text{Ac} + \_\_$$
$$^{225}_{89}\text{Ac} \rightarrow \_\_ + {}^{4}_{2}\text{He}$$
$$\_\_ \rightarrow {}^{217}_{85}\text{At} + {}^{4}_{2}\text{He}$$
$$^{217}_{85}\text{At} \rightarrow \_\_ + {}^{4}_{2}\text{He}$$

69. Write a partial decay series for Th-232 undergoing the following sequential decays: α, β, β, α.

70. Write a partial decay series for Rn-220 undergoing the following sequential decays: α, α, β, α.

Half-Life

71. Suppose you have a 1000-atom sample of a radioactive nuclide that decays with a half-life of 2.0 days. How many radioactive atoms are left after 10 days?

72. Iodine-131 is often used in nuclear medicine to obtain images of the thyroid. If you start with 1.0×10^9 I-131 atoms, how many are left after approximately one month? I-131 has a half-life of 8.0 days.

73. A patient is given 0.050 mg of technetium-99m, a radioactive isotope with a half-life of about 6.0 hours. How long until the radioactive isotope decays to 6.3×10^{-3} mg?

74. Radium-223 decays with a half-life of 11.4 days. How long will it take for a 0.240 mol sample of radium to decay to 1.50×10^{-2} moles?

75. One of the nuclides in spent nuclear fuel is U-234, an alpha emitter with a half-life of 2.44×10^5 years. If a spent fuel assembly contains 2.80 kg of U-234, how long will it take for the amount of U-234 to decay to less than 0.10 kg?

76. One of the nuclides in spent nuclear fuel is U-235, an alpha emitter with a half-life of 703 million years. How long would it take for the amount of U-235 to reach one-eighth of its initial amount?

77. A radioactive sample contains 1.55 g of an isotope with a half-life of 3.8 days. How much (in grams) of the isotope will remain after 11.4 days?

78. A 58 mg sample of a radioactive nuclide is administered to a patient to obtain an image of her thyroid. If the nuclide has a half-life of 12 hours, how much of the nuclide remains in the patient after 4.0 days?

79. Each of the following nuclides is used in nuclear medicine. List them in order of most active (most number of decay events per second) to least active (least number of decay events per second).

Nuclide	*Half-Life*
P-32	14.3 days
Cr-51	27.7 days
Ga-67	78.3 hours
Sr-89	50.5 days

80. Each of the following nuclides is used in nuclear medicine. List them in order of most active (most number of decay events per second) to least active (least number of decay events per second).

Nuclide	*Half-Life*
Y-90	64.1 hours
Tc-99m	6.02 hours
In-111	2.8 days
I-131	8.0 days

Radiocarbon Dating

81. A wooden boat discovered just south of the Great Pyramid in Egypt had a carbon-14 content of approximately 50% of that found in living organisms. How old is the boat?

82. A layer of peat beneath the glacial sediments of the last ice age had a carbon-14 content of 25% of that found in living organisms. How long ago was this ice age?

83. An ancient skull has a carbon-14 content of 1.563% of that found in living organisms. How old is the skull?

84. A mammoth skeleton has a carbon-14 content of 12.50% of that found in living organisms. When did the mammoth live?

Fission and Fusion

85. Write a nuclear reaction for the neutron-induced fission of U-235 to form Xe-144 and Sr-90. How many neutrons are produced in the reaction?

86. Write a nuclear reaction for the neutron-induced fission of U-235 to produce Te-137 and Zr-97. How many neutrons are produced in the reaction?

87. Write a nuclear equation for the fusion of two H-2 atoms to form He-3 and one neutron.

88. Write a nuclear equation for the fusion of H-3 with H-1 to form He-4.

Cumulative Problems

89. The fission of U-235 produces 3.2×10^{-11} J/atom. How much energy does it produce per mole of U-235? per kilogram of U-235?

90. The fusion of deuterium and tritium produces 2.8×10^{-12} J for every one atom of deuterium and one atom of tritium. How much energy is produced per one mole of deuterium and one mole of tritium?

91. Bi-210 is beta emitter with a half-life of 5.0 days. If a sample contains 1.2 g of Bi-210, how many beta emissions would occur in 5.0 days?

92. Po-218 is an alpha emitter with a half-life of 3.0 minutes. If a sample contains 55 mg of Po-218, how many alpha emissions would occur in 6.0 minutes?

93. If a person living in a high radon area is exposed to 0.400 rem of radiation from radon per year, and his total exposure is 0.585 rem, what percentage of his total exposure is due to radon?

94. An X-ray technician is exposed to 0.020 rem of radiation at work. If her total exposure is the national average (0.36 rem), what fraction of her exposure is due to on-the-job exposure?

Highlight Problems

95. Closely examine the following diagram representing the alpha decay of sodium-20 and draw in the missing nucleus.

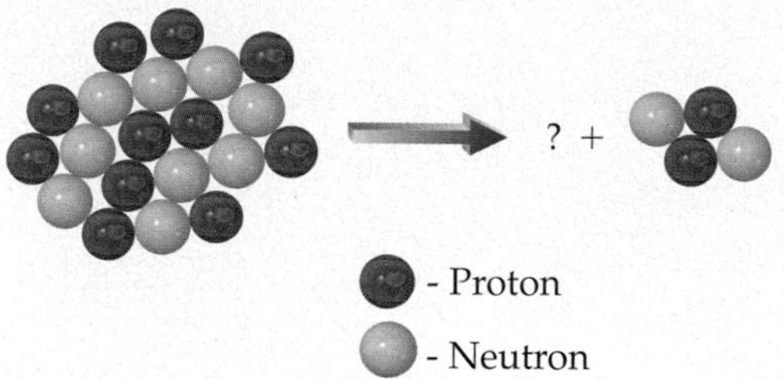

96. Closely examine the following diagram representing the beta decay of fluorine-21 and draw in the missing nucleus.

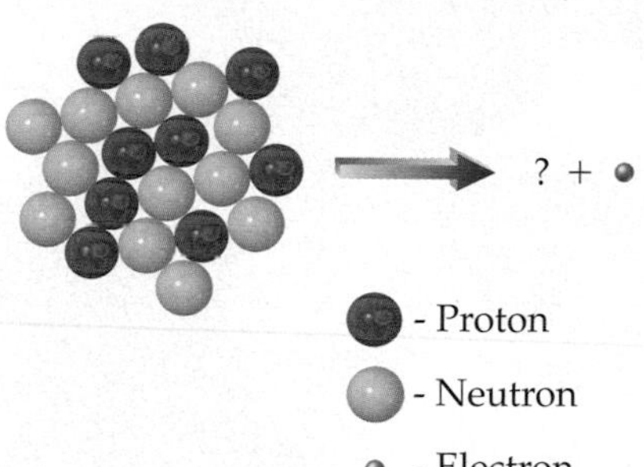

97. Closely examine the following diagram representing the positron emission of carbon-10 and draw the missing nucleus.

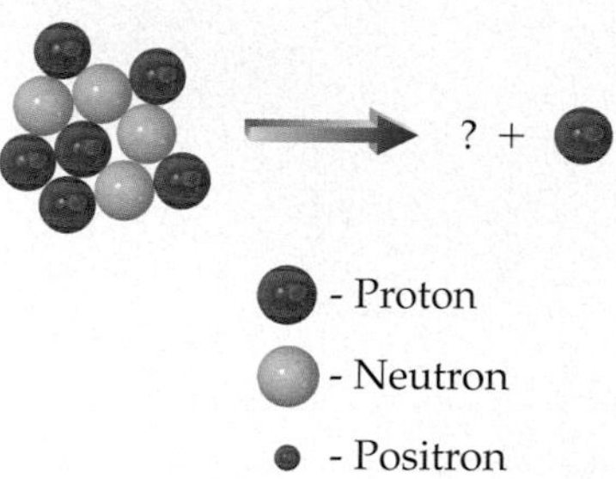

98. A radiometric dating technique uses the decay of U-238 to Pb-206 (the half-life for this process is 4.5 billion years) to determine the age of the oldest rocks on earth and by implication the age of the Earth itself. The oldest uranium-containing rocks on earth contain approximately equal numbers of uranium atoms and lead atoms. Assuming the rocks were pure uranium when they were formed, how old are the rocks?

Answers to Skillbuilder Exercises

Skillbuilder 17.1 $^{216}_{84}\text{Po} \rightarrow\, ^{212}_{82}\text{Pb} +\, ^{4}_{2}\text{He}$
Skillbuilder 17.2 $^{228}_{89}\text{Ac} \rightarrow\, ^{228}_{90}\text{Th} +\, ^{0}_{-1}\text{e}$
Skillbuilder Plus p. 616 $^{235}_{92}\text{U} \rightarrow\, ^{231}_{90}\text{Th} +\, ^{4}_{2}\text{He}$
$^{231}_{90}\text{Th} \rightarrow\, ^{231}_{91}\text{Pa} +\, ^{0}_{-1}\text{e}$ $^{231}_{91}\text{Pa} \rightarrow\, ^{227}_{89}\text{Ac} +\, ^{4}_{2}\text{He}$

Skillbuilder 17.3 $^{22}_{11}\text{Na} \rightarrow\, ^{22}_{10}\text{Ne} +\, ^{0}_{+1}\text{e}$
Skillbuilder 17.4 7.00×10^{-3} moles
Skillbuilder 17.5 No, the carbon-14 content suggests that the scroll is from very recent times.

APPENDIX 1
Physical Constants, Conversion Factors, and Geometric Formulas

A1.1 Fundamental Physical Constants

Atomic mass unit	1 amu	$= 1.660\,539 \times 10^{-27}$ kg
	1 g	$= 6.022\,142 \times 10^{23}$ amu
Avogadro's number	N_A	$= 6.022\,142 \times 10^{23}$/mol
Electron charge	e	$= 1.602\,176 \times 10^{-19}$ C
Gas constant	R	$= 8.314\,472$ J/(mol · K)
		$= 0.082\,058\,2$(L · atm)/(mol · K)
Mass of electron	m_e	$= 5.485\,799 \times 10^{-4}$ amu
		$= 9.109\,382 \times 10^{-31}$ kg
Mass of neutron	m_n	$= 1.008\,665$ amu
		$= 1.674\,927 \times 10^{-27}$ kg
Mass of proton	m_p	$= 1.007\,276$ amu
		$= 1.672\,622 \times 10^{-27}$ kg
Pi	π	$= 3.141\,592\,653\,6$
Planck's constant	h	$= 6.626\,069 \times 10^{-34}$ J · s
Speed of light in vacuum	c	$= 2.997\,924\,58 \times 10^{8}$ m/s

A1.2 Useful Conversion Factors

Length: SI unit = meter (m)

- 1 m = 39.37 in
- 1 in = 2.54 cm (exactly)
- 1 mile = 5280 ft = 1.609 km
- 1 angstrom (Å) = 10^{-10} m

Mass: SI unit = kilogram (kg)

- 1 kg = 2.205 lb
- 1 lb = 16 oz = 453.6 g
- 1 ton = 2000 lb
- 1 metric ton = 1000 kg = 1.103 tons
- 1 g = 6.022×10^{23} atomic mass units (u or amu)

Energy: SI unit = joule (J)

- $1\ J = 1\ N \cdot m$
- $1\ cal = 4.184\ J$ (exactly)
- $1\ L \cdot atm = 101.33\ J$
- 1 megaelectron volt (MeV) $= 1.602 \times 10^{-13}\ J$
- 1 amu = 931.5 MeV

Pressure: SI unit = pascal (Pa)

- $1\ Pa = 1\ N/m^2$
- $1\ bar = 10^5\ Pa$
- 1 atm $= 1.01325 \times 10^5$ Pa (exactly)
  = 1.01325 bar = 760 mmHg
  = 760 torr (exactly)

Volume: SI unit = cubic meter (m^3)

- $1\ L = 1000\ cm^3 = 1.057$ qt (U.S.)
- 1 gal (U.S.) = 4 qt = 8 pt
  = 32 fluid ounces
  = 3.785 L

Temperature: SI unit = Kelvin (K)

- K = °C + 273.15
- °C = (5/9)(°F − 32°)
- °F = (9/5)(°C) + 32°

A1.3 Useful Geometric Formulas

Perimeter of a rectangle $= 2l + 2w$

Circumference of a circle $= 2\pi r$

Area of a rectangle $= l \times w$

Area of a triangle = (1/2)(base × height)

Area of a circle $= \pi r^2$

Area of a sphere $= 4\pi r^2$

Volume of a sphere $= (4/3)\pi r^3$

Volume of a cylinder or prism = area of base × height

APPENDIX 2

Solubility Rules for Ionic Compounds

Compounds That Are Mostly Soluble

Compounds containing Li^+, Na^+, K^+, or NH_4^+
Compounds containing NO_3^- and $C_2H_3O_2^-$
Compounds containing Cl^-, Br^-, and I^-
(Exception: When these ions pair with Ag^+, Hg_2^{2+}, or Pb^{2+}, the compounds are insoluble.)
Compounds containing SO_4^{2-}
(Exception: When SO_4^{2-} pairs with Sr^{2+}, Ba^{2+}, Pb^{2+}, and Ca^{2+}, the compounds are insoluble.)

Compounds That Are Mostly Insoluble

Compounds containing OH^- and S^{2-}
(Exceptions: When these ions pair with Li^+, Na^+, K^+, or NH_4^+, the compounds are soluble.
When S^{2-} pairs with Ca^{2+}, Sr^{2+}, or Ba^{2+}, the compounds are soluble.
When OH^- pairs with Ca^{2+}, Sr^{2+}, or Ba^{2+}, the compounds are slightly soluble*.)
Compounds containing CO_3^{2-} and PO_4^{3-}
(Exception: When these ions pair with Li^+, Na^+, K^+, or NH_4^+, the compounds are soluble.)

APPENDIX 3

Mathematics Review

A3.1 Basic Algebra

In chemistry, you often have to solve an equation for a particular variable. For example, suppose you want to solve the following equation for V.

$$PV = nRT$$

To solve an equation for a particular variable means you should isolate that variable on one side of the equation. The rest of the variables or numbers will then be on the other side of the equation. To solve the above equation for V, simply divide both sides by P.

$$\frac{PV}{P} = \frac{nRT}{P}$$

$$V = \frac{nRT}{P}$$

The Ps cancel and we are left with an expression for V. For another example, consider solving the following equation for °F.

$$°\mathrm{C} = \frac{(°\mathrm{F} - 32)}{1.8}$$

First, eliminate the 1.8 in the denominator of the right side by multiplying both sides by 1.8.

$$(1.8)\ °\mathrm{C} = \frac{(°\mathrm{F} - 32)}{\cancel{1.8}}(\cancel{1.8})$$

$$1.8\ °\mathrm{C} = (°\mathrm{F} - 32)$$

Then eliminate the -32 on the right by adding 32 to both sides.

$$1.8\ °\mathrm{C} + 32 = (°\mathrm{F} - 32) + 32$$
$$1.8\ °\mathrm{C} + 32 = °\mathrm{F}$$

We are now left with an expression for °F.

In general solve equations by following these guidelines:

- Cancel numbers or symbols in the denominator (bottom part of a fraction) by multiplying by the number or symbol to be canceled.
- Cancel numbers or symbols in the numerator (upper part of a fraction) by dividing by the number or symbol to be canceled.
- Eliminate numbers or symbols that are added by subtracting the same number or symbol.
- Eliminate numbers or symbols that are subtracted by adding the same number or symbol.

- Whether you add, subtract, multiply, or divide, **always perform the same operation to both sides of a mathematical equation.**

For a final example, solve the following equation for x.

$$\frac{67x - y + 3}{6} = 2z$$

Cancel the 6 in the denominator by multiplying both sides by 6.

$$(\not{6})\frac{67x - y + 3}{\not{6}} = (6)2z$$

$$67x - y + 3 = 12z$$

Eliminate the +3 by subtracting 3 from both sides.

$$67x - y + 3 - 3 = 12z - 3$$
$$67x - y = 12z - 3$$

Eliminate the $-y$ by adding y to both sides.

$$67x - y + y = 12z - 3 + y$$
$$67x = 12z - 3 + y$$

Cancel the 67 by dividing both sides by 67.

$$\frac{\not{67}x}{\not{67}} = \frac{12z - 3 + y}{67}$$

$$x = \frac{12z - 3 + y}{67}$$

FOR PRACTICE **Using Algebra to Solve Equations**

Solve each of the following for the indicated variable.

a) $P_1V_1 = P_2V_2$; solve for V_2

b) $\frac{V_1}{T_1} = \frac{V_2}{T_2}$; solve for T_1

c) $PV = nRT$; solve for n

d) K = °C + 273; solve for °C

e) $\frac{3x + 7}{2} = y$; solve for x

f) $\frac{32}{y + 3} = 8$; solve for y

Answers:

a) $V_2 = \frac{P_1V_1}{P_2}$

b) $T_1 = \frac{V_1T_2}{V_2}$

c) $n = \frac{PV}{RT}$

d) °C = K − 273

e) $x = \frac{2y - 7}{3}$

f) $y = 1$

A3.2 Mathematical Operations with Scientific Notation

Writing numbers in scientific notation is covered in detail in section 2.2. Briefly, a number written in scientific notation consists of a **decimal part,** a number that is usually between 1 and 10, and an **exponential part,** 10 raised to an **exponent,** n.

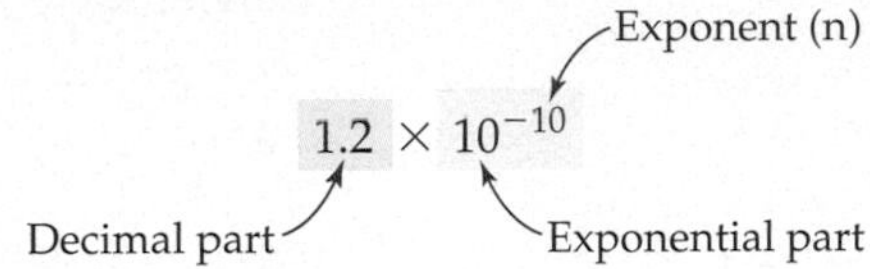

Each of the following numbers is written in both scientific and decimal notation.

$1.0 \times 10^5 = 100{,}000$ $\quad$ $1.0 \times 10^{-6} = 0.000001$

$6.7 \times 10^3 = 6700$ $\quad$ $6.7 \times 10^{-3} = 0.0067$

Multiplication and Division

To multiply numbers expressed in scientific notation, multiply the decimal parts and add the exponents.

$$(A \times 10^m)(B \times 10^n) = (A \times B) \times 10^{m+n}$$

To divide numbers expressed in scientific notation, divide the decimal parts and subtract the exponent in the denominator from the exponent in the numerator.

$$\frac{(A \times 10^m)}{(B \times 10^n)} = \left(\frac{A}{B}\right) \times 10^{m-n}$$

Consider the following example involving multiplication.

$$\begin{aligned}(3.5 \times 10^4)(1.8 \times 10^6) &= (3.5 \times 1.8) \times 10^{4+6} \\ &= 6.3 \times 10^{10}\end{aligned}$$

Consider the following example involving division.

$$\begin{aligned}\frac{(5.6 \times 10^7)}{(1.4 \times 10^3)} &= \left(\frac{5.6}{1.4}\right) \times 10^{7-3} \\ &= 4.0 \times 10^4\end{aligned}$$

Addition and Subtraction

To add or subtract numbers expressed in scientific notation, rewrite all the numbers so that they have the same exponent, then add or subtract the decimal parts of the numbers. The exponents remained unchanged.

$$\begin{array}{r} A \times 10^n \\ \pm B \times 10^n \\ \hline (A \pm B) \times 10^n \end{array}$$

Notice that the numbers *must have* the same exponent.

Consider the following example involving addition.

$$\begin{array}{r} 4.82 \times 10^{7} \\ +3.4 \times 10^{6} \\ \hline \end{array}$$

First, express both numbers with the same exponent. In this case, we rewrite the lower number and perform the addition as follows:

$$\begin{array}{r} 4.82 \times 10^{7} \\ +0.34 \times 10^{7} \\ \hline 5.16 \times 10^{7} \end{array}$$

Consider the following example involving subtraction.

$$\begin{array}{r} 7.33 \times 10^{5} \\ -1.9 \times 10^{4} \\ \hline \end{array}$$

First, express both numbers with the same exponent. In this case, we rewrite the lower number and perform the subtraction as follows:

$$\begin{array}{r} 7.33 \times 10^{5} \\ -0.19 \times 10^{5} \\ \hline 7.14 \times 10^{5} \end{array}$$

FOR PRACTICE **Mathematical Operations with Scientific Notation**

Perform each of the following operations.

a) $(2.1 \times 10^{7})(9.3 \times 10^{5})$

b) $(5.58 \times 10^{12})(7.84 \times 10^{-8})$

c) $\dfrac{(1.5 \times 10^{14})}{(5.9 \times 10^{8})}$

d) $\dfrac{(2.69 \times 10^{7})}{(8.44 \times 10^{11})}$

e) $\begin{array}{r} 1.823 \times 10^{9} \\ +1.11 \times 10^{7} \\ \hline \end{array}$

f) $\begin{array}{r} 3.32 \times 10^{-5} \\ +3.400 \times 10^{-7} \\ \hline \end{array}$

g) $\begin{array}{r} 6.893 \times 10^{9} \\ -2.44 \times 10^{8} \\ \hline \end{array}$

h) $\begin{array}{r} 1.74 \times 10^{4} \\ -2.9 \times 10^{3} \\ \hline \end{array}$

Answers:

a) 2.0×10^{13}
b) 4.37×10^{5}
c) 2.5×10^{5}
d) 3.19×10^{-5}
e) 1.834×10^{9}
f) 3.35×10^{-5}
g) 6.649×10^{9}
h) 1.45×10^{4}

A3.3 Logarithms

The log of a number is the exponent to which 10 must be raised to obtain that number. For example, the log of 100 is 2 because 10 must be raised to the 2nd power to get 100. Similarly, the log of 1000 is 3 because 10 must be raised to the 3rd power to get 1000. The logs of several multiples of 10 are shown below.

$\log 10 = 1$
$\log 100 = 2$
$\log 1{,}000 = 3$
$\log 10{,}000 = 4$

The log of a number smaller than one is negative because 10 must be raised to a negative exponent to get a number smaller than one. For example, the log of 0.01 is −2 because 10 must be raised to −2 to get 0.01. Similarly, the log of 0.001 is −3 because 10 must be raised to −3 to get 0.001. The logs of several fractional numbers are shown below.

$\log 0.1 = -1$
$\log 0.01 = -2$
$\log 0.001 = -3$
$\log 0.0001 = -4$

The logs of numbers that are not multiples of 10 can be computed on your calculator. See your calculator manual for specific instructions.

Inverse Logarithms

The inverse logarithm or invlog function (sometimes called antilog) is exactly the opposite of the log function. For example, the log of 100 is 2 and the inverse log of 2 is 100. The log function and the invlog function undo one another.

$\log 1000 = 3$
$\text{invlog } 3 = 1000$
$\text{invlog } (\log 1000) = 1000$

The inverse log of a number is simply 10 rasied to that number.

$\text{invlog } x = 10^x$
$\text{invlog } 3 = 10^3 = 1000$

The inverse logs of numbers can be computed on your calculator. See your calculator manual for specific instructions.

FOR PRACTICE **Logarithms and Inverse Logarithms**

Perform each of the following operations.

a) $\log 1.0 \times 10^5$
b) $\log 59$
c) $\log 1.0 \times 10^{-5}$
d) $\log 0.068$
e) $\text{invlog } 7.0$

f) invlog 1.44
g) invlog −6.0
h) invlog −0.250
i) invlog (log 88)

Answers:

a) 5.00
b) 1.77
c) −5.00
d) −1.17
e) 1×10^7
f) 28
g) 1×10^{-6}
h) 0.56
i) 88

APPENDIX 4

Glossary

absolute zero The coldest temperature possible. Absolute zero (0 K or −273 °C or −459 °F) is the temperature at which molecular motion stops. Lower temperatures do not exist.

acid A molecular compound that dissolves in solution to form H^+ ions. Acids have the ability to dissolve some metals and will turn litmus paper red.

acid rain Acid precipitation in the form of rain; created when fossil fuels are burned, releasing SO_2 and NO_2, which then react with water in the atmosphere to form sulfuric acid and nitric acid.

acid–base reaction A reaction that forms water and other products, usually including a salt.

acidic solution A solution containing a concentration of H_3O^+ ions greater than 1.0×10^{-7} (pH < 7).

activation energy Amount of energy that must be absorbed by reactants in their ground states to reach the transition state so that a reaction can occur; an energy hump that normally exists between the reactants and products.

activity series of metals A listing of metals (and hydrogen) in order of decreasing activity, decreasing ability to oxidize, and decreasing tendency to lose electrons.

actual yield Amount of product actually produced by a chemical reaction.

addition polymer A polymer formed by the addition of monomer units one to another without the elimination of any atoms.

alcohol Organic compound containing an -OH functional group bonded to a carbon atom and having the general formula *R*—OH.

aldehyde Organic compound with the general formula *R*CHO.

alkali metals The Group 1A elements, which are highly reactive metals.

alkaline battery A dry cell employing half-reactions that use a base. Allows longer battery life.

alkaline earth metals The Group 2A elements, which are fairly reactive metals.

alkaloids Organic bases found in plants.

alkanes Hydrocarbons in which all carbon atoms are connected by single bonds. Noncyclic alkanes have the general formula C_nH_{2n+2}.

alkene Hydrocarbon that contains at least one double bond between carbon atoms. Noncyclic alkenes have the general formula C_nH_{2n}.

alkyl group In an organic molecule, any group containing only singly bonded carbon atoms and hydrogen atoms.

alkyne Hydrocarbon that contains at least one triple bond between carbon atoms. Noncylclic alkynes have the general formula C_nH_{2n-2}.

alpha (α) radiation Occurs when an unstable nucleus emits an energetic particle composed of 2 protons and 2 neutrons.

alpha (α)-helix The most common secondary protein structure. The amino acid chain is wrapped into a tight coil in which the side chains extend outward. The structure is maintained by hydrogen bonding interactions between NH and CO groups along the peptide backbone of the protein.

alpha particle A form of radiation consisting of two protons and two neutrons and represented by the symbol ^4_2He.

amine Organic compound that contains nitrogen and has the general formula NR_3 where *R* may be an alkyl group or a hydrogen atom.

amino acid A molecule containing an amine group, a carboxylic acid group, and an *R* group (also called a side chain). Amino acids are the building blocks of proteins.

amorphous Type of solid matter where atoms or molecules do not have long-range order (e.g. glass and plastic).

amphoteric In Brønsted-Lowry terminology, a substance that can act as an acid or a base.

anion A negatively charged ion.

anode The electrode where oxidation occurs in an electrochemical cell.

aqueous solution A homogeneous mixture of a substance with water.

aromatic ring A ring of carbon atoms containing alternating single and double bonds. Another name for the benzene ring.

Arrhenius acid Substance that produces H^+ ions in aqueous solution.

Arrhenius base Substance that produces OH^- ions in aqueous solution.

Arrhenius, Svante Swedish chemist who proposed molecular definitions of acids and bases.

atmosphere (atm) The simplest unit of pressure; 1 atm is defined as the average pressure at sea level.

atom Smallest identifiable unit of an element.

atomic element An element that exists in nature with single atoms as the basic unit.

atomic mass A weighted average of the masses of each naturally occurring isotope of an element; atomic mass is the average mass of the atoms of an element.

atomic mass unit (amu) The common unit to express the masses of protons and neutrons. 1 amu = 1.66×10^{-24} g

atomic number (Z) The number of protons in the nucleus of an atom.

atomic size The atomic size of an atom is determined by how far the outermost electrons are from the nucleus.

atomic solid Solids whose composite units are individual atoms. (e.g. diamond, C; iron, Fe)

atomic theory States that all matter is composed of tiny particles called atoms.

Avogadro, Amadeo A professor of physics; the number of atoms in a mole (6.022×10^{23}) is called Avogadro's number in his honor.

Avogadro's law The volume (V) of a gas and the amount of the gas in moles (n) are directly proportional.

balanced equation A chemical equation in which the numbers of each type of atom on both sides of the equation are equal.

base A molecular compound that dissolves in solution to form OH^- ions. Bases have a slippery feel and turn litmus paper blue.

base chain The longest continuous chain of carbon atoms in an organic compound.

basic solution A solution containing a concentration of OH^- ions greater than 1.0×10^{-7}. (pH > 7).

Becquerel, Antoine-Henri French scientist who accidentally discovered radioactivity in 1896 while studying X-rays.

bent The molecular geometry in which 3 atoms are not in a straight line. This geometry occurs when the central atoms contain 4 electron groups (2 bonding and 2 nonbonding) or 3 electron groups (2 bonding and 1 nonbonding).

benzene (C_6H_6) A particularly stable organic compound with six carbon atoms and alternating single and double bonds in a ring structure.

beta (β) radiation Occurs when an unstable nucleus emits an electron.

beta (β)-pleated heet A common pattern in the secondary structure of proteins. The protein chain is extended (as opposed to coiled) and forms a zigzag pattern. The peptide backbones of neighboring chains interact with one another through hydrogen bonding to form zigzagged sheets.

beta particle A form of radiation consisting of an energetic electron and represented by the symbol ${}_{-1}^{0}e$.

binary acid An acid containing only hydrogen and a nonmetal.

binary compound A compound containing only two different kinds of elements.

biochemistry The study of the chemical substances and processes that occur in plants, animals, and microorganisms.

Bohr model A model for the atom in which electrons travel around the nucleus in circular orbits at specific, fixed distances from the nucleus.

boiling point The temperature at which the vapor pressure of a liquid is equal to the pressure above it.

boiling point elevation The increase in the boiling point of a solution caused by the presence of the solute.

bonding pair Electrons that are shared between two atoms in a chemical bond.

bonding theory A model that predicts how atoms bond together to form molecules.

Boyle's law The volume (V) of a gas and its pressure (P) are inversely proportional.

branched alkane An alkane composed of carbon atoms bonded in chains containing branches.

Brønsted-Lowry acid A proton (H^+ ion) donor.

Brønsted-Lowry base A proton (H^+ ion) acceptor.

buffer A solution that resists pH change by neutralizing added acid or added base.

Calorie (Cal) Energy unit equivalent to 1000 little *c* calories.

calorie (cal) The amount of energy required to raise the temperature of 1 g of water by 1 °C.

carbohydrate Polyhydroxyl aldehydes or ketones or their derivatives, containing multiple –OH groups and often having the general formula $(CH_2O)_n$.

carbonyl group A carbon atom double bonded to an oxygen atom.

carboxylic acid Organic compound with the general formula *R*COOH.

catalyst A substance that increases the rate of a chemical reaction but is not consumed by the reaction.

cathode The electrode where reduction occurs in an electrochemical cell.

cation A positively charged ion.

cell The smallest structural unit of living organisms that has the properties associated with life.

cell membrane The perimeter of the cell that holds the contents of the cell together.

cellulose A common polysaccharide composed of repeating glucose units linked together.

Celsius (°C) A temperature scale often used by scientists. On this scale, water freezes at 0 °C and boils at 100 °C. Room temperature is approximately 25 °C.

chain reaction A self sustaining chemical or nuclear reaction yielding energy or products that cause further reactions of the same kind.

charge A fundamental property of protons and electrons. Charged particles experience forces such that like charges repel and unlike charges attract.

Charles's law The volume (V) of a gas and its temperature (T) expressed in kelvin are directly proportional.

chemical bond Sharing or transfer of electrons to attain stable electron configurations among the bonding atoms.

chemical change A change in which matter changes its composition.

chemical energy Energy associated with chemical changes.

chemical equation Represents a chemical reaction; the reactants are on the left side of the equation and the products are on the right side.

chemical formula A way to represent a compound. At a minimum, the chemical formula indicates the elements present in the compound and the relative number of atoms of each element.

chemical properties Those that a substance can display only through changing its composition.

chemical reaction The process by which one or more substances transform into different substances via a chemical change. Chemical reactions often emit or absorb energy.

chemical symbol A one or two letter abbreviation for an element. Chemical symbols are listed directly below the atomic number on the periodic table.

chemistry The science that seeks to understand what matter does by studying what atoms and molecules do.

chromosome A biological structure containing genes and located within the nucleus of a cell.

codon A sequence of three bases that acts as a code for one amino acid.

colligative properties Physical properties of solutions that depend upon the number of solute particles present but not the type of solute particles.

collision theory Theory of reaction rates; collision theory states that effective collisions between reactant molecules must occur in order for the reaction to occur.

color change One evidence of a chemical reaction, involving the change in color of a substance before and after a reaction.

combined gas law Combines Boyle's law and Charles's law; it is used to calculate how a property of a gas (*P, V,* or *T*) changes when two other properties are changed at the same time.

combustion reaction A reaction in which a substance reacts with oxygen, emitting heat and forming one or more oxygen containing compounds.

complementary base In DNA, a base capable of precise pairing with only one other base.

complete ionic equation A chemical equation showing all the species as they are actually present in solution.

complex carbohydrate A carbohydrate composed of many repeating saccharide units.

compound A substance composed of two or more elements in fixed definite proportions.

compressible Property of a gas. Gases are compressible because, in the gas phase, atoms or molecules are separated by large distances.

concentrated solution A solution containing large amounts of solute.

condensation A physical change in which a substance is converted from its gaseous form to its liquid form.

condensation polymer A class of polymers that expel atoms, usually water, during their formation or polymerization.

condensed structural formula A shorthand way to write a structural formula.

conjugate acid–base pair In Brønsted-Lowry terminology, two substances related to each other by the transfer of a proton.

conservation of energy, law of States that energy can neither be created nor destroyed. The total energy is constant and cannot change; it can be transferred but not created.

conservation of mass, law of States that in a chemical reaction matter is neither created nor destroyed.

constant composition, law of States that all samples of a given compound have the same proportions of their constituent elements.

conversion factor Used to convert between two separate units; a conversion factor is constructed from any two quantities known to be equivalent.

copolymer Polymers that are composed of two different kinds of monomers and result in chains composed of alternating units rather than a single repeating unit.

core electrons The electrons that are not in the outermost principal shell of an atom.

corrosion The oxidation of metals (e.g. rusting of iron).

covalent atomic solid An atomic solid, such as diamond, which is held together by covalent bonds.

covalent bond The bond that results when two nonmetals combine in a chemical reaction. In a covalent bond, the atoms share their electrons.

critical mass The mass of uranium or plutonium required for a nuclear reaction to be self sustaining.

crystalline Type of solid matter with atoms or molecules arranged in well-ordered, three-dimensional array with long-range, repeating order. (e.g. salt and diamond)

Curie, Marie Sklodowska Polish-born French chemist who was a codiscoverer of radioactivity and also codiscovered elements (radium and polonium).

cytoplasm In a cell, the region between the nucleus and the cell membrane.

Dalton, John Developed the atomic theory, which explained the law of conservation of mass, as well as other laws and observations of the time by proposing that all matter was composed of small, indestructible particles called atoms.

Dalton's law of partial pressure The sum of the partial pressures of each of the components in a gas mixture equals the total pressure.

daughter nuclide The nuclide product of a nuclear decay.

decimal part One part of a number expressed in scientific notation.

decomposition A reaction in which a complex substance decomposes to form simpler substances; $AB \rightarrow A + B$.

Democritus Greek philosopher who theorized that matter was ultimately composed of small, indivisible particles he called *atomos* or atoms.

density (*d*) A fundamental property of materials that differs from one substance to another. The units of density are those of mass divided by those of volume, and most commonly expressed in g/cm^3, g/mL, or g/L.

derived unit A unit formed from other units.

dilute solution A solution containing small amounts of solute.

dimer A molecule formed by the joining together of two smaller molecules.

dipeptide Two amino acids linked together via a peptide bond.

dipole moment A measure of the separation of charge in a bond or in a molecule.

diprotic acid An acid containing two ionizable protons.

disaccharide A carbohydrate that can be decomposed into two simpler carbohydrates.

dispersion force The default intermolecular force present in all molecules and atoms. Dispersion forces are caused by fluctuations in the electron distribution within molecules or atoms.

displacement A reaction in which one element displaces another in a compound, $A + BC \rightarrow AC + B$.

dissociation In aqueous solution, the process in which a solid ionic compound separates into its ions.

disubstituted benzene A benzene in which two hydrogen atoms have been substituted with an atom or group of atoms.

DNA (deoxyribonucleic acid) Long chainlike molecules that occur in the nucleus of cells and act as blueprints for the construction of proteins.

dot structure A drawing that represents the valence electrons in atoms as dots; represents a chemical bond as the sharing or transfer of electron dots.

double bond When two electron pairs are shared between two atoms. In general, double bonds are shorter and stronger than single bonds.

double displacement A reaction in which two elements or groups of elements in two different compounds exchange places to form two new compounds, $AB + CD \rightarrow AD + CB$.

dry cell Ordinary battery (voltaic cell); does not contain large amounts of liquid water.

duet The name for the two electrons corresponding to a stable Lewis structure in hydrogen and helium.

dynamic equilibrium In a chemical reaction, the condition in which the rate of the forward reaction equals the rate of the reverse reaction.

electrical current The flow of electronic charge. Examples are electrons flowing through a wire or ions flowing through a solution.

electrical energy Energy associated with the flow of electrical charge.

electrochemical cell A device that creates electrical current from a redox reaction.

electrolysis Process in which electrical current is used to drive an otherwise nonspontaneous redox reaction.

electrolytic cell An electrochemical cell used for electrolysis.

electromagnetic radiation Type of energy that travels through space at a constant speed of 3.0×10^8 m/s (186,000 miles/s) and exhibits both wave-like and particle-like behavior. Light is a form of electromagnetic radiation.

electromagnetic spectrum A spectrum which includes all wavelengths of electromagnetic radiation.

electron Negatively charged particle that occupies most of the atom's volume but contributes almost none of its mass.

electron configuration Shows the occupation of orbitals by electrons for a particular element.

electron geometry The geometrical arrangement of the electron groups in a molecule.

electron group A general term for a lone pair, single bond, or multiple bond in a molecule.

electron spin A fundamental property of all electrons that causes them to have magnetic fields associated with them. The spin of an electron can either be oriented up $(+1/2)$ or down $(-1/2)$.

electronegativity The ability of an element to attract electrons within a covalent bond.

element A substance that cannot be broken down into simpler substances.

emission spectrum Spectrum associated with emission of electromagnetic radiation by elements or compounds.

empirical formula A formula for a compound that gives the smallest whole number ratio of each type of atom.

empirical formula molar mass Sum of the molar masses of all the atoms in an empirical formula.

endothermic A process that absorbs heat energy.

endothermic reaction A chemical reaction that absorbs energy from the surroundings.

endpoint The point in a reaction at which the reactants are in exact stoichiometric proportions.

energy The capacity to do work.

English system Common unit system used in the United States.

enzymes Biological catalysts that increase the rates of biochemical reactions; enzymes are abundant in living organisms.

equilibrium constant (K_{eq}) The ratio, at equilibrium, of the concentrations of the products raised to their stoichiometric coefficients divided by the concentrations of the reactants raised to their stoichiometric coefficients.

equivalent The stoichiometric proportions of elements and compounds in a chemical equation.

ester Organic compound with the general formula *RCOOR*.

ester linkage The type of bond that joins glycerol to fatty acids with the general formula —COO—.

ether Organic compound with the general formula *ROR*.

evaporation Process in which molecules in a liquid, undergoing constant random motion, acquire enough energy to overcome attractions with neighbors and fly into the gas phase.

excited state An unstable state for an atom or molecule in which energy has been absorbed but not reemitted. An electron has moved from ground state into a higher energy orbital.

exothermic Describes processes that release heat energy.

exothermic reaction A chemical reaction that gives off energy to the surroundings.

experiment A test that looks for other observable predictions of a theory and can also be used to validate theories.

exponent A number that represents the number of times a term is multiplied by itself. For example, in 2^4 the exponent is 4 and represents $2 \times 2 \times 2 \times 2$.

exponential part One part of a number expressed in scientific notation; it represents the number of places the decimal point has moved.

Fahrenheit (°F) The temperature scale that is most familiar in the United States; water freezes at 32 °F and boils at 212 °F.

family (of elements) Elements that have similar outer electron configurations and therefore similar properties. Families occur in vertical columns in the periodic table.

family (of organic compounds) A group of organic compounds with the same functional group.

fatty acid A type of lipid; a fatty acid is a carboxylic acid with a long hydrocarbon tail.

Fermi, Enrico Italian physicist who played an important role in the development of nuclear fission. In collaboration with Szilard, he constructed the first nuclear reactor.

film badge dosimeter Badges used to measure radiation exposure that consist of photographic film held in a small case that is pinned to clothing.

fission, nuclear The process in which a heavy nucleus is split into nuclei of smaller masses and energy is emitted.

formula mass Average mass of the molecules (or formula units) that compose a compound.

formula unit Basic unit of ionic compounds; formula units are the smallest electrically neutral collection of cations and anions that compose the compound.

freezing point depression The decrease in the freezing point of a solvent caused by the presence of a solute.

frequency Number of wave cycles or crests that pass through a stationary point in one second.

fuel cell Voltaic cell in which the reactants are constantly replenished.

functional group A characteristic set of atoms that characterize a family of organic compounds.

fusion, nuclear The combination of light atomic nuclei to form heavier ones; fusion emits large amounts of energy.

galvanic (voltaic) cell An electrochemical cell that spontaneously produces electrical current.

gamma radiation High energy, short wavelength electromagnetic radiation.

gamma rays The shortest wavelength and most energetic form of electromagnetic radiation. Each ray consists of an energetic photon emitted by an atomic nucleus and represented by the symbol ${}^{0}_{0}\gamma$.

gas In gaseous matter, atoms or molecules are separated by large distances and are free to move relative to one another.

gas evolution reaction Reaction that occurs in liquids and forms a gas as one of the products.

gas formation One evidence of a chemical reaction is that agas forms when two substances are mixed together.

Geiger-Müller counter A radioactivity detector consisting of a chamber filled with argon gas that discharges electrical signals when high energy particles pass through it.

gene A sequence of codons within a DNA molecule that codes for a single protein. Genes vary in length from hundreds to thousands of codons.

genetic material The inheritable blueprint for making organisms.

glycogen A type of polysaccharide; has a structure similar to starch, but the chain is highly branched.

glycolipid A biological molecule composed of a nonpolar fatty acid and a hydrocarbon chain and a polar section composed of a sugar molecule such as glucose.

glycoside linkage The link between monosaccharides in a polysaccharide.

ground state The state of an atom or molecule in which the electrons occupy the lowest possible energy orbitals available

group (of elements) Elements that have similar outer electron configurations and therefore similar properties. Groups occur in vertical columns in the periodic table.

Hahn, Otto German chemist credited as one of the discoverers of nuclear fission.

half-cell Compartment in which the oxidation or reduction half-reaction occurs in a galvanic or voltaic cell.

half-life The time it takes for one-half of the parent nuclides in a radioactive sample to decay to the daughter nuclides.

half-reaction Either the oxidation part or the reduction part of a redox reaction.

halogens The Group 7A elements, which are very reactive nonmetals.

heat absorption One evidence of a chemical reaction, involving the intake of energy.

heat capacity The quantity of heat energy required to change the temperature of a given amount of a substance by 1 °C.

heat emission One evidence of a chemical reaction, involving the evolution of thermal energy.

heat of fusion The amount of heat required to melt one mole of a solid at its melting point with no change in temperature.

heat of vaporization The amount of heat required to vaporize one mole of liquid at its boiling point with no change in temperature.

heterogeneous mixture A mixture, such as oil and water, that has two or more regions with different compositions.

homogeneous mixture A mixture, such as salt water, that has the same composition throughout.

human genome All of the genetic material of a human being; the total DNA of a human cell.

Hund's rule States that when filling orbitals of equal energy, electrons will occupy empty orbitals singly before pairing with other electrons.

hydrocarbon A compound that contains only carbon and hydrogen atoms.

hydrogen bond A strong dipole–dipole interaction between molecules containing hydrogen directly bonded to a small, highly electronegative atom, such as N, O, or F.

hydrogenation The chemical addition of hydrogen to a compound.

hydronium ion The H_3O^+ ion. Chemists often use $H^+(aq)$ and $H_3O^+(aq)$ interchangeably to mean the same thing—a hydronium ion.

hypothesis A theory before it has become well established. A tentative explanation for an observation or scientific problem that can be tested by further investigation.

hypoxia A shortage of oxygen in the tissues of the body.

ideal gas law Combines the four properties of a gas—pressure *(P)*, volume *(V)*, temperature *(T)*, and number of moles *(n)* in a single equation showing their interrelatedness. $PV = nRT$ (R = ideal gas constant).

indicator A substance that changes color with acidity level, often used to detect the endpoint of a titration.

infrared (IR) light The fraction of the electromagnetic spectrum that lies between visible light and microwaves. Infrared light is invisible to the human eye.

insoluble Does not dissolve in water.

instantaneous dipole A type of intermolecular force resulting from transient shifts in electron density within an atom or molecule.

intermolecular forces Attractive forces that exist between molecules.

International System (of Units) The most convenient system for science measurements; it is based on the metric system. The set of standard units agreed upon by scientists throughout the world.

ion A charged particle.

ion product constant (K_w) The product of the H_3O^+ ion concentration and the OH^- ion concentration in an aqueous solution. At room temperature, $K_w = 1.0 \times 10^{-14}$.

ionic bond The bond that results when a metal and a nonmetal combine in a chemical reaction. In an ionic bond, the metal transfers one or more electrons to the nonmetal.

ionic compound A compound formed between a metal and one or more nonmetals.

ionic solid Solid compounds composed of metals and nonmetals joined by ionic bonds.

ionization The forming of ions.

ionization energy The energy required to remove an electron from an atom in the gaseous state.

ionizing power The ability of radiation to ionize other molecules and atoms.

isomers Molecules with the same molecular formula but different structures.

isoosmotic Solutions having osmotic pressures equal to bodily fluids.

isotope scanning The use of radioactive isotopes to identify disease in the body.

isotopes Atoms with the same number of protons but different numbers of neutrons.

Joule, James English scientist who demonstrated that energy could be converted from one type to another as long as the total energy was conserved. The SI unit of energy is the joule (J), named after James Joule.

Kekulé Fredrich August German chemist who played a central role in elucidating many early organic chemical structures, especially benzene.

Kelvin (K) The temperature scale that avoids negative temperatures by assigning 0 K to the coldest temperature possible, absolute zero. Absolute zero (−273 °C or −459 °F) is the temperature at which molecular motion stops. The size of the kelvin degree is identical to the Celsius degree.

ketone Organic compound with the general formula *RCOR*.

kilogram (kg) The SI standard unit of mass.

kilowatt-hour (kWh) A unit of energy equal to 3.6 million Joules.

kinetic energy Energy associated with the motion of an object.

kinetic molecular theory A simple model for gases that predicts the correct behavior for most gases under many conditions.

Lavoisier, Antoine French chemist who studied combustion and made careful measurements on mass changes during chemical reactions. Lavioisier discovered the law of conservation of mass.

Le Châtelier's principle States that when a chemical system at equilibrium is disturbed, the system shifts in a direction that minimizes the disturbance.

lead-acid storage battery Automobile batteries consisting of six electrochemical cells wired in series. Each cell produces 2 volts for a total of 12 volts.

Lewis structure A drawing that represents chemical bonds between atoms as shared or transferred electrons; the valence electrons of atoms are represented as dots.

Lewis theory A simple theory for chemical bonding involving diagrams with bonds between atoms as lines or dots. In this theory, atoms bond together to obtain octets (8 valence electrons).

Lewis, G.N. An American chemist who developed a simple theory for chemical bonding, now called Lewis theory.

light emission One evidence of a chemical reaction involving the giving off of electromagnetic radiation.

limiting reactant The reactant that limits the amount of product formed in a chemical reaction.

linear The molecular geometry of a molecule containing 2 electron groups (2 bonding groups, and no lone pairs).

linearly related When two variables are plotted one against the other and the graph produces a straight line.

lipid Cellular component that is insoluble in water, but extractable with nonpolar solvents.

lipid bilayer Structure formed by lipids in the cell membrane.

liquid In liquid matter, atoms or molecules are packed close to each other (about as closely as in a solid) but are free to move around and by each other.

logarithmic scale A scale involving logarithms. A logarithm entails an exponent that indicates the power to which a number is raised to produce a given number. (e.g. the logarithm of 100 to the base 10 is 2)

lone pair Electrons that are only on one atom in a Lewis structure.

main-group elements Groups 1A–8A on the periodic table. These groups have properties that tend to be predictable based on their position in the periodic table.

Manhattan Project Top-secret U.S. endeavor whose main goal was to build an atomic bomb before the Germans did. The project was led by physicist J.R. Oppenheimer and culminated on July 16, 1945, when the world's first nuclear weapon was successfully detonated.

mass The mass of an object is a measure of the quantity of matter within it.

mass number (A) The sum of the number of neutrons and protons in an atom.

mass percent composition (or mass percent) The percentage, by mass, of each element in a compound.

matter Anything that occupies space and has mass. Matter exists in three different states: solid, liquid, and gas.

Meitner, Lise German chemist credited as one of the discoverers of nuclear fission.

melting point The temperature at which a solid turns into a liquid.

Mendeleev, Dmitri A Russian chemistry professor who studied the known elements and developed the periodic law. Mendeleev then organized the known elements in a table, which is the foundation of today's periodic table.

metallic atomic solid An atomic solid, such as iron, which is held together by metallic bonds that, in the simplest model, consist of positively charged ions in a sea of electrons.

metallic character Elements become more metallic as you move from right to left on the periodic table. Metallic elements tend to lose electrons in chemical reactions.

metalloids Those elements that fall in the middle right side of the periodic table and have properties between metals and nonmetals.

metals Those elements that tend toward the bottom left side of the periodic table and tend to lose electrons in chemical reactions.

meter (m) The SI standard unit of length.

metric system Common unit system used throughout most of the world.

microwaves The part of the electromagnetic spectrum that is between the infrared region and the radio wave region. Microwaves are efficiently absorbed by water molecules and can therefore be used to heat water-containing substances.

millimeter of mercury (mm Hg) A unit of pressure that originates from the method used to measure pressure with a barometer. Also called a *torr.*

miscibility The ability for two liquids to mix without separating into two phases, or the ability of one liquid to mix with (dissolve in) another liquid.

mixture A substance composed of two or more different types of atoms or molecules combined in variable proportions.

molality (m) A common unit of concentration of a solution defined as number of moles of solute per kilogram of solvent.

molar mass The mass of one mole of atoms of an element. An element's molar mass in grams per mole is numerically equivalent to the element's atomic mass in amu.

molar solubility The solubility of a substance in units of moles per liter (mol/L).

molar volume The volume occupied by one mole of gas under standard temperature and pressure conditions (22.4 L).

molarity (M) A common unit of the concentration of a solution defined as moles of solute per liter of solution.

mole A large number, 6.022×10^{23}, used when dealing with small particles such as atoms or molecules. A mole of any element has a mass in grams that is numerically equivalent to its atomic mass in amu.

molecular compound A compound formed between two or more nonmetals. Molecular compounds have distinct molecules as their simplest identifiable units.

molecular element An element that does not normally exist in nature with single atoms as the basic unit. Instead, these elements usually exist as diatomic molecules—two atoms of that element bonded together—as their basic units.

molecular equation A chemical equation showing the complete, neutral formulas for every compound in a reaction.

molecular formula A formula for a compound that gives the specific number of each type of atom in a molecule.

molecular geometry The geometrical arrangement of the atoms in a molecule.

molecular solid Solids whose composite units are molecules.

molecule Two or more atoms grouped in well-defined clusters. A molecule is the smallest identifiable unit of a molecular compound.

monomer An individual repeating unit that makes up a polymer.

monoprotic acid An acid containing only one ionizable proton.

monosaccharide A carbohydrate that cannot be decomposed into simpler carbohydrates.

monosubstituted benzene A benzene in which one of the hydrogen atoms has been substituted with another atom or group of atoms.

m-RNA (ribonucleic acid) Long chainlike molecules that occur throughout cells and act as blueprints for the construction of proteins.

net ionic equation An equation that shows only the species that actually change during a reaction.

neutral solution A solution in which the concentrations of H_3O^+ and OH^- are equal (pH = 7).

neutralization When an acid and a base are mixed, the $H^+_{(aq)}$ from the acid combines with the $OH^-_{(aq)}$ from the base to form $H_2O_{(l)}$.

neutron A nuclear particle with no electrical charge and nearly the same mass as a proton.

nitrogen narcosis An increase in nitrogen concentration in bodily tissues and fluids that results in feelings of drunkeness.

noble gases The Group 8A elements, which are chemically inert gases.

nonbonding atomic solid An atomic solid, such as solid xenon, which is held together by relatively weak dispersion forces.

nonelectrolyte solution A solution containing a solute that dissolves as molecules; therefore, the solution does not conduct electricity.

nonmetals Those elements that tend toward the upper right side of the periodic table and tend to gain electrons in chemical reactions.

nonpolar molecule A molecule that does not have a dipole moment (charge separation).

nonvolatile A compound is said to be nonvolatile if it does not vaporize easily.

normal boiling point The boiling point of a liquid at a pressure of 1 atmosphere.

normal alkane (or n-alkane) Alkane composed of carbon atoms bonded in a straight chain with no branches.

nuclear equation Represents the changes that occur during radioactivity and other nuclear processes.

nuclear radiation The energetic particles emitted from the nucleus of an atom when it is undergoing a nuclear process.

nuclear theory of the atom States that most of the atom's mass and all of its positive charge is contained in the nucleus. Most of the volume of the atom is empty space occupied by negatively charged electrons.

nucleic acid Biological molecules, such as deoxyribonucleic acid (DNA) and ribonucleic acid (RNA), that act as the templates from which proteins are made.

nucleotide An individual unit composing nucleic acids.

nucleus (of a cell) The part of the cell that contains the genetic material.

nucleus (of an atom) Small core containing most of the atom's mass and all of its positive charge. The nucleus is made up of protons and neutrons.

observation The first step in the scientific method. An observation must measure or describe something about the physical or natural world.

octet The number of electrons, eight, around atoms with stable Lewis structures.

octet rule States that an atom will give up, accept, or share electrons in order to achieve a filled outer electron shell, which usually consists of 8 electrons.

Oppenheimer, J.R. Director of the Manhattan Project, the U.S. endeavor to build the world's first nuclear weapon.

orbital diagram An electron configuration in which electrons are represented as arrows in boxes corresponding to orbitals of a particular atom.

orbital Region around the nucleus of an atom where an electron is most likely to be found.

organic chemistry The study of carbon-containing compounds and their reactions.

organic molecule Molecule containing carbon and hydrogen and possibly nitrogen, oxygen, or sulfur.

osmosis The flow of solvent from a lower concentration solution through a semipermeable membrane to a higher concentration solution.

osmotic pressure The pressure produced on the surface of a semipermeable membrane by osmosis or the pressure required to stop osmotic flow.

oxidation The gain of oxygen, the loss of hydrogen, or the loss of electrons (the most fundamental definition).

oxidation state (or oxidation number) Number that can be used as a mechanical aid in writing formulas and balancing equations. It is computed for each element based on the number of electrons assigned to it in a scheme where the most electronegative element is assigned all of the bonding electrons.

oxidation–reduction (redox) reaction Reaction in which electrons are transferred from one substance to another.

oxidizing agent In a redox reaction, the substance being reduced. Oxidizing agents tend to gain electrons easily.

oxyacid An acid containing hydrogen, a nonmetal, and oxygen.

oxyanion An anion containing oxygen. Most polyatomic ions are oxyanions.

oxygen toxicity The result of increased oxygen concentration in bodily tissues.

parent nuclide The original nuclide in a nuclear decay.

partial pressure The pressure due to any individual component in a gas mixture.

pascal (Pa) The SI unit of pressure, defined as 1 newton per square meter.

Pauli exclusion principle States that no more than two electrons can occupy an orbital and that the two electrons must have opposite spins.

penetrating power The ability of a radioactive particle to penetrate matter.

peptide bond The bond between the amine end of one amino acid and the carboxylic acid end of another amino acid. Amino acids link together via peptide bonds to form proteins.

percent natural abundance The percentage amount of each isotope of an element in a naturally occurring sample of the element.

percent yield In a chemical reaction, the percentage of the theoretical yield that was actually attained.

period A horizontal row of the periodic table.

periodic law States that when the elements are arranged in order of increasing relative mass, certain sets of properties recur periodically.

periodic table An arrangement of the elements in which atomic number increases from left to right and elements with similar properties appear in columns called families or groups.

permanent dipole A separation of charge resulting from the unequal sharing of electrons between atoms.

perpetual motion machine A device that continuously produces energy without the need for energy input.

pH scale A scale used to quantify acidity or basicity. A pH of 7 is neutral; a pH lower than 7 is acidic; and a pH greater than 7 is basic. The pH is defined as follows: $pH = -\log [H_3O^+]$

phenyl group The term for a benzene ring when other substituents are attached to it.

phospholipid A lipid with the same basic structure as a triglyceride, except that one of the fatty acid groups is replaced with a phosphate group

phosphorescence The slow, long-lived emission of light that sometimes follows the absorption of light by some atoms and molecules.

photon A particle of light or a packet of light energy.

physical change A change in which matter does not change its composition even though its appearance might change.

physical properties Those properties that a substance displays without changing its composition.

polar covalent bond A covalent bond between atoms of different electronegativities. Polar covalent bonds have a dipole moment.

polar molecule A molecule with polar bonds that add together—they do not cancel each other—to form a net dipole moment.

polyatomic ion An ion that is composed of a group of atoms with an overall charge.

polymer A molecule with many similar units, called monomers, bounded together in a long chain.

polypeptide A short chain of amino acids joined by peptide bonds.

polysaccharide Long, chainlike molecule composed of many adjacent monosaccharide units. Polysaccharides are a type of polymer.

positron A nuclear particle which has the mass of an electron, but carries a +1 charge.

positron emission Occurs when an unstable atomic nucleus emits a positron. In positron emission, a proton is transformed into a neutron.

potential energy The energy of a body that is associated with its position.

precipitate An insoluble product formed through the reaction of two solutions containing soluble compounds.

precipitation reaction A reaction that forms a solid or *precipitate* upon mixing two aqueous solutions.

prefix multipliers Used by the SI system with the standard units. These multipliers change the value of the unit by powers of 10.

pressure The push (or force) exerted per unit area by gaseous molecules as they collide with the surfaces around them.

primary protein structure The sequence of amino acids in a protein's chain. Primary protein structure is maintained by the covalent peptide bonds between individual amino acids.

principal quantum number Indicates the shell that an electron occupies.

principal shell The shell indicated by the principal quantum number.

products The final substances produced in a chemical reaction; represented on the right side of a chemical equation.

proof For alcoholic beverages, the proof is twice the percent alcohol content by volume. For example, an eighty proof liquor contains 40% alcohol.

properties The characteristics we use to distinguish one substance from another.

protein A biological molecule composed of a long chain of amino acids joined by peptide bonds. In living organisms, proteins serve many varied and important functions.

proton Positively charged particle inside the nucleus of an atom. A proton's mass is 1 amu.

Proust, Joseph A chemist who formally stated the idea that elements combine in fixed proportions to form compounds and established the law of constant composition.

pure substance A substance composed of only one type of atom or molecule.

quantification The assigning of a number to an observation; allows an individual to specify a quantity or property precisely.

quantum (*pl.* **quanta)** Precise amount of energy contained in a photon; the difference in energy between two atomic orbitals.

quantum number (*n*) An integer that specifies the energy of an orbital. The higher the quantum number n is, the greater the distance between the electron and the nucleus and the higher its energy.

quantum mechanical model The foundation of modern chemistry; explains how electrons exist in atoms and how those electrons affect the chemical and physical properties of elements.

quarternary structure In a protein, the way that individual chains fit together to compose the protein. Quaternary structure is maintained by interactions between amino acids on the different chains.

radio waves The longest wavelength and least energetic form of electromagnetic radiation. Radio waves are used extensively as a medium for communication and broadcasting.

radioactive A substance that emits tiny, invisible, energetic particles from changes in the nuclei of its composite atoms.

radioactivity The emission of tiny, invisible particles from the unstable nuclei of atoms. Many of these particles can penetrate matter.

radiocarbon dating A technique used to estimate the age of fossils and artifacts through the presence of natural radioactivity in our environment.

radiotherapy Treatment of disease with radiation, such as using gamma rays to kill rapidly dividing cancer cells.

random coil The name given to an irregular pattern of a secondary protein structure.

rate of a chemical reaction (reaction rate) The amount of reactant that changes to product in a given period of time. Also defined as the amount of a product that forms in a given period of time.

reactants The initial substances in a chemical reaction, represented on the left side of a chemical equation.

recrystallization A technique to purify a solid; involves dissolving the solid in a solvent at high temperature, creating a saturated solution, then cooling the solution to cause the crystallization of the solid.

reducing agent In a redox reaction, the substance being oxidized. Reducing agents tend to lose electrons easily.

reduction The loss of oxygen, the gain of hydrogen, or the gain of electrons (the most fundamental definition).

rem Stands for *roentgen equivalent man;* a weighted measure of radiation exposure that accounts for the ionizing power of the different types of radiation.

resonance structures Two or more Lewis structures that are necessary to describe the bonding in a molecule or ion.

reversible A reaction that is able to proceed in both the forward and reverse directions.

RNA (ribonucleic acid) Long chainlike molecules that occur throughout cells and act as blueprints for the construction of proteins.

Rutherford, Ernest Physicist who explained that the mass and positive charge of an atom must all be concentrated in a small space called the nucleus; discovered the nuclear theory of the atom.

salt An ionic compound that usually remains dissolved in a solution after an acid–base reaction has occurred.

salt bridge An inverted, U-shaped, tube that contains a strong electrolyte. Used to complete the circuit in an electrochemical cell, allowing the flow of ions between the two half-cells.

saturated fat A triglyceride composed of saturated fatty acids. Saturated fat tends to be solid at room temperature.

saturated hydrocarbon A hydrocarbon that contains no double or triple bonds between the carbon atoms.

saturated solution A solution that holds the maximum amount of solute under the solution conditions. If additional solute is added to a saturated solution, it will not dissolve.

scientific law A statement that summarizes past observations and predicts future ones. Scientific laws are usually formulated from a series of related observations.

scientific method The way that scientists learn about the natural world. The scientific method consists of observation, law, theory, and experimentation.

scientific notation Used to write very big or very small numbers, often containing many zeros, more compactly. A number written in scientific notation consists of a decimal part, an exponential part, and an exponent.

scintillation counter A device used to detect radioactivity in which radioactive particles pass through a material that emits ultraviolet or visible light in response to excitation by the radioactive particles. The light is detected and turned into an electrical signal.

second (s) The SI standard unit of time.

secondary protein structure Short-range periodic or repeating patterns often found along protein chains. Secondary protein structure is maintained by interactions between amino acids that are fairly close together in the linear sequence of the protein chain or adjacent to each other on neighboring chains.

semiconductor Compound or element that possesses intermediate electrical conductivity that can be changed and controlled.

semipermeable membrane A membrane that selectively allows some substances to pass through but not others.

SI units The most convenient system of units for science measurements; it is based on the metric system. The set of standard units agreed upon by scientists throughout the world.

Significant digits (figures) The non-place-holding digits in a reported measurement; they represent the precision of a measured quantity.

simple carbohydrate A monosaccharide or disaccharide.

simple sugar A monosaccharide or disaccharide.

single bond A chemical bond in which one electron pair is shared between two atoms.

solid Phase of matter in which atoms or molecules pack close to each other in fixed locations.

solid formation One evidence of a chemical reaction involving the formation of a solid.

solubility The amount of a compound, usually in grams, that will dissolve in a certain amount of solvent.

solubility rules A set of empirical rules used to determine whether or not an ionic compound is soluble.

solubility-product constant (K_{sp}) The equilibrium expression for a chemical equation that represents the dissolving of an ionic compound in solution.

soluble Dissolves in solution.

solute The minority component of a solution.

solution A homogenous mixture of two or more substances.

solution map A visual outline employed in this book that shows the strategic route required to solve a problem.

solvent The majority component of a solution.

specific heat capacity (or specific heat) The heat capacity of a substance expressed in joules per gram degree celsius (J/g °C).

spectator ions Ions that do not participate in a reaction; they appear unchanged on both sides of a chemical equation.

standard temperature and pressure (STP) Conditions often assumed in calculations involving gases [$T = 0$ °C (273 K) and $P = 1$ atm].

starch A common polysaccharide composed of repeating glucose units.

states of matter The three forms in which matter can exist: solid, liquid, and gas.

steroid A biological compound containing a 17-carbon 4-ring system.

stock solution A concentrated form in which solutions are often stored.

stoichiometry The numerical relationship between chemical quantities in a balanced chemical equation. Stoichiometry allows us to predict the amounts of products that form in a chemical reaction based on the amounts of reactants.

Strassmann, Fritz Chemist credited as one of the discoverers of nuclear fission.

strong acid An acid that completely ionizes in solution.

strong base A base that completely dissociates in solution.

strong electrolyte A substance whose aqueous solutions are good conductors of electricity.

strong electrolyte solution A solution containing a solute that dissociates into ions; therefore, the solution conducts electricity well.

structural formula A two-dimensional representation of molecules that not only shows the number and type of atoms, but also how the atoms are bonded together.

sublimation A physical change in which a substance is converted from its solid form directly into its gaseous form.

subshell In quantum mechanics, specifies the shape of the orbital and is represented by different letters (s, p, d, f).

substituent An atom or group of atoms that has been substituted for a hydrogen atom in an organic compound.

substitution reaction A reaction in which one or more atoms are replaced by one or more different atoms.

supersaturated solution A solution holding more than the maximum amount of solute. The solute will normally precipitate from (or come out of) a supersaturated solution.

surface tension The tendency of liquids to minimize their surface area that results in a "skin" on the surface of the liquid.

synthesis A reaction in which simpler substances combine to form more complex substances, $A + B \rightarrow AB$.

temporary dipole A type of intermolecular force resulting from transient shifts in electron density within an atom or molecule.

terminal atom An atom that is located at the end of a molecule or chain.

tertiary structure A protein's structure that consists of the large-scale bends and folds due to interactions between the R groups of amino acids that are separated by large distances in the linear sequence of the protein chain.

tetrahedral The molecular geometry of a molecule containing 4 electron groups (4 bonding groups, and no lone pairs).

theoretical yield The amount of product that can be made in a chemical reaction based on the amount of limiting reactant.

theory A theory describes the underlying reasons for observations and laws. It is a model for the way nature is and predicts behavior that extends well beyond the observations and laws from which it was formed.

Thomson, J.J. English physicist who discovered the electron. Thomson discovered that electrons were negatively charged, that they were much smaller and lighter than atoms, and that they were uniformly present in many different kinds of substances.

titration Laboratory procedure used to determine the amount of a substance in solution. In a titration, a reactant in a solution of known concentration is reacted with another reactant in a solution of unknown concentration until the reaction reaches the endpoint.

torr A unit of pressure named after Evangelista Torricelli. Also called a *millimeter of mercury*.

Torricelli, Evangelista Italian physicist who invented the barometer and has the pressure unit (torr) named in his honor.

transition metals The elements in the middle of the periodic table whose properties tend to be less predictable based simply on their position in the periodic table. Transition metals lose electrons in their chemical reactions, but do not necessarily acquire noble gas configurations.

triglyceride A fat or oil; a tryglyceride is a tri-ester composed of glycerol with three fatty acids attached.

trigonal planar The molecular geometry of a molecule containing 3 electron groups, 3 bonding groups, and no lone pairs.

trigonal pyramidal The molecular geometry of a molecule containing 4 electron groups, 3 bonding groups, and 1 lone pair.

triple bond A chemical bond consisting of three electron pairs shared between two atoms. In general, triple bonds are shorter and stronger than double bonds.

Type I compound A compound where the charge is implied for the metal; they contain metals that always form cations with the same charge.

Type II compound A compound where the charge is not implied for the metal; they contain metals that forms cations with different charges.

ultraviolet (UV) light The fraction of the electromagnetic spectrum that is between the visible region and the X-ray region. UV light is invisible to the human eye.

units Previously agreed upon quantities used to report experimental measurements. Units are vital in chemistry.

unsaturated fat (or oil) Name for triglyceride when the fatty acids that compose it are unsaturated. Unsaturated fats tend to be liquids at room temperature.

unsaturated hydrocarbon A hydrocarbon that contains one or more double or triple bonds between its carbon atoms.

unsaturated solution A solution holding less than the maximum amount of solute under the solution conditions. If additional solute is added to an unsaturated solution, it will dissolve.

valence electrons The electrons in the outermost principal shell of an atom; they are involved in chemical bonding.

valence shell electron pair repulsion (VSEPR) A theory that allows the prediction of the shapes of molecules based on the idea that electrons—either as lone pairs or as bonding pairs—repel one another.

vapor pressure The partial pressure of a vapor in dynamic equilibrium with its liquid.

vaporization The phase transition between a liquid and a gas.

viscosity A manifestation of intermolecular forces; viscosity is the resistance of a liquid to flow.

visible light The fraction of the electromagnetic spectrum that is visible to the human eye. The visible region is bracketed by wavelengths of 400 nm (violet) and 780 nm (red) and contains all wavelengths in between.

vital force A mystical or supernatural power that was believed to be employed by living organisms to allow them to produce organic compounds.

vitalism Belief that living things contain a nonphysical "force" that allows them to synthesize organic compounds.

volatile The tendency to vaporize easily.

voltage Potential difference between two electrodes; the driving force that causes electrons to flow.

volume A measure of space. Any unit of length, when cubed, becomes a unit of volume.

wavelength Distance between adjacent wave crests in a wave.

weak acid An acid that does not completely ionize in solution.

weak base A base that does not completely dissociate in solution.

weak electrolyte A substance whose aqueous solutions are poor conductors of electricity

Wöhler, Fredrich German chemist who contributed greatly to the development of organic chemistry and was the first to synthesize an organic compound in the laboratory.

X-rays The portion of the electromagnetic spectrum that lies between the ultraviolet (UV) region and the gamma ray region. X-rays can penetrate substances that normally block light and are often used in medical diagnosis.

APPENDIX 5
Answers to Odd-Numbered Exercises

Note: Answers in the Questions section are written as briefly as possible. Student answers may vary and still be correct.

Chapter 1

Questions

1. Soda fizzes due to the interactions between carbon dioxide and water under high pressures. At room temperature, carbon dioxide is a gas and water is a liquid. Through the use of pressure, the makers of soda force the carbon dioxide gas to dissolve in the water. When the can is sealed, the solution remains mixed. When the can is opened, the pressure is released and the carbon dioxide molecules escape in bubbles of gas.
3. Chemists study molecules and interactions at the molecular level to learn about and explain macroscopic events. Chemists attempt to explain why ordinary things are as they are.
5. Chemistry is the science that seeks to understand what matter does by studying what atoms and molecules do.
7. The scientific method is the way chemists investigate the chemical world. The first step consists of observing the natural world. Later observations can be combined to create a scientific law, which summarizes and predicts behavior. Theories are models that strive to explain the cause of the observed phenomenon. Theories are tested through experiment. When a theory is not well established, it is sometimes referred to as a hypothesis.
9. A law is simply a general statement that summarizes and predicts observed behavior. Theories seek to explain the causes of observed behavior.
11. To say "It is just a theory" makes it seem as if theories are easily discardable. However, many theories are very well established and are as close to truth as we get in science. Established theories are backed up with years of experimental evidence, and they are the pinnacle of scientific understanding.
13. The atomic theory states that all matter is composed of small, indestructible particles called atoms. John Dalton formulated this theory.

Problems

15. **a.** observation **b.** theory
    c. law **d.** observation
17. **a.** All atoms contain a degree of chemical reactivity. The larger the size of an atom, the higher the chemical reactivity of that atom.
    b. There are many correct answers. One example is: Conceivably, when the size of an atom is increased, the surface area of the atom is also increased; an atom with a greater surface area is more likely to react chemically.

Chapter 2

Questions

1. Without units, the results are unclear and it is hard to keep track of what each separate measurement entails.
3. Often scientists work with very large or very small numbers that contain a lot of zeros. Scientific notation allows these numbers to be written more compactly, and the information is more organized.
5. Zeros count as significant digits when they are interior zeros (zeros between two numbers) and when they are trailing zeros (zeros after a decimal point). Zeros are **not** significant digits when they are leading zeros, which are zeros to the left of the first nonzero number.
7. For calculations involving only multiplication and division, the result carries the same number of significant figures as the factor with the fewest significant figures.
9. When rounding numbers, one must round down if the last (or left-most) digit dropped is four or less, or one must round up if the last (or left-most) digit is five or more.
11. The basic SI unit of length is the meter. The kilogram is the SI unit of mass. Lastly, the second is the SI unit of time.
13. For measuring a Frisbee, the unit would be the meter and the prefix multiplier would be *centi-*. The final measurement would be in centimeters.
15. **a.** 2.42 cm **b.** 1.69 cm
    c. 21.58 cm **d.** 20.85 cm
17. Units act as a guide in the calculation and are able to show if the calculation is off track. The units must be followed in the calculation, so that the answer is correctly written and understood.
19. A conversion factor is a quantity used to relate two separate units. They are constructed from any two quantities known to be equivalent.
21. The solution map for converting grams to pounds is:

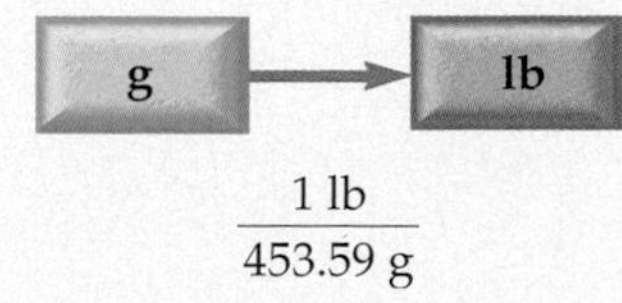

$$\frac{1 \text{ lb}}{453.59 \text{ g}}$$

23. The solution map for converting meters to feet is:

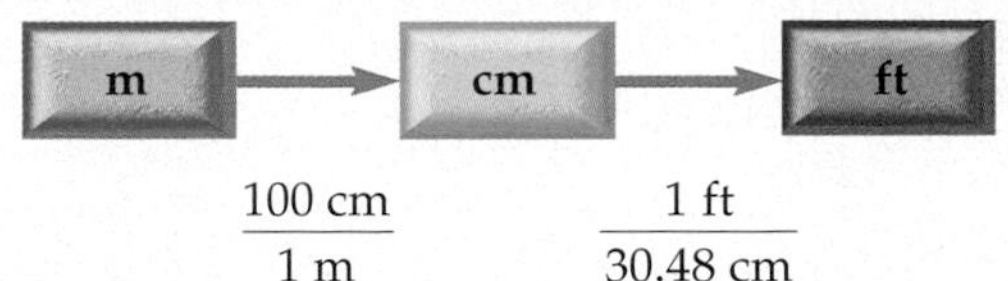

$$\frac{100\text{ cm}}{1\text{ m}} \qquad \frac{1\text{ ft}}{30.48\text{ cm}}$$

25. The density of a substance is the ratio of its mass to its volume. Density is a fundamental property of materials and differs from one substance to another. Density can be used to relate two separate units, thus working as a conversion factor. Density is a conversion factor between mass and volume.

Problems:

27. **a.** 3.2667×10^7 **b.** 1.193×10^6
c. 1.8175×10^7 **d.** 4.81×10^5
29. **a.** 7.461×10^{-11} m **b.** 1.57×10^{-5} mi
c. 6.32×10^{-7} m **d.** 1.5×10^{-5} m
31. 8.74×10^{-4}; 34,560,000,000; 1×10^8; 9,800,000,000,000
33. **a.** 602,200,000,000,000,000,000,000
b. 0.00000000000000000016 C
c. 299,000,000 m/s
d. 344 m/s
35. **a.** 3,890,000,000 **b.** 0.00059
c. 8,680,000,000,000 **d.** 0.0000000786
37. **a.** 54.9 mL **b.** 48.7 °C
c. 5.553 mL **d.** 46.83 °C
39. **a.** 0.00608 **b.** 540,802
c. 0.000000000000000005 **d.** 0.009090
41. **a.** 4 **b.** 4
c. ambiguous **d.** 6
43. **a.** correct **b.** 3
c. 7 **d.** correct
45. **a.** 343.0 **b.** 0.009651
c. 3.526×10^{-8} **d.** 1,127,000,000 or 1.127×10^9
47. **a.** 2.3 **b.** 2.4
c. 2.3 **d.** 2.4
49. **a.** 42.3 **b.** correct
c. correct **d.** 0.0456
51. **a.** 0.054 **b.** 0.619
c. 1.2×10^8 **d.** 6.6
53. **a.** 4.22×10^3 **b.** correct
c. 3.9969 **d.** correct
55. **a.** 200.6 **b.** 41.4
c. 183.3 **d.** 1.22
57. **a.** correct **b.** 1.0982
c. correct **d.** 3.53
59. **a.** 3.9×10^3 **b.** 632
c. 8.93×10^4 **d.** 6.34
61. **a.** 3.15×10^3 **b.** correct
c. 2.0 **d.** correct
63. **a.** 2.14×10^3 g **b.** 6.172 m
c. 1.316×10^{-3} kg **d.** 25.6 mL
65. **a.** 57.2 cm **b.** 38.4 m
c. 0.754 km **d.** 61 mm
67. **a.** 16 in **b.** 91.2 ft
c. 6.2 mi **d.** 8478 lb
69. 1.1×10^3 g
71. 50 min
73. 4.7×10^3 cm^3
75. **a.** 1.0×10^6 m^2 **b.** 1.0×10^{-6} m^3
c. 1.0×10^{-9} m^3
77. **a.** 2.15×10^{-4} km^2 **b.** 2.15×10^4 dm^2
c. 2.15×10^6 cm^2
79. 1.49×10^6 mi^2
81. 11.4 g/cm^3, lead
83. 1.26 g/cm^3
85. Yes, the density of the crown is 19.3 g/cm^3
87. **a.** 463 g **b.** 3.7 L
89. **a.** 3.38×10^4 g (gold); 5.25×10^3 g (sand)
b. Yes, the mass of the bag of sand is different than the mass of the gold vase; thus, the weight-sensitive pedestal will sound the alarm.
91. 10.6 g/cm^3
93. 0.098 lb/in^3
95. 2.5×10^5 lbs
97. 30 km/L
99. 108 km; 47.2 km
101. 9.1×10^{10} g/cm^3

Chapter 3

Questions

1. Matter is defined as anything that occupies space and possesses mass. It can be thought of as the physical material that makes up the universe.
3. The three states of matter are solid, liquid, and gas.
5. In a crystalline solid, the atoms/molecules are arranged in geometric patterns with repeating order. In amorphous solids, the atoms/molecules do not have long-range order.
7. The atoms/molecules in gases are not in contact with each other and are free to move relative to one another. The spacing between separate atoms/molecules is very far apart. A gas has no fixed volume or shape; rather it assumes both the shape and the volume of the container it occupies.
9. A mixture is two or more pure substances combined in variable proportions.
11. Pure substances are those composed of only one type of atom or molecule.
13. A mixture is formed when two or more pure substances are mixed together; however a new substance is not formed. A compound is formed when two or more elements are bonded together and form a new substance.

15. In a physical change the composition of the substance does not change, even though its appearance might change. However, in a chemical change, the substance undergoes a change in its composition.
17. Energy is defined as the capacity to do work.
19. Some different kinds of energy are kinetic, potential, electrical, chemical, and thermal.
21. Three common units for temperature are kelvin, Celsius, and Fahrenheit. They differ in how they define zero and in the size of the respective degrees.
23. Heat capacity is the quantity of heat energy required to change the temperature of a given amount of the substance by 1 °C.
25. $°F = \frac{9}{5}(°C) + 32$

Problems

27. **a.** element **b.** element
c. compound **d.** compound
29. **a.** homogeneous **b.** heterogeneous
c. homogeneous **d.** homogeneous
31. **a.** pure substance - element
b. mixture - homogeneous
c. mixture - heterogeneous
d. mixture - heterogeneous
33. **a.** chemical **b.** physical
c. physical **d.** chemical
35. physical – colorless; odorless; gas at room temperature; one liter has a mass of 1.260 g under standard conditions; mixes with acetone
chemical – flammable; polymerizes to form polyethylene
37. **a.** chemical **b.** physical
c. chemical **d.** chemical
39. **a.** physical **b.** chemical
41. 210 kg
43. **a.** yes **b.** no
45. 15.1 g of water
47. **a.** 7.77 cal **b.** 2.35×10^3 J
c. 8.53×10^3 cal **d.** 1.20 kJ
49. **a.** 9.0×10^7 J **b.** 0.249 Cal
c. 1.31×10^{-4} kWh **d.** 1.1×10^4 cal
51. 3.44×10^9 J
53. 8.4×10^6 J
55. **a.** 100 °C **b.** -3.2×10^2 °F
c. 298 K **d.** 310 K
57. −62°C, 2.1×10^2 K
59. 159 K, −173 °F
61. 1.0×10^4 J
63. 8.7×10^5 J
65. 58 °
67. 31 °
69. 1.0×10^1 °C
71. 0.24 J/g °C; silver
73. 2.2 J/g °C
75. 49 °C
77. 70.2 J
79. 1.7×10^4 kJ
81. 67 °C
83. 6.0 kWh
85. 6.1 g of fuel
87. **a.** pure substance **b.** pure substance
c. pure substance **d.** mixture
89. physical change
91. Small temperature changes in the ocean have a great impact on global weather because of the high heat capacity of water.
93. **a.** Sacramento is further inland than San Francisco, so Sacramento is not as close to the ocean. The ocean water has a high heat capacity and will be able to keep San Francisco cooler in hot days of summer. However, Sacramento is away from the ocean in a valley, so it will experience high temperatures in the summer.
b. San Francisco is located right next to the ocean, so the high heat capacity of the seawater keeps the temperature in the city from dropping. In the winter, the ocean actually helps to keep the city warmer, compared to an inland city like Sacramento.

Chapter 4

Questions

1. It is important to understand atoms because they compose matter and the properties of atoms determine matter's properties. The atom is the key to connecting the microscopic world with the macroscopic world.
3. Democritus theorized that matter was ultimately composed of small, indivisible particles called atoms. Upon dividing matter, one would find tiny, indestructible atoms.
5. Rutherford's gold foil experiment involved sending positively charged alpha particles through a thin sheet of gold foil and detecting if there was any deflection of the particles. He found that most passed straight through, yet some particles showed some deflection. This result contradicts the plum-pudding model of the atom because the plum-pudding model does not explain the deflection of the alpha-particles.
7.

Particle	Mass (kg)	Mass (amu)	Charge
Proton	1.67262×10^{-27}	1	+1
Neutron	1.67493×10^{-27}	1	0
Electron	0.00091×10^{-27}	0.00055	−1

9. Matter is usually charge neutral due to protons and electrons having opposite charges. If matter were not charge neutral, many unnatural things would occur, such as objects repelling or attracting each other.

11. A chemical symbol is a unique one- or two-letter abbreviation for an element. It is listed below the atomic number for that element on the periodic table.
13. Mendeleev noticed that many patterns were evident when elements were organized by increasing mass; from this he formulated the periodic law. He also organized the elements based on this law and created the basis for the periodic table being used today.
15. The periodic table is organized by listing the elements in order of increasing atomic number.
17. Nonmetals have varied properties (solid, liquid, or gas at room temperature); however, as a whole they tend to be poor conductors of heat and electricity and they all tend to gain electrons when they undergo chemical changes. They are located toward the upper right side of the periodic table.
19. Each column within the main group elements in the periodic table is labeled as a family or group of elements. The elements within a group usually have similar characteristics.
21. An ion is an atom or group of atoms that has lost or gained electrons and has become charged.
23. **a.** ion charge = +1 **b.** ion charge = +2 **c.** ion charge = +3 **d.** ion charge = −2 **e.** ion charge = −1
25. The percent natural abundance of isotopes is the relative amount of each different isotope in a naturally occurring sample of a given element.
27. Isotopes are noted in this manner, $^{A}_{Z}X$. X represents the chemical symbol, A represents the mass number, and Z represents the atomic number.

Problems

29. **a.** Correct.
b. False; different elements contain different types of atoms according to Dalton.
c. False; one cannot have 1.5 hydrogen atoms; combinations must be in simple, whole-number ratios.
d. Correct.
31. **a.** Correct.
b. False; most of the volume of the atom is empty space occupied by tiny, negatively charged electrons.
c. False, the number of negatively charged particles outside the nucleus equals the number of positively charged particles inside the nucleus.
d. False, the majority of the mass of an atom is found in the nucleus.
33. a, b, d
35. b, d
37. approximately 1.8×10^3 electrons
39. **a.** 27 **b.** 77 **c.** 92 **d.** 14 **e.** 4
41. **a.** 25 **b.** 47 **c.** 79 **d.** 82 **e.** 16
43. **a.** C, 6 **b.** N, 7 **c.** Na, 11 **d.** K, 19 **e.** Cu, 29
45. **a.** gold, 79 **b.** silicon, 14 **c.** nickel, 28 **d.** zinc, 30 **e.** tungsten, 74
47.

silver	Ag	47
gallium	Ga	31
arsenic	As	33
astatine	At	85

49. **a.** metal **b.** metal **c.** nonmetal **d.** nonmetal **e.** metalloid
51. a, d, e
53. a, b
55. c, d
57. a, b, e
59. **a.** halogen **b.** noble gas **c.** halogen **d.** neither **e.** noble gas
61. **a.** 6A **b.** 3A **c.** 4A **d.** 4A **e.** 5A
63. b, oxygen; it is in the same group or family.
65. b, chlorine and fluorine; they are in the same family or group.
67.

Na	1A	alkali metals	metal
He	8A	noble gases	nonmetal
I	7A	halogens	nonmetal
Sr	2A	alkali earth metals	metal

69. **a.** e^- **b.** O^{2-} **c.** $2e^-$ **d.** Cl^-
71. **a.** −2 **b.** +3 **c.** +4 **d.** −1
73. **a.** 19 protons, 18 electrons **b.** 16 protons, 18 electrons **c.** 38 protons, 36 electrons **d.** 24 protons, 21 electrons
75. **a.** False; Ti^{2+} has 22 protons and 20 electrons.
b. True
c. False; Mg^{2+} has 12 protons and 10 electrons
d. True
77. **a.** Rb^+ **b.** K^+ **c.** Al^{3+} **d.** O^{2-}
79. **a.** 3 electrons lost **b.** 1 electron lost **c.** 1 electron gained **d.** 2 electrons gained
81.

Ca	Ca^{2+}	18	20
Te	Te^{2-}	54	52
In	In^{3+}	46	49
Sr	Sr^{2+}	36	38

83. **a.** Z = 1, A = 2 **b.** Z = 24, A = 51 **c.** Z = 20, A = 40 **d.** Z = 73, A = 181
85. **a.** $^{16}_{8}O$ **b.** $^{19}_{9}F$ **c.** $^{23}_{11}Na$ **d.** $^{27}_{13}Al$

87. **a.** $^{60}_{27}Co$ **b.** $^{22}_{10}Ne$
c. $^{131}_{53}I$ **d.** $^{244}_{94}Pu$

89. **a.** 11 protons, 12 neutrons
b. 88 protons, 178 neutrons
c. 82 protons, 126 neutrons
d. 7 protons, 7 neutrons

91. 6 protons, 8 neutrons, $^{14}_{6}C$

93. 85.47 amu

95. **a.** 49.31% **b.** 78.92 amu

97. 121.8 amu, Sb

99. 7.8×10^{17} electrons

101.

Symbol	Number of Protons	Number of Neutrons	A (Mass Number)	Natural Abundance
Sr-84 or $^{84}_{38}Sr$	38	46	83.9134 amu	0.56%
Sr-86 or $^{86}_{38}Sr$	38	48	85.9093 amu	9.86%
Sr-87 or $^{87}_{38}Sr$	38	49	86.9089 amu	7.00%
Sr-88 or $^{88}_{38}Sr$	38	50	87.9056 amu	82.58%

Atomic mass of Sr = 87.62 amu

103.

Symbol	Z	A	No. of Protons	No. of Electrons	No. of Neutrons	Charge
Fe	*26*	*58*	*26*	24	*32*	2+
Cu	*29*	*53*	*29*	*27*	24	2+
P	15	*31*	*15*	15	16	*0*
O	*8*	16	*8*	*10*	*8*	2−

105. 153 amu, 52.2%

107. **a.** Nt-304 = 72%, Nt-305 = 4%, Nt-306 = 24%
b.

120
Nt
304.5

Chapter 5

Questions

1. Yes; when elements combine with other elements, a compound is created. Each compound is unique and contains properties different from those of the elements that compose it.

3. The law of constant composition states that all samples of a given compound have the same proportions of their constituent elements. Joseph Proust formulated this law.

5. The more metallic element is generally listed first in a chemical formula.

7. An atomic element is one that exists in nature with a single atom as the basic unit. A molecular element is one that exists as a diatomic molecule as the basis unit. Molecular elements include H_2, N_2, O_2, F_2, Cl_2, Br_2, and I_2.

9. The systematic name can be directly derived by looking at the compound's formula. The common name for a compound acts like a nickname and can only be learned through familiarity.

11. The block that contains the elements for Type II compounds is known as the transition metals.

13. The basic form for the names of Type II ionic compounds is to have the name of the metal cation first, followed by the charge of the metal cation, and finally the base name of the nonmetal anion with *-ide* attached to the end.

15. For compounds containing a polyatomic anion, the name of the cation is first, followed by the name for the polyatomic anion. Also, if the compound contains both a polyatomic cation and a polyatomic anion, one would just use the names of both polyatomic ions.

17. The form for naming molecular compounds is to have the first element preceded by a prefix to indicate the number of atoms present. This is then followed by the second element with its corresponding prefix and *-ide* placed on the end of the second element.

19. To correctly name a binary acid one must begin the first word with *hydro-*, which is followed by the base name of the nonmetal plus *-ic* added on the end. Finally, the word *acid* follows the first word.

21. To name an acid with oxyanions ending with *-ite*, one must take the base name of the oxyanion and attach *-ous* to it; the word *acid* follows this.

Problems

23. Yes; the ratios of sodium to chlorine in both samples were equal.

25. 2.06×10^3 g

27. NBr_3

29. **a.** Fe_2O_3 **b.** PCl_3
c. PCl_5 **d.** Ag_2O

31. **a.** 4 **b.** 4
c. 6 **d.** 4

33. **a.** magnesium, 1; chlorine, 2
b. sodium, 1; nitrogen, 1; oxygen, 3
c. calcium, 1; nitrogen, 2; oxygen, 4
d. strontium, 1; oxygen, 2; hydrogen, 2

35. **a.** atomic **b.** molecular
c. molecular **d.** atomic

37. **a.** molecular **b.** ionic
c. ionic **d.** molecular

39. helium → single atoms
CCl_4 → molecules
K_2SO_4 → formula units
bromine → diatomic molecules

41. **a.** formula units **b.** single atoms
c. molecules **d.** molecules

43. **a.** ionic, Type I **b.** molecular
c. molecular **d.** ionic, Type II

45. a. SrO b. Na_2O
c. Al_2S_3 d. $SrBr_2$
47. a. $KC_2H_3O_2$ b. K_2CrO_4
c. K_3PO_4 d. KCN
49. a. K_3N, K_2O, KF b. Ba_3N_2, BaO, BaF_2
c. AlN, Al_2O_3, AlF_3
51. a. cesium chloride b. strontium bromide
c. potassium oxide d. lithium fluoride
53. a. chromium(II) chloride
b. chromium(III) chloride
c. tin(IV) oxide
d. lead(II) iodide
55. a. Type II, chromium(III) oxide
b. Type I, sodium iodide
c. Type I, calcium bromide
d. Type II, tin(II) oxide
57. a. barium nitrate b. lead(II) acetate
c. ammonium iodide d. potassium chlorate
e. cobalt(II) sulfate f. sodium perchlorate
59. a. $CuBr_2$ b. $AgNO_3$
c. KOH d. Na_2SO_4
e. $KHSO_4$ f. $NaHCO_3$
61. a. sulfur dioxide b. nitrogen triiodide
c. bromine pentafluoride d. nitrogen monoxide
e. tetranitrogen tetraselenide
63. a. CO b. S_2F_4
c. Cl_2O d. PF_5
e. BBr_3 f. P_2S_5
65. a. PBr_5 phosphorus pentabromide
b. P_2O_3 diphosphorus trioxide
c. SF_4 sulfur tetraflouride
d. correct
67. a. chlorous acid b. hydroiodic acid
c. sulfuric acid d. nitric acid
69. a. H_3PO_4 b. HBr c. H_2SO_3
71. a. 47.02 amu b. 110.98 amu
c. 153.81 amu d. 148.33 amu
73. PBr_3, Ag_2O, PtO_2, $Al(NO_3)_3$
75. a. ionic, cobalt(III) cyanide
b. acid, chromic acid
c. ionic, potassium chloride
d. molecular, dinitrogen tetrahydride
77. a. calcium nitrite b. potassium oxide
c. phosphorus trichloride d. correct
79. a. $Sn(SO_4)_2$ 310.8 amu b. HNO_2 47.02 amu
c. $NaHCO_3$ 84.01 amu d. PF_5 125.97 amu
81. a. platinum(IV) oxide 227.1 amu
b. dinitrogen pentoxide 108.02 amu
c. aluminum chlorate 277.33 amu
d. phosphorous pentabromide 430.47 amu
83. a. molecular element b. atomic element
c. ionic compound d. molecular compound
85. $NaOCl$; $NaOH$; $CaCO_3$; $NaHCO_3$; $Ca(PO_4)_2$; $NaAl(SO_4)_2$

Chapter 6

Questions

1. Chemical composition lets us determine how much of a particular element is contained within a particular compound.
3. There are 6.022×10^{23} atoms in 1 mole of atoms.
5. One mole of any element has a mass equal to its atomic mass in grams.
7. a. 30.974 g b. 195.08 g
c. 12.011 g d. 51.996 g
9. Each element has a different atomic mass number. So, the subscripts that represent mole ratios cannot be used to represent the ratios of grams of a compound. The grams per mole of one element always differs from the grams per mole of a different element.
11. a. 11.19 g H $\equiv$ 100 g H_2O
b. 53.29 g O $\equiv$ 100 g $C_8H_{12}O_6$
c. 84.12 g C $\equiv$ 100 g C_8H_{18}
d. 52.14 g C $\equiv$ 100 g C_2H_5OH
13. The empirical formula gives the smallest whole number ratio of each type of atom. The molecular formula gives the specific number of each type of atom in the molecule. The molecular formula is always a multiple of the empirical formula.
15. The empirical formula mass of a compound is the sum of the masses of all the atoms in the empirical formula.

Problems

17. 3.6×10^{24} atoms
19. a. 2.0×10^{24} atoms b. 5.8×10^{21} atoms
c. 1.38×10^{25} atoms d. 1.29×10^{23} atoms
21. a. 72.7 dozen b. 6.06 gross
c. 1.74 reams d. 1.45×10^{-21} moles
23. 0.356 mol
25. a. 2.36×10^{-2} mol b. 0.571 mol
c. 0.477 mol d. 4.9×10^{-3} mol
27. 8.07×10^{18} atoms
29. 8.44×10^{22} atoms
31. a. 8.80×10^{22} atoms b. 4.87×10^{23} atoms
c. 2.84×10^{22} atoms d. 7.02×10^{23} atoms
33. 2.6×10^{21} atoms
35. 1.61×10^{25} atoms
37. 6.20×10^{21} molecules
39. a. 1.2×10^{23} molecules
b. 1.21×10^{24} molecules
c. 3.5×10^{23} molecules
d. 6.4×10^{22} molecules
41. 0.10 mg
43. 9.4 mol Cl

45. d, 3 mol O
47. a. 3.8 mol C b. 0.546 mol C
c. 19.6 mol C d. 183 mol C
49. a. 32 g b. 43 g
c. 31 g d. 19 g
51. a. 1.4×10^3 kg b. 1.4×10^3 kg
c. 2.1×10^3 kg
53. 84.8% Sr
55. 36.1% Ca; 63.9% Cl
57. 10.7 g
59. 6.6 mg
61. a. 63.65% b. 46.68%
c. 30.45% d. 25.94%
63. a. 39.99% C; 6.73% H; 53.28% O
b. 26.09% C; 4.39% H; 69.52% O
c. 60.93% C; 15.37% H; 23.69% N
d. 54.48% C; 13.74% H; 31.78% N
65. Fe_3O_4, 72.4% Fe
67. NO_2
69. a. NiI_2 b. $SeBr_4$ c. $BeSO_4$
71. C_2H_6N
73. a. C_3H_6O b. $C_5H_{10}O_2$ c. $C_9H_{10}O_2$
75. P_2O_3
77. NCl_3
79. C_4H_8
81. a. C_6Cl_6 b. C_2HCl_3 c. $C_6H_3Cl_3$
83. 2.43×10^{23} atoms
85. 1.67×10^{21} molecules
87. a. CuI_2: 20.03 % Cu; 79.97% I
b. $NaNO_3$: 27.05% Na; 16.48% N; 56.47% O
c. $PbSO_4$: 68.32% Pb; 10.57% S; 21.10% O
d. CaF_2: 51.33% Ca; 48.67% F
89. 1.8×10^3 kg rock
91. 59 kg Cl
93. 1.1×10^2 g H
95. $C_4H_6O_2$
97. $C_{10}H_{14}N_2$
99. a. 1×10^{57} atoms per star
b. 1×10^{68} atoms per galaxy
c. 1×10^{77} atoms in the universe
101. $C_{16}H_{10}$

Chapter 7

Questions

1. A chemical reaction is the change of one or more substances into different substances, for example burning wood, rusting iron, and protein synthesis.
3. The main evidences of a chemical reaction are a color change, the formation of a solid, the formation of a gas, the emission of light, and the emission or absorption of heat.
5. a. gas b. liquid
c. solid d. aqueous
7. a. reactants: 4 Ag, 2 O, 1 C
products: 4 Ag, 2 O, 1 C
balanced: yes
b. reactants: 1 Pb, 2 N, 6 O, 2 Na, 2 Cl
products: 1 Pb, 2 N, 6 O, 2 Na, 2 Cl
balanced: yes
c. reactants: 3 C, 8 H, 2 O
products: 3 C, 8 H, 10 O
balanced: no
9. If a compound dissolves in water than it is soluble. If it does not dissolve in water, it is insoluble.
11. When ionic compounds containing polyatomic ions dissolve in water, the polyatomic ions usually dissolve as intact units.
13. The solubility rules are a set of empirical rules for ionic compounds that were deduced from observations on many compounds. The rules help us determine whether particular compounds will be soluble or insoluble.
15. The precipitate will always be insoluble; it is the solid that forms upon mixing two aqueous solutions.
17. A complete ionic equation shows the reactants and products as they are actually present in solution. The net ionic equation shows only the species that actually change during the reaction. Spectator ions are shown in the complete ionic equation but are not shown in the net ionic equation. An example of each:
Complete ionic equation:

$$Na^+(aq) + OH^-(aq) + H^+(aq) + NO_3^-(aq) \rightarrow H_2O(l) + Na^+(aq) + NO_3^-(aq)$$

Net ionic equation:

$$H^+(aq) + OH^-(aq) \rightarrow H_2O(l)$$

19. An acid is a compound characterized by its sour taste, its ability to dissolve some metals, and its tendency to form H^+ ions in solution. A base is a compound characterized by its bitter taste, its slippery feel, and its tendency to form OH^- ions in solution.
21. A redox reaction is a reaction involving the transfer of electrons. An example of a redox reaction is the rusting of iron: $4\,Fe(s) + 3\,O_2(g) \rightarrow 2\,Fe_2O_3(s)$
23. You can classify chemical reactions by either 1) the type of chemistry occurring during the reaction, such as acid–base chemistry verses precipitation chemistry, or 2) by studying what happens to atoms during the reaction. We use both methods for classifying reactions so we can study the similarities and differences between reactions.
25. Two examples of synthesis reactions:

$$2\,Na(s) + Cl_2(g) \rightarrow 2\,NaCl(s)$$
$$CaO(s) + CO_2(g) \rightarrow + CaCO_3(s)$$

27. Two examples of single-displacement reactions:

$$2\,Na(s) + 2\,HOH(l) \rightarrow 2\,NaOH(aq) + H_2(g)$$
$$Zn(s) + CuCl_2(aq) \rightarrow ZnCl_2(aq) + Cu(s)$$

Problems:

29. a. Yes; there is a color change showing a chemical reaction.
b. No; the state of the compound changes, but no chemical reaction takes place.
c. Yes; there is a formation of a solid in a previously clear solution.
d. Yes; there is a formation of a gas when the yeast was added to the solution.
31. Yes; a chemical reaction has occurred; for the presence of the bubbles is evidence for the formation of a gas.
33. Yes; a chemical reaction has occurred. We know this due to the color change of the hair.
35. a. $2\,Cu(s) + S(s) \rightarrow Cu_2S(s)$
b. $2\,SO_2(g) + O_2(g) \rightarrow 2\,SO_3(g)$
c. $4\,HCl(aq) + MnO_2(s) \rightarrow 2\,H_2O(l) + Cl_2(g) + MnCl_2(aq)$
d. $2\,C_6H_6(l) + 15\,O_2(g) \rightarrow 12\,CO_2(g) + 6\,H_2O(l)$
37. a. $Mg(s) + 2\,CuNO_3(aq) \rightarrow 2\,Cu(s) + Mg(NO_3)_2(aq)$
b. $2\,N_2O_5(g) \rightarrow 4\,NO_2(g) + O_2(g)$
c. $Ca(s) + 2\,HNO_3(aq) \rightarrow H_2(g) + Ca(NO_3)_2(aq)$
d. $2\,CH_3OH(l) + 3\,O_2(g) \rightarrow 2\,CO_2(g) + 4\,H_2O(g)$
39. $2\,Na(s) + 2\,H_2O(l) \rightarrow H_2(g) + 2\,NaOH(aq)$
41. $2\,SO_2(g) + O_2(g) + 2\,H_2O(l) \rightarrow 2\,H_2SO_4(aq)$
43. $V_2O_5(s) + 2\,H_2(g) \rightarrow V_2O_3(s) + 2\,H_2O(l)$
45. $C_{12}H_{22}O_{11}(aq) + H_2O(l) \rightarrow 4\,CO_2(g) + 4\,C_2H_5OH(aq)$
47. a. $3\,N_2H_4(l) \rightarrow 4\,NH_3(g) + N_2(g)$
b. $3\,H_2(g) + N_2(g) \rightarrow 2\,NH_3(g)$
c. $Cu_2O(s) + C(s) \rightarrow 2\,Cu(s) + CO(g)$
d. $H_2(g) + Cl_2(g) \rightarrow 2\,HCl(g)$
49. a. $BaO_2(s) + H_2SO_4(aq) \rightarrow BaSO_4(s) + H_2O_2(aq)$
b. $2\,Co(NO_3)_3(aq) + 3\,(NH_4)_2S(aq) \rightarrow Co_2S_3(s) + 6\,NH_4NO_3(aq)$
c. $Li_2O(s) + H_2O(l) \rightarrow 2\,LiOH(aq)$
d. $Hg_2(C_2H_3O_2)_2(aq) + 2\,KCl(aq) \rightarrow Hg_2Cl_2(s) + 2\,KC_2H_3O_2(aq)$
51. a. $4\,Al(s) + 3\,O_2(g) \rightarrow 2\,Al_2O_3(s)$
b. correct
c. $2\,NiS(s) + 3\,O_2(g) \rightarrow 2\,NiO(s) + 2\,SO_2(g)$
d. $Fe_2O_3(s) + 3\,CO(g) \rightarrow 2\,Fe(s) + 3\,CO_2(g)$
53. $C_6H_{12}O_6(aq) + 6\,O_2(g) \rightarrow 6\,CO_2(g) + 6\,H_2O(l)$
55. $2\,NO(g) + 2\,CO(g) \rightarrow N_2(g) + 2\,CO_2(g)$
57. a. soluble; Na^+ and NO_3^-
b. soluble; Pb^{2+} and $C_2H_3O_2^-$
c. insoluble
d. soluble; NH_4^+ and S^{2-}
59. AgCl; $BaSO_4$; $CuCO_3$; Fe_2S_3
61. Soluble: CaS, LiOH, K_2S, NH_4Cl, K_2CO_3, *Na_2S, SrS*
Insoluble: Hg_2Cl_2, $Cu_3(PO_4)_2$, $PbSO_4$, $CaSO_4$, AgCl, *Hg_2I_2, $PbCl_2$*
63. a. $NH_4Cl(aq) + AgNO_3(aq) \rightarrow AgCl(s) + NH_4NO_3(aq)$
b. NO REACTION
c. $CrCl_2(aq) + Li_2CO_3(aq) \rightarrow CrCO_3(s) + 2\,LiCl(aq)$
d. $3\,KOH(aq) + FeCl_3(aq) \rightarrow Fe(OH)_3(s) + 3\,KCl(aq)$
65. a. $Na_2CO_3(aq) + Pb(NO_3)_2(aq) \rightarrow PbCO_3(s) + 2\,NaNO_3(aq)$
b. $K_2SO_4(aq) + Pb(CH_3CO_2)_2(aq) \rightarrow PbSO_4(s) + 2\,KCH_3CO_2(aq)$
c. $Cu(NO_3)_2(aq) + BaS(aq) \rightarrow CuS(s) + Ba(NO_3)_2(aq)$
d. NO REACTION
67. a. correct
b. NO REACTION
c. correct
d. $Pb(NO_3)_2(aq) + 2\,LiCl(aq) \rightarrow PbCl_2(s) + 2\,LiNO_3(aq)$
69. K^+, $C_2H_3O_2^-$
71. a. $Ag^+(aq) + NO_3^-(aq) + K^+(aq) + Cl^-(aq) \rightarrow AgCl(s) + K^+(aq) + NO_3^-(aq)$
$Ag^+(aq) + Cl^-(aq) \rightarrow AgCl(s)$
b. $Ca^{2+}(aq) + S^{2-}(aq) + Cu^{2+}(aq) + 2\,Cl^-(aq) \rightarrow CuS(s) + Ca^{2+}(aq) + 2\,Cl^-(aq)$
$Cu^{2+}(aq) + S^{2-}(aq) \rightarrow CuS(s)$
c. $Na^+(aq) + OH^-(aq) + H^+(aq) + NO_3^-(aq) \rightarrow H_2O(l) + Na^+(aq) + NO_3^-(aq)$
$H^+(aq) + OH^-(aq) \rightarrow H_2O(l)$
d. $6\,K^+(aq) + 2\,PO_4^{3-}(aq) + 3\,Ni^{2+}(aq) + 6\,Cl^-(aq) \rightarrow Ni_3(PO_4)_2(s) + 6\,K^+(aq) + 6\,Cl^-(aq)$
$3\,Ni^{2+}(aq) + 2\,PO_4^{3-}(aq) \rightarrow Ni_3(PO_4)_2(s)$
73. $Hg_2^{2+}(aq) + 2\,NO_3^-(aq) + 2\,Na^+(aq) + 2\,Cl^-(aq) \rightarrow Hg_2Cl_2(s) + 2\,Na^+(aq) + 2\,NO_3^-(aq)$
$Hg_2^{2+}(aq) + 2\,Cl^-(aq) \rightarrow Hg_2Cl_2(s)$
75. a. $2\,Na^+(aq) + CO_3^{2-}(aq) + Pb^{2+}(aq) + 2\,NO_3^-(aq) \rightarrow PbCO_3(s) + 2\,Na^+(aq) + 2\,NO_3^-(aq)$
$Pb^{2+}(aq) + CO_3^{2-}(aq) \rightarrow PbCO_3(s)$
b. $2\,K^+(aq) + SO_4^{2-}(aq) + Pb^{2+}(aq) + 2\,CH_3CO_2^-(aq) \rightarrow PbSO_4(s) + 2\,K^+(aq) + 2\,CH_3CO_2^-(aq)$
$Pb^{2+}(aq) + SO_4^{2-}(aq) \rightarrow PbSO_4(s)$
c. $Cu^{2+}(aq) + 2\,NO_3^-(aq) + Ba^{2+}(aq) + S^{2-}(aq) \rightarrow CuS(s) + Ba^{2+}(aq) + 2\,NO_3^-(aq)$
$Cu^{2+}(aq) + S^{2-}(aq) \rightarrow CuS(s)$
d. NO REACTION
77. $HCl(aq) + KOH(aq) \rightarrow H_2O(l) + KCl(aq)$
$H^+(aq) + OH^-(aq) \rightarrow H_2O(l)$
79. a. $2\,HCl(aq) + Ba(OH)_2(aq) \rightarrow 2\,H_2O(l) + BaCl_2(aq)$

b. $H_2SO_4(aq) + 2\,KOH(aq) \rightarrow 2\,H_2O(l) + K_2SO_4(aq)$
c. $HClO_4(aq) + NaOH(aq) \rightarrow H_2O(l) + NaClO_4(aq)$

81. a. $HBr(aq) + NaHCO_3(aq) \rightarrow H_2O(l) + CO_2(g) + NaBr(aq)$
b. $NH_4I(aq) + KOH(aq) \rightarrow H_2O(l) + NH_3(g) + KI(aq)$
c. $2\,HNO_3(aq) + K_2SO_3(aq) \rightarrow H_2O(l) + SO_2(g) + 2\,KNO_3(aq)$
d. $2\,HI(aq) + Li_2S(aq) \rightarrow H_2S(g) + 2\,LiI(aq)$

83. b and d are redox reactions; a and c are not.

85. a. $2\,C_2H_6(g) + 7\,O_2(g) \rightarrow 4\,CO_2(g) + 6\,H_2O(g)$
b. $2\,Ca(s) + O_2(g) \rightarrow 2\,CaO(s)$
c. $2\,C_3H_8O(l) + 9\,O_2(g) \rightarrow 6\,CO_2(g) + 8\,H_2O(g)$
d. $2\,C_4H_{10}S(l) + 15\,O_2(g) \rightarrow 8\,CO_2(g) + 10\,H_2O(g) + 2\,SO_2(g)$

87. a. double displacement
b. synthesis or combination
c. single displacement
d. decomposition

89. a. synthesis
b. decomposition
c. synthesis

91. a. $2\,Na^+(aq) + 2\,I^-(aq) + Hg_2^{\ +}(aq) + 2\,NO_3^-(aq) \rightarrow Hg_2I_2(s) + 2\,Na^+(aq) + 2\,NO_3^-(aq)$
$Hg_2^{\ +}(aq) + 2\,I^-(aq) \rightarrow Hg_2I_2(s)$
b. $2\,H^+(aq) + ClO_4^-(aq) + Ba^{2+}(aq) + 2\,OH^-(aq) \rightarrow 2\,H_2O(l) + Ba^{2+}(aq) + ClO_4^-(aq)$
$H^+(aq) + OH^-(aq) \rightarrow H_2O(s)$
c. NO REACTION
d. $2\,H^+(aq) + 2\,Cl^-(aq) + 2\,Li^+(aq) + CO_3^{2-}(aq) \rightarrow H_2O(l) + CO_2(g) + 2\,Li^+(aq) + 2\,Cl^-(aq)$
$2\,H^+(aq) + CO_3^{2-}(aq) \rightarrow H_2O(l) + CO_2(g)$

93. a. NO REACTION
b. NO REACTION
c. $K^+(aq) + HSO_3^-(aq) + H^+(aq) + NO_3^-(aq) \rightarrow H_2O(l) + SO_2(g) + K^+(aq) + NO_3^-(aq)$
$H^+(aq) + HSO_3^-(aq) \rightarrow H_2O(l) + SO_2(g)$
d. $Mn^{3+}(aq) + 3\,Cl^-(aq) + 3\,K^+(aq) + PO_4^{3-}(aq) \rightarrow MnPO_4(s) + 3\,K^+(aq) + 3\,Cl^-(aq)$
$Mn^{3+}(aq) + PO_4^{3-}(aq) \rightarrow MnPO_4(s)$

95. a. acid–base; $KOH(aq) + HC_2H_3O_2(aq) \rightarrow H_2O(l) + KC_2H_3O_2(aq)$
b. gas evolution; $2\,HBr(aq) + K_2CO_3(aq) \rightarrow H_2O(l) + CO_2(g) + 2\,KBr(aq)$
c. synthesis; $2\,H_2(g) + O_2(g) \rightarrow 2\,H_2O(l)$
d. precipitation; $2\,NH_4Cl(aq) + Pb(NO_3)_2(aq) \rightarrow PbCl_2(s) + 2\,NH_4NO_3(aq)$

97. $3\,CaCl_2(aq) + 2\,Na_3PO_4(aq) \rightarrow Ca_3(PO_4)_2(s) + 6\,NaCl(aq)$
$3\,Ca^{2+}(aq) + 6\,Cl^-(aq) + 6\,Na^+(aq) + 2\,PO_4^{3-}(aq) \rightarrow Ca_3(PO_4)_2(s) + 6\,Na^+(aq) + 6\,Cl^-(aq)$
$3\,Ca^{2+}(aq) + 2\,PO_4^{3-}(aq) \rightarrow Ca_3(PO_4)_2(s)$
$3\,Mg(NO_3)_2(aq) + 2\,Na_3PO_4(aq) \rightarrow Mg_3(PO_4)_2(s) + 6\,NaNO_3(aq)$
$3\,Mg^{2+}(aq) + 6\,NO_3^-(aq) + 6\,Na^+(aq) + 2\,PO_4^{3-}(aq) \rightarrow Mg_3(PO_4)_2(s) + 6\,Na^+(aq) + 6\,NO_3^-(aq)$
$3\,Mg^{2+}(aq) + 2\,PO_4^{3-}(aq) \rightarrow Mg_3(PO_4)_2(s)$

99. Ca^{2+} and Cu^{2+} were present in the original solution.
1st: $Cu^{2+}(aq) + SO_4^{2-}(aq) \rightarrow CuSO_4(s)$
2nd: $Ca^{2+}(aq) + CO_3^{2-}(aq) \rightarrow CaCO_3(s)$

101. Figure 7.9 chemical
Figure 7.10 physical

Chapter 8

Questions

1. Reaction stoichiometry is very important to chemistry. It gives us a numerical relationship between the reactants and products that allows chemists to plan and carry out chemical reactions to obtain products in the desired quantities,
e.g., How much CO_2 is produced when a given amount of C_8H_{10} is burned?
How much $H_2(g)$ is produced when a given amount of water decomposes?

3. 1 mol $N_2 \equiv$ 2 mol NH_3
3 mol $H_2 \equiv$ 2 mol NH_3

5. 1 mol $Cl_2 \equiv$ 2 mol NaCl

7. mass A $\rightarrow$ moles A $\rightarrow$ moles B $\rightarrow$ mass B (A = reactant, B = product)

9. The limiting reactant is the reactant that limits the amount of product in a chemical reaction.

11. The actual yield is the amount of product actually produced by a chemical reaction. The percent yield is the percentage of the theoretical yield that was actually attained.

13. d

Problems

15. a. 1 mol C b. 0.5 mol C
c. 2 mol C d. 1 mol C

17. a. 2.6 mol NO_2 b. 11.6 mol NO_2
c. 8.90×10^3 mol NO_2 d. 2.012×10^{-3} mol NO_2

19. a. 2.96 mol HCl
b. 2.96 mol H_2O
c. 0.740 mol Na_2O_2
d. 0.987 mol SO_3

21. **a.** 1.2 mol $PbO(s)$, 1.2 mol $SO_2(g)$
    b. 0.80 mol $PbO(s)$, 0.80 mol $SO_2(g)$
    c. 7.9 mol $PbO(s)$, 7.9 mol $SO_2(g)$
    d. 5.3 mol $PbO(s)$, 5.3 mol $SO_2(g)$

23.

mol N_2H_4	mol N_2O_4	mol N_2	mol H_2O
4	2	6	8
6	3	9	12
3	1.5	4.5	6
5.8	2.9	8.7	11.6
13.4	6.7	20.1	26.8
15	7.5	22.5	30

25. $2\ C_4H_{10}(g) + 13\ O_2(g) \rightarrow 8\ CO_2(g) + 10\ H_2O(g)$
    32 mol O_2
27. **a.** $Pb(s) + 2\ AgNO_3(aq) \rightarrow Pb(NO_3)_2(aq) + 2\ Ag(s)$
    b. 19 mol $AgNO_3$
    c. 56.8 mol Ag
29. **a.** 0.213 g O_2 **b.** 0.385 g O_2
    c. 96 g O_2 **d.** 3.4×10^{-4} g O_2
31. **a.** 3.0 g NaCl **b.** 3.2 g $CaCO_3$
    c. 3.0 g MgO **d.** 2.3 g NaOH
33. **a.** 8.9 g Al_2O_3, 9.7 g Fe **b.** 3.0 g Al_2O_3, 3.3 g Fe
35. **a.** 2.3 g HCl **b.** 4.3 g HNO_3
    c. 2.2 g H_2SO_4
37. 123 g H_2SO_4, 2.53 g H_2
39. **a.** 2 mol A **b.** 1.8 mol A
    c. 4 mol B **d.** 40 mol B
41. **a.** 1.5 mol C **b.** 3 mol C
    c. 3 mol C **d.** 96 mol C
43. **a.** 1 mol K **b.** 1.8 mol K
    c. 1 mol Cl_2 **d.** 14.6 mol K
45. **a.** 1.3 mol MnO_3 **b.** 4.8 mol MnO_3
    c. 0.107 mol MnO_3 **d.** 27.5 mol MnO_3
47. **a.** 1.0 g F_2 **b.** 10.5 g Li
    c. 6.79×10^3 g F_2
49. **a.** 1.3 g $AlCl_3$ **b.** 24.8 g $AlCl_3$
    c. 2.17 g $AlCl_3$
51. 74.6%
53. CaO; 25.7 g $CaCO_3$; 75.5%
55. O_2; 5.07 g NiO; 95.9%
57. Pb^{+2}; 262.7 g $PbCl_2$; 96.1%
59. 1.5 g HCl
61. 3 kg CO_2
63. 4.7 g Na_3PO_4
65. 469 g Zn
67. salicylic acid ($C_7H_6O_3$); 2.71 g $C_9H_8O_4$; 74.1%
69. NH_3; 120 g CH_4N_2O; 72.9%
71. b; the loudest explosion will occur when the ratio is 2 hydrogen to 1 oxygen, for that is the ratio that occurs in water.
73. 2×10^{13} kg CO_2; 150 years

Chapter 9

Questions

1. Both the Bohr model and the quantum mechanical model for the atom were developed in the early 1900's. These models serve to explain how electrons are arranged within the atomic structure and how the electrons affect the chemical and physical properties of each element.
3. Light travels at a speed of 3.0×10^8 m/s. The speed of light is extremely fast, for a flash of light from the equator could circle our earth 7 times in one second.
5. A blue object appears blue because when white light is present the object absorbs all wavelengths of light except blue, which it reflects.
7. Wavelength and frequency are inversely related; the longer the wavelength, the lower the frequency, and vice versa.
9. X-rays pass through many substances that block visible light and are therefore used to image bones and organs.
11. Ultraviolet light contains enough energy to damage biological molecules, and excessive exposure increases the risk of skin cancer and cataracts.
13. Microwaves can only heat things containing water, and therefore the food, which contains water, becomes hot, but the plate does not.
15. The Bohr model is a representation for the atom in which electrons travel around the nucleus in circular orbits with a fixed energy at specific, fixed distances from the nucleus.
17. The Bohr orbit describes the path of an electron as an orbit or trajectory (a specified path). A quantum mechanical orbital describes the path of an electron using a probability map.
19. The subshells are *s*, which contains 2 electrons; *p*, which contains 6 electrons; *d*, which contains 10 electrons; and *f*, which contains 14 electrons.
21. The Pauli exclusion principle states that separate orbitals may hold no more than 2 electrons, and when 2 electrons are present in a single orbital, they must have opposite spins. When writing electron configurations, the principle means that no box can have more than 2 arrows, and the arrows will point in opposite directions.
23. [Ne] represents $1s^22s^22p^6$
    [Kr] represents $1s^22s^22p^63s^23p^64s^23d^{10}4p^6$
25. 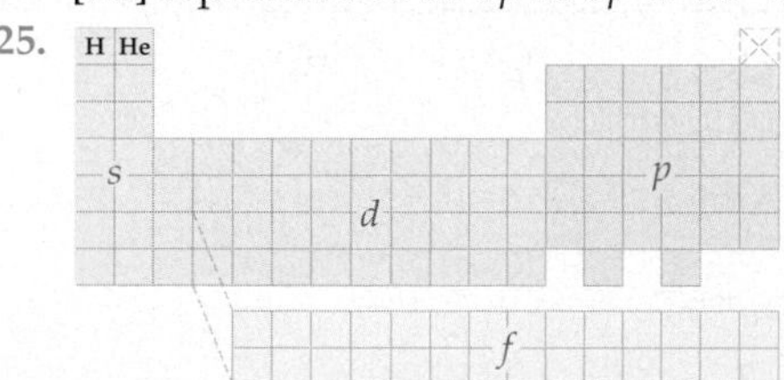

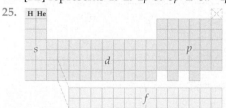

27. Group 1 elements form +1 ions because they lose one valence electron in the outer *s* shell to obtain a noble gas configuration. Group 7 elements form −1 ions

because when they gain an electron to fill their outer *p* orbital to obtain a noble gas configuration.

Problems

29. infrared
31. radiowaves < microwaves < infrared < ultraviolet
33. gamma, ultraviolet, or X-rays
35. **a.** X-rays **b.** X-rays **c.** microwaves
37. energies, distances
39. $n = 6 \rightarrow n = 2$ 410 nm
$n = 5 \rightarrow n = 2$ 434 nm
41.

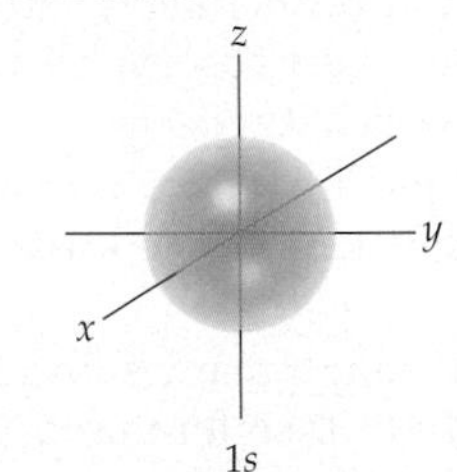

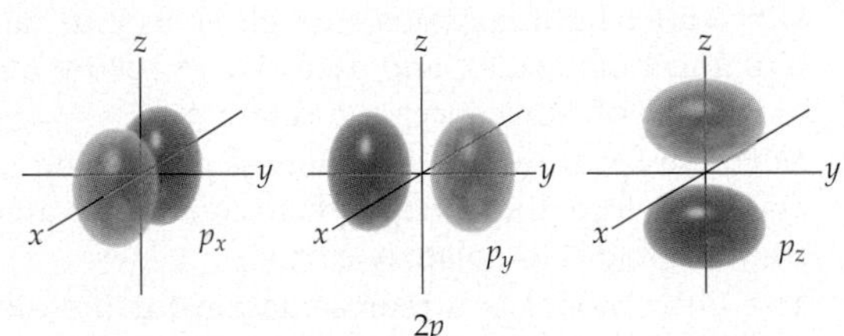

2p

The 2*s* and 3*p* orbitals are bigger than the 1*s* and 2*p* orbitals.

43. Electron in the 2*s* orbital
45. $2p \rightarrow 1s$
47. **a.** $1s^22s^22p^2$ **b.** $1s^22s^22p^63s^1$
c. $1s^22s^22p^63s^23p^6$ **d.** $1s^22s^22p^63s^23p^2$
49. **a.**

↑↓	↑↓	↑↓ ↑ ↑
1s	2s	2p

b.

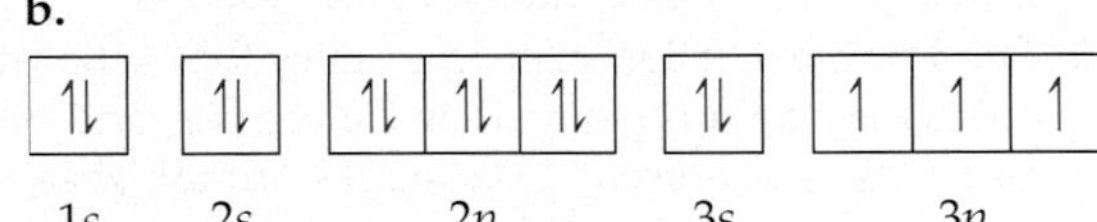

c.

↑↓	↑↓	↑↓ ↑↓ ↑
1s	2s	2p

d.

↑↓	↑↓	↑↓ ↑↓ ↑↓	↑↓
1s	2s	2p	3s

51. **a.** [Ar] $4s^23d^{10}4p^1$
b. [Ar] $4s^23d^{10}4p^3$
c. [Kr] $5s^1$
d. [Kr] $5s^24d^{10}5p^2$
53. **a.** [Ar] $4s^23d^2$
b. [Ar] $4s^23d^3$
c. [Ar] $4s^13d^5$
d. [Ar] $4s^23d^5$
55. Valence electrons are underlined
a. $1s^2\underline{2s^22p^1}$
b. $1s^2\underline{2s^22p^3}$
c. $1s^22s^22p^63s^23p^64s^23d^{10}4p^6\underline{5s^2}4d^{10}\underline{5p^3}$
d. $1s^22s^22p^63s^23p^6\underline{4s^1}$
57. **a.** 6 **b.** 6
c. 7 **d.** 1
59. **a.** ns^1 **b.** ns^2
c. ns^2np^3 **d.** ns^2np^5
61. **a.** [Ne] $3s^23p^1$ **b.** [He]$2s^2$
c. [Kr] $5s^24d^{10}5p^1$ **d.** [Kr] $5s^24d^2$
63. **a.** [Ar] $4s^23d^{10}4p^3$ **b.** [Xe] $6s^2$
c. [Ar] $4s^23d^8$ **d.** [Xe] $6s^24f^{14}5d^{10}6p^3$
65. **a.** 2 **b.** 3
c. 5 **d.** 6
67. **a.** 8 **b.** 8
c. 18 **d.** 18
69. **a.** Al **b.** S
c. Ar **d.** Mg
71. **a.** Cl **b.** Ga
c. Fe **d.** Rb
73. **a.** Na **b.** Ge
c. cannot tell **d.** P
75. Pb < Sn < Te < S < Cl
77. **a.** In **b.** Si
c. Pb **d.** C
79. F < S < Si < Ge < Ca < Rb
81. **a.** Sr **b.** Bi
c. cannot tell **d.** As
83. S < Se < Sb < In < Ba < Fr
85. Alkali metals have the general electron configuration of ns^1. If they lose their one *s* electron, they will obtain the electron configuration of a noble gas. This loss of an electron will give the alkali metal a +1 charge.
87. **a.** $1s^22s^22p^63s^23p^6$
b. $1s^22s^22p^63s^23p^6$
c. $1s^22s^22p^63s^23p^6$
d. $1s^22s^22p^63s^23p^64s^23d^{10}4p^6$
89. Metals tend to form positive ions because they tend to lose electrons. Elements on the left side of the periodic table have only a few extra electrons, which they will lose to gain a noble gas configuration. Metalloids tend to be elements with 3 to 5 valence electrons; they could lose or gain electrons to obtain a noble gas configuration. Nonmetals want to gain electrons to fill their almost full valence shell, so they tend to form negative ions and are on the right side of the table.
91. **a.** Can only have 2 in the *s* shell and 6 in the *p* shell: $1s^22s^22p^63s^23p^3$.
b. There is no 2*d* subshell: $1s^22s^22p^63s^23p^2$.
c. There is no 1*p* subshell: $1s^22s^22p^3$.
d. Can only have 6 in the *p* shell: $1s^22s^22p^63s^23p^3$.
93. Bromine is highly reactive because it reacts quickly to gain an electron and obtain a stable valence shell.

Krypton is a noble gas because it already has a stable valence shell.

95. 660 nm
97. 8 min, 19 sec
99. The quantum-mechanical model provided the ability to understand and predict chemical bonding, which is the basic level of understanding of matter and how it interacts. This model was critical in the areas of lasers, computers, semiconductors, and drug design. The quantum-mechanical model for the atom is considered the foundation of modern chemistry.
101. Ultraviolet light is the only one of these three types of light that contains enough energy to break chemical bonds in biological molecules.

Chapter 10

Questions

1. Bonding theories predict how atoms bond together to form molecules and also predict what combinations of atoms form molecules and what combinations do not. Likewise, bonding theories explain the shapes of molecules, which in turn determine many of their physical and chemical properties.
3. Ne: $1s^2 2s^2 2p^6$, 8 valence electrons
   Ar: $1s^2 2s^2 2p^6 3s^2 3p^6$, 8 valence electrons
5. According to Lewis theory, a chemical bond is the sharing or transfer of electrons to attain stable electron configurations among the bonding atoms.
7. The Lewis structure for potassium has 1 valence electron while the Lewis structure for monatomic chlorine has 7 valence electrons. From these structures we can determine that if potassium gives up its one valence electron to chlorine, K^+ and Cl^- are formed; therefore the formula must be KCl.
9. Double and triple bonds are shorter and stronger than single bonds.
11. You determine the number of electrons that go into the Lewis structure of a molecule by summing the valence electrons of each atom in the molecule.
13. The octet rule is not sophisticated enough to be correct every time. For example, some molecules that exist in nature have an odd number of valence electrons, and thus will not have octets on all their constituent atoms. Some elements tend to form compounds in nature in which they have more (sulfur) or less (boron) than 8 valence electrons.
15. VSEPR theory predicts the shape of molecules using the idea that electron groups repel each other.
17. **a.** 180° **b.** 120° **c.** 109.5°
19. Electronegativity is the ability of an element to attract electrons within a covalent bond.
21. A polar covalent bond is a covalent bond that has a dipole moment.
23. It is important to know if a molecule is polar or nonpolar in order to determine how it will react with other molecules.

Problems

25. $1s^2\underline{2s^2 2p^3}$, ·N̈: (underlined electrons are the ones included)
27. **a.** K· **b.** Al: **c.** ·P̈: **d.** :Är:
29. :Ẍ: Halogens tend to gain one electron in a chemical reaction.
31. M: Alkaline earth metals tend to lose two electrons in a chemical reaction.
33. **a.** $[:\ddot{Cl}:]^-$ **b.** $[:\ddot{Se}:]^{2-}$ **c.** Na^+ **d.** Mg^{2+}
35. **a.** $Na^+[:\ddot{F}:]^-$ **b.** $Ca^{2+}[:\ddot{O}:]^{2-}$ **c.** $[:\ddot{Br}:]^- Sr^{2+}[:\ddot{Br}:]^-$ **d.** $K^+[:\ddot{O}:]^{2-}K^+$
37. **a.** SrSe **b.** $BaCl_2$ **c.** Na_2S **d.** Al_2O_3
39. **a.** $Li^+ [:\ddot{F}:]^-$ **b.** $Li^+ [:\ddot{O}:]^{2-} Li^+$ **c.** $Li^+ [:\ddot{N}:]^{3-} Li^+$ (with a third Li^+ above)
41. **a.** $Cs^+ [:\ddot{Cl}:]^-$ **b.** $Ba^{2+} [:\ddot{O}:]^{2-}$ **c.** $[:\ddot{I}:]^- Ca^{2+} [:\ddot{I}:]^-$
43. **a.** H:P̈:H (with H above P) **b.** :C̈l:S̈:C̈l: **c.** :F̈:F̈: **d.** H:Ï:
45. **a.** Ö=Ö **b.** :C≡O: **c.** H:Ö:N̈=Ö ⟷ H:Ö=N̈:Ö: **d.** Ö=S̈:Ö: ⟷ :Ö:S̈=Ö
47. **a.** H:C≡C:H **b.** $H_2C=CH_2$ (each C bonded to two H) **c.** H–N̈=N̈–H **d.** H₂N̈:N̈H₂ (H H over :N:N: over H H)
49. **a.** :N≡N: **b.** S̈=Si=S̈ **c.** H:Ö:H **d.** I:N̈:I (with I above N)
51. **a.** Ö=Se:Ö: ⟷ :Ö:Se=Ö
    b. $[CO_3]^{2-}$ resonance structures: three structures, each with one C=O double bond and two C–O single bonds, connected by ⟷

c. $[:\ddot{C}l:\ddot{O}:]^-$

d. $[:\ddot{O}:\ddot{C}l:\ddot{O}:]^-$

53. a. $\left[\begin{array}{c} :\ddot{O}: \\ :\ddot{O}:\ddot{P}:\ddot{O}: \\ :\ddot{O}: \end{array}\right]^{3-}$

b. $[:C\equiv N:]^-$

c. $[\ddot{O}=N:\ddot{O}:]^- \longleftrightarrow [:\ddot{O}:N=\ddot{O}]^-$

d. $\left[\begin{array}{c} :\ddot{O}: \\ :\ddot{O}:\ddot{S}:\ddot{O}: \end{array}\right]^{2-}$

55. a. $\begin{array}{c} :\ddot{C}l: \\ :\ddot{C}l:B:\ddot{C}l: \end{array}$

b. $\ddot{O}=\ddot{N}:\ddot{O}: \longleftrightarrow :\ddot{O}:\ddot{N}=\ddot{O}$

c. $\begin{array}{c} H \\ H:\ddot{B}:H \end{array}$

57. a. 4 b. 4
c. 4 d. 2

59. a. 3 bonding groups, 1 lone pair
b. 2 bonding groups, 2 lone pairs
c. 4 bonding groups, 0 lone pairs
d. 2 bonding groups, 0 lone pairs

61. a. tetrahedral
b. trigonal planar
c. linear
d. trigonal planar

63. a. 109.5° b. 120°
c. 180° d. 120°

65. a. linear, linear
b. trigonal planar, bent
c. tetrahedral, bent
d. tetrahedral, trigonal pyramidal

67. a. 180° b. 120°
c. 109.5° d. 109.5°

69. a. linear, linear
b. trigonal planar, bent
c. tetrahedral, trigonal pyramidal

71. a. trigonal planar b. bent
c. trigonal planar d. tetrahedral

73. a. 1.2 b. 1.8 c. 2.8

75. a. polar covalent b. ionic
c. pure covalent d. polar covalent

77. a. polar b. nonpolar
c. nonpolar d. polar

79. a. nonpolar b. polar
c. nonpolar d. polar

81. a. nonpolar b. polar
c. nonpolar d. polar

83. a. $1s^22s^22p^63s^23p^6\underline{4s^2}$, Ca: (underlined electrons are the ones included)

b. $1s^22s^22p^33s^23p^6\underline{4s^2}3d^{10}\underline{4p^1}$, $\dot{G}a$:

c. [Ar] $\underline{4s^2}3d^{10}\underline{4p^3}$, $\cdot\dot{\underset{\cdot}{A}}s$:

d. [Kr] $\underline{5s^2}4d^{10}\underline{5p^5}$, $:\ddot{\underset{\cdot\cdot}{I}}:$

85. a. ionic, $K^+[:\ddot{S}:]^{2-}K^+$ b. covalent, $\begin{array}{c} H \\ :\ddot{F}:\ddot{C}=\ddot{O} \end{array}$

c. ionic, $Mg^{2+}[:\ddot{S}e:]^{2-}$ d. covalent, $\begin{array}{c} :\ddot{B}r: \\ :\ddot{B}r:\ddot{P}:\ddot{B}r: \end{array}$

87. $\begin{array}{c} \ddot{O} \\ \| \\ :\ddot{C}l:C:\ddot{C}l: \end{array}$, polar, $\begin{array}{c} O \\ | \\ C \\ Cl \quad\quad Cl \end{array}$

89. $\begin{array}{c} \ddot{O} \\ H \; \| \\ H:\ddot{C}:C:\ddot{O}:H \\ H \end{array}$ $\quad$ $\begin{array}{c} H \\ H\cdots C \\ H \quad C=O \\ O \\ H \end{array}$

91. $H:\ddot{C}l: + Na^+[:\ddot{O}:H]^- \longrightarrow$

$H:\ddot{O}:H + Na^+[:\ddot{C}l:]^-$

93. a. $K^+[:\ddot{O}:H]^-$

b. $K^+\left[\begin{array}{c} :\ddot{O}: \\ \ddot{O}=\ddot{N}:\ddot{O}: \end{array}\right]^- \longleftrightarrow K^+\left[\begin{array}{c} \ddot{O} \\ \| \\ :\ddot{O}:\ddot{N}:\ddot{O}: \end{array}\right]^- \longleftrightarrow$

$K^+\left[\begin{array}{c} :\ddot{O}: \\ :\ddot{O}:\ddot{N}=\ddot{O} \end{array}\right]^-$

c. $Li^+[:\ddot{I}:\ddot{O}:]^-$

d.

$Ba^{2+}\left[\begin{array}{c} \ddot{O} \\ \| \\ C \\ :\ddot{O}: \; :\ddot{O}: \end{array}\right]^{2-} \longleftrightarrow Ba^{2+}\left[\begin{array}{c} :\ddot{O}: \\ C \\ :\ddot{O} \quad\quad \ddot{O}: \end{array}\right]^{2-} \longleftrightarrow$

$Ba^{2+}\left[\begin{array}{c} :\ddot{O}: \\ C \\ :\ddot{O} \quad\quad \ddot{O}: \end{array}\right]^{2-}$

95. a. $[:\ddot{O}:\ddot{O}:]^-$ b. $[:\ddot{O}:]^-$

c. $:\ddot{O}:H$ d. $\begin{array}{c} H \\ H:\ddot{C}:\ddot{O}:\ddot{O}: \\ H \end{array}$

97. a. H_2Se, H–Se–H (bent structure)

b. structure is correct

c. PCl_3, Cl–P(–Cl)–Cl (trigonal pyramidal structure)

d. structure is correct

Chapter 11

Questions

1. Pressure is the push (or force) exerted per unit area by gaseous molecules as they collide with the surfaces around them.
3. The longest straw that could work correctly is about 10.3 m long.
5. Gases are compressible due to the large amount of empty space between their molecules. Gases assume the shape and volume of their container because the atoms or molecules are in constant motion. Gases have low densities because they are mostly empty space.
7. The main units used to measure pressure are Pa, atm, mm Hg, torr.
9. Boyle's law from the perspective of kinetic molecular theory states that if the volume of a gas container is decreased, the same number of gas particles is crowded into a smaller volume, causing more collisions with the walls of the container and therefore increasing the pressure.
11. When an individual is more than a couple of meters underwater, the air pressure in the lungs is greater than the air pressure at the water's surface. If a snorkel were used, it would move the air from the lungs to the surface, making it very difficult to breath.
13. Charles's law from the perspective of kinetic molecular theory states that if the temperature of a gas is increased, the particles move faster and will collectively occupy more space.
15. The combined gas law is $\frac{P_1V_1}{T_1} = \frac{P_2V_2}{T_2}$. It is useful when more than one variable of a gas changes at the same time.
17. Avogadro's law from the perspective of kinetic molecular theory states that if the numbers of particles increase, they occupy more volume.
19. The ideal gas law is most accurate when the volume of gas particles is small compared to the space between them. It is also accurate when the forces between particles are not important. The ideal gas law breaks down at high pressures and low temperatures. This breakdown occurs because the gases are no longer acting according to the kinetic molecular theory.
21. Dalton's law states that the sum of the partial pressures in a gas mixture must equal the total pressure. $P_{tot} = P_A + P_B + P_C + \ldots$
23. Deep-sea divers breathe helium with oxygen because helium, unlike nitrogen, does not have physiological effects under high-pressure conditions. The oxygen concentration in the mixture is low to avoid oxygen toxicity.
25. Vapor pressure is the partial pressure of a gas in with its liquid. Partial pressure increases with increasing temperature.

Problems

27. **a.** 1.16 atm **b.** 1.33 atm **c.** 1.005 atm **d.** 0.971 atm
29. **a.** 1.7×10^3 torr **b.** 36 mm Hg **c.** 1.28×10^3 mm Hg **d.** 834.2 torr
31. **a.** 0.832 atm **b.** 632 mm Hg **c.** 12.2 psi **d.** 8.43×10^4 Pa
33. **a.** 809.0 mm Hg **b.** 1.065 atm **c.** 809.0 torr **d.** 107.9 kPa
35. 5.7×10^2 mm Hg
37. 1.8 L
39. 4.8 L
41. 58.9 mL
43. 5.7 L
45. 4.33 L
47. 1.95×10^3 mm Hg
49. 0.76 L
51. 877 mm Hg
53. 3.0 L
55. 2.1 mol
57. 1.42 mol
59. 0.23 mol
61. 44.0 g/mol
63. 4.00 g/mol
65. 712 torr
67. 10.7 atm
69. 700 mm Hg
71. 0.87 atm N; 0.25 atm O
73. 0.34 atm
75. **a.** 56 L **b.** 1.3×10^2 L **c.** 732 L **d.** 9.2×10^2 L
77. **a.** 59.1 L **b.** 30.0 L **c.** 72.1 L **d.** 0.125 L
79. 0.350 g
81. 42 L
83. 33 L H_2; 16 L CO
85. 8.83 L
87. 12.6 g
89. 0.5611 g
91. $V = \frac{nRT}{P}$

$$= \frac{1.00\ \text{mol}\left(0.0821 \frac{\text{L}\cdot\text{atm}}{\text{mol}\cdot\text{K}}\right)(273\ \text{K})}{1.00\ \text{atm}} = 22.4\ \text{L}$$

93. 27.9 g/mol
95. C_4H_{10}
97. 0.827 g
99. 0.128 g
101. 0.935 L
103. **a.** SO_2, 0.0127 mol **b.** 65.8%
105. **a.** NO_2, 24.0 g **b.** 61.7%
107. c; with the aid of Boyle's law, we can see that an increase in density relates directly to an increase in pressure. Thus c, with the highest density, has the greatest pressure.
109. 22.8 g

Chapter 12

Questions

1. The bitter taste is due to the interaction of molecules with receptors on the surface of specialized cells on the tongue.
3. Intermolecular forces are what living organisms depend on for many physiological processes. Intermolecular forces are also responsible for the existence of liquids and solids.
5. The magnitude of intermolecular forces relative to the amount of thermal energy in the sample determines the state of the matter.
7. Properties of solids:
   a. Solids have high densities in comparison to gases.
   b. Solids have a definite shape.
   c. Solids have a definite volume.
   d. Solids may be crystalline or amorphous.
9. Properties of solids explained
   a. Solids have high densities in comparison to gases.
   The solid particles are in close contact, whereas gas particles are not.
   b. Solids have a definite shape.
   The particles of a solid are in fixed positions.
   c. Solids have a definite volume.
   The particles of a solid are in close contact.
   d. Solids may be crystalline or amorphous.
   A crystalline solid is a well-ordered three-dimensional array of solid particles.
11. Surface tension is the tendency of liquids to minimize their surface area. Molecules at the surface have few neighbors to interact with via intermolecular forces.
13. Evaporation is a physical change in which a substance is converted from its liquid form to its gaseous form. Condensation is a physical change in which a substance is converted from its gaseous form to its liquid form.
15. Evaporation below the boiling point occurs because molecules on the surface of the liquid experience fewer attractions to the neighboring molecules and can therefore break away. At the boiling point, evaporation occurs faster due to more of the molecules having sufficient thermal energy to break away (including internal molecules).
17. Acetone has weaker intermolecular forces than water. Acetone is more volatile than water.
19. Vapor pressure is the partial pressure of a gas in dynamic equilibrium with its liquid. It increases with increasing temperature and also increases with decreasing strength of intermolecular forces.
21. A steam burn is worse than a water burn at the same temperature (100 °C), because when the steam condenses on the skin, it releases large amounts of additional heat.
23. As the first molecules freeze, they release heat, making it harder for other molecules to freeze without the aid of a refrigeration mechanism, which would draw heat out.
25. Dispersion forces are the default intermolecular force present in all molecules and atoms. Dispersion forces are caused by fluctuations in the electron distribution within molecules or atoms. Dispersion forces are the weakest type of intermolecular force and increase with increasing molar mass.
27. Hydrogen bonding is an intermolecular force and is sort of a super dipole–dipole force. Hydrogen bonding occurs in compounds containing hydrogen atoms bonded directly to fluorine, oxygen, or nitrogen.
29. Molecular solids as a whole tend to have low to moderately low melting points; however, strong molecular forces can increase their melting points relative to each other.
31. Ionic solids tend to have much higher melting points than molecular solids.
33. Water is unique for a couple reasons. Water has a low molar mass, yet it is still liquid at room temperature and has a relatively high boiling point. Unlike other substances, which contract upon freezing, water expands upon freezing.

Problems

35. The 55 mL of water in a dish with a diameter of 12 cm will evaporate more quickly because it has a larger surface area.
37. Acetone feels cooler while evaporating from one's hand, for it is more volatile than water and evaporates much faster.
39. The ice's temperature will increase from −5 °C to 0 °C, where it will then stay constant while the ice completely melts. After the melting process is complete, the water will continue to steadily rise in temperature until it reaches room temperature (25 °C).

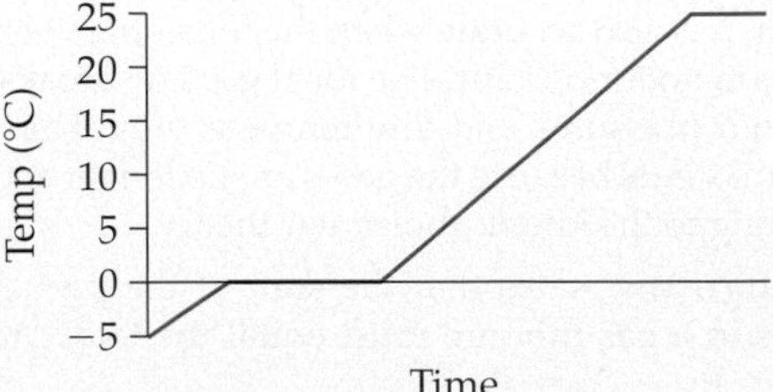

41. Allowing 0.05 g of 100 °C steam to condense on your hand would cause a more severe burn than spilling 0.05 g of 100 °C water on your hand. When the steam condenses on the skin, it releases large amounts of additional heat.
43. The −8 °C ice chest will cause the water in the watery bag of ice to freeze, which is an exothermic process. The freezing process of the water will release heat, and the temperature of the ice chest will increase.
45. The ice chest filled with ice at 0 °C will be colder after a couple hours than the ice chest filled with water at 0 °C. This is because the ice can absorb additional heat as it melts.
47. Denver is a mile above sea level and thus has a lower air pressure. Due to this low pressure, the point at which the vapor pressure of water equals the external pressure will occur at a lower temperature.
49. 78.4 kJ
51. 6.8 kJ
53. 6.25 kJ
55. 371 g
57. 9.65 kJ
59. 15.9 kJ
61. **a.** dispersion
    b. dispersion
    c. dispersion, dipole–dipole
    d. dispersion, dipole–dipole, hydrogen bonding
63. **a.** dispersion, dipole–dipole
    b. dispersion, dipole–dipole, hydrogen bonding
    c. dispersion
    d. dispersion
65. d, because it has the highest molecular weight
67. CH_3OH, due to strong hydrogen bonding
69. NH_3, because it has the ability to form hydrogen bonds
71. These two substances are not miscible, for H_2O is polar and $CH_3CH_2CH_2CH_2CH_3$ is nonpolar.
73. **a.** atomic **b.** molecular
    c. ionic **d.** atomic
75. **a.** molecular **b.** ionic
    c. molecular **d.** molecular
77. C. LiCl(*s*); it is an ionic solid and possesses ionic bonds resulting in a higher melting point.
79. **a.** Ti(*s*); Ti is a covalent atomic solid and Ne is a nonbonding atomic solid.
    b. H_2O(*s*); while both are molecular solids, water has strong hydrogen bonding.
    c. Xe(*s*); both are nonbonding atomic solids, but Xe has a higher molar mass.
    d. NaCl(*s*); NaCl(*s*) is an ionic solid and CH_4(*s*) is molecular solid.
81. **a.** 1.9×10^4 J **b.** 19 kJ
    c. 4.6×10^3 cal **d.** 4.6 Cal
83. 2.7 °C
85. 88.1 g
87. **a.** H:Se:H (with lone pairs on Se); bent; dispersion, dipole–dipole
    b. :O=S–O: (with lone pairs); bent; dispersion, dipole–dipole
    c. $CHCl_3$ (H above C, Cl on left, right and below, each with lone pairs); tetrahedral; dispersion, dipole–dipole
    d. :O=C=O: ; linear; dispersion
89. $Na^+[:F:]^-$ $Mg^{2+}[:O:]^{2-}$ MgO has a higher melting point, because the magnitude of the charges on the ions is greater.
91. The molecule in the interior has the most neighbors. The molecule on the surface is more likely to evaporate. In three dimensions, the molecules that make up the surface area have the least amount of neighbors and are more likely to evaporate than the molecules in the interior.
93. a. Yes, if the focus is on the melting of icebergs. No, the melting of an ice cube in a cup of water will *not* raise the level of the liquid in the cup; when the ice cube melts, the volume of the water created will be less than the volume of the initial ice cube.
    b. Yes, for the ice sheets that sit on the continent of Antarctica are above sea level and if some melted the water created would be added to the ocean without decreasing the amount of ice below the ocean's surface.

Chapter 13

Questions

1. A solution is a homogeneous mixture of two or more substances. Some examples are air, seawater, soda water, and brass.
3. In a solution, the solvent is the majority component of the mixture and the solute is the minority component. For example, in a seawater solution, the water is the solvent and the salt content is the solute.
5. Solubility is the amount of the compound, usually in grams, that will dissolve in a specified amount of solvent.
7. In solutions with solids, soluble ionic solids form strong electrolyte solutions, while soluble molecular solids form nonelectrolyte solutions. Strong electrolyte solutions are solutions containing solutes that dissociate into ions, for example, $BaCl_2$ and NaOH.
9. Recrystallization is a common way to purify a solid. In recrystallization, enough solid is put into high-temperature water until a saturated solution is created. Then the solution cools slowly, and crystals result from the solution. The crystalline structure tends to reject impurities, resulting in a purer solid.
11. The bubbles formed on the bottom of a pot of heated water (before boiling) are dissolved air coming out of the solution. These gases come out of solution because the solubility of the dissolved nitrogen and oxygen decreases as the temperature of the water rises.

13. The higher the pressure above a liquid, the more soluble the gas is in the liquid.
15. A dilute solution is one containing small amounts of a solute relative to solvent and a concentrated solution is one containing large amounts of solute relative to solvent.
17. Molarity is a common unit of concentration of a solution defined as moles of solute per liter of solution.
19. The boiling point of a solution containing a nonvolatile solute is higher than the boiling point of the pure solvent. The melting point of the solution, however is lower.
21. Molality is a common unit of concentration of a solution expressed as number of moles of solute per kilogram of solvent.
23. Water tends to move from lower concentrations to higher concentrations and when the salt water is being passed through the human body, the salt content draws the water out of the body, causing dehydration.

Problems

25. c, d
27. **a.** solute: salt, solvent: water
    b. solute: sugar, solvent: water
    c. solute: soda, solvent: water
29. **a.** hexane **b.** water **c.** ethyl ether
31. ions, strong electrolyte solution
33. unsaturated
35. recrystallization
37. At room temperature water contains some dissolved oxygen gas; however, the boiling of the water will remove dissolved gases.
39. Under higher pressure the gas (nitrogen) will be more easily dissolved in the blood. To reverse this process, the diver should ascend to relieve the pressure.
41. **a.** 8.11% **b.** 10.3% **c.** 6.37%
43. 13%
45. 9.6 g
47. **a.** 2.7 g **b.** 13.7 mg **c.** 0.99 kg
49. **a.** 3.3×10^3 g
    b. 77 g
    c. 15 g
51. 1.6×10^2 g
53. 1.3×10^3 g
55. 1.4×10^4 mL
57. **a.** 0.52 M **b.** 0.263 M **c.** 0.199 M
59. **a.** 0.149 M **b.** 0.461 M **c.** 3.00×10^{-2} M
61. 0.60 M
63. **a.** 1.8 mol **b.** 0.38 mol **c.** 0.238 mol
65. **a.** 0.39 L **b.** 0.056 L **c.** 0.10 L
67. 4.8 g
69. 19 g
71. 8.8 L
73. 0.38 M
75. Dilute 0.045 L of the stock solution to 2.5 L.
77. 6.0×10^2 mL
79. 17.7 mL
81. **a.** 0.025 L **b.** 0.020 L **c.** 1.03 L
83. 4.45 mL
85. 0.373 M
87. 1.2 L
89. **a.** 1.0 m **b.** 3.92 m **c.** 0.52 m
91. 1.28 m
93. **a.** −1.6 °C **b.** −2.70 °C
    c. −8.9 °C **d.** −4.37 °C
95. **a.** 100.060 °C **b.** 100.993 °C
    c. 101.99 °C **d.** 101.11 °C
97. −1.27 °C, 100.348 °C
99. 2.28 M, 12.3%
101. 0.43 L
103. 319 mL
105. 0.17 L
107. −3.22 °C, 100.886 °C
109. −1.6 °C, 100.48 °C
111. **a.** Water will flow from left to right.
     b. Water will flow from right to left.
     c. Water won't flow between the two.
113. 3×10^4 L

Chapter 14

Questions

1. Sour gummy candies are coated with a white powder that is a mixture of citric acid and tartaric acid. The combination of these two acids creates the sour taste.
3. Acids have a sour taste; acids dissolve many metals; and acids turn litmus paper red.
5. Carboxylic acids are organic compounds with the general formula R—COOH. Some examples are acetic acid ($HC_2H_3O_2$) and formic acid ($HCHO_2$).
7. Alkaloids are organic bases found in plants.
9. The Arrhenius definition of an acid is a substance that produces H^+ ions in aqueous solution. An example:

   $HCl(aq) \rightarrow H^+(aq) + Cl^-(ag)$

11. The Brønsted-Lowery definition states that an acid is a proton donor and a base is a proton acceptor. The following is an example of a chemical equation demonstrating this definition:

    $HCl(aq) + H_2O(l) \rightarrow H_3O^+(aq) + Cl^-(aq)$

13. An acid–base neutralization reaction occurs when an acid and a base are mixed and the $H^+(aq)$ from the acid combines with the $OH^-(aq)$ from the base to form $H_2O(l)$. An example follows.

    $HCl(aq) + KOH(aq) \rightarrow H_2O(l) + KCl(aq)$

15. $2\,HCl(aq) + K_2O(s) \rightarrow H_2O(l) + 2\,KCl(aq)$
17. A titration is a laboratory procedure where a reactant in a solution of known concentration is reacted with another reactant in a solution of unknown concentration until the reaction has reached the endpoint. The

endpoint is the point at which the reactants are in exact stoichiometric proportions.

19. A strong acid is one that will completely dissociate in solution, while a weak acid does not completely dissociate in solution.

21. Monoprotic acids (such as HCl) contain only one hydrogen ion that will dissociate in solution, while diprotic acids (such as H_2SO_4) contain two hydrogen ions that will dissociate in solution.

23. Yes, pure water contains H_3O^+ ions. Through self-ionization, water acts as an acid and a base with itself; water is amphoteric.

25. **a.** $[H_3O^+] > 1.0 \times 10^{-7}$ M; $[OH^-] < 1.0 \times 10^{-7}$ M
b. $[H_3O^+] < 1.0 \times 10^{-7}$ M; $[OH^-] > 1.0 \times 10^{-7}$ M
c. $[H_3O^+] = 1.0 \times 10^{-7}$ M; $[OH^-] = 1.0 \times 10^{-7}$ M

27. A buffer is a solution that resists pH change by neutralizing added acid or added base.

29. The cause of acid rain is the combustion of fossil fuels.

31. Acid rain damages structures made out of metal, marble, cement, and limestone, as well as harming and possibly killing aquatic life and trees.

Problems

33. a. acid; $HNO_3(aq) \rightarrow H^+(aq) + NO_3^-(ag)$
b. base; $KOH(aq) \rightarrow K^+(aq) + OH^-(ag)$
c. acid; $HC_2H_3O_2(aq) \rightarrow H^+(aq) + C_2H_3O_2^-(ag)$

35.

	B-L Acid	B-L Base	Conj. Acid	Conj. Base
a.	HBr	H_2O	H_3O^+	Br^-
b.	H_2O	NH_3	NH_4^+	OH^-
c.	HNO_3	H_2O	H_3O^+	NO_3^-
d.	H_2O	C_5H_5N	$C_5H_5NH^+$	OH^-

37. a, c

39. **a.** Cl^- **b.** HSO_3^-
c. CHO_2^- **d.** F^-

41. **a.** NH_4^+ **b.** $HClO_4$
c. H_2SO_4 **d.** HCO_3^-

43. **a.** $HI(aq) + NaOH(aq) \rightarrow H_2O(l) + NaI(aq)$
b. $HBr(aq) + KOH(aq) \rightarrow H_2O(l) + KBr(aq)$
c. $2\ HNO_3(aq) + Ba(OH)_2(aq) \rightarrow 2\ H_2O(l) + Ba(NO_3)_2(aq)$
d. $2\ HClO_4(aq) + Sr(OH)_2(aq) \rightarrow 2\ H_2O(l) + Sr(ClO_4)_2(aq)$

45. **a.** $2\ K(aq) + 2\ HBr(aq) \rightarrow H_2(g) + 2\ KBr(aq)$
b. $Ca(aq) + 2\ HBr(aq) \rightarrow H_2(g) + CaBr_2(aq)$
c. $Sr(aq) + 2\ HBr(aq) \rightarrow H_2(g) + SrBr_2(aq)$
d. $2\ Rb(aq) + 2\ HBr(aq) \rightarrow H_2(g) + 2\ RbBr(aq)$

47. **a.** $MgO(aq) + 2\ HI(aq) \rightarrow H_2O(l) + MgI_2(aq)$
b. $K_2O(aq) + 2\ HI(aq) \rightarrow H_2O(l) + 2\ KI(aq)$
c. $Rb_2O(aq) + 2\ HI(aq) \rightarrow H_2O(l) + 2\ RbI(aq)$
d. $CaO(aq) + 2\ HI(aq) \rightarrow H_2O(l) + CaI_2(aq)$

49. **a.** 0.1400 M **b.** 0.138 M
c. 0.08764 M **d.** 0.182 M

51. 0.1018 M

53. **a.** strong **b.** weak
c. strong **d.** weak

55. **a.** $[H_3O^+] = 2.5$ M **b.** $[H_3O^+] = 1.2$ M
c. $[H_3O^+] < 0.25$ M **d.** $[H_3O^+] < 2.25$ M

57. **a.** strong **b.** weak
c. strong **d.** weak

59. **a.** $[OH^-] = 0.88$ M **b.** $[OH^-] < 0.88$ M
c. $[OH^-] = 0.88$ M **d.** $[OH^-] = 1.55$ M

61. **a.** acidic **b.** acidic
c. neutral **d.** basic

63. **a.** 6.7×10^{-6} M, basic **b.** 1.1×10^{-6} M, basic
c. 4.5×10^{-9} M, acidic **d.** 1.4×10^{-11} M, acidic

65. **a.** 3.7×10^{-3} M, acidic **b.** 4.0×10^{-13} M, basic
c. 9.1×10^{-5} M, acidic **d.** 3.0×10^{-11} M, basic

67. **a.** basic **b.** neutral
c. acidic **d.** acidic

69. **a.** 7.77 **b.** 7.0
c. 5.66 **d.** 3.13

71. **a.** 2.8×10^{-9} M **b.** 5.89×10^{-12} M
c. 1.3×10^{-3} M **d.** 6.0×10^{-2} M

73. **a.** 7.28 **b.** 6.41
c. 3.86 **d.** 13.0

75. **a.** 1.8×10^{-10} M **b.** 3.4×10^{-2} M
c. 3.2×10^{-13} M **d.** 1.8×10^{-6} M

77. various answers

79. c, d

81. $HCl(aq) + NaF(aq) \rightarrow HF(aq) + NaCl(aq)$
$HCl(aq) + KC_2H_3O_2(aq) \rightarrow HC_2H_3O_2(aq) + KCl(aq)$

83. 50.0 mL

85. 0.16 L

87. 65.2 g

89. 60.0 g/mol

91. **a.** 2.6, acidic **b.** 12, basic
c. 8.0, basic **d.** 1.71, acidic

93.

$[H_3O^+]$	$[OH^-]$	pH	Acidic or Basic
1.0×10^{-4}	1.0×10^{-10}	4.00	acidic
5.5×10^{-3}	1.8×10^{-12}	2.26	acidic
3.1×10^{-9}	3.2×10^{-6}	8.50	basic
4.8×10^{-9}	2.1×10^{-6}	8.32	basic
2.8×10^{-8}	3.5×10^{-7}	7.55	basic

95. **a.** $[H_3O^+] = 0.0150$ M
$[OH^-] = 6.67 \times 10^{-13}$ M
pH = 1.82
b. $[H_3O^+] = 1.5 \times 10^{-3}$ M
$[OH^-] = 6.7 \times 10^{-12}$ M
pH = 2.8
c. $[H_3O^+] = 9.77 \times 10^{-4}$ M
$[OH^-] = 1.02 \times 10^{-11}$ M
pH = 3.01
d. $[H_3O^+] = 0.0878$ M
$[OH^-] = 1.14 \times 10^{-13}$ M
pH = 1.06

97. **a.** $[OH^-] = 0.15$ M
$[H_3O^+] = 6.7 \times 10^{-14}$ M
pH = 13

b. $[OH^-] = 3.0 \times 10^{-3}$ M
$[H_3O^+] = 3.3 \times 10^{-12}$ M
pH = 11

c. $[OH^-] = 9.6 \times 10^{-4}$ M
$[H_3O^+] = 1.0 \times 10^{-11}$ M
pH = 11

d. $[OH^-] = 8.7 \times 10^{-5}$ M
$[H_3O^+] = 1.1 \times 10^{-10}$ M
pH = 9.9

99. **a.** weak **b.** strong
c. weak **d.** strong

101. approximately 8 times more concentrated

Chapter 15

Questions

1. The two general concepts involved in equilibrium are sameness and changelessness.
3. The rate of a chemical reaction is the amount of reactant that changes to product in a given period of time.
5. By controlling reaction rates, chemists can control the amount of a product that forms in a given period of time and have control of the outcome.
7. The two factors that influence reaction rates are concentration and temperature. The rate of a reaction increases with increasing concentration. The rate of a reaction increases with increasing temperature.
9. In a chemical reaction, dynamic equilibrium is the condition in which the rate of the forward reaction equals the rate of the reverse reaction.
11. Because the rate of the forward and reverse reactions is the same at equilibrium, the relative concentrations of reactants and products becomes constant.
13. The equilibrium constant is a measure of how far a reaction goes, and this constant is significant because it is a way to quantify the concentrations of the reactants and products at equilibrium.
15. A small equilibrium constant shows that a reverse reaction is favored and that when equilibrium is reached, there will be more reactants than products. A large equilibrium constant shows that a forward reaction is favored and that when equilibrium is reached, there will be more products than reactants.
17. No, the particular concentrations of reactants and products at equilibrium will not always be the same for a given reaction—they will depend on the initial concentrations.
19. Various answers depending on the answer for question number 12.
21. Decreasing the concentration of a reactant in a reaction mixture at equilibrium causes the reaction to shift to the left.
23. Decreasing the concentration of a product in a reaction mixture at equilibrium causes the reaction to shift to the right.
25. Increasing the pressure of a reaction mixture at equilibrium if the product side has fewer moles of gas particles than the reactant side causes the reaction to shift to the right.
27. Decreasing the pressure of a reaction mixture at equilibrium if the product side has fewer moles of gas particles than the reactant side causes the reaction to shift to the left.
29. Increasing the temperature of an exothermic reaction mixture at equilibrium causes the reaction to shift left, absorbing some of the added heat. Decreasing the temperature of an exothermic reaction mixture at equilibrium causes the reaction to shift right, releasing heat.
31. $K_{sp} = [A^{2+}][B^-]^2$
33. The solubility of a compound is the amount of the compound that dissolves in a certain amount of liquid, and the molar solubility is the solubility in units of moles per liter.
35. Two reactants with a large K_{eq} for a particular reaction might not react immediately when combined because of a large activation energy, which is an energy hump that normally exists between the reactants and products. The activation energy must be overcome before the system will undergo a reaction.
37. No, a catalyst does not affect the value of the equilibrium constant; it simply lowers the activation energy and increases the rate of a chemical reaction.

Problems

39. **a.** $K_{eq} = \dfrac{[N_2O_4]}{[NO_2]^2}$ **b.** $K_{eq} = \dfrac{[NO]^2[Br_2]}{[BrNO]^2}$
c. $K_{eq} = \dfrac{[H_2][CO_2]}{[H_2O][CO]}$ **d.** $K_{eq} = \dfrac{[CS_2][H_2]^4}{[CH_4][H_2S]^2}$

41. **a.** $K_{eq} = \dfrac{[Cl_2]}{[PCl_5]}$ **b.** $K_{eq} = [O_2]^3$
c. $K_{eq} = \dfrac{[H_3O^+][F^-]}{[HF]}$ **d.** $K_{eq} = \dfrac{[NH_4^+][OH^-]}{[NH_3]}$

43. $K_{eq} = \dfrac{[H_2]^2[S_2]}{[H_2S]^2}$

45. **a.** reactants **b.** products
c. reactants **d.** both

47. 0.0394
49. 1.79×10^{-5}
51. 0.0987
53. 0.82
55. 0.119
57.

T(K)	$[N_2]$	$[H_2]$	$[NH_3]$	K_{eq}
500	0.115	0.105	0.439	1.45×10^3
575	0.110	0.25	0.128	9.6
775	0.120	0.140	0.00439	0.0584

59. **a.** shift right
b. shift left

c. shift right

61. a. no effect b. shift left
c. shift left d. shift right

63. a. shift right b. shift left

65. a. no effect b. no effect

67. a. shift right b. shift left

69. a. shift left b. shift right

71. a. no effect b. shift right
c. shift left d. shift right
e. no effect

73. a. $CaSO_4(s) \rightleftharpoons Ca^{2+}(aq) + SO_4^{2-}(aq)$
$K_{sp} = [Ca^{2+}][SO_4^{2-}]$
b. $AgCl(s) \rightleftharpoons Ag^+(aq) + Cl^-(aq)$
$K_{sp} = [Ag^+][Cl^-]$
c. $CuS(s) \rightleftharpoons Cu^{2+}(aq) + S^{2-}(aq)$
$K_{sp} = [Cu^{2+}][S^{2-}]$
d. $FeCO_3(s) \rightleftharpoons Fe^{2+}(aq) + CO_3^{2-}(aq)$
$K_{sp} = [Fe^{2+}][CO_3^{2-}]$

75. $K_{sp} = [Fe^{2+}][OH^-]^2$

77. 7.0×10^{-11}

79. 1.35×10^{-4}

81. 7.04×10^{-5}M

83. 2.61×10^{-3}M

85.

Compound	[Cation]	[Anion]	K_{sp}
$SrCO_3$	2.4×10^{-5}	2.4×10^{-5}	5.8×10^{-10}
SrF_2	1.0×10^{-3}	2.0×10^{-3}	4.0×10^{-9}
Ag_2CO_3	2.6×10^{-4}	1.3×10^{-4}	8.8×10^{-12}

87. 3.3×10^{-7}

89. 534 g

91. b, c, d

93. 1.13×10^{-18} M, 5.4×10^{-16} g

95. e

97. 35.5 L

Chapter 16

Questions

1. A fuel-cell electric vehicle is an automobile running on an electric motor that is powered by hydrogen. The fuel cells use the electron-gaining tendency of oxygen and the electron-losing tendency of hydrogen to force electrons to move through a wire, creating the electricity that powers the car.

3. **a.** Oxidation is the gaining of oxygen and reduction is the losing of oxygen.
b. Oxidation is the loss of electrons and reduction is the gain of electrons.
c. Oxidation is an increase in oxidation state and reduction is a decrease in oxidation state.

5. A reducing agent is the substance being oxidized that causes the reduction of the other substance.

7. Good reducing agents have a strong tendency to *lose* electrons in reactions.

9. The oxidation state of a monotomic ion is equal to its charge.

11. For an ion, the sum of the oxidation states of the individual atoms must add up to *the charge of the ion*.

13. In a redox reaction, an atom that undergoes an increase in oxidation state is *oxidized*. An atom that undergoes a decrease in oxidation state is *reduced*.

15. When balancing redox equations, the number of electrons lost in the oxidation half-reaction must *equal* the number of electrons gained in the reduction half-reaction.

17. When balancing aqueous redox reactions occurring in acidic media, hydrogen is balanced using *H^+ ions*.

19. The metals at the top of the activity series are the most reactive.

21. The metals at the bottom of the activity series are least likely to lose electrons.

23. If the metal is listed above H_2 on the activity series, it will dissolve in acids such as HCl or HBr.

25. Oxidation occurs at the *anode* of an electrochemical cell.

27. The salt bridge joins the two half-cells or completes the circuit—it allows the flow of ions between the two half-cells.

29. The common dry cell battery does not contain large amounts of liquid water and is composed of a zinc case that acts as the anode. The cathode is a carbon rod immersed in a moist paste of MnO_2 that also contains NH_4Cl. The anode and cathode reactions occur to produce a voltage of about 1.5 volts.
anode reaction: $Zn(s) \rightarrow Zn^{2+}(aq) + 2\,e^-$
cathode reaction: $2\,MnO_2(s) + 2\,H_2O(l) + 2\,e^- \rightarrow 2\,MnO(OH)(s) + 2\,OH^-(aq)$

31. Fuel cells are like batteries, but the reactants are constantly replenished. The reactants constantly flow through the battery, generating electrical current as they undergo a redox reaction.
anode reaction:
$2\,H_2(g) + 4\,OH^+(aq) \rightarrow 4H_2O(l) + 4\,e^-$
cathode reaction:
$O_2(g) + 2\,H_2O(l) + 4\,e^- \rightarrow 4\,OH^-(aq)$

33. Corrosion is the oxidation of metals; the most common is rusting of iron.
oxidation:
$2\,Fe(s) \rightarrow 2\,Fe^{2+}(aq) + 4\,e^-$
reduction:
$O_2(g) + 2\,H_2O(l) + 4\,e^- \rightarrow 4\,OH^-(aq)$
overall:
$2\,Fe(s) + O_2(g) + 2\,H_2O(l) \rightarrow 2\,Fe(OH)_2(s)$

Problems

35. a. H_2 b. Al c. Al

37. a. Sr is oxidized, O_2 is reduced.
b. Ca is oxidized, Cl_2 is reduced.
c. Mg is oxidized, Ni^{2+} is reduced.

39. a. Sr is the reducing agent, O_2 is the oxidizing agent.
b. Ca is the reducing agent, Cl_2 is the oxidizing agent.

c. Mg is the reducing agent, Ni^{2+} is the oxidizing agent.

41. b (Fe_2), d (Cl_2)

43. a (K), c (Fe)

45. a. N_2 is oxidized and the reducing agent. O_2 is reduced and the oxidizing agent.
b. C is oxidized and the reducing agent. O_2 is reduced and the oxidizing agent.
c. Cl is oxidized and the reducing agent. Sb is reduced and the oxidizing agent.
d. K is oxidized and the reducing agent. Pb^{2+} is reduced and the oxidizing agent.

47. a. 0 b. +2
c. 0 d. 0

49. a. Na: + 1; Cl:−1 b. Ca: + 2; F:−1
c. S: + 4; O:−2 d. H: + 1; S:−2

51. a. +2 b. +4 c. +1

53. a. C: + 4; O:−2 b. O: − 2; H:+1
c. N: + 5; O:−2 d. N: + 3; O:−2

55. a. +1 b. +3
c. +5 d. +7

57. a. Sb + 5→ + 3, reduced
Cl − 1→0, oxidized
b. C + 2→ + 4, oxidized
Cl0 → − 1, reduced
c. N + 2→ + 3, oxidized
Br0 → − 1, reduced
d. H0 → + 1, oxidized
C + 4→ + 2, reduced

59. Na is the reducing agent. H is the oxidizing agent.

61. a. $3\,K(s) + Cr^{3+}(aq) \rightarrow Cr(s) + 3\,K^+(aq)$
b. $Mg(s) + 2\,Ag^+(aq) \rightarrow Mg^{2+}(aq) + 2\,Ag(s)$
c. $2\,Al(s) + 3\,Fe^{2+}(aq) \rightarrow 2\,Al^{3+}(aq) + 3\,Fe(s)$

63. a. reduction, $5\,e^- + MnO_4^-(aq) + 8\,H^+(aq) \rightarrow Mn^{2+}(aq) + 4\,H_2O(l)$
b. oxidation, $2\,H_2O(l) + Pb^{2+}(aq) \rightarrow PbO_2(s) + 4\,H^+(aq) + 2\,e^-$
c. reduction, $10\,e^- + 2\,IO_3^-(aq) + 12\,H^+(aq) \rightarrow I_2(s) + 6\,H_2O(l)$
d. oxidation, $SO_2(g) + 2\,H_2O(l) \rightarrow SO_4^{2-}(aq) + 4\,H^+(aq) + 2\,e^-$

65. a. $PbO_2(s) + 4\,H^+(aq) + 2\,I^-(aq) \rightarrow I_2(s) + Pb^{2+}(aq) + 2\,H_2O(l)$
b. $5\,SO_3^{2-}(aq) + 6\,H^+(aq) + 2\,MnO_4^-(aq) \rightarrow 5\,SO_4^{2-}(aq) + 2\,Mn^{2+}(aq) + 3\,H_2O(l)$
c. $S_2O_3^{2-}(aq) + 4\,Cl_2(g) + 5\,H_2O(l) \rightarrow 2\,SO_4^{2-}(aq) + 8\,Cl^-(aq) + 10\,H^+(aq)$

67. a. $ClO_4^-(aq) + 2\,H^+(aq) + 2\,Cl^-(aq) \rightarrow ClO_3^-(aq) + Cl_2(aq) + H_2O(l)$
b. $3\,MnO_4^-(aq) + 24\,H^+(aq) + 5\,Al(s) \rightarrow 3\,Mn^{2+}(aq) + 5\,Al^{3+}(aq) + 12\,H_2O(l)$
c. $Br_2(aq) + Sn(s) \rightarrow Sn^{2+}(aq) + 2\,Br^-(aq)$

69. c, Pb

71. b, Cu^{2+}

73. b, Al

75. b, c

77. Fe, Cr, Zn, Mn, Al, Mg, Na, Ca, K, Li

79. Al

81. a. $2\,Al(s) + 6\,HCl(aq) \rightarrow 2\,Al^{3+}(aq) + 6\,Cl^-(aq) + 3\,H_2(g)$
b. no reaction
c. $Pb(s) + 2\,HCl(aq) \rightarrow Pb^{2+}(aq) + 2\,Cl^-(aq) + H_2(g)$
d. $2\,Cr(s) + 6\,HCl(aq) \rightarrow 2\,Cr^{3+}(aq) + 6\,Cl^-(aq) + 3\,H_2(g)$

83.

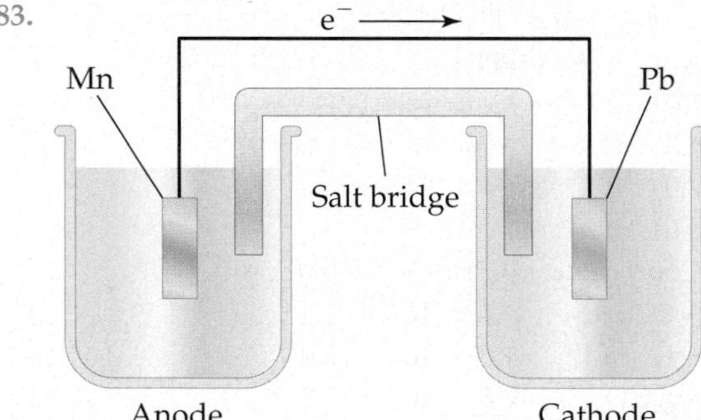

85. d

87. $Zn(s) + 2\,MnO_2(s) + 2\,H_2O(l) \rightarrow Zn(OH)_2(s) + 2\,MnO(OH)(s)$

89.

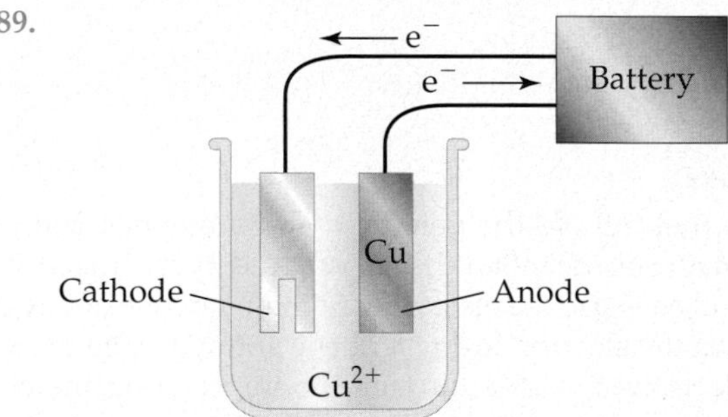

anode: $Cu(s) \rightarrow Cu^{2+}(aq) + 2\,e^-$
cathode: $Cu^{2+}(aq) + 2\,e^- \rightarrow Cu(s)$

91. a, Zn; c, Mn

93. a. redox; Zn is oxidized; Co is reduced.
b. not redox
c. not redox
d. redox; K is oxidized; Br is reduced.

95. 34.9 mL

97. 0.054 mol

99. a. $2\,Cr(s) + 6\,HI(aq) \rightarrow 2\,Cr^{3+}(aq) + 6\,I^-(aq) + 3\,H_2(g)$, 98 mL HI

b. $2\ Al(s) + 6\ HI(aq) \rightarrow$
$2\ Al^{3+}(aq) + 6\ I^{-}(aq) + 3\ H_2(g)$, 68 mL HI

c. no

d. no

101. 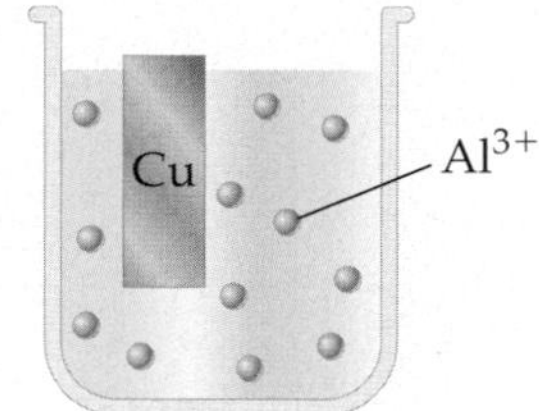

103. Many of the Zn atoms on the electrode would become Zn^{2+} ions in solution. Many Ni^{2+} ions in solution would become Ni atoms on the electrode.

Chapter 17

Questions

1. Radioactivity is the emission of tiny, invisible particles by disintegration of atomic nuclei. Many of these particles can pass right through matter. Atoms that emit these particles are radioactive.
3. Antoine-Henri Becquerel first discovered radioactivity while he was studying X-rays. He came across the idea when his photographic plate became exposed even in complete darkness, concluding that the uranium atoms he was using were constantly emitting some tiny particle.
5. Marie Curie further investigated uranic rays, and she discovered two new elements. Seeing these rays in numerous elements, she changed their name from uranic rays to radioactivity.
7. X: chemical symbol, used to identify the element.
   A: mass number, which is the sum of the number of protons and number of neutrons in the nucleus.
   Z: atomic number, which is the number of protons in the nucleus.
9. Alpha radiation occurs when an unstable nucleus emits a small piece of itself composed of 2 protons and 2 neutrons. The symbol for an alpha particle is ${}^{4}_{2}He$.
11. Alpha particles have high ionizing power and low penetrating power compared to beta and gamma particles.
13. When an atom emits a beta particle, its atomic number increases by one, because it now has an additional proton.
15. Gamma radiation is electromagnetic radiation, and the symbol for a gamma ray is ${}^{0}_{0}\gamma$.
17. Gamma particles have low ionizing power and high penetrating power compared to alpha and beta particles.
19. When an atom emits a positron, its atomic number decreases by one, because it now has one less proton.
21. A nuclear equation represents the changes that occur during radioactivity and other nuclear processes. For a nuclear equation to be balanced, the sum of the atomic numbers on both sides of the equation must be equal and the sum of the mass numbers on both sides of the equation must be equal.
23. A film-badge dosimeter is a badge that consists of photographic film held in a small case that is pinned to clothing. It is used to monitor a person's exposure to radiation. The more exposed the film has become in a given period of time, the more the person has been exposed to radioactivity.
25. In a scintillation counter, the radioactive particles pass through a material that emits ultraviolet or visible light in response to excitation by radioactive particles. The light is detected and turned into an electrical signal.
27. The half-life is the time it takes for one-half of the parent nuclides in a radioactive sample to decay to the daughter nuclides. One can relate the half-life of objects to find their radioactive decay rates.
29. The decaying of uranium in the ground is the source of radon in our environment. Radon increases the risk of lung cancer because it is a gas that can be inhaled.
31. All living organisms contain a uniform amount of carbon-14 because when an organism is alive, carbon-14 is repeatedly incorporated into the body tissue through carbon dioxide and plants. The decay rate of carbon-14 and the continuous formation of it into the body create an equilibrium of carbon-14 in a living organism.
33. We know that carbon-14 dating is accurate because it can be checked against objects whose ages are known from other methods. The age limit for which carbon-14 dating is useful is up to 50,000 years. The carbon-14 amount in older objects is too small to measure accurately.
35. Nuclear fission is the process in which a heavy nucleus is split by a bombardment of neutrons into nuclei of smaller masses and energy is emitted. Meitner, Strassmann, and Hahn discovered it when they were repeating Fermi's experiments and examining the products from bombarding uranium with neutrons.
37. Critical mass is the mass of uranium or plutonium required for a nuclear reaction to be self-sustaining.
39. The process of nuclear fission can generate heat, which then can be used to boil water and create steam, which turns a turbine and generates electricity.
41. Nuclear electricity generation creates a lot of energy for the relative small mass of fuel. Furthermore, a nuclear power plant generates no air pollution and no greenhouse gases.
43. No, a nuclear reactor cannot detonate the way a nuclear bomb can, because the uranium fuel used in electricity generation is not sufficiently enriched in U-235 to produce a nuclear detonation.
45. Modern nuclear weapons use both fission and fusion. In the hydrogen bomb, a small fission bomb is detonated first to create a high enough temperature for the fusion reaction to proceed.

47. Radiation can affect the molecules in living organisms by ionizing them.
49. Lower doses of radiation over extended periods of time can increase cancer risk damaging DNA. Occasionally a change in DNA can cause cells to grow abnormally and to become cancerous.
51. The main unit of radiation exposure is the rem, which stands for *roentgen equivalent man*. The average American is exposed to 1/3 of a rem of radiation per year.
53. Isotope scanning can be used in the medical community to detect and identify cancerous tumors. Likewise, isotope scanning can produce necessary images of several different internal organs.

Problems

55. ${}^{207}_{82}\text{Pb}$
57. 81 protons, 126 neutrons
59. **a.** beta particle
  b. neutron
  c. gamma ray
61. **a.** ${}^{234}_{92}\text{U} \rightarrow {}^{230}_{90}\text{Th} + {}^{4}_{2}\text{He}$
  b. ${}^{230}_{90}\text{Th} \rightarrow {}^{226}_{88}\text{Ra} + {}^{4}_{2}\text{He}$
  c. ${}^{226}_{88}\text{Ra} \rightarrow {}^{222}_{86}\text{Rn} + {}^{4}_{2}\text{He}$
  d. ${}^{222}_{86}\text{Rn} \rightarrow {}^{218}_{84}\text{Po} + {}^{4}_{2}\text{He}$
63. **a.** ${}^{214}_{82}\text{Pb} \rightarrow {}^{214}_{83}\text{Bi} + {}^{0}_{-1}\text{e}$
  b. ${}^{214}_{83}\text{Bi} \rightarrow {}^{214}_{84}\text{Po} + {}^{0}_{-1}\text{e}$
  c. ${}^{231}_{90}\text{Th} \rightarrow {}^{231}_{91}\text{Pa} + {}^{0}_{-1}\text{e}$
  d. ${}^{227}_{89}\text{Ac} \rightarrow {}^{227}_{90}\text{Th} + {}^{0}_{-1}\text{e}$
65. **a.** ${}^{11}_{6}\text{C} \rightarrow {}^{11}_{5}\text{B} + {}^{0}_{+1}\text{e}$
  b. ${}^{13}_{7}\text{N} \rightarrow {}^{13}_{6}\text{C} + {}^{0}_{+1}\text{e}$
  c. ${}^{15}_{8}\text{O} \rightarrow {}^{15}_{7}\text{N} + {}^{0}_{+1}\text{e}$
67. ${}^{241}_{94}\text{Pu} \rightarrow {}^{241}_{95}\text{Am} + {}^{0}_{-1}e$
  ${}^{241}_{95}\text{Am} \rightarrow {}^{237}_{93}\text{Np} + {}^{4}_{2}\text{He}$
  ${}^{237}_{93}\text{Np} \rightarrow {}^{233}_{91}\text{Pa} + {}^{4}_{2}\text{He}$
  ${}^{233}_{91}\text{Pa} \rightarrow {}^{233}_{92}\text{U} + {}^{0}_{-1}\text{e}$
69. ${}^{232}_{90}\text{Th} \rightarrow {}^{228}_{88}\text{Ra} + {}^{4}_{2}\text{He}$
  ${}^{228}_{88}\text{Ra} \rightarrow {}^{228}_{89}\text{Ac} + {}^{0}_{-1}\text{e}$
  ${}^{228}_{89}\text{Ac} \rightarrow {}^{228}_{90}\text{Th} + {}^{0}_{-1}\text{c}$
  ${}^{228}_{90}\text{Th} \rightarrow {}^{224}_{88}\text{Ra} + {}^{4}_{2}\text{He}$
71. 31 atoms
73. 18 hrs
75. 1.2×10^{6} yrs
77. 0.194 g
79. Ga-67 > P-32 > Cr-51 > Sr-89
81. 5,730 yrs
83. 34,380 yrs
85. ${}^{235}_{92}\text{U} + {}^{1}_{0}\text{n} \rightarrow {}^{144}_{54}\text{Xe} + {}^{90}_{38}\text{Sr} + 2{}^{1}_{0}\text{n}$; 2 neutrons
87. ${}^{2}_{1}\text{H} + {}^{2}_{1}\text{H} \rightarrow {}^{3}_{2}\text{He} + {}^{1}_{0}\text{n}$
89. per mole $= 1.9 \times 10^{13}$J
  per kg $= 8.2 \times 10^{13}$J
91. 1.7×10^{21} β emissions
93. 68.4%
95. nucleus with 9 protons and 7 neutrons
97. nucleus with 5 protons and 5 neutrons

PHOTO CREDITS

FM Page vii: Professor Nivaldo Jose Tro. **Chapter 1** Page 1: Richard Megna/Fundamental Photographs. Page 3 (Bl): Richard Megna/Fundamental Photographs. Page 3 (Br): Richard Megna/Fundamental Photographs. Page 4 (C): Richard Megna/Fundamental Photographs. Page 4 (B): Kevin Fleming/CORBIS. Page 5 (T): The Metropolitan Museum of Art. Page 5 (B): CORBIS. Page 6: IBM Almaden Research Center. Page 7 (T): Andy Sacks/Getty Images Inc. - Stone. Page 7 (B): Science Photo Library/Photo Researchers, Inc. **Chapter 2** Page 12: Kim Steele/Getty Images, Inc. Page 14 (Cl): Richard Megna/Fundamental Photographs. Page 14 (Cc): Richard Megna/Fundamental Photographs. Page 14 (Cr): Richard Megna/Fundamental Photographs. Page 14 (B): Richard Megna/Fundamental Photographs. Page 15 (T): Richard Megna/Fundamental Photographs. Page 15 (C): Richard Megna/Fundamental Photographs. Page 17: NASA Headquarters. Page 22: Richard Megna/Fundamental Photographs. Page 23 (Cl): National Institute of Standards and Technology. Page 23 (Cc): Bureau International des Poids et Mesures. Page 23 (Cr): Dr. Donald Sullivan/National Institute of Standards and Technology. Page 23 (B): Richard Megna/Fundamental Photographs. Page 24: Richard Megna/Fundamental Photographs. Page 33: Litespeed Titanium. Page 40: Richard Megna/Fundamental Photographs. Page 52 (T): NASA/JPL/California Institute of Technology Archives. Page 52 (C): NASA/Science Photo Library/Photo Researchers, Inc. Page 52 (B): Chandra X-ray Observatory/M. Weiss/Harvard-Smithsonian Center for Astrophysics. Page 53: Litespeed Titanium. **Chapter 3** Page 56 (Ta): John A. Rizzo/Getty Images, Inc. Page 56 (Tb): Jim Wehtje/Getty Images, Inc. Page 56 (Bl): IBM Almaden Research Center. Page 56 (Br): IBM Almaden Research Center. Page 57 (l): David Chasey/Getty Images, Inc. Page 57 (c): Getty Images, Inc. - Photodisc. Page 57 (r): Mark Downey/Getty Images, Inc. Page 58: Albert Copley/Visuals Unlimited. Page 59 (Cl): Richard Megna/Fundamental Photographs. Page 59 (Clc): Richard Megna/Fundamental Photographs. Page 59 (Crc): Imagination Photo Design. Page 59 (Cr): Kip Peticolas/Fundamental Photographs. Page 59 (B): Adalberto Rios/Getty Images, Inc. Page 60 (T): Ryan McVay/Getty Images, Inc. Page 60 (C): Martial Colomb/Getty Images, Inc. Page 61: Getty Images, Inc. Page 62: Getty Images, Inc. Page 64 (l): Richard Megna/Fundamental Photographs. Page 64 (r): Richard Megna/Fundamental Photographs. Page 65: Jess Alford/Getty Images, Inc. Page 72: Russell Illig/Getty Images, Inc. Page 73: C Squared Studios/Getty Images, Inc. Page 88: Don Chambers, Ph. D. Page 88: Spike/Getty Images, Inc. **Chapter 4** Page 92 (Tl): Siede Preis/Getty Images, Inc. Page 92 (Tr): Corbis Digital Stock. Page 92 (B): University of Pennsylvania, Van Pelt Library. Page 93: IBM Almaden Research Center. Page 96: StockTrek/Getty Images, Inc. Page 97 (T): Richard Megna/Fundamental Photographs. Page 97 (B): Jeremy Woodhouse/Getty Images, Inc. - Photodisc. Page 98 (l): Mel Curtis/Getty Images, Inc. Page 98 (r): John A. Rizzo/Getty Images, Inc. Page 100 (T): Charles D. Winters/Photo Researchers, Inc. Page 100 (B): Mark Marten/Photo Researchers, Inc. Page 101: Novosti/Science Photo Library/Photo Researchers, Inc. Page 103: Getty Images, Inc. - Photodisc. Page 104 (Ct): John A. Rizzo/Getty Images, Inc. Page 104 (Cb): Michael Dalton/Fundamental Photographs. Page 104 (Bt): Richard Megna/Fundamental Photographs. Page 104 (Bb): Ed Degginger/Color-Pic, Inc. Page 105 (Tl): Richard Megna/Fundamental Photographs. Page 105 (Tr): Ed Degginger/Color-Pic, Inc. Page 105 (Cl): Ed Degginger/Color-Pic, Inc. Page 105 (Cr): Tom Bochsler/Pearson Education/PH College. Page 109: D Falconer/Getty Images, Inc. - Photodisc. Page 112: U.S. Department of Energy. **Chapter 5** Page 127: Richard Megna/Fundamental Photographs. Page 128 (Tl): Richard Megna/Fundamental Photographs. Page 128 (Tr): C Squared Studios/Getty Images, Inc. Page 128 (B): Nancy R. Cohen/Getty Images, Inc. Page 129 (l): JoLynn E. Funk/JoLynn E. Funk. Page 129 (c): Sami Sarkis/Getty Images, Inc. Page 129 (r): Russell Illig/Getty Images, Inc. Page 133 (T): M. Freeman/Getty Images, Inc. - Photodisc. Page 133 (B): Charles D. Winters/Photo Researchers, Inc. Page 134 (T): Ken Karp/Omni-Photo Communications, Inc. Page 134 (B): Dr. E.R. Degginger/Bruce Coleman Inc. Page 141 (l): Richard Megna/Fundamental Photographs. Page 141 (r): Michael Dalton/Fundamental Photographs. Page 145 (l): Will & Deni McIntryre/Photo Researchers, Inc. Page 145 (r): Gunter Marx Photography/CORBIS. Page 158 (Tl): Getty Images Inc. - Hulton Archive Photos. Page 158 (Tr): Michael Dalton/Fundamental Photographs. Page 158 (B): Richard Megna/Fundamental Photographs. Page 159 (a): Charles D. Winters/Photo Researchers, Inc. Page 159 (b): Richard Megna/Fundamental Photographs. Page 159 (c): Richard Megna/Fundamental Photographs. Page 159 (d): Richard Megna/Fundamental Photographs. Page 161 (Tl): Richard Megna/Fundamental Photographs. Page 161 (Tr): Richard Megna/Fundamental Photographs. **Chapter 6** Page 163: Charles O'Rear/CORBIS. Page 164 (T): Joseph P. Sinnot/Fundamental Photographs. Page 164 (Bl): Richard Megna/Fundamental Photographs. Page 164 (Br): Richard Megna/Fundamental Photographs. Page 165: Richard Megna/Fundamental Photographs. Page 166: Nancy R. Cohen/Getty Images, Inc. Page 168 (T): Richard Megna/Fundamental Photographs. Page 168 (Cl): Kristen Brochmann/Fundamental Photographs. Page 168 (Cr): Kristen Brochmann/Fundamental Photographs. Page 172 (C): Siede Preis/Getty Images, Inc. Page 172 (Bl): G.K. & Vikki Hart/Getty Images, Inc. Page 172 (Bc): C Squared Studios/Getty Images, Inc. Page 177: NASA/Tom Pantages. Page 178: Richard Megna/Fundamental Photographs. Page 200 (Tl): Jeff Maloney/Getty Images, Inc. Page 200 (Tc): Getty Images, Inc. - Photodisc. Page 200 (Tr): NASA/Goddard Space Flight Center. Page 200 (B): NASA/Johnson Space Center. **Chapter 7** Page 204: Richard Megna/Fundamental Photographs. Page 205 (T): Richard Megna/Fundamental Photographs. Page 205 (Bl): Richard Megna/Fundamental Photographs. Page 205 (Br): Ryan McVay/Getty Images, Inc. Page 206 (Cl) Tom Bochsler/Pearson Education/PH College. Page 206 (a): Richard Megna/Fundamental Photographs. Page 206 (b): Richard Megna/Fundamental Photographs. Page 206 (c): Richard Megna/Fundamental Photographs. Page 206 (d): Jeff J. Daly/Fundamental Photographs. Page 206 (e): Richard Megna/Fundamental Photographs. Page 206 (Bl): Getty Images, Inc. Page 209: Sami Sarkis/Getty Images, Inc. Page 215 (l): Richard Megna/Fundamental Photographs. Page 215 (r): Richard Megna/Fundamental Photographs. Page 221 (C): Don Tremain/Getty Images, Inc. Page 221 (B): Michael Dalton/Fundamental Photographs. Page 223: Richard Megna/Fundamental Photographs. Page 224 (l): Richard Megna/Fundamental Photographs. Page 224 (r): PhotoEdit. Page 228 (l): Richard Megna/Fundamental Photographs. Page 228 (c): Richard Megna/Fundamental Photographs. Page 228 (r): Richard Megna/Fundamental Photographs. Page 229: Richard Megna/Fundamental Photographs. Page 244: Fundamental Photographs. Page 245 (l): Richard Megna/Fundamental Photographs. Page 245 (r): Richard Megna/Fundamental Photographs. **Chapter 8** Page 248 (l): Kristen Brochmann/Fundamental Photographs. Page 248 (lc): Kristen Brochmann/Fundamental Photographs. Page 248 (rc): Kristen Brochmann/Fundamental Photographs. Page 248 (r): Kristen Brochmann/Fundamental Photographs. Page 249 (Tl): Kristen Brochmann/Fundamental Photographs. Page 249 (Tr): Kristen Brochmann/Fundamental Photographs. Page 249 (Cr): Kristen Brochmann/Fundamental Photographs. Page 251: Cheryl Sheridan. Page 256 (l): Kristen Brochmann/Fundamental Photographs. Page 256 (c): Kristen Brochmann/Fundamental Photographs. Page 256 (r): Kristen Brochmann/Fundamental Photographs. Page 264 (a): Richard Megna/Fundamental Photographs. Page 264 (b): Richard Megna/Fundamental Photographs. Page 264 (c): Richard Megna/Fundamental Photographs. Page 264 (d): Richard Megna/Fundamental Photographs. Page 274: Nancy R. Cohen/Getty Images, Inc. Page 275: Stephen Frisch/Stock Boston. Page 249 (Cl): Kristen Brochmann/Fundamental Photographs. **Chapter 9** Page 277: AP/Wide World Photos. Page 278 (l): Princeton University Library. Page 278 (r): American Institute of Physics/Emilio Segre Visual Archives. Page 280: Rob Gage/Getty Images, Inc. - Taxi. Page 282: Larry Mulvehill/Photo Researchers, Inc. Page 283 (l): Bob Richards/Sierra Pacific Innovations Corporation. Page 283 (r): IBob Richards/Sierra Pacific Innovations Corporation. Page 284 (T): D Falconer/Getty Images, Inc. - Photodisc. Page 284 (Bl): Fundamental Photographs. Page 284 (Br): Fundamental Photographs. Page 285: Fundamental Photographs. Page 287: SOHM, JOE/Photo Researchers, Inc. Page 288: AP/Wide World Photos. Page 293: Jim Wehtje/Getty Images, Inc. Page 316: Hisham F. Ibrahim/Getty Images, Inc. Page 317: Karsh/Woodfin Camp &

Associates. **Chapter 10** Page 336: Kip Peticolas/Fundamental Photographs. Page 341 (l): Richard Megna/Fundamental Photographs. Page 341 (r): Richard Megna/Fundamental Photographs. Page 352: Cheryl Sheridan. **Chapter 11** Page 363: Steve Cole/Getty Images, Inc. Page 370 (T): Larry Brownstein/Getty Images, Inc. Page 370 (B): Richard Megna/Fundamental Photographs. Page 376: Getty Images, Inc. - Photodisc. Page 392: K Knudson/Getty Images, Inc. - Photodisc. Page 405: Donald Johnston/Getty Images Inc. - Stone. **Chapter 12** Page 409: Michael Dalton/Fundamental Photographs. Page 410: John A. Rizzo/Getty Images, Inc. Page 411: Richard Megna/Fundamental Photographs. Page 411: Lois Ellen Frank/CORBIS. Page 412 (Tl): ANGELINA LAX/Photo Researchers, Inc. Page 412 (Tr): Richard Megna/Fundamental Photographs. Page 412 (C): NASA/Johnson Space Center. Page 413 (a): Richard Megna/Fundamental Photographs. Page 413 (b): Richard Megna/Fundamental Photographs. Page 413 (c): Richard Megna/Fundamental Photographs. Page 414: Michael Dalton/Fundamental Photographs. Page 417: Richard Megna/Fundamental Photographs. Page 419: Charles D. Winters/Photo Researchers, Inc. Page 423: Richard Megna/Fundamental Photographs. Page 428 (l): Tony Freeman/PhotoEdit. Page 428 (c): Richard Megna/Fundamental Photographs. Page 428 (r): Paul Silverman/Fundamental Photographs. Page 429 (T): Kristen Brochmann/Fundamental Photographs. Page 429 (Bl): Dr. Mark J. Winter. Page 429 (Bc): Getty Images, Inc. - Photodisc. Page 429 (Br): Bryan Peterson Corbis/Stock Market. Page 431 (T): Kristen Brochmann/Fundamental Photographs. Page 431 (B): Lawrence Migdale/Stock Boston. **Chapter 13** Page 446 (a): Richard Megna/Fundamental Photographs. Page 446 (b): Richard Megna/Fundamental Photographs. Page 446 (c): Richard Megna/Fundamental Photographs. Page 448: Richard Megna/Fundamental Photographs. Page 449: Richard Megna/Fundamental Photographs. Page 452: C.J. Allen/Stock Boston. Page 464: Mark Burnett/Stock Boston. Page 466: E.R. Degginger/Photo Researchers, Inc. Page 469: Steve Cole/Getty Images, Inc. Page 481 (T): M. Freeman/Getty Images, Inc. - Photodisc. Page 481 (B): Culligan International Company. **Chapter 14** Page 484 (l): Page 484 (r): Richard Megna/Fundamental Photographs. Page 485: Richard Megna/Fundamental Photographs. Page 486: Getty Images, Inc. - Photodisc. Page 487 (C): Richard Megna/Fundamental Photographs. Page 487 (B): Richard Megna/Fundamental Photographs. Page 492 (T): Paul Silverman/Fundamental Photographs. Page 492 (B): Richard Megna/Fundamental Photographs. Page 494 (l): Richard Megna/Fundamental Photographs. Page 494 (r): Richard Megna/Fundamental Photographs. Page 495 (a): Richard Megna/Fundamental Photographs. Page 495 (b): Richard Megna/Fundamental Photographs. Page 495 (c): Richard Megna/Fundamental Photographs. Page 498 (a): Richard Megna/Fundamental Photographs. Page 498 (b): Richard Megna/Fundamental Photographs. Page 499 (a): Richard Megna/Fundamental Photographs. Page 499 (b): Richard Megna/Fundamental Photographs. Page 509: David M. Martin, M.D./Science Photo Library/Photo Researchers, Inc. Page 513 (l): NYC Parks Photo Archive/Fundamental Photographs. Page 513 (r): Kristin Brochmann/Fundamental Photographs. **Chapter 15** Page 547: Getty Images, Inc. Page 551 (l): Richard Megna/Fundamental Photographs. Page 551 (c): Richard Megna/Fundamental Photographs. Page 551 (r): Richard Megna/Fundamental Photographs. Page 554: Astrid & Hanns-Frieder Michler/SPL/Photo Researchers, Inc. **Chapter 16** Page 573: Alexandra Winkler/Reuters NewMedia Inc./CORBIS. Page 575 (Tl): Santokh Kochar/Getty Images, Inc. Page 575 (Tr): Ryan McVay Getty Images, Inc. Page 575 (B): Richard Megna/Fundamental Photographs. Page 581: Ranald Mackechnie/Getty Images Inc. - Stone. Page 586 (Tl): Richard Megna/Fundamental Photographs. Page 586 (Tr): Richard Megna/Fundamental Photographs. Page 586 (l): Richard Megna/Fundamental Photographs. Page 588 Richard Megna/Fundamental Photographs. Page 589: Richard Megna/Fundamental Photographs. Page 591 (Tl): Kim Steele/Getty Images, Inc. Page 591 (Tr): Edmond Van Hoorick/Getty Images, Inc. Page 595: Draeger Safety, Inc. Page 596: Runk/Schoenberger/Grant Heilman Photography, Inc. Page 597: Imprints Photography. Page 606: IdaTech Corp. **Chapter 17** Page 611: American Institute of Physics/Emilio Segre Visual Archives. Page 612: Richard Megna/Fundamental Photographs. Page 624: AP/Wide World Photos. Page 625: UPI/CORBIS. Page 626: StockTrek/Getty Images, Inc. Page 627: Pete Saloutos/Corbis/Stock Market. Page 631: Larry Mulvehill/Photo Researchers, Inc.

INDEX

D

O

P

Q

R

S

T

Tro, Introductory Chemistry, Accelerator CD

LICENSE AGREEMENT

YOU SHOULD CAREFULLY READ THE TERMS AND CONDITIONS BEFORE USING THE CD-ROM PACKAGE. USING THIS CD-ROM PACKAGE INDICATES YOUR ACCEPTANCE OF THESE TERMS AND CONDITIONS.

Prentice Hall, Inc. provides this program and licenses its use. You assume responsibility for the selection of the program to achieve your intended results, and for the installation, use, and results obtained from the program. This license extends only to use of the program in the United States or countries in which the program is marketed by authorized distributors.

LICENSE GRANT

You hereby accept a nonexclusive, nontransferable, permanent license to install and use the program ON A SINGLE COMPUTER at any given time. You may copy the program solely for backup or archival purposes in support of your use of the program on the single computer. You may not modify, translate, disassemble, decompile, or reverse engineer the program, in whole or in part.

TERM

The License is effective until terminated. Prentice-Hall, Inc. reserves the right to terminate this License automatically if any provision of the License is violated. You may terminate the License at any time. To terminate this License, you must return the program, including documentation, along with a written warranty stating that all copies in your possession have been returned or destroyed.

LIMITED WARRANTY

THE PROGRAM IS PROVIDED "AS IS" WITHOUT WARRANTY OF ANY KIND, EITHER EXPRESSED OR IMPLIED, INCLUDING, BUT NOT LIMITED TO, THE IMPLIED WARRANTIES OR MERCHANTABILITY AND FITNESS FOR A PARTICULAR PURPOSE. THE ENTIRE RISK AS TO THE QUALITY AND PERFORMANCE OF THE PROGRAM IS WITH YOU. SHOULD THE PROGRAM PROVE DEFECTIVE, YOU (AND NOT PRENTICE-HALL, INC. OR ANY AUTHORIZED DEALER) ASSUME THE ENTIRE COST OF ALL NECESSARY SERVICING, REPAIR, OR CORRECTION. NO ORAL OR WRITTEN INFORMATION OR ADVICE GIVEN BY PRENTICE-HALL, INC., ITS DEALERS, DISTRIBUTORS, OR AGENTS SHALL CREATE A WARRANTY OR INCREASE THE SCOPE OF THIS WARRANTY. SOME STATES DO NOT ALLOW THE EXCLUSION OF IMPLIED WARRANTIES, SO THE ABOVE EXCLUSION MAY NOT APPLY TO YOU. THIS WARRANTY GIVES YOU SPECIFIC LEGAL RIGHTS AND YOU MAY ALSO HAVE OTHER LEGAL RIGHTS THAT VARY FROM STATE TO STATE.
Prentice-Hall, Inc. does not warrant that the functions contained in the program will meet your requirements or that the operation of the program will be uninterrupted or error-free. However, Prentice-Hall, Inc. warrants the CD-ROM(s) on which the program is furnished to be free from defects in material and workmanship under normal use for a period of ninety (90) days from the date of delivery to you as evidenced by a copy of your receipt.
The program should not be relied on as the sole basis to solve a problem whose incorrect solution could result in injury to person or property. If the program is employed in such a manner, it is at the userís own risk and Prentice-Hall, Inc. explicitly disclaims all liability for such misuse.

LIMITATION OF REMEDIES

Prentice-Hall, Inc.ís entire liability and your exclusive remedy shall be: 1. the replacement of any CD-ROM not meeting Prentice-Hall, Inc.'s "LIMITED WARRANTY" and that is returned to Prentice-Hall, or 2. if Prentice-Hall is unable to deliver a replacement CD-ROM that is free of defects in materials or workmanship, you may terminate this agreement by returning the program.
IN NO EVENT WILL PRENTICE-HALL, INC. BE LIABLE TO YOU FOR ANY DAMAGES, INCLUDING ANY LOST PROFITS, LOST SAVINGS, OR OTHER INCIDENTAL OR CONSEQUENTIAL DAMAGES ARISING OUT OF THE USE OR INABILITY TO USE SUCH PROGRAM EVEN IF PRENTICE-HALL, INC. OR AN AUTHORIZED DISTRIBUTOR HAS BEEN ADVISED OF THE POSSIBILITY OF SUCH DAMAGES, OR FOR ANY CLAIM BY ANY OTHER PARTY. SOME STATES DO NOT ALLOW FOR THE LIMITATION OR EXCLUSION OF LIABILITY FOR INCIDENTAL OR CONSEQUENTIAL DAMAGES, SO THE ABOVE LIMITATION OR EXCLUSION MAY NOT APPLY TO YOU.

GENERAL

You may not sublicense, assign, or transfer the license of the program. Any attempt to sublicense, assign or transfer any of the rights, duties, or obligations hereunder is void. This Agreement will be governed by the laws of the State of New York. Should you have any questions concerning this Agreement, you may contact Prentice-Hall, Inc. by writing to:

ESM Media Development
Higher Education Division
Prentice-Hall, Inc.
1 Lake Street
Upper Saddle River, NJ 07458

Should you have any questions concerning technical support, you may write to:

New Media Production
Higher Education Division
Prentice-Hall, Inc.
1 Lake Street
Upper Saddle River, NJ 07458

YOU ACKNOWLEDGE THAT YOU HAVE READ THIS AGREEMENT, UNDERSTAND IT, AND AGREE TO BE BOUND BY ITS TERMS AND CONDITIONS. YOU FURTHER AGREE THAT IT IS THE COMPLETE AND EXCLUSIVE STATEMENT OF THE AGREEMENT BETWEEN US THAT SUPERSEDES ANY PROPOSAL OR PRIOR AGREEMENT, ORAL OR WRITTEN, AND ANY OTHER COMMUNICATIONS BETWEEN US RELATING TO THE SUBJECT MATTER OF THIS AGREEMENT.

MINIMUM SYSTEM REQUIREMENTS

MACINTOSH

- Processor: 133 Mhz (266 Mhz or higher recommended)
- RAM: 96 MB RAM (128 MB or higher recommended)
- Operating System: Power PC System, OS 8.5 (OS 9 or higher recommended)
- Hardware: 4x CD-ROM (12x or higher recommended)
- Monitor Resolution: 1024 x 768 pixels
- Software: A browser (IE5.5 or higher) with the QuickTime plugin (v5 or higher) and the proper shockwave plugin (8.5 or higher is necessary to view the animations.)

WINDOWS 98, ME, NT, XP and 2000

- Processor: 150 Mhz (266 Mhz or higher recommended)
- RAM: 96 MB RAM (128 MB or higher recommended)
- Hardware: 4x CD-ROM (12x or higher recommended)
- Monitor Resolution: 1024 x 768 pixels
- Software: A browser (IE5.5 or higher) with the QuickTime plugin (v5 or higher) and the proper shockwave plugin (8.5 or higher is necessary to view the animations.)